International Scientific Practical Conference on Breakthrough Technologies and Communications in Industry (BTCI 2018)

IOP Conference Series: Materials Science and Engineering
Volume 483

Volgorad, Russia
20 - 21 November 2018

Editors:

Elena Yu. Malushko
Nikolay L. Shamne

ISBN: 978-1-5108-8480-9
ISSN: 1757-8981

Printed from e-media with permission by:

Curran Associates, Inc.
57 Morehouse Lane
Red Hook, NY 12571

Some format issues inherent in the e-media version may also appear in this print version.

Printed by Curran Associates, Inc. (2019)

For permission requests, please contact the Institute of Physics at the address below.

Institute of Physics
Dirac House, Temple Back
Bristol BS1 6BE UK

Phone: 44 1 17 929 7481
Fax: 44 1 17 920 0979

techtracking@iop.org

Additional copies of this publication are available from:

Curran Associates, Inc.
57 Morehouse Lane
Red Hook, NY 12571 USA
Phone: 845-758-0400
Fax: 845-758-2633
Email: curran@proceedings.com
Web: www.proceedings.com

Table of contents

Volume 483

The I International Scientific Practical Conference "Breakthrough Technologies and Communications in Industry"

20–21 November 2018, Volgograd, Russian Federation

Accepted papers received: 4 January 2019
Published online: 20 March 2019

Preface

Papers

Abstract

International Scientific Practical Conference "Breakthrough technologies and communications in industry" (BTCI2018)

PREFACE

The conference was held on 20-21 November, 2018 at Volgograd state university (Volgograd, Russian Federation) and provided a forum place for discussion for PhD students, teachers, scientists, researchers from research institutes and industrial enterprises from Volgograd, Moscow, Irkutsk and other cities from Russia and abroad (Sweden, Germany, China).

The conference served as a platform for the researchers to discuss and carry out a research on current production and communication technology in industry and production department. The aim of the conference was to present the innovative approaches to developing convergent solutions in engineering and technology, and also in interdisciplinary scientific researches. As the venue bore interdisciplinary character, the scientists and experts also had an opportunity to cover the applied aspects like training, oral communication, technical translation, interaction and others.

Key topics of the conference covered priority research areas of technical, communication, educational and other related sciences and took into account global trends in social and industrial development. BTCI2018 topics are built on the fundamentals of sustainability, with vigorous attention to innovative technologies for producing and testing permanent joints, modern technologies for producing and processing materials in industry, automation, informatization and management in the enterprise, production communication and ethics, environmental protection, safety and labor protection in enterprises, breakthrough technologies and technology for subsoil development, development of methods to stimulate innovation and investment activity of enterprises.

IOP Conf. Series: Materials Science and Engineering **483** (2019) 011001 doi:10.1088/1757-899X/483/1/011001

EDITORS

Elena Yu. Malushko

Volgograd State University - Federal State Autonomous Educational Institution of Higher Education " Volgograd State University " (VolSU)

E-mail: e.malushko@volsu.ru

Nikolay L. Shamne

Volgograd State University - Federal State Autonomous Educational Institution of Higher Education " Volgograd State University " (VolSU)

E-mail: nikolay.shamne@volsu.ru

IOP Conf. Series: Materials Science and Engineering **483** (2019) 011002 doi:10.1088/1757-899X/483/1/011002

Peer review statement

All papers published in this volume of *IOP Conference Series: Materials Science and Engineering* have been peer reviewed through processes administered by the proceedings Editors. Reviews were conducted by expert referees to the professional and scientific standards expected of a proceedings journal published by IOP Publishing.

Electronic course development for future engineers training

Zh V Smirnova[1], O I Vaganova[1], M N Gladkova[2], M L Gruzdeva[1], S. N Kaznacheeva[1]

[1]Minin Nizhny Novgorod State Pedagogical University, st Ulyanova 1, Nizhny Novgorod 603004, Russia
[2] Tyumen Higher Military Engineering Command School named Marshal Engineering Troops A I Proshlyakova, Ministry of Defense of the Russian Federation, L. Tolstoy str. 1, 625001,Tyumen, Russia

Abstract. Due to the necessity of implementing the requirements of the Federal State Educational Standard, there is a need to use electronic information in educational environment. The article considered creation of an electronic course in the context of blended learning, since this approach seems to be one of the most effective, bringing better results. The educational process in this case is based on integration and blending of traditional and e-learning technologies. Materials were placed in the electronic educational environment Moodle. Objective: to identify features of the development of an electronic course for future engineers. A training scheme how to blend traditional model and e-learning is introduced. The recommendations on the selection of educational material are provided. For the 1st year students whose major is engineering and construction control of the teacher and is extremely important. Thus, an electronic course of «Basics of Geology» was developed to combine blended learning and traditional approach. Special recommendations for creating such a course are developed. A scenario for the development of the training course consisting of several stages is proposed. We have analyzed the implementation of this course into practice.

1 Introduction

Today, there is a tendency in the world to use electronic educational resources, which is caused by the need to increase mobility and independence of students. Electronic information in educational environment in Russia should be used according to the Federal State Educational Standard.

In 2016, the Government of Russia established the Project "Modern Digital Educational Environment in Russia". Its goal was to create conditions for improving the quality and expanding the possibilities of continuing education for all citizens through the development of the digital educational space. Note that the "National Open Education Platform" is being implemented within the framework of this project.

The implementation of the "Modern Digital Educational Environment in the Russian Federation" shows that the number of training platforms is growing, there are several of them. Online courses are numerous. To choose a suitable course is quite difficult. Therefore, higher educational institutions should pay attention to the development of internal university courses in specific disciplines. With the transition to the federal state educational standards of a new generation, the time of classroom pressure has decreased and the number of independent work of students has increased [1]. This trend may adversely affect students whose first year training requires constant practice and teacher control [2]. We have identified future engineers for whom the e-course will allow implementing the principle of virtual academic mobility by providing high-quality educational content (statistical texts with a large number of illustrations and explanations, video lectures with demonstrations of technical processes in dynamics, files with additional information and buildings, final tests), while there is a necessity to provide teachers as well.

2 Methodology

IOP Conf. Series: Materials Science and Engineering **483** (2019) 012001 doi:10.1088/1757-899X/483/1/012001

We considered the tools of electronic courses and developed special recommendations for its creation. The practice of using the course of the discipline «Fundamentals of Geology» is analyzed.

The developed e-course was tested for students whose major is «Construction». According to the students' achievements who were trained the help of this course, a survey was conducted in which students answered the following questions: «Did the course contribute to better learning?», «Would you like to use this course in further education?» 80% of students answered positively. In addition, final testing showed higher results of than before the introduction of the course. The students noted the advantage of being able to control the level of their achievements at any stage of the study of the course, as well as the positive effect of constant access to necessary materials.

3 Results and discussions

Experts emphasize the use of:

- traditional learning (without using any electronic technologies);
- blended learning (Blended-learning, part of the courses implemented in the network, combined learning) [3];
- full online learning (does not provide for full-time interaction between the teacher and students).

We will consider the creation of an electronic course in the framework of blended learning, where it serves as an auxiliary element [4].

The educational process in this case is based on the integration of traditional and e-learning technologies [5]. E-learning can take up to 80% of the time. With the reduction of classroom activities, they are replaced by the electronic environment [6].

A well-organized electronic training course is not only a place for storing contents, but also a means for interaction between a teacher and a student, as well as a means of accompanying the educational process [7].

Course development should not begin with finding the right content and developing it, but with defining the planned learning outcomes. Next, the teaching strategy is determined scenarios of interaction between students and teachers for maximum involvement of the first ones in the educational process. Selection and structuring of educational content occurs at the last stage [8].

The organization of the course in an electronic environment occurs so that the order of the key components of the educational process becomes different. The traditional teaching model assumes that at the beginning the presentation and explanation of materials in the classroom conditions and the performance of tasks and the consolidation of the material goes to extracurricular work, and verification and control again takes place in the classroom, then using e-learning, the process begins with the problem statement. Then the student independently acquaints himself with the educational material located on the e-learning course and with the materials of open educational resources predetermined by the teacher [9]. The student conducts self-control understanding of the content. Thus, the activity of the student begins with independent work, but takes place in an environment with the participation of the teacher. In the classroom, a teacher helps a student clarify understanding and discuss assignments [10]. Then there is a demand to provide the automated control of learning outcomes [11].

We recommend the following scenario development course. At the first stage the following actions are performed:

- this is the design of learning outcomes. It is necessary to formulate the results of training in the discipline as a whole, and then structure it into separate thematic sections [12];
- development of assessment tools (a system for assessing the achievement of initially planned results by students) [13];
- this is a selection of educational content.

The teacher needs to select educational resources on the Internet that could be used on the course (audio-video materials, graphics, electronic textbooks) [14]. The collected content is best structured into subsections with intermediate control in the form of testing [15]. Note that the time to study the subsection should be no more than 15 minutes. 45-60 minutes are allocated for mastering disciplinary knowledge [16].

The second stage is devoted to the placement of content according to the requirements for the structure of sections of the electronic course in the electronic educational environment Moodle [17]. We speak about this environment because it has a number of advantages over the others and is most often used by higher educational institutions [18]. The structure of the e-course in Moodle should have the following elements:

- a section where information is reflected throughout the course as a whole, where a glossary, a forum, a brief description of the course can be located;

Figure 1 presents a section on the subject «Fundamentals of Geology» for students in the direction of training «Vocational training», the profile «Construction».

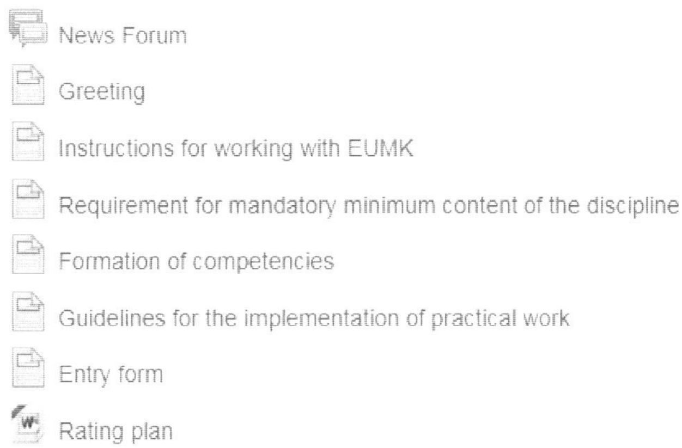

Figure 1. Elements of the electronic course on the discipline «Fundamentals of Geology», located on the platform «Moodle»

The teacher provides information using the tools of the platform at his discretion [19]. The main thing is that the student can understand what this discipline is for and how to approach the performance of tasks. Future engineers, thus, receive full awareness of the discipline and competencies that they can acquire after passing it.

- theme sections (there should also be educational materials on topics) [20];

For the engineering discipline «Fundamentals of Geology» one of these sections is as follows.

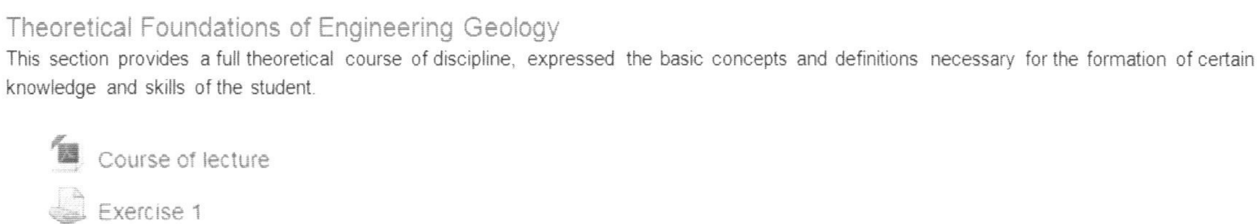

Figure 2. Section on the discipline «Fundamentals of Geology» - section with materials of the final assessment event

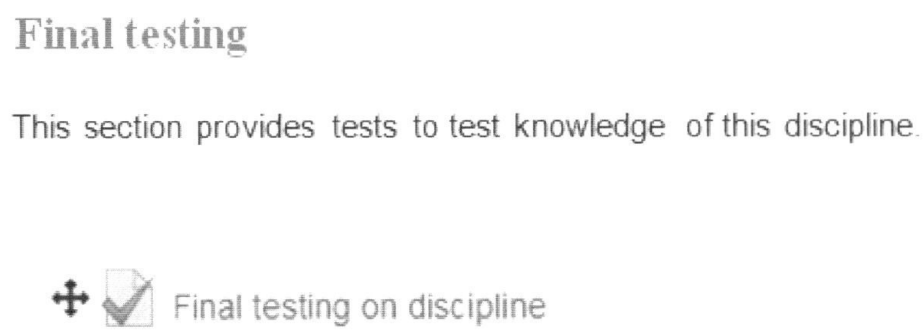

Figure 3. Final testing on the electronic platform Moodle on the discipline «Fundamentals of Geology»

Training material should be selected according to the following criteria:

- ensuring the disclosure of all the themes of the program of the academic discipline while maintaining intra-subject and interdisciplinary communication;
- reliability of materials;
- compliance of materials with the current state of the studied science.

At the same time it is necessary to observe the methodical principles

- output of information for better memorization and assimilation of material by students;
- the sequence of presentation of information (educational material is organized in such a way that when studying new topics briefly repeated previous ones);
- the use of problematic style of presentation in the preparation of tasks;
- the construction of tasks should be made taking into account the future profession of students. In our case of engineers, for example, to establish methods of dealing with the negative influence of geological processes and phenomena on the engineering and geological properties of soils or engineering structures;
- design materials in the same style;
- the use of hypertexts in educational material (hyperlinks can be made to various illustrations, multimedia information - audio recordings and video sequences, additional materials, educational and scientific literature).

For each assignment, guidelines for their implementation are developed and indicators and evaluation criteria are specified [21].

To increase the validity of the assessment of learning outcomes for the assignment, it is recommended to use a certain period of time between attempts [22]. For example, the second attempt may be granted no earlier than 24 hours.

The third stage is approbation. At this stage, various kinds of errors are identified. The course can be tested by colleagues to identify critical comments, as well as by the students themselves [23].

The e-course on the discipline «Fundamentals of Geology» was originally introduced in addition to the traditional teaching of the discipline for a group of engineering-construction students of the 1st course to identify the degree of their learning of the educational material. According to the results of mastering the discipline with the use of this course, a survey was conducted in which students answered the following questions: «Did the course contribute to better learning?», «Would you like to use this course in further education?» 80% of students answered positively. In addition, final testing showed higher results of mastering the discipline than before the introduction of the course. The students themselves point out the advantages of being able to control the level of mastering the material at any stage of the study of the discipline, the positive effect of the ability to constantly update the content and the quick availability of the necessary materials.

4 Conclusions

In the process of work, we highlighted the features of the development of an electronic course for future engineers in the context of blended learning. The presented recommendations will help teachers to navigate in creating a course. The results of a survey conducted after the students graduated from the course «Basics of Geology» allowed to conclude that there is a necessity to introduce an e-course as an addition to traditional engineering education.

E-courses allow you to make future engineers more mobile, independent, give them the freedom to manifest their creative component and as a result help to form a graduate who is ready to carry out their professional activities.

References

[1] Smirnova ZH V, Vaganova O I, Trutanova A V 2017 Itogovaya gosudarstvennaya attestatsiya kak sposob kompleksnoy otsenki kompetentsiy *Karelian Scientific Journal* **6(3)** 74-77

[2] Tsyplakova S A, Grishanova M N, Korovina E A, Somova N M 2016 Theoretical bases of designing of educational systems *Azimuth of Scientific Research: Pedagogy and Psychology* Vol 5 N **1 (14)** 131-133

[3] Vaganova O I, Koldina M I, Trutanova A V 2017 Development content of professional pedagogical education in the conditions of realization of competence approach *Baltic Humanitarian Journal* T 6 **2 (19)** 97-99

[4] Vaganova O I, Smirnova ZH V, Trutanova A V 2017 Organization of research activities of bachelor of professional education in electronic form *Azimuth of Scientific Research: Pedagogy and Psychology* **6(3)** 239-241

[5] Yashin S N, Yashina N I, Ogorodova M V, Smirnova Z V, Kuznetsova S N, Paradeeva I N 2017 On the methodology for integrated assessment of insurance companies' financial status *Man in India* **97 (9)** 37-42

[6] Bicheva I B, Filatova O M 2017 Formation of the teacher-leader in the educational process of the university *Vestnik of Minin University* **3 (20)** 5

[7] Bogorodskaya O V, Golubeva O V, Gruzdeva M L, Tolsteneva A A, Smirnova Z V 2018 Experience of approbation and introduction of the model of management of students' Independent work in the university *Advances in Intelligent Systems and Computing* **622** 387-397

[8] Bulaeva M N, Vaganova O I, Koldina M I, Lapshova A V, Khizhnyi A V 2017 Preparation of Bachelors of Professional Training Using MOODLE *In International conference on Humans as an Object of Study by Modern Science* 406-411

[9] Vaganova O I, Ilyashenko L K 2018 The main directions of implementation technologies of student-centered education in high school *Vestnik of Minin University* vol 6 **3** 2

[10] Garina, E P, Kuznetsov V P, Egorova A O, Romanovskaya E V, Garin A P 2017 Practice in the application of the production system tools at the enterprise during mastering of new products *Contributions to Economics* **9783319606958** 105-112

[11] Garina E P, Kuznetsov V P, Romanovskaya E V, Andryashina N S, Efremova A D 2018 Research and generalization of design practice of industrial product development (by the example of domestic automotive industry) () *Quality* **19 (S2)** 135-140

[12] Iltaldinova E Yu, Filchenkova I F, Frolova S V 2017 Peculiarities of the organization of postgraduate support of graduates of the targeted training program in the context of supporting the life cycle of the teacher's profession *Vestnik of Minin University* **3 (20)** 2

[13] Ilyashenko L K, Prokhorova M P, Vaganova O I, Smirnova Z V, Aleshugina E A 2018 Managerial preparation of engineers with eyes of students *International Journal of Mechanical Engineering and Technology* **9 (4)** 1080-1087

[14] Ilyashenko L K, Smirnova Z V, Vaganova O I, Prokhorova M P, Abramova N S 2018 The role of network interaction in the professional training of future engineers *International Journal of Mechanical Engineering and Technology* **9 (4)** 1097-1105

[15] Ilyashenko L K, Vaganova O I, Smirnova Z V, Gruzdeva M L, Chanchina A V 2018 Structure and content of the electronic school-methodical complex on the discipline "mechanics of soils, foundations and foundations" *International Journal of Mechanical Engineering and*

Technology **9 (4)** 1088-1096

[16] Kutepov M M, Vaganova O I, Trutanova A V 2017 Possibilities of health-saving technologies in the formation of a healthy lifestyle *Baltic Humanitarian Journal* **6(3)** 210-213

[17] Kuznetsov V P, Romanovskaya E V, Egorova A O, Andryashina N S, Kozlova E P 2018 Approaches to developing a new product in the car building industry *Advances in Intelligent Systems and Computing* **622** 494-501

[18] L K Ilyashenko 2018 Pedagogical Conditions of Formation of Communicative Competence of Future Engineers in the Process of Studying Humanitarian Disciplines *International Journal of Civil Engineering and Technology* 9(3) 607-616

[19] Markova S M, Sedykh E P, Tsyplakova S A, Polunin V Y 2017 Perspective Trends of Development of Professional Pedagogics as a Science *In International conference on Humans as an Object of Study by Modern Science* 129-135

[20] Smirnova Zh V, Krasikova O G 2018 Modern tools and technologies for assessing learning outcomes *Vestnik of Minin University* Vol 6 **3** 9

[21] Potashnik Y S, Garina E P, Romanovskaya E V, Garin A P, Tsymbalov S D 2018 Determining the value of own investment capital of industrial enterprises *Advances in Intelligent Systems and Computing* 6**22** 170-178

[22] Smirnova Zh V, Mukhina M V, Kutepova L I, Kutepov M M, Vaganova O I 2018 Organization of the research activities of service majors trainees *Advances in Intelligent Systems and Computing* **622** 187-193

[23] Vaganova O I, Gladkov A V, Trutanova A V 2017 Formation of professional competencies of bachelors in the conditions of e-learning *Baltic Humanitarian Journal* vol 6 **2 (19)** 190-193

Current requirements for assessing the results of student training

O I Vaganova[1], Zh V Smirnova[1], N S Abramova[2], J M Tsarapkina[3], L MBazavlutskaya[4]

[1]Minin Nizhny Novgorod State Pedagogical University, ul. Ulyanova,1, Nizhny Novgorod, (603004), Russia
[2] Tyumen Higher Military Engineering Command School named Marshal Engineering Troops A.I. Proshlyakova, Ministry of Defense of the Russian Federation, L. Tolstoy str., 1, (625001),Tyumen, Russia
[3]Russian State Agrarian University - Moscow Timiryazev Agricultural Academy, Russia
[4] Federal State Budgetary Educational Institution of Higher Education "South-Ural State Humanitarian-Pedagogical University. Address Russia, 454080, Russian Federation, Ural Federal District, Chelyabinsk Region, Chelyabinsk, Lenin Ave., 69

Abstract. The article offers an overview of current requirements for assessing the results of student training. To do this, the authors consider the use different evaluation tools established by the Federal State Educational Standard of Higher Education to assess the educational activities of students. The article reveals the tasks of evaluation tools, their structural elements, and the documents they must comply with. The authors have proposed recommendations for the development of assessing students' knowledge (major "The use of units and the operation of engineering electrical equipment"). The article also considers principles that should guide the preparation of assessment tools. As an example, the authors developed a structure for evaluating the report on the course "Engineering", revealing the expected results and offering a detailed scheme to follow.

1 Introduction

Due to changes in educational paradigm the requirements for university students' training have also been altered. The necessity for these transformations was also determined by the needs of the state and society. Today, employers require graduates who are fully prepared for professional activity, ready to act in non-standard conditions and to solve tasks quickly.

In the system of higher education there are some measures by means of which the quality of education is regulated, allowing training of competent bachelors. This includes the quality of objectives, educational and occupational standards, norms, and the quality of educational resources.

If these measures are implemented, students' training can be considered as being carried out at a high level.

Among these requirements, one should pay attention to the way students are trained at a higher educational institution in order society and manufacturing requirements. At the same time, a qualitative assessment of training results should be provided [1].

Today, professionals in the service industry are in demand. The service sector, usually in industrialized countries, accounts for more than 70% of the revenues that form the country's gross domestic product; in Russia, this figure is at the level of 60%. Therefore, highly qualified personnel in a service sector are so in demand. To ensure it, a higher education institution should conduct a qualitative assessment of training results.

The Feral state Educational Standard emphasizes the need to use funds of appraisal funds to certify students for compliance of their achievements with the requirements of the main educational corriculum.

2 Methodology

In the article we consider current requirements for assessing the results of student training. In a higher educational institution, based on the requirements of the Federal State Educational Standard, assessment funds for students are used for evaluating students' activities, which are developed for each discipline. Taking into account the competence approach, they help not only to manage the process of acquiring knowledge and skills by students, but also to check the formation of competencies. We have proposed recommendations for the development of appraisal tools in the specialty "Use of divisions and operation of electrical engineering tools", for the course "Introduction to engineering", and also a structure of report evaluation for the specified course is presented [2]. As practice shows, these recommendations help to evaluate the preparation of students more objectively and efficiently.

3 Results and discussions

Assessment tools is an integral part of the normative and methodological support of the system for evaluating the quality of students 'mastering of the basic professional educational curriculum of higher education, which is designed to ensure an objective assessment of the quality of students' training in the implementation of the educational process at the University [3].

Appraisal tools contain monitoring materials designed to assess the achieved results. They are created for each discipline and module for the final certification [4].

The tasks of the appraisal tools include:

- managing the process of acquiring learners of knowledge and skills, checking the formation of competencies established by the Federal State Educational Standard for a specific area of training [5];
- management of the achievement of the objectives of the implementation of the basic professional educational program of higher education, defined as a set of general cultural, universal, general professional and professional competencies of graduates [6];
- assessment of students' achievements in the process of studying the discipline;
- ensuring the compliance of learning outcomes with the objectives of future professional activities [7].

Assessment funds are developed and implemented by higher education institutions [8].

This procedure should be guided by the following principles:

- validity (the evaluation tool should check exactly the results that were planned) [9];
- the degree of mastery of competencies should be evaluated, that is, all tasks and criteria for evaluating their implementation should be expressed in formulations that reflect their readiness for a particular type of activity [10];
- assessment of the degree of preparation for future professional activity should not be replaced by an assessment of the level of possession only of special professional competencies [11].
- objectivity (use of tools that ensure the reliability of assessment results) [12];
- openness of the assessment procedure (both students and examiners should have common criteria for evaluating performance);
- timeliness (maintenance of developing feedback) [13].

When developing appraisal funds, its compliance should be considered:

- The federal state educational standard of higher education in the relevant area of training [14];
- the main professional educational program of higher education;
- curriculum;
- work program [15];
- rating plan;
- educational technologies used in teaching the discipline.

Appraisal funds must have the following structural elements:

- title page;
- a passport containing the levels of formation of each competence;
- a set of evaluation tools;

- test and examination materials (examination tickets, questions, assignments, criteria and rating scale) [16].

Each assessment tool consists of three parts:

- the first part is organizational-methodical (with indication of the formed competencies and decomposition at the levels “knows, knows, owns”) [17];
- the second part is the content (a set of test and measurement materials aimed at checking the competencies mentioned in the first part) [18];
- the third part is the criterion-evaluation (indicates the criteria, procedures and scale of assessment, here limits are set that allow you to accept or reject the task performed by the student [19].

As we have said, the country today is extremely in demand engineering personnel, therefore, we will offer recommendations for the development of appraisal tools in the specialty "Use of units and operation of electrical engineering engineering", in the discipline "Introduction to engineering".

The student must achieve the following results:

- be able to develop technical specifications;
- be able to develop technical proposals;
- be able to develop draft, technical and working projects. [20];
- know the regulatory legal acts [21].

These provisions should be reflected in the passport [22].

As we have already mentioned, an integral part of appraisal funds is a set of appraisal funds. Next, we will present the structure of the assessment report or message for students whose major is "The use of units and the operation of engineering electrical equipment":

- formed competencies [23];
- labor activities;
- indicators of the achievement of labor actions and competencies;
- The subject of the discipline: "The design process";
- Approximate topics of reports;
- criteria and assessment procedure (rating scale). Here is the teacher's explanation of the procedure for preparing the report, the estimation algorithm is presented.

The report is prepared once per semester.

The rating is set according to the following criteria: the degree of completeness of the material disclosure, the degree of compliance with the professional standard.

One point is assigned for each completed item:

- the content of the report corresponds to the stated topic;
- the studied theoretical position was formulated, and its place in the theory should be determined;
- marked the range of concepts and terms that will be needed to describe the studied position;
- a description and comparison of examples of the use of the studied situation in Russia and abroad;
- in the report, semantic parts are highlighted and the logic of reasoning is visible when moving from one part to another;
- the student freely expounds the material of the report, communicates with the audience;
- the report is accompanied by a multimedia presentation, the presentation lasts for 10-15 minutes;
- The report traces interim and final conclusions;
- in the report there is a link to sources, authors of studies;
- The student clearly answers the questions asked on the topic of the report.

The maximum score for the report is 5 points. The report will be counted if the student receives at least 3 points.

4 Conclusions

The article reviewed the current requirements for assessing the results of student training. It is determined that the evaluation takes place on the basis of evaluation funds which are developed for each discipline.

We have proposed recommendations for the development of appraisal tools for students whose major is "Application of units and operation of electrical engineering tools". The presented structure of report assessment in the course provides detailed description of the expected results and gives a thorough algorithm of actions. These recommendations help to take a more systematic approach to the development of appraisal tools as a result of which students are trained and evaluated more objectively and efficiently.

References

[1] Bicheva I B, Filatova O M 2017 Formation of the teacher-leader in the educational process of the university *Vestnik of Minin University* **3 (20)** 5

[2] Bogorodskaya O V, Golubeva O V, Gruzdeva M L, Tolsteneva A A, Smirnova Z V 2018 Experience of approbation and introduction of the model of management of students' Independent work in the university *Advances in Intelligent Systems and Computing* **622** 387-397

[3] Bulaeva M N, Vaganova O I, Koldina M I, Lapshova A V, Khizhnyi A V 2017 Preparation of Bachelors of Professional Training Using MOODLE *In International conference on Humans as an Object of Study by Modern Science* 406-411

[4] Fedorov A A, Paputkova G A, Ilaltdinova E Y, Filchenkova I F, Solovev M Y 2017 Model for employer-sponsored education of teachers: Opportunities and challenges *Man in India* **97 (11)** 101-114

[5] Garina E P, Kuznetsov V P, Egorova A O, Romanovskaya E V, Garin A P 2017 Practice in the application of the production system tools at the enterprise during mastering of new products *Contributions to Economics* (**9783319606958**) 105-112

[6] Garina E P, Kuznetsov V P, Romanovskaya E V, Andryashina N S, Efremova A D 2018 Research and generalization of design practice of industrial product development (by the example of domestic automotive industry) *Quality* **19 (S2)** 135-140

[7] Iltaldinova E.Yu., Filchenkova IF, Frolova S.V. 2017 Peculiarities of the organization of postgraduate support of graduates of the targeted training program in the context of supporting the life cycle of the teacher's profession *Vestnik of Minin University* **3 (20)** 2

[8] Ilyashenko L K, Prokhorova M P, Vaganova O I, Smirnova Z V, Aleshugina E A 2018 Managerial preparation of engineers with eyes of students *International Journal of Mechanical Engineering and Technology* **9 (4)** 1080-1087

[9] Ilyashenko L K, Smirnova Z V, Vaganova O I, Prokhorova M P, Abramova N S 2018 The role of network interaction in the professional training of future engineers *International Journal of Mechanical Engineering and Technology* **9 (4)** 1097-1105

[10] Ilyashenko L K, Vaganova O I, Smirnova Z V, Gruzdeva M L, Chanchina A V 2018 Structure and content of the electronic school-methodical complex on the discipline "mechanics of soils, foundations and foundations" *International Journal of Mechanical Engineering and Technology* **9 (4)** 1088-1096

[11] Kutepov M M, Vaganova O I, Trutanova A V 2017 Possibilities of health-saving technologies in the formation of a healthy lifestyle *Baltic Humanitarian Journal* **6(3)** 210-213

[12] Kuznetsov V P, Romanovskaya E V, Egorova A O, Andryashina N S, Kozlova E P 2018 Approaches to developing a new product in the car building industry *Advances in Intelligent Systems and Computing* **622** 494-501

[13] L K Ilyashenko 2018 Pedagogical Conditions of Formation of Communicative Competence of Future Engineers in the Process of Studying Humanitarian Disciplines *International Journal of Civil Engineering and Technology* **9(3)** 607-616

[14] Markova S M, Sedykh E P, Tsyplakova S A, Polunin V Y 2017 Perspective Trends of Development of Professional Pedagogics as a Science *In International conference on Humans as an Object of Study by Modern Science* 129-135

[15] Potashnik Y S, Garina E P, Romanovskaya E V, Garin A P, Tsymbalov S D 2018 Determining

the value of own investment capital of industrial enterprises *Advances in Intelligent Systems and Computing* **622** 170-178

[16] Smirnova Zh V, Mukhina M V, Kutepova L I, Kutepov M M, Vaganova O I 2018 Organization of the research activities of service majors trainees *Advances in Intelligent Systems and Computing* **622** 187-193

[17] Smirnova ZH V, Gruzdeva M L, Krasikova O G 2017 Open electronic courses in the educational activities of the university *Vesknik of Minin University* **4(21)** 3

[18] Smirnova Zh V, Krasikova O G 2018 Modern tools and technologies for assessing learning outcomes *Vestnik of Minin University* Vol 6 **3** 9

[19] Smirnova ZH V, Vaganova O I, Trutanova A V 2017 Final state certification as a way to comprehensive assessment of competences *Karelian Scientific Journal* vol 6 **3(20)** 74-77

[20] Tsyplakova S A, Grishanova M N, Korovina E A, Somova N M 2016 Theoretical bases of designing of educational systems *Azimuth of Scientific Research: Pedagogy and Psychology* Vol 5 **1 (14)** 131-133

[21] Vaganova O I, Koldina M I, Trutanova A V 2017 Development content of professional pedagogical education in the conditions of realization of competence approach *Baltic Humanitarian Journal* T 6 **2 (19)** 97-99

[22] Vaganova O I, Smirnova ZH V, Trutanova A V 2017 Organization of research activities of bachelor of professional education in electronic form *Azimuth of Scientific Research: Pedagogy and Psychology* **6(3)** 239-241

[23] Yashin S N, Yashina N I, Ogorodova M V, Smirnova Z V, Kuznetsova S N, Paradeeva I N 2017 On the methodology for integrated assessment of insurance companies' financial status *Man in India* **97 (9)** 37-42

IOP Conf. Series: Materials Science and Engineering **483** (2019) 012003 doi:10.1088/1757-899X/483/1/012003

Practice-oriented approach implementation in vocational education

Zh V Smirnova[1], O I Vaganova[1], D A Loshkareva[2], Konyaeva E A[3], M N Gladkova[4]

[1]Minin Nizhny Novgorod State Pedagogical University, ul. Ulyanova,1, Nizhny Novgorod, 603004, Russia
[2] Nizhny Novgorod State University of Architecture and Civil Engineering, Ilyinskaya str., 65, Nizhny Novgorod, 653950, Russia
[3]South Ural State Humanitarian and Pedagogical University, Lenina 69, Chelyabinsk, Chelyabinsk Oblast, 454080 Russian Federation
[4] Tyumen Higher Military Engineering Command School named Marshal Engineering Troops A.I. Proshlyakova, Ministry of Defense of the Russian Federation, L. Tolstoy str., 1, 625001, Tyumen, Russia

Abstract. The article is devoted to consideration of a practice-oriented approach implementation in a higher educational institution. The essence of this approach and possibility of its implementation in bachelors' training is revealed. If the selected methods are introduced into practice students' training will be more successful. Method of specific situations (case-study); educational games; project method that fully meets the requirements of the competence approach are under consideration. As a result of the analysis of bachelors' final qualifying works inNizhny Novgorod State Pedagogical University named after Kozma Minin (major «Vocational training in industry)» for 2016-2017, the authors showed their compliance with the requirements of the Federal State Educational Standard. It is very significant for students who are going to work in various fields of industries. This justifies the need to use practice-oriented approach in training, which helps to teach graduates to be ready for professional activity.

1 Introduction

With the changing needs of the modern labor market, the state and society as a whole, there has been a change in the educational paradigm, leading to global transformations in the field of higher education. Competence-based approach, designed to train a mobile, creative, competent graduate who is able to solve emerging problems in a non-standard way has become an integral part of the learning process. Professionalism of future bachelors, in accordance with the requirements of the new Federal State Educational Standard, should be formed in conditions as close as possible to the future field of activity, that is, there should be enough practice in the students' life to improve their skills and more quickly adapt to the real work process in the future. Therefore, there is a need to use a practice-oriented approach. The use of practice-oriented methods in the classroom is becoming increasingly relevant. The main task of the teacher is not to transfer a certain amount of knowledge and skills, but to prepare the student in such a way that he can put into practice the obtained theoretical knowledge and be ready for certain types of activities.

2 Methodology

The framework of competence-based approach forms the need for student autonomy. In teaching pedagogical disciplines, it is especially important that future university graduate feels confident in his actions and is ready for the successful implementation of certain types of activities. And in order for the student to maximize their capabilities, it is necessary to build his practical skills using practice-oriented approach. The practice-oriented approach is an approach aimed at bringing the activities of the educational organization closer to the needs of real life, allowing you to create conditions for the purposeful training of students to be the competitive. According to P. Obraztsov and T. Dmitriyenko, the introduction of a practice-oriented approach should contribute to the formation of students' personal qualities that are important for their future professional activity, as well as knowledge and skills that will ensure that they perform their professional duties successfully. These authors consider it necessary to update the content of education by enhancing its practical orientation. However it is impossible to achievewithout studying theoretical basics [4]. The development of this issue can be found in works of V.S. Bezrukova, A.A. Kirsanov, N.E. Kuznetsova, O.S. Zaitseva, E.N. Ilina, E.I. Meshcheryakova.

We have presented an analysis of the graduation qualification thesis of bachelors in 2016-2017. We divided them into 2 groups.Sample size - more than 100 works. The study allowed determining the compliance of these final works with the requirements of the Federal State Educational Standard on their practical orientation. Among the evaluation criteria were also highlighted the professional competencies and the development of specific activities and proposals for a certain problem.

Among practice-oriented methods successfully implemented in higher educational institutions, we have identified method of specific situations (case-study); educational games and project method.

3 Results and discussions

R.A. Azhimullaeva speaks about practical focus of education as one of the fundamental didactic principles. According to her, the teacher pays great attention to the connection between theory and practice and brings the learning environment to future professional conditions [1].

E.I. Meshcheryakova believes that the use of a practice-oriented approach is a sequence of targeted pedagogical procedures, operations and techniques that together constitute an integrated didactic system, realizing which the students have professionally significant qualities and competencies necessary for successful implementation of professional activities [2].

It is important to note some signs of practice-oriented learning:

- use of problem-based learning methods, project method;
- organization of training in a group, in a team [3];
- integration of academic courses in such a way as to create a holistic view of the future professional field of activity among students [4].

From this it follows that the essence and the main idea of practice-oriented learning is to create such conditions for future bachelors in which they can get work experience, acquire necessary competences and become prepared for real professional activity, having skills of productive thinking, interaction with each other, critical thinking, solving complex problems based on the analysis of production situations.

As the leading pedagogical principles we selected the following:

- focus on practical training;
- taking into account the needs of the labor market;
- consideration of the requirements of the competence approach [5];
- focus on the competitiveness of the future bachelor [6].

As noted by G.A. Tyryginoy, the development of the educational program in this case should take place not only in the audience, but also in real professional conditions.

The possibilities of practice-oriented approach in the teaching of disciplines in higher educational institutions are revealed with the help of special methods. We have identified the following:

- the method of specific situations (case-study);
- didactic games;
- project method [7].

Building a practice-oriented learning process based on data from the methods will allow the future bachelors to come as close as possible to their profession, and will enable them to build a holistic learning process [8].

Content for students using the case method is presented in the form of small problems, and students acquire knowledge and necessary skills in the process of active work to develop solutions. The distinctive features of the case method are:

- the teacher's creation of a problem situation based on facts from real life [9];
- the ability to find an alternative solution [10];
- the possibility of group evaluation of decisions;
- organization of an individual approach to each student, taking into account his needs [11];
- providing the learner with freedom in learning;
- the formation of the students' skills in working with information [12].

The use of cases allows students to develop skills in analysis and critical thinking, to combine theory and practice, to articulate their point of view clearly [13]. In addition, the case method fully meets the requirements of the competence-based approach of student self-development of knowledge.

Students gain the skill of self-presentation in the process of presenting their answer to the task, learn to formulate questions and answer arguments [14].

Business game (development of real future professional activity through game imitation, recreation of basic types of behavior). During the game, students also perform activities that are close to professional ones. Students analyze them and find alternative solutions [15]. So students gain experience in finding effective ways to solve problems or out of difficult situations. As a result, the assimilation of social and professional experience is activated and the conditions are created for the actualization and transformation of knowledge and skills acquired in the process of learning that are associated with future professional activities [16].

Students can prepare essays, reports, after which a discussion is organized. Here the project method can be used, which successfully helps to implement a practice-oriented approach [17].

The theme of the project for a student should be close for the better development of his general and professional skills, creative thinking [18]. At the stage of developing and organizing a project implementation plan, students, with the help of a teacher, set a goal, draw up an action plan, determine sources of information and distribute functions among working groups, choose the form of project presentation [19].

At the project implementation stage, students select and structure the material in accordance with the plan, prepare for the presentation [20].

The project presentation stage is characterized by the presentation of the work carried out within the project. Students learn to build their ideas logically, to present them in a short form, to speak in public [21].

Next is the stage at which the evaluation and analysis of results is important. Discussing the results, the disadvantages and advantages of the project work are revealed [22]. The teacher sums up, summarizes the results and gives final assessment of the project [23].

Speaking about practice-oriented approach at the university, it is necessary to note such an obligatory element as final qualification works, which, according to the basic requirement of the Federal State Educational Standard, should have practical orientation.

We have presented the analysis of the final qualification works of bachelors in 2016-2017. We divided them into 2 groups. Criteria for evaluating the work was practical orientation to a specific problem existing in an organization or industry, the development of specific measures and proposals to solve the problem, possession of professional competencies. The list of current problems studied by students in the framework of final qualifying works is quite wide and varied.

In the students' works of the second group, attention was paid to the marketing analysis of specific enterprises, which includes:

- study of the demand for products, markets;
- assessment of the competitiveness of products and finding ways to improve its level;
- assessment of production efficiency.

Students were offered recommendations to improve the performance of enterprises:

- the organizational structure of the marketing service has been developed. The first group of students developed a provision "On the service of marketing", as well as job descriptions for the head of the marketing service and specialists on the basis of current legislation. Also, students were engaged in identifying the shortcomings of legislation in various fields and offered options for their completion and improvement of laws.

Student developments and proposals help companies and enterprises to develop, and this means that the activities of students have real practical benefits.

As a result, 91% of the students' work respond the above criteria, since the defence of the graduation qualification thesis passed on the assessment of "good" and "excellent". At the same time 87% of the students' work of the first group and 63% of the students' work of the second group received practical implementation. As a result of project activities, students develop the necessary practical skills, working with various sources of information, independently choosing the content relevant to them, compile a presentation, logically build their speech. Thanks to this method, it is easier for students to master these or other competencies.

4 Conclusion

The use of practice-oriented technologies is very important for students to acquire demanded competences because students\ adaptation to real working conditions depends on practice. The result of using practice-oriented approach in training is a bachelor who is able to effectively apply his practical competences in practice. The essence of the practice-oriented approach is to approximate the conditions of an educational institution to professional ones in order to shape the competitiveness of future professionals through gaining work experience and acquiring the necessary competencies. Practice-oriented approach successfully complements competence-based approach, thereby creating an environment for the formation of independent, mobile, productive students.

The study proves the fact that graduates have a sufficiently high proficiency not only in theoretical material, but also in practical skills for solving professional problems, as well as in-depth and comprehensive knowledge of the training profile.

Practice-oriented approach reflects basic processes of the modernization of education, where the basis is training of competent bachelors able to quickly adapt to the conditions of specific practical activities.

References

[1] Fedorov A A, Paputkova G A, Ilaltdinova E Y, Filchenkova I F, Solovev M Y Model for employer-sponsored education of teachers: Opportunities and challenges (2017) Man in India 97 (11) 101-114

[2] Garina E P, Kuznetsov V P, Egorova A O, Romanovskaya E V, Garin A P 2017 Practice in the application of the production system tools at the enterprise during mastering of new products *Contributions to Economics* **9783319606958** 105-112

[3] Garina E P, Kuznetsov V P, Romanovskaya E V, Andryashina N S, Efremova A D 2018 Research and generalization of design practice of industrial product development (by the example of domestic automotive industry) *Quality* **(S2)** 135-140

[4] Iltaldinova E Yu, Filchenkova I F, Frolova S V 2017 Peculiarities of the organization of postgraduate support of graduates of the targeted training program in the context of supporting the life cycle of the teacher's profession *Vestnik of Minin University* **3 (20)** 2

[5] Ilyashenko L K, Prokhorova M P, Vaganova O I, Smirnova Z V, Aleshugina E A 2018 Managerial preparation of engineers with eyes of students *International Journal of Mechanical Engineering and Technology* **9 (4)** 1080-1087

[6] Ilyashenko L K, Smirnova Z V, Vaganova O I, Prokhorova M P, Abramova N S 2018 The role of network interaction in the professional training of future engineers *International Journal of Mechanical Engineering and Technology* **9 (4)** 1097-1105

[7] Ilyashenko L K, Vaganova O I, Smirnova Z V, Gruzdeva M L, Chanchina A V 2018 Structure

and content of the electronic school-methodical complex on the discipline "mechanics of soils, foundations and foundations" *International Journal of Mechanical Engineering and Technology* **9 (4)** 1088-1096

[8] Kutepov M M, Vaganova, O I, Trutanova A V 2017 Possibilities of health-saving technologies in the formation of a healthy lifestyle *Baltic Humanitarian Journal* **6(3)** 210-213

[9] Kuznetsov V P, Romanovskaya E V, Egorova A O, Andryashina N S, Kozlova E P 2018 Approaches to developing a new product in the car building industry *Advances in Intelligent Systems and Computing* **622** 494-501

[10] L K Ilyashenko 2018 Pedagogical Conditions of Formation of Communicative Competence of Future Engineers in the Process of Studying Humanitarian Disciplines *International Journal of Civil Engineering and Technology* **9(3)** 607-616

[11] Markova S M, Sedykh E P, Tsyplakova S A, Polunin V Y 2017 Perspective Trends of Development of Professional Pedagogics as a Science *In International conference on Humans as an Object of Study by Modern Science* 129-135

[12] Smirnova Zh V, Krasikova O G 2018 Modern tools and technologies for assessing learning outcomes *Vestnik of Minin University* Vol 6 **3** 9

[13] Potashnik Y S, Garina E P, Romanovskaya E V, Garin A P, Tsymbalov S D 2018 Determining the value of own investment capital of industrial enterprises *Advances in Intelligent Systems and Computing* **622** 170-178

[14] Smirnova Zh V, Mukhina M V, Kutepova L I, Kutepov M M, Vaganova O I 2018 Organization of the research activities of service majors trainees *Advances in Intelligent Systems and Computing* **622** 187-193

[15] Smirnova ZH V, Gruzdeva M L, Krasikova O G 2017 Open electronic courses in the educational activities of the university *Vestnik of Minin University* **4(21)** 3

[16] Prokhorova M P, Semchenko A A 2018 Involving of trainees-future teachers of professional training in project activities in the discipline *Vestnik of Minin University* vol. 6 **2** 6

[17] Tsyplakova S A, Grishanova M N, Korovina E A, Somova N M 2016 Theoretical bases of designing of educational systems *Azimuth of Scientific Research: Pedagogy and Psychology* Vol. 5 **1 (14)** 131-133

[18] Vaganova O I, Koldina M I, Trutanova A V 2017 Development content of professional pedagogical education in the conditions of realization of competence approach *Baltic Humanitarian Journal* T 6 **2 (19)** 97-99

[19] Vaganova O I, Smirnova ZH V Trutanova A V 2017 Organization of research activities of bachelor of professional education in electronic form *Azimuth of Scientific Research: Pedagogy and Psychology* **6(3)** 239-241

[20] Yashin S N, Yashina N I, Ogorodova M V, Smirnova Z V, Kuznetsova S N, Paradeeva I N 2017 On the methodology for integrated assessment of insurance companies' financial status *Man in India* **97 (9)** 37-42

[21] Bicheva I B, Filatova O M 2017 Formation of the teacher-leader in the educational process of the university *Vestnik of Minin University* **3 (20)** 5

[22] Bogorodskaya O V, Golubeva O V, Gruzdeva M L, Tolsteneva A A, Smirnova Z V 2018 Experience of approbation and introduction of the model of management of students' Independent work in the university *Advances in Intelligent Systems and Computing* **622** 387-397

[23] Bulaeva M N, Vaganova O I, Koldina M I, Lapshova A V, Khizhnyi A V 2017 Preparation of Bachelors of Professional Training Using MOODLE *In International conference on Humans as an Object of Study by Modern Science* 406-411

IOP Conf. Series: Materials Science and Engineering **483** (2019) 012004 doi:10.1088/1757-899X/483/1/012004

Solid-state modeling and basic training of specialists in the field of mechanical engineering

V V Telegin[1], I V Telegin[2], A V Kirichek[3]

[1] Head of department, Lipetsk State Technical University, Lipetsk, 398055, Russia
[2] Associate Professor, Lipetsk State Technical University, Lipetsk, 398055, Russia
[3] Professor, Bryansk State Technical University, Bryansk, 241035, Russia

E-mail: vv.telegin@yandex.ru, igor.v.telegin@gmail.com, avk.57@ya.ru

Abstract. One of the main elements of basic training for modern machine-building enterprises is modern 3D-modelling technology. Currently, it is not only the construction of solid-state models of mechanical systems, but also their support – accumulation and analysis of data for the working efficiency of a product during the entire period of its operation. The article is devoted to forming of the initial level of knowledge, experience and practical skills necessary to work with models designed to manage the products life cycle. Different methods of teaching 3D modeling in Autodesk software environment are also considered in the article. Attention is focused on the interrelation between these methods and traditional methods of construction.

Introduction

The design process of mechanical engineering products is currently based on creating their 3D models [1, 2]. These models are used to perform calculations, simulate the normal operation of the product, and analyze various options for its improvement [3, 4]. On the basis of 3D models, technologies for manufacturing parts [5, 6] (or other components) of a product and assembly operations are being developed.

In recent years, there appear more and more widespread technologies, which are based on the process for development an information model in parallel with the real product (a 3D-model is its component) to manage the life cycle of this product (PLM – Product Lifecycle Management) [7].

A well–known analogue is BIM-technology in construction (Building Information Model – an information model of a building) [8]. The essence of PLM-technology is creating, filling and changing the structure of information data that determine the condition of the product throughout its life cycle, from reasoning the development to its liquidation (GOST P 53791-2010).

Modern and traditional design methods

At first glance it would seem obvious that traditional design, which is based on the methods of descriptive geometry and engineering graphics, as well as theoretical mechanics, material resistance, theory of mechanisms and machines, etc. is in the past [9]. But this is by no means always the case [10]. Computer technology can dramatically improve design performance and accuracy, analyze and find a selection of solutions that meet the specified criteria. However, it only remains for the engineer to determine these criteria and formulate a design problem. Its solution depends on the engineer's competence which is formed, among other things during the process of studying traditional design methods [11].

New technologies, including PLM technologies, do not always simply replace the obsolete ones, rather they are based on the technologies that existed before. The educational process in a higher educational institution is limited in time. If we talk about the discipline "Engineering graphics", which is discussed in this article, we must not use the principle "first master, what was before, and then study 3D-modeling" [10]. In that event this principle is wrong not only because the development of the discipline will take more time, but because the approach itself is in error. According to the authors, outdated tools should be replaced with the new ones, but the volume and the content of the theory is worth retention if possible. It is desired not just to save, but to remake it in terms of 3D modeling tools, FSES (Federal State Educational Standard) requirements and professional standards for machine-building enterprises, specialized software products, i.e., of such companies as Autodesk (USA), SolidWorks Corporation (an independent division of Dassault Systemes – France), Ascon (Russia) and others. The content of the revised course "Engineering graphics", obviously, will be different from the traditional one [12], but it will preserve its basic theoretical foundations [13].

The content and methods of teaching the basic course of engineering graphics

A student who has mastered the course of engineering graphics must:

- have theoretical knowledge to work with three-dimensional objects;
- be able to use them to create the design documentation of engineering products;
- have practical skills of working with 2D-graphics and 3D-modeling in modern software products.

In this statement, the problem of shaping the course "Engineering graphics" is difficult and time-consuming. The Department of engineering graphics of LSTU has been working in this field for many years. The first results were obtained, tested and published in 2013 – 2014.

The structure of the discipline "Engineering graphics", currently studied by first-year technical students of the Institute of mechanical engineering of LSTU is the following:

1st term: 1 hour of lecture and 2 hours of practice, consultation, unsupervised work. Only 3-4 credits, depending on the training line.

2nd term: 1 hour of lecture and 2 hours of practice, consultation, unsupervised work. As in the first term 3-4 credits depending on the training line. It is planned to replace an hour of lectures with an additional hour of practical training.

The software is Autodesk AutoCAD and Inventor Professional. The choice of Autodesk software products is due to the following reasons:

- recognized leadership in the Russian market, along with the companies Ascon and SolidWorks Corporation, popularity and widespread distribution;
- similarity of interfaces and principles of operation to the software of other companies;
- a high level of capabilities of Autodesk Inventor Professional (and SolidWorks) not only in the field of 3D-modeling and development of design documentation in accordance with the USDD (Unified System of Design Documentation), but also in the field of complex calculations (strength, fatigue, and others), both on the basis of traditional methods (the disciplines "Material resistance", "Theory of mechanisms and machines", "Machine parts") and finite element methods;
- extensive import and export capabilities (3D models created in Autodesk Inventor are exported to SolidWorks and vice versa virtually with no loss);
- the ability provided by Autodesk to install licensed free software not only in computer classrooms of educational institutions, but (***and most importantly***) on students' home computers.

Educational process control during practical classes and lectures is carried out by means of LiteManager program [14]. The following features of the program can improve the quality and efficiency of the educational process:

- demonstration of educational material and methods of working with software (AutoCAD, Inventor) from the teacher's computer to the computer of each student;
- real-time tracking of each student's activity;
- intervention in the student's work with software (AutoCAD, Inventor), e.g. in the process of 2D, 3D-modeling or making drawings.

Methodological materials required to perform graphic works in Autodesk AutoCAD and Inventor Professional software products can be found in the cloud and are used during practical classes and when working independently on home computers. The Internet provides access to the necessary literature, primarily to the texts of state standards.

The theoretical part of the discipline "Engineering graphics" includes the study of the basics of descriptive geometry (1st term) and engineering drawing (2nd term).

There are three individual graphic works on the topics of positional and metric problems for points, lines and planes, faceted surfaces and surfaces of revolution in the 1st term. Each of the works is performed in AutoCAD environment using both methods of descriptive geometry and 3D modeling. In addition, the first 8 – 10 out of 36 hours of practical training provided by the curriculum in this term, are assigned for the acquisition of practical skills in the space of the AutoCAD model. It is extremely important that a student, who was trained in the 1st term, could perceive the surrounding objects as a result of operations of unification, subtraction and intersection of their surfaces (bodies). The student must also know the methods of performing these operations.

The 2nd term is entirely devoted to the theory and practice of making design documentation in accordance with the standards of USDD for engineering products in Autodesk Inventor Professional. This term involves the following graphic works: a 3D model and drawing for a part (template "Part"); a 3D model and a drawing of bolted joints (templates "Assembly" and "Schema"); reading of an assembly drawing or building a 3D model of a product according to its assembly drawing; designing a reducer shaft. The first two graphic works are the foundations of the theory and practice of drawing. The third and the fourth works improve the knowledge and practical skills. One of the most important results of training must be the awareness that while making a 3D-model of the product, i.e. its drawing, the student makes this product using a computer and the appropriate software.

The control of students' mastering the material studied. It is assumed that all graphic works performed by students, for the most part, are carried out during practical classes. Using LiteManager allows you to save and analyze, if necessary, the results of the student's work at each lesson. The system of online testing (https://i-exam.ru/) is widely used to track how well the students mastered the theoretical material. During the term, students are offered about five tests on the topics under study in the system of Internet simulators and on the home computer as part of independent work. The final exam or test includes testing, usually in the system of FEPE (Federal Exam in Professional Education).

Figure 1 shows a 3D model of a shaft made as a part of the final graphic work.

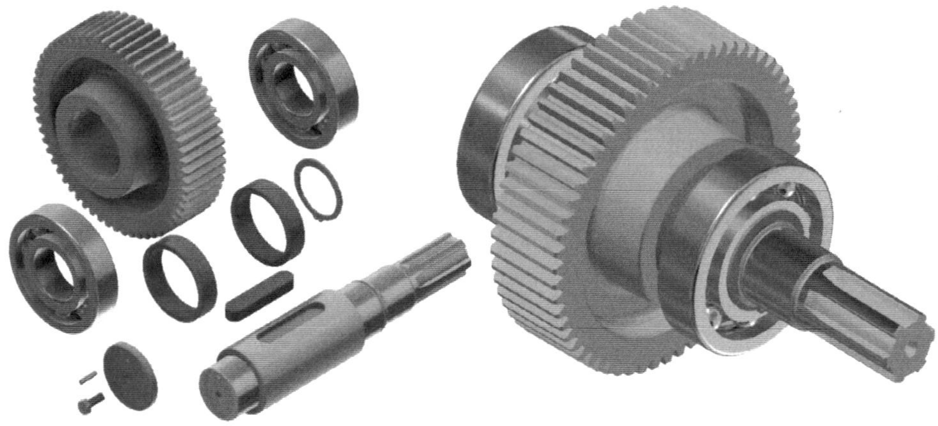

Figure 1. A design task

Initial design data include standard parts of the construction (bearings, a key, a washer, a screw, a pin, a ring), the availability and location of the shaft elements (chamfers, fillets, grooves for the grinding wheel output, center holes), the geometry parameters of standard shaft elements (slots), the gear parameters (a module, the number of teeth, the angle of teeth, the profile angle, the displacement coefficient, the facet width and the hub length). The values of the missing sizes are set by the student independently.

In the context of this work a student develops a 3D model of an assembly (figure 1), its drawing (not given in the article) and a drawing of a shaft. A fragment of this drawing is shown in figure 2.

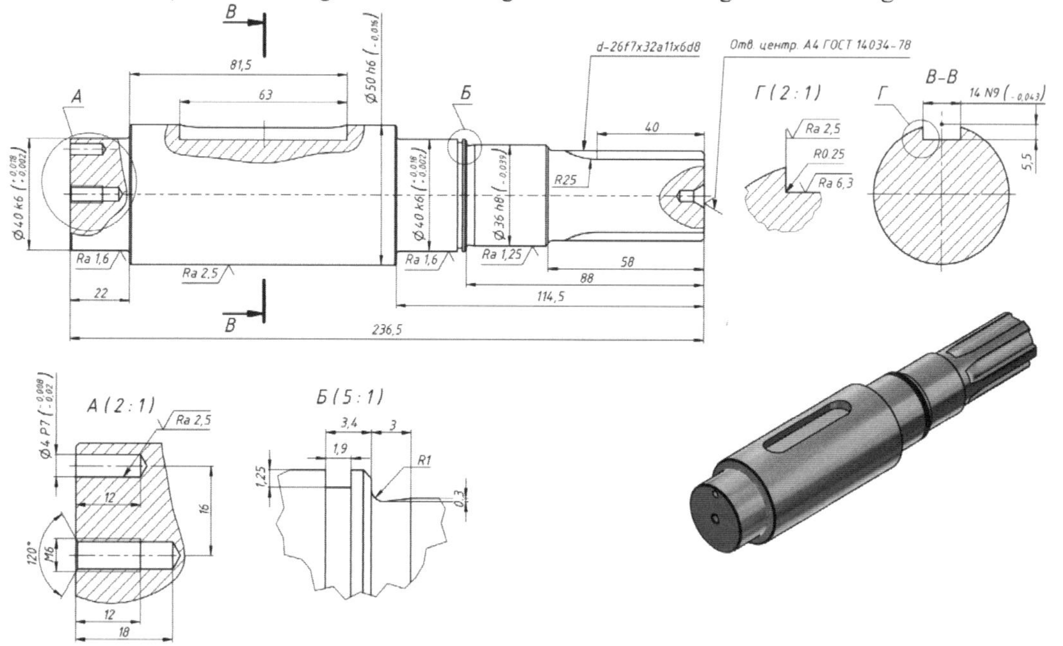

Figure 2. A fragment of a shaft drawing

Conclusion

There is an obvious need to establish a balance between the methods of 3D-modeling and the traditional methods based on the propositions of descriptive geometry, drawing and design which were developed in the last century. The authors believe that the theoretical basis for the training of engineers in the field of mechanical engineering must contain the experience gained by the previous generations of designers, technologists and other experts. The practical component, of course, is the modern methods of design.

When assessing the content of the course of engineering graphics presented in this article, it should be kept in mind that this is only the beginning of training in the field of design. In future the students will have such disciplines as "Machine Parts", "Fundamentals of Interchangeability" and others.

References

[1] Telegin V V, Kozlov A M, Sakalo V I 2017 Solid Modeling and Dynamic Analysis of the Mechanisms of Press-forging Machines *Procedia Engineering* **206** 1258-1263

[2] Heifets A L, Loginovsky A N, Butorina I V, Vasilyev V N 2012 *Engineering 3D-computer graphics: Textbook for bachelors. 2nd ed.* (Moscow: Yurayt Publishing House) p 464

[3] Telegin V and Kozlov A 2016 Computer realization of research into the dynamics of mechanical systems *IOP Conf. Series: Materials Science and Engineering* **124** (2016) 012101

[4] Telegin I V and Telegin V V (2018) Mathematical model of the part shape and automation on its basis of the process of developing the drawing parameters of the axisymmetric forging *The Turkish Online Journal of Design, Art and Communication* TOJDAC ISSN: 2146-5193 458-464

[5] Telegin I V, Volodin I M, Zolotukhin P I 2018 The mathematical modeling for assessing the effectiveness of hot forging extruded round in plan forgings on crank presses *International Journal of Engineering & Technology* **7** (2.2) 30-34

IOP Conf. Series: Materials Science and Engineering **483** (2019) 012004 doi:10.1088/1757-899X/483/1/012004

[6] Kozlov A M, Telegin V V, Kozlov A A 2018 The influence of roll working surface topography on wear resistance and quality of cold rolled grain-oriented steel *Journal of Chemical Technology and Metallurgy* **53** 5 943-949
[7] Product Lifecycle Management 2010 *TAdvise.State. Buisness. IT*
[8] Grakhov V P, Mokhnachev S A, Ishtryakov A H 2015 The Development of BIM-design as an element of competitiveness *Modern problems of science and education* **1** (1)
[9] Seregin V I, Senchenkova L S, Surkova N G, Beloborodova T L 2017 Electronic geometric modeling in the structure of the course "Engineering graphics" *International journal* Part 1 **08** (62) 88-93
[10] Polubinskaya L G 2017 Computer graphics in the course "Engineering graphics" *VII International Internet Conference "Quality problems of graphic training of students at a technical university: traditions and innovations" KGP-2017*
[11] Telegin V V, Telegin I V, Stepanov A S 2018 3D-Forming and Autodesk Autocad at the Initial Stage of Engineering Training of Specialists in Technical Areas *International Journal of Engineering and Technology (UAE)* **7** (3.5) 1-3
[12] Korolev Yu I, Ustyuzhanina S Yu 2011 *Engineering graphics: Textbook for universities. Standard of the third generation* (St. Petersburg.: Peter) p 464
[13] Gordon V O, Sementsov - Ogievsky M A 1988 *A course of descriptive geometry: Tutorial. 23rd ed, Rev.* ed Yu B Ivanova (Moscow: Nauka) p 272
[14] Yakhnovets D A 2011 Certificate of state registration of computer program №2011618072. Litemanager. October 14th, 2011

Organizational specificity of batch production enterprises of processing industry

H-Ch Brauweiler, V V Kurchenkov, L V Ponomareva

Western Saxon University of Applied Sciences, Scheffelstr 39, R. 6.304, D-08066, Zwickau, Germany.
Volgograd State University, pr.Universitetsky, 100, 400062, Volgograd, Russia

E-mail: kurchenkov@volsu.ru

Abstract. This article deals with the approaches to the management of batch production in enterprises of processing industries. The advantages of batch production in comparison with piece production and continuous production are proved in the text. The authors show the mechanism of parallel or consecutive production line management. The comparison of key parameters of batch production in foreign and Russian enterprises is carried out in the article. The authors describe similarity and difference of batch production at the enterprises which are characterized by the discontinuity of production cycle with an exact calculation of temporal parameters of the stages of direct labor process. This comparison is made according to the methodology of similarity determination of the elements of production management in its various types and primarily in batch and piece production. Three variants of productive process modernization of the batch production are analyzed in the text of the article. Their integrating efficiency is assessed using such parameters as annual production volume of output, production cycle continuity, level of prime equipment load.

1. Introduction

The improvement of production systems under the conditions of dynamically developing economy and increase of their flexibility and high market adaptability become an important issue nowadays. Various approaches to modernization of production systems today vary significantly and consist of new managerial techniques that improve management of separate operating processes by means of the elimination of “extra” costs of production and by means of the production range expansion at industrial enterprises [1]. Furthermore, elimination of risks and introduction of risk management are an additional important factor and a part of aforementioned new managerial techniques [2].

Modernization of models of continuous and batch production acquires considerable significance under the existing conditions. In this respect it is necessary to analyze the possibilities of the use of batch production techniques at the enterprises of process industries. It is indispensable to assess their efficiency with respect to the strategic goals of enterprises. These are: introduction of new technologies, growth of the flexibility of processes for the satisfaction of quickly changing demand, diversification in the limited period of time. Simultaneously, introduction of new forms of production management is to be correlated with the expenses for their development and the integration within already working production lines.

At the modern stage of development high technologies require considerable investments, long cycle of introduction and on the contrary, the competition demands the fastest production of the wide range of goods [3]. Nowadays flexible production systems capable of changing over to new goods manufacturing in the short time with fractional modernization and readjustment of the equipment emerges full blown [4].

In this respect integration of manufacturing of various types of goods on a single technological line gains particular significance [5]. The use of batch production contributes to this issue to a great extent. Actually batch production is characterized by a relatively simple way of the innovations’ introduction into the operating process. And general-purpose equipment which is typical for batch production is the most flexible for readjustment as it accelerates new technology developments assimilation. At the same time production management can be carried out with fewer costs that has a positive impact on the terms of the return on investment. Meanwhile the batch production makes the system more flexible relative to the fluctuations of the market demand, and to the continuous production systems.

Depending on the technology used at a specific enterprise it is possible to use hybrid models of production organization which combine the advantages of continuous and batch methods [6]. Indeed as a rule, the processing is carried out with the use of continuous flow of materials and sippliers according to specified tasks. However if an enterprise expects to diversify the output and to minimize the investments into the installation of new equipment, it is possible to organize a discontinuous flow of batch process (multi-item changing flow line). In particular, such an approach can be implemented in enterprises specializing in the production of construction materials.

2. Results and discussion

The use of such models can be analyzed on the example of the enterprises producing the construction bricks. The major part of the brick making plants existing in the South of Russia have a continuous automated (completely or partially) production line manufacturing only one type of commodity. “Volzhskiy kirpichniy zavod” (Volzhskiy brick making plant) is one of such factories which can be taken as an example or a prototype. This model is economically viable by the decrease in the production cycle, reduction in expenses per unit of output and by other factors. But market circumstances require the growth of the range of output for the capture of new market segments [7]. For instance it is reasonable to branch out into a new type of production i.e. facing bricks together with the production of construction bricks or ceramic brick M150. At the classical model of industrial management it will require the construction of a new production line that in its turn will be combined with high investments [8] or shutdown or readjustment of the existing line that will make a simultaneous production of various types of production impossible.

The classical technological pattern of brick production is presented in Figure 1. The process includes the placement of raw material into the feeding unit, its dosage and directing into low and high roll mills for stone and metal separation. Then comes the flow into the mixing unit for making the clay sufficiently humid, pressing and cutting of clay bricks, its further drying and burning. When it is not possible to organize a continuous synchronized process of new commodity manufacturing, it is indispensable to introduce the batch method of production what will require a partial modification or modernization of the production technology line [9].

Such a modernization can be carried out in various variants. In the first case, in order to save the investments, production can be organized in such a way that batches are produced in a continuous way together with the readjustment of equipment for every product. Another variant implies mixing up of the batches of raw materials of various products and its incessant processing until a definite stage of process is achieved. After an intermediary storage (it requires modernization of the equipment), the process keeps up for various batches with the readjustment of the same equipment. The third variant implies the purchase of additional equipment for the simultaneous processing of batches of various products at different stages of production [10].

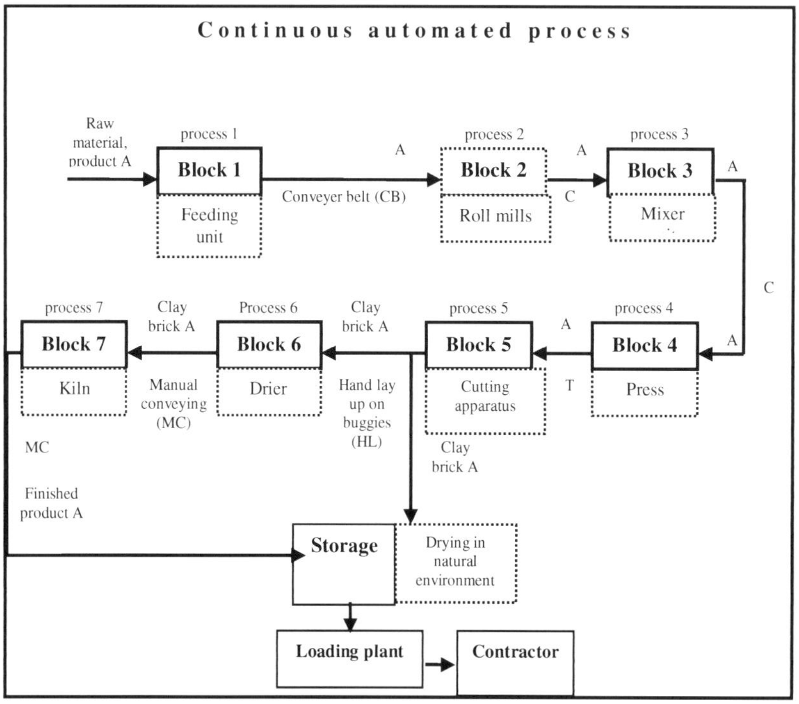

Figure 1. Flow chart of the continuous automated process of the production of construction brick at an enterprise.

Figure 2 presents the first variant of the modernization of the process which implies the consecutive (or in connection with the incoming orders) alternation of the batches of different products (ceramic and facing bricks) with the readjustment of every type of equipment (mixing unit, press, cutting apparatus, water pump). Batch B comes into processing, after that batch A goes partially into the storage where the drying of the clay brick in the natural environment takes place and partially it goes into the tunnel kiln for gas drying.

The main tasks of the production management at the given type of modernization come down to the determination of batch sizes for various products and the period of their alternation, calculation of the time of the start of the batch production, time of the technological cycle taking into account the period of readjustment of the equipment. The advantages of the method are minimal investments and diversification of output. Additional expenditures are connected with the increase in complexity of planning and labor costs.

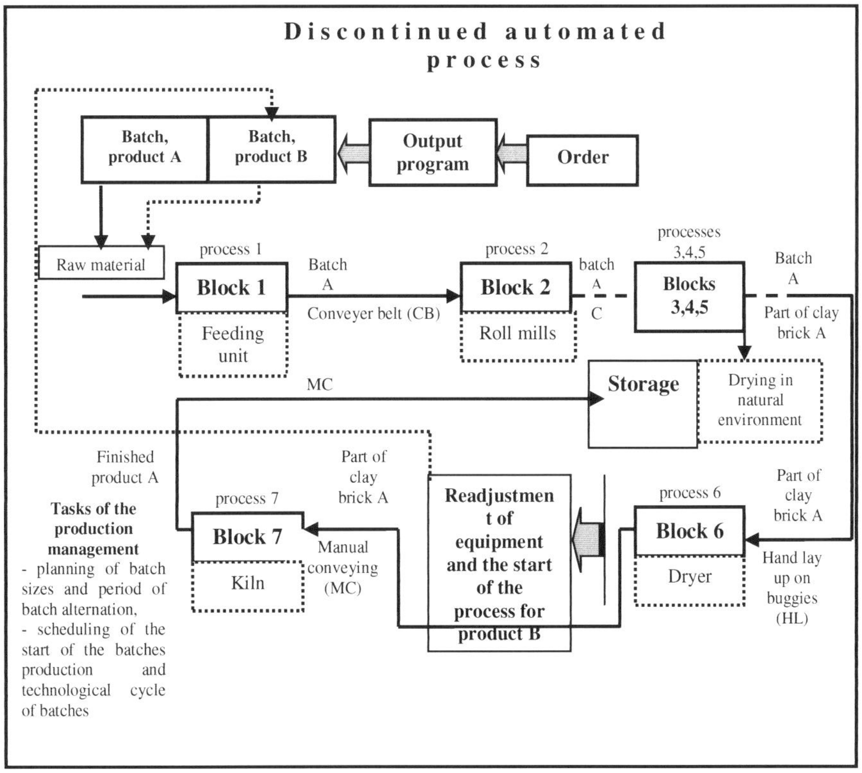

Figure 2. The first variant of the production modernization of two types of products with the use of batch methods**.**

In figure 3 the second variant of the modernization of the production process of the construction brick is shown.

The given variant implies the equipping of the clay mixer with an additional section for an intermediary reservation (and if necessary for an additional mixing up) of some parts of raw material. Before this stage the raw material for products A and B is mixed and goes into the processing simultaneously. After sending batch A of clay brick for drying and for readjustment of the equipment the processing of the batch B starts. The batch of clay bricks comes from the section of the intermediary reservation in the mixer. The process can be organized in smaller batches which reduces the storage costs per unit of output.

The main tasks of the production management are: determination (calculation) of optimal batch sizes and the period of technological cycle taking into account the time of raw material waiting and also coordination of the start of output of batches of various products at the stage of burning. The advantage of the method is the investment saving (modernization of one single installation and the decrease of the units of equipment for readjustment) with the opportunity of increasing the flexibility of the process. The disadvantages of this method are the increase of planning costs, lengthening of the technological cycle due to the waiting time of raw materials in the expectation of the release of the equipment (during processes 3 and 6).

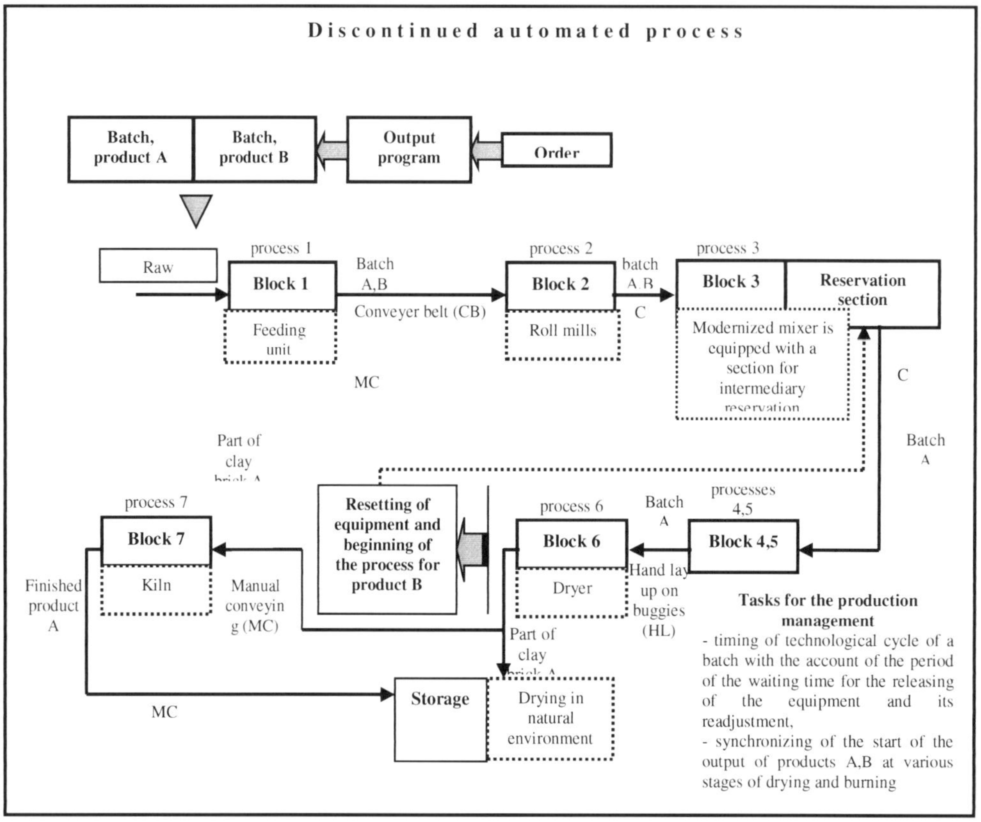

Figure 3. The second variant of modernization of production of two types of products with the use of batch methods.

Figure 4 shows the third variant of the production modernization of the brick manufacturing at an enterprise. It comes down to the purchase of additional equipment and specifically of the presses and cutting apparatuses for the organization of a synchronized parallel processing of various products. Raw material for products A and B is processed together and gets into the mixer equipped with the output channels into various devices (press and cutting apparatuses). Further the process is going in a parallel way until the burning stage at which the waiting time of the raw material of one of the batches in the expectation of the release and reset of the kiln takes place.

The mentioned method is the most investment intensive but at the same time allows organizing a parallel or simultaneous processing (at almost all stages) of two products at relatively low costs. The main tasks of the production organization are reduced to the determination of the optimal batch sizes, period of technological cycle of the batches' processing taking into account the time of waiting and readjustment of the equipment [11].

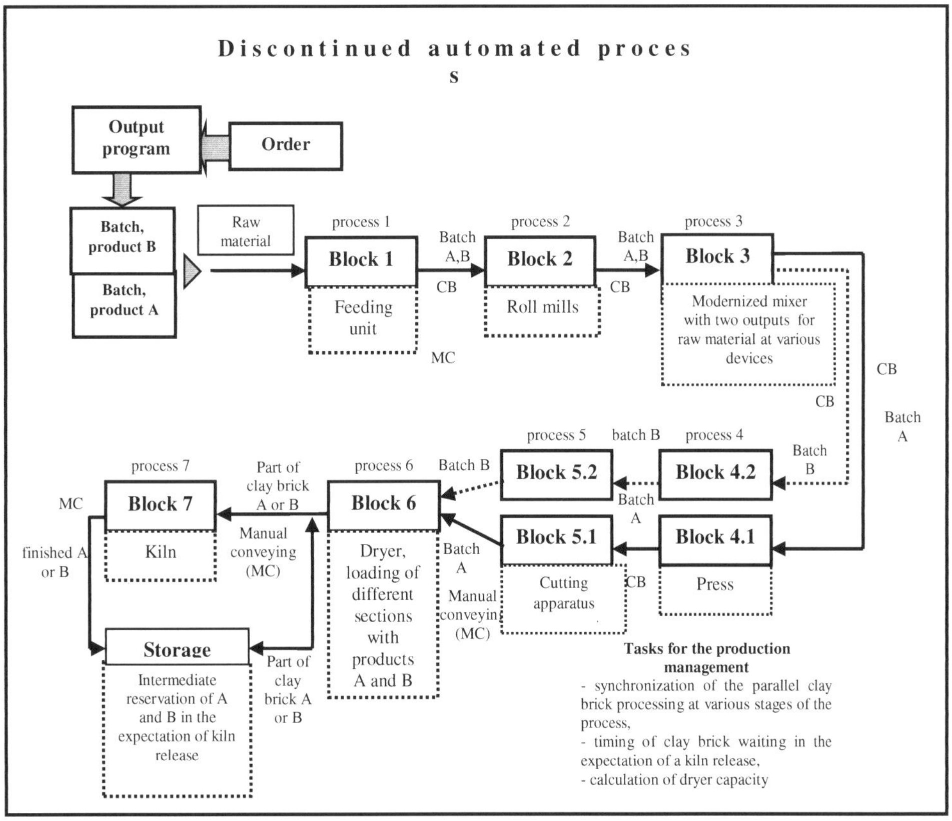

Figure 4. The third variant of modernization of the production of two types of product with the use of batch methods.

It is possible to assess the efficiency of modernization of the production system on the basis of the comparison of three analyzed types taking into account such criteria as flexibility of the process and output differentiation, expenditures on production management, length of the production cycle, level of the equipment load, costs per unit of output (Table 1).

Table 1. Assessment of the efficiency dynamics of the production system after the introduction of batch methods.

Criteria	**Indices of the efficiency of a production system**		
Variants of the modernization	**The first**	**The second**	**The third**
Flexibility of the process	Low	Middle	High
Differentiation of the output	Middle	Middle	Middle
Production management costs	Low	High	Middle
Production cycle length	High	Middle	Middle
Machine load	Middle	Middle	High
Costs per unit of output	Middle	Middle	Middle

The assessment of the suggested variants of modernization which is shown in the table is made on the basis of average quantitative indices chosen for the efficiency assessment and is presented as a qualitative assessment at threshold limit values.

3. Conclusions

On the basis of the data mentioned above, the conclusion can be made that the first variant of the modernization process provides the achievement of the strategic goals of an enterprise in the sphere of increase in output range with the minimum investment. The expenditures for the production management at this variant are the lowest. However economic efficiency of the process is low as the production cycle turns out to be the longest one.

The second variant of changes allows increasing of flexibility of the process at the expense of a partial modernization of devices i.e. relatively small capital investments. The mentioned type of changes significantly increases the complexity of the production management as it requires calendar scheduling for small batches of various products including the waiting time at two stages of the process.

The third variant is characterized by the highest capital intensity as it implies the purchase of additional equipment. At the same time it has the highest flexibility and the economic efficiency of the process allowing organizing a continuous flow of raw material for two products almost at all stages of production. On this basis this variant can become more preferable.

Thus it is worth mentioning again that the batch production is characterized by flexible technology, universal equipment and relative simplicity of the introduction of innovations into the process [12]. It meets the market requirements in the output of a variable, quickly changing production and it can be integrated into the existing process without significant investments. Meanwhile the parameters of the batch production process (general-purpose equipment, relatively small volume of production, discontinuity of the process and so on) influence the flexibility of the process and variety of the output, lengthen the production cycle and require additional expenditures for the industrial management, including the costs of the full load of the machinery of the enterprise.

References

[1] Glinnikov M 2004 Proizvodstvennoe planirovanie: podkhody, resheniya *Director* **6**
[2] Brauweiler H-Ch 2018 Risikomanagement in Unternehmen 2 Auflage Springer
[3] Brauweiler H-Ch 2015 Claimmanagement Springer
[4] Sinitsa L M 2004 Organizatsiya proizvodstva Minsk. UP "IVC Minfina" 426
[5] Singo S 2006 Study of the production system of Toyota from the point of view of the production management Moscow IKSI Publ 192-198
[6] Lobov F M 2003 Effective production management Phoenix Publ 265 p
[7] Kurchenkov V V, Fetisova O V, Orlova A A, Gladkaya E A 2017 The importance of the regions' economic activity evaluation in Russia 2012-2016 *Revista Galega de Economia* **26** **(3)** 33-44
[8] Mazur I I, Shapiro V D 2001 Restructuring of enterprises and companies Moscow Ekonomika Publ 86-90
[9] Brauweiler H-Ch 2002 Innovationen im peripheren Raum. Eine Analyse am Beispiel der Granzraume Sud- und Ostsachsens. Mit einem Geleit von Matthias Kramer. Wiesbaden: Dt.Univ Verlag 141 p.
[10] Ovshinov S A 2010 Development of the system of the management and planning of batch production at the enterprises of processing industry *Extended abstract of Cand. Sci. (Econ.) Dissertation* 24 p
[11] Kurchenkov V V Ovshinov S A 2011 Determination of an optimal variant of the production modernization at industrial enterprises *Vestnik Bulletin of the Volgograd Institute of Business. Business. Education. Law* **3(16)** 43 – 50
[12] Stefan A 2001 Schedule in batch production Hamburg. Deutscher Universitaet Verlag 504 p

IOP Conf. Series: Materials Science and Engineering **483** (2019) 012006 doi:10.1088/1757-899X/483/1/012006

The analysis of the impact of technological processes of hot forging on the dynamics of the crank press

I V Telegin[1], A M Kozlov[2], V I Sakalo[3]

[1]Associate Professor, Lipetsk State Technical University, Lipetsk, 398055, Russia
[2]Professor, Lipetsk State Technical University, Lipetsk, 398055, Russia
[3]Professor, Bryansk State Technical University, Bryansk, 241035, Russia

E-mail: igor.v.telegin@gmail.com, kam-48@yandex.ru

Abstract. The article deals with a dynamic model of a crank hot stamping press, which allows evaluating the effect of elastic torsional, bending deformations and tensile strains of compression of its elements, taking into account gaps in the kinematic pairs, depending on the technological process implemented on this press. Given that there are always several options to realize a technological process of hot volume forging, designed for a specific part, the proposed method allows you to choose the best option in terms of the working conditions of the press.

1. Introduction
In the foreseeable future hot volume forging (HVF) using crank hot stamping presses (CHSP) will remain one of the main manufacturing methods high performance forgings. The key element of their cost is metal and hardware investments. The classical method of making forgings of a given size is to tighten the flash bridge conditions to increase its width and to reduce the height. At the same time, the stamping force increases, and the operating conditions of the equipment deteriorate, especially due to the oscillatory (dynamic) processes in the system combining shafts, a connecting rod, a slide and a press frame [1, 2, 3].

The obvious means of improving reliability and durability of the press is to create new technological schemes of HVF which make it possible to minimize the dynamic processes in the CHSP by changing the technological loads graphs and their magnitude [5, 6].

2. A mathematical model
A mathematical model of the press is a system of ordinary, nonlinear, piecewise continuous differential equations of the second order, describing the motion of the mass system, interconnected by kinematic and elastic-dissipative bonds of a dynamic model (figure 1). In developing a dynamic model of a GHSP maximum loads were assumed to occur at the end of the working stroke of the slider. In this case the rotating component of the force acting from the slider to the main shaft of the press is minimal. During this time interval, the press elements (a frame, a connecting rod and a slider) are mainly under tensile-compression deformation, and the main shaft is working for bending. For the rest of the time, the main shaft loading circuit is torsion and bending.

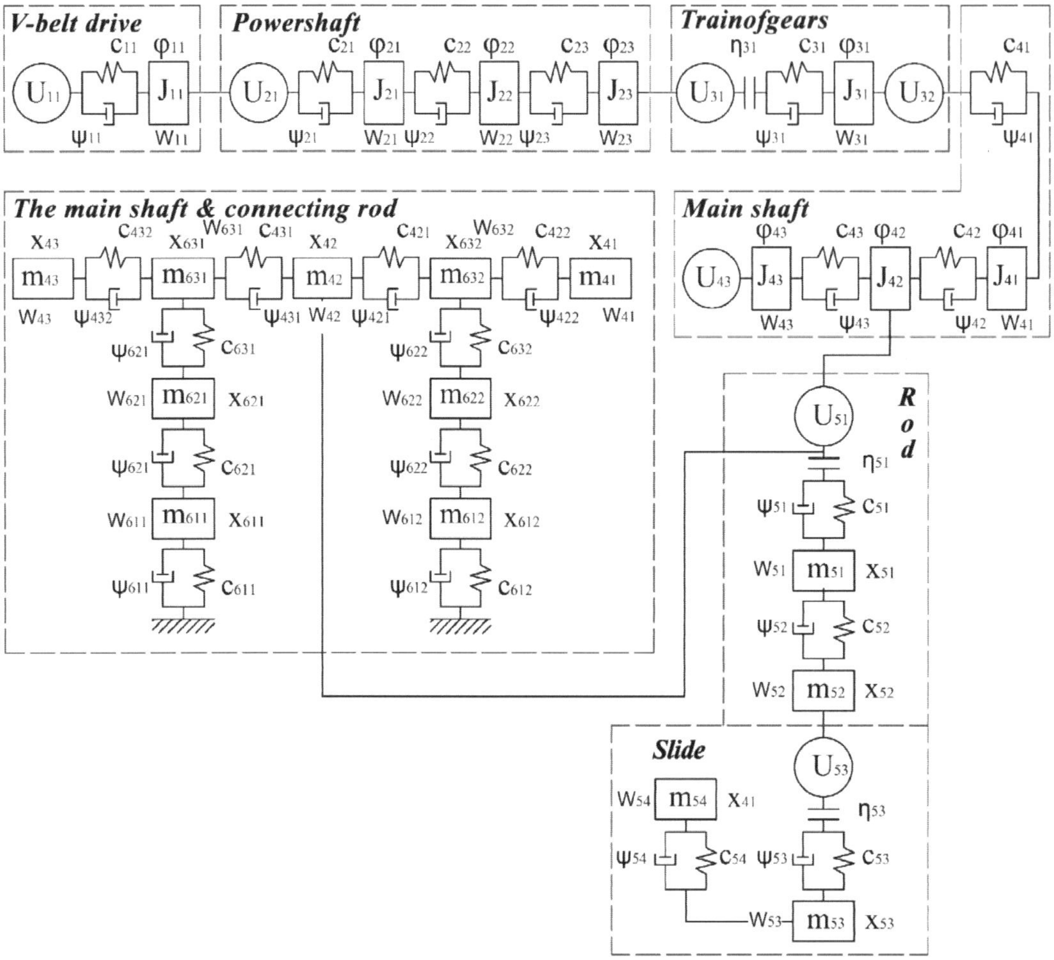

Figure 1. A dynamic model of a crank hot-stamping press.
J_{ij} – moment of inertia, m_{ij} – mass, C_{ij} – stiffness, ψ_{ij} – coefficient of dissipation, η_{ij} – gap,
U_{ij} – function of position, W_{ij} – external load of the mass j of the i-th mechanism.

The method of constructing a mathematical model is based on the application of the Lagrange equation of the second kind, and, for example, the equations of masses motion for a crank-slider mechanism (figure 1) can be written as:

$$\begin{cases} m_{51}\ddot{x}_{51} = -c_{51}(x_{51} - U_{51}) - b_{51}(\dot{x}_{51} - U'_{51}\cdot\dot{\varphi}_{42}) + \Omega_{56} \\ \qquad + c_{52}(x_{52} - x_{51}) - b_{52}(\dot{x}_{52} - \dot{x}_{51}) + W_{51} \\ m_{52}\ddot{x}_{52} = -c_{52}(x_{52} - x_{51}) - b_{52}(\dot{x}_{51} - \dot{x}_{51}) + \\ \qquad + c_{53}(x_{53} - U_{53}) - b_{53}(\dot{x}_{53} - U'_{53}\cdot\dot{x}_{52}) + W_{52} \\ m_{53}\ddot{x}_{53} = -c_{53}(x_{53} - U_{53}) - b_{53}(\dot{x}_{53} - U'_{53}\cdot\dot{x}_{52}) + \\ \qquad + c_{54}(x_{54} - x_{53}) - b_{54}(\dot{x}_{54} - \dot{x}_{53}) + W_{53} \\ m_{54}\ddot{x}_{54} = -c_{54}(x_{54} - x_{53}) - b_{54}(\dot{x}_{54} - \dot{x}_{53}) + W_{54} \end{cases}, \qquad (1)$$

$$\begin{aligned} \Omega_{45} &= c_{51}(x_{51} - U_{51}) + b_{51}(\dot{x}_{51} - U'_{51}\cdot\dot{\varphi}_{42}) \\ \Omega_{56} &= c_{51}(x_{42} - x_{51} + U_{51}) + b_{51}(\dot{x}_{42} - \dot{x}_{51} + U'_{51}\cdot\dot{\varphi}_{42}) \end{aligned}, \qquad (2)$$

$$
\begin{aligned}
&U_{51} = R \cdot (1 - \cos(\varphi_{42}) + L - \sqrt{L^2 - 2R \cdot \sin(\varphi_{42})}, \\
&U'_{51} = R \cdot \sin(\varphi_{42}) \cdot \left(1 + \frac{R}{\sqrt{L^2 - 2R \cdot \sin(\varphi_{42})}} \right), \\
&U_{53} = x_{52},\ U'_{53} = 1.
\end{aligned} \tag{3}
$$

In equations (1 – 3) $x_{5n}, \dot{x}_{5n}, \ddot{x}_{5n}$ are linear displacements, velocities and accelerations of masses of the crank mechanism (n = 1 ... 4); m_{5n}, c_{5n}, b_{5n} are masses and elastic-dissipative characteristics of the model; W_{5n} are external loads: friction forces, a balancer of the slider, a processing load; Ω_{45} is the impact of the connecting rod of the press on its main shaft; Ω_{56} is the impact of the system "frame – main shaft" on the connecting rod; R is the radius and L is the length of the connecting rod, respectively.

To assess the level of dynamic processes in a crank-slider mechanism of the press, we will use the criteria (4):

$$
K_s = \frac{P_{cp}}{Ps_{cp}},\quad K_m = \frac{P_{\max}}{Ps_{\max}}, \tag{4}
$$

where P_{cp}, P_{Scp}, P_{max} and P_{Smax} are an average dynamic, an average static, the maximum absolute value of the dynamic and static loads in the investigated element of the mechanism, calculated for the time interval $[T_1, T_2]$ respectively. The coefficient K_m determines a relative level of the maximum load in the investigated press mechanism and is an indicator of strength. The value of K_S depends on the amplitude, the frequency, the rate of vibration damping and is an indicator that determines wear and a fatigue failure. The values of an average and the maximum dynamic loads, for example, in the press rod can be found from the solution to its equations of motion according to the dependencies (5):

$$
P_{cp} = \frac{1}{T_2 - T_1} \int_{T_1}^{T_2} \left| c_{51}(x_{51} - U_{51}) \right| \cdot dt,\quad P_{\max} = \max \left| c_{51}(x_{51} - U_{51}) \right|_{T_1}^{T_2}. \tag{5}
$$

One of the modern trends of designing CHSP processes is the formation of pre-transition technological projections that on reducing the forging efforts allow a forging with minimal stamping radii to be made, and machining allowances to be reduced (figure 2).

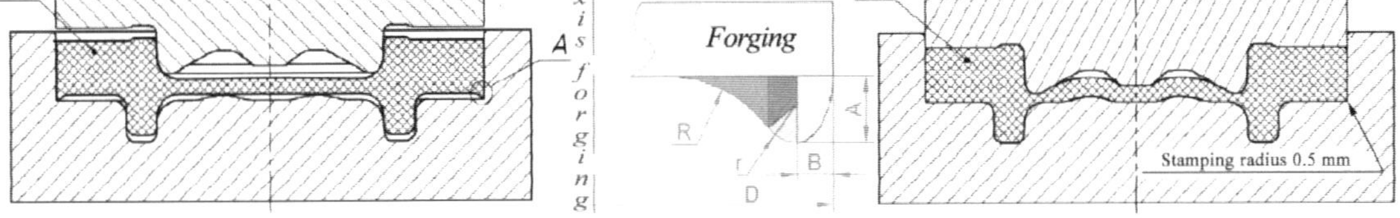

Figure 2. Stamping round forgings (the final transition):
1 – half-finished product, formed during the preliminary forming transition, 2 – finished forging.

The mathematical model (6, 7), setting the configuration of the projection (figure 2) at $y(x)$ is:

$$y=\begin{cases}-\dfrac{A}{B}\cdot\sqrt{\left|(x-0.5\cdot D+B)^2-B^2\right|}, & 0\le x-0.5\cdot D+B\le B\\ r-A-\sqrt{\left|(x-0.5\cdot D+B)^2-r^2\right|}, & -r\cdot\sin\alpha\le x-0.5\cdot D+B\le 0\\ -R+\sqrt{\left|(x+(r+R)\cdot\sin\alpha-0.5\cdot D+B)^2-R^2\right|}, & \begin{cases}x-0.5\cdot D+B\le -r\cdot\sin\alpha\\ x-0.5\cdot D+B\ge -(r+R)\cdot\sin\alpha\end{cases}\end{cases} \quad (6)$$

In equations (6) the angle in radians is $\alpha=\arccos\left(\dfrac{R-A+r}{r+R}\right)$.

$$\begin{cases}\overline{\alpha}=90-\dfrac{180\cdot\alpha}{\pi};\quad 20°\le\overline{\alpha}<90°\\ 0.6\le\dfrac{4\cdot A^2}{\pi\cdot A\cdot B+2\cdot(r^2-R^2)\cdot\alpha+4\cdot(A-r)\cdot(r+R)\cdot\sin\alpha+(r+R)^2\cdot\sin 2\alpha}\le 1.2\end{cases} \quad (7)$$

Expression (7) determines the conditions of defect-free stamping at the final transition of a semi-finished product with a projection obtained at the preliminary transition.

3. Results and discussion

In the context of the influence of geometric parameters of a tool for hot forging extruded round in plan forgings on the dynamics of a crank press, the following tasks were solved:

- The search for the parameter values A, B, R and r (figure 2), allowing to minimize the maximum static force P_{Smax} at the final transition when forming a stamping radius not exceeding 0.5 mm on the outer side of a forging.
- The calculation of the criteria for K_m, K_S, P_m, and P_s at the obtained values A, B, R, and r.
- The study of the impact of changes of the values A and R on the criteria K_m and K_S. Here, taking into account the results of problem №1, $B = 0.5 \cdot A$, $r = B$ are adopted.

To calculate dynamic characteristics of a press, special software [7] was used. This software is based on dynamic (figure 1) and mathematical (1) – (5) models, as well as the software of the company QuantorForm Ltd. – QForm 8. Its aim is to simulate the HVF process and calculate technological loads [3, 5, 8, 9].

The object of study is forging, manufactured on the CHSP K8542. The press capacity is 16 MN to fit the technological scheme presented in figure 2.

Problem №1 – the solution by the method of coordinate descent with limits (7). The result is P_{Smax} = 10.78 MN, P_{Scp} =2.18 MN at A = 3.0 mm; B = 1.5 mm; R = 3.5 mm; r = 1.5 mm.

Problem №2. The result is K_m = 1.05, K_S = 1.27, P_m = 11.32 MN and P_s = 2.78 MN.

Problem №3. The variation range of the parameters A and R is ≈ 10%. The results are outlined in table 1.

Table 1. Values of the criteria K_m/K_S at different values of the parameters A and R.

		Parameter A, mm						
		2.4	**2.6**	**2.8**	**3.0**	**3.2**	**3.4**	**3.6**
Parameter R, mm	**3.0**	1.06/1.27	1.06/1.28	1.05/1.26	1.06/1.26	1.07/1.27	1.06/1.27	1.04/1.27
	3.2	1.05/1.28	1.06/1.28	1.05/1.27	1.05/1.26	1.06/1.27	1.06/1.27	1.04/1.28
	3.4	1.05/1.28	1.04/1.29	1.04/1.28	1.06/1.26	1.05/1.26	1.06/1.28	1.06/1.29
	3.5	1.05/1.29	1.04/1.30	1.04/1.28	**1.05/1.27**	1.06/1.28	1.06/1.29	1.07/1.29
	3.6	1.05/1.29	1.05/1.30	1.04/1.28	1.05/1.28	1.06/1.28	1.05/1.29	1.07/1.29
	3.8	1.05/1.30	1.05/1.31	1.04/1.29	1.06/1.30	1.06/1.29	1.06/1.30	1.08/1.30
	4.0	1.04/1.30	1.06/1.31	1.04/1.29	1.06/1.30	1.05/1.29	1.06/1.31	1.08/1.32
	4.2	1.04/1.31	1.06/1.32	1.04/1.30	1.06/1.31	1.06/1.30	1.07/1.32	1.09/1.34

Table 1 shows different zones of values of the parameters *A* and *R*. Yellow ink (the right top corner) marks the appearance of stamping defects; brown ink (the bottom left corner) marks the stamping radius exceeding the established value (0.5 mm); green ink marks close to optimum values of varied parameters.

4. Conclusion

The article presents a mathematical model of plastic deformation of a forging made on a crank hot stamping press. It allows estimating the level of dynamic loads in the press elements, depending on geometric parameters of the tool used to form the forging. The efficiency of this model is checked by hot forging extruded round in plan forgings on crank presses. As a result close to optimum values of the tool geometric parameters were found.

Acknowledgments
This research was financially supported by the Ministry of Science and Education of Russia (project No. 11.9505.2017/8.9).

References

[1] Volodin I M 2006 *Modeling the processes of hot volume forging: Monograph* (Moscow: Mashinostroenie vol 1) p 253
[2] Semenov E I et al 1986 *Forging and stamping. Reference book in 4 volumes* (*Hot stamping* vol 2) ed E I Semenov (Moscow: Mashinostroenie) p 592
[3] Svistunov V A 2008 *Forging and stamping equipment. Crank presses: Tutorial* (Moscow: MSIU) p 698
[4] Vlasov V I, Borzykin A Ya, Bukin-Batyrov I K et al 1982 *Crank forging machines* ed V I Vlasov (Moscow: Mashinostroenie) p 424
[5] Telegin V and Kozlov A 2016 Computer realization of research into the dynamics of mechanical systems *IOP Conf. Series: Mater. Sci. Eng.* **124** (2016) 012101
[6] Volodin I M, Logunov G S, Martyugin V S et al 2014 Patent RF №2509620. A method of manufacturing hollow flange forgings No. 2011135292/02. Announced 23.08.2011. Publ. 27.02.13 Byul. No. 6 of the *Invention, utility models* 2014 Bulletin No 8
[7] Telegin V V, Kozlov A M, Sakalo V I 2017 Solid Modeling and Dynamic Analysis of the Mechanisms of Press-forging Machines *Procedia Engineering* **206** 1258–1263
[8] Telegin I V and Telegin V V (2018) Mathematical model of the part shape and automation on its basis of the process of developing the drawing parameters of the axisymmetric forging *The Turkish Online Journal of Design, Art and Communication* TOJDAC ISSN: 2146-5193 458–464
[9] Vakalov A A 2005 *The application of computer modelling in in the development of hot forming processes of blade blanks*

IOP Conf. Series: Materials Science and Engineering **483** (2019) 012007 doi:10.1088/1757-899X/483/1/012007

Creation of institutional conditions and mechanisms of entrepreneurship development in the region (using as the example the Volgograd region)

M M Guzev[1], M V Ledeneva[1], G G Egorov[1], L V Shamray-Kurbatova[2]

[1] Volzhskiy Humanitarian Institute, Volzhskiy Branch of the Volgograd State University, ul. 40 Let Pobedy, 11, Volzhskiy, Volgograd region, 404133, Russia
[2] Private Educational Institution of Higher Education «Volgograd Institute of Business», ul. Kachintsev, 63, Volgograd, 400010, Russia

e-mail: ledeneva.mv@vgi.volsu.ru

Abstract. The transfer of the Russian Federation to the market economy had a negative impact on the industry of the Volgograd region and determined the process of deindustrialization. For the reconstruction of the industry since 2013 to present day the institutional sphere of small and medium sized entrepreneurship (SME) development within the government program "Economic development and innovative economy" is being created. The program supposed the foundation of centers for entrepreneurial support, business incubators, microfinance centers, regional guarantee funds, engineering centers, prototyping centers, centers for export support and other elements of infrastructure in the regions of the Russian Federation. The article studies the process of the creation of institutional conditions and mechanisms of entrepreneurship development in the Volgograd region, the statistical data on the governmental support of small and medium sized enterprises from the point of view of the created elements of institutional environment. Despite significant efforts on the creation of institutional sphere, the tendency which is observed today is the escape of business from the Volgograd region and transfer to more successful regions of the Russian Federation. This fact is connected first of all with a considerable reduction of incomes and the purchasing ability of population. The authors analyzed indices of the development of small and medium sized enterprises in dynamics. The conclusion about negative trends in this sphere is made. The innovative development of a region is a driver of the industrial development needs special conditions. In particular they are industrial parks, cluster potential of the territory and also new mechanisms of innovations' encouragement i.e. innovative vouchers. For the moment the efforts are made in the implementation of the project in the chemical and pharamaceutical industrial cluster.

1. Introduction

An important role in the support of small and medium sized enterprises belongs to the government which creates the institutional environment for entrepreneurial development at the federal and regional levels. The international and national science has accumulated a number of researches carried out by academic scientists, social organizations of entrepreneurs, expert and consulting agencies which are aimed at the study of the measures that should be taken for entrepreneurial development.

However a constant optimization and improvement of forms and ways of governmental support of small and medium sized enterprises is needed. On of the mechanisms is bench marketing which allows using and adjusting the foreign practice and the regional experience of other regions. In this respect it is important to study the forms and mechanism of entrepreneurship encouragement, removal of administrative barriers, improvement of laws for business in the Volgograd region.

The creation of institutional conditions of the support of small and medium sized enterprises in the region is an important factor of its sustainable social and economic development. In Russia the programs of federal and regional level in this sphere are realized with effort as it is connected with the imperfectness of both laws and administrative management. The institutional prerequisites and measures of the entrepreneurship encouragement in the region are also determined by the condition of health care system, educational system, standard of living and quality of life of the population, focus on the social policy and other factors.

2. Results and discussion

At the present time serious measures of the support of small and medium sized enterprises are taken at the federal level of the Russian Federation. The federal law № 209-FZ dated 24.07.2007 "On development of small and medium sized enterprises in the Russian Federation" specify the creation in the Russian Federation the development of the infrastructure for the support of SMEs which is accomplished within the federal program of the Russian Federation "Economic development and innovative economy" for the period 2013-2020. The measures of the government support of SMEs' include the reduction of administrative burden of the SMEs, development of special taxation regimes, development of the net of regional organizations of the SMEs' support (centers for entrepreneurial support, business incubators, centers for export support, engineering centers, microfinance organizations, guarantee funds and so on). Only in 2017 7,5 billion rubles were allocated from the federal budget for the governmental support of SMEs from 82 regions of the Russian Federation including the Volgograd region (124,4 mln rubles) [1]. In 2018 408,76 million rubles were allocated from the reserve fund of the Government for the co financing of capital investments into the construction of a business incubator (Kaluga region), technological park (Lipetsk region), Tutaev industrial park "Master" (Yaroslavl region), industrial park "Safonovo" (Smolenks region), industrial park "Aurora" (Republic of Dagestan).

During 2011-2013 in Russia the institute of the regulation assessment was developed for the prevention of the appearance of new barriers for business and the reduction of the existing impediments developed. In the result a new model of departmental law making was created which is focused on the analysis by the business community of laws suggested by federal authorities. However the impact of business community on the measures taken by government is limited [2].

Over recent years in the Volgograd region the infrastructure of the support for SMEs within the implementation of the federal program "Economic development and innovation economy" and corresponding regional programs was in general created. In the region there exist: center for entrepreneurial support, business incubator, regional engineering center, engineering center at the Volgograd State Technical University, regional microfinance center, regional guarantee fund. In 2018 the center for innovations in the social sphere was established. It will assist the social entrepreneurs and socially oriented non profit organizations. In 2020 the prototyping center is planned for opening.

In 2018 the regional Fund for Micro Financing of Entrepreneurship gives loans to SMEs at 7,75% interest year what is almost twice as low than interest rates in commercial banks. The guarantee fund gives the guarantees on bank loans for about 70% from the amount of the credit.

The center for support of entrepreneurship provides free advisory services concerning finance planning, marketing, patent and license support, legal support, labour legislation. In this center the representatives of SMEs can fill in the claim forms, references and other procedural papers with the help of legal consultants.

The indices which characterize the institutional environment and measures for support of SMEs are presented in Table 1.

Table 1. Governmental support of SMEs in the Volgograd region from the budget of the Ministry for Economic Development

Year	Volume of subsidies from the state budget allocated for the governmental	Number of loans provided against the guarantee from the guarantee fund	The amount of loans provided against the guarantee from the guarantee fund, thousand rubles	Number of SMEs (residents of the business incubator)	Total number of the employees hired by the SMEs (residents of the business incubator)

IOP Conf. Series: Materials Science and Engineering **483** (2019) 012007 doi:10.1088/1757-899X/483/1/012007

	support of SMEs, thousand rubles				
2015	297498	96	902205	15	51
2016	152000	120	1064816	15	68

Source: Federal Service for State Statistics

The Volgograd region is one of ten pilot areas where the educational project of the Public Company "Russian Export Center" is being realized. In 2016 the Center for Export Support Coordination was established for the provision of informational, analytical, consulting and organizational support of foreign economic activity of SMEs, contribution to the attraction of investments and exit of export focused SMEs into international markets. The center of coordination of export support of the Volgograd region provides informational, analytical, advisory and organizational support of the foreign economic activity of SMEs, assistance to the investment raising, support at the exit of firms into foreign markets. During the training course offered by the center for export support the entrepreneurs were taught to negotiate business, close transaction with foreign firms, studied the customs rules of delivery, fiscal issues of the activity, specificity of online trade for exporters. Summarizing the results of training within the educational project Public Company "Regional Export Center", 131 employees from 40 companies obtained the educational certificates. Besides the Center for Export Support helps SMEs prepare the papers for the registration of companies with foreign investment. In 2017 the Public Company "Regional Export Center" organized the points of information for the enterprises of the Volgograd region at the largest Russian and international exhibitions and also business missions in Volgograd and abroad.

Since 2017 the informational source "Business navigator" was founded in the Volgograd region. With the help of this source a potential entrepreneur can get the access to the information about the existing market niches for business, all types of support for SMEs and specialized financial and credit resources in a "one window mode". The web portal contributes to the creation of an approximate business plan and gives the information about competitors.

In the Volgograd region the non profit partnership "Volgograd Center for Technology Transfer" also functions. The organization mentioned above is the representative of the Fund for Support of SMEs in the scientific and technological sphere of the Volgograd region and operator of the project "Encouragement of youth innovative entrepreneurship in the Russian Federation" in the Volgograd region. Its main tasks are: establishment of small innovative enterprises, provision of measures in the identification of perspective R&D works directed on a further profit making among students and young researchers, search of partners in the commoditization of developments; monitoring of the R&D results funded by the programs "Participant of youth innovation competition" (UMNIK) and "Start" of the Fund of the assistance to the development of SMEs in scientific and technological sphere.

By 2020 the creation of a chemical and pharmaceutical cluster is expected to be finished. The goal of the creation of the chemical and pharmaceutical cluster is the provision of population with strategically important, innovative and high class drugs what will implement the conception of national security. The chemical and pharmaceutical cluster of the Volgograd region will consist from the educational, scientific, manufacturing and social functional areas. The main participants of the cluster will become the leading scientific and research organizations of the Volgograd region: Volgograd state medical university, Volgograd scientific and research medical center at the Administration of the Volgograd region, research institute of pharmacology at the Volgograd state medical university and also the largest chemical enterprises of the region (PC "Khimprom", PC "Kaustik" and PC "Volzhskiy orgsintez").

The agreement between the administration of the Volgograd region and the company RUSAL was reached on the creation of a new industrial cluster. The idea of the project of the Volgograd technological valley specifies the establishment of new enterprises next to the production of primary aluminum with the opportunity of the direct delivery of the liquid metal. The production of major conversion goods is planned. They are final goods including construction materials, automobile components, structures and constructions using the aluminum roll sections. It is reasonable to include into the clusters the SMEs which could carry out research and development work upon the order of large companies.

The integrated program "Creation of technological parks in the sphere of high technologies" started to realize in the Volgograd region. The attempts to open a technological park were made but they were unsuccessful. It's an open secret that both the procedures of use of new technologies and also the program complexes for their realization also assume considerable time for its establishment, further normal functioning and identification of inevitable problems and drawbacks of legal and organizational character.

The Federal Law "On the provision of governmental and municipal services" fixed as the main ones the principles connected with the realization of the citizens of their rights in the sphere of entrepreneurship: the rights of claimant and the duty of the organ providing governmental and municipal services, set of definite requirements on the interaction with claimants; order of payment; interdepartmental informational communication between the representatives of various authorities which is determined in administrative regulations; ways of agreements' settlement what is one of the most considerable innovations in the regional policy connected with the development of local entrepreneurship. Further their optimization is needed.

For the moment the existing administrative regulations connected with the provision of governmental services present in its essence the static documents which are in reality static papers which are approved by the departmental act of a corresponding federal authority and as a rule contain significant excess information. The federal bodies of executive power notice the necessity of the optimization of both the structure of administrative regulation and the legal basis of the provision of governmental services and functions of the government administration. The provisions mentioned above specify the order of informational and legal advisory interaction which is written in the corresponding papers duplicating the existing stipulations of the regulation.

Thus in the region a significant work in the creation of institutional conditions and mechanisms of entrepreneurship development was carried out. At the same time SMEs do not have a corresponding impact on its economic development. Besides we can even state the tendency of the escape of business from the Volgograd region.

The dynamics of the number of SMEs (Table 2) shows that their total number practically does not change and fluctuates at the number of 74-75 thousand. The number of the recorded small firms in the Volgograd region tends to reduce from 2010 to the present moment. By the middle of the year 2018 their number reduced to 2526. The number of micro firms during this period grew what suggests that some part of small enterprises out of necessity reduce costs under the influence of the overall stagnation and move into the group of "micro firms" [3]. It is worth mentioning that not all registered enterprises do business in reality. Thus according to the data of continuous reading of small and medium sized firms in 2015 only 18715 small enterprises or 59,6% from their total number worked in practice [4]. The negative dynamics shows also the number of sole proprietors which in fact dido business in the reference year. The number of the employed in small and medium sized firms in 2010-2016 reduced by one third. A sharp increase of the number of SMEs in 2016 was due to the growth of prices for their production. It is worth mentioning that according to the data of the Federal Service for State Statistics the inflation rate in 2015 made up 12,9%.

The main reasons of the observed outflow of business from the Volgograd region became the reduction of already low real disposable incomes of the population and the business number what lead to demand reduction for goods and services and the profitability of enterprises. According to the Federal Service of State Statistics the Volgograd region in 2016 had the 73th rank in the Russian Federation. Among the regions which are attractive for life and doing business the natives of Volgograd mention the Krasnodar krai, Belgorod region, republic of Tatarstan, Moscow and Saint Petersburg [5]. These regions are also affected by the crisis but at a lesser degree than the Volgograd

IOP Conf. Series: Materials Science and Engineering **483** (2019) 012007 doi:10.1088/1757-899X/483/1/012007

region. Thus for instance the disposable real incomes of the population in the Volgograd region in 2016 made up 88,9% from the rate of 2015 what is one of the worst indices in the regions of the Russian Federation (larger rates of real incomes drop are observed in the Republic of Tyva). After all labour resources, human capital are a decisive factor of entrepreneurial development.

Table 2. Indices of SMEs' development in the Volgograd region in 2010–2016

Year	Number of SMEs without micro firms)	Number of micro firms	Number of middle sized enterprises	Number of working sole proprietors	Business sales in current prices, mln. rubles	Number of the employed in SMEs, thousand persons
2010	3729	22974	407	47119	316263,9	353367
2011	3560	21329	290	49053	283409,2	289396
2012	3794	23354	219	47241	305900,5	277924
2013	3354	24658	217	46889	318035,2	268841
2014	3414	25247	216	44945	383490	265766
2015	3081	28332	238	37522	551434,5	264252
2016	2820	30932	181	41425	413380,3	239143

Source: the data of the Federal Service for State Statistics

The main problems in the development of human capital in the Volgograd region are: low salaries, social vulnerability especially of health care professionals and education workers; poor accessibility of health care services provided by the government for the population. A low standard of living does not allow the population using the services of private clinics. There are problems of youth employment after professional training, especially higher education. There is poor integration of the interests of employers, professional educational institutions and students in particular.

The problems of the Volgograd region and other regions of the Russian Federation which were mentioned above are the result of the so called "post communist changes" of 1990s which lead to a crisis in science and education, "brain drain", reduction of high tech industries, reduction of life standard of population and the failure of management institutes [6].

The main indicators of human capital reduction of a territory are the outflow of enrollees, young specialists and labour force of the working age from the Volgograd region into other regions. The Volgograd region lags considerably upon the budgetary expenditures per capita, average nominal payable salaries, new housing supply, number of provided mortgage loans, natural population growth, total volume of dwelling units calculated per head of population.

Another reason of the outflow of business from the Volgograd region became the pressure on business in the form of office audits, growth of number of court suits, cases of invoicing blocks of entities and sole proprietors, lack of attention of regional authorities to the problems of SMEs. Thus for instance the Volgograd region passed the regional law on the trade regulation (in the support of the Federal Law № 381-FZ dated December, 28th 2009 "On the foundations of the government regulation of trade in the Russian Federation) although the need was obvious considerably earlier.

Further a constraining factor for business development is the uncertainty of the economic situation and the tax burden which is assessed by SMEs as a heavy one.

Finally it is a frequent change of regional leaders (and their teams), lack of unity in the local elite, incompetency, lobbying of group interests which also does not contribute to the development of the entrepreneurship in the region.

The coordination of legal acts between regional and local authorities in the sphere of entrepreneurial support in the taxation remains a problem. Thus according to article 72, paragraph 10 of the Constitution of the Russian Federation the establishment of common principles of taxation is a joint competence of federal and regional authorities of the Russian Federation. Article 132 of the Constitution of the Russian Federation says that the local authorities have the right of tax and duties' assessment. Certainly the tax powers of regional and local authorities are limited by the federal tax law. Thus the regions of the Russian Federation and municipal authorities in an order and limits fixed by the Tax Code of the Russian Federation determine tax rates, the order and terms of payment of regional and local taxes unless this taxation specificity is not written in the Tax Code of the Russian Federation. Besides in the order and limits specified in the Tax Code of the Russian Federation, the public and legal organs mentioned above can determine the specificity of the tax base, tax exemptions, foundations and the order of their use according to regional and local taxes correspondingly [7].

3. Conclusion

The first order task of the creation of institutional conditions and mechanisms of entrepreneurial development in the region should become the leading out of industrial complex of the Volgograd region at a higher technological level, its modernization, identification of "growing points", increase of the output of science intensive production. The most perspective from this point of view are chemical industry, machine construction as the most science intensive industries. As the research shows the entrepreneurship is rather an opportunity in the future in the countries driven by innovations than a better way of earning a living in the countries focused on the efficiency [8], [9].

On the basis of the commercialization of R&D of higher educational institutions and the scientific research of the region at the support of the Administration of the Volgograd region of small and medium sized innovative startup companies via the financing of the infrastructural support of SMEs of the region (Volgograd regional business incubator, Center of Engineering of the Volgograd region, Regional guarantee fund, Regional microfinance center) it is possible to develop new technology intensive enterprises in the sphere of informational technologies, pharmaceuticals, chemical industry and so on.

The opening of specialized industrial grounds (industrial parks) equipped with engineering infrastructure, approach roads, technological parks, development of sectoral regional clusters with the creation on their basis of innovative territorial clusters will contribute to the attraction of investments into the region. It is necessary to pay attention to the creation of optimal conditions for existence and development of firms (residents) from various regions, creation of additional tools and services. Good results which the Kaluga region shows where achieved thank to the PC "Corporation of development of the Kaluga region". With the help of this private public partnership 12 industrial parks were created. In this region 99 investment projects are implemented, an industrial production special economic zone "Kaluga" functions.

In order to encourage the innovation activity of SMEs by means of their cooperation with public and private providers of R&D services it is indispensable to introduce the innovative vouchers in the Volgograd region. An innovative voucher is a security (certificate) confirming the right of the owner, small or medium sized enterprise to get R&D services from a higher educational institution or a specialized consulting company within the nominal value of the voucher for the solution of problems of enterprise [10].

The public and private partnership within the mechanism of creation of industrial parks and the implementation of cluster potential of a territory will contribute to the creation of conditions for an innovative development of the region. The Volgograd region has all the necessary prerequisites for the development of agricultural, pharmaceutical, metallurgical, petrochemical clusters which will be a factor of encouragement of SMEs. In any case this fact is confirmed by the experience of other countries.

It is necessary to improve further the administrative regulations. The informational coordination of governmental authorities and municipal authorities is expected to ensure the right of getting such information both upon request and within the realization of corresponding powers of an authority, i.e. the diffusion of information among a definite audience. Such a situation will allow providing a full scale access to the systems of informational interaction of entrepreneurs as consumers of governmental services.

One of the forms of a successful and perspective provision of federal and municipal services in the Russian Federation influencing the entrepreneurship development became the "one window" system.

Unfortunately Russia does not provide the "one window" system for all forms of entrepreneurial financial reporting. The financial reporting is presented to various authorities and moreover it is duplicated. That's why the introduction of such a governmental service is extremely important for Russia.

It is reasonable to include the obligatory entrepreneurship lessons into the school curriculum and to offer business trainings in higher educational institutions in dependence on the specialization of the faculty what was suggested by the All Russian public organization of small and medum sized entrepreneurship "Opora Rossii" (Support of Russia) [2].

For the development of entrepreneurship in the region the analysis of risks arising on the basis of legal regulation in budgetary and fiscal areas matters. The mentioned situation is connected with the powers of the regions of the Russian Federation within local administrative independence. Frequently when analyzing the territorial risks of budget sphere a special accent is made on an unjustified growth of the tax burden on the budgets of the regions of the Russian Federation and the budgets of municipalities for the provision of their balanced character.

The general legal mechanism in the prevention of appearance of territorial risks in the budgetary sphere can be presented by the following measures. These are legal regimes connected with the implementation of the so called budgetary immunity admitting the levying of budgetary funds only on the basis of Act of court except the cases set by the Budgetary Code of the Russian Federation (Article 239), provision of the opportunity of creation of reserve funds of the regions of the Russian Federation within their budgets for the financial provision of their expenditures in case of an insufficiency of budget profit (Article 81.1 of Budgetary Code of the Russian Federation) and also the creation of reserve funds of executive authorities of the Russian Federation and municipal authorities for a financial provision of unpredictable expenditures (Article 81 of Budgetary Code of the Russian Federation). All these measures can contribute to the improvement of institutional conditions of the entrepreneurship development in the Volgograd region.

References

[1] Decree of the Government of the Russian Federation № 129-p dated January, 28th 2017. Consulting system "Consultant +". Date of reference 01.09.17.

[2] Faktory stimulirovaniya predprinimatel'stva v Rossii i v mire [Factors of encouragement of entrepreneurship in Russia and in the world. 2014]. Obshcherossiiskaya obshchestvennaya organizatsiya malogo i srednego predprinimatel'stva «Opora Rossii» [All Russian non governmental organization of small and medium sized business "Support of Russia"]. Moscow. 184 p.

[3] Ledeneva M V, Parfenova M V, Vasilieva S I, Shamray-Kurbatova L V. 2017 The effectiveness of the government support to small and medium-sized businesses in the Volgograd region including an innovative entrepreneurship. *International Journal of Advanced Biotechnology and Research*. Vol-8 **2** pp. 687–697

[4] Volgogradskaya oblast' v tsifrakh [Volgograd region in numbers. 2016]. Short reference book. Territorial organ of the Federal Service for State Statistics in the Volgograd region. Vogogradstat, 2017. 268 p

[5] Vederina E. Business outflows from Volgograd into other regions Available at: https://v1.ru/text/business/88844162813952.html

[6] Loginova E, Plaksunova T, Litvinova A, Litvinov E, Parfenova M, Talalaeva N 2017 Determinants of Innovational Development of Economy of Modern Russia *Journal of Applied Economic Sciences.* Vol XII **2 (48)** 356–373

[7] Omelekhina N V 2018. Financial risks of regional and local level: legal identification and species variety *Financial law* **6** 11–17

[8] Ştefănescu D, Maior P 2012 What Influences more the Entrepreneurship in European Countries: Opportunity or Necessity? *"Ovidius" University Annals, Economic Sciences Series* Vol XII. **2** 224–229

[9] Kelley D, Bosma N, Amorós J E, *GEM 2010 Global Report*, retrieved from http://www.gemconsortium.org/docs/download/266. Date of reference 01.09.17.
[10] Ledeneva M V, Parfenova M V, Guzev M M 2018 Innovative vouchers and strategy of their introduction in the Volgograd region. *Advances in Economics, Business and Management Research* **39** 210–215

Transport nominations in cross-cultural communication training

A A Isakova

Industrial University of Tyumen, 38, Volodarskogo str., Tyumen, 625000, Russia

IsakovaAA@yandex.ru

Abstract. The article is devoted to studying of transport nomination as a cross-cultural communicative problem. The solution of this problem requires a new interdisciplinary synthesis, i.e. use of knowledge of various sciences such as linguistics in our case. Our research deals with the analysis of transport nominations, transport advertising and brand adaptations in language. The language of transport nominations has a multicultural basis. It is a proceeding unit due to changes of geopolitical, social, and economic priorities. Some processes of globalization, cross-cultural communication and environment influence on adaptation of transport nominations in Russian communicative area. Brand nominations have cross-cultural features, national character and particular passion. All of them become a part of our life and students have to cope with issues of correct pronunciation and translation, global marked language units and foreign or national elements in training of functioning processes of transport brand commodity

1. Introduction

The purpose of education is a formation of certain personality qualities and an acquisition of specific skills. The problems of higher technical education create on the basis of student vocational training. These problems are specified in the purposes of individual disciplines and the formation of relevant competencies. The postmodernism epoch contributed to the change of the education paradigm and its ideal having openness to cross-cultural dialogue in the context of national and world values. A person is forced to accept the rules of his “game” getting into new conditions of a new genre. The solution of this problem requires a new interdisciplinary synthesis, i.e. use of knowledge of various sciences such as linguistics in our case. The successful person is one who is more able to quickly switch the "code" and "lives" in the "genre" [1].

The modern linguistics is anthropocentric science. It opens new ways of studying of psychology and thinking through language which contains rich material for the analysis of specific and national peculiarities, cross-cultural and interlingual contacts. Facts of linguistic reality, new concepts and objects of the surrounding world develop science and technological progress [2].

The relevance of the following research lies in the fact of the growing interest in the processes of transport nomination and its modeling abilities in the multicultural space of modern Russia.

The research is focused on identification of some tendencies of modern Russian development which are reflected in processes of interlingual contacts. It is certainly interesting for modern linguistics which seeks to describe the national peculiarities of the linguistic picture of the world in detail, to form a holistic view of the structure and factors of the nominative language system, to identify and characterize multidimensional relationships of language units.

The research of transport nominations requires knowledge of trademarks and brands.

The language side of trademarks has so far been practically ignored, it is explained by the limited scope of their functioning, the artificiality of education, and the specificity of semantics. In general, it should be noted that numerous foreign trademark manuals and manuals refer to publications in the field of economics. Special attention is paid to the consideration of the legal aspects of their functioning.

IOP Conf. Series: Materials Science and Engineering **483** (2019) 012008 doi:10.1088/1757-899X/483/1/012008

Recently, the term "brand" has become widely used. The difference is quite significant between a trademark and a brand. Firstly, the "concept of a trademark" remains highly specialized and they are operated by lawyers most often. The word "brand" is widely known. Secondly, a brand cannot be registered, but a trademark is possible to register. Brand can't be registered, but it can be estimated.

Brand means a financial success of any company. It is a well-known trademark in consciousness of consumers. It is a winning result of the company, i.e. a brand is only a successful trademark. Thus, the brand is a guarantee of quality. That is Qualities in assortment: Swedish, German, Spanish, English, American, Russian, Japanese, Chinese etc.

Theoretical value of the research is focused on identifying some trends in the development of modern Russian language which are reflected in the processes of interlingual contacts. The national characteristics of the world linguistic picture, structure and factors of the nominative language system and cross-cultural relationships of linguistic units are interesting for modern linguistics and pedagogics.

The subject of our research is mental information by J. Fodor [3] and cognitive theory by R. Jackendoff [4].

The following techniques and methods were used to solve methodological and methodological issues of the study: a descriptive method, a method of observation and study of experience, etc. The methodological basis of our research is a philosophical concept of culture and morality, communicative, interethnic and ecological culture, etc. V. Belyaev believes cultural approach is one of the modern approaches in pedagogy. We fully share the opinion of the scientist that "a person becomes the heir of culture in the process of education and upbringing" [5]. The Saransk Cultural School of N. I. Voronina enjoys special prestige, their research devotes to integration of science, culture, education and cross-cultural regional policy [6]. The famous Tyumen author researches the key features of the current socio-cultural situation in the Tyumen region, which is determined both by the all-Russian processes and specific local factors [7]. There are many curriculums, university courses in this area.

The environment is constantly reflected in the mind of a person as a carrier of language, culture, spirituality, and life experience. The principle of consciousness understanding is developed in the works of A. N. Leontiev, G. F. V. Hegel, E. F. Tarasov, I. A. Sternin, Yu. A. Sorokin and others. Scientists consider language as a symbolic key to culture, a guide to social reality and instructions for understanding culture. Yu.A. Sorokin emphasizes the multi-layered language consciousness. The correlation of consciousness and mental processes of the cognitive-communicative system is a continuous interrelated process in which life features appear (ecosphere, social landscape, typical economic work, activity, worldview, etc.) [8].

The great value is given to problems of creation of commodity brands by the foreign scientists claiming that the unsuccessful name can lead to a market failure, and successful name can lead to bring millions. (K. Buhler, J. Praninskas, T Matumcs, A Room).

2. Transport nomination in context of cross-cultural communication training

Our research is devoted to the analysis of the transport nominations, transport brand advertising and trademarks adaptations in language. It has a quite practical value in training of functioning processes of brand commodity and translation, understanding and correct pronunciation during the text working. A special function of advertising texts is penetration and adaptation of foreign names of goods in Russian language.

The scientific approach requires knowledge of commodity naming and understanding of the potential buyer consciousness. Special techniques are used during creation of commodity brand. It allows using verbal (all levels of language) and non-verbal units in order to influence on the consciousness of the consumer and to achieve the desired result. All the variety of existing techniques is focused on the image of the product according to the customer desires and expectations. Moreover, at present there is a need to analyze the processes of the emergence and structural-semantic characteristics of transport brands, the laws of formation and development of nominative language units.

The world of transport isn't heartless and cold, it is penetrated by those feelings help the person to be guided with the world of subjects and objects. Besides, transport flows influence ecological safety [10] and the vital activity of modern person [11]. This plan of the technology existence has a definite influence on the physical world: some names become more popular, others do not, and therefore some vehicles "die", others continue to live transforming into new forms.

The study of language contacts of different nations ("the world in the language") is also presented in the advertising texts of cars:

- Daewoo car advertising "I love Uzbeks, they start up well" - correlation with Eastern temperament,
- Mercedes car "Real German quality",
- Opel Corsa "Germans are driving for permanent residence in Russia"

The advertising field actively exploits ethnic stereotypes of the perception of a particular people:

For example,

- advertising of the BMW car (It shows wide and clean German roads in the video. Voice of the announcer says behind the scenes: wide roads, wide possibilities)
- Impeccable Japanese concentration. Mitsubishi Lancer. A real Japanese drive.

Each transport nomination has a history of assimilation in Russian and an area of distribution and functional loading.

Sometimes good knowledge of other languages and history of creation is necessary to understand a name meaning of this or that car. For example:

Maybach (Germany) - a car. Wilhelm Maybach was a talented designer and a creator of his own car.

MAN (Germany) – a truck, bus. The history of MAN Nutzfahrzeuge is associated with the development of diesel engines for trucks and railway rolling stock. It was renamed in honor of the first founder of this company in 1987 whose name is Heinrich Mandeville. Three first letters of a surname have been taken during creation of commodity brand.

Mercedes-Benz (Germany) – a car. Mercedes Jelinek was a daughter of a rich Austrian businessman. when she was 11 only she had demanded from the father that cars which he intended to buy had her name.

It should be noted that some transport nominations are translated literally into Russian. For example:

Volkswagen means "people's car" in German.

Audi is a car. The company "Audi" was founded in 1909 by August Horch. It has begun its life since famous firm Horch. In 1909 the talented inventor August Horch built a very unsuccessful 6-cylinder engine which almost brought the company to the brink of bankruptcy. Then he was expelled from his own company. But Horch immediately founded another company nearby. He started to produce cars named by Horch again. His companions sued Horch in court. And Augustus Horch had to change the Latinized version of the former name. The word "horch" means "listen" in German. This is a real fact of using interlanguage connections.

The active position of the creators of transport advertising texts is embodied in a conscious language-creating process, in a targeted selection of a large number of techniques. The creators form and consolidate a positive image of goods or service for consumers. They try to use the national peculiarities of the country where the car will be sold. Application of various emotional and expressional units is allocated to units of verbalization of linguistic and cultural competence [12]. They are case phenomena. They are the main components of national cognitive base. Moreover, case names reflect the values, cultural experience and mentality.

Popular phrases of feature films, quotations of literary works, lines of famous songs, proverbs, sayings, popular expressions, phraseological units, famous statements of public figures are used in the language of advertising of transport brands.

IOP Conf. Series: Materials Science and Engineering **483** (2019) 012008 doi:10.1088/1757-899X/483/1/012008

For example,

the BMW car "I didn't think the future would come so soon,

MERCEDES-BENZ The sheet music staff winds between trees, it becomes the road through which the car travels. Antonio Vivaldi, The Four Seasons, Summer.

FORD Fiesta "Smart Fiesta with voice control features! The stupid Robodog never hears the owner".

MITSUBISHI Outlander "If Darwin was right other cars would simply die ".

Transport nominations are primarily linguistic signs of the world picture created by a certain people in a certain historical period. Such attitude gives rise to the flourishing of the "precedence" phenomenon because time has advanced the following postulate: "there is maximum of emotions and meaning in the minimum amount of time". Case names are understood as individual names associated primarily with well-known artwork, historical figure, political figure, geographical reality which have a certain value for the peoples.

These names constitute cultural stereotypes, cultural symbols. They can be references to the state name (such as: Nissan, ARO, Iran Khodro, etc.), the names of historical regions and various significant geographical objects that are symbols of the country (the Ural motorcycle, Tatra cars, Aston Martin, Valday cars, Vauxhall, Otosan, Shanghai, Scania, the passenger ship Altai, cars Volga, Oka, Isuzu, etc.), names of mythological creatures, deities, semi-mythical and mythical heroes (Lada cars mean the Old Russian deity of love, Mazda means the Zoroastrian Supreme deity of light, Mercury means ancient Roman god of trade Mercury, Svyatogor means the epic athlete, the yacht Andrew the Apostle, the plane Ilya Muromets is a national fantastic hero, SsangYong means two dragons). Transport can be called in honor of heroic people and famous persons of the past (the Prince Vladimir cars, Yury Dolgorukiy, Kalita, Lincoln, the Nette steamship, passenger ships Sergey Lazo, Prokopy Pleshcheev, Pyotr Bereznitsky); names of the cities (cars: Moskvich, Zhiguli, Hino, Lublin, Pontiac, Prague, cranes: Uglich, Ivanovo, Husqvarna motorcycle, etc.).

Analysis of transport brands shows their nomination as follows: percentage, i.e. more correct.

Table 1. The principle of nominations.

	The principle of nominations	%
1	Persons, i.e. they're called in honor of the creator of a vehicle (for example, the Ford car is called in honor of a creator Henry Ford, Porsche is called in honor of a designer Ferdinand Porsche, Foden is called in honor of E.R. Foden);	51.2
2	Objects, i.e. they're called by the name of the settlement where the first car was created (for example, Otosan cars are in honor of the Otosan-Uchi district, MAZ means Minsk Automobile Plant, Komatsu means Komatsu city on the west coast of Japan);	45.7
3	Actually identifying objects (for example, Caterpillar tracked vehicles means "caterpillar", Nissan cars is "Ni" from Nihon - Japan, "san" from "Sangio" - industry, Volkswagen means "people's car " in German).	3.1

We understand modern Russian mentality as a kind of unity which is not a national system but it's an original synthesis of pagan and Christian, Western European and Eastern, Slavic and Asian, Catholic and Orthodox, capitalist and socialist. This fact determines the peculiarities of the development of the Russian multi-human personality. In this regard, the fact of the nomination of Russian cars is very interesting. "to love the motherland", patriotism is considered an integral feature of the Russian people character.

It can be traced even in the names of various national vehicles (Volga, Zhiguli, KAMAZ, Volzhanin, etc.). Russian vehicles are named either in honor of the heroes or in honor of a specific territory of the Russian Federation (Ilya Muromets, Ivanovets, KAMAZ). Moreover, advertisers create advertising slogans to promote pride of "their native side" and to stimulate Russians in order to buy domestic cars. For example:

UAZ Patriot. True, Reliable friend.

GAS. Russian cars

Ural Strong cars for strong Russia.

So-called "soft" advertising is mainly used in advertising Russian cars. "Soft" advertising spreads information about the product and also it is directed to creating the favorable atmosphere around these goods, to calling for a sense of pride in the Russian car. In the construction of there are no imperative constructions in such advertising text. This is often emotional advertising playing on symbolism, deep motives.

- the Lada Kalina car. Our city has apparently fallen in love.
- the tractor "BELARUS". They could not win the Russian winter but it is won by THE BELARUS TRACTOR.
- KAMAZ - King of the roads. Tanks are not afraid of dirt!

Besides there are many facts of adaptation of foreign transport names in Russia. For example, the Mitsubishi transport brand. Mitsubishi is one of the largest and oldest machine-building concerns in Japan. This word means three diamonds. It has come from a logo of the company chosen by Ivasaki Yataro. There are two ways of this word pronouncing in the daily speech: [mitsubisi] and [mɪtˌsu :ˈ bi :ʃ i] in Russia. Both options have taken the place in media and advertising. Pronunciation [mitsubisi] is correct, it is used by producers of this car and the Japanese advertisers.

It can be assumed that the use of the pronunciation of the brand affects the advertising and cognitive consciousness of potential car owners. Our students have interviewed the leaders of the Mitsubishi dealerships and we found that residents of Moscow and the European part of Russia use the pronunciation [mɪtˌsuːˈbiːʃi] more often. They pronounce this word using English transcription.

Our task was to determine the percentage of using pronunciation of the brand by the Tyumen citizens. Linguistic methods were used to solve the problem. The use of these methods in describing the functioning of a language does not differ from similar tools in the humanitarian [12] and technical [13] sciences. However use of methods of measurement of language realization allows to modify significantly a concept of language system and a possibility of its functioning.

Residents of Tyumen often use the Japanese version of the pronunciation of this car as a percentage, i.e. more correct.

Table 2. Frequency of Mitsubishi using in Tyumen.

	Absolute frequency	Relative frequency
[mɪtsuːˈbɪʃɪ]	57	0.38
[mitsubisi]	93	0.62

We turned to the computer program of the Russian National Corpus in order to confirm or refute the data. The corpus contains both written texts (fiction, memoirs, journalism, scientific, religious literature, everyday printed matter), as well as recordings of oral texts (public speaking and private conversations). Corpus articles are dated 1994-2017. The popularity of the Mitsubishi brand is quite high, the program issued a figure of 86 only in the main body and 254 uses in the newspaper body, all texts are devoted to analyzing the brand and do not contain the advertising slogans.

The pronunciation "mitsubisi" is used 87% in the National Corpus of the Russian language, it's more often than "mitsubishi". And nevertheless, advertisers continue to use the last one.

We dare to suggest that this incident is associated with the processes of globalization and the influence of the English language. By the way, this research was very interesting and useful for students of Tyumen industrial university.

3. Conclusion

The received results allow to confirm to confirm the earlier hypotheses, to expand and supplement the developments on the problem of psycholinguistics, language contacts and intercultural communication. The results of an individual nomination are first generalized in the speech practice of society, then they enter a common language for all members of society and become facts of public consciousness [14]. Advertising has a significant impact on the modern person's worldview. During our life transport advertising gets into all spheres of life, filling it with meanings and images. Influencing the cognitive system of people, the advertised trademarks become part of intercultural communication. It influences the cognitive system of people. Transport brands become a part of cross-cultural communication.

Thus, the language of transport, transport nominations has a multicultural basis, it a proceeding unit due to changes of geopolitical, social, and economic priorities. Some processes of globalization, cross-cultural communication and environment influence on adaptation of transport nominations in Russian communicative area. Brand nominations have cross-cultural features, national character and particular passion. All of them become a part of our life and students have to cope with issues of correct pronunciation and translation, global marked language units and foreign or national elements in training of functioning processes of transport brand commodity. Therefore, this problem demands a particular testing, a training curriculum and teaching at the university.

References

[1] Khairullina N G, Sadykova H N, Ustinova O V, Gadzhigasanova N S and Tretyakova O V 2016 *European Journal of Science and Theology* **4** (12) 97-110.

[2] Isakova A A and Isakov A A 2018 *Astra Salvensis* **6** 782-794

[3] Fodor J A 1975 *The language of thought* (Cambridge: Harvard university press) 75

[4] Jackendoff R 1992 *Languages of the mind* (London: MIT press) 34

[5] Belyaev V I 1999 *Pedagogika* **6** 19

[6] Voronina N I 2014 *Mordovia University Bulletin* **4** 189-196

[7] Bazylev V N, Bel'chikov YU A, Leont'ev A A and Sorokin YU A 2013 *Rossijskaya yusticiya* **4** 43.

[8] Isakova A A 2017 *Integration of education* **1**(21) 46-54

[9] Andrianova E V, Arbitajlo I YA, Ahmedzyanova R M, Dashina YA N Kombarova T V, Korepanov G S, Pecherkina I F, Stepanova M V, Tarasova A N, Hvoshchin A S, Hudyakova M V, Davydenko V A and Romashkina G F 2015 *Social and cultural dynamic potrait of Tyumen region* (Tyumen: TSU)

[10] Kolesov V I, Danilov O F and Petrov A I 2017 *IOP Conf: Earth and Environmental Science* **19** 012059.

[11] Karnauhov V N 2013 *Problems of trafic safety in Russia VI Tyumen Conf* 75-79.

[12] Baranov A N 2001 *Vvedenie v prikladnuyu lingvistiku* (Moscow: EHditorial URSS) 360

[13] Artamonov E I, Balabanov A V, Romakin V A, Smirnov S V 2012 *CAD/CAM/PDM 12 Conf* 12-17

[14] Ufimceva N V, Tarasov E F 2009 *Voprosy psiholingvistiki* **10** 21

IOP Conf. Series: Materials Science and Engineering **483** (2019) 012009 doi:10.1088/1757-899X/483/1/012009

Improvement of procedural approach to the production management at industrial enterprises

A V Litvinova[1], A V Gorbunova[2] and E O Litvinov[1]

[1] Volzhskiy Humanitarian Institute, Volzhskiy Branch of the Volgograd State University, ul. 40 Let Pobedy, 11, Volzhskiy, Volgograd region, 404133, Russia
[2] Volzhskiy Polytechnical Institute, Branch of the Volgograd State Technical University, ul. Engelsa, 42a, Volzhskiy, Volgograd region, 404121, Russia

e-mail: litvinova_av@mail.ru

Abstract. The importance of the process approach in the management of industrial enterprises is shown. The peculiarities of the implementation of the process approach in the pipe metallurgical industry, content of the processes and criteria of their efficiency are studied. An important problem for the implementation of the process approach is connected with the absence of a clear and reliable assessment of quality management processes. The method of the efficiency assessment of managerial decisions taken during the process is suggested and tested. It is done by means of use of a correcting ratio reflecting the managerial quality upon every criterion of the process and the calculation of an integrated index of the process efficiency on its basis. It is determined that the developed methodology allows assessing objectively the dynamics of the process, purposefully changing the characteristics of every process in the needed direction, interpreting definitely the condition of all processes of an enterprise taking into account the quality of the management.

1. Introduction

In the course of development of theory and practice of management the vertical, hierarchical, strictly functional systems of management at industrial enterprises were replaced by a managerial concept which analyzed an enterprise as a set of the most important processes from the point of view of its efficiency which provides the creation of rigid horizontal connections at all managerial levels.

The comparative analysis of functional and process approaches to management made by A.V. Kladov [2, P.147-148] showed that the process approach improves the way of division of labour existing in the enterprise between separate functional subdivisions. This approach integrates the fulfilled functions into a total efficiency of organizational management thank to flexibility, adaptability, dynamism and transparency. From the point of view of O.V. Koltsov and V.I. Menshchikov [3, P.117] the process approach to management in contrast to functional approach directs the enterprise to the best final result of its activity.

The concept of process approach was finally formulated in 1980s of the last century and it was firstly written into the standards of the International Organization for Standardization (ISO) within quality management systems.

The main element of the process approach is generally meant a periodically repeated, adjusted consequence of actions which can be easily formalized [4, P.258]. According to the terminology of ISO Standard 9001-2000 «Systems of quality management. Requirements» a process is the activity using the resources and managed with the aim of the transformation of exits into entries. According to a justified opinion of V.G. Chebotarev and A.I. Gromov [1, P.14] the processes are considered as fundamental characteristics of an enterprise as an organizational system and the notion "enterprise management" for the moment is inseparable from the notion "management by business processes".

The ISO standards determine the necessity of the identification by the organizations of the content of their processes and ways of management of these processes but they do not present the strict requirements to their content and documentary formalization. The types of processes recommended by ISO that can be used by the organizations generally include four large groups: management processes, main production processes, implementation, monitoring, control and management.

For the documentation of the process approach the industrial enterprises as a rule use typical block schematic diagrams of processes which include such elements as the entry and the exit of the process, resources of the process, owner of the process, suppliers and consumers of the process, criteria of the process (qualitative or quantitative ones) which characterize the efficiency of a process. The use of a typical structure of processes according to the opinion of A. V. Kladov [2, P.146] allows joining the processes into a cyclical structure which is founded on the comparison of the result of every process and the reaction to the fulfillment of requirements.

An important impact on the creation of the structure and content of processes accompanying the activity of organizations has scientific research. For instance, A.G. Shapovalov [5, P.237] suggested to use the processes which characterize the introduction of new technological solutions in an organization. Ya. V. Grebnev [6, P.64] studied the opportunity of use of the process determining the degree of self organization in the course of management by organizational changes at the enterprises, process of the efficiency assessment of organizational changes and so on. A.V. Litvinova and A.V. Gorbunova [7] recommended the inclusion into the criteria of process assessment the criteria "Costs on process quality" for the reduction of the volume of useless (connected with the overuse of resources) of costs which can arise in the course of the realization of business processes.

Despite the availability of a number of methodological approaches to the organization of the process approach at industrial enterprises every from them is independent in the methods of the systematization of processes in their activity and determination of the elements of their processes. The differences mentioned above are clearly observed not only between enterprises but also between entire industries.

The pipe and metallurgical industry which used the process approach in compliance with the norms of ISO standards in early 1990 of the XX century had a considerable success in this sphere. The orientation to ISO standards specifically and a high level of the competitiveness achieved due to this fact contributed significantly to overcoming by Russian metallurgical enterprises of crisis tendencies typical for the Russian economy recently.

According to the results of 2017 the production index of metallurgical industry of the country made up 96,4% (in 2016 it was 97,7%) and the tube production in 2017 remained the leader of the industry. Thus the reduction of steel production, rolled steel was accompanied by a considerable growth of the volumes of the production of conductor weldless tubes (by 19,8%).

The implementation of oil and gas projects, recovery of industrial production in the world market is observed today. The large scale programs of technological modernization, strengthening of the ruble exchange rate, trend for import replacement existing today contributed to the recovery of pre crisis positions of national tube and metallurgical enterprises. For instance in February 2017 "Gazprom" company announced the purchase of almost 600 thousand tons of large diameter tubes for the gas pipeline "Sila Sibiri" (Power of Siberia). The suppliers of tubes became six leading tube producing companies of the country [8,9].

The leaders of pipe and tube metallurgical production in Russia are PC "Vyksunskiy metallurgical plant", PC "Chelabyinsk pipe rolling plant", PC "Pervouralskiy new pipe plant". PC "Northern pipe plant", PC "Novosibirskiy metallurgical plant named after Kuzmin", PC "Izhorskiy tube plant", PC "Uraltrubprom", PC "Almetievskiy tube plant", PC "Gazpromtrubinvest". The leading enterprises of pipe and tube production in the Volgograd region are PC "Volzhskiy tube plant" (part of the group PC "Tube metallurgical company"), Closed Corp. "Tube plant "Prophil-Akras"" in Volzhskiy.

The Closed Corporation "Tube plant "Prophil-Akras"" named after V.V. Makarov specializes on the production of electro welded pipes, gas pipelines and heating systems, pipes of square, rectangular, oval, plane oval section. Also it focuses on the production of various constructions. The annual volume of production reaches more than 120 thousand tons. The plant is one of top three in Russia upon the production of shaped tube and has the 7th rank upon the volume of production of small diameter pipes. The plant has an active policy in quality management using the process approach.

The goal of the research is the study of peculiarities of the implementation of the process approach at the enterprises of pipe and tube production, identification of problems and development of measures directed on its efficiency growth.

2. Materials and Methods (Model)

The analysis of the implementation of process approach in the Closed Corporation "Pipe plant "Prophil-Akras" was founded on the study of the composition and the content of process and on the efficiency criteria used in an enterprise.

The enterprise uses a typical for the tube metallurgical industry approach to the systematization of processes (Table 1).

All processes in the enterprise are divided into three groups: basic, managerial and supporting ones. The efficiency of every process is characterized by the corresponding criteria. When developing the composition and content of criteria at the enterprise the management assumes that when the number of criteria grows the safety and reliability of the assessment of every process increases. However a general drawback for all processes at the enterprise is clearly observed. It is the absence of efficiency assessment of managerial decisions taken in the course of their realization.

Meanwhile in terms of the standard ISO 9001-2015 «Explanatory text. Requirements. Guide for introduction» the process approach is directed on the systematic definition and management of processes and ensuring of interaction in a way that helps to reach the targeted results in compliance with the policy and strategic direction of the firm's activity. The achievement of this goal in the opinion of the authors is possible only on the basis of the growth of process management qualities and the systematic assessment of the efficiency level of the taken managerial decision from the point of view of their timeliness, operational efficiency, justification, flexibility, objectivity, multivariance, consistency of the goals set earlier and taken decisions, availability of the implementation mechanism. It is obvious that the absence of the efficiency assessment of managerial decisions taken for the management under the conditions of a quickly changing internal and external environment of the enterprise prevents achieving the goals of the process approach.

It is worth mentioning that in a series of scientific research devoted to the problems of the implementation of process approach at industrial enterprises [7,10,11,12], a serious attention is paid to the improvement of qualitative assessment of efficiency processes. However the research allowing to assess the efficiency of managerial decisions taken during their realization do not exist. At the same time E. M. Malkova and E.V. Konovalova [12] fairly assume that the main effect from the introduction of process management is the growth of the manageability of business processes and as a result the reduction of costs for their realization, growth of operational efficiency and competitiveness of the enterprise. From this point of view the assessment of the process management becomes a first order task.

Table 1. Processes in Closed Corporation "Tube plant "Prophil-Akras"" named after V.V. Makarov

Process	Criteria
	Main processes
Determination, planning and analysis of requirements of consumers and production	1. Gathering of demands for the beginning of the planned period (month) 2. Provision of production in compliance to the demands for the end of the planned period (month) 3. Satisfaction of consumers 4. Fulfillment of the plan of the shipped production, share of exports in the total volume of shipment
Purchase of basic materials	1. Provision of the demand for production in basic materials and compliance of purchased materials to set requirements 2. Delays in the work of electric welding pipe workshop 3. Volume of inappropriate basic material 4. Qualification of suppliers of main materials

Purchase of additional materials	1. Provision of the need of production in supplementary materials and compliance of the purchased materials to set requirements 2. Qualification of suppliers of supplementary materials
Production of goods	1. Implementation of production schedule 2. Implementation of norms for unsuitable production in the total volume of production 3. Technological delays
Storage and shipment of final production	1. Implementation of shipment schedule 2. Remarks on the storage conditions of final product
	Managerial processes
Activity planning	1. Implementation of expenditures budget 2. Profitability of shipped production 3. Implementation of profit budget 4. Implementation of profit plan
Organization and analysis of the activity	1. Assessment of quality management system from the point of view of top management 2. Achievement of goals set in the sphere of quality 3. Production discipline observing
	Supporting processes
Infrastructural management	1. Delays of mechanical equipment 2. Delays of electrical equipment 3. Implementation of schedules of the equipment repairing 4. Implementation of check schedules of measurement and equipment control 5. Implementation of schedule of building and construction inspection
Personnel management	1. Satisfaction of the demand in personnel 2. Implementation of personnel training plan 3. Turnover of employees
Production environment management	1. Realization of measures on labour safety 2. Traumatism level

For the efficiency assessment of managerial decisions taken in the course of the process it is suggested to adjust every process criteria for a correcting coefficient reflecting the degree of achievement of the acknowledged in theory and practice of management indices of managerial decisions efficiency (Table 2). The assessment should be made taking into account the understanding of quality management historically existing at the given enterprise, experience and qualification of specialists engaged in the process approach or in first turn the owners of process. It also should be accompanied by a detailed analysis of deviations which caused the reduction of process efficiency [13].

Table 2. Values of correction coefficients which determine the efficiency of managerial decisions taken in the course of processes

Process efficiency characteristics	Values of coefficient of efficiency of managerial decisions K_i
Timeliness, operational efficiency, justification, flexibility, objectivity, multivariance, consistency of the goals set and taken decisions, availability of the mechanism of implementation of managerial decisions	1.0
Justification, flexibility, objectivity but insufficient operational efficiency, timeliness and flexibility of managerial decisions	0.8
Insufficiency of justification of taken managerial decisions	0.6
Flexibility, objectivity and operational efficiency of taken managerial decisions but absence of a real mechanism of its realization	0.4
Managerial decisions contradict the goals set and decisions taken	0.2
Taken managerial decision did not provide the achievement of the set goals of the process	0

It is suggested to carry out the efficiency assessment of managerial decisions taken in the course of the process by means of the aggregation of the values of all the criteria into an integral index of the process efficiency. This integrated index should take into account the correction of every criteria for a coefficient which determines the degree of managerial decisions' efficiency:

$P = \sum d_i * K_i \ / \ n,$

where P – integral coefficient of process efficiency;
i – cardinal number of the process criterion (i=1,..,n);
n – number of criteria;
d_i – factual value of the process criteria, % from the normative value;
K_i – coefficient determining the efficiency of managerial decisions taken in the course of process.

3. Results and Discussion

The methodical approaches to the calculation and determination of normative values of process criteria used in Closed Corporation "Pipe plant "Prophil-Akras" named after V.V. Makarov are presented on the example of the process "Purchase of basic materials" (Table 3). The goal of this process is to provide the production with the basic material (metal) in compliance with technical norms and delivery terms.

Table 4 shows an adjusted description of the process "Purchase of basic materials", its planned and factual values of their criteria taking into account the coefficients which determine the efficiency of managerial decision taken in the course of the process. In order to specify the normative values of process efficiency we used as a target the maximally admissible normative value of every criterion within the limits of an accessible band of values. It allows identifying the maximal deviation of the efficiency coefficient of managerial decisions during the process.

Table 3. Composition, content and methods of calculation of criteria of the process "Purchase of basic materials"

№	Criterion	Measurement unit	Method of calculation	Normative value	Rank of importance
1	Provision of the need of metallic production and correspondence of purchased materials to requirements set by laws on purchases	%	Ratio of factual supply of metal (tons) to the need in purchase of metal according to production schedule (tons)	100 (±5)	5
2	Percentage of delays in the work of electric welding pipe workshop	%	Ratio of nominal time for technological process to the time of delays in the work of the workshop	0 (±0)	3
3	Volume of inappropriate metal	%	Ratio of the volume of the sorted metal (tons) to factual supply of metal (tons)	2 (+0;-0,5)	4
4	Qualification of suppliers of main materials	%	Ratio of the number of qualified (assessed upon five criteria and obtained more than 2,5 grades in 3 grade scale) of suppliers to the total number of suppliers	50 (-25,+50)	5

Table 4. Values of the criteria of the process "Purchase of basic materials" with the account of the coefficient of the managerial decisions efficiency

Criterion	Meas-	Value of criterion	Value of	Factual

	urement unit	normative	factual	deviation (+/–)	Percentage of achievement of a normative value	coefficient of managerial decisions efficiency	value of criteria with the account of the coefficient of efficiency
Provision of the need of metallic production	%	100 (±5)	98	-7	93.3	1	93.3
Percentage of delays in the work of electric welding pipe workshop	%	0 (±0)	1.8	-1.8	98.2	0.4	39.3
Volume of inappropriate metal	%	2 (+0,-0,5)	2.8	-1.3	71.4	0.6	42.8
Qualification of suppliers of basic materials	%	50 (-25,+50)	79	+54	126.0	1	126.0

The data in Table 4 show that at the provision of production with metal (coiled stock), the number of delays in the work of electric welding pipe workshop and the volume of inappropriate metal, there exists the non fulfillment of the maximal normative values of process criteria. At the same time a low efficiency of managerial decisions taken during the process is observed. The managerial decisions are assessed according to two criteria: percentage of delays of work of electric welding pipe workshop and the volume of inappropriate metal. A high efficiency is observed upon the following criteria: qualification of suppliers of basic materials and the provision of production with metal. The specificity of the metallurgical production is that the purchase of metal (strips, iron plate) is made in advance of factual production and it is the main condition of provision of the continuity of a technological process. Despite the fact that for the beginning of production process the provision with basic material made up 98%, the managerial decisions which are taken in time which are directed on the delivery of metal from the store room for the provision of continuity of the technological process and the implementation of the plan allowed assessing maximally high the efficiency of managerial decisions taken in compliance with the criteria "Provision of demand for metal production".

The results of the calculation of an integral index of process efficiency "Purchase of basic materials" upon the data of the factual value (percentage from the normative value) of its criteria is presented in Table 5.

Table 5. Results of the calculation of an integral index of efficiency process "Purchase of basic materials"

Variants of calculation of an integral index of efficiency process	Value of integral index P, %
Calculation on the basis of normative values of criteria	100.0
Calculation on the basis of factual values of criteria without the use of efficiency coefficient of managerial decisions	97.2
Calculation on the basis of factual values of criteria with the use of coefficients of managerial decisions efficiency	75.3

The data of Table 5 show that the value of the integral index calculated on the basis of factual values of process criteria with the use of the efficiency coefficient of managerial decisions are 21,9% lower than the value of the integral index calculated without this coefficient. The value of the integral index with factual values is lower than the integrated index which is calculated on the basis of normative values of criteria by 24,7%. The reasons which caused such deviations should be identified by means of the analysis of parameters determining the quality of managerial decisions within every process. They deal with specific problems: economic issues (cost of raw material supplies, material, volume of store of final goods and so on), production (specificity of technology and organization of production, regularity of supplies, functioning of equipment and workers) and so on. The identification of the most important reasons of unfavourable deviations in the value of process criteria and development of measures directed on the elimination of deviation reasons will allow channeling the process under analysis into a stable condition.

4. Conclusion

The method of the calculation of an integral index of the process efficiency allows improving the process approach to the management which is used at the enterprises of tube metallurgical industry due to the following reasons:

- aggregation of heterogeneous, with various dynamics, process criteria into an integral index of its efficiency;
- objective assessment of dynamics of process productivity with the account of the quality of managerial decisions taken in the process of its implementation and as a result there should be a targeted change of parameters of every process in the required direction;
- clear interpretation of the condition of all the processes at enterprises presented into an aggregated view for every moment of time within the whole system of an enterprise's management, forecasting of their further development, optimization of the structure and content;
- implementation of variants of calculation of process management at an enterprise in dependence on its external and internal environment and real resource provision.

The suggested methodology has a universal character and can be used at industrial enterprises of various forms and types of activity.

References

[1] Chebotarev V G, Gromov A I 2010 *Business –informatics* **1(11)** 14-21
[2] Kladov A V 2011 *Vestnik Voronezhskogo gosudarstvennogo tekhnicheskogo universiteta* Vol 7 **1** 145-148
[3] Koltsova O V, Menshchikova V I 2008 *[Bulletin of Tambov University. Series. Humanitarian sciences]*. **5(61)** 113-118
[4] Davudov M A 2007 *Journal of Saint Petersburg university of economics and finance* **4(52)** 257-259
[5] Shapovalov A G 2010 Analysis of approaches to the process of scientific and organizational projecting *Science and contemporaneity* **5-3** 235-239
[6] Grebnev Ya V 2014 Important problems of economic sciences **41-1** P. 57-64
[7] Litvinova A V Gorbunova A V 2011 Development of theoretical and methodical approaches to the systematization and assessment of expenditures on quality (Volgograd: Volgograd state university of architecture and civil engineering) 164 p
[8] Review of black metallurgy market. First half year. Research center of "Deloit" company in CIS 24 p
[9] Obzor rynka chernoi metallurgii. Pervoe polugodie 2018 g. Review of black metallurgy market. First half year. Research center of "Deloit" company in CIS 24 p
[10] Trisch R, Gorbenko E, Dotsenko N, Kim N and Kiporenko A. 2016 *Eastern-European Journal of Enterprise Technologies* **4/3(82)** 18-24
[11] Popova E M 2011 *Bulletin of Southern Russian State Technical University. Novocherkassk*

polytechnical institute. Series: Social and economic science **2** 204-209
[12] Malkova E M Konovalova E V 2014 *Humanitarian, economic and social sciences* **3** 288-297
[13] Udalov F E, Voronov N A, Yagunova N A 2015 In the collection of articles *Increase of managerial, economic, social, innovative and technical potential of enterprises, industries and economic complexes.* Collection of articles. VII International scientific and practical conference 73-76

IOP Conf. Series: Materials Science and Engineering **483** (2019) 012010 doi:10.1088/1757-899X/483/1/012010

Professional translators' ethics in workplace communication

V A Mityagina, A P Naumova

Volgograd State University, 100 Universitetsky Ave., Volgograd, 400062, Russia

E-mail: anaumova9@volsu.ru

Abstract. The article deals with the history of the concept of translation ethics in foreign and Russian literature. The understanding of this concept from different points of view is analysed. On the basis of the content of electronic discussion platforms (a group in the social network and professional forums) specific characteristics of translators' workplace communication on ethical issues are highlighted.

1. Introduction

The ethics of interaction in different areas of professional life became a relevant subject of study in different research paradigms, because the importance of communicative competences and the ability to be a member of a team is growing exponentially: the post-industrial era increased the role of a person who organises and implements an interactive process. Globalization as a space for international contacts in different areas of society's life revealed new aspects of mediated communication, translator ethics and ethics of translation began to receive a new understanding, primarily in the framework of translation studies, which has become a truly integral branch of the Humanities.

2. Research

One of the most relevant, very successful and concise definitions of the concept of "ethics of translation" belongs to the Austrian translation scholar Peter Sandrini. He published it in his article on translation studies in the "Lexicon of the Humanities": "The ethics of translation has repeatedly become the subject of discussion, but in general it depends on theoretical principles of the translation concept; its clear model still remains an object of study. Translation supports and makes it possible to interact across language and cultural boundaries with communication partners who belong to a foreign culture. Any text and any translate is a product of a certain culture: the production of texts and their reception are determined by current standards and parameters. They depend on point in time and differ in different cultures. Translation creates texts, manipulates texts to make them available to new recipients" [1].

The social determinism of translator ethics is a culture of translation – a historically established, self-regulating system that is a part of culture in general, dialectically related to attitudes towards translation practice. From the point of view of the Austrian scholar in translation studies Erich Prunč, this is a set of socially established, regulated and adjustable standards, conventions, presuppositions, values and patterns of behaviour common to all those involved in translational processes [2, 3, 4, 5].

As a social construct, the culture of translation is a fact of regulating interests of participants and functions of individuals and institutions involved in translation. Actors are not only translators, but also all those involved in the process - authors initiating translation - translators - addressees-recipients of translation. The culture of translation may reflect, first of all, actual provisions of translation studies, and then the actions of translators will have an instructive character. A democratically well-designed ethics of translation involves responsible handling of interests and potential of partners in interaction. This way of forming translation ethics is reflected by the term of loyalty ("Loyalität" in German) introduced by Christiane Nord. This term implies fidelity to the original, responsibility and professionalism of the translator towards the author, customer and to addressees [6, 7, 8].

It is worth pointing out that some researchers consider the term "loyalty" insufficiently comprehensive, because it lacks moral correction of the translator as an ethically active integral person [9, 10]. Therefore, a universal focus of loyalty of translators in relation to other partners on interaction needs to be replaced by a multidimensional loyalty of communicating actors to each other.

Translators performing their responsibilities within the scope of instructive, equivalence-oriented translation studies should possess the competence to deal creatively with a conflict of interests and take responsibility for making a decision in favour of a certain partner on interaction. This responsibility implies solidarity with other translators, as well as respecting provisions of the professional code of ethics [11, 12]. Thus, the framework conditions for implementing ethics of translation are transformed in a subjective manner in the reflected ethical behaviour of the translator. Victor Shadrin makes an attempt to present the ethical concept of translator's activity and identifies four basic models of translation ethics: representative ethics, service ethics, communicative ethics and normative ethics [13]. Rather schematic modelling of ethical parameters of professional behaviour of translators made by this researcher does not allow to present a set of actions performed in the overall scope of their characteristics. Therefore, the determinism of ethics by actual theoretical principles of translation studies remains the main issue.

Along with the Scopos theory, a significant role in the formation of current ethics of translation was also played by descriptive paradigm. It considered, in particular, the role of "patronage" - a person or institution controlling the actions of the translator [14]. Undoubtedly, these ideas led to the development of translation standards, which are based on a system of social values, public and individual expectations regarding communicative behaviour and the choice of translator's decisions in a given situation. [15].

Andrew Chesterman sets the norms quite in the classical spirit of providing the exact similarity to the original and minimizing misunderstandings in translation [16, 17], and suggests using the oath of Jerome in the translation activity by analogy with the Hippocratic oath:

1. I swear to keep this Oath to the best of my ability and judgement. [Commitment]
2. I swear to be a loyal member of the translators' profession, respecting its history. I am willing to share my expertise with colleagues and to pass it on to trainee translators. I will not work for unreasonable fees. I will always translate to the best of my ability. [Loyalty to the profession]
3. I will use my expertise to maximize communication and minimize mis-understanding across language barriers. [Understanding]
4. I swear that my translations will not represent their source texts in unfair ways. [Truth]
5. I will respect my readers by trying to make my translations as accessible as possible, according to the conditions of each translation task. [Clarity]
6. I undertake to respect the professional secrets of my clients and not to exploit clients' information for personal gain. I promise to respect dead-lines and to follow clients' instructions. [Trustworthiness]
7. I will be honest about my own qualifications and limitations; I will not accept work that is outside my competence. [Truthfulness]
8. I will inform clients of unresolved problems, and agree to arbitration in cases of dispute. [Justice]
9. I will do all I can to maintain and improve my competence, including all relevant linguistic, technical and other knowledge and skills. [Striving for excellence] [18].

It is worth noting that attempts to offer such consolidated standardizing texts were made by many scholars in translation. Taking into account the position of translators "in a special intercultural space, at the crossroads of cultures", Anthony Pym formulates five principles for translator ethics:

1. Translators are responsible for their product as soon as they accept to produce it.
2. Translators are responsible for the probable effects of their translations.
3. Translator ethics need not involve deciding between two cultures.

4. Transaction costs should not exceed the total benefits ensuing from the corresponding cooperative interaction.
5. Translators, insofar as they are more than simple messengers, are responsible for the capacity of their work to contribute to long-term stable, cross-cultural cooperation [19].

In Russian translation studies, the conceptual solution regarding translator ethics has not been elaborated yet, but many theorists with vast professional experience gave different versions of rules, postulates and advice. It should be noted that all of them were published in modern Russia: Ryurik Minyar-Beloruchev wrote about the etiquette of the interpreter [20], Andrey Chuzhakin and Pavel Palazhchenko formulated 10 rules of ethics for interpreters [21], and Mikhail Zwilling, back in 1964, published "Tips for simultaneous interpreters and those who would like to become such", in which 10 humorous precepts for simultaneous interpreters were given [22]. Forty years later these tips were published in Bridges, translators and interpreters' journal, and M. Zwilling noted some inaccuracies and errors committed by him while translating. These tips first appeared in German at bulletin board of MGIMO University.

It can be stated that these publications, as well as the work of Gennady Miram [23], reflected the tradition of ironic attitude of the 90s to the definition of rules and norms that were replaced by the postulates of translation ethics proposed by Irina Alekseeva in her textbook [24]. These rules are addressed to both interpreters and translators, they are included in the Chapter "Ethics of the Translator" and based on the author's understanding of such categories as moral principles of the translator, standards of professional behaviour of the translator, professional suitability and professional requirements, knowledge of technical support of translation, legal and social status of the translator [25]. The postulates formulated by I. Alekseeva cover various aspects of the translation process and its actor, let us compare:

- The translator must take care of his health, as the quality of translation depends on his physical condition.
- In translation, the translator must follow the rules of its execution, the correct attitude to the customer. [26].

Such different "responsibilities" of the translator are undoubtedly related to multidimensional and complex character of the contexts of both interpretation and translation. The wording of the rules governing the behaviour of the translator, is determined by constant throughout the entire history of translation requirements of equivalence and adequacy. Therefore, in the discourse on ethics we note the following requirements: "to follow the rules of its design", as well as "with the help of professional actions known to him always strive to convey as much as possible the invariant of the source text, focusing on functional dominants of the original", "has no right to change the meaning and composition of the text during translation, reduce it or expand it, if the additional task to adapt, select, add, etc. is not set by the customer" [27].

The dependence of studying translator ethics on his work consisting in creating the translation text is noted by many researchers, and the factor of the personality of the translator receives a proper assessment [28, 29, 30, 31].

In modern translation studies, ethical problems of translation activity require new understanding, because the role of ethical codes, which are developed by both large translation associations and individual translation agencies, increases significantly against the background of large-scale professional communication of translators via electronic platforms. The problems of ethical regulation of translator's behaviour in workplace interactions represent one of the main discussion topics.

Communicative interaction of translators on business issues is a communication about their professional activities, ways of its implementation, improvement and optimization. In this respect, the concept of workplace communication of translators is synonymous with the concept of professional discourse of translators. The electronic form of communication, in particular, within the framework of various discussion Internet sites, provides for a prompt resolution of topical issues faced by translators in the course of their professional activities.

Now, let us analyse the content of the forums of the largest translation resource ProZ.com in English and the "Code of Ethics" group in Russian on Facebook social network to identify the specifics of translation ethics and its application in practice. The subject of messages on the indicated resources is similar and raises the following questions:

- interaction with customers (forms, terms, conditions of remuneration; transfer of copyright, characteristics of employment contracts with translators, methods of conflict management, disclosure and dissemination of information, etc.);
- the behaviour of the translator in the course of translation job (cancellation of the order on the eve of the event; use of the signal button during simultaneous translation; withholding information in court translation, distribution of tips between a guide-translator and a driver, etc.);
- interaction with colleagues (transferring an order to another translator if it is impossible to carry it out by himself; claiming to execute an order of a colleague, etc.);
- discussion of the ethical code of the translator (creation of a national ethical code; discussion of already existing codes of industry organizations, translation agencies and communities).

The topics presented reflect the vectors of interaction between communicants of translation interaction (translator-customer, translator-translator), as well as the target audience of ethical standards in translation – members of translation community (in a broader sense – participants of the translation market). The diversity of the issues discussed correlates with the fact that today translation is an independent emerging branch of the economy with numerous important, pressing issues that need to be resolved for its successful development. In his work "Translation as a Profession", Daniel Gouadec very aptly highlights the basic rules of professional ethics of the translator and the contexts that every bona fide translator should follow (*Basic rules. Vis-à-vis the work provider, client. In the course of a translation/localisation job. With regard to payment. Vis-à-vis colleagues or fellow translators. Vis-à-vis partners*) [32]. This classification is explicitly proved by practical examples of discussion in forums.

Discussing ethical aspects of translation on electronic platforms is also a way of training prospect professionals. Thus, the analysed data contain the following messages:

- novice translators asking their senior colleagues to share their experience: *Greetings for everybody. I am wondering if you can help me with your experience I am a novice and I would like to know some examples in real life at doing this task of interpreting regarding to ethics where you have to make a decision. Case studies where you have to decide what to do. It does not matter what kind of task was requested by your client* [https://www.proz.com/forum/translation_theory_and_practice/273961-ethical_dilemmas.html];
- students in translation: *Dear colleagues! I would like to ask those of you who have worked / are working as an interpreter with heads of state, Prime Ministers, Ministers, etc., to fill out a short questionnaire on neutrality ... The research is anonymous; I will have access to the received data exclusively. The results of the research will be presented in the thesis "Neutrality of the Interpreter in Interpreting the First Persons", the electronic version of which will soon be available in the repository of graduation papers of Charles University (Prague, Czech Republic)* [https://www.facebook.com/groups/TranslatorsCodex/]. – Translation from Russian into English is made by the authors.

Temporal characteristic allows to classify messages into two categories: urgent and non-urgent. The first are situations in which it is urgently necessary to get a piece of advice from colleagues to adequately perform a professional task. For example, calculating characters in order to get payment for the translation of the Power Point presentation; choosing a translator's style of clothing when working with customers from the Middle East, etc. The second are the so-called "case-studies", when the translator discusses the actions taken after the fact. For example, whether to agree in the future to translate a document of poor typographical quality, on which the text for translation is not accurately displayed; whether translators can publish photos from events at which they worked in social networks, etc.

3. Conclusion

The analysis of theoretical works allowed to establish that ideas regarding the development of translation ethics evolved over several stages. Initially, the ethics of translation was interpreted as a reflection of the culture of society as a whole. Later, the culture / ethics of translation is considered as a social entity that takes into account the interests of partners on interaction (authors-initiators of translation, translators and recipients of translation). Being in the framework of equivalence-oriented translation, translators acquire a new competence to smooth out conflicts of interest, which directly implies following the provisions of the professional code based on the system of social values. Ethical principles of the translator as a participant of intercultural communication are also formulated. In Russian translation studies, the first ideas about translation ethics found their expression in the form of advice and recommendations to interpreters. Subsequently, postulates addressed to both translators and interpreters were formulated. Analysis of the content of professional group and forums allows us to identify several areas of discussion on ethical issues (relations «translator↔employer/customer/client», relations «translator↔translator», relations «translator↔novice translator/student», professional behaviour of the translator, code of ethics). This proves the fact that translation ethics today is not only a reflection of the culture of society, but also it represents an important tool for optimizing translation activities and developing the industry in general.

References

[1] Sandrini P 2011 *Lexikon der Geisteswissenschaften. Sachbegriffe - Disziplinen – Personen,* ed H Reinalter and P Brenner (Wien: Böhlau) pp 1195–1100
[2] Prunč E 1997 *TextConText 11* **NF 1.1997** 99–127
[3] Prunč E 2002 *TextConText 15* **NF 5/2** 283–299
[4] Prunč E 2002 *Translation zwischen Theorie und Praxis. Innsbrucker Ringvorlesungen zur Translationswissenschaft I,* ed L N Zybatow (Frankfurt am Main et al.: Lang) pp 139–166
[5] Prunč E 2003 *La direccionalidad en traducción e interpretación: perspectivas teóreticas, profesionales y didácticas,* ed D Kelly, A Martin et al. (Granada: Atrio) pp 67–89
[6] Nord Ch 1989 *Lebende Sprachen* **24** 100-105
[7] Nord Ch 1997 *Translating as a Purposeful Activity. Functionalist Approaches Explained* (Manchester: St. Jerome Publishing) p 154
[8] Nord Ch 2001 *The Translator* **7(2)** 185–202
[9] Prunč E 1997 *TextConText 11* **NF 1.1997** 99–127
[10] Chesterman A 1997 *Memes of Translation: The Spread of Ideas in Translation Theory* (Amsterdam and Philadelphia: John Benjamins Publishing Co.) vii, p 219
[11] Pym A 1992 *Translation and Text Transfer: An Essay on the Principles of Intercultural Communication* (Frankfurt am Main: Peter Lang GmbH, Internationaler Verlag der Wissenschaften) p 228
[12] Chesterman A 1997 *Memes of Translation: The Spread of Ideas in Translation Theory* (Amsterdam and Philadelphia: John Benjamins Publishing Co.) p 181
[13] Shadrin V I 2015 *Translation and Comparative Linguistics* **11** 13–16
[14] Hermans T 2009 *The Routledge Companion to Translation Studies* (London & New York: Routledge) p 94
[15] Hermans T 2009 *The Routledge Companion to Translation Studies* (London & New York: Routledge) p 95
[16] Chesterman A 1997 *Memes of Translation: The Spread of Ideas in Translation Theory* (Amsterdam and Philadelphia: John Benjamins Publishing Co.) vii, p 219

[17] Chesterman A 1997 *Translation as Intercultural Communication: Selected papers from the EST Congress, Prague 1995* (Amsterdam and Philadelphia: John Benjamins Publishing Co.) pp 147–157
[18] Chesterman A 2001 *The Translator* **7(2)** 139-154, 152–153
[19] Pym A 1997 *Pour une éthique du traducteur* (Arras: Artois Presses Université, Presses de l'Université d'Ottawa) pp 136–137
[20] Miniar-Beloruchev R K 1999 *How to become a translator?* (Moscow: Gotika) pp 86–93
[21] Chuzhakin A P and Palazhchenko P R 1999 *The World of Translation, or Eternal Search for Mutual Understanding* (Moscow: R.Valent) pp 13–14
[22] Zwilling M Ya 1964 *Translator's Notebooks* **2** 121–122
[23] Miram G E 1999 *Profession: Translator* (Kiev: Nika-Centre) p 160
[24] Alekseeva I S 2004 *Introduction to Translation Studies* (Moscow: Academia Publishing Centre) pp 29–33
[25] Alekseeva I S 2004 *Introduction to Translation Studies* (Moscow: Academia Publishing Centre) pp 26–42
[26] Alekseeva I S 2004 *Introduction to Translation Studies* (Moscow: Academia Publishing Centre) p 33
[27] Alekseeva I S 2004 *Introduction to Translation Studies* (Moscow: Academia Publishing Centre) pp 29–30
[28] Van Wyke B 2013 *The Routledge Handbook of Translation Studies* (London & New York: Routledge) pp 548–560
[29] Inghilleri M 2009 *The Routledge Encyclopedia of Translation Studies* (London & New York: Routledge) pp 100–104
[30] Inghilleri M 2013 *Interpreting Justice: Ethics, Politics and Language* (New York: Routledge) p 170
[31] Hermans T 2009 *The Routledge Companion to Translation Studies* (London & New York: Routledge) pp 93–105
[32] Gouadec D 2007 *Translation as a Profession* (Amsterdam and Philadelphia: John Benjamins Publishing Co.) pp 241–245

IOP Conf. Series: Materials Science and Engineering **483** (2019) 012011 doi:10.1088/1757-899X/483/1/012011

Sustainable production as the dominant value of environmental discourse

E B Pavlova, E V Terentyeva

People Friendship University of Russia, 6, Miklouho-Maclay St, Moscow, 117198, Russia
Volgograd State University, 100 Universitetskij Ave, Volgograd, 400062, Russia

E-mail: ev_terenteva@volsu.ru

Abstract. The environmental policy of modern enterprises aims to reduce the anthropogenic impact on the environment, to preserve natural resources, to create safe jobs, and to produce environmentally friendly products. All these elements combined are included in the concept of "sustainable production". The formation of the value bases of this concept is carried out in the process of communication which becomes the main way of introducing ideas of sustainable production into the consciousness of people, so this determines the relevance of our study.
The purpose of the present study is to establish key concepts that represent the idea of sustainable production as the dominant value. To achieve it, the rationale for distinguishing the industrial environmental discourse within the broader concept of "environmental discourse" was substantiated. Based on discourse analysis and content analysis of English texts of websites of companies engaged in production, as well as publications in high-ranking English-language journals on production ecology, the basic values of industrial environmental discourse and its key concepts were established.
The results show that sustainable production is the dominant value of modern industrial environmental discourse. This concept is verbalized in various naming units and correlates with the broader concept of environmental discourse – "sustainable development".

1. Introduction

Environmental communication can be distinguished into a separate type on the basis of at least two criteria: the topic of communication and the presence of the agent. In the topic of environmental communication in addition to the production and technical aspects associated with the presence of engineering and technological solutions to prevent and / or reduce pollution of nature, the social dimension plays a significant role. It is associated with the social responsibility of humanity for the state of the environment; therefore, environmental communication serving the sphere of production reveals an axiological component. The foregoing determines the relevance of the study, the purpose of which is to establish value bases that realize the idea of sustainable production in modern communication on environmental issues.

Russian scientists have begun to discuss the concept of "environmental communication" relatively recently, primarily in connection with the consideration of environmental media discourse [1]. American scientists have been developing environmental communication studies as an interdisciplinary scientific field based on the achievements of sociology, ecology, environmental management, and communication theory since the mid-1980s [2]. However, despite more than thirty years of scientific discussion about what should be understood by environmental communication, there has not been a unified approach to the interpretation of this concept among scientists yet. According to A. J. Flor, environmental communication is a set of communication strategies and methods that regulate the rational use of natural resources and compliance with the requirements of environmental protection [3]. M. Meisner interprets eco-communication as the process of information exchange on environmental issues, which brings together its interpretation with a broad understanding of environmental discourse [4]. Such an interpretation is quite appropriate, since discourse and communication can be equated if the language is studied in its natural habitat, that is, in the process of communication. R.J. Cox sees pragmatic and constitutional functions in eco-communication. According to the author, the essence of the pragmatic function of environmental communication is the implementation of education, prevention, persuasion and cooperation in the process of environmental activities. The constitutional function is realized in the process of forming the environmental awareness of the society. The constitutional function is aimed to introduce the systems of ecological values, ideologies, and views into people's consciousness through communication [5]. It is important to emphasize that the constitutional function is relevant for industrial environmental discourse, which can be considered as part of environmental discourse. Environmental discourse is a very wide communicative phenomenon. It is represented by such types as scientific, popular science, journalistic, legal, artistic, educational discourses, and discourse of environmental non-commercial advertising. Without a doubt, it is possible to distinguish the industrial type that serves the sphere of working relations, ensuring the verbalization of the environmental policy of modern enterprises. In this article, the industrial environmental discourse is used to refer to the communication on environmental topics between social actors generated by working relations.

2. Methodology

The study of industrial environmental discourse was based on advances in the field of discourse (K. Aijmer [6], D. Bolinger [7], J.P. Gee [8], M. Jung [9], D. Schiffrin [10], M. Stubbs [11]), ecological discourse (R. Alexander, A. Stibbe [12]; S. Chen [13], A. Stibbe [14]), ecolinguistics (A. Fill [15], E. Haugen [16], N.L. SHamne [17], S.V. Steffensen [18]), axiology (M. Reale [19], M. Rokeach [20], S. H. Schwartz [21]).

Standard methods of discourse analysis were used. Content analysis was also used, which was based on the study of relevant environmental publications in well-established scholarly journals over the past 5 years and environmental publications posted on the websites of well-known international environmental organizations and leading enterprises and holdings of various industries (manufacturing, mining, agricultural, construction) (DEA Deutsche Erdoel AG). Special attention was paid to the websites of WIPRO, TCS, Tech Mahindra Ltd., Essar Oil Ltd и Larsen and Toubro Ltd. These companies are among the global leaders in green manufacturing technology.

The classification of values by M. Rokeach [20] which proposes the division of human values into terminal (personal and social) and instrumental (moral values and competence values) was used to analyze the value content of texts. Based on the analysis of axiologically oriented argumentation, including the mention of the most important value concepts [22], key concepts relevant to the industrial environmental discourse were identified.

3. Results and discussion

The growing rates of production are accompanied by increasing depletion of natural resources and increasing anthropogenic load on nature as a whole. Humanity has already faced the growing scale of natural hazards and disasters and should understand in the near future the long-term consequences of the intensive production activities of the previous decades. Under such circumstances, maintaining the relative balance of social and ecological systems is becoming increasingly problematic. One of the ways out of this situation is to implant a system of environmental values in the minds of people. These values are actively created in the environmental communication.

3.1. Values of environmental communication through the prism of M. Rokeach's classification

The typology of the values that underlie environmental communication is correlated with so-called basic values, which are not a kind of complement to economics or politics. The basic values are the generalized ideas of people about the norms of behaviour based on historical experience and containing the cultural signs of each ethnic group separately and of humanity as a whole. To analyze the value content of the ecological discourse texts, the classification of values proposed by M. Rokeach, who divided the values into terminal (basic) values-goals and instrumental values-means, was used. The researcher considered the fact that values affect the functioning of a number of social phenomena studied by various social sciences and humanities. M. Rokeach divided terminal values into personal and social, and among instrumental values he distinguished values of morals and values of competence. In connection with the given typology of values, it should be noted that this division is relative, since the distinction between terminal and instrumental values may become blurred in different situations due to the fact that the same value can act as both terminal and instrumental. The values of industrial environmental communication correspond in general to terminal (security, harmony, comfortable life) and instrumental (responsibility, helpfulness) values according to M. Rokeach's classification. The environmental policy of modern enterprises aims to reduce the anthropogenic load on the environment, to preserve natural resources, to create safe jobs, and to produce environmentally friendly products that ensure a comfortable, safe life. All these elements combined correspond to the terminal and instrumental values listed above and are included in the concept of "sustainable production".

3.2. Terminal and instrumental values of industrial environmental discourse

The present study focuses on the analysis of nominations that realize the idea of sustainable production in industrial environmental discourse. As already mentioned, we distinguish industrial environmental discourse taking into account two criteria: 1) industrial environmental topics (sustainable production, sustainable materials and technologies, recycling) and 2) composition of subjects or agents of discourse. Representatives of business, industrial, mining, agricultural, and construction companies can act as subjects of industrial environmental discourse. In contrast to the industrial type, environmental discourse demonstrates a wider thematic diversity and a wider composition of subjects, including governments, environmental governmental and non-governmental organizations, communities of scientists, experts, eco-activists, advocates of environmentalism philosophy, eco-journalists. Environmental discourse verbalizes terminal values (fairness, nonviolence, safety, equality, partnership), as well as instrumental values (donation, helpfulness, responsibility). These values are reflected in slogans of environmental organizations and texts written by leading experts, scientists, eco-activists.

In the industrial environmental discourse, the most important terminal value is verbalized by such word combination as Protection & Safety. The concept of Sustainability is an instrumental value that ensures the preservation of the biosphere as a condition for survival of mankind.

There are several meanings of the lexeme sustainability in modern dictionaries of the English language:

1) the quality of being able to continue over a period of time

2) environment: the quality of causing little or no damage to the environment and therefore able to continue for a long time

3) the idea that goods and services should be produced in ways that do not use resources that cannot be replaced and that do not damage the environment [23].

Many authors argue that there is no common and unified understanding of what "Sustainability" is. One of the reasons behind the large number of definitions is the many different interpretations of the "sustainability" concept: e.g., seeing sustainability as an environmental initiative; as a goal or a process; as an integration of different aspects; or as a compromise between pillars, etc. Researchers claim that the large number of terms and definitions in the sustainable manufacturing research field is a barrier to sharing knowledge, particularly between academia and industry [24]. The most common understanding is the following: Sustainability – meeting the needs of society in ways that can continue indefinitely into the future without damaging or depleting natural resources. In short, meeting present needs without compromising the ability of future generations to meet their own needs [25]. However, it also needs to be defined more clearly. Let us give an example posted on a resource for manufacturers and industry experts on key manufacturing topics in the USA, which clarifies the conceptual component of the term "Sustainability" by comparing it with the term "Going Green": "Going green" is not the same thing as sustainability, albeit they are related. Although the terms are often used interchangeably, green is more frequently associated with a singular product or process. Examples include improving a specific operation so that it does not harm the environment or creating a product made entirely out of recycled materials. Sustainability is typically more associated with an organization's holistic approach; it takes the entire production process and logistics into consideration. For example, you may purchase a green product made out of recycled goods. However, if that product was made overseas, and environmentally harmful methods were used to transport that product to the United States, this would not be adhering to sustainable principles. In the manufacturing world, it is advantageous to focus on both green and sustainability [26].

For manufacturers and managers of the industrial sector, the same resource reveals the potential value of the concept of Sustainability, which is concisely reflected in the heading "Boost Workforce Morale and Innovations": Sustainability improvements are a collaborative effort. When employees work together to identify and implement green and sustainable initiatives, it fosters a culture of teamwork and continuous improvement. Employees work harder when they are engaged and have a sense of pride in their company. By internally communicating the importance of changes and the impact they are having on the business and environment, manufacturers will positively influence their corporate culture. Sustainability can also ignite innovation. For example, if you challenge your engineers and machinists to reduce material scraps or recycle more waste during the manufacturing process, it often leads to additional ideas for operational improvements [27].

3.3. Key concept of industrial environmental discourse

The instrumental value of Sustainability is concretized for the industrial environmental discourse by the key concept of Sustainable production. In environmental texts, this concept acts as the dominant value which unites such lexical units with the common semantics as "natural environment", "long time", "minimization of damage". These lexical units are related to the meanings of the lexeme "sustainable":

1) able to continue over a period of time

2) environment: causing little or no damage to the environment and therefore able to continue for a long time [28].

The key concept of Sustainable production verbalizes the notion that "until now, all human activities have been based on the paradigm of unlimited resources and unlimited world's capacity for regeneration; from now on, the awareness of the termination of this assumption means that all related behavioural models must be changed" [29].

In industrial environmental discourse, the linguistic representation of the key concept of "Sustainable production" is achieved through synonymy and metaphor which reveal its conceptual, figurative and evaluative components. The concept of "Sustainable production" is verbalized by such synonymous terminological combinations as Sustainable manufacturing, Sustainable manufacture, Manufacturing sustainability, Eco-Friendly Production, Environmentally Friendly Production, Sustainable and eco-friendly production, Sustainable approach to manufacturing.

Metaphorization is present in such nominations as Green production, Green manufacturing, Clean production. In the industrial environmental discourse, there is a desire to distinguish the meaning of synonymous terminological combinations, for example: "The use of the term green production rather than sustainable manufacture emphasises operational processes economic effectiveness within industry and their inter-relationship with the society. Green production is implicitly linked to product design as innovations in product and process transform the overall manufacturing system" [30].

The key concept of Sustainable production becomes the basis for the creation of a number of terminological nominations which includes hyphenated compounds (eco-friendly, environment-friendly), words in the semantic field of "management", "industrial production": Sustainable & environment-friendly industrial production; Eco-friendly Production Systems; Sustainable Materials and Technologies, Eco-friendly oil production, Water and Energy Sustainability, Sustainable approach to manufacturing, Sustainable construction, Green production technology, Green manufacturing technology; Sustainable management, Sustainable strategies.

The idea of sustainable production in the industrial environmental discourse contributes to a change in thinking based on sustainable manufacturing values, which become crucial for sustainable production and sustainable management [31]. Moreover, the change in values that accompanies green production also challenges conventional thinking on production operations design and management [30].

4. Conclusion

The paper concludes that industrial environmental discourse can be distinguished within the framework of environmental discourse on the basis of two criteria, such as environmental topics and the composition of subjects or agents of discourse. It also establishes terminal (Protection & Safety) and instrumental (Sustainability) values and shows that the dominant value of environmental discourse is the concept of Sustainable production which is verbalized in various naming units.

In this way, value potential of industrial environmental discourse, focused on the dominant value “sustainable production”, promotes the creation of environmentally friendly technologies, which allows us to look to the future of the planet and humanity with a certain degree of optimism.

References

[1] SHarkova E A 2014 Environmental communication as the informational and the communactive component of environmental political process. *Vlast'* **7** 43–48.
[2] Littlejohn S W, Foss K A 2003. Encyclopedia of Communication Theory (New Mexico: Sage Reference Publishing)
[3] Flor A G 2003 Environmental Communication (Quezon City: University of the Philippines Diliman)
[4] Meisner M 2009 What is environmental communication? Retrieved December 7. http://www.esf.edu/ecn/whatisec.htm
[5] Cox R 2010 Environmental communication and the public sphere (Los Angeles [etc.]: SAGE Publications)
[6] Aijmer K 2004 Pragmatic markers in spoken interlanguage *Nordic Journal of English Studies* **3(1)** 173–190.
[7] Shamne N L, Shishkina E V 2017 German popular scientific medical online media: structural and functional aspects. *Vestnik Volgogradskogo gosudarstvennogo universiteta-seriya 2-Yazykoznanie.* **16(2)** 143-151
[8] Gee J P 1996 Social Linguistics and Literacies: Ideologies in Discourses (London) 97–121
[9] Jung M 1996 Linguistische Diskursgeschichte, Öffentlicher Sprachgebrauch. Praktische und historische Perspektiven. Georg Stötzel zum 60. Geburtstag gewidmet. K. Böke, M. Jung, M. Wengeler (Hrsg.) (Opladen: Westdeutscher Verlag)
[10] Schiffrin D 1994. Approaches to Discourse (Cambridge, MA: BlackwellPublishers Inc.) 1994
[11] Stubbs M 1983 Discourse Analysis: the Sociolinguistic Analysis of Natural Language (Oxford: Blackwell)
[12] Alexander R & Stibbe A 2014 From the analysis of ecological discourse to the ecological analysis of discourse. *Lang. Sci.* A. **41** 104–10
[13] Chen S 2016 Language and ecology: A content analysis of ecolinguistics as an emerging research field *Ampersand.* **3** 108–16
[14] Stibbe A 2015 Ecolinguistics: Language, Ecology, and the Stories We Live by (London: Routledge)
[15] Fill A 2018 Introduction. The Routledge Handbook of Ecolinguistics. Routledge 29–41.
[16] Haugen E 1972 The Ecology of Language (Stanford University Press)
[17] Shamne N L 2010 EHkolingvistika: teoreticheskie i prikladnye aspekty. Language, person,

society. Sankt-Peterburg 125–136.
[18] Steffensen S V, Fill A 2013 Ecolinguistics: the state of the art and future horizons. *Lang. Sci.* **41**. pp. 6–25.
[19] Kushneruk S P 2018 Peculiarities of the official communication in professional communities. *Vestnik Volgogradskogo gosudarstvennogo universiteta-seriya 2-Yazykoznanie.* **17(2)** 170-173
[20] Rokeach M 1973 The Nature of Human Values (N.Y).
[21] Schwartz S H 1992 Universals in the Content and Structure of Values: Theoretical Advances and Empirical Tests in 20 Countries, In: M. Zanna, Ed., Advances in Experimental Psychology, (Academic Press: Orlando), pp. 1-65
[22] Valeeva N G, Ivanova N V 2011 Procedural issues of discourse analysis. *Vestnik RUDN, seriya Voprosy obrazovaniya: yazyki i special'nost'.* **3** 20–24
[23] Cambridge Dictionary: https://dictionary.cambridge.org
[24] Moldavska A, Welo T 2017 The concept of sustainable manufacturing and its definitions: A content-analysis based literature review *Journal of Cleaner Production.* **166** 744–55.
[25] Green Technology: green-technology.org
[26] NUSTBlogs:https://www.nist.gov/blogs/manufacturing-innovation-blog/five-benefits-embracing-sustainability
[27] NISTBlogs: https://www.nist.gov/blogs/manufacturing-innovation-blog/five-benefits-embracing-sustainability-and-green-manufacturing
[28] Cambridge Dictionary: https://dictionary.cambridge.org
[29] Garetti M, Taisch M 2012 Sustainable manufacturing: trends and research challenges. *Production Planning and Control.* **23(2-3)** 83–104
[30] Baines T, Brown S, Benedettini O, Ball P 2012. Examining green production and its role within the competitive strategy of manufacturers. *Journal of Industrial Engineering and Management.* **5(1)** 53–87.
[31] Abdul Rashid S H, Evans S, Longhurst P 2008 A comparison of four sustainable manufacturing strategies. *International Journal of Sustainable Engineering.* **1(3)** 214–29.

IOP Conf. Series: Materials Science and Engineering **483** (2019) 012012 doi:10.1088/1757-899X/483/1/012012

Translators Training for Joint Venture Employment

A V Prostov[1], O I Popova[1], E V Stepanova[2]

[1]Volgograd State University, 400062, Universitetskiy av., 100, Volgograd, Russia,
[2] Volgograd Institute of Management, branch of Russian Academy of National Economy and Public Administration, 400131, Gagarin street, building 8, Volgograd, Russia,

E-mail: [1]sacha_prost@mail.ru, [2]olgapopovavlg@yandex.ru, [3]stepanova.volg@mail.ru

Abstract. The paper deals with some issues of professional translator and interpreter training for employment in joint Russian and French enterprises. The key factors determining the need to develop the problem is changing political and economic situation, increased dependence on high quality translation and localisation for purposes of international business and higher requirements to personnel providing translation services in the context of globalization and the active introduction of digital technologies into economic, social, managerial processes in the global market in countries with different levels of economic development. At the same time, profound changes in translator and interpreter training have occurred over the past 20 years which is related both to globalization in the area of professional education, accessibility and relevance of various methods and technologies for teaching translation personnel, as well as to growing demands from business employers determined by the increasing complexity of economic activity. Scientific and technological context requires from mediator and translators new expertise and adaptive skills due to global trends of mass production including localisation of joint ventures operations and websites. The authors of the paper describe the principles of translators and interpreters training based on specific texts use, analyse advantages of "mixed" group format classes and compare them to traditional teaching approaches. The suggested methods are based on the Interpretive Theory of Translation developed by the French Translation School (ESIT); specific exercise series have been compiled for development of native and source language skills accompanied by relevant background knowledge and corresponding methodology. The authors have proved the effectiveness of the mixed group classed format conducted with the use of the developed complex of training exercises for the need of business environment helping to overcome linguistic, cultural and economic differences as well as to enhance comprehension within multinational companies through internationalization, standardization and localization.

1. Introduction

Over the last few years, the political and economic situation in the world has changed dramatically. Due to the sanctions imposed by the United States and the European Union against Russia, there has been a trend of withdrawal of international business from Russia, however, a number of large enterprises, including French, still preferred to stay in our country. Fortunately, nothing has changed in the partnership between Russian and European universities.

In this context, it is appropriate to recall the profound changes that have occurred recently in the modern educational space, affecting both the whole concept of higher education and certain areas of development of each Russian university. Russia's participation in the Bologna process and, as a result, trends in curriculum internationalization, prospects for international cooperation and student mobility have not spared most federal and regional universities in particular, Volgograd State University and Volgograd Institute of Management, branch of Russian Academy of National Economy and Public Administration. Increasingly, international students specifically French, German or English-speaking are nowadays present in translation classes together with Russian students, which mark a new stage in the Russian University's awareness of its place in global educational processes and provide new opportunities for effective training of translators for international business, in particular, for joint

Russian-French enterprises. It should be stipulated that mixed groups format of various types of translation classes requires the use of specific exercise and training complexes different from those used in single national groups aimed at enhancing understanding, native and source language skill as well as at building up relevant background knowledge [1].

Speaking about the traditional training of professional translation and interpretation personnel for the production sphere it is necessary to keep in mind the work with specific technical texts increasing cognitive and procedural awareness, moreover, training should involve all types of translation, both written and oral. Precisely the work with technical texts in the classroom for translation and interpretation is fundamental, as it involves mastering technical or business terminology which the translator will certainly face at the multinational company.

2. Materials and methods

The authors suggest particular methodology considering existing academic goals, strategies of international business integration, programs of digitalization of training technologies, intercultural communication and expertise on the basis of the existing Russian state training standards and curriculum. We have developed the didactic training model aimed at enhancing professional translation and intercultural communication skills, selecting appropriate translation strategies and technologies of the mixed group format experience. We argue that the methods applied to various types of texts in a range of potential professional communicative situations leads to the formation of translation competence.

In order to prepare a student for an oral analysis of such texts, it is necessary to develop the following skills: the isolation of meaning, a rapid information analysis, visualization, and an analysis of fragments of the basic ideas. When working with technical texts, all these skills are necessary, which does not contradict the use of exact terms. Any term has its meaning regardless of the context and situation. These are lexemes that retain the meaning on the speech and text levels like that they have on the language level, and thus impose on the interpreter a mandatory search for the equivalents. That is, on the one hand, the translator uses equivalents regardless of the original language, and on the other hand, recodes the term. In the methodological context of the Paris school, "future translator must learn to recognize concepts that are hidden behind the terms and only then look for the right equivalents in the target language" [2]. In preparation for the interpretation of special texts, students first need to study the texts in their own language. It is obvious that they do not use the method of "fishing" technical terms for further search in the dictionary while working with a technical text. Before you start looking up the correspondent term in another language, you need a full understanding of it in the original language. Here, dictionaries and encyclopedias are our best assistants [3]. As for transcoding, it is the easiest and most accessible component of interpretation, when it would seem sufficient to know a couple of terms of the original/the corresponding term of the target language. At the same time, we should not forget about the role of context in determining the value, given that the context can significantly change it. Work with context should be given a lot of time in the classroom.

In addition, consultations with experts play an important role in the work with special texts. In this regard, interesting is the experience of the Leonardo da Vinci Institute (Brussels, Belgium). The students of this university prepare to translate technical texts in collaboration with experts in a particular industry. Moreover, the mechanism of such interaction is very simple: the future interpreter him-/herself addresses to the organization, arranges a meeting. The experts, as a rule, do not refuse to give advice and they do it for free [4]. By the way, the university has long opened a Terminology center of European significance, which is constantly updated with new terms created by students in the process of translation practice. Being impressed by this kind of "capitalist" relationship, the authors of these lines decided to take a similar experiment in the mid-90s in post-Soviet Russia. Everything turned out well, except the fact that only those people whom the teacher knew personally agreed to give the students advice.

The Taking into account the fact that in addition to linguistic and translation training future professionals need to have competence in intercultural communication, conducting joint classes on consecutive and simultaneous translation with speakers of different languages and cultures is of greatest interest. The fundamental principle on which all training is based is the use of interpretative translation theory.

The founders of the higher school of interpreters (ESIT) at the Sorbonne – Paris III – and the authors of the interpretative theory of translation M. Lederer and D. Seleskovitch insist on the allocation of three levels of transfer based in his research on the practice of translation and translation conferences. M. Lederer talks about the level of language (translation of words out of context), the level of speech (the meaning of words is updated, that is, clarified by a specific context) and, finally, about the level of the text (the text contains additional extra linguistic information, takes into account the background knowledge of the translator and the information that appears when reading the text) [1].

Thus, the authors of the interpretative theory of translation emphasize the dominant role of extra linguistic factors in recreating the meaning of a statement in another language. It is this provision that causes the greatest controversy when discussing the translations made. The explanation in this case can be the use of different sources of information on the background of the problem [5].

Very painful are the topics related to the past military conflicts, in particular, to the First and Second World Wars. Russian students have a rather vague idea of the reasons why the First world war was unleashed, they do not know who were Russia's allies in that war, they know very little about the conditions of the (in many respects humiliating) peace. For many years, the ideology of "victorious socialism" did not allow to remember that war. As a result, in Russia almost no one remembers the date of November 11 – the day of the armistice. On the contrary, for Belgian and French students this day is well known since it is a holiday in their countries.

It is very difficult to translate texts on the World War II. If it is clear with Belgium– it fought on the side of Germany, it is not so clear with France. It was the only one of the four victorious countries occupied by the Wehrmacht and all years of the occupation it was ruled by the pro-fascist regime of Marshal Petain. On the other hand, the organization of the Resistance movement, the role of Charles de Gaulle... To understand the original text, it is necessary to "isolate" its meaning (consisting of linguistic forms and including non-verbal elements of the situation), and in order to translate, it is not enough to restore in another language the "linguistic forms" that are part of the text. To translate a text at the meaning level, you need to restore its meaning, so the translation at the text level does not coincide with the translation either at the language level or at the speech level. In this context, the difference in the semantic translation of the texts devoted to the two wars among the French, Belgians and Russians is quite understandable.

Unfortunately, the theme of war continues to be relevant (a vivid example is the situation in Ukraine). It is quite difficult for French-speaking students to understand the information they receive from the Western media, from conversations with their relatives living in Europe, from Russian sources. In this situation, students are forced to move to another level of translation – the level of speech, preferring not to use background knowledge, even to the detriment of the quality of the translation. In this regard, training of Russian and international students in mixed groups has many advantages compared to classes in traditional groups. First, the French-speaking student is always ready to help and explain to his Russian-speaking "fellow" unknown lexical units, that is, to act as a "live" explanatory dictionary [6].The mission is not only very noble, but also mutually beneficial: on the one hand, Russian students have the opportunity to deepen their knowledge of the French language, learning to perceive unfamiliar vocabulary through an understandable context. On the other hand, French students act in an unfamiliar role of linguists-theorists, analyzing well-known, at first glance, but such difficult with a more detailed consideration of the concept.

3. Results

It should be noted that this analytical activity is quite difficult for both Russian and international students. The main difficulty is not the content of the texts themselves, but a heavy official style with a huge number of subordinate clauses, with "empty" vocabulary and clichés familiar to native speakers, but having almost no sense and difficult to translate [6]. As an example, we can cite the materials of any forum, conference on economic/social issues in Russian and French.

At the workplace, the translator will necessarily face the problem of translating realities, so the experience of training in mixed groups is indispensable in preparing to work with this kind of vocabulary, as well as with similar at first glance concepts. For example, the concept of "agglomeration" (*agglomération*, f) was widely discussed at the consecutive translation class (topic – international and regional forums). First we had to go to the definitions. The business dictionary of the Russian language defines it as "a concentrated location of settlements, which are interconnected not only on a territorial basis, but also by various economic, social and cultural ties." [7].

The French dictionary defines agglomeration as "a collection of settlements that make up a village, town or city, regardless of administrative boundaries. Agglomeration consists of the main city and the urbanized area around it, which is its continuation (its suburbs)" (authors' translation). (*Une agglomération est un ensemble d'habitations qui constituent un village, un bourg ou une ville, indépendamment des limites administrative. Les agglomérations sont constituées d'une ville principale et de la zone urbanisée qui l'entoure de manière continue (sa banlieu).* The discussion revolved around what is considered the agglomeration of Volgograd. At first glance, everything is clear: this is the "millionaire city", the satellite town of Volzhsky, Krasnoslobodsk (on the opposite bank of the Volga), the worker's settlement of Gorodishche. However, a French student, who knows both the surroundings of Volgograd and Paris suburbs, could not agree to put an equal sign between the agglomerations of Paris and Volgograd.

The thing is that the union between the French capital and suburbs (*banlieue*) is much closer than between large and suburban cities in Russia. So in France, one city from another can only be separated by a crossroads. Nevertheless, after heated debates, a consensus was established, and all parties were satisfied, first of all, with the new knowledge gained not from stories or TV shows, but directly from the residents of this country.

In the course of this dispute was also raised the question of the coincidence of the scope of the concepts "*cite*" and the "*worker's settlement*". According to the definition, these are full equivalents, but to the surprise of Russian – speaking students in the French version there is a connotative meaning that is absent in Russian. For the French, *Cité* is a small town somewhere near Paris, inhabited by Arabs, where it is not safe to be at any time of the day.

It is worth noting that the presence itself of international students at the University makes it possible to prepare for classes in translation outside the classroom. We are talking about the so-called "work in a group". This format is quite common in translation universities in Paris and Brussels. The meaning of this work is to independently work out the skills of translation cursive outside the mandatory classroom. The fact that such classes are attended by native speakers of different languages, makes it possible to better control the quality of translation, allows you to clarify the nuances of translation directly from a native speaker.

Of course, there are some difficulties in such mixed groups. For example, French students often complain that they do not have time to write down some interesting translations quickly understanding what they are talking about. Similarly, when translating French-speaking students need more time to understand the content of the text, which inevitably causes a failure of the pace of the lesson.

A certain drawback can be considered psychological stress experienced by Russian students at the beginning of joint training – the mere thought that now will have to say something in French in the presence of French-speaking students, causes severe psychological stress.

However, this disadvantage is easy to turn into an advantage, if you remember that every time the translator has to overcome him-/herself, "stepping over" through fear and uncertainty.

Training in sight interpretation, which is an aspect of interpretation, plays an important role in the training of a professional interpreter. It is important to note that work on this type of translation takes a different place in the training of interpreters in different translation schools.

In the Soviet school and then in the Russian school, the translation is traditionally included in the structure of the course and state exams of the first and second foreign languages.

Exercises for sight interpretation contribute to the development of mental skills: the speed of understanding and extracting ideas from the text; the ability to perceive the text as a whole; synthesis skills, etc.

3.1. Written translation

The course is crucial for formation of both translation and intercultural communication expertise in the international mixed group classes. In the Russian federal state standard it is included to the variable part of the professional cycle. In mixed student groups the instruction is conducted in parallel with the course "Theory of Translation", "Abstracting and Annotation in Interpretation" and "Recording Technique". Students acquire skills in business and technical translation using specific texts of various types and genres for written communication [8]. We should stipulate that the first two years of studies are primarily aimed at source language skills development: translation and interpretation as types of professional mediation are introduced in the first semester of the third year. Being the first practical translation course in the profile "Translation and Translation Studies", written translation in mixed groups is founded on the idea that from the very first steps students become aware that the word-for-word translation does not provide efficient communication but has to be interpretative and imply the creation of the similarly cognitive and pragmatic effects of the source and target texts [9]. Referring to the framework curriculum, the training includes theory of translation and practical skills taught using different specific texts, traditional exercises and digital technologies [10].

3.2. Sight interpretation

Sight interpretation transforms written communication in oral. You must first look through the text in the original language, and then give it orally in the desired language. The format of the text received in the target language depends on its pragmatic orientation and communicative situation. Given that this type of translation is often in demand at the enterprise in order to quickly get acquainted with the incoming documentation or correspondence, the customer often does not need a full translation. In this case, the translator must make an abstract in the target language.

In the classroom for this type of sight interpretation one of the most important skills that a student must gain is the compression of the text. Working out the skills of speech compression is traditionally carried out at several levels: at the level of word-groups, at the level of phrases and at the level of super-phrasal units. In this case, two forms of exercise are used: the sight interpretation and translation at the hearing. In the classroom, it is possible to work with a modified version of this exercise and use only the development of speech compression skills at the level of text or text fragment. And students are previously informed about the conditional situation of communication: the characteristic of the enterprise, the summary of negotiations, the characteristic of participants are given. Sessions of exercises at the preparatory stage are about 7-10 minutes... understanding and extracting ideas from the text; the ability to perceive the text as a whole; synthesis skills, etc.

4. Discussions

Several other exercises are required when developing skills for a full sight interpretation, which is necessary in such communicative situations as, for example, a report or a press release on the results of the economic forum, etc. In this case, training involves several stages.

Select a French text of 200-300 words. As a rule, it is authentic, representing a part of the indictment or press release drawn up at the end of the economic forum. The student has 3-5 minutes to read the text and 5-7 minutes to translate. It is desirable to record the translation on a voice recorder. To achieve greater efficiency in learning to translate from a sheet of paper it is necessary to follow several steps.

The first stage is a preview reading (2-3 minutes). At this stage, the student has the opportunity: to identify the type of text and understand the main content; to highlight key words; to identify possible problems; to fix in writing any possible equivalent for key words; to try to divide the text into fragments long enough to be able to analyse and understand them and short enough to then translate them orally. These fragments may include a number of sentences or a whole paragraph.

The second stage is to record the transfer to an electronic device. The third stage involves the comparison of the translation with the original text. The next stage – the analysis and search for solutions to the difficulties – involves analyzing the mistakes and trying to understand their cause (lack of knowledge of vocabulary, inability to find an equivalent, loss of the beginning of the phrase, etc.), finding ways to solve them (allowed the use of search engines and dictionaries).

The next step is to discuss possible variants of translation cursive with further fixation of the recorded translation using the recording technique.

And at last, students submit the final translation taking into account all the corrected mistakes, starting from the second stage. In conclusion, the post-translation analysis of the finished translation text is carried out.

The work in simultaneous translation classes is built in a slightly different way. At the preliminary stage of teaching simultaneous translation, the Russian school offers special exercises aimed at overcoming the difficulties associated with this type of translation, for example, listening with counting. For listening there is a text in a foreign language, one student rhythmically counts aloud, then any of the listeners transmits the main content of the heard text in the target language [2]. The French translation school uses similar exercises. The only difference is that students pronounce three-digit numbers in descending order (for example, 234, 233, 232, etc.) therefore, there are probably no differences between French-speaking and Russian-speaking students.

The training of translators and interpreters at each stage is conducted on the basis of specific types of texts used selected from simple to more difficult. All the exercises are of three main types corresponding to the three steps of translation which include pre-translation text analysis, translation itself and checking/correcting of the translation.

Initially, students form research competence learning to retrieve information and to acquire additional background, cultural and factual data to fully understand the source text and to render it fully into the target language [11]. Digital technologies ensure effective information retrieval results. The skill is also accompanied by selection of suitable translation strategy. In the course of preliminary research glossaries of specific terms and cultural background data are compiled by students [12]. They focus on ensuring adequate impact of the target text similar to that of the source text itself due to understanding of cultural background in the mixed groups.

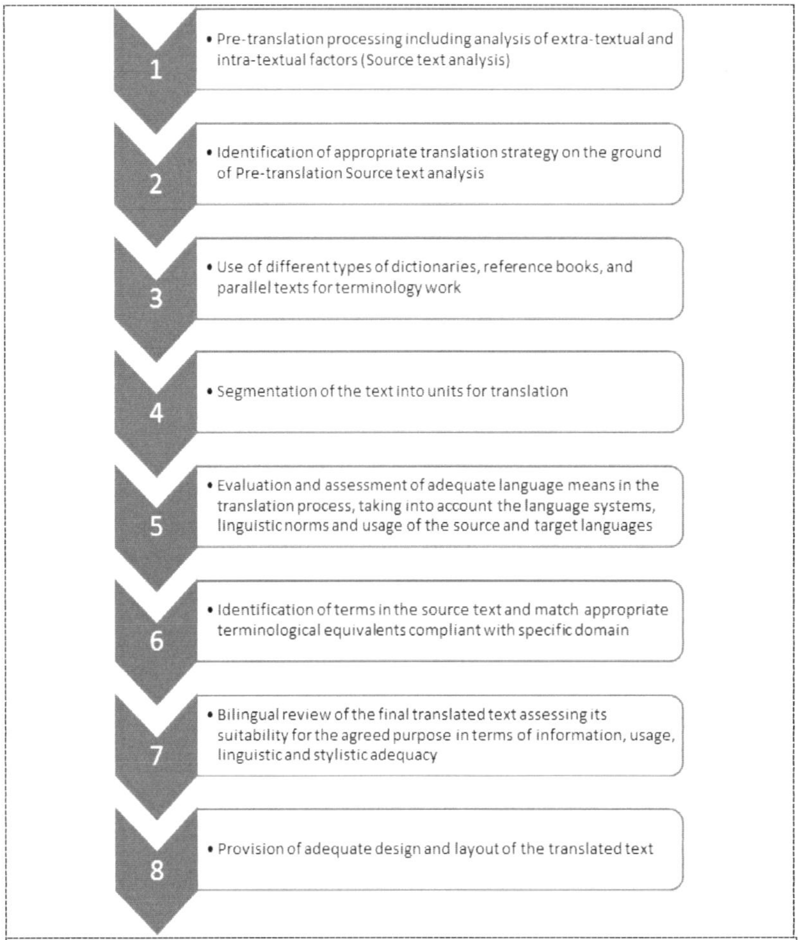

Figure 1. Didactic Strategy in Translation Training).

However, the basic principle of the interpretive theory of translation: "one should translate not words but thoughts, not the letter but the meaning" [3]. In this regard, among the preparatory exercises of undoubted interest are exercises "report-comment" and "fairy tale". The first involves working with simple texts in the language "B" (first foreign language) and "C" (second foreign language). The student is required to tell, describe, what was heard in the language "A" (native language). And the theme (plot) in both languages should be similar. Thus, attention is drawn to different ways of expressing the same events in different languages. As for the second type of exercise, in this case, you can take a fairy tale or a story well-known in both cultures, for example, "Cinderella" or "Red riding hood". Russian-speaking students closely monitor the action of the French fairy tale, but instinctively use in the translation purely Russian phrases, familiar to them from childhood. In addition, the translation is often carried out "ahead of schedule" – translated what has not been said. This is due to the fact that the story is familiar from childhood – the student does not even translate, but rather retells what he/she has long known. As practice shows, such carelessness is fraught with some danger. Thus, during the simultaneous translation of the fairy tale "Red riding hood" Russian-speaking students did not notice some changes in the plot, such as the fact that Red riding hood was not carrying cakes with jam, as traditionally described in the fairy tale in Russian, but croissants and a bottle of wine. The fact that the wolf did not eat the grandmother, but locked her in a trunk, and that it was not three woodcutters who released her, but one hunter, also did not confuse the students. French trainees, on the contrary, translated the content very accurately, despite the existence of a large number of versions of this fairy tale.

5. Conclusion

Therefore, the training of future translators to work at joint ventures includes several complementary aspects. First of all, it is work with special texts related to various industries. Working with these texts involves possession of all types of translation – written and oral (sight translation, consecutive and simultaneous translation). Special attention should be paid to the problems of intercultural communication in order to better understand each other within an international team. In this regard, classes in "mixed" groups (e.g. French/Russian) seem to be the most effective.

Specific authentic educational material adopted to mixed group needs, simulation of typical translation situations of intercultural communication, ability to determine the type of the source text, its pragmatic purpose and potential receptor are essential in translators training in Russian higher education institutions. The authentic text used for training during translation classes include business correspondence, (letters of inquiry, complaint or offer), official documents (invitations, condolences congratulation, etc.), corporate documentation and papers of individuals (birth certificates, Certificate of Birth, of Secondary Education, etc.), documents of business and multinational corporation (statutes, agreements of federal and local government bodies using digital technologies.

References

[1] Lederer M 1994 *La Traduction aujourd'hui* (Paris: Hachette) 52-66
[2] Alekseeva I S 2000 *Vocational training of a translator: studies. manual for translators and teachers* (SPb: Institute of Foreign Languages) p 79
[3] Selescovitch D, Lederer M 2002 Pédagogie raisonnée de l'interprétation *Communautés européennes: Didier Erudition* p 169
[4] Stepanova E V, Buletova N E 2018 Legal regulation of internationalization in Russia public institutes of higher education: a case of Volgograd Institute of Management *Volgograd: Legal Concept* vol 17 **1** 104-112
[5] Stepanova E V 2015 Translation studies: didactic training model of intercultural communication, experts in higher education institutions of Russia *EDAMBA 2015 International Scientific Conference for Doctoral Students and Post-Doctoral Scholars The Era of Science Diplomacy: Implications for Economics, Business, Management and Related Disciplines Conference Proceedings University of Economics in Bratislava, Vydavatel'stvo EKONÓM* 831-838
[6] Korolkova S A, Novozhilova A A et al. 2017 Tourist Discourse: Strategies and Problems of Translation of Hotel Sites *Bulletin of the Volgograd State University* vol 16 **3** 80-89
[7] Tomareva I G, Marianina L A et al. 2016 Communicative ability in teaching a foreign language (on the example of English and German) *Kazan Science* **6** 50-51
[8] Gulyaeva E V 2017 Pragmalinguistic Characteristics of Global English (based on the Nobel lectures) *Bulletin of Volgograd State Pedagogical University* **8 (121)** 93-97.
[9] Orlova E R, Buletova N E 2016 Paradoxes of the service economy and the role of the state in strategic management of their effects *International journal of applied engineering research* **11 13** 8024-8030
[10] Demushina O N 2018 Electronic communications as a channel of interaction between government and society at the regional level *Proc. Int. Conf. Strategic communications in the modern world* **1** 471-475
[11] Usacheva A N, Sidorovich T S 2014 Interpreting for a foreign delegation (English) *Volgograd State University* 75
[12] Buletova N E, Zlochevsky I A 2018 Application of effective ranking method in the construction of the ratings (for example GDP per capita) *Journal of Engineering and Applied Sciences* **13 7** 1725-1731

IOP Conf. Series: Materials Science and Engineering **483** (2019) 012013 doi:10.1088/1757-899X/483/1/012013

Specificity of intercultural communication and documents translation in aluminium industry

Alexandra Usacheva, Victoria Skrotskaya

Volgograd State University, Universitetsky Prospect 100, Volgograd 400062, Russian Federation

usacheva.alex@yandex.ruбviktoriyaskrotskaya@mail.ru

Abstract. The paper is devoted to studying specificity of technical translation in aluminium industry. It is based on our on-site translation experience at the aluminium plant. The authors analyse the specificity of oral and written communication modes between technical specialists of different cultural backgrounds and languages. The study provides some important patterns of aluminium industry installation peculiarities referring to translating phone negotiations, meetings, reports, official letters, acts and manuals.

1. Introduction

Intercultural communication in the workplace, for any branch of industry as well as for aluminium industry, sets a particular portion of translation restrictions in face-to-face and technology-mediated interaction. "The workplace is a highly meaningful context for intercultural communication where persons who come from different countries, identify with different ethnic groups or speak different languages get to collaborate and develop relationships with one another [1]". Whenever a translator's job is conducted within the framework of their delegated duties, it should be done under a number of conditions defined either by the norms of business communication, or by the standards designed for a particular branch of industry, or even for a definite plant or company. There is also some specificity regarding a certain type of communication, i.e. oral speech mode and written speech mode.

Communication modes can vary greatly in aluminium industry. As we were to moderate communication between subcontractors and supervisors, translation was mostly required during the process of supervised installation. In terms of oral speech mode, we would like to consider the following list of situations which the interpreter is sure to face:

- Phone conference/ negotiations
- High-level meetings
- Scheduled/ routine meetings
- On-site communication

In terms of written speech mode, there is a list of documents that are most likely to be translated:

- Manuals
- Official letters
- Acts
- Reports

Nowadays, the biggest amount of aluminium production equipment is supplied by foreign producers, and the supplier's party delegates their staff responsible for correct assembly and installation of the purchased equipment. In our case, two *Furnace Tending Assemblies as well as two anode handling systems* have been purchased by the aluminium smelter and it is extremely important to provide smooth communication between the two parties – the Buyer and the Seller. While supervisors usually have some experience of working with a translator, it is an absolutely new concept for subcontractors which results in serious challenges for an interpreter working on-site.

IOP Conf. Series: Materials Science and Engineering **483** (2019) 012013 doi:10.1088/1757-899X/483/1/012013

2. Discussion

2.1. Oral speech mode

Within the conditions stated above for oral speech mode, the interpreter should provide either simultaneous or consecutive translation.

During phone conferences or negotiations, consecutive interpreting should be provided. The main challenge the interpreter may face, is the official phone conversation protocol. On the one hand, the main function of the interpreter is to convey the idea of a speaker, but on the other hand, in real life, it is the interpreter who makes a call and starts a conversation. The interpreter should know the phone conversation protocol, because in spite of the fact that the top-level managers are supposed to know official courtesies, in most cases it becomes the job of the interpreter to make sure that the communication will go smoothly.

First of all, there is a great difference between the norms and standards in Russia and the norms and standards in Europe. When the Russian party says, "Send us the document", we translate, "We kindly ask you to send us the document". Another difference is that the managers from some European countries feel much more relaxed about the standards. They say, "Hi" and "Bye-bye", while Russians stick to more formal and official equivalents like, "Good afternoon" and "Goodbye".

Another issue we need to take care of is the transfer of names. "Proper names create special problems for translators [2]." In many cases, so much communication takes place, and with so many different countries and companies that the managers forget *who is who* and what is the right way to pronounce names properly. Since nowadays English is the language of official communication, and all the negotiations are conducted in English, the interpreters working with this language have to face the problem of names written in languages they do not know. For example, French names present a considerable difficulty for the Russian-English translator because some letters are not pronounced in the French language, while others are pronounced differently like in such names as *Aloïs* or *Marceau*. To cope with this challenge, it is highly recommended to know the basic reading rules of the language and culture we communicate with. Another solution is to give a phone-call to your peer specialising in the source language though it is not always possible.

Working for a big company, you inevitably have to talk with representatives of a large number of companies producing equipment for various stages of aluminium production. It means that the translator should be able to get familiar with new products in almost no time, as well as to rely a lot on their proposition when it is necessary to get into technical details. The translator should be able to match the terms pronounced by one party with those pronounced by the other party – in this kind of translation success comes when you understand what is discussed by the parties. However, it is necessary to add that since both sides are interested in conveying their ideas they try to do their best to explain profoundly what they mean. In many cases before the conversation starts, the interpreter is explained which issue will be discussed within the framework of a pre-planned phone-call.

The protocol for high-level meetings differs from the one we use for phone negotiations. First of all, not everything is translated as most of the scheduled high-level meetings are attended by the same managers representing both parties.

Let's consider one more important issue – translation mode. In many cases it is required to combine simultaneous and consecutive translation. When general matters are discussed, it is preferable to translate simultaneously in order to save time, but when it comes to some technical matters the clients prefer to be translated consecutively thus making sure that the meaning is conveyed accurately and precisely. They also make a lot of sketches and drawings which allow the interpreter to clearly see what exactly is being discussed. Time is the main factor. Actually an average meeting is held for an hour or two. It means that non-stop simultaneous translation continues for at least 30-40 minutes exceeding the approved standards for a conference interpreter [3, 4].

Everyday meetings are less formal and continue for 10-20 minutes. They may be held both on the site and in the office, but normally include the discussion of the planned work or the work that was done the day before. These meetings more resemble face-to-face friendly communication between two people rather than official negotiations. Thus, the language that both communicants use is quite informal and plain. The major type of translation in this case is consecutive [5].

Intercultural communication at the workplace can be various [6, 7]. In most cases we work with the subcontractors and supervisors. While supervisors usually have some experience of working with a translator, for the subcontractors it is an absolutely new skill that some of them cannot master instantly. There are some problems that any interpreter comes across while working on the site as "the workplace is not a mere container for encounters and relations. It is a social space that prescribes a specific orientation to relationships and actions, and suffuses interactions with considerations of professional and organisational identification, and with economic and political implications" [6]

Many subcontractors have problems with expressing their thoughts clearly. They start somewhere in the middle and don't bring their thoughts to any logical conclusion. They also tend to use professional slang instead of choosing proper technical terms for naming processes and objects. For example, in terms of working on assembling the anodes handling systems a subcontractor may come to the supervisor and say, "We have a problem with the hopper". The real problem is that we have more than two hoppers in the system. But an even bigger problem is that when the supervisor asks to elaborate a little, the subcontractor usually answers, "You know, that one," while being at the supervisor's container. Another example is when the subcontractor uses a slang word and then cannot explain what that is. Mechanics also like to give proper names or professional slang names to their tools, for instance, a hammer can be called "Masha", and a water level is just a "blue stick".

They do not name the objects at all. When you see the equipment sometimes it is easier to just point at the objects and say "this" or "that", but the problem is that when the clients get inside the equipment there is not always enough space for the interpreter to get in. It means that you have to be outside but keep translating. If you don't see the objects that are pointed at, it is highly probable that translating even simultaneously you wouldn't be able to keep up with the speed of the speaker that means that when they point in the right direction you are still translating about the left.

Supervisors are not native speakers. Since almost all the technical terminology is in English, it is preferable to work with an English interpreter, however, it can also set some minor difficulties like lack of vocabulary or a specific accent of a client.

The translator cannot write anything down. Working on the site means working in a noisy, dirty and hectic environment. The people working there are pressed with time and cannot spend an hour discussing issues at hand. They need fast and efficient communication with foreign specialists. Thus, the translator doesn't have time to make notes, put down numbers and figures. They should translate right away converting every spoken word from one language into another.

2.2. *Written speech mode*

While the translation of technical manuals normally doesn't present any difficulties for the experienced translator, it can be tricky for the one who is new in this job. Manuals very often contain an abundance of specific industry terminology as well as accepted well-known terms used in a particular supplier context. For example, in the context of crane building, *exhauster*, *blower* and *vacuum pump* denote one and the same thing. "Translation of manuals requires not only necessary translation skills, but also certain tools that make such translation consistent and clear to the user. The key ones include computer-assisted translation software. Such software or platforms enable smooth and efficient translation based on previously translated texts. With such software translators can use glossaries. These are specialist thematic dictionaries compiled based on previous translation projects. Such glossaries are particularly useful when it comes to translation of manuals" [8].

Another issue with manuals is that they are supposed to be written either in the language of the country where a company is based, or in the language which is used by this company's employees at the proficiency level high enough to communicate fluently. In many cases the translator has to deal with an English text with another language grammar, structure, some words, and even mistakes. For instance: *It is used to move the bundle previously and made the two other bundle. The telemeter who is given the absolute position, etc.*

Official letters do not pose much of a problem for the translator, except for specific terms common for technical translation in general. "Professionals, especially those in industry, are busy people… Many disciplines and specialties have a special language all their own. Technical terms are a helpful shorthand when you're communicating within the profession" [9].

Acts, on the other hand, turn out to be quite debatable [10]. One can find a lot of various documents with a wide range of text types nominated *Acts* in the Russian language. That is why the Russian word *Act* can be translated as *act*, *official notification,* and *report*. However, at some enterprises there are their own standards to be followed. At Volgograd Aluminium Smelter, for instance, all the Acts are translated as *certificates*. It is extremely confusing for the recipients of such a document since this title does not reflect its real content. Another peculiarity that should be taken into consideration is the way the company presents itself. For example, everyone knows that the type of a business entity should not be translated with the help of a foreign equivalent. However, some of the companies do that traditionally: *ОП ООО "РУСАЛ ИТЦ"* is translated as *"RUSAL ETC" LLC"*, even though LLC is a US-specific form of a private limited company.

Technical reports prepared by supervisors present a completely unique piece of document. There is one common problem in many non-English-speaking countries: "In these Oman based companies, translations are produced by different social agents (e.g. translators, managers, employees, chairpersons) who play a significant role in setting up new domestic laws, such as the Capital Market Authority (CMA), applying international standards and principles (e.g. IFRS), and introducing new terminology to Arab audiences" [11]. In our case, the two main types of reports are daily reports and weekly reports. Daily reports include the amount of work done during the day with some comments, the amount of manpower present on the site, as well as some complaints such as lack of tools or equipment. There is also a separate column for the work planned for the next working day, comments on the required tools, and the amount of manpower to conduct the job. The report is written in two languages, namely English and Russian, and signed by two parties, the client and the contractor.
Weekly reports consist mainly of the information already mentioned in daily reports; however, the structure is a little bit different. A weekly report includes such data as average manpower per CW, summary of meetings, executed works, achieved progress plus total progress, quality control, acceptance and miscellaneous. This document is also prepared in two languages and signed by both, the client and the contractor. No matter how simple it may seem, even this kind of documents sometimes rise intercultural conflicts: "the periodic engineering report can become a source of conflict when North American engineers collaborate with colleagues abroad" [12], [13].

Taking into consideration the information stated above we come to the conclusion that the main peculiarities of technical translation (written speech mode) at the aluminium smelter consist of the translation standards approved by the smelter, special terminology that include colloquial and slang words, terminology typical for a particular company as well as the national specificity of the source text prepared by non-native speakers.

3. Conclusion

The terminology of the aluminium industry becomes a prominent feature of the technical translation in terms of this industry. However, the peculiarities of intercultural communication that were discussed in the present paper are characteristic not only of the technical translation in the aluminium industry, but also of technical translation in general. That is why we sum up the most important findings both for interpreting and translating activities.

3.1. For the oral speech mode

They are the following:

- norms and standards of the official phone communication in Russia differ from the norms and standards in Europe;
- ability to read foreign names partially defines the success of the phone communication;
- knowledge of the high-level meeting protocol is essential for the provision of smooth communication;
- lack of appropriate terminology on the job site puts the interpreter in the situation when they should be able to navigate quickly and freely in the framework of the discussed topic.

3.2. For the written speech:

They are the following:

- diversity of terms naming one and the same object or process forces the translator to juggle the multi-variation of terminology;
- translation standards approved by the smelter restrict the translation variation that can be potentially used by the translator;
- terminology typical for a particular company puts the translator into the situation, where they have to get familiar with the words usage in the limited time;
- national specificity of the source text prepared by the non-native speakers requires the adaptation of the structure in the translated text.

References

[1] Lahti M, Valo M 2017 Intercultural Workplace Communication. *Oxford Research Encyclopedia of Communication.* from
[2] Parianou A 2007Translating Proper Names: a Functionalist Approach. *A Journal of Onomastics* **55** 407-416.
[3] Benoît Kremer, Klaus Ziegler 2017New ISO standards relating to conference interpreting. *aiic.net*.
[4] Oliver Pospiech, Almute Löber, Angelika Eberhardt, Anja Rütten, Julia Böhm, Klaus Ziegler, Angela Drosser 2015 How to make a living as a conference interpreter: Part 2 – Best practice for pro. *aiic.net*
[5] Daniel Gile, Consecutive vs. Simultaneous: Which is more accurate? Interpretation Studies, No. 1, December 2001, pages 8-20. (c) 2001 by the Japan Association for Interpretation Studies
[6] Lahti Malgorzata Communicating interculturality in the workplace. Jyväskylä: University of Jyväskylä, 2015, 70 p.
[7] Yates Lynda 2010. Speech act performance in workplace settings. In Alicia Martinez-Flor & Esther Usó-Juan (eds.), Speech act performance: Theoretical, empirical and methodological issues, 109–126 (Amsterdam: John Benjamins)
[8] How to translate manuals? https://summalinguae.com/technical-translation-manuals/
[9] Bly, Robert W 2003Webster's New World Letter Writing Handbook (Webster's New World) 408 p
[10] Byrne J 2006 Technical translation: Usability strategies for translating technical documentation. Technical Translation: Usability Strategies for Translating Technical Documentation 1-280
[11] Khalfan S Al Obaidani *Translating Terminology in Business Annual Reports* (English-Arabic)
[12] Kushneruk S P 2018 Peculiarities of the official communication in professional communities. *Vestnik Volgogradskogo gosudarstvennogo universiteta-seriya 2-Yazykoznanie.***17(2)** 170-173
[13] Luzón María 2005 Genre Analysis in Technical Communication. *In IEEE Transactions on Professional Communication* **48** 1-11

Professionally-related communication: sociolinguistic monitoring of most frequently used words of youth sociolect

A A Petrova[1], N A Sytina[1], O Ergunova[2]

[1]Volgograd State University, 100 Prospekt Universitetskij, Volgograd, 400062, Russia

[2]The Ural State University of Economics, Department of regional and municipal economics and management Ural Federal University named after the first President of Russia B.N.Yeltsin, Department of Integrated Marketing Communications and Branding Yekaterinburg, Russia forstud@vstu.ru

n.sytina@volsu.ru

Abstract. The current article deals with functioning of youth sociolect in professionally-related and everyday communication. It relies on the concept of professional discourse. A special attention is given to professional interaction, based on professional interests. The field study is developed to investigate the usage of youth sociolect words, involving macro-sociolinguistic monitoring of frequency occurrence of English slang words in production sphere. The initial stage of the field study included the questionnaire development, defining the boundaries of monitoring, the number of lexical items under monitoring, the place of monitoring and the age of respondents. The second step in the project was monitoring. Finally, the results of monitoring were analyzed. The study was carried out through on-line questionnaire, face-to-face and remote interviewing, administered to respondents both in the production sphere and other spheres of activity. Native and non-native speakers from different countries and regions participated in the on-line questionnaire. The monitoring of frequency usage demonstrates that a limited number of lexical units of English youth slang are used in professional activity; the other part of lexicon presents a diverse picture. The article gives a number of diagrams and graphs to provide statistics confirming priority and universal use of English words of youth slang.

1. Introduction

As is known, the typology of social communication is based on the stratification of society and professional activity, with industrial specific communication features in economy, politics and social sphere as well [1]. Moreover, there are such spheres as everyday communication, professional, business communication, sport business communication and etc. It is assumed that in our modern era of communication, professional communication, which is connected with production relations, processes and other components of production activity, is ever more relevant as we spend most of our hours at workplace and occupation provides us with livelihood opportunities [2]. Thus, it makes possible to speak about professional discourse as an interaction of people united by professional interests. Professional discourse has received scholarly attention depending on the respective research approach: strategic planning in communicative management of production companies [3], corporate communication at enterprise in terms of communication backgrounds of business discourse [4], professional oath [5] and etc. However, there is no empirical research on professional communication in the domestic communicative studies [cf. 6].

The purpose of this study is to investigate the frequency use of youth words, taking into consideration the fact that youth language penetrates in all spheres of communication including the professional language. This idea is confirmed by findings of Degaltseva which provide evidence for the fact that most jargon words used by bank employees are the words of youth and computer jargons [7, p 66]. Moreover, the frequency of word use of the youth sociolect in professional and everyday communication will be compared in this study. As it deals with the processes of society and language interaction, this article provides understanding of youth language as a variety of social and group dialects; thus, in this respect, the term "youth sociolect" is used. The youth sociolect is a multidimensional phenomenon including such parameters as language, social status, culture and age; at the same time it performs differentiating as well as integrating functions [8]. The evolution of the youth sociolect can be proved by its concept and theme development, increase and renewal of lexical items and phraseology. Currently, young people demonstrate different sociolects depending on a sphere of communication [9]. This allows us to suggest "common youth jargon" which it is understood by all young people regardless their social background, occupation and professional activity and is being actively used in conversational interaction [10].

The study focuses on lexical units of youth sociolect, codified in the Dictionary of Youth Slang, based on English, German, French and Russian languages [11]. Therefore, the frequency usage of English words from this dictionary, used in professionally-related and everyday spheres of communication, is studied [12].

Related research topics include sociolinguistic studies, studies on evaluation, studies on pragmatic and youth language.

The study is based on the methods adopted in the area of macro-and microsociolinguistics: field study, questionnaire, face-to-face and on-line interviewing, statistical data processing as well as methods of scientific description: observation, collection of information, interpretation and classification [13].

2. Discussion of results

This work is thus designed to investigate the use of youth sociolect in professional as well as everyday communication. The starting point for the analysis involved the development of the pilot project for the field study aimed at macro-sociolinguistic monitoring of the frequency use of English youth slang words. It included the questionnaire development, the boundaries of monitoring, the number of lexical items under monitoring, the place of monitoring and the age of respondents; the second step of the project included monitoring. Finally, the results of monitoring were analyzed [14].

Native and non-native speakers (they speak English) participated in the Internet on-line survey, the 16 - 44 age range. The survey was carried out on such websites as Google Forms and Survio. The respondents were from different spheres of activity (Professional employee, student English teacher, Employee, Industrial Engineer, Part-time employee, 3D Artist, HR employee, Assistant, Foreign Education Consultant, Freelancer), as well as different countries of residence (Poland, United Kingdom, Russia, Viet Nam, Yemen, Thailand, Belgium, Croatia, Bulgaria, Macedonia, Turkey, Germany, Algeria, U.S.A., Greece, Egypt, Indonesia, Korea).

The questionnaire includes a sample of 20 slang words from the dictionary, listed in alphabetical order.

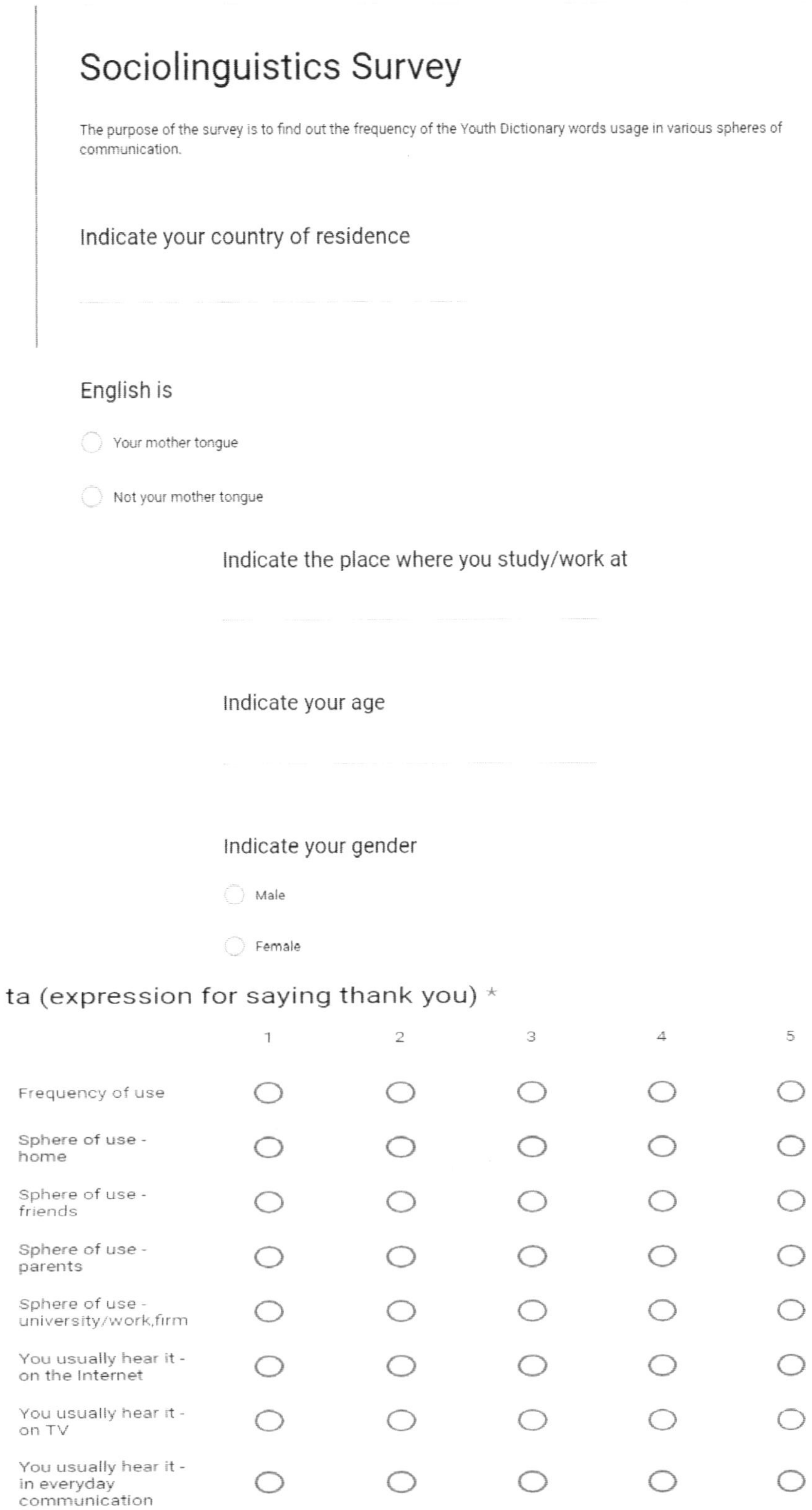

Sociolinguistics Survey

The purpose of the survey is to find out the frequency of the Youth Dictionary words usage in various spheres of communication.

Indicate your country of residence

English is

- Your mother tongue
- Not your mother tongue

Indicate the place where you study/work at

Indicate your age

Indicate your gender

- Male
- Female

ta (expression for saying thank you) *

	1	2	3	4	5
Frequency of use	○	○	○	○	○
Sphere of use - home	○	○	○	○	○
Sphere of use - friends	○	○	○	○	○
Sphere of use - parents	○	○	○	○	○
Sphere of use - university/work,firm	○	○	○	○	○
You usually hear it - on the Internet	○	○	○	○	○
You usually hear it - on TV	○	○	○	○	○
You usually hear it - in everyday communication	○	○	○	○	○

Figure 1. The example of the questionnaire on the website Google Forms:

IOP Conf. Series: Materials Science and Engineering **483** (2019) 012014 doi:10.1088/1757-899X/483/1/012014

The study examined 200 lexical units. 155 informants were included in the survey. Let us consider some examples.

2.1. Field study 1.
The questionnaire includes the following slang words [11, pp 92-96].

1. *Ya* • yes; you; your
2. *To yack* • to talk, especially uninterruptedly and idly; gab; chatter
3. *Yada* • et cetera (popularized by the Jerry Seinfeld tv show)
4. *Yahoo* • exclamation of delight; "hooray"; a weird person
5. *Yard* •a thousand dollars
6. *Yas* • used to express high levels of approval; often times the word is drawn out to add emphasis!
7. *Yawn* • something that is very boring
8. *Yay!* • an exclamation of happiness
9. *Yeah* • conversational form of "yes"
10. *Yeet* • used when the speaker is excited about the situation
11. *Yeezy* • rapper Kanye West
12. *Yes man* • a person who always agrees with superiors
13. *Yikes!* • an expression of surprise or astonishment
14. *YMMV* • short for "your mileage may vary"
15. *Yo* • used to get someone's attention; used as a prefix to a greeting
16. *Yob* • a loutish, uncultured person; back-slang; derived from the word boy
17. *To yoink* • to steal
18. *YOLO* • acronym for "you only live once"
19. *Yonks* • a long time
20. *Yo-yo* • a stupid person; someone who doesn't know what he's doing

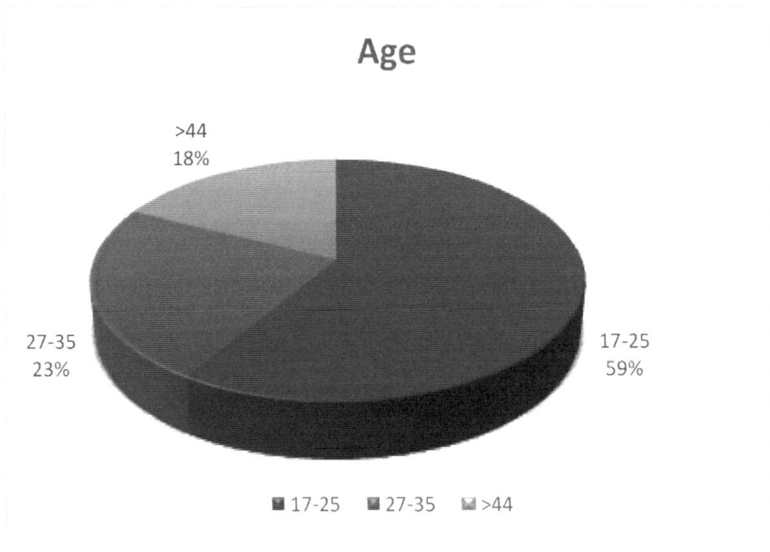

Figure 2. Age of respondents.

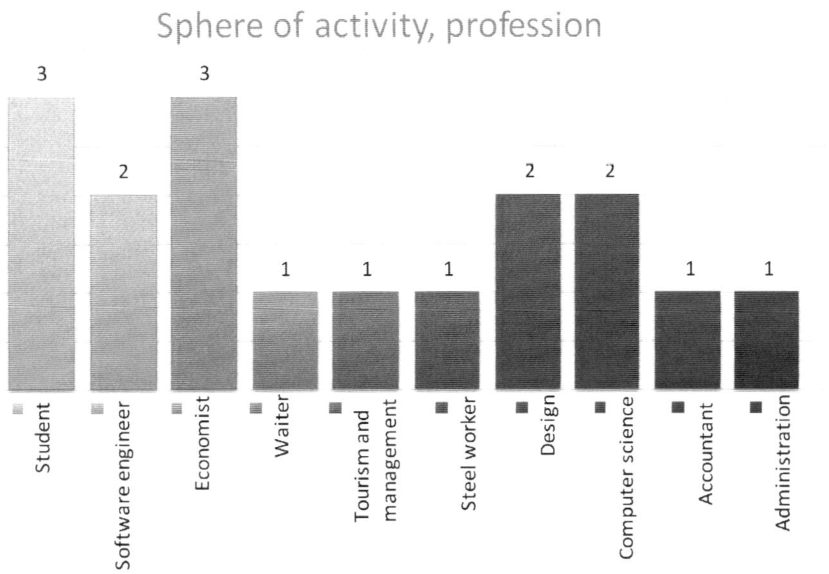

Figure 3. Sphere of activity, profession.

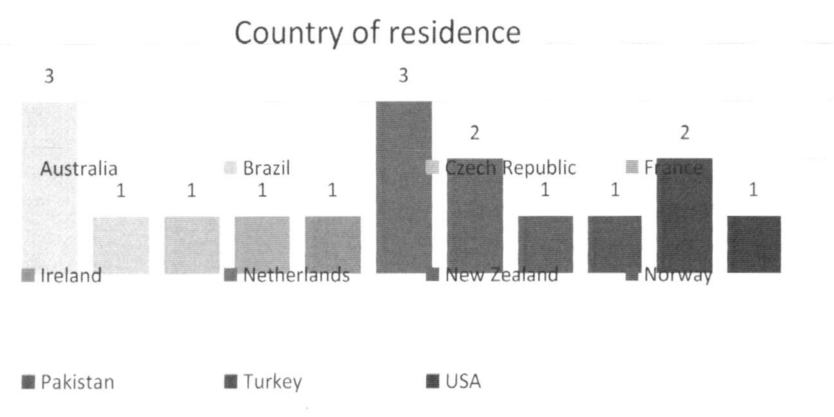

Figure 4. Country of residence.

As to the language competence, there are 5 native speakers and 12 are non-native speakers.

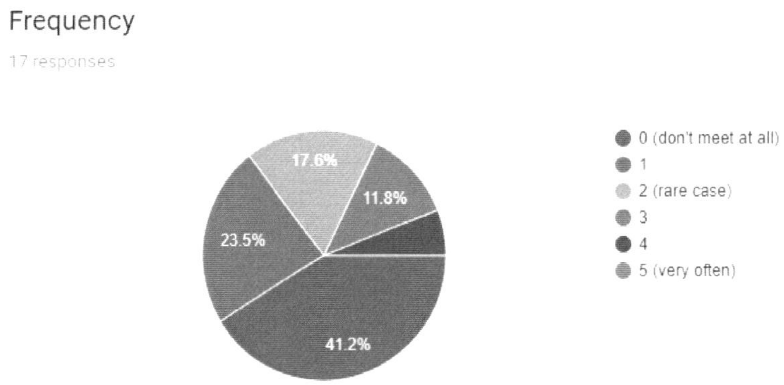

Figure 5. The results on lexical unit 2. (To yack - to talk, especially uninterruptedly and idly; gab; chatter)

The frequency of the word usage (from 0 to 5 where 0- do not meet at all, 5-very often) demonstrates the results: "0: do not meet at all" - 7 informants, "1" - 4 informants, "2: rare case" - 3 informants, "3" - 2 informants, "4" - 1 informant and "5: very often" - 0 informant.

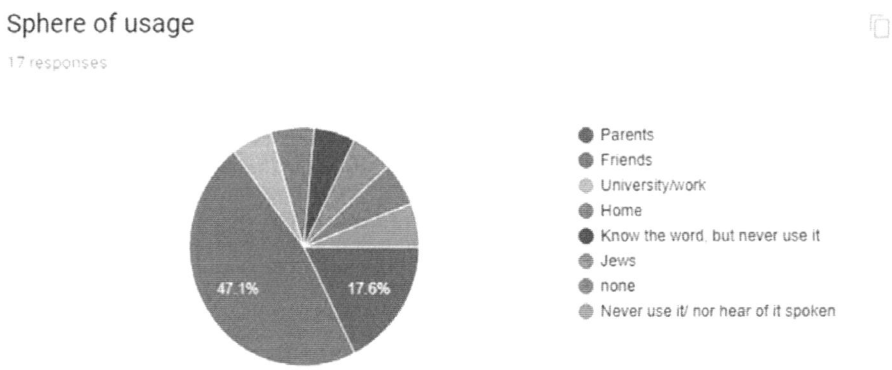

Figure 6. Sphere of usage.

According to the sphere of usage, the results are: "parents" - 3 informants, "university/work" - 1 informant, "friends" - 8 informants, "home" - 1 informant, "other: I know the word, but never use it" - 1 informant , "other: Jews" - 1 informant , "other": none" - 1 informant and "other": never use it/nor hear of it spoken" - 1 informant .

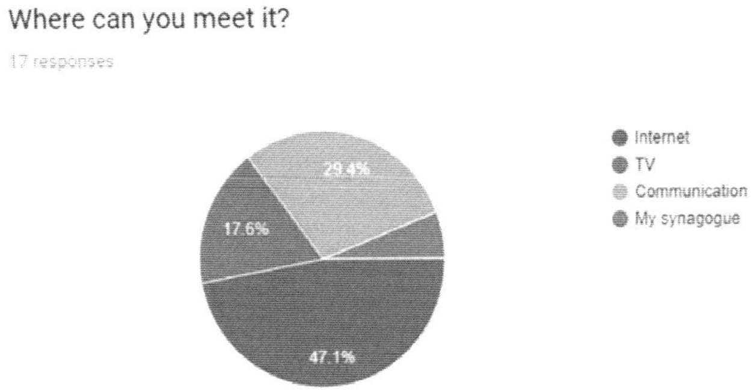

Figure 7. Where you can meet it.

The respondents meet the word: "Internet" - 8 informants; "TV" - 3 informants; "Communication" - 5 informants; "Other": my synagogue" -1 informant

According to the results of field study 1, almost all the words from the above list are popular not only with young people [14]. The most frequently used words are lexical units expressing exclamation and agreement, for instance, *ya, yeah, yay!, yikes*! and etc. In particular, the word *yeah* is most frequently used slang word. It is worth noting that all 17 informants meet it very often as among friends as well in professional communication [15].

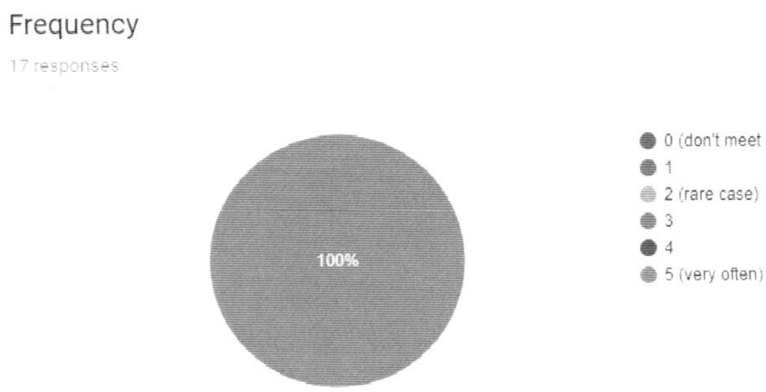

Figure 8. The results on lexical unit 9. (Yeah - conversational form of "yes")

According to the frequency usage, all 17 informants use the word very often.

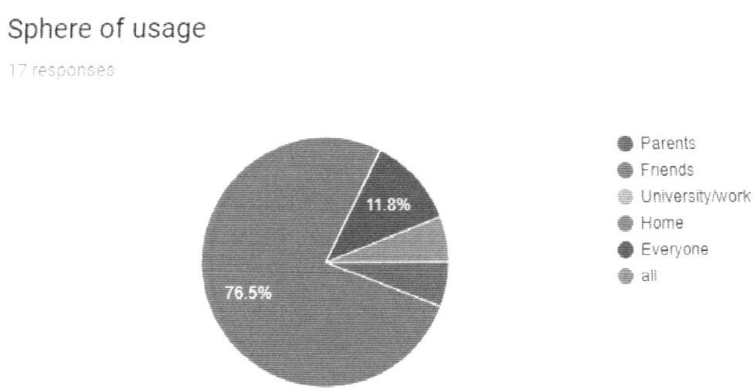

Figure 9. Sphere of usage.

As to the spheres of usage, the results as follows: "Parents" -1 informant, "University/work" - 0 informant, "Friends" - 13 informants, "Home" - 0 informant and "Other: all" - 3 informants. Where you can meet: "Internet" - 4 informants, "TV" - 0 informant, "Communication" – 11 informants and "other: all" - 2 informants.

Thus, the most frequent areas where the informants use slang words are *friends*, the *Internet and professional activity*. According to the field study, the frequency of word use is as follows: every communication is 86% and professionally-related communication is 14%. The less popular word is *yet*. 13 respondents out of 17 have not met it before. The abbreviations YOLO and YMMV are not widely used as well. Let us consider some examples.

Example 1. *Yeezy* is a rap-singer Kanye West. As this proper word characterizes both a man and a subculture, such as hip-hop, it is an expected result that the greater part of the respondents does not know this word. The most popular spheres of usage are the Internet and friends; one of the informants reports that he has heard this word on the radio.

Example 2. A remarkable fact that has emerged is that the word "Yes man" is mainly used in such spheres of activity as *work* and *university*.

Example 3. The word "*Yo*" is used to attract attention and as a greeting. The word is quite popular among the respondents. All 17 informants report that communication *with friends, the Internet and professional activity* are the most popular spheres of this word usage.

2.2. Field study 2.

8 native speakers from Great Britain and United States participated in the survey as well from Lebanon, Nigeria (English is the official language) and Germany (respondent is from Great Britain). Moreover, 7 non-native speakers from Germany, Portugal, Croatia, Mexico, Columbia and Costa Rika were also included in the study. With regard to gender, 7 informats are females and 8 are males.

Table 1. Representation of respondents (language, country, place of study/work, age and gender).

No.	English	Country	Place of study/work	Age	Gender
3	Non-native speaker	Germany	University Oldenburg	28	Female
5	Non-native speaker	Mexico	Mexico City Intl. Airport	29	Male
6	Native speaker	UK	Self employed	26	Male
7	Native speaker	Germany	door2door	22	Male
8	Native speaker	United States	Restaurant	25	Female
13	Native speaker	Costa Rica	Government	35	Female
14	Native speaker	UK	Education	27	Male
15	Non-native speaker	Croatia	Volgograd State Technical University	23	Female

Thus, the findings of the study indicate that most frequently used words from the list are "*OK*" и "*Oh my god!*". The phrases "*On the heavy*" and "*On the beam*" are less frequently used. A part of respondents reports that they do not know about 50% of words.

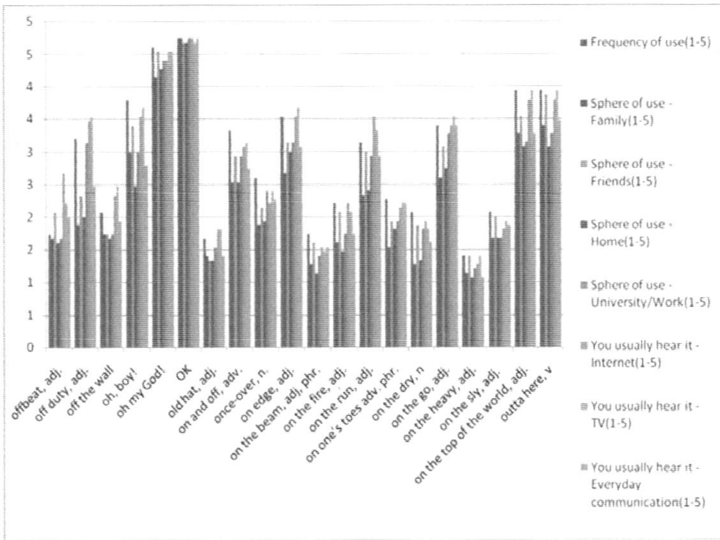

Figure 10. Frequency of word use.

2.3. Field study 3.
17 respondents were included in this study, mostly native speakers.

The respondents hear most words from the given list from their friends (11): *dinosaur, dorky, drag queen, duck soup, dude, dynamite, fam, fangirl, flake out, fleek, to be on fleek.* 6 words from the list are used in communication: *deadbeat, deck, dirt, fairdinkum, fomo, front*. At working place/study they hear the following words more often (5): *dampsquib, deadpan, dodgy, dunno, flatoutlike a lizarddrinking*. Only 2 words are heard from parents: *dicey, dinero*. The word "dirty" is the only word which is mostly used at home. The words used more frequently are *fam, dinero*. The less used word is *dragqueen.*

Conclusion
Thus, the results of the survey indicate that 20% of 200 analyzed words are most frequently used by the speakers both in professionally-related and every day communication; 80% of words present a diverse picture: some lexical units are not used/ heard /identified; some of the words are used/ heard /identified in rare cases. In some cases respondents (mostly native speakers) identified a word but marked it as obsolete. For instance, the slang word "*backhander*":

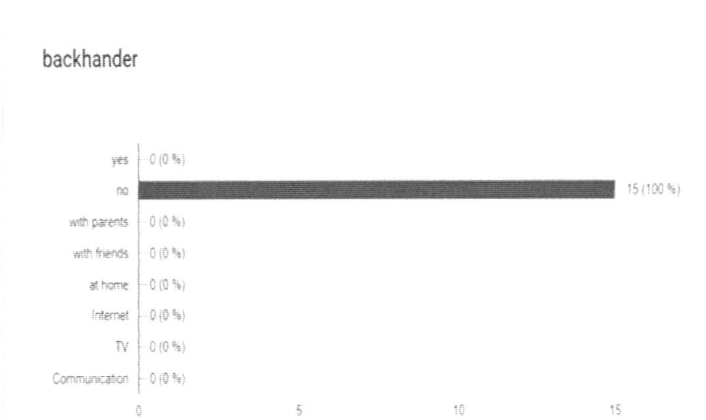

Figure 11. The results on lexical unit "*backhander*".

The lexical unit "backhander" is reported as an "unknown word". However, a native speaker reports that the word "bribe" is used instead of "backhander" in his state [15].

Out of 20 % most frequently used youth slang words, about 7-8% of lexical units are regularly used in the production spheres, such as Mexico City Intl. Airport, Restaurant, Industrial Engineer, Professional employee and etc. [16].

In conclusion, the present article makes evident the relevance of studying youth language in the field of professional communication [17]. It should be noted that this kind of research is a rare case, mainly, slang words manifestation has given attention in fiction, mass-media, the Internet, in other words, multi-media sources. Therefore, this pilot study is an attempt to cover global and macro-sociological space of English slang words function in professionally-related and everyday communication. Finally, it indicates the necessity to supply monitoring of words usage in productive sphere by providing particular themes with transcripts.

References

[1] Gnatyuk O L 2010 *Bases of the theory of communication* (Moscow: Knorus) 256

[2] Tyunyukova E V 2006 *Communication strategy of the organization* (Novosibirsk) 223

[3] Poskochinova O G The specific features of formation of communication strategies at production companies 2014 *J. Management consultation* **12** 52

[4] Kerer K A The specific communication features of production discourse (the film "Practice" as an example) 2018 *Science J.of VSU: Linguistics and international communication* **2** 128-131

[5] Lizunkov V, Politsinskaya E, Malushko E, Kindaev A, Minin M 2018 Population of the world and regions as the principal energy consumer *International journal of energy economics and policy* **8 (3)** 250-257

[6] Zalyaeva E O, Solnyshkina M I and Sirazieva Z N Pragmalinguistic parameters of a professional aoth as a performative speech genre 2016 *Turkich on-line J. of Design Art and Communication* **6**, 2530-36

[7] Ergunova O T, Lizunkov V G, Malushko E Yu, Marchuk V I, Ignatenko A Yu 2017 Forming the system of strategic innovation management at the high-tech engineering enterprises *IOP Conference Series: Materials Science and Engineering* **177 (1)** 012046

[8] Schmitt R and Pustički N 2015 Zur Konstruktion einer amüsanten Unterbrechung während einer argumentativen Auseinandersetzung *Ko-Konstruktionen in der Interaktion. Die gemeinsame Arbeit an Äußerungen und anderen sozialen Ereignissen* 183-208

[9] Degaltseva A V Some specific features and causes for using jargon words in electronic business communication 2013 *Proc. Int. Conf. on Stratification of national language in modern Russian society* 61-66

[10] Pavlova E E and Seliphonova E D 2015The specific features of youth slang of modern Englhish language (sms-wring as an example) *Proc. Int. Conf., LV, In the world of science and art: issues on phylology, art and culture* **12 (55)** 65 -72

[11] Fedotov N G, Moiseev A V, Syemov A A, Lizunkov V G, Kindaev A Y 2017 New Methods of Three-Dimensional Images Recognition Based on Stochastic Geometry and Functional Analysis *IOP Conference Series: Materials Science and Engineering* **177 (1)** 012047

[12] Malushko E, Maletina O, Lizunkov V, Tsybaneva V 2016 Use of virtual learning system for educating students with disabilities and special needs *International Multidisciplinary Scientific Conferences on Social Sciences and Arts* 481-487

[13] Cajo ST and Bahlo N "Oh that which is sweet"- Spontaneous Conversations. A communicative Genre in Terms of Sociability and Purpose Orientation 2016 *Deutsche Sprache* **44** (1) 77-96

[14] Anishchenko O A The evolution of youth sociolect and its development perspectives 2015 *Proc. Int. Conf.* 5th 3-11

[15] *Dictionary of Youth Slang (English, German, French and Russian languages)* 2017 ed Shamne N L, Rebrina L N and et al (Volgograd: VolSU) p 381
[16] Shamne N L, Shishkina E V 2017 German popular scientific medical online media: structural and functional aspects *Vestnik Volgogradskogo Gosudarstvennogo Universiteta-seriya 2-yazykoznanie* **16 (2)** 143-151
[17] Serebryakova S V, Milostivaya A I 2017 Semantic emergence as a translation problem *Vestnik Volgogradskogo Gosudarstvennogo Universiteta-seriya 2-yazykoznanie* **16 (3)** 48-57

Top 100 Occupations in Demand at Modern Russian Enterprises: Interaction and Communication in the 21st century

N V Yudina[1], O A Seliverstova[2]

[1]Vladimir Branch of Financial University under the Government of the Russian Federation, 1 Tikhonravova Street, Vladimir, 600037, Russia
[2]Vladimir State University, 87 Gorky Street, Vladimir, 600000, Russia

dr.yudina@mail.ru

Abstract. Today labour market reflects hi-tech and innovative economic development of the XXI century. The system of professional occupations is subject to the effect of various factors, prestige being one of them. The paper analyses 100 occupations highly demanded by Russian enterprises. These nominations represent the synergetic nature of professional sphere in the life of society involving major advances in science and engineering. The paper offers subject-based classification of prospective occupations and provides characteristic of their communicative potential.

1. Introduction

Occupations nominations or job titles are lexical units which can be referred to as markers of industrial and economic spheres. Over the recent years, job titles have been under study by researchers in various fields: some presenting a socioeconomic classification of job titles with regard to health issues [1], others undertaking a gender study [2], still others offering a language modelling approach to discovering novel labour market occupations [3]. Considered together the studies above provide evidence of the high value of further research focused on jobs and occupations, and their titles, since these nouns join together labour and social spheres, and specify an individual's position in the society. In the Russian language there are over 8,000 lexical units for job titles, position titles, and occupations. This variety accounts for scholarly novelty of our paper which seeks to reveal nominations of the most highly demanded jobs in the XXI century.

2. Main part

The current economic situation in Russia affects the conditions of HR policies and determines its directions and prospects. Noteworthy is the view the RF President Vladimir Putin shared in relation to prospective jobs, mentioning in this regard "experts in such fields as big numbers and their processing, artificial intelligence, aerial, submersible, and ground vehicles" [4].

Research into focus areas is made by a number of federal and regional enterprises at the government initiative. Thus comprehensive data about the most demanded and prospective jobs in the form of reference resource can be found on the official web-site of the Ministry of Labour and Social Protection [5].

The work on compiling of Directory of Occupations has been carried out since 2015 as a joint project of the Ministry of Labour and Social Protection and the National Council and involves Union of Industrialists and Euntrepreneurs, Chamber of Commerce and Industry, all-russian non-governmental orgatization of small and medium business Opora Rossi, Federation of Independent Trade Unions of Russia, Agency for Strategic Initiatives to facilitate new projects, Agency for Professional Associatons and Regular labour Force "World Skills Russia", National Agency for Skills Development, as well as educational establishments and unions. Presently, the Directory of

Occupations is a current regularly updated official resource approved by the order of the Ministry of Labor of Russia dated November 02, 2015 № 832 "On Approval of The Directory of Occupations Demanded In The Labor Market, New and Promising Occupations, Including Those Requiring Secondary Vocational Education " (amended by the order of February 10, 2016 № 46).

Directory of Occupations specifies the most prestigious and popular jobs at today's Russian enterprises from the legally approved 1600 fields of labour activity. All professions are divided into larger groups in accordance with activities performed. Thus, the portal describes 40 groups of professions. Let us consider some of the thematic groups. In this paper the authors undertake to compile their own classification of popular occupations.

We think it appropriate to distinguish 100 lexical units productive in the twenty-first century from 1600 names of the professional activities. The criteria for identifying relevant nominations are their being included into The All-Russian Classifier of Occupations and Economic Activities, into directories structure, atlases of jobs, as well as into regional lists (the "Top region" project), and their frequency in the modern Russian language (according to The National Corpus of the Russian language).

In the course of analysis of the legal grounds, official lists and codifiers, and types of work, we arrived at the conclusion that most nominations belong to the sphere of industrial production. Some sources, namely The All-Russian Classifier of Vocational Occupations, Positions of Employees and Tariff Categories [6], Atlas of Popular Professions, Directory of Occupations Of the Russian Ministry of Labour [7] do not distinguish the sphere of military-professional activity. However, the view of emerging work spheres will be incomplete without nominations of military ranks and positions.

The list of military personnel occupations can be found in the Federal law of 28.03.1998 No. 53-FZ (as amended 03.08.2018) "Concerning Military Duty and Military Service" [8], also in the "OK 010-2014 (ISCO-08) The All-Russian Classifier of Occupations (adopted and signed into law by the order of Rosstandart of 12.12.2014 No. 2020-St) [9].

Detailed examination of the information presented in the codifiers, classifiers of labour activity, regulated by the normative legal acts has let us identify the most productive lexical units that reflect the current situation in the labor market.

As a result of analysis, occupations in demand were classified with regard to activity field. Thus, "a set of generalized labour functions of similar nature, results, and working conditions" [10: 8] is referred to as occupation field. Occupations are grouped with regard to their integration of nature, implying the same purpose, objects, processes involved, and instruments of labour [ibid: 22].

The sequence of groups and nominations of occupation are listed on the basis of the quantitative principle of the nominations to be included. The most productive in the XXI century are 10 types of professional activity, namely:

1) social activities (health, education, physical education and sports), 16 nominations, for example: obstetrician, nursery teacher, doctor, instructor in physical education, nurse, baby sitter, speech pathologist, teacher of high school and college, pharmacist, psychologist, social worker, athlete, coach, teacher, pharmacist, medical assistant;

2) activities in the field of public services (13 nominations): car mechanic, receptionist, bartender, driver, maid, designer, pastry chef, fashion designer, landscape designer, waiter, hairdresser, cook, seamstress;

3) activities in the field of industry (instrument-making, mechanical engineering, robotics, intersectional industrial activity), 12 nominations, e.g.: engineer-mechatronics, laboratory chemist, mobile robot technician, adjuster-repairman of equipment, equipment operator, mechanic, specialist in additive technologies, technician in composite materials, technician-designer, turner, miller, electrician;

4) activities in the field of culture, art, media, publishing (11 nominations), namely: actor, performing artist, presenter, sound engineer, art critic, correspondent, cameraman, editor, Director, photographer, choreographer;

5) administrative-and-managerial and official activities, including the field of law (10 nominations), e.g.: attorney, executive, clerk, director, chief, notary, secretary, forensic investigator, bailiff, and lawyer;

6) activities in the field of architecture, topography, construction and housing (9 nominations), e.g.: architect, cadastral engineer, mason, painter, plumber, locksmith, cartographer, topographer, plasterer;

7) activities in the field of agriculture, forestry, fish farming (8 nominations), e.g.: agriculturist, veterinarian, winemaker, huntsman, breeder, animal technician, ichthyologist, forester;

8) activities in the field of financial, management and marketing activities (8 nominations), e.g.: actuary, auditor, business analyst, accountant, dealer, manager, insurance broker, SSM-specialist;

9) activities in the field of information and communication technologies (6 nominations), namely: database administrator, programmer, WEB and multimedia applications developer, network and system administrator, information security technician, technical writer;

10) military-professional activity (6 nominations), namely: admiral, general, captain, major, colonel, private.

The above classification does not claim to be complete and will be supplemented in the course of our further work.

We would like to dwell on a group of occupations related to industrial activity (instrument-making, mechanical engineering, robotics, industrial activity). Nominations in this group most fully represent the synergetic nature of industrial activity. Thus, such lexical units as engineer and technician are part of the composite names of the mechatronics engineer, and design technician. Moreover, the classification can be supplemented by such complex lexemes as logistics engineer, metallurgical engineer, mechanical engineer, microchip engineer, software engineer, process engineer, and many others.

3. Results and Discussion

Our observations show that each field features clearly distinguished nuclear nominations, most susceptible to transformation to produce more complex word-building, lexical, and syntactic structures. For example, from the group of names describing activities in the social sphere, the noun doctor is productive. The latter is a hyperonym in relation to a composite and two-component nominations like military doctor, resuscitator doctor, sanitary doctor, and many others [11: 29-33]. Alternatively, the Russian word "артист" (thematic group-activities in the field of culture, art, media, publishing) is a unified designation of representatives of different occupations in the sphere of art, eg.: It stands for "theatre actor", "film actor", "orchestra musician", "dance group performer", "choir singer", "solo singer", "circus performer", and others [11: 13-14]. Finally, noteworthy is high frequency of the nomination manager in media language [12: 80]. It features productivity at the lexical and word-building levels, resulting in neologisms, previously unrecorded in reference books and codifiers.

It should be noted that a number of new nominations is "the representation of well-known productive models with a key word" [13: 60-64]. The nomination of technical writer (activities in the field of information and communication technologies) is interesting since the common perception of a writer in the minds of native speakers is associated with the field of art. The occupation of a technical writer appeared in the XXI century as a result of transformation of responsibilities. It refers to a specialist engaged in the "development of technical documentation and methodological support of products in the field of information technology" [14, 15].

4. Results and conclusions

Occupations referring to the field of industrial activities are currently the most promising [16]. The field of industry covers such areas as aircraft, automotive, nuclear industry, rocket, and space industry, shipbuilding. Automation and robotics in production make the occupations of an engineer, a mobile robot technician, an equipment repair technician, and a design technician more and more popular, prestigious and, most importantly, universal [17]. Noteworthy is the group of nominations referring to activities in the field of information and communication technologies. These are especially relevant in the era of information and digital society. It is for a reason that most occupations in the field of industrial, information and communication activity are listed in the Top Region ratings of various regions in the Russian Federation [18].

In the 21st century, the boundaries between industries are getting "blurred" and occupations in demand are at the intersection of different fields. To conclude, we would like to note that professional activity has integrative nature due to industrial, scientific, and technological progress [19]. The study of emerging occupations of the XXI century opens up new horizons for further study of the nomination system of occupations and labour fields, as well as for the study of industrial ethics and communication in the modern world.

References

[1] Schyllert Ch, Andersson M, Hedman L, Ekström M, Backman H, Lindberg A & Rönmark E 2018 *Job titles classified into socioeconomic and occupational groups identify subjects with increased risk for respiratory symptoms independent of occupational exposure to vapour, gas, dust, or fumes*, *European Clinical Respiratory Journal*, 5:1, DOI: https://doi.org/10.1080/20018525.2018.1468715

[2] Foster S 2018 *Job titles need to reflect modern times and go gender neutral, British Journal of Nursing* 27:7, pp 417-417

[3] Marrara S, Pasi G, Viviani M, Cesarini M, Mercorio F, Mezzanzanica M, and Pappagallo M 2017 *A language modelling approach for discovering novel labour market occupations from the web. In Proceedings of the International Conference on Web Intelligence (WI '17)*. (New York, NY, USA) pp 1026-1034. DOI: https://doi.org/10.1145/3106426.3109035

[4] Karnaukhova N, Gordeev V 2018 *Putin declared the most demanded jobs of the future* [Electronic resource]. Access mode: https://www.rbc.ru/politics/02/09/2018/5b8b00859a79471448f0fded (access date: 10.11.2018)

[5] *Directory of Occupations* [Electronic resource]. Access mode: http://spravochnik.rosmintrud.ru/about (access date: 1.11.2018)

[6] Malushko E, Maletina O, Lizunkov V, Tsybaneva V 2016 Use of virtual learning system for educating students with disabilities and special needs *International Multidisciplinary Scientific Conferences on Social Sciences and Arts* 481-487

[7] *The All-Russian Classifier of Vocational Occupations, Positions of Employees and Tariff Categories* [Electronic resource]. Access mode: http://professions.org.ru (access date: 10.11.2018)

[8] *Atlas of Popular Professions, Directory of Occupations Of the Russian Ministry of Labour* [Electronic resource]. Access mode: http://spravochnik.rosmintrud.ru (access date: 10.11.2018)

[9] 2018 *Federal law of 28.03.1998 No. 53-FZ "Concerning Military Duty and Military Service"* (as amended 03.08.2018) [Electronic resource]. Access mode: http://base.garant.ru free (access date: 9.11.2018)

[10] Shamne N L, Shishkina E V 2017 German popular scientific medical online media: structural and functional aspects *Vestnik Volgogradskogo Gosudarstvennogo Universiteta-seriya 2-yazykoznanie* **16 (2)** 143-151

[11] Ferreira R, Lizunkov V G, Politsinskaya E V 2017 Formation of entrepreneurial competencies of university graduates in conditions of transition to the universities of the third generation *Novosibirsk State Pedagogical University Bulletin* **7 (6)** 195-211

[12] *The All-Russian Classifier of Occupations* (adopted and signed into law by the order of Rosstandart of 12.12.2014 No. 2020-St) [Electronic resource]. Access mode http://www.consultant.ru/document/cons_doc_LAW_177953/ (access date: 9.11.2018)

IOP Conf. Series: Materials Science and Engineering **483** (2019) 012015 doi:10.1088/1757-899X/483/1/012015

[13] 2016 *Formation of the system of professional qualifications* (Moscow: Pero Publishing House)
[14] Gorbunova M V, Kirilyuk E V and Oreshkina A P 2011 *1000 professions* (Rostov-on-the-Don)
[15] Isaeva N V 2009 *Comound Nominations of People's Occupations in Advertisements* (Moscow: Vestnik MGOU) pp 78-81
[16] Ergunova O T, Lizunkov V G, Malushko E Yu, Marchuk V I, Ignatenko A Yu 2017 Forming the system of strategic innovation management at the high-tech engineering enterprises *IOP Conference Series: Materials Science and Engineering* **177 (1)** 012046
[17] Golovanova E I 2015 *New Words for People's Occupations and Positions in Russian* (Chelyabinsk) pp 60-4
[18] Serebryakova S V, Milostivaya A I 2017 Semantic emergence as a translation problem *Vestnik Volgogradskogo Gosudarstvennogo Universiteta-seriya 2-yazykoznanie* **16 (3)** 48-57
[19] Hardware and Software Complex *National Register of Professional Standards* [Electronic resource]. Access mode: http://profstandart.rosmintrud.ru/obshchiy-informatsionnyy-blok/natsionalnyy-reestr-professionalnykh-standartov/index .php?ELEMENT_ID=50435

IOP Conf. Series: Materials Science and Engineering **483** (2019) 012016 doi:10.1088/1757-899X/483/1/012016

Need Analysis of College-Educated Labor Resources in the Russian Arctic

K S Zaikov[1], N A Kondratov[1], M Yu Kuprikov[2], N M Kuprikov[2] and E Tamitskaya[1]

[1]Northern (Arctic) Federal University named after M.V. Lomonosov, 17 Embankment of The Northern Dvina, Arkhangelsk, 163002, Russia
[2]Moscow Aviation Institute (National Research University), 4 Volokolamsk str., Moscow, 125993, Russia

E-mail: n.kondratov@narfu.ru

Abstract. In the first half of the 20th century due to the changed economic conditions, growing interest of foreign countries and their corporate sector in the Arctic, the implementation of large-scale investment projects in mining and processing of minerals, the use of transport routes in the Russian Arctic, the issue of human resources provision is becoming more topical. The paper presents the requirement analysis of college-educated labor resources in the sub-federal units of the Russian Arctic in the current (until 2020) and medium-term (until 2022) perspective using sociological and analytical methods. Highly qualified personnel and their advanced technical solutions are considered as the main conditions for the adaptation of Arctic communities and economic systems to environmental changes and new requirements of world technological paradigms. The study results can be recommended for the federal and regional executive authorities for formulating policy in the academic space management, targeted college-educated personnel training for the real sector of economy of the Russian Arctic. The paper can be useful for enterprises operating in the Arctic macro-region, for developing a system for monitoring and evaluating staffing needs and introducing such a system into the management of the socio-economic development of the Arctic zone of Russia.

1. Introduction

Humanity has been actively developing the Arctic for the last 80 years. Several economic models (American, European, and Russian) were formed during this period. They are associated by the fact that almost a third of industry production is created in the materials sector. In 2016, Arctic GDP was about $ 300 million. Manufacturing and service industries in the Arctic region have never been significant. Examples include metalworking and mechanical engineering (shipbuilding and shiprepairing) and food industry. Food industry has been developed in populated places on the coast of the Arctic Ocean based on indigenous peoples' fishery and deer farming.

Russian Arctic in the early 21st century turns into an independent object of public policy and management. Several-trillion-dollar natural resources potential as well as technological and production potential, which participates in the production of 10–12 % of GDP and creates about 25 % of Russian export [1], are concentrated in this macro-region. In 2008–2013, the President of Russia approved the Basic Principles of State Policy of the Russian Federation in the Arctic for the period up to 2020 and further perspective and the Development Strategy of the Arctic Zone of the Russian Federation and national security protection for the period until 2020. In order to develop natural resources effectively, state efforts are focused on elements, the progressive use of which will provide a synergistic effect in the implementation of the full range of priorities for the development of the Arctic zone of the Russian Federation. One of the elements is labor resources with knowledge of trends of Arctic environmental, socio-ecological and socio-economic systems, which can effectively participate in investment (often international) projects on the natural resources and transport development (first of all, the Northern Sea Route, NSC) in the Arctic.

2. Research methodology

Development of human resources in higher education and graduates staffing level in the Arctic economy branches are of great interest in the scientific community both in Russia and abroad at the state level. The specific features of labor resources of the Russian Arctic, such as nomadic life, mobility, rotational work, a higher number of people working in the natural resource service (environmental management) than in the southern and central regions of the country, single-industry towns with the only city-forming enterprise, a large proportion of people employed in state and municipal enterprises, gender differentiation of activities into "male" and "female" are identified in the Development Strategy of the Arctic Zone of the Russian Federation. The most important challenge for the Arctic labor market in the long-term prospects (after 2025) will be the need for quantitative preservation and qualitative improvement of labor and human resources in the Arctic zone of Russia for benefit of increasing labor productivity, including among indigenous minorities. This raises the question of modernization of the education system as a supplier of highly qualified personnel. Obviously, the universities that will accumulate new knowledge about the Arctic natural and socio-economic systems will become the catalyst for these processes; the educational programs will be linked to the demands of enterprises and government bodies in the municipalities of the Arctic zone of Russia. Government efforts will be aimed at securing the population in the Far North and the Arctic zone, attracting young professionals, maintaining the system of social guarantees.

The factors influencing the population dynamics in the Arctic region, the specifics of manpower training, the direction of state policy modernization in relation to the population of the North, are updated in the works of specialists of the Kola Scientific Center of the Russian Academy of Sciences [2]. The issue is also described in publications of Heleniak T. [3], Semenova A. and Popelnitskaja I.M. [4], Povarich I.P. and Harchenko A.P. [5], Belonozhko M.L., Silin A.N. and Barbakov O.M. [6], Giltman M.A. [7], Gorokhov A., Zaikov K., Kondratov N., et al. [8]. Klyukina E.S. [9] analyzes the quality of the population (health and the educational level) on the example of the Murmansk Oblast to determine resource potential and to select regional development scenarios. Sinenko P.V. [10] notes that the socio-economic situation in the north of Russia has been characterized by destructive processes for a long time, which is reflected in a decrease in living standards of the population, an outflow of young qualified personnel, low labor productivity, poor state of infrastructure and housing and communal services. Ivanova M.V. and Klyukina E.S. [11] note that by 2018, the subjects of the Russian Arctic will begin to experience a shortage of labor resources. According to increasing mobility of population of the economically active age, new qualification requirements in "old" industries, growing struggle for personnel are the main problems in the labor market. Based on a sociological study and analysis of statistical information reflecting the state of housing and social infrastructure in the subjects of the Arctic zone of Russia, Pavlenko V.I. and Kutsenko S.Yu. form an idea of human life and activities in the Arctic macroregion and the consequent possibility of securing labor resources in areas with harsh climatic resources [12].

The demographic potential of a specific territory of the Arctic zone of the Russian Federation, most often in the Murmansk Oblast, is considered in publications. The works of scientists contain generalized data on the personnel potential of the subjects of the North; issues of labor requirement for enterprises and executive authorities, personnel training in the system of higher education of the Arctic zone of Russia are practically not covered [13, 14, 15].

In 2016, the Northern (Arctic) Federal University named after M.V. Lomonosov (NArFU, Arkhangelsk) studied the problem to provide valuable insights of labor resources of the Russian Arctic. The scientists assessed the current (currently observed) and medium-term (until 2020–2022) needs of employers engaged in economic activities in the subjects of the RF Arctic zone in the labor resources of various levels of higher education (Bachelor's, specialist's, Master's programs and postgraduate education). Evaluation and forecast of labor requirement were carried out on the basis of data provided by 50 enterprises, including both great (over 5,000 employees) and large (from 1,000 to 5,000 people).

3. Results and discussion

The collected data demonstrate that the current requirement of college-educated labor resources in enterprises and organization in the Russian Arctic is 6,198 people, including specialists of a bachelor degree level – 3,680, Master's degree level – 1,145, specialist's degree level – 1,364, postgraduate education – 9 people. An idea current labor requirement is presented in Figure 1.

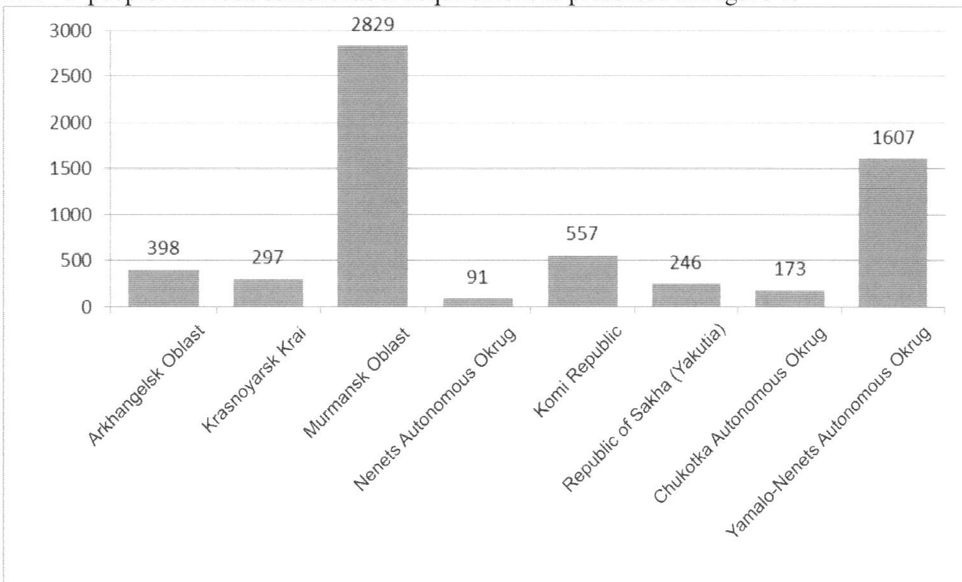

Figure 1. Current requirement of college-educated labor resources in the Russian Arctic

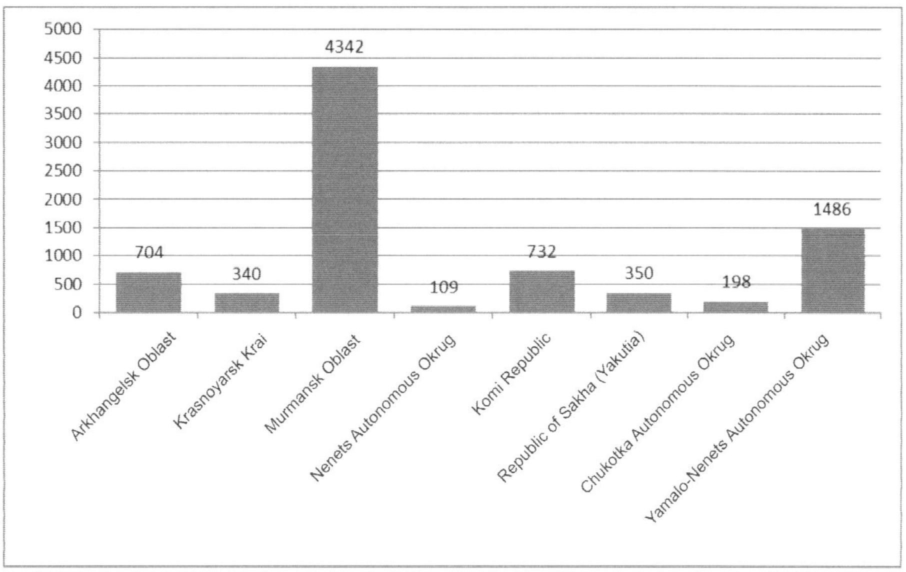

Figure 2. Medium-term requirement of college-educated labor resources in the Russian Arctic

Medium-term requirement of college-educated labor resources in the Russian Arctic subjects is 8,261 people, including specialists of a bachelor degree level – 4,658, Master's degree level – 1,753, specialist's degree level – 1,836, postgraduate education – 14 people. The idea of a medium-term need for labor resources is presented in Figure 2.

The greatest current and medium-term need for labor resources is observed in the Murmansk Oblast (2829 and 4342 respectively), the Yamalo-Nenets Autonomous Okrug (1607 and 1486) and the urban district Vorkuta in the Komi Republic (557 and 732). The lowest figures are found in the Nenets Autonomous Okrug (9 and 109), the Republic of Sakha (Yakutia) (246 and 350) and the Chukotka Autonomous Okrug (173 and 198).

The greatest current recruitment needs are observed in the manufacturing industry, including mechanical engineering (1,100 people), social sphere (including education and health care) (2,184 people), and mining (592 people) (Figure 3).

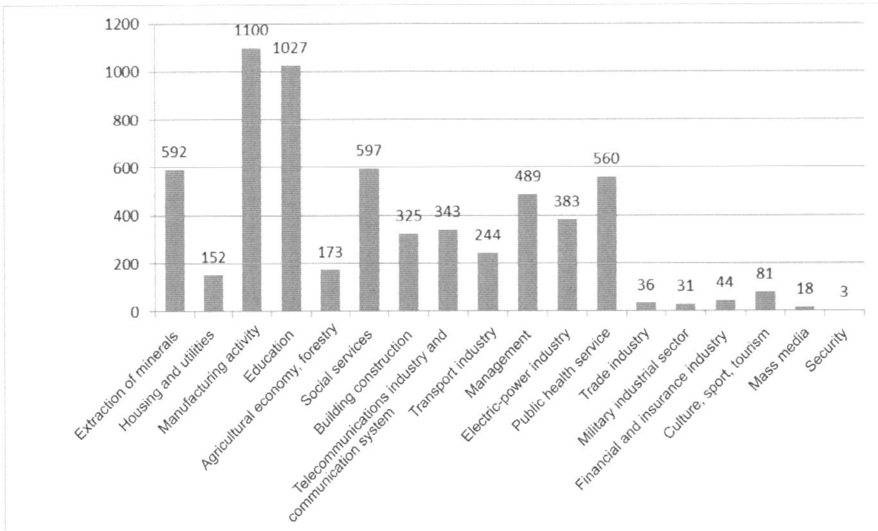

Figure 3. Current requirement of college-educated labor resources in the Russian Arctic by branches of the economy

High demand for labor resources not related to raw materials industries can be explained by the diversification of the sectoral structure of the economy in the most populated and economically developed regions of the Arctic zone – Murmansk and Arkhangelsk Oblasts. More than 1 million people live there, and conditions for large-scale investment projects for the extraction of minerals and transport infrastructure development in the Russian Arctic are created.

Analysis of the medium-term need for college-educated personnel demonstrates the growth of demand compared to the current one. The most popular will be specialties for the manufacturing and engineering industries (1800 people), mining (1280 people), communications (659), social sphere (1653 people) (Figure 4). The projected increase in staffing needs in the extractive industries and in the field of communication may be associated with the beginning of implementation of mining projects and modernization of the NSR after 2022.

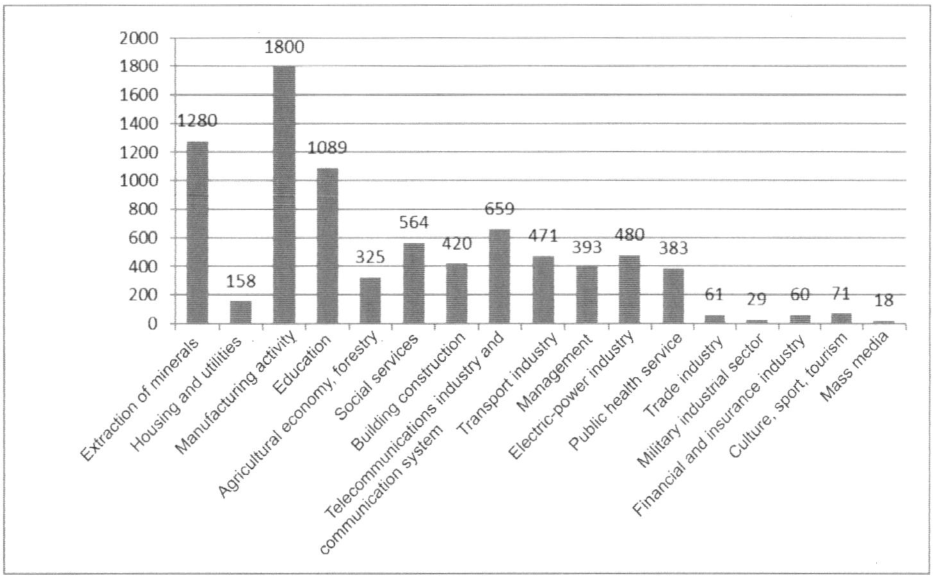

Figure 4. Medium-term requirement of college-educated labor resources in enterprises in the Russian Arctic by branches of the economy

4. Conclusion

In the course of the study, the current and medium-term need for college-educated labor resources for enterprises and organizations engaged in economic activities in the Arctic zone of the Russian Federation is identified. The greatest requirement of labor resources will be typical for manufacturing, metalworking and mechanical engineering, electric power industry, mining, communication and services (including education and health). This fact is discovered by using the sociological method. A significant impact on the formation of a medium-term requirement (until 2022) may have the beginning of the implementation of deferred projects for the natural resources development in the Russian Arctic and their delivery to the consumer.

The results of the study have scientific and practical value. The presented materials develop an understanding of needs for college-educated labor resources in enterprises and organizations located in the Arctic zone of the Russian Federation. Comparison of the results of this study with expert-analytical materials of the UArctic, the Arctic Council, the working group on education and science of the Barents Euro-Arctic Council can be carried out only in a meaningful context and on a structural basis. The research materials were tested at the All-Russian and international scientific conferences organized at NArFU in 2016–2018: “Monitoring and Assessment of the Development of the Arctic Zone”, “Training for the Arctic: from Problems to Finding Solutions”, “Arctic is a National Megaproject: Human Resourcing and Scientific Support”. The collected data allowed improving the information-analytical and methodological decision-making process for the development of state policy in the field of training and the development of education in the Arctic zone of the Russian Federation. Using the survey materials and their analysis, the portal of the National Arctic Research and Education Consortium (established in 2016 on the basis of NArFU) functions. This media displays universities and research and educational centers providing training in the Arctic zone of the Russian Federation, a list of educational programs implemented on their basis, data on the number of students, as well as employers in the region.

References

[1] Zaikov K S, Kalinina M R, Kondratov N A and Tamitskiy A M 2017 *Economic and social changes: facts, trends, forecast* **10 vol 3** 59–77

IOP Conf. Series: Materials Science and Engineering **483** (2019) 012016 doi:10.1088/1757-899X/483/1/012016

[2] Selin V S, Vasilyev V V and Shirokova L N 2011 *Russian Arctic: Geography, Economy, Zoning* (Apatity, Kola Scientific Center RAS) 190
[3] Heleniak T 2009 *Polar Geography* **32** 31–60
[4] Semenova A and Popelnitskaja I M 2017 *Journal of Siberian Federal University. Humanities & Social Sciences* **10 vol 12** 1905–15
[5] Povarich I P and Harchenko A P 2015 *Tomsk State University Journal* **400** 253–60
[6] Belonozhko M L, Silin A N and Barbakov O M 2018 *International Journal of Ecological Economics and Statistics* **39 vol 1** 132–142
[7] Giltman M A 2017 *Journal of the New Economic Association* **3 (35)** 103–124
[8] Gorokhov A, Zaikov K, Kondratov N, Kuprikov M, Kuprikov N and Tamickij A 2018 *European Journal of Contemporary Education* **7(3)** 485–497
[9] Klyukina E S 2015 *The North and the Market: Forming the Economic Order* **4 (47)** 14–23
[10] Sinenko P V 2016 *Economics: Yesterday, Today and Tomorrow* **6 vol 12** 18–25
[11] Ivanova M V and Klyukina E S 2017 *Proc. 7th Intern. Sociological Grushin Conf. "Towards the Future. Prediction in Sociological Research"* ed. by A.V. Kuleshov (Moscow, Russia All-Russ. Public Opinion Research Center Publ.) 1497–1501
[12] Pavlenko V I and Kutsenko S Yu 2018 *Human Ecology* **2** 52–8
[13] Petrov A N, Zbeed S O and Kavin F A 2018 *Arctica i Sever* **30** 5 – 22
[14] Zamjatina N Yu and Piljasov A N 2018 *Arctica i Sever* **31** 5 – 27
[15] Govorova N V 2018 *Arctica i Sever* **31** 52 – 61

IOP Conf. Series: Materials Science and Engineering **483** (2019) 012017 doi:10.1088/1757-899X/483/1/012017

Communicative aspects of professional training of personnel

I S Dikarev, S A Korobov, S B Tokareva

[1]Volgograd State University, Universitetskiy av, 100, 400062, Volgograd, Russia

econmanag@volsu.ru

Abstract. In the context of rising costs of human resources, the effective use of human capital becomes an important component of productive investment in the future. The economic feasibility of investing in training, reproducing and supporting cadres proved by the authors of the theory of human capital is reflected in the active cooperation of enterprises with the education system. The effectiveness of this cooperation largely depends on the established channels of communication and trust between business, government and the institute of education. In this regard, the task of studying the communicative aspects of professional training of personnel at the federal and regional levels is challenging. The article presents an analysis of the legal field of educational activities. The results of the study of the regional specifics of communications of enterprises with the authorities and educational institutions in the field of professional training of personnel of the Volgograd region, obtained using the monitoring data of the Committee on Industry and Trade of the Administration of the Volgograd Region, are presented. According to the study, professional standard plays a key role in ensuring an efficient conjunction between production and education. The production shapes the requirements for the training level and content in particular.

1. Introduction

The authors of the theory of human capital (T. Schultz [1], G.S. Becker [2], [3]) proved the equivalence of investments in education, the accumulation of industrial experience, health, awareness and investment in the acquisition and creation of new equipment and technologies. Russian researchers also come to the conclusion that education costs have high rates of profitability not only for the individual and society as a whole, but also for a particular enterprise, because the knowledge of workers can replace and save resources (materials, electricity, working time, production space and etc.) [4].

Objective trends in the development of industrial production and the experience of leading state and private companies indicate that the success of an organization is largely determined by a competent personnel policy that ensures the recruitment of personnel, the adaptation of new employees and the management of professional growth of employees. In this regard, it is beneficial for an enterprise to focus on «the quality of a person as a participant in production» [5]. The social order of a particular enterprise for training can be implemented through cooperation with educational institutions in the form of targeted recruitment; opening of the basic departments; involvement of employees of enterprises as employers in the development and implementation of professional educational programs.

At the institutional level, the most important communication channel in the process of the training of personnel is the joint work of the executive authorities, employers' associations, professional communities, educational institutions and other interested organizations in the development of professional standards. The system of professional standards is designed, on the one hand, to become a tool for the modernization of the economy (due to consistency with state educational standards); on the other hand, to guarantee graduates of educational institutions, when applying for a job, a position and level of salary that corresponds to their qualifications and labor functions.

The availability of professional standards is an objective condition for streamlining labor relations, and clearness in matters of qualification requirements for employees of various professions serves as a preventive measure against possible violations of the labor rights of citizens. In the absence of such benchmarks in the labor market, it is impossible to achieve a unified approach among employers to determine the level of education, knowledge, competencies and job responsibilities of workers employed by various organizations.

Additional opportunities for the development of external communication of industrial enterprises are opened up by the branch in Government Relations (GR) in Russia, which makes communication between business, government and society public, legitimate and civilized. Thus, on the one hand, it ensures the coordination of public, state and corporate goals; on the other hand, a system of social communication technologies is being developed, which ensures the creation and maintenance of direct and indirect interactions of enterprises with government bodies and local self-governing authorities This creates a positive business environment.

A strategic pathway for building partnerships based on social trust between enterprises, authorities and the institute of education is the involvement of business associations as intermediaries of business associations [6]. In countries with market economies, they are an effective institution for coordinating the external relations of enterprises; create conditions for the formation of a socially-oriented market economy; participate in the formation of the legal environment of entrepreneurship. Business associations provide a full-fledged communication of enterprises with authorities and institutions with a lack of their own resources, allow any company or firm to gain access to making power decisions, regardless of their resource capabilities. Studies show that in economies with a low level of development of market infrastructure institutions, business associations participate in the production of not only "corporate", but also "public goods", and also ensures «coordination between firms in the development and maintenance of industry standards» [7]. Most often, enterprises become members of all-Russian industry-specific business associations, but there are also a lot regional business associations. In Russia and China, examples of successful business associations representing the interests of entrepreneurs in relations with the authorities are the chambers of commerce organized according to the territorial principle. For example, the Chamber of Commerce and Industry of Russia includes many territorial chambers of commerce, the members of which are both commercial and non-profit organizations of the regions.

Business associations have an institutional status, which is explained by their access to the system of personnel training. So, regional chamber of commerce and industry is actively involved in the professional and public accreditation of educational programs in universities, which indicates the recognition of the graduates' training quality. The graduates' skills meet professional standards, the requirements of the labor market for professionals and employees of the relevant profile.

2. Methods

The analysis of theoretical literature was carried out using the methods of interpretation and categorical analysis. In the study of the legislative framework were used comparative legal and logical-legal methods, as well as the method of legal interpretation. When studying the regional specifics of enterprise communications with government authorities and educational institutions in the field of professional training of personnel, a statistical description of data for 2016–2017 taken from official sources was used.

3.Results and Discussion

The interests of the company require that the introduction of new equipment and technologies be preceded by the training of competent personnel. However, since the 1990s, in Russia there has been a reduction in the volume of training of workers directly at work, the destruction of the institution of mentoring. As a result, at domestic enterprises, the training of professional workers has ceased to be an obligatory stage of education, training and professional development of personnel. The lack of full-fledged professional training directly in the production of young people who have received basic training in secondary and higher educational institutions gets Russian enterprises in a tight corner. In particular, they do not have the opportunity to apply the Japanese experience of hiring, when applying for a job, the decisive factors are not the graduate's professional skills, but his ability to adapt and learn during the production process in order to bring the professional level to the level required by the company. At the regional level, the target solution could be the best solution to this problem, but enterprises do not always have the finances to do so. Therefore, they prefer interaction at the institutional level, which implies the participation of business and entrepreneurs as employers in the development of professional and educational standards and in the implementation of educational programs. By participating in the development of professional standards, associations of employers, professional communities and other interested organizations can indirectly «program», predetermine the content of future federal state educational standards for relevant professions and specialties. However, this relation has a reverse, negative side: the primacy of the professional standard in relation to the standard educational sometimes leads in practice to the fact that the absence of the first (professional) standard hinders the adoption of the second.

It is important that associations of employers, professional communities and other interested organizations have the opportunity not only to join the development of professional standards, but also to initiate this work. The rules for the development and approval of professional standards, approved by Government Decree No. 23 of January 22, 2013, provide the possibility of developing professional standards both at the expense of the federal budget and at the own expense of developers. At the same time, the draft of professional standard is mandatory to be discussed with representatives of employers, professional communities, trade unions (their associations) and other interested organizations.

The analysis of the current legislation revealed a number of problems that act as an objective obstacle both for ensuring unhindered communication of representatives of industry and the educational sphere in general, and for the "synchronization" of professional and educational standards, especially. In particular, there is no uniform approach to the understanding of the term "qualification" in labor and education legislation. While in art. 195.1 of the Labor Code of the Russian Federation, the qualification of an employee is defined as «the level of knowledge, skills, professional skills and experience of the employee» according to art. 2 of the Federal Law of December 29, 2012 No. 273-FZ «On Education in the Russian Federation» qualification is «the level of knowledge, skills and competencies that characterizes the preparedness to perform a certain type of professional activity». As it is rightly noted in literature, this is not one and the same concept, but two: qualification «according to work» and qualification "according ti education", the first relates to the employee, and the second to the graduate of an educational organization [8]. To eliminate this conflict, the legislator should take a unified approach to the understanding of qualifications, correct the relevant standard and in the future adhere to the chosen approach when formulating the provisions of regulatory legal acts.

In the Volgograd region, the institutional body responsible for the development of personnel policy in the sphere of production is the Committee on Industry and Trade of the Administration of the Volgograd Region (hereinafter the Committee) [9]. Carrying out the implementation of the regional state program «Development of the industry of the Volgograd region and increasing its competitiveness for 2014–2020» in accordance with federal strategic directions [10], the Committee monitors the current and future needs of industrial enterprises in the region annually.

In 2017, according to information received from 118 enterprises of Volgograd and the Volgograd region, the need for personnel increased almost one and a half times compared to 2016 and amounted to about 20 thousand people until 2023 [11]. The monitoring covered major cities of the region (Volgograd, Volzhsky, Kamyshin, Kotelnikovo), where the industry is actively developing and there are enterprises where major investment projects are being implemented: OOO EuroChem VolgaKaliy, VGAZ-SUAL branch of SUAL group, JSC Kaustik, OOO Kamyshin Textile, OOO Volgabas, Voltayr-prom company, etc.

A feature of the monitoring conducted in 2017 was that the data were collected taking into account training areas and needs of enterprises in personnel. This made it possible to compare the education of graduates with the required for the enterprise education for a particular profession. Based on the needs of industrial enterprises, propositions were sent to the regional base of working specialties for industry in the Volgograd region, 30 key industrial enterprises and investment projects in the industry, as well as about 30 of the most popular and promising training areas were identified.

The monitoring allowed to determine the need of industrial enterprises of the Volgograd region in the period for 2017-2023 (by branch of production):

- production of machinery and equipment (more than 8 thousand people);
- mining operations (more than 2.5 thousand people);
- chemical production (more than 2.5 thousand people);
- metallurgical production (more than 2 thousand people);
- consumer industry (more than 2 thousand people);
- production and distribution of electricity, gas and water (more than 1 thousand people).

Let us compare this calculation with the dynamics of personnel training for industrial enterprises of the Volgograd region by institutions of secondary vocational education in the region (Table 1).

Table 1. Training of personnel in institutions of secondary vocational education for industry in the region [12].

Groups of training	Number of graduates in 2016	Recruitment in 2017
Machinery	946	1200
Chemical engineering	176	275
Metallurgical engineering	43	250
Consumer industry	87	156

Thus, the trend of training of production personnel is positive, although not sufficient for the full realization of the needs of enterprises of the Volgograd region in the field of light industry.

The proposals sent by the Committee to the regional base of working specialties were taken into account when approving the list of popular and promising professions and specialties requiring secondary vocational education by the Committee of Education and Science of the Volgograd Region (decree of the Committee of Education and Science of the Volgograd Region of April 24, 2017 No. 50).

Modern strategies for attracting young people to education for the purpose of further employment for industrial enterprises of the region reflect the ideas of the creators of the human capital theory, according to which a person is making predominantly rational economic decisions on his future and career (when choosing education and profession) considerations.

It is this «economic approach to social issues» (G.S. Becker) that underlies the vocational guidance work with students of educational institutions. According to this approach, ideas and motives generally obey the economic pattern of behavior and can be described in terms of supply and demand, competition, etc. Thus, the main emphasis in providing industrial enterprises with personnel is placed on increasing the prestige of industrial occupations.

To this end, the annual regional forum «Education» is held in the Volgograd region, within which meetings of representatives of various industrial enterprises and students of educational institutions of the region take place [13]. Also held an annual regional competition «Engineer of the Year». In 2017, the competition was held in 4 nominations: «Technological engineer»; «Design Engineer»; «Mechanical Engineer»; «Software Engineer» [14].

Volgograd region is actively involved in the all-Russian action «Week without turngates». The action is a set of activities aimed at informing career guidance of the leading enterprises of the Volgograd region and popularizing professions and specialties that are in demand in industrial production, acquainting students of institutions of general, secondary and higher professional education with modern enterprises and with professions of industrial production.

In 2017, 25 industrial enterprises of the Volgograd region and more than 1.000 students from educational institutions in the region took part in the action. In particular, excursions were organized and students of educational institutions familiarized themselves with the work of innovators of production and modern technologies, such enterprises of the region as Redaelli SSM company, OJSC OAO Volgograd Radio-Technical Equipment Plant, Voltayr-Prom company, Volzhsky Pipe plant Volgogradneftemash, OOO VOLMA and others. In addition, tours of the WorldSkills Russia, an all-Russian Championship are organized and conducted annually in the Volgograd Region [15].

4.Conclusions

According to the results of the study conducted on the example of the Volgograd region, the following conclusions can be drawn.

1. When solving the problem of professional training of personnel, enterprises in the region prefer the institutional level of interaction with government bodies and educational institutions. This is reflected in the participation of business associations namely the territorial Chamber of Commerce in the accreditation of professional educational programs; in the participation of industry associations of employers in the development of professional standards, which play a key role in ensuring an effective interaction between production and the education system; participation of enterprises as employers in the implementation of professional educational programs.

2. A study of current legislation revealed the problem of «synchronization» of professional and educational standards. To solve it, it is necessary to develop a unified approach to the understanding of qualifications and adjust the regulatory framework to reflect this. A non-institutional reserve for solving the problem of professional training of personnel is career guidance, within which communication should be based on the patterns of human capital formation.

References

[1] Schultz T W The Economic Importance of Human Capital in Modernization (1993) Education Economics **1 (1)** 13-19

[2] Becker G. Growing human capital investment in China compared to falling investment in the United States (2012) *Journal of Policy Modeling* **34 (4)** 517-524

[3] Becker G.S. Health as human capital: Synthesis and extensions (2007) *Oxford Economic Papers* **59 (3)** 379-410

[4] Lutz W, Cuaresma J C, Sanderson W (2008) Economics: The demography of educational attainment and economic growth *Science* **319 (5866)** 1047-1048

[5] Schultz T W On investing in specialized human capital to attain increasing returns (1988) The state of development economics 339-356

[6] Yakovlev A A, Zudin A Yu, Golikov V V (2011) Business associations and their role in the processes of modernization in *Russia Social Sciences and modernity* **3** 25-35

[7] Doner R, Schneider B R (2000) Business associations and economic development: why some associations contribute more than others *Business and Politics* **2(3)** 261-288

[8] Kurushina V A, Kurushina E V, Zemenkova M Y (2018) Technologies of polytechnic education in global benchmark higher education institutions *IOP Conference Series: Materials Science and Engineering* **357 (1)**

[9] Committee of industry and trade of the Volgograd region. Access mode: http://promtorg.volgograd.ru/

[10] About the state program of the Volgograd region «Development of the industry of the Volgograd region and increase of its competitiveness». Access mode: http://docs.cntd.ru/document/460218749

[11] Committee of industry and trade of the Volgograd region. Access mode: http://promtorg.volgograd.ru/current-activity/promyshlennost/obespechenie-kadrami-promyshlennykh-predpriyatiy/

[12] Committee of education, science and youth policy of the Volgograd region. Orders. Access

mode: http://obraz.volgograd.ru/norms/acts/672/
[13] Committee of education, science and youth policy of the Volgograd region. XIII Volgograd regional forum «Education-2017». Access mode: http://obraz.volgograd.ru/current-activity/cooperation/news/137294/
[14] Committee on labor and employment of the Volgograd region. Annual Volgograd regional competition «Engineer of the year». Access mode: http://ktzn.volgograd.ru/other/trud-i-zanyatost-naseleniya/sotsialnoe-partnerstvo/professionalnye-konkursy/ezhegodnyy-volgogradskiy-oblastnoy-konkurs-inzhener-goda/
[15] Committee of industry and trade of the Volgograd region. Industrial enterprises of the region are invited to participate in the Action «Week without turnstiles» Access mode: http://promtorg.volgograd.ru/current-activity/cooperation/news/182024/

Network support for personnel training: evaluation component

N V Zolotykh[1], A V Chernyaeva[1], T U Shevchenko[2]

[1]Volgograd State Agricultural University, 26 University Aven., Volgograd 400002, Russia
[2]Volgograd State Socio-Pedagogical University, 27 Lenina Aven., Volgograd 400066, Russia

E-mail: shefdtj@yandex.ru

Abstract. The article is devoted to the evaluation component of graduate personnel training. Professional training of graduates of agricultural sector is a set of indicators (theoretical knowledge in the field of agricultural education and the ability to apply this knowledge in new and unusual situations), providing subsequent socio-professional adaptation and potential career growth of graduates of agricultural universities, taking into account the requirements of the modern labor market. There are four components in the evaluation of graduate professional training at Agricultural universities for those who has completed a certain level of education (bachelor), which is seen in the readiness and ability to be successful and productive in socio-economic activities, taking into account possible social risks: professional and personal potential, agroecological culture, basic and innovative technologies. On the material of research an attempt is made to reflect one of the fragments of network support for professional formation of production personnel at Agricultural University.

1. Introduction

Currently, the system of professional agricultural education is going through a complicated period of adaptation in the context of socio-economic reforms, innovation and technological developments. Higher education institutions of the Ministry of Agriculture of the Russian Federation should be more mobile, flexible and be more responsive to the needs of the modern labor market. Rapidly changing priorities replace the vector of assessment of professional qualities of production personnel to the stage of preparation of a graduate at Agricultural University. This aspect implies the search for the evaluation components of the graduate, providing an effective result of the education, the most adapted to the real conditions of agricultural industry.

Modern didactics in issues of students' professional training has been confirmed in the network interaction of subsystems of teaching staff and students in order to ensure the constant development of the student's personality in accordance with the social demand and production requirements [1;2;3]. Network support for professional personnel assumes to accept what is offered to graduates. In particular, paradoxical induction utilizes what a graduate is able to do or doing.

At Agricultural University network support is carried out through the strategy of development of agricultural education until 2030. Its aim is to teach students pedagogical strategies that allow to organize a professional activity for the production, processing and sale of agricultural products, as well as the development of rural areas for the purpose of personal self-actualization of students in the educational process.

At present, there is no common view on the elaboration of pedagogical measuring materials for the assessment of educational achievements on the basis of monitoring the formation and determination of the level of graduates' professional training at Agricultural University, forms of its organization and didactic technologies [4;5;6].

The platforms for the implementation of the results of scientific work among the staff of Volgograd State Agricultural University in the framework of network support were: Federal State Budgetary Education Institution of Higher Education 'Volgograd State Agricultural University', Federal State Budgetary Education Institution of Higher Education 'Bashkir State Agricultural University', Federal State Budgetary Education Institution of Higher Education 'Dagestan State Agricultural University', Federal State Budgetary Education Institution of Higher Education 'Kostroma State Agricultural University', Federal State Budgetary Education Institution of Higher Education 'Voronezh State Agricultural University N.A. Emperor Peter the Great', Federal State Budgetary Education Institution of Higher Education 'Don State Agricultural University'.

2. Results and Discussion

The evaluation component of the graduate at Agricultural University is provided by a set of modern requirements for the professional training of the graduate [7]. At the same time, integrated groups of indicators have the greatest potential in terms of training for professional personnel: professional and personal capability, agroecological culture, basic and innovative technologies. The level of the expert's professional activity in agrarian profile is defined by degree of indicators' formation.

The content and procedural aspects of training should be taken into account when the evaluation component of the graduate professional training at Agricultural University is realized. The modeling of the content of professional training of the graduate is based on the recourses of pedagogical support and includes the components of educational standards: programs and course material [8;9;10]. In this case, the modeling of the professional training refers to the cooperative activity of the teacher and the student where different forms of training are divided into three stages: empirical, theoretical and practical.

The basic information is systematized and summarized within the groups of basic and innovative technologies, for the purpose of invariant preparation and the fastest graduate's adaptation to the conditions of the entry into professional activity.

Pedagogical support for the graduates professional training is regulated by internal documents of the University, related to the requirements of the general management and structural departments of the University.

The assessment of graduate professional readiness at Agricultural University lies in the personal-developing sphere. It contributes to the activation and realization for production personnel potential. A unique holistic person is in the focus who is ready to self-development and self-improvement. As a result, each position of the evaluation component creates a certain set of favorable conditions for a complex assessment of graduates professional readiness.

The process of the personality's professionalization in terms of training at Agricultural University is associated with a consistent strengthening and deeping of personal and value-semantic orientations of the graduate's consciousness: (professional and personnel potential, agroecological culture, basic and innovative technologies) [11]. These changes are the internal conditions for the professional training's assessment.

In order to evaluate the graduate professional training, the experimental program should be organized. It, in turn, must be a part of the process of professional training; the diagnostics of indicators of integrated groups also should be implemented. These conditions are external factors in the process of assessment of graduates' professional training at Agricultural University.

The professional training is a complex system that includes external and internal factors that consistently form the levels: "Human" – "Agroecological resourse" – "Technologies" – "Innovations". A specific foundation of the integrated structure of the graduate professional training at Agricultural University is the formation and development of his professional and personal potential (the level of "Human"), responsibility for his vital activity, especially in relation to nature, as the main source of agricultural production.

The rational use of natural resources (the level of" Nature") will preserve the ecological balance of the biological basis and agricultural production. The formation of agroecological culture of the future specialist is one of the main agricultural teaching and training, regadless of the education field.

Having such a "foundation", it is necessary to form a technological base of professional education of the specialist, a scientific elaboration of technological programs and strategies in the sphere of agriculture (the level of "Technology") are meant here. Developing, testing and implementing innovative technologies (the level of the "Innovations") should be in agricultural production in perspective.

To sum up, the graduate professional training at Agricultural University involve two subjective sense-forming groups, they are professional and personnel potential and agroecological culture. The process of being formed of these two groups allows to move to the mainly objective level of technological readiness.

The analysis of works has enabled the researchers to use the following conclusion on the essence of the notion "the level of the professional training of a graduate". The presence of a set of formed indicators is meant, which provide the subsequent social and professional adaptation and potential career move of the graduate at Agricultural University taking into account the requirements of the modern labour market in the conditions of intensification of agricultural production.

Identifying the levels of the graduate's professional training, we base upon the methodological consideration of V.G. Afanasyev. He highlights the emergence of a new system based on disparate components at the initial level, the certain unity on the middle level and the replacing one system with another as a result of changes and the development of a new system at higher levels [12]. This scientist's position helped us to identify 4 groups of graduates of Agricultural University with low (critical) level, middle (acceptable) level, high (the best possible) level and advanced (scientific) level of professional training's formation of graduates of Agricultural University. Let us briefly discuss the generalized characteristics of each level. The obtained results are generalized in the characteristics related to different levels of pedagogical competence. Characteristics of the levels of formation of graduates' professional training can be presented in the following form:

2.1 Low (critical) level

Indicators of formation.

The first integrated group-professional and personnel potential.

Graduate:

- doesn't show adaptive abilities, socially not oriented, has some difficulties in making decisions and in personification of professional interaction;
- has some difficulties in behavior regulation (volitional effort is not easy);
- communicative potential does not allow establishing meaningful communication;
- not oriented on the moral normativity of his own personnel and professional actions.

The second integrated group – agroecological culture.

Graduate:

- doesn't express ability to use agroecological knowledge in separate spheres of life;
- didn't take part in the activity, that controlled the state of the environment and the compliance with ecological regulations of production and land use;
- does not use environmentally sound technologies of agricultural production during the practice;
- doesn't search for ways to preserve the quality of natural potential and ways of its reproduction while doing practical tasks;
- didn't take part in a monitoring study or agroecological assessment or expertise.

The third integrated group – basic technologies

Graduate:

- didn't take part in any regional seminars, conferences, contests, or competitions;
- did not pass the professional retraining program;
- didn't take part in the student's work of agro teams;

- didn't receive an invitation for employment.

The fourth integrated group – innovative technologies.

Graduate:

- didn't take part in special elective course or program, scientific project;
- didn't participate in any international contests, competitions or exhibitions;
- formulates the issue of diploma project, which is mainly theoretical (includes the results of research, but does not rely on the real needs of agricultural enterprises);
- doesn't have any patents for invention or utility model, computer program.

2.2. Middle (acceptable) level

Indicators of formation.

The first integrated group - professional and personnel potential.

Graduate:

- shows adaptive abilities sitationally, is socially oriented, has no difficulties in making decisions and in personification of professional interaction;
- has no difficulties in behavior regulation (is oriented on the social norms);
- communicative potential is oriented on the personnel communication;
- is oriented on the moral normativity of his own personnel and professional actions partially;

The second integrated group – agroecological culture

Graduate:

- expresses ability to use agroecological knowledge in separate spheres of life;
- took part in the activity, that controlled the state of the environment and the compliance with ecological regulations of production and land use;
- uses environmentally sound technologies of agricultural production during the practice partially;
- sometimes searches for some ways to preserve the quality of natural potential and ways of its reproduction while doing practical tasks;
- took part in a monitoring study or agroecological assessment or expertise.

The third integrated group – basic technologies.

Graduate:

- took part in any regional seminars, conferences, contests, or competitions;
- passed the professional retraining program;
- took part in the student's work of agro teams;
- received an invitation for further employment.

The fourth integrated group – innovative technologies.

Graduate:

- took part in special elective course or program, scientific project;
- didn't participate in any international contests, competitions or exhibitions;
- formulates the issue of diploma project, which is mainly scientific and practical (includes the results of research, relies on the real needs of agricultural enterprises);
- filed a patent application for invention or utility model, computer program.

2.3. High (the best possible) level

Indicators of formation.

The first integrated group - professional and personnel potential.

Graduate:

- shows adaptive abilities, is socially oriented, has no difficulties in making decisions and in personification of professional interaction;
- has no difficulties in behavior regulation (is oriented on cooperative activity);
- communicative potential is oriented on the professional interaction;
- is oriented on the moral normativity of his own personnel and professional actions.

The second integrated group – agroecological culture

Graduate:

- expresses ability to use agroecological knowledge in separate spheres of life;
- takes part in the activity, that controlled the state of the environment and the compliance with ecological regulations of production and land use;
- uses environmentally sound technologies of agricultural production during the practice;
- regularly searches for some ways to preserve the quality of natural potential and ways of its reproduction while doing practical tasks;
- takes part in a monitoring study or agro-ecological assessment or expertise.

The third integrated group – basic technologies.

Graduate:

- took the first prizes in some regional seminars, conferences, contests, or competitions;
- passed the professional retraining program during some semesters;
- took part in the student's work of agro teams several times;
- received an invitation for further employment according to the results of practice.

The fourth integrated group – innovative technologies.

Graduate:

- took part in special elective course or program, scientific project;
- won prize-places in some international contests, competitions or exhibitions;
- formulates the issue of diploma project with his scientific guider, which is mainly scientific applied (includes the results of research, relies on the real needs of agricultural enterprises);
- has a patent application for invention or utility model, computer program.

2.4. Advanced (scientific) level

Indicators of formation.

The first integrated group - professional and personnel potential.

Graduate:

- shows the highest level of adaptive abilities, is socially oriented, has no difficulties in making decisions and in personification of professional interaction;
- has no difficulties in behavior regulation (is oriented on a result);
- communicative potential is of high level (has the leading positions in personnel and professional communication);
- is completely oriented on the moral normativity of his own personnel and professional actions.

The second integrated group – agroecological culture

Graduate:

- expresses ability to use agroecological knowledge in separate spheres of life regularly;
- takes part in the activity, that controlled the state of the environment and the compliance with ecological regulations of production and land use regularly;
- uses environmentally sound technologies of agricultural production regularly during the practice;
- offers some ways to preserve the quality of natural potential and ways of its reproduction while doing practical tasks;
- takes part in a monitoring study or agro-ecological assessment or expertise regularly.

The third integrated group – basic technologies.

Graduate:

- won prize-places often in some regional seminars, conferences, contests, or competitions;
- passed some professional retraining program during training;
- took part in the student's work of agro teams regularly;
- received an invitation from some households for further employment according to the results of practice.

The fourth integrated group – innovative technologies.

Graduate:

- takes part in special elective courses or program, scientific project regularly;

IOP Conf. Series: Materials Science and Engineering **483** (2019) 012018 doi:10.1088/1757-899X/483/1/012018

- won prize-places in some international contests, competitions or exhibitions several times;
- formulates the issue of diploma project, which is mainly scientific (includes the results of research, relies on the real needs of agricultural enterprises, where he tests his research and tries to implement it);
- has some patents application for invention or utility model, computer program.

Statistical data has required essential changes in the construction of the entire assessment process for graduate's professional training at Agricultural University, including a computer program for integrated statistical evaluation and its experimental approbation.

The ability to achieve the maximum possible unity of social demand and natural potential in the agricultural sector can be considered as a sign of high assessment of the level of formation of the graduate's professional training at Agricultural University. Being aware of that fact that the environmental protection is the only means of preserving human beings as a species and the very formation of professional readiness is not possible without their personal participation is also considered as a criterion of high level of professional training.

The principles of systematization, science and mathematical modeling are the basis of efficiency's estimate of the graduate's professional training. They were realized in theoretical and practical approach and in some requirements of scientific substantiation, criterions of measurement procedures.

The scientific research has been carried out since February 2018 and contains quantitative statement of the issue under the study. The assessment's integral index of the level of vocational training's formation was calculated with the help of developed integral mathematical statistical method. The control comparative diagnostics was carried out among 100 students of different departments: 35.03.06 "Agroingeering", 35.03.03 "Agrochemistry and Agriculture soil science" (specialty "Agroecology"), 36.03.02 "Zootechni".

Database methods of assessing the graduate's level of professional training at Agricultural University were developed in phpMyAdmin, by using Database Management System (DBMS) application. It is a web-interface for system administration DBMS MySQL. All this is illustrated in figure 1.

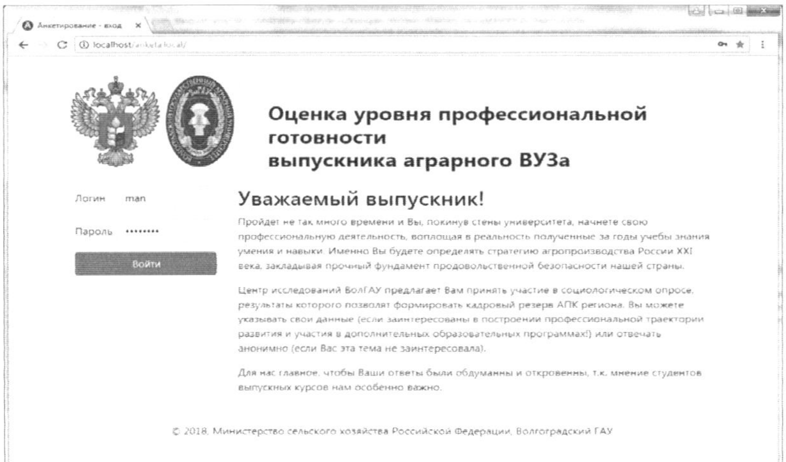

Figure 1. System of screen forms

The program provides a selection of prepared formats *.doc or *.docx questionnaire options. The paragraph of the questionnaire template text is a structural element in the automated questionnaire generation via the interface. It includes the rules of registration of structural elements of the questionnaire and takes into account 4 types of them: the beginning of the questionnaire, a question, an answer, a number, a title and a weight of index, appropriated to the group of questions. The program makes it possible to generate the following types of reports: 1. summary report; 2. the year report; 3. universities reports; 4. education field reports; 5. detailed report.

IOP Conf. Series: Materials Science and Engineering **483** (2019) 012018 doi:10.1088/1757-899X/483/1/012018

The detailed report provides a selection of questionnaire on the maximum set of parameters: 1. title of questionnaire; 2. university; 3. education field; 4. specialty; 5. year of admission; 6. completed questionnaire or not; 7. date and the beginning of questionnaire (optional).

Private groups' indicators are determined by taking into account the sample from the general database filtered by the specified parameters (kind of questionnaire, university, education field and others). Particular indicators characterize the average index for each of the categories.

Calculation algorithm of integrated parameter of the level of professional training was carried out according to the following:

$$I = \sum ai \times(\sum bj\ ni\ j=1\ ni\ )m\ i=1\ , \quad (1)$$

where

m – the count of question groups in the questionnaire;

ni – the count of questions in group;

i – index of questions' group

j – index of the question in a group;

ai – significance of the indicator corresponding to the group of questions;

bj – score of "j" question.

The values of the parameters 'ai' and 'bj' can scale the value of the integral index 'I' to the necessary range of values. It should be noted that significance indicators a_i,, can be adjusted only by system administrator. The following values are accepted:

a_1 = 0,5 – part of the indicator's contribution of "Basic technologies" to the integral;

a_2 = 0,3 – part of the indicator's contribution «Innovative technologies»;

a_3= 0,1 – part of the indicator's contribution «Agroecological culture»;

a_4 = 0,1 – part of the indicator's contribution «Professional and personnel potential».

The integral index is calculated according to these coefficients. A detailed report of the results is shown in the figure 2.

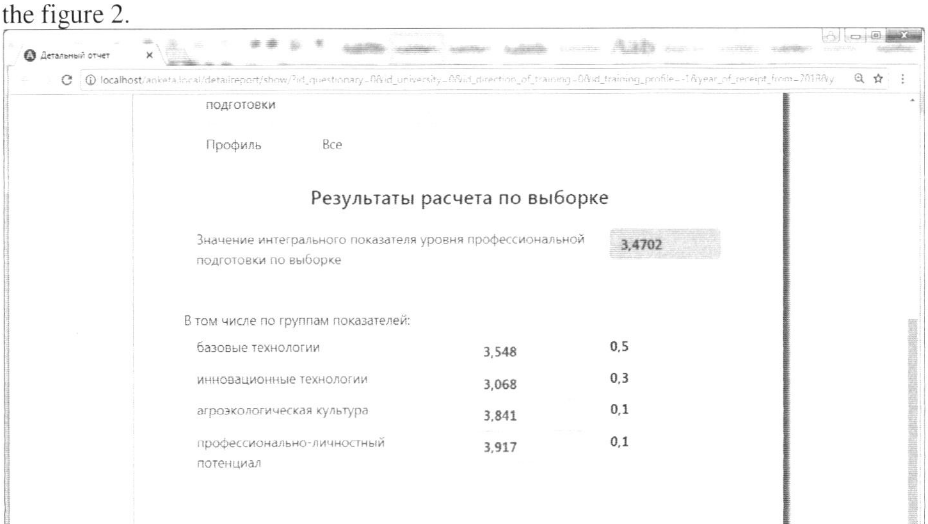

Figure 2. A detailed report: calculation results

The main page in the mode of the survey organizer provides an opportunity for the respondent to fill in not a required string "Notice", the results of the questionnaire are saved. The value of the string is displayed on the page "Results" in the administrator mode.

3. Conclusion

The evaluation component of the graduate professional training at Agricultural University, presented in the scientific research, and built on its web-based interface of diagnostic assessment's module showed a high effectiveness in solving the tasks of network support.

Assessment of the level of personnel training makes it possible to identify peculiar differences in the ability of critical evaluation, presentation and results' research defence, in using methodological, general scientific and special knowledge on the stage of studying at the University. The results of the study presented above may be of practical use for various specialists at Agricultural Universities.

Prospects for further research are obvious for us in the assessment enhancement for graduate professional training at Agricultural Universities and this research ideas and principle extension for the various student training.

References

[1] Rozhdestvenskya M G 2014 *The World of Science, Culture, Education* **2 (45)** 116-119
[2] Aleksandrova E A 2016 *Modern Problems of Science and Education* **6**
[3] Kajl U A 2008 *Bulletin of Buryat State University* **1** 21-24
[4] Lobok A M 2008 *School technology* **1** 49–59
[5] Zolotykh N V, Chernyaeva A V 2014 *Proceedings of Nizhnevolzskiy Agrouniversity Complex: Science and Higher Vocational education* **4 (36)** 222-226
[6] Mogilevkin E A 2012 *HR-tools: practical assessment. How to identify employees who can give the maximum result: educational and practical guide* (Saint-Petersburg: Rech) p 320
[7] Volodarskya I A 2010 *Russian Scientific Journal* **14** 48-56
[8] Zaitseva G A 2014 *Bulletin* **3 (30)** 54–57
[9] Chuchalin A I, Epihin A V, Muratova E A 2013 *Higher Education in Russia* **1** 13-20
[10] Zvyagdinskiy V I 2006 *Dictionary of Pedagogy* (Moscow: Academia) p 352
[11] Isakova E K, Lazarenko D V 2014 Rusnauka, Science-Pedagogy **9**
[12] Afanasyev V G 2018 *System and society* (Moscow: Lenand) p 242-244
[13] Safronofa E M, Beylinson L S, Zolotykh N V, Shevchenko T U 2016 *International Journal of Enviromental & Science Education* **11 (16)** 8968-8980
[14] Shevchenko T U, Slyshkin G G, Moseyko A A and Korovina K G 2018 *SHS Web of Conferences* **50** 01173

IOP Conf. Series: Materials Science and Engineering **483** (2019) 012019 doi:10.1088/1757-899X/483/1/012019

Artificial intelligence and digital transformations in the society

A V Lavrentyeva, A A Dzikia, A E Kalinina, D P Frolov, E A Akhverdiev and A S Barakova

[1]Volgograd State University, pr-t Universitetsky, 100, 400062, Volgograd, Russia

E-mail: oponir@volsu.ru, lavra.ne@mail.ru

Abstract. The article describes the global trends in the digital economy development. The role of artificial intelligence for the formation of the institutional environment of ecosystems is highlighted. Tendencies of society's institutional changes in connection with digital transformation of economy are described. New organizational forms of cooperation and the role of artificial intelligence for the formation of effective methods of organizations management that seek to build a strategy of interaction in the digital environment are described. The problems of transition to a new level of post-institutional economy development are highlighted. The paper describes the boundaries of the digital economy and the trends of changes in the institutional environment

1. Introduction

The challenges of the 21st century are associated with a sharp impetus to the development of digital economy, globalization and the increasing speed of various changes. Global trends that affect all spheres of public life are transforming traditional ways of life, turning industrial sectors and legislation upside down. Moreover, these processes do not take place on a local scale, but on the territory of all countries and affect each of us, as the digital economy has already become one of the determinants of globalization. This process has developed since the beginning of the industrial revolution in the 17th century, but began to develop actively since the robotisation in the 21st century. Today we are on the threshold of transition to the sixth technological mode and this process is accelerating every year. The speed of innovation is growing exponentially, and new technologies such as the Internet of things, artificial intelligence, and robotics allow routine tasks to be automated. As these technologies become more widely accepted, efficiency will no longer be a key difference for enterprises. Instead, today's business leaders are looking for 25-30% faster performance and productivity improvements. They achieve this by combining the Internet of things, artificial intelligence and other digital technologies.

2. Artificial intelligence in management

Progress in information and communication technologies (ICT), in which information and knowledge are becoming available to more people, change the institutional character of the society. Societies are becoming ecosystems mediated by various digital media platforms, digital services and technologies that enable surveillance, data collection and analytics. This progress and changes in technology are bound to change the way societies are organized and how their members interact with each other. Massive amounts of data are generated every moment from a growing number of sources. Companies recognize that the data they own and the way they use it can give them a competitive advantage. Big data and business analytics also challenge existing businesses and well-established companies. However, there is a limited understanding of how organizations need to change to embrace these technological innovations, and the business changes they entail, which can transform businesses and societies.

Over the past few years, competition has increased and the system of effective management of organizations that seek to build a strategy of interaction with different users by analysing large amounts of data, using business analytics and the so-called smart environments and attracting the attention of stakeholders at the decision-making stage has changed. With the help of artificial intelligence and modern information systems of data analysis, organisations collect information about the preferences and requests of users, needs and opportunities at a particular time, build forecasts for the demand of goods and services depending on many factors, increasing the efficiency of organisations, reducing the cost of production of unclaimed goods and the cost of advertising budgets. Existing technologies allow processing large volumes of high-speed, complex, temporal data for its collection, storage, distribution, management and analysis [25]. The Internet generates and offers a large amount of data every second and companies are often found to be loaded with this data obtained from various fields (social media, landing pages, Internet search engines, etc.) [11]. At the same time, this data shows the behaviour of customers (existing and potential), competition trends, prices, costs and other important indicators for business. Moreover, these processes do not have an extensive nature in relation to specific sectors of the economy, but are characterized by pluralism and show their worth even in the field of health. Nowadays, big data technology allows us to take into account the preferences and interests of Internet users, build individual behaviour through complex digital stages and without the efforts of people, offer and independently orient them in the digital space. These processes are one of the determinants of the development of artificial intelligence.

Speaking about process automation and artificial intelligence (hereinafter – AI), it is necessary to understand the multiplicity of complex algorithmic methods that allow machines and computers to simulate human actions [14]. With the help of these technologies, it is possible today to streamline and simplify a large number of tasks that are important in the process of creating a product. In this context, there is a significant facilitating role of big data for artificial intelligence, the work of which is provided by information about interactions, preferences of people, and the history of cyclical actions. Today, it is predicted that the stimulation of AI growth will be carried out in three ways:

1. creating a virtual workforce,
2. increasing existing skills and workforce in the digital economy,
3. innovations used in AI work.

It is worth noting that the automation of processes with elements of artificial intelligence is already being implemented in various areas of business.

In this part, we are interested in the issue raised above – about the methods and measures of regulation of the digital economy. Representatives of politics, business, and lawmakers fluctuate between two positions. The first ones propose to provide a free market without violating the free movement of information, to abandon data control and only provide legal certainty of the sphere. The latter are on the position of the need to control data and information, attempts to regulate the maximum possible types of digital legal relations to ensure confidence in digital trade and the legality of the existence of such sectors of economy. This problem is currently the most urgent, each state and state association solve them in their own way. Big data may be one of the most significant technological failures in business and academic ecosystems in recent years [18].

While the vast majority of executives say that achieving business sustainability is a priority for them, companies are also seen as responsible for multiple societal challenges with social, environmental and economic consequences [15, 19]. According to the report of McKinsey & Company, which was released in 2015, 45% of professions in the next 15-20 years will be replaced by robots or programmes. The volume of displaced money according to experts' forecasts will be more than 20 tron dollars only in the US economy.

There is a growing need for accessible, networked ecosystems and data management standards that provide greater access to information in society. Investing in such infrastructures allows innovation and digitization of urban services and the introduction of a wide variety of technological ecosystems. Digital ecosystems are now an integral part of many areas (e.g. business, healthcare, transport, finance). But there remains an important question of how to make data available to a large number of stakeholders and at the same time keep them safe to accelerate sustainable development of society.

In the ecosystems of the 21st century, none of their participants can be considered in isolation. All participants in this system must actively interact and cooperate with each other forming new chains of connections and teams among themselves, to create new knowledge, innovation, technology, the interdisciplinary nature of the discipline of artificial intelligence can be a driving force that will give meaning to big data and improve the relationship between stakeholders and data. Digitalization of services and transformation of modern business models are needed to accelerate the creation of sustainable societies. New digital business models will not only be more accurate and efficient, but will also go beyond economic needs and address social issues that generate shared value that affects companies, organizations, consumers and the public at large. Big data plays a key role in this transformation and brings them together from different sources, shares them with different stakeholders and analyses them in different ways, enabling digital transformation and the creation of sustainable societies. Comprehensive analysis and access to large amounts of data and relationships allow the development of new business models that will provide new solutions that will benefit all participants in the ecosystem. As a result, political, economic and legal disputes arise on the issues of the actual non-legal existence of the digital economy on the international platforms and on the platforms of legislative bodies of various countries. The risks stemming from this situation range from the abuse of intellectual property to digital fraud in the financial and credit sphere, to hacker attacks on power grid units and commissioned or independent cyberterrorists capable of paralysing public institutions.

New organizational forms of cooperation are emerging, such as joint communities, peer-to-peer organizations, as well as holocracy and other forms of self-government. Thanks to digital technologies, they are transforming large industries and public sectors, such as IT, transport, hotels and housing, scientific discoveries, health, military, and even changing the way knowledge and labour are generated and organised.

The literature mentions a variety of matching terms and concepts for collaborative organisational forms, such as I-form, virtual organisation [15], C-form [20], market management [8], community [10], meta-organisation [13], and actor-oriented organisations [8]. Echoing the calls for research and the theory of adaptive, self-organizing and innovative organisations [5], there is a need for systematic discussion by scientists and practitioners, to analyse and design collaborative organizational forms, and to understand the dynamism and heterogeneity demonstrated by modern organizations. In addition, organisations combine and match mechanisms from different forms into new, rapidly developing hybrids, some of which are radically different from the usual hybrids of the market hierarchy known from the classical literature of the economic organisation [22]. We need better understand the distinctive design features of collaborative organizational forms, including new collaborative hybrids. In particular, we need a coherent framework to describe, analyse and conceptualize organisational forms of cooperation and how they relate to conventional projects.

How to effectively organise and manage production activities in new organisations is the main issue. This applies to key issues such as decision-making, the choice between buying different products, ownership and distribution of resources, contracting, knowledge sharing and other forms of cooperation. Most studies on comparative analysis of organisational forms in the literature of economic organisation were based on two ideal forms - markets and hierarchies - and combinations of the two, called hybrids. With the advent of knowledge and the networked economy, the time for market hierarchy is becoming increasingly inefficient. The importance of trust and cooperation in the knowledge economy is increasing, Adler suggested the community as the third ideal type, which is superior to co-production of knowledge and with trust to view as the primary coordination mechanism [2].

Currently, legislators from different countries and the digital business community find it difficult to come to common understanding. The former advocate legal regulation of all aspects of the new economy in order to avoid undermining the constitutional and economic foundations, while the latter

IOP Conf. Series: Materials Science and Engineering **483** (2019) 012019 doi:10.1088/1757-899X/483/1/012019

take the position of free regulation, which will not interfere with the development of innovation, digital trade and new ways and forms of relations between people and organisations.

In this regard, in our opinion, one of the most important issues arises – the issue of the legal boundaries of the digital economy, which should ensure a balance between stimulating the development of a new way of economic relations and the security of citizens and states, as risks and threats acquire completely new forms.

It is worth noting that the time of the so-called digital revolution in the economy, scares the public in terms of the rapid development of the digital economy in a fairly short time and its deep inclusion in almost all spheres of public life [23], which fundamentally transform the relationship [13]. In addition, lawmakers understand that modern technological developments are so diverse and unpredictable that developments under soft (free) regulation can expose the inefficiency of the legal system and expose public institutions to even greater risk, as the limits of the emergence of innovative developments, as well as the speed of their implementation and use on an industrial scale, will grow exponentially [9].

In the United States "Digital Economy Agenda" was adopted in 2015, which covers 4 areas of the digital economy (free and open Internet, trust and security on the Internet, Internet accessibility, innovation and new technologies in the digital economy) [17]; in the UK a number of programmes are developed and implemented, for example, "UK Digital Strategy"[24]; in India there is a program "Digital India", adopted in 2014 and aimed at turning the country into a "digital society and knowledge economy" [7]; in Germany in 2011 a strategy called "Industry 4.0" was adopted, which involves a full transition to Internet production by 2030 [19]; in China in 2015 the "Internet plus" concept was adopted [4].

In Russia in 2017, the "Digital Economy of the Russian Federation" Programme (hereinafter - the "Digital Economy of the RF" Programme) was adopted. The programme of the government of the Russian Federation consists of the five areas:

1. legal regulation,
2. personnel and education,
3. research competence and technological background,
4. information infrastructure,
5. information security.

Legal regulation is the principal direction for the functioning and development of the digital economy. In this direction, according to the "Digital Economy of the RF" Programme, it is planned to solve the following tasks:

- creation of a permanent mechanism for managing changes and competencies in the field of legal regulation of the digital economy (control over the state of the regulatory framework, rapid response to changes);
- removal of key legal restrictions and creation of new legal institutions, allowing to implement priority measures for the formation of the digital economy;
- formation of complex legislative regulation (bringing to a common conceptual denominator of basic regulations and reflection of the general principles of regulation of the digital economy);
- adoption of measures aimed at the development and promotion of activities in the digital environment;
- formation of policy for the digital economy development in the territory of the Eurasian Economic Union;
- creation of a methodological basis for the development of competencies in the field of regulation of the digital economy (adoption of methodological documents on the development of competencies of lawyers, civil servants and other persons involved in the preparation, adoption and application of legal acts in the field of digital economy).

All these tasks should be implemented in the bosom of the existence and development of those technologies that have been disclosed earlier, and which in turn form the basis of the digital economy. In the legal community, the discussion and evaluation of the programme mentioned by us has become particularly relevant in recent years. Many of the researchers consider it successful and complete, others are sceptical, pointing to the possibility of improvement. However, we proceed from the position that the existence of such a document, implemented by the state, is already a progress, as the society, represented by the state bodies, has realised the need to develop provisions aimed at the development of the digital economy.

At the same time, there is a danger that in the process of creating an institutional and legal system regulating the ecosystem of the new digital environment, a situation may arise in which the transition and functioning of the state in the new economic realities will be difficult. We are talking about the correct definition of the legal boundaries of the digital economy in the Russian Federation.

3. Conclusion

As noted earlier, the digital economy is currently based on four general technologies that have great potential for development. All of them are already radically changing the traditional way of life in various spheres of human activity. Speaking about the problems arising in the digital economy existence system, it is necessary to highlight some important ones.

First, the development of basic digital legal concepts. Depending on how accurately, affordably and professionally this or that concept is built, the functioning of the entire regulatory system and the effectiveness of its action will depend. For example, such concepts as "digital economy", "blockchain", "artificial intelligence", "smart contracts", "cloud electronic signature", "biometric personal data", "cyber physical system", "crowd funding", "cryptocurrency" and many others require legal definition [25]. As we can see, the digital economy is actively changing the legal plane, introducing new legal proto-institutions. In this regard, some lawyers talk about the existence of the so-called "digital jurisprudence"[1].

Secondly, such an array of new concepts, which are likely to emerge in the next few years, points to the problem of a temporal character of existing economic mechanisms. This situation implicitly leads to the fact that the traditional approaches and norms that exist today will no longer be enough. For example, the mechanisms of formation of the price of a product, its cost, the sale of digital products, the composition of the delivery set and other economic mechanisms are already gaining a new appearance. However, both traditional and specified processes do not arise at the same time, but are created purposefully, therefore, the main task is to create them, after which the regulatory framework should be modified [21]. Today, the countries are already facing the task of legal regulation of the full cycle of digital products. In addition, it is necessary to take into account the specifics of such products – its intangible form. For jurisprudence, there is a huge task to streamline these relations, because, for example, it is not clear how the transfer of rights to products from one owner to another will be carried out. This also applies to lease, a pledge, etc. In addition to the issues raised, there is a need to determine the legality and legal force of these transactions in Russia. For example, most large foreign companies enter into essentially illegal or sham M&A agreements, to which Western law enforcers do not respond. This policy is based on the principle of the impossibility of full regulation of the digital economy on the one hand and the provision of freedom in the development of its activities in the global sense on the other. Indeed, an objective view of existing large companies and how they profit raise some questions that do not yet have a single recognised solution. For example, Facebook and Google, registered in the United States, pay income tax in the United States, although the activities through which their income is generated, is conducted in all countries where there is Internet. How can the tax issue be handled delicately or should the situation be left untouched?

Thirdly, there are questions about the preservation of the existing order of labour relations. The digital economy highlights the creative and intellectual work, which entails the non-normality of the working schedule, the independence of the employee from the employer, special conditions for the organization of work. All these and other aspects of the changes should be taken into account in the labour legislation in order to harmonize and stabilize labour in the digital economy. To date, the idea of transition to paperless interaction of workers and employers, the creation of electronic employment record books, etc. is proposed.

Fourth, the issue how to identify the subjects of legal relations and the rules of processing, storage and collection of data. Today, there is a need to change the federal legislation in terms of unification of the identification requirements of the subject, as well as to expand the possibilities and methods of such identification. There are proposals to ensure the possibility of identification and authentication by any technically possible means: electronic (mobile) signature, subscriber number of an individual, biometric data. In this regard, it is possible to regulate the status of electronic documents. For example, the conclusion of contracts in electronic form, through correspondence between two representatives of organsations or ordinary individuals, can reach a new level. Today, there is some uncertainty in connection with the impossibility of a reliable definition of the person who sent the message or expressed the will to commit legally significant actions. In some court systems, in order to preserve and ensure the stability of business turnover, the names in the "header" of the messages are accepted as the signature of electronic agreements.

The need to change the rules of data collection, storage and processing arise in connection with the use of information by legal entities and individuals for the development of digital relations. The evolution of economic relationships dictates the removal of legal constraints that impede the rapid and efficient use of data. Therefore, the immediate task of the legislator is to develop rules for prompt and non-discriminatory access to data. Changing this area will entail the problem of the principles of disclosure and protection of various types of secrets, which will also require its legal solutions in terms of respect for the rights of all parties of legal relations.

Fifth, the novelties of the institutional and legal system of economic relations determine the need to change the principles of state systems. Already today, there is talk about the possibility of using blockchain technology for state information bases: tax, property, car owners, etc. In addition, the United States has already launched a new form of electronic lawmaking, which provides access to ordinary users to discuss bills [3]. The results of the digital economy development, as we can see, greatly improve the political, social and legal life of society. To this end, we will change the procedure by changing the rules of providing electronic evidence, expanding the practice of conducting trials using video communication systems, the development of the mechanism of electronic interaction of the parties to the trial.

Finally, it is worth noting no less important financial and legal problem of objects of legal relations. The most obvious examples are the question of the recognition subject to the right of cryptocurrencies, robots, communications, information, etc. It is interesting that in 2017 in Belgium, a robot named Fran Pepper became the first humanoid, which was included in the population register. A similar example has recently occurred in Saudi Arabia, where the humanoid robot Sofia was given citizenship. These facts are frightening to modern society, but in our opinion to make such drastic action in respect of machinery in general, is premature. It is necessary to proceed from the fact that machines with artificial intelligence still do not have full capacity, that is, for the actions of machines with AI should be responsible persons who own or control them. In this case, the closest to our legal system is the way to draw an analogy of machines with AI with legal entities that under certain conditions could be endowed with a limited legal personality.

It is obvious that the boundaries of the digital economy we have outlined are not exhaustive. The experience of European and Asian regulation of the digital economy shows that in the process of law-making there are previously unaccounted problems that need to be solved. However, the above analysis shows that the problem of institutional and legal support of a new way of economic life, affecting the political and social life of society, is an urgent topic at the moment. It should be emphasised that today it is necessary to effectively build an institutional environment and legislation, not just regulating the economy, but contributing to its innovative development. In this regard, there are three models of regulation [12]:

1. A traditional model of control and supervisory functions of the state (possibly through a regulated body of experts)
2. A self-regulation model
3. A mixed (compromise) model

The international community and authoritative researchers prefer the third model as the most effective and harmonious one. In addition, we must not forget that one of the main consequences of the digital economy is globalization. This fact necessitates the resolution of some important regulatory issues at the interstate level. However, as the practice of several years shows, compromise can be achieved with some difficulties. In the opinion of some researchers, this situation is due to the desire of the state to maintain a leading position and dominance in the markets of the digital economy. This is another reason why the Russian expert community, together with the legislator, should take all possible measures for a rapid and effective transition from the traditional to the electronic (digital) economy. At the same time, this process should not be rigid, but should contain an element of easy regulation in order to create conditions for the development of a competitive economy.

4. Acknowledgement
The work was supported by Russian Science Foundation (project №18-78-10075.).

References
[1] Andreeva L A 2017 K voprosu o «tsifrovoj» yurisprudentsii. *Aktual'nye problemy yurisprudentsii: sb. st. po mater. I mezhdunar. nauch.-prakt. konf* **1(1)** 32-41
[2] Adler P S, Kwon S W, Heckscher C 2008 Professional work: the emergence of collaborative community. *Organ Sci* **19** 359–76
[3] Cass R S 2014 Democratizing Regulation, Digitally, Simposium Regulating the digital economy, *Democracy a Journal of ideas.* [Electronic Resource] https://democracyjournal.org/magazine/34/democratizing-regulation-digitally
[4] 2015 China unveils Internet Plus action plan to fuel growth http://english.gov.cn/policies/latest_releases/2015/07/04/content_281475140165588.htm
[5] Child J, McGrath R G 2001 Organizations unfettered: organizational form in an information-intensive economy *Acad Manag* **44** 1135–48
[6] Demil B, Lecocq X 2006 Neither market nor hierarchy nor network: the emergence of bazaar governance *Organ Stud* **27**1447–66
[7] Digital India 2014 A programme to transform India into digital empowered society and knowledge economy. *Press Information Bureau*. New Delhi: Government of India, Cabinet, http://pib.nic.in/newsite/PrintRelease.aspx?relid=108926
[8] Fjeldstad Ø D, Snow C C, Miles R E and Lettl C 2012 The architecture of collaboration. [Electronic Resource]. *Strat. Mgmt. J* **33** 734-50
[9] Gulati R, Puranam P, Tushman M 2012 Meta-organization design: rethinking design in interorganizational and community contexts. *Strategic Manage J* **33** 571–86.
[10] Heckscher C C, Adler P S at al 2006 The firm as a collaborative community: reconstructing trust in the knowledge economy (Oxford University Press, Oxford)
[11] Hoffman A 2018 The next phase of business sustainability. *Stanford Social Innovation Review* (Spring)
[12] Johns N 2015 Regulating the Digital Economy. *Observer Research Foundation* **6** 2
[13] Lakhani K R, Iansiti M 2017 The Truth About Blockchain *Harvard Business Review*, Jan-Feb Issue
[14] Mirjana Stankovic, Ravi Gupta, Bertrand Andre Rossert, Gordon I. Myers and Marco Nicoli 2017 Exploring Legal, Ethical and Policy Implications of Artificial Intelligence, White Paper of the Global Forum on Law *Justice and Development*
[15] Newman D, Data Analysts and Scientists More Important Than Ever for the Enterprise, Digitalistman, 28 April 2017 http://www.digitalistmag.com/author/daniel-newman.
[16] Porter M E, Kramer M R 2019 Creating shared value. In: Lenssen G, Smith N (eds) Managing sustainable business, vol 89 (Springer, Dordrecht) 327–50
[17] Pritzker Penny U.S. unveils Digital Economy Agenda that supports new tech, worker participation https://www.linkedin.com/pulse/today-commerce-launches-digital-attaché-

program-address-pritzker?trk=eml-b2_content_ecosystem_digest-recommended_articles-73-null&midToken=AQHvNG0sBYtAzg&fromEmail=fromEmail&ut=2vvXcZ4-89T781.

[18] Roffe P, Seuba X 2017 Current Alliances in International Intellectual Property Lawmaking: *The Emergence and Impact of Mega-Regionals*, A CEIPI-ICTSD Publication Series, September, https://www.researchgate.net/publication/319469614_The_Regulatory_Framework_for_Digital_Trade_in_the_Trans-Pacific_Partnership_Agreement

[19] Safiullin A R 2017 Industriya 4. 0 i prioritety razvitiya ehkonomiki i obshhestva v Germanii *Vestnik UlGTU* 3 (79)

[20] Seidel M-DL, Stewart K J 2011 An initial description of the C-form. *Res Sociol Organ* **33** 37–72.

[21] Semenov E H Kak postroit' tsifrovuyu ehkonomiku v Rossii: shest' glavnykh problem https://centerforpoliticsanalysis.ru/position/read/id/kak-postroit-tsifrovuju-ekonomiku-v-rossii-shest-glavnyh-problem.

[22] Shah S K 2006 Motivation, governance, and the viability of hybrid forms in open source software development. *Manag Sci* **52** 1000–14

[23] TechAmerica Foundation's Federal Big Data Commission, 2012, quoted in Amir Gandomi and Murtaza Haider, Beyond the Hype: Big Data Concepts, Methods and Analytics, *International Journal of Information Management* Vol 3 2015 p 137-44 at p 138

[24] UK Digital Strategy, Available at: https://www.gov.uk/government/publications/uk-digital-strategy/uk-digital-strategy

[25] Vaipan V A 2017 Osnovy pravovogo regulirovaniya tsifrovoj ehkonomiki *Pravo i ehkonomika* **11**

IOP Conf. Series: Materials Science and Engineering **483** (2019) 012020 doi:10.1088/1757-899X/483/1/012020

Assessing competitiveness of innovational systems

S A Korobov, V O Moseyko, V S Epinina, E Y Marusinina, S I Korobova

Volgograd State University, Universitetskiy av., 100, 400062, Volgograd, Russia

econmanag@volsu.ru

Abstract. Russian economy is undergoing the most difficult and perhaps most dramatic stage of its development after the period of market relations formation. This situation creates respective competitiveness imperatives thus drawing increased attention to competition not only on the international level but also between national regions. This work studies an approach to assessing competitiveness of innovative socioeconomic systems based on I. Adizes' methodology. For assessing competitiveness of innovational systems, the authors suggest to apply the integral index of socioeconomic system innovational competitiveness and individual indexes of innovation competitiveness in short-term and long-term perspectives.

1. Introduction

Russian economy is currently undergoing the most difficult and perhaps most dramatic stage of its development after the period of market relations formation. Difficulties in most economic areas are caused directly by sanctions imposed by leading Western countries or by their consequences. As a result, certain complications related to promoting products on international markets and maintaining high level of competitiveness on the internal market arise only due to the impossibility to use the resource potential of Western economies. All these processes create respective competitiveness imperatives and consequently draw increased attention to competition not only on the international level the but also between national regions

Existing economic practices indicate insufficient stability of Russian economy that is mostly focused on using natural resources. This determines the relevance of invigorating innovational activities in all spheres of national economy. Innovational development of Russia should become the main condition for national economic growth and increase of economic stability and competitiveness. All these factors make achieving high competitiveness level of various economic systems the primary objective. Its fulfillment requires innovational development and reforms of the national economy.

2. Methods

This work studies a methodological approach to assessing competitiveness of innovational socioeconomic systems (hereinafter – systems) of different levels or, to be more precise, competitiveness of systems in relation to their innovational development. Innovation-based competitiveness can be defined as innovational competitiveness [1]. Thus, this work provides an approach to assessing innovational competitiveness. The method's objectivity makes it universally applicable for assessing innovational competitiveness of enterprises, industries and regions.

Several works involving the authors of this article offer to implement the method for assessing competitiveness of socioeconomic systems [2, 3] based on I. Adizes' methodology [4].

In order to assess competitiveness of socioeconomic systems, these works offer to apply the integral competitiveness index. It includes individual indexes of innovation competitiveness in short-term and long-term perspectives. The index is estimated using four values demonstrating the level of functionality, systematicity, proactivity and harmony.

Individual index of socioeconomic system competitiveness in short-term perspective is formed by two indicators estimating levels of functionality, and systematicity.

Individual index of competitiveness in long-term perspective is also formed by two indicators estimating levels of socioeconomic system's proactivity and harmony.

Integral and individual competitiveness indexes are suggested to be assessed as the geometric means of the product of specific indicators comprising these indexes [5].

3. Result and Discussion

First of all, it must be pointed out that the notion of innovational system is relative because it is applicable not only to specific systems and is not opposed to other, "non-innovational" systems. The notion of innovational system is applicable to all socioeconomic systems involved in entrepreneurial activities because all of these systems are innovational to a certain extent. According to J. Schumpeter, entrepreneurship always includes innovation that could be aimed at designing new products, applying new production technologies, creating new types of organizations, exploring new raw materials and market opportunities. Thus, socioeconomic systems must be distinguished not according to the principle "Is this system innovational or not?" but to the principles like "To what extent is this system innovational?" or "How much is this system more innovational than others?"

We may assume when comparing two systems that the system with a higher competitiveness level could have lower levels of innovation perception and implementation and, consequently, has a lower innovational level [6,7].

Table 1 lists indicators of functionality, systematicity, proactivity and harmony of socioeconomic systems and gives examples of economic values that could be used for interpreting these indicators.

Table 1. Indicators forming system competitiveness indexes.

Indicator	Methodological content of assessment indicator	Economic values interpreting the indicator
Functionality indicator	Short-term efficiency, meeting customers' demands.	Sales volume. Organizations' turnover. Organizations' turnover depending on economic activity type.
Systematicity indicator	Short-term efficiency – economic efficiency of resources supply.	Gross value added in relation to the employment. Gross value added in relation to the population number.
Proactivity indicator	Provision of innovational activities, initiation of changes, adaptation to new challenges and opportunities.	Innovational products volume. Volume of scientific developments funding. Capital assets investments.
Harmony indicator	Provision of integrational cooperation. Integrational interrelation level.	Personnel number. The number of enterprises and organizations depending on economic activity type. The number of regional enterprises and organizations.

Functionality indicator demonstrate short-term efficiency of the socioeconomic system meaning mostly the customers' satisfaction. The most applicable economic values for this indicator are related to the sales. High sales volume indicates attractiveness of products, i.e. their competitive advantages determined by customers' attractiveness conditions. Higher sales indicate higher efficiency and functionality of the system due to its ability to successfully perform its functions. Economic values enabling to interpret this indicator include sales volume, organizations' turnover and organizations' turnover depending on economic activity type.

Systematicity indicator should demonstrate the systems' short-term efficiency as well, meaning primarily its economic efficiency in relation to usage of various resources and factors during the production process. The most applicable values for interpreting this indicator determine traditional production efficiency. The more systematized the production process is, the more efficiently it uses resources and factors, thus demonstrating that this product has cost-related competitive. The general systematicity level is maintained in the socioeconomic system by implementation of management, budget, audit, analysis, control and other procedures during production. In our opinion, economic values enabling to interpret this indicator to a certain extent may include gross value added in relation to the employment or gross value added in relation to population number.

Traditionally, gross value added is understood as output of products and services minus intermediate consumption. It seems evident that high total cost of the system's products and services combined with low cost of used resources could be the cause of high organizational systematicity level [8].

Thus, individual index of socioeconomic system competitiveness in the short-term perspective defined as the geometric mean of the product of functionality and systematicity indicators reflects efficiency of production and market sales of products and services, i.e. consumer- and cost-related attractiveness of the system.

Proactivity indicator is supposed to demonstrate the systems' long-term efficiency meaning mostly provision. The most applicable economic values for this indicator define the system's activities on proactive provision of innovative advantages for their product in full accordance with system's entrepreneurial behavior in the market environment. Such economic values enabling the indicator's interpretation may include innovational products volume, volume of scientific developments funding and investments in capital assets for systems of different levels.

Harmony indicator is supposed to demonstrate the socioeconomic systems' long-term efficiency meaning mostly provision of integrational cooperation within the system and cooperation between its structural elements. The most applicable values for this indicator demonstrate integrity between the system's elements. Harmony of the system is determined by integrative attributes of its elements understood as their willingness to form coalitions [9].

The integrity of the system meaning the interrelation of its elements increases its organic ability to adapt to external and internal changes. High adaptive quality allows the system to give to its products competitive advantages attracting consumers. These advantages can be consumer-, cost-, innovation related, etc. Comparing two systems functioning with the same efficiency, the one that is more differentiated, internally and externally interrelated and has higher number of internal elements has higher harmony level. High structural differentiation of the system's economy enables to minimize the negative impact of crisis factors. In our opinion, economic values that allow to interpret this indicator to a certain extent include the number of the enterprise's personnel, the number of enterprises and depending on economic activity type and the number of regional enterprises and organizations.

Thus, individual index of competitiveness in the short-term perspective can be estimated as the geometric mean of the product of proactivity and harmony indicators. The index demonstrates the socioeconomic system's ability for proactive provision of innovational advantages and the integrity of its elements.

The integral index of socioeconomic system competitiveness is estimated basing on all four indicators. Comparing systems using this index enables to provide the integral assessment of their competitiveness.

This work recommends to use an approach to assessing innovational systems' competitiveness or assessing systems depending on their innovational activities. This approach is based on methodology developed by I. Adizes.

The algorithm of this approach is demonstrated in Figure 1.

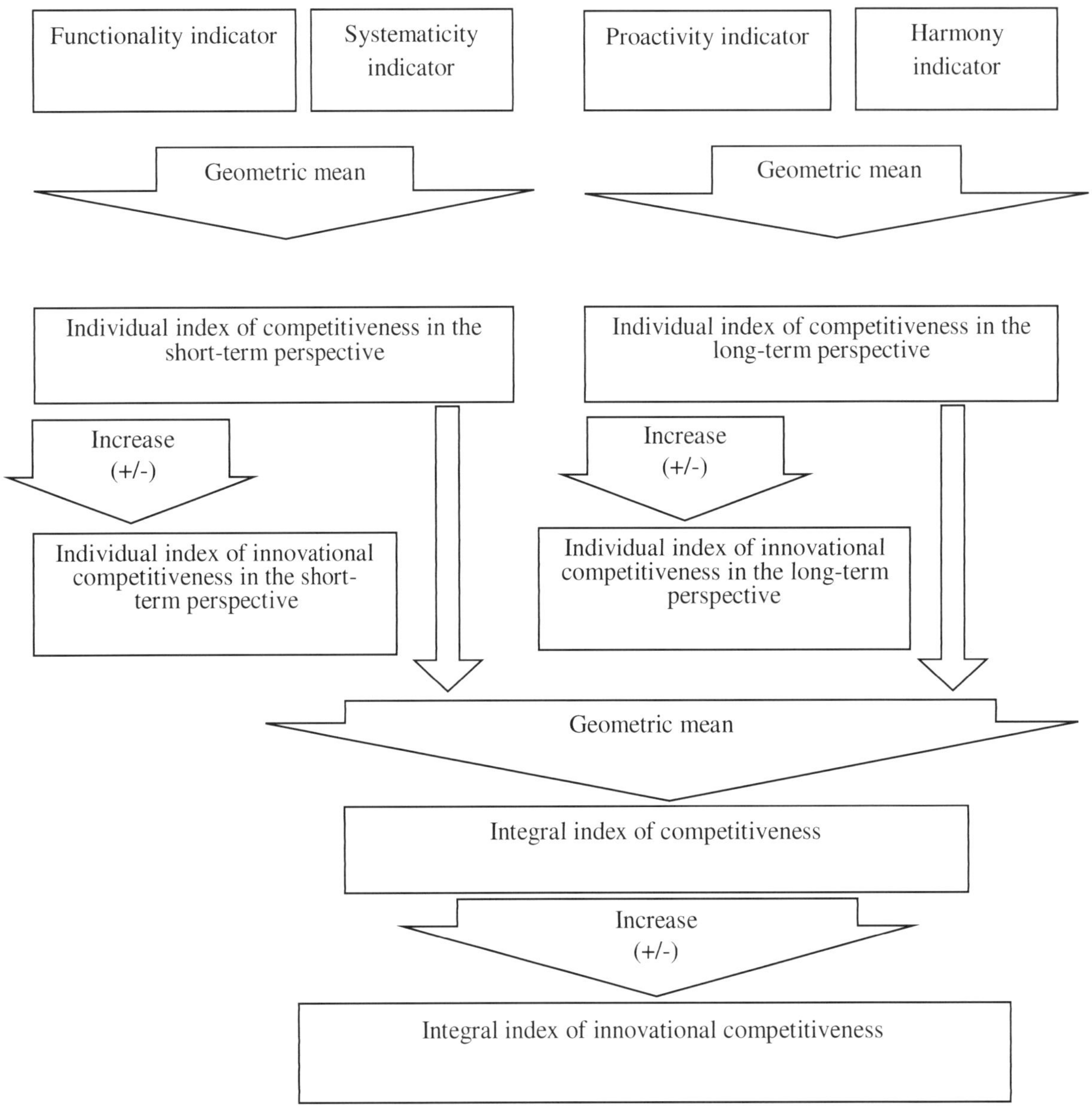

Figure 1. Formation of integral indexes of innovational competitiveness of socioeconomic systems

The principal difference of this competitiveness assessment method from approaches studied in other works is that final indexes of innovational competitiveness are estimated basing on changes in "traditional" competitiveness indexes.

This approach has a classical mechanics analogy. A body moves with a certain velocity. But in case of non-uniform motion, in order to estimate the velocity and the distance we need to know not only the object's initial speed, but also its acceleration defined as change of speed with time [10].

4. Conclusion

For analyzing the socioeconomic systems' innovational competitiveness, we recommend to estimate it using parameters (competitiveness indexes) assessing its initial competitiveness.

Indeed, in the market economy it is possible for a company that initially has lower competitiveness compared to another one to eventually overcome this distance and hold leading positions. Higher speed of competitiveness growth has a more important role than competitiveness value itself. Such speed can be determined by better perception to innovations and consequently to innovational development. Thus, socioeconomic systems can be assessed according to their competitiveness level and their innovational competitiveness level.

Individual indexes of innovational competitiveness of socioeconomic system in short-term and long-term perspectives are assessed basing on the increase (or decrease) of its respective individual indexes of competitiveness. The final index of innovational competitiveness is estimated basing on the increase (or decrease) of respective competitiveness index.

5.Acknowledgments

The reported research was funded by Russian Foundation for Basic Research and the government of Volgograd region, grant № 18-410-340009

References

[1] Schumpeter J A (2010) Capitalism, socialism and democracy 1-442

[2] Moseiko V O, Korobov S A, Frolov D P 2015 Socio-Economic Systems' Competitiveness Assessment Method *Asian Social Science* **11 (20)** 1-8

[3] Adizes I 1979 Organizational passages-Diagnosing and treating lifecycle problems of organizations *Organizational Dynamics* **8 (1)** 3-25

[4] Adizes I, Cudanov M, Rodic D 2017 Timing of proactive organizational consulting: Difference between organizational perception and behavior *Amfiteatru Economic* **19 (44)** 232-248

[5] Moseiko V O, Korobov S A, Tarasov A V 2015 Cognitive modelling in decision-making: resource-factor analysis potential *Creat. Econ* **9(5)** 629–644

[6] Anholt S 2007 Competitive Identity: the New Brand Management for Nations *Cities and Regions* p. 160

[7] Camagni R 2002 On the concept of territorial competitiveness: sound or misleading? *Urban Studies* **39(13)** 2395-2411

[8] Korobov S A, Moseiko V O, Marusinina E Y, Novoseltseva E G, Epinina V S 2017 The substance of a rational approach to entrepreneurship socio-economic development *Contributions to Economics* 207-223

[9] Huovari J, Kangasharju A, Alanen A 2001 Constructing an index for regional competitiveness. *Pellervo Economic Research Institute Working* **44** 23

[10] Harihayati T, Lubis R, Atin S, Widianti U D 2018 The company's performance assessment using balanced scorecard *IOP Conference Series: Materials Science and Engineering* 407 (1)

Development of entrepreneurial skills in the innovation sector

S A Korobov[1], V O Moseyko[1], V S Epinina[1], E Y Marusinina[1], S I Korobova[1]

[1]Volgograd State University, Universitetskiy av., 100, 400062, Volgograd, Russia

econmanag@volsu.ru

Abstract. The innovation sphere could not flourish without the entrepreneurial skills development of people, who participate in the production relations in this sphere. To possess skills in the innovation spere means to provide a new approach to the work with additional solutions to a challenge your enterprise currently faces. The paper examines the particularities of enterpreneurial skills' shaping as the main condition for development of the innovation sphere. The following entrepreneurial skills are considered: the ability of a person to carry out economic activities, which is effected through the economic gene of production; the ability of a person to entrepreneurial arbitration; a person's ability to undertake actions, defined as entrepreneurial spirit; the ability of a person to integrate with the internal and external business environment.

1. Introduction

Due to western economic sanctions against Russia the development of the innovation sphere of the Russian economy is an important task which solution will improve national production driven by modern achievements in science and technology, increase the competitiveness of economic entities at both the federal and the state levels, as well as the level of the economy as a whole.

Today one may state that the expected breakthrough in the field of industrial use of high-tech developments, inventions and technologies went wrong. As a result, selling of high-tech products on the market is fragile in the domestic economy. The lack of positive dynamics from 2010 to 2016 in terms of the number of large and medium-sized enterprises that carried out innovative activities should be considered as one of the trends of the period under review [1].

All this determines the relevance of any research addressing the development of the innovation sphere.

An entrepreneur leads all production processes, including those taking place in the innovation sphere. Therefore, the entrepreneurial skills development of people who participate in the production relations is crucial for an economic success. These skills are abilities and competencies that form the perfect mix of unique knowledge and values that make companies competitive. The aim of this paper is to analyse the particularities of enterpreneurial skills' shaping as one of the main conditions for development of the innovation sphere.

2. Methods

In accordance with the resource-oriented approach, a business entity, like any socio-economic system, manufactures a product that can be conceived of as the production function Q = F (A, T, M, O, Inf, Ins), where Q is the manufactured product , A is a human factor, T is a technical-technological factor; M stands for a material and resource factor, O stands for an organizational factor, Inf means an informational and Ins - institutional factors of product creation [2]. In accordance with the resource-based view, resources are given the major role in helping companies to achieve higher organizational performance.

3.Results and Discussion

Let us consider some fundamental thesis according to the behavior of an individual entrepreneur. We assume that the principles of behavior of an individual entrepreneur are also characteristic for all business entities. Indeed, according to P. Drucker, "Entrepreneurship is based on the same principles, regardless of whether the entrepreneur is a functioning large institution or an individual who opens his enterprise. It makes almost no difference whether the entrepreneur is operating in the production sector or in the service sector, in the state or another sector. The operational principles of entrepreneurial organizations and the problems they face in many respects are similar [3].

1. The ability of a person to carry out economic activities through the implementation of the economic gene of production.

Proposed and introduced into scientific circulation by Professor Oleg V. Inshakov the notion of an economic gene for the production (ecogene of a product) is essential for understanding the genesis of any economic activity. Obviously, the concept of an ecogenic product is important in the theoretic substantiation of the main features of entrepreneurial activity in the innovation sphere. The legitimacy of the use of a resource-based view to the innovation sphere of management is confirmed by the conclusions of J. Schumpeter. According Schumpeter entrepreneurs deal with the implementation of "new combinations of production factors" [4].

Engaging in an economic activity, a person always makes as a result of s/he activity one or another product. Handling with the future product as a kind of model, a businessman develops his entrepreneurial skills. These skills involve designing, creating, shaping of the production process of a future product, using the necessary production factors. This corresponds to the logic of the modern theory of factors of production, that identifies entrepreneurship as a controlling force organizing the process of combining all the resources involved in the production of goods and services [5].

Obviously, the process of "improving" the ecogenic product is nothing more than innovation, which, according to Schumpeter, can be aimed at creating a new product, using new production technologies, creating new types of organizations, developing new sources of raw materials and new markets.

2. The ability of a person to provide an entrepreneurial activity of arbitration.

The purpose of the functioning of any business entity (including individual entrepreneurs) is to extract profits, primarily due to the difference in the prices of purchase and sale.

In essence, we consider activity of arbitration as an activity of a business entity aimed at obtaining entrepreneurial profits due to the difference between the production resources it acquired on the market and the goods produced by this entity from these resources in the framework of the relevant production process and then offered to them again on the market. An entrepreneur, offering a ready-to-use product on the market, acts as the owner of resources – the "price-taker", and when acquiring resources on the market, the entrepreneur appears to be a consumer – the "price-searcher". Pricing, which consists in the constant implementation of entrepreneurial assessments, is the most initial element of entrepreneurial behavior, aimed at the constant search for opportunities for arbitration, that is, for making a profit [6].

According to P. Drucker, the entrepreneurial function is characteristic of human activity in general, it "is not a specific property of a particular group of people or a class of people; it is inherent in any activity and burdens any actor".

And, finally, entrepreneurship according to P. Drucker, "is more a behavioral than a personal feature", that is, the ability characteristic of the behavior of any individual, and not exclusively for individuals.

The arbitrage behavior of an entrepreneur can be identified in "all transactions aimed at making profit, even in those enterprises that process certain means and resources through the production process into ready-to-use products".

Therefore, the ability to arbitrate should be considered as an exertion by any person of entrepreneurial behavior in a market environment that is developed and reinforced by the individual in the form of entrepreneurial skill.

Obviously, while offering its product on the market, the entrepreneur is looking for new markets and for new resources. Therefore, entrepreneurial arbitration should be viewed as an innovative ability of an entrepreneur, which, according to P. Drucker is aimed at creating the development of new sources of raw materials and new markets.

3. The ability of a person to undertake actions, defined as entrepreneurial spirit.

A significant role in the implementation of entrepreneurial activity is played by such a person's ability as entrepreneurial spirit. It's an attitude and approach to thinking that actively seeks out change, rather than waiting to adapt to change. It's a mindset that embraces critical questioning, innovation, service and continuous improvement. This ability is typical of each individual.

"Entrepreneurial spirit is a quality of an individual involved in a business activity manifested as energy, resourcefulness, ingenuity, practical thinking and efficiency in solving problems, primarily due to the ability to effectively solve tasks, persistence, leadership, change management, networkbuilding, and strategic thinking. Without entrepreneurial spirit, success in entrepreneurial activity is impossible".

Each person in different situations for themselves seeks to take various steps aimed, in general, at improving his or her condition or position. This striving is accompanied in practice by the manifestation on the part of an individual of such personal qualities as purposefulness, independence, decisiveness, initiative, energy, resourcefulness, ingenuity, riskiness, sharpness, efficiency, practicality, etc. Differences or boundaries between these concepts, which are mostly of special psychological qualities, are quite blurred, but, nevertheless, each of them has its own individual semantic content, expressing a certain, quite imaginable, human quality and all of them, in aggregate, integrally constitute shape of such a thing as the entrepreneurial nature of the individual. Therefore, enterprise should be presented as a collective definition, combining the meaning and meaning of many qualitative concepts typical for an individual.

All of these personal qualities characterize different facets of entrepreneurship, which, from an ordinary person, basically "do" an entrepreneur, form his personal potential. Moreover, being present in each person in varying degrees, and in varying degrees, they "make" him an entrepreneur. When definitions are used in everyday life - "highly" enterprising person or "light" enterprising person, then, first of all, they imply a degree of expression of the general level of all these qualities in an individual.

It is obvious that an entrepreneur, showing in his activity an entrepreneurial activity developed and embedded in the relevant skills, accompanies his work with innovative activity. Innovation and enterprise are equal and mandatory conditions that determine the economic development of any business entity. Thus, the development according to P. Drucker consists of sudden bursts of entrepreneurial and innovative energy [7].

4. The ability of a person to build relations with internal and external environment.

This quality, characteristic of business entities, is determined by their ability to build or organize relationships with the partners. You can build or organize relations with the internal environment - within the enterprise (organization, company, project). The result of this process is the harmonization of relations and interests of all employees of the enterprise and, as a result, the successful organization of all internal business processes.

Relations with the external environment are relations of a business entity (enterprises, organizations, firms) about their production activities with other parties to contractual relations operating in the external environment. The result of this process is also the harmonization of relations and interests with all the entities with which the enterprise (organization, firm) interacts: customers, suppliers, partners, intermediaries, government officials, competitors, contractors, external shareholders, etc. As a result, the successful organization of all business processes for interaction [8].

Much attention is paid to ensuring and forming relationships within the business environment in the methodology of a well-known scientist and practice of I. Adizes [9, 10]. Adizes called the process of ensuring a relationship of mutual understanding and trust between the members of a business community integration. At the same time, the main outcome of the integration process of a business entity, both inside and outside itself, is effectiveness in the long term.

It is obvious that by building or organizing relations in the course of its production activity, an entrepreneur is always looking for new forms for such processes. Therefore, the ability to build relationships should be considered as an innovative quality of an entrepreneur, developed and consolidated in him as a skill.

4. Conclusion

Entrepreneurial skills are formed, consolidated and, finally, realized by separate individuals due to certain abilities, characteristic to each individual in varying degrees. These separately identified abilities determine his general entrepreneurial ability, as well as his ability to engage in entrepreneurship in the innovation sphere. This is, first of all, the ability of a person to do business, which can be represented as his ability to form an ecogene of a product, using production factors. To form an ecogenic product means to realize the process of production of the product, using the appropriate factors of production.

Further, for carrying out entrepreneurial activity in the innovation sphere, an entrepreneur shows, in varying degrees for each person, the ability to arbitrate - the desire to constantly carry out entrepreneurial assessments in order to derive benefit from the difference between prices for acquired resources – conditions and products produced by an entrepreneur of relevant production factors. Thanks to the ability of people to arbitrate, entrepreneurs also put together, consolidate and, finally, implement in practice relevant skills.

Another important ability, which is also peculiar in varying degrees to each person, which is formed of skills, becomes permanent and also manifests itself in the entrepreneur in the implementation of his entrepreneurial activity, this enterprise is the ability to do something. This quality of the entrepreneur constantly encourages him to carry out innovative and innovative actions, steps and actions in the framework of business activities.

And finally, an entrepreneur, in order to carry out his activities, manifests in the innovation sphere in the form of entrenched skills, the characteristic ability of people to varying degrees for building or organizing relationships in the course of their production activities.

Thus, for the successful business activity in the innovation sphere, it is necessary to pay attention to development and consolidation of specific entrepreneurial skills based on the individual's abilities: to carry out economic activities through the realization of the economic gene, to master entrepreneurial arbitration, to undertake actions, and finally build relations with the internal and external environment. Development of entrepreneurial skills is seen as vital to promoting innovation, competitiveness and economic growth of the region.

5. Acknowledgments

The reported research was funded by Russian Foundation for Basic Research and the government of Volgograd region, grant № 18-410-340009.

References

[1] Korobov S A, Moseiko V O, Marusinina E Y, Novoseltseva E G, Epinina V S 2017 The substance of a rational approach to entrepreneurship socio-economic development *Contributions to Economics* 207-223
[2] Inshakov O V 2006 Evolutional economy and economic genetics *VI International Symposium on Evolutional economy* (Moscow: IE RAN) 45-58
[3] Drucker P F, Noel J L 1986 Innovation and entrepreneurship: Practices and principles *Journal of Continuing Higher Education* **34 (1)** 22-23
[4] Schumpeter J A 2010 Capitalism, socialism and democracy 1-442
[5] Makarov V L, Kleiner G B 2007 Microeconomy of knowledge (Moscow CJSC Publishing House «Economcs»)
[6] Dolan E G 2006 Science, public policy, and global warming: Rethinking the market-liberal position *Cato Journal* **26 (3)** 445-468

[7] Drucker P F 1987 Social innovation-Management's new dimension *Long Range Planning* **20 (6)** 29-34

[8] Harihayati T, Lubis R, Atin S, Widianti U D 2018 The company's performance assessment using balanced scorecard *IOP Conference Series: Materials Science and Engineering* **407 (1)**

[9] Adizes I 1979 Organizational passages-Diagnosing and treating lifecycle problems of organizations *Organizational Dynamics* **8 (1)** 3-25

[10] Adizes I, Cudanov M, Rodic D 2017 Timing of proactive organizational consulting: Difference between organizational perception and behavior *Amfiteatru Economic* **19 (44)** 232-248

Managing access to enterprise information based on the mandatory model

A S Ksenofontov, I I Mamuchiev, L A Moskalenko

Federal State Budgetary Educational Institution of Higher Education «Kabardino-Balkarian State University named after H.M. Berbekov», 173 Chernyshevsky Street, Nalchik, 360004, Russia

E-mail: misteribragim@yandex.ru

Abstract. The article describes the main aspects of working with Oracle Label Security (OLS), discusses the OLS capabilities for managing access to corporate data. The access to information is controlled by the OLS policy, which has a standard set of components: Labels, Levels, Compartments (optional security levels), Groups (optional security levels), Policy. The proposed methodology complies with the requirements of the Government of the Russian Federation "Digital Economy of the Russian Federation" program among the main end-to-end digital technologies, which include big data.

1. Introduction

The purpose of the work is to study the main aspects of the work of Oracle Label Security and apply the use of data labels to limit access to enterprise information using OLS.

For the past 30 years, Oracle has been a leader in developing advanced data protection solutions that enable the security of sensitive information. Oracle Label Security (OLS) is part of Oracle's in-depth approach to security and is considered the most advanced information access control solution based on information classification. This ability is crucial to ensuring the "need-to-know" principle for accessing data and combining information. Combining information not only reduces the cost, but also increases the efficiency of analysis and the effectiveness of decision-making [1].

The need for the most sophisticated means of controlling access to sensitive information is becoming increasingly important as companies address emerging security requirements with regard to data fusion and confidentiality. In particular, maintaining separate databases (DB) for highly sensitive customer data is costly and incurs unnecessary administrative costs. OLS guarantees the ability to label information with a data label or a classification of information. This feature allows the database to know, in essence, what information is sensitive, and allows sharing secret information in the same table as the entire data set, without compromising security (Figure 1).

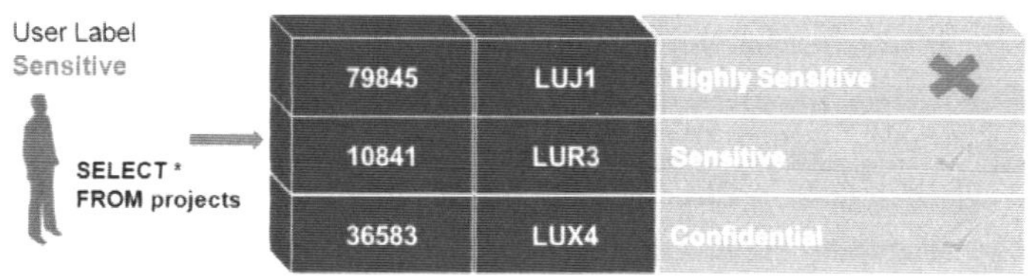

Figure 1. Access Control OLS.

2. Access to secret data

Access to sensitive information is controlled by matching the data label with the requesting user or security label. A custom label or security permission can be viewed as an extension of standard privileges and database roles. OLS is used in the database under the application level, providing reliable protection and eliminating the need for complex application representations [2].

The classification of information plays a vital role not only in adhering to the need-to-know principle, but also in reliably combining highly sensitive information. Historically very sensitive information is stored in physically separate systems. But this approach has limited the ability to perform advanced analysis and business intelligence. OLS provides the most advanced and flexible solutions for classifying information [3].

OLS applies information at the row level to provide access control that restricts users to information that they are allowed to access. This allows organizations to control their own operating and storage costs, allowing information with different degrees of sensitivity to be combined in the same database [4].

A characteristic feature of the OLS function is the multi-level access policies that allow you to store access control attributes in clusters.

Inside the policy, data access levels are defined: "highly sensitive", "sensitive" and "confidential" information or other sets of levels: "top secret", "secret", "confidential" and "unclassified". In addition to these levels, there are four more: "trade secret", "proprietary", "company confidential confidentiality" and "public domain" [5]. A user with a higher level of access has access to information marked with its level and levels below.

OLS is a commercial option for Oracle Database Enterprise Edition [6].

The OLS policy has a standard set of components:

– Labels. Labels for data and users, as well as permissions for users and software units, regulate access to marked protected objects. Labels consist of the following elements:

– Levels. Levels indicate the type of sensitivity you wish to assign to a line, for example, sensitive or highly sensitive.

Optional security levels:

- Compartments. Information may have the same level (public, confidential and secret), but may belong to different projects within the company, for example, ACME Merger and IT Security. Departments present projects in this example that help establish more accurate access controls. They are most often used in public institutions.
- Groups. Groups establish companies that possess or acquire access to information, for example, in the UK, USA, Asia, and Europe. Groups are used in both commercial and state environments and are often used instead of departments because of their flexibility.

– Policy. Policy - the name associated with labels, rules, authorizations and protected tables [7].

3. Mandate access model

Mandatory access control (mandatory access control, MAC, sometimes translated as compulsory access control) - delimiting the access of subjects to objects, based on the assignment of a confidentiality label for information contained in objects, and issuing official permits (access) to subjects to access information of this level confidentiality.

According to the requirements of the FSTEC, mandatory access control or "access labels" are considered to be the key difference between the protection systems of the State Secrets of the Russian Federation of the senior classes 1B and 1B from the junior classes of the protection systems on the classical division of rights in the access matrix.

The main purpose of the BCH is to prevent information leaks from objects with a high level of access to objects with a low level of access.

For systems with the BCH, the security check task is algorithmically solvable. Such systems are characterized by higher reliability. The disadvantage of the BCH is a higher complexity of implementation [8].

There are requirements for a mandatory mechanism, which are as follows:

1. Each subject and object of access must be associated with classification labels reflecting their place in the corresponding hierarchy. Subjects and objects should be assigned classification levels using these labels. These labels should form the basis of the mandatory principle of access control.

2. The security system when entering new information into the system should require and receive from the authorized user the classification labels of this information. When authorized to add a new subject to the list of users, a classification mark must be assigned to it. External classification labels (subjects, objects) must exactly correspond to internal labels (inside the protection system).

3. The protection system should implement the mandatory principle of access control for all objects with explicit and implicit access from any of the subjects:

a) a subject can read an object only if the hierarchical classification in the classification level of the subject is not less than the hierarchical classification in the classification level of the object (Figure 2).

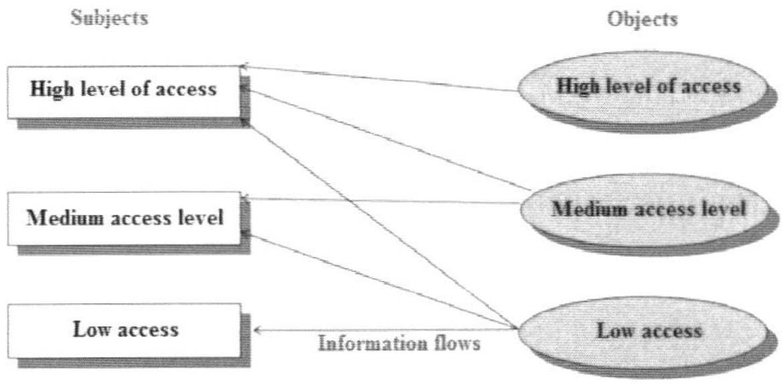

Figure 2. Reading information.

b) the subject realizes writing to the object only if the classification level of the subject in the hierarchical classification is not greater than the classification level of the object in the hierarchical classification (Figure 3) [9].

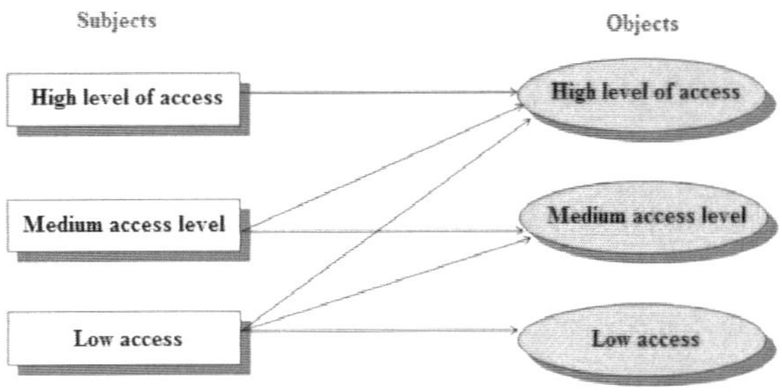

Figure 3. Recording information.

There is also a place to consider the Bella LaPadula model - one of the first security models - and subsequently the most frequently used one. The model was developed by David Bell and Leonardo LaPadula to simulate computer operation. To illustrate the model, consider a system of two files and two processes (Figure 4). One file and one process are unclassified, the other file and process are secret.

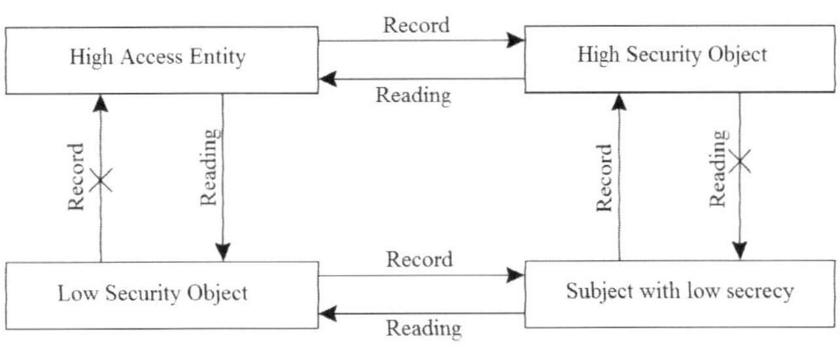

Figure 4. Illustration of the Bella LaPadula model.

The implementation of the mandatory rules of access control should take into account the possibility of tracking, changing the classification levels of subjects and objects by specially selected subjects.

Advantages:

– With the help of multi-level models, a significant simplification of the administration task is permissible. Moreover, this applies both to the initial setting of the demarcation access policy (such a high level of detail in the setting of subject-object relationship is not required), and to further incorporation into the administration scheme of new objects and access subjects.

– The most important advantage is that the user cannot fully control access to the resources that he forms.

– Such a system prohibits a user or process with a certain level of trust from gaining access to information, processes or devices of a more protected level.

Disadvantages:

Separately taken categories of the same level are equivalent, which in most cases leads to redundancy of access rights for certain subjects within the limits of the corresponding levels [9].

4. Differentiation of access to database resources

We describe for each group of users the rights of access to each table. Access rights must be distributed so that for each database object there is at least one user who has the right to add and delete data from the object. The rights are shown in table 1.

Table 1. Access rights to database resources for user groups

Tables	User groups (roles)						
	A	B	C	D	E	F	G
Employee	SIUD	SIUD	S	SIU	SIUD	-	S
Department	SIUD	SIUD	S	SIU	SIUD	-	S
Location	SIUD	SIUD	S	SIU	SIUD	-	S
Job	SIUD	SIUD	S	SIU	SIUD	-	S
Customer	SIUD	SIUD	-	-	-	-	S
Sales_order	SIUD	SIUD	-	-	-	-	S
Item	SIUD	SIUD	-	-	-	-	S
Product	SIUD	SIUD	S	S	S	S	S
Price	SIUD	SIUD	S	S	S	S	S

Where: S - data reading (select); I - add data (insert); U - data modification (update); D - delete data (delete), A - Database_administrator, B - Director, C - Employees, D - Bookkeeping, E - Human_Resources_Department, F - Client, G - Special_privileges.

Rights are assigned by the DBA (or security administrator, if the system is complex and there are several administrators).

Practical implementation of mandatory access control to data by means of Oracle Label Security: 1) creation of users; 2) giving users the right to create a session with the server; 3) provision of administrator rights and the right to create tables and unlimited table space; 4) the creation of tables; 5) establishing links between relational database tables; 6) filling in tables with data; 7) creating OLS policies; 8) adding labels on lines; 9) change of policy on enforcement during reading; 10) the issuance of the right to read the protected OLS table without any specific level of tolerance; 11) view table entries through users.

5. Conclusions

As part of this work, the following tasks were performed: 1) analysis of the subject area, database design at the logical level; 2) implemented a database in the environment of Oracle Database Enterprise Edition; 3) selected user groups, their functions, tasks and requests; 4) accounts of user groups are created and administered on the basis of the mandatory access model; 5) information is marked with a data label or information classification; 6) the mandatory access restriction model has been adapted for individual records of the table; 7) an algorithm has been developed for differentiating access according to the mandate model; 8) secured security policy in multi-user database systems; 9) access control to database resources has been implemented by means of the OLS function; 10) formed a database security administrator's guide.

The server database contains 9 tables with the following number of records:

Employee table - 32 entries. Department table - 11 entries. Location table - 4 entries. Job table - 6 entries. Customer table - 33 records. Table Sales_Order - 100 records. Table Item - 271 entries. Product table - 31 entries. Price table - 58 records.

Selected 7 groups of users and transferred to the platform Oracle Database 12c Enterprise Edition Release 12.2.0.1.0 - 64bit Production, each filled with 7 accounts.

The accounts are object and operator-administered on the basis of the mandatory model.

A database security administrator's guide has been developed, which describes how to install and configure the operation mode in Oracle Database 12c Enterprise Edition Release 12.2.0.1.0 - 64bit Production and activate the Oracle Label Security feature.

Testing of user groups access to database resources was carried out using a specific example.

When using the mandatory access restriction model, the BOP is fully controlled by the server security administrator.

The proposed methodology meets the requirements of the Government of the Russian Federation "Digital Economy of the Russian Federation" program among the main end-to-end digital technologies, which include big data.

References

[1] Oracle Label Security. Technical White Paper. ORACLE WHITE PAPER | MARCH 2018 [Electronic resource] Access mode: http://www.oracle.com/technetwork/wp-dbsec-ols-201702-3634252.pdf

[2] An Oracle White Paper June 2013. Oracle Label Security with Oracle Database 12c [Electronic resource] Access mode: http://www.oracle.com/technetwork/database/options/label-security/label-security-wp-12c-1896140.pdf

[3] An Oracle White Paper March 2009. Oracle Label Security in Government and Defense Environments [Electronic resource] Access mode: http://www.oracle.com/us/products/database/database-govdef-label-security-twp-066569.pdf

[4] Oracle Label Security 18c [Electronic resource] Access mode: http://www.oracle.com/technetwork/database/options/label-security/index-084797.html

[5] Oracle Database Release 12.2. Administrator's Guide. 2 Understanding Data Labels and User Labels [Electronic resource] Access mode: https://docs.oracle.com/en/database/oracle/oracle-database/12.2/olsag/understanding-data-labels-and-user-labels.html#GUID-2C0383D3-4AA5-4263-B938-827E2CCC40C0

[6] Oracle Label Security. First acquaintance [Electronic resource]. - Access mode: https://habr.com/post/185946/

[7] Oracle Label Security. Administrator's Guide 12c Release 2 (12.2). May 2017 [Electronic resource] Access mode: https://docs.oracle.com/en/database/oracle/oracle-database/12.2/olsag/introduction-to-oracle-label-security.html#GUID-ACEE3C58-A7CF-4548-976C-7D1C9F79A047

[8] Security database systems. 2. Access control [Electronic resource] Access mode: https://ppt-online.org/68872

[9] Access control models: discrete, mandatory, role [Electronic resource] Access mode: http://gman1990.ru/articles.php?article_id=83

IOP Conf. Series: Materials Science and Engineering **483** (2019) 012023 doi:10.1088/1757-899X/483/1/012023

The statistical researches of flexible manufacturing system's efficiency

E N Malyshev, I A Zenkina and V A Fedorov

Kaluga Branch of the Bauman Moscow State Technical University, 2 Bazhenova Street, Kaluga, 248000, Russia

E-mail: en.malishev@yandex.ru

Abstract. Currently engineering production is mainly medium-sized. The parts quantity in batch depends on the customer's requirements (for example, just-in-time delivery) and differs from each other significantly – a range from one to thousand in number. Similar designed products are made by means of flexible manufacturing systems (FMSs). These systems are elaborate and expensive. Therefore, their application needs to be proved. The statistical research allows establishing effective limits of system parameters. The statistical method of Monte Carlo is used for research and allows simulating variation parameters. In article the FMS for production of electric wires contacts is considered. The quantity of such contacts types is estimated in tens. One of important efficiency indicators of FMS is the machine flexibility, which shows system switching ease. The basic element's operation of a manufacturing system and time consumption on its operation on one product is specified. The modelling results are described in the graph's bar in article. The statistical researches method allowed to estimate flexible manufacturing systems efficiency taking batch deviation parts quantity and the changeover need.

1. Introduction

The competitive enterprises face challenges in responding to customer order quickly because the parts quantity in batch is various [1, 2]. Design variability and batches quantity requires the use of statistical modeling methods [3, 4]. Simulation of FMSs function allows establishing limits where flexible production systems are effective [5, 6].

At present, the key characteristics of FMSs effectiveness are: the number of products variations that should be manufactured by this system, and the relative time on FMSs changeover [7]. The research presented below shows the parts quantity in batch also has a significant impact on the FMSs effectiveness. The product quantity of each kind may be varying [8]. The research is conducted by the statistical modeling method.

Figure 1 presents a block-diagram of the FMS for production of electric wires contacts. Figure 2 presents some contacts. The quantity of such contacts types is estimated in tents and the enterprises producing such connectors prefer the flexible production which is quickly adapting by product's design variety, volume and terms of production to the consumer.

Basic elements (BEj) of the simulated system are: hook hopper feeder $(BE1)$, loading chute $(BE2)$, slide damper $(BE3)$, boring mill turret $(BE4)$ and back gate $(BE5)$.

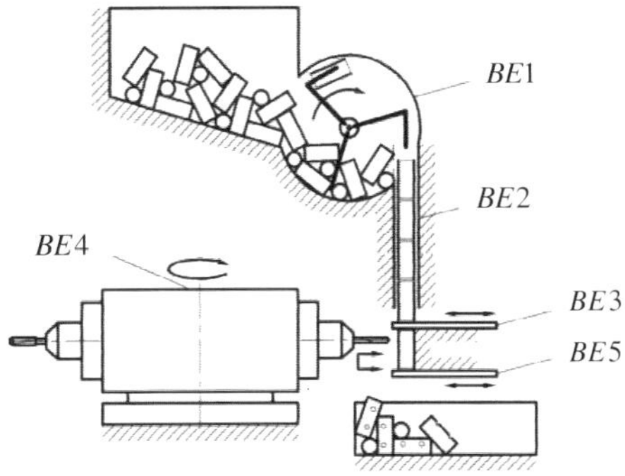

Figure 1. FMS for production of electric wires contacts.

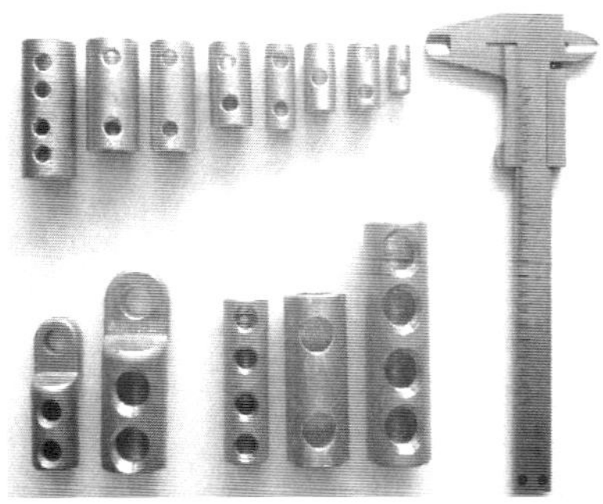

Figurer 2. Electric wires contacts.

2. Research object model

One of important efficiency indicators of FMS is the machine flexibility, which shows system switching ease. This machine flexibility is defined as K_{MF},

$$K_{MF} = \frac{T_{CH}}{T_{PM} + T_{SR}}, \tag{1}$$

where T_{CH} = time of system changeover, T_{PM} = time of product manufacturing, T_{SR} = time of system restoration and reparation which aren't connected in system changeover, $(T_{PM} + T_{SR})$ = system's usage time.

It is generally accepted that a flexible manufacturing systems are high-flexibility level's systems if $K_{MF} \le 0.1$, average-flexibility level's systems if $0.1 < K_{MF} \le 0.2$ and low-flexibility level's systems if $0.2 < K_{MF}$.

Table 1 shows the main elements of the production system and the time of its work ($t_{op.BEj}$) on one product.

Table 1. A basic element's operation of a manufacturing system and its operation time.

Basic element, BEj	The basic element's operation (system's usage time)	Time, $t_{op.BEj}$ (s.)
$BE1$	Adjusting the position and transfer of each single product from feeder to a loading chute. Operating diameter of feeder is 400 mm; quantity of hooks – one; angular speed of hook's rotation – 23 rpm	2.6
$BE2$	Movement of part on a loading chute from feeder to the slide damper. Chute's length – 2000 mm.	0.6
$BE3$	Movement of finish part from loading chute to the machining position (slide damper application).	0.4
$BE4$	Boring mill turret's operation (drilling two holes and thread cutting)	19.7
$BE5$	Movement of product from machining position to the container (back gate application)	0.4

In table 2 the operations content and time consumption for FMSs elements changeover are specified.

Table 2. An operations content and time consumption for FMSs elements changeover.

Basic element, BEj	The operations content for FMSs elements changeover	Time, $t_{ch.BEj}$ (min.)
$BE1$	Adjustment of hook's working radius	11.80
$BE2$	Not applicable	0.00
$BE3$	Position's adjustment of the slide damper and machining	1.05
$BE4$	Retooling (drill, tap)	7.41
	Adjustment of the CNC-program	1.84
$BE5$	Position's adjustment of the machining and back gate	1.05

After the first product was transferred to chute, the next products are passed from feeder, until the chute is filled. Time of transfer is written down by $t_{op.BE1}$. The chute is filled quickly because machining time ($t_{op.BE4}$) exceeds transfer time ($t_{op.BE1}$).

A product leaves a loading chute through time period equal to the sum of machining time $t_{op.BE4}$ and time of consecutive application of slide damper and back gate application ($t_{op.BE3}+t_{op.BE5}$).

Thus, $i-$ batch production time by N_i-products can be calculated (figure 3) if FMSs elements are ready and there is no operation's break:

$$T_{PMi}=t_{op.BE1}+t_{op.BE2}+\left(t_{op.BE3}+t_{op.BE4}+t_{op.BE5}\right)N_i\,. \tag{2}$$

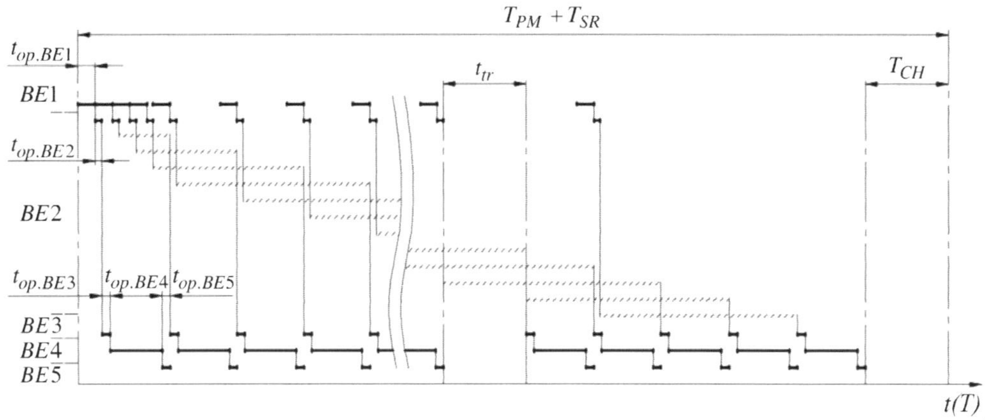

Figure 3.
Batch production time

The time of system restoration and reparation T_{SR} are mainly tool's repairs and can be calculated:

$$T_{SR}=\frac{t_{op.BE4}N_i}{T}t_{tr}\,, \tag{3}$$

where $i =$ batch number ($i=1,...,n$), $T =$ tool life, t_{tr} = one tool's repair time.

Therefore, as for $i-$ batch, the machine flexibility of FMS can be calculated:

$$K_{MFi}=\frac{t_{ch.BE1}+t_{ch.BE2}+t_{ch.BE3}+t_{ch.BE4}+t_{ch.BE5}}{T_{PMi}=t_{op.BE1}+t_{op.BE2}+\left(t_{op.BE3}+t_{op.BE4}+t_{op.BE5}\right)N_i+\dfrac{t_{op.BE4}N_i}{T}t_{tr}} \tag{4}$$

As for n consecutive batches, the Machine flexibility of FMS can be calculated:

$$K_{MF} = \frac{\sum_{i=1}^{n} K_{MFi} N_i}{\sum_{i=1}^{n} N_i} \tag{5}$$

3. Statistical research

The dependences analysis has set the time consumption for FMSs elements changeover (T_{CH}) and parts quantity in batch (N_i) influence the Machine flexibility most significantly. The statistical method of Monte Carlo is used for a research and its method allows simulating variation of the parameters [9, 10].

The pseudorandom values (N_i) were generated in the range from 50 to 500 in numbers products based on uniform probability distribution for $n = 100$ of consecutive processed batches parts.

The feeder changeover need was modeled by replacement of feature $(t_{ch.BE1})$ on the feature $(k_{ch.i} t_{ch.BE1})$, where $k_{ch.i}$ = factor of feeder changeover need for $i-$ parts production: $k_{ch.i} = 1$ if changeover is applicable and $k_{ch.i} = 2$ if changeover is not applicable.

The factor P_{ch} of changeover need probability (changeover is necessary for all production batches and $k_{ch.i} = 1$) was set at four standards:

– $P_{ch} = 0.00$ – changeover isn't necessary for all production batches (high-flexibility feeder) and $k_{ch.i} = 0$ for all production batches to simulation model;

– $P_{ch} = 0.33$ – changeover is necessary for 33 % of all production batches (average-flexibility feeder) and $k_{ch.i} = 1$ for 33 % of random production batches;

– $P_{ch} = 0.66$ – changeover is necessary for 66 % of all production batches (low-flexibility feeder) and $k_{ch.i} = 1$ for 66 % of random production batches;

– $P_{ch} = 1.00$ – changeover is necessary for all production batches (in-flexible feeder) and $k_{ch.i} = 1$ for all production batches to simulation model.

Statistical researches for each level have been made a 100 times that has allowed simulating a significant volume of situations with variable of parts quantity in batch and FMSs changeover quantity. The modeling results are presented in the bar graph, depicted in figure 4. The bar graphs relating separately to various changeover need probability standard are spline integrated.

From histograms it is visible that the FMS can be the high-flexibility level's system if $P_{ch} = 0.00$ $(K_{MF} = 0.089 \pm 0.015)$. The FMS can be the average-flexibility level's system if $P_{ch} = 0.33$ $(K_{MF} = 0.121 \pm 0.022)$, $P_{ch} = 0.66$ $(K_{MF} = 0.152 \pm 0.028)$ and $P_{ch} = 1.00$ $(K_{MF} = 0.186 \pm 0.027)$.

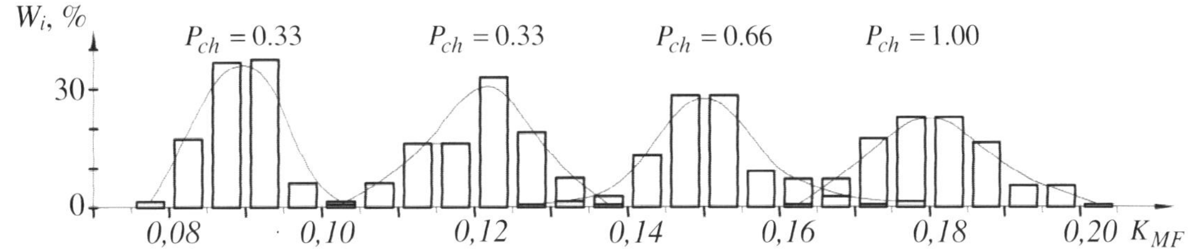

Figure 4. Influence of a changeover need's probability factor on machine flexibility with parts quantity in batch in the range from 50 to 500.

Figure 5 presents a modeling results bar graph is shown for the enterprise initial state. The parts quantity in batch has varied from 200 to 500, all FMSs elements were changeovers when a new size/design product began to be made $P_{ch} = 1.00$ $\left(K_{MF} = 0.143 \pm 0.012\right)$.

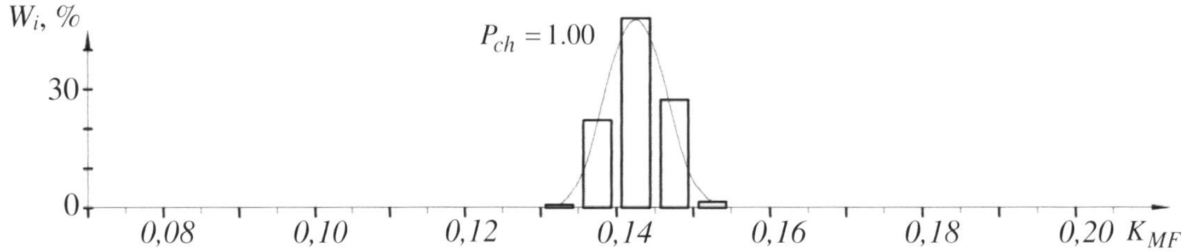

Figure 5. Influence of a changeover need's probability factor on machine flexibility with parts quantity in batch in the range from 200 to 500.

The modeling results are determined in accord with probability belief of 90% and with supposition that the constructed distributions are centrally symmetric. Highest precision in results assessment wasn't required at this stage.

4. Conclusions and recommendations

1. The statistical researches method has allowed estimating flexible manufacturing system's efficiency with due regard for parts quantity in batch deviation and the changeover need.

2. The parts quantity in batch varied has significant impact on FMSs efficiency. The FMSs with parts quantity in batch from 200 to 500 by indicator of machine flexibility can be estimated by average-flexibility level's system, but with quantity from 50 to 500 – by low-flexibility level's system.

3. The executed modeling has shown the application of low-flexibility feeder allows to downgrade of machines flexibility index almost on ~25 % in comparison with in-flexible feeder, the application of average-flexibility feeder – to downgrade by ~50 %, the application of high-flexibility feeder – to downgrade ~100 %. The use of high-flexible devices is the perspective direction of FMSs creation.

References

[1] Kevin W 1998 *European Journal of Operational Research* **109(3)** 587–98
[2] Yumin H, Ram R, Milton L S and Kathryn E S 2015 *International Journal of Production Research* **53(17)** 5106–17
[3] Helton J C, Johnson J D and Oberkampf W L 2004 *Reliability Engineering & System Safety* **85(1–3)** 39–71
[4] Jeong K C and Kim Y D 1998 *International Journal of Production Research* **36** 2609–26
[5] Helton J C 1994 *Risk Analysis* **14(4)** 483–511
[6] Saygin C, Chen F F and Singh J 2001 *The International Journal of Advanced Manufacturing Technology* **18(10)** 755-63
[7] He Y, Smith M L and Dudek R A 2010 *Robotics and Computer-integrated Manufacturing* **26** 671–6
[8] Weng M X, Wu Z, Qi G and Zheng L 2008 *International Journal of Production Research* **46** 2197–213
[9] Kiviet J F 2011 *Foundations and Trends® in Econometrics* **5(1)** 1–181
[10] Janssen H 2013 *Reliability Engineering & System Safety* **109** 123–32

IOP Conf. Series: Materials Science and Engineering **483** (2019) 012024 doi:10.1088/1757-899X/483/1/012024

Syntactic features of scientific articles on materials science

O O Kandrashkina[1], El Vl Revina[1]

[1]Samara State Technical University, 244 Molodogvardeyskaya street, Samara, 443100, Russia

petergrifon@mail.ru

Abstract. The paper gives an outline of the syntactic features of the texts. The purpose of the research is to define grammatical forms and syntactic structures observed in scientific articles from the journal "Advanced Engineering Materials" (2018). The article also presents the analysis of verbal constructions and phrases in scientific articles on materials science. The tense forms used in the articles under investigation are also considered. The analysis has revealed the gerund to be the most frequently used verbal form. The main functions of the gerund in the articles are subject and adverbial modifier. The infinitives are not so frequent as gerunds and are used in the function of adverbial modifier of purpose mostly. The participles in the articles fulfill the function of an attribute either in pre- or post-position. The other syntactic features include passive voice forms of present simple, past simple and present perfect simple tenses.

1. Introduction

Language used in science describes and interprets the world, projecting the objective character of scientific knowledge. Scientific and technical communication can be defined as a process of gathering, organizing, presenting and refining information. Persuasion is also a part of such communication which appeals to objectivity to convince an audience. It is also a process inevitably shaped by its contexts, and which is improved when it recognizes its contexts.

Effective scientific and technical communication is an integral part of all professions and fields of study in an industrialized society [1]. The qualities attributed to effective scientific and technical communication, such as accuracy, clarity, precision and thoroughness, correspond to the understanding of what makes science and technology work effectively.

The main purpose of any scientific writing is undoubtedly to acquaint the public with researcher` findings. Scientific publications serve as a ground for academic communication and exchange of opinions [2]. Any publication should therefore cover the background, the procedure and results and meaning of the research. Scientific words differ from ordinary and literary words since they do not accumulate emotional associations and implications. Scientific language is supposed to be more direct, free from alternative, and much less artistic than literary language. The language of science is characterized by impersonal style [3]. It is generally considered that scientific or technical texts are quite neutral, as they contain a lot of terms and do not have many stylistic expressive features. However technical and scientific texts possess their own peculiarities despite the factual and objective information they represent.

The main distinguishing feature between the scientific and literal text is that any scientific text do not create any aesthetic effect. It purely states the facts, supporting and proving them with experimental data. So the words in scientific writing do not convey any connotative or symbolic meaning, it is primarily terms and words in their direct meaning.

While analyzing scientific texts, one should pay attention to the general style features of scientific writing, such as clarity, precision and logical sequence of ideas and viewpoints presented. Integrity is the other essential characteristic of any scientific article [4]. Each thought should be logically expressed and supported by evidence. The logical structure of writing allows the reader to follow and grasp the text with ease.

IOP Conf. Series: Materials Science and Engineering **483** (2019) 012024 doi:10.1088/1757-899X/483/1/012024

The scientific writing has its own peculiarities. Syntactic features of scientific texts have been studied and defined. Being aware of these features helps scientists write their paper correctly and to convey the results of the research. It is also of great importance for non-speakers of English, because foreign scientists should be able to understand the peculiarities of the English syntax in scientific articles to gain the main idea and get to know the research itself, its results and meaning for the scientific field.

Theoretical discussions of the grammatical features of academic writing can be found in works by scholars from systemic functional linguistics [5]. Its approach to academic language regards nominalization as a pivotal linguistic device that distinguishes formal technical writing from everyday spoken interaction and as a major source of syntactic complexity in writing. Nominalization, i.e. verbs and adjectives converted to nouns through derivation, is a powerful device for packing and integrating information. Nominals substitute the clauses where the same thing could be expressed. As for grammatical constructions and forms used in technical writing, the most common are known to be the Passive Voice, the Participle, the Infinitive, the Gerund. Scientific and technical writing is also characterized by the abundance of terms and specialized terminology [6, 7]. Sentence structures are presented by complex or compound-complex sentences with two or more subordinate clauses joined syndetically or asyndetically.

2. Non-finite verb forms

Non-finite verb forms are considered to possess the characteristics of a verb, as well as other parts of speech. So, the Participle has a double nature of a verb and an adjective; the Gerund and the Infinitive combine the characteristics of a verb with those of a noun . Verbals can only be part of the predicate and they must always be in connection with the finite form of the verb. The verbals can form predicative constructions, consisting of a nominal expressed by noun or pronoun and a verbal expressed by a participle, a gerund or an infinitive [8]. In such way the verbal is connected with the nominal element by means of predication.

In contrast with the finite forms, the tense distinctions of the verbals are relative showing if the action is simultaneous with the action expressed by the finite verb or prior to it.

Verbals can fulfill different syntactic functions. The Participle can be an attribute, adverbial modifier, a predicative, part of Complex Object. The Gerund can function in the sentence as a subject, a predicative, an object, an attribute, an adverbial modifier, be a part of a compound verbal predicate and of a complex object. The Infinitive can serve as a subject, predicative, an object or as an adverbial modifier. Moreover, infinitives, participles, gerunds and verbless clauses represent four general types of nonfinite subordinate clauses.

2.1 Participle and Gerund

We analyzed articles from the scientific journal "Advanced Engineering Materials". The following examples from the articles show how non-finite verb forms are used and what their syntactic functions in the sentence are. First, we consider the syntactic functions of the participle:

While the effective dielectric permittivity being improved by adding the ceramic nanofillers into polymer matrix, the electric field may concentrate near the nanofiller–matrix interfaces, and make these regions weak points in the breakdown process.

Absolute Participial construction is a secondary clause that modifies the whole meaning of the main clause. In the sentence a Nominative Absolute Participial construction functions as an adverbial modifier of time. In the construction the participle **being improved** stands in predicate relation to a noun **permittivity** in the common case.

The examples given below show The Past Participle serving as an attribute:

Two special microstructures **filled** with BaTiO3 nanofibers and nanoparticles at the same volume fraction of 10% are considered.

Based on the switching mechanism, commonly studied resistive switching devices can be generally divided into two groups: **devices based on the migration** and redistribution of metal cations such as

Ag or Cu ions, and **devices based on the migration** of anions such as oxygen ions, **leading** to redistribution of oxygen vacancies (VOs).

In the sentence there are three forms of Participles II and one form of Participle I. All the Past Participles, the first detached attribute in pre-position and the others in post-position, fulfill the function of attribute. The Present Participle functions as an adverbial modifier of result. Moreover, parallel constructions that modify the subject *devices* aim to point out two different devices operating on different principles.

By directly changing the properties of the storage material itself, the device also directly modulates the signals **propagating** through it, thus **allowing** logic functions **to be performed** at the same physical locations as the memory element, **leading** to significant speed and energy benefits.

The example presents a variety of verbal forms with -ing ending. So it is quite difficult to define each form at first sight. The first verbal with preposition *by directly changing* is a gerund in the function of adverbial modifier of manner. The Participle I *propagating* as an adverbial modifier of manner modifies the noun in post-position. The construction "participle + passive infinitive" with conjunctive adverb *thus* functions as adverbial modifier of result. Such clauses are very common in formal written English. The participle I *leading* is an adverbial modifier of result. The abundance of participle clauses make the simple sentence more extended without being a complex one.

It should be noted that a prepositional phrase that includes - by + a gerund as an adverbial modifier of manner is quite common in scientific articles. It makes it easier for the authors to show the methods of analysis or to explain the chain of experimental studies. More examples of gerund in this function are given below:

It is also noted that hollow structuring of the electrode materials **has been proved** as a promising avenue **to boost** the electrochemical performance **by effectively alleviating** the stress-induced structural variation during long-term electrochemical reactions.

The shell number and composition can be precisely controlled **by simply tuning** the amount of VOT **used** in the reaction systems, **thus allowing** the preparation of triple-shelled Co3O4@Co3V2O8 nanoboxes.

The gerund functioning as a subject is also frequently used in scientific articles.

Therefore, **developing** highly efficient and robust HER electrocatalysts to lower the overpotentials in alkaline media is highly desirable for electrochemical water splitting.

Being transparent for light-emitting diodes (LEDs) significantly expands their applications **by displaying** visual information on objects **without affecting** their original appearance and functionality.

In the sentences the Gerunds at the beginning are subjects, the second prepositional form is an adverbial modifier of manner and the Gerund after preposition without is an adverbial modifier of condition. The next example also illustrates the subject expressed by the gerund form.

2.2 Infinitive

Infinitives in the sentence can also fulfill different syntactical functions. The examples given below examples illustrates the usage of the infinitive phrases in the function of adverbial modifier of purpose:

To reveal the spatial distribution of different elements in the T-Co3O4@Co3V2O8 NBs, elemental mapping analysis is carried out on a representative T-Co3O4@Co3V2O8 nanobox under scanning transmission electron microscopy (STEM) mode.

To obtain further insight into the formation process of multishelled Co_3O_4@Co_3V2O_8 nanoboxes, syntheses with different VOT concentrations are carried out.

Afterward, the assynthesized particles are annealed in air **to obtain a variety of hollow structures**.

To investigate the protective role of the ETLA_x during the top electrode deposition, we exposed the ETLA_x-coated QD films to Ar plasma (30 W, 13 Pa).

... have been proposed and explored for years in order **to improve the energy density.**_

In aim **to regulate** these factors for better catalytic performance, surface reorganization **was regarded** as a promising way to modify the electrocatalysts.

Attributive function of the infinitive can also be observed in scientific papers but its cases are not so common. The following sentences illustrate the function of infinitive as an attribute.

When coupling with vanadium element **to form cobalt vanadates** (such as Co3V2O8), the electrochemical performance of these ternary TMOs can be largely improved.

Another effort **to improve the EL performance** is structural engineering of QDs.

3. Other grammatical verb forms

Along with non-finite verb forms, finite verb forms in different grammatical categories are widely used by scientists and researchers. As for The Passive Voice forms, they allow to create an objective tone in scientific papers. The passive voice is also necessary when the doer of the action (the subject) is less important than the action itself or the result of the action. Therefore, the Passive voice constructions are involved in creating an impersonal tone in scientific discourse. This grammatical verb form seems to reduce the involvement of the researcher who actually carried out different actions: formulating the assumptions, collecting the data, conducting experiments, providing interpretation of the findings, drawing conclusions [9, 10]. Resent studies on passive constructions in scientific writing show that many researches argue the prevailing role f passive forms over active forms. The passive forms seem to have lost their significance for scientific discourse in written form. Nowadays there is a tendency among linguists and experts to consider active verb forms more appropriate in scientific writing.

The analysis has revealed the simple and perfect passive forms used in scientific articles, such as present simple passive and present perfect simple passive forms. Passives are typically used in main clauses if the sentence is a simple one. Rarely the passive forms are observed in relative clauses. The present passive forms are widely used for the description of the procedures:

Then, this phase-field model **is employed** to perform a high throughput calculation.

Finally, an artificial sandwich microstructure **is designed and optimized**.

Present perfect simple passive forms are found to referring to previous studies, experiments or results achieved in the given area of study. These forms are usually modified by an adverb, e.g. **have been extensively studied, have also been performed, has been extensively carried out, have been successfully demonstrated.** These forms are observed in Introduction section of the paper.

Past Simple Active or Passive verb forms are rarely observed due to the fact that current research should reflect the actual and present state of events and the latest findings in given field of science. Past simple forms are used mainly when researchers want to highlight previous experimental results or about certain last studies in their field.

To examine the effect of the ETLA x on the carrier dynamics, the TRPL spectra of QDs in Tr-QLEDs **were measured** under applied biases.

CoP nanowire arrays grown on carbon cloth (CoP NWs/CC) **were synthesized** by the phosphorization of CoOHF nanowire arrays.

Therefore, the above results clearly suggest that the CoP nanowires supported on carbon clothes **were** successfully **synthesized**.

We **prepared** a series of alloyed core/ shell CdSe/ZnS QDs by growing ZnS shells of different thicknesses.

4. Sentence types

In scientific writing different sentence types can be observed. If we look at the above-mentioned examples with non-finite verb forms in the sentences, we can notice that most of the sentences are simple ones, expanded by phrases or constructions. A simple sentence can be very short, consisting of a noun for the subject and a verb for the predicate, but in scientific writing it can have further extensions at the sentence initial, middle, and final positions. Simple sentences can be expanded by additional phrasal segments from the beginning, middle, and at the end by such phrases like noun phrases, adjective phrases, adverb phrases, verb phrases, infinitive phrases, gerundive phrases, participle phrases, participial phrases, and appositive phrases [11]. The examples with nonfinite clauses show sentences extended from the beginning or from the end.

To verify the function of surface reorganization by oxygen plasma engraving, all the pristine and treated CoP NWs/CC are compared in alkaline HER as electrocatalysts.

To further understand the reason for outstanding HER activity of CoP NWs 40s/CC after surface reorganization, the systematic characterizations were performed **to show the homojunction**.

In such a way, sentences are expanded without any clauses, dependent or independent. The main independent clause usually occupies the middle part of the whole sentence, extended by participle, infinitive or gerund clauses. However, complex and compound sentence types remain a distinctive feature of the scientific papers.

Complex sentences with subordinate clauses in scientific writing are connected syndetically, with subordinate conjunctions. They usually contain two clauses. The dependent clause is essential clause in scientific papers.

It can be seen that the volume fraction of the breakdown phase does not increase **until** the electric field reaches 140 kV mm^{-1}. **When** the applied electric field is greater than 140 kV mm^{-1}, the volume fraction of the breakdown phase begins to increase. **When** the applied electric field reaches 200 kV mm^{-1}, the volume fraction of the breakdown phase gets saturated, indicating that the polymer nanoparticle composite is totally broken down. **In contrast**, the polymer nanofiber composite is not totally broken down **until** the applied electric field reaches 225 kV mm^{-1} (B point on Figure 1e).

Complex sentences with subordinate clauses are mainly used for the description of the experiment. In the abovementioned example complex sentences illustrate the influence of the electric field on the material under different parameters. The adverbial clauses with subordinate conjunctions **when** and **until** express cause/effect relations between the clauses and denote the outcomes of the conducted experiment. The attributive relative clauses introduced by the relative pronoun **which** are primarily Object relative clauses in the articles:

Taking CoP nanowire arrays grown on carbon cloth (denoted as CoP NWs/CC) as an example, the homostructure CoOx/CoP interface was precisely-controlled constructed by the adjustment of **plasma treatment time, which** is visually verified by the X-ray absorption fine structure (XAFS).

The current recorded with the surface reorganization CoP catalyst (CoP NWs 40s/CC) shows an outstanding HER performance for −0.146 V at 100 mA cm^{-2} versus the reversible hydrogen **electrode which** possesses the outstanding alkaline HER performance in reported electrocatalysts

The vertical phase segregation maximizes the morphological continuity of semiconductor **component, which provides** the effective pathways for charge transport.

Compound sentences in the articles under investigation are often made up of two clauses.

This indicates the surface oxidation of CoP occurred **and** the chemical valence of cobalt becomes higher after the surface reorganization.

As a result, the device lifetime with ETLA_0 is very short, less than 1 h at 1500 cd m^{-2}, **and** the device lifetime variation between devices is large.

A unique solid-state battery system was designed **and** its working mechanism is illustrated in **Scheme 1**.

The most frequently used coordinating conjunctions are **and** and **but**, whereas the subordinating conjunctions are more varied and they are mainly used to define sequence and cause/effect relations between the sentences. The examples of the conjunctions from the articles are **when**, **after**, **before**, **until**, **if**, **while**.

The sentence types mentioned above are appeared to prove the syntactic complexity of scientific articles. In other words hypotaxis along with parataxis are indispensable features of any research paper.

5. Conclusion

Scientists and researches make use of a variety of syntactic structures in differing proportions. Syntax of research papers are characterized by a number of peculiarities that have been studied. The purpose of the article has been to reveal syntactic features in the material science articles. The use of non-finite verb forms, The Passive forms, the impersonal character of writing, different sentence types have been observed as the essential traits of scientific articles. The analysis of technical articles allows us to define the syntactic functions of the verbals. The Participle is used primarily as an attribute in pre-position or post-position. The Gerund functions as an adverbial modifier of manner and as a subject. The Infinitive is observed, expressing an adverbial modifier of purpose. It also used to modify a proceeding noun as an attribute. The passive verb forms are mainly observed in the past simple, present simple and present perfect tenses. Although most linguists consider active verb forms more As for sentence types, the most common are simple sentences extended by infinitive phrases, gerundive phrases, participle phrases. Complex sentences are found to be preferable for describing experimental procedure or for showing the cause/effect relations of different phenomena types of machinery employed or different devices. It also indicates the behavior of a tested object or phenomenon under different conditions. The above-mentioned features tend to cluster together as characteristic of scientific writing. The result of the current study will be immensely useful for the novice researchers of science, because it can help them to write their research papers with grammatical accuracy and organize their findings in a systematic manner. Moreover, being aware of syntactic features in scientific articles it is of great importance to scientists who are not native English speakers, as it help them to get the main point of the article and to understand the text more clearly.

References

[1] Collier J and Toomey D 1997 *Scientific and Technical Communication: Theory, Practice, and Policy* (London: SAGE Publications, Inc.) p 5

[2] Rabinowitz H and Vogel S 2009 *The Manual of Scientific Style: A Guide for Authors, Editors, and Researchers* (London: Academic press) pp. 8-11

[3] Jameel A 2012 *Eng. Lang. and Liter. Studies* **2(1)** 49-52

[4] Glasman-Deal H *Science Research Writing for Non-Native Speakers of English* (London: Imperial College Press) pp.1-3

[5] Liming L and Lan L 2016 *The J. of Asia TEFL* **1** 48-65

[6] Halliday M and Martin J *Writing Science* 2014 (Abingdon: Routledge) p 78-9

[7] Koulaidis V, Dimopoulos K and Sklaveniti Sp 2001 *Proc. of the Learn. Conf. Learn. for the Future* ed Bill Cope Mary Kalantzis (Common Ground Publishing Pty Ltd) pp. 7-11

[8] Huddleston R and Pullum G 2002 *The Cambridge Grammar of the English Language* (Cambridge: Cambridge University Press) pp. 1171-85

[9] Livnat Z 2010 *Lidil Revue de linguistique et de didactique des langues* **41** 109-10

[10] Leong P A 2014 *J. of Sci. Communic.* **13** 1-16

[11] Mehmet D 2013 *Procedia Social and Behav. Sci.* **70** 560-566

IOP Conf. Series: Materials Science and Engineering **483** (2019) 012025 doi:10.1088/1757-899X/483/1/012025

Optimization model of motivational process as a basis of management decisions

T L Bezrukova[1], S S Kirillova[1], Kuznetsov S A[1], Pecherskaya O A[1]

[1]Voronezh State University of Forestry and Technologies named after G.F. Morozov, 394087, 8, Timiryazeva str., Voronezh, Russian Federation

bezrukova_t_l@mail.ru

Abstract. Management decision making is based on an assessment of employee performance and is an essential management practice for any organization, because employee progress is important for improving the efficiency and development of each organization. The study is aimed at identifying the dependencies which take into account all the factors contributing to the improvement of quality and productivity of labor, in other words, identifying the features of motivational process building that contributes to effective management. The assessment has been carried out using correlation analysis of indicators for assessing the interrelationships of the motivational process factors to create an optimization model as the basis for making adequate management decisions using the example of furniture industry enterprises. The optimization model proposed in the article connects motivation of employees with the effectiveness of organization in making effective management decisions.

1. Introduction

In modern conditions, the issues of motivational process optimization in industrial enterprises are particularly important, since management decisions are largely dependent on how well the staff is satisfied with the work and existing working conditions. Saving money on motivating and incentives negatively affects the development of any enterprise. [1]

The efficiency of industrial enterprises is determined by the goals and interests of management, and the goals and interests of employees. If the main task of a manager is making profit and success in the market, then the main task of an employee is to get the maximum reward for his/her work and realization of personal interests. [2] At the same time, recognizing the personal interests of employees and upgrading the efficiency of the work are serious problems in motivational process optimization which is a basis for management decisions.

Issues of optimization of the motivational process are one of the most significant problems. These issues are of great importance for Voronezh furniture industry enterprises. To date, enough attention has been given to the problems of motivation and many papers have been published on this topic. [3] A large number of existing theories of motivation are divided into two categories: content and process ones. The theories of A Maslow, D McClelland, F Herzberg, C Alderfer and others were wide-spread among the content theories, the theories of K Levin, V Vrum, S Adams, Porter-Lawler, D MacGregor and others were wide-spread among the process theories. Among the works of domestic scientists studies of the motivational process by V A Yadov, A G Zdramomyslov, V P Rozhin, A N Leontyev, N F Naumova, I F Belyaeva and others should be noted. [4]

However, the issues of motivational process optimization as the basis of management decisions are poorly developed. Therefore, the development of an optimization model of motivational process as the basis of management decisions (on the example of furniture industry) is quite relevant.

2. Theoretical, Informational, Empirical, and Methodological Grounds of the Research

Various factors influence personnel motivation, but material factor has the highest ranking, as a result of which material motivational effect is of particular relevance. There are many definitions of motivation. Motivation refers to the instinctive component of human behavior. Mental concept of the human will does not significantly affect motivation. It is also believed that motivation is the cause of bodily reactions of the organism in response to external influences. Unconscious behavior of people is a reaction to an external stimulus. [5] Motivation arises with changes in internal needs of the body, and its formation is influenced by expectations about the future. The nature of labor norms and values acquired by an individual is of great importance for the formation of labor motivation.

For competent management decision-making in an enterprise, it is necessary to identify specific needs that motivate its employees with an offer of optimal remuneration that would help the employee to meet these needs, bearing in mind personal characteristics of individual's needs. [6]

Process motivation theories primarily rely on the perception and cognition of people. [7] Therefore, the task of a competent leader, according to these theories, should be expressed in effective time distribution of rewards and punishments with provision of continuous reinforcement of the employee.

The main attention should be paid on the analysis of factors which form the basis of motivation. Based on several surveys, we have obtained the following results of the most significant factors in the work activity of furniture industrial enterprises in the Voronezh Region. Analysis of wide range of managers and employees of furniture industry enterprises, from small to large ones, has been conducted. The analysis has shown that their indicators differ incomparably.

Labor efficiency (LE) is a function that establishes a relationship between input and output values through parameters. Input value parameters (x) make production output: x_1– financial resources spent on material incentives; x_2– cost of production; x_3 – level of motivation of top, middle and lower managers (from 0 to 1); x_4 – level of workers' motivation (from 0 to 1); x_5 – number of motivated workers; x_6 – number of unmotivated workers; x_7 – wages fund; x_8 – wages fund of motivated workers. Input value parameters (a) make an assessment of the enterprise activity, i.e. profit: a_1 – labor efficiency of motivated workers; a_2 – labor efficiency of unmotivated workers; a_3 – percent of defective goods (level of complaints).

It is reasonable to start with correlation analysis and use the data to create an optimization model of the motivation process on the example of furniture industry enterprises.

3. Results

Correlation analysis is considered to be the most natural and mathematically sound method for assessing the relationship between the indicators of the studied system. [8]. Pearson correlation coefficients for pairs of indicators of furniture industry enterprises have been calculated using the following formulas:

$$r_{xy} = \frac{\sum_{i=1}^{n}\left[(x_i - \bar{x})\cdot(y_i - \bar{y})\right]}{(n-1)\cdot\sigma_x\cdot\sigma_y}, \tag{1}$$

Where x_i and y_i – values of indicators of industrial enterprises, between which the correlation for the i-th quarter is determined; r_{xy} – correlation coefficient of x and y indicators (hereinafter *r* abbreviation is used); n – total number of quarters for all surveyed enterprises for which x and y are known; *x* and *y* – assessments of average values of the indicators determined by the formulas

$$\bar{x} = \frac{1}{n}\sum_{i=1}^{n} x_i; \qquad \bar{y} = \frac{1}{n}\sum_{i=1}^{n} y_i; \tag{2}$$

σ_x and σ_y – standard deviations of the indicators, calculated as follows:

$$\sigma_x = \sqrt{\frac{\sum_{i=1}^{n}(x_i - \bar{x})^2}{n-1}}; \qquad \sigma_y = \sqrt{\frac{\sum_{i=1}^{n}(y_i - \bar{y})^2}{n-1}}. \tag{3}$$

Further analysis should take into account that the Pearson correlation coefficient [9] may take values from –1 to 1. The more pronounced relationship between the indicators, the greater the absolute value of the correlation coefficient is. Nature of relationship indicators can be judged on the sign of the correlation coefficient: when $r > 0$ the relationship is direct one (with an increase in x indicator, y indicator also increases), when $r < 0$ the relationship is inverse one (with an increase in x indicator, y indicator decreases) (formula 4). When calculating the correlation coefficients, their statistical significance is evaluated at the same time, that is, the sufficiency of the initial data for establishing the correlation with a given level of confidence. The condition of the statistical significance of the correlation coefficient with p level is as follows:

$$|r_{xy}|\sqrt{\frac{n-2}{1-r_{xy}^2}} > t_{1-p/2}^{n-2}, \tag{4}$$

Where $t_{1-p/2}^{n-2}$ – critical value of the Student's coefficient which is given for a number of degrees of freedom (n - 2) and level of significance p. Further, only statistically significant correlations with the level $p = 0.05$ (correlations are reliable when the value is not less than 95 %) are considered.

In this paper, the matrix of correlation coefficients and significance levels are calculated using STATISTICA 10 program.

Table 1 has been prepared for the correlation analysis, in which indicators for each of the seven furniture industry enterprises are filled over five years. Thus, the size of the analyzed table is 35 rows (7 organizations × 5 years) with 16 columns. The correlation coefficients are presented in table 1, where all calculated coefficients are indicated above the main diagonal, and coefficients for only statistically significant correlations are below the main diagonal. The numbers of furniture industrial enterprises of the Voronezh Region which are shown in the cells of the table: 1 - Anstrem, LLC, PC, 2 - Furniture of Chernozemye, LLC, CC, 3 - Somovo-Mebel, LLC, 4 - Gefest-Mebel, LLC, 5 - Lux-Hall, LLC 6 - Furniture for Office, LLC, 7 - Grafskaya Kitchen, LLC, and for which correlation between these indicators is statistically significant. A minus sign is placed after the enterprise number for inverse correlations. Cells are darkened if the most number of enterprises (four or more) have statistically significant correlations of these indicators.

The more numbers in the cells of table 1, the more common is the dependence of the corresponding pair of indicators for furniture industry enterprises. The most common are the correlations that are simultaneously found in five or more of the seven enterprises (the darkened cells in table 1).

From non-obvious correlations (disregarding the effect of time t), the following ones can be distinguished: with an increase in the size of material incentives C_m, revenue R and production cost PC increases; with an increase in the wage fund W_f, revenue R and production cost PC increase; increase in labor productivity of unmotivated workers LE_{uw} leads to an increase in revenue R and the cost of production PC.

However, statistically significant correlations of the total correlation matrix (table 2) are in good agreement with general correlations in table 1.

IOP Conf. Series: Materials Science and Engineering **483** (2019) 012025 doi:10.1088/1757-899X/483/1/012025

Table 1. General correlations for the analyzed furniture enterprises.

Indicators	Indicators														
	t	R	LE_w	N_w	R	C_m	PC	K_m	K_w	N_{mw}	N_{uw}	W_f	W_{fmw}	LE_{mw}	LE_{uw}
t	1.00														
R	1-, 5-, 7-	1.00													
LE_w	4	3	1.00												
N_w	3-, 7-	4, 7		1.00											
R	7-	3, 6-	1, 2, 3, 7		1.00										
C_m	1-, 3-, 4-, 5-, 6-, 7-	1, 4, 5, 6, 7		1, 3, 4, 7	6-	1.00									
PC	1-, 5-, 7-	1, 2, 3, 4, 5, 6, 7	3	4, 7	3, 6-	1, 4, 5, 6, 7	1.00								
K_m	5-, 6	7		3, 7	5	3, 7	7	1.00							
K_w			1		1			5	1.00						
N_{mw}	5-, 7-	4, 5, 7	1	2, 4, 7		4, 7	4, 7	3, 5, 7	1, 5, 6	1.00					
N_{uw}	1-, 3-, 7-	1, 4, 7	5-	3, 4, 7		1, 3, 4, 7	1, 4, 7	7		4, 7	1.00				
W_f	1-, 3-, 4-, 5-, 6-, 7-	1, 4, 5, 6, 7		1, 3, 4, 7	6-	1,0	1, 4, 5, 6, 7	3, 7		4, 7	1, 3, 4, 7	1.00			
W_{fmw}	4-, 5-, 7-	4, 5, 7		2, 3, 4, 7		2, 3, 4, 7	4, 7	3, 5, 7	5,	3, 4, 5, 7	4, 7	2, 3, 4, 7	1.00		
LE_{mw}		4, 6, 7-		4, 5, 7-	6-	4, 6, 7-	4, 6, 7-	1-, 3-	1-	1-, 3-, 4, 7-	4, 7-	4, 6, 7-	3-, 4, 7-	1.00	
LE_{uw}	5-	2, 4, 5, 6, 7-		4, 7-	6-	4, 5, 6, 7-	2, 4, 5, 6, 7-		5	4, 5, 7-	3-, 4, 7-	4, 5, 6, 7-	4, 5, 7-	4, 6, 7	1.00

Table 2. Correlation matrix for the seven analyzed enterprises.

Indicators	Indicators														
	t	R	LE_w	N_w	R	C_m	PC	K_m	K_w	N_{mw}	N_{uw}	W_f	W_{fmw}	LE_{mw}	LE_{uw}
t	1.00	-0.12	0.12	-0.15	0.08	-0.22	-0.14	-0.09	0.00	-0.05	-0.20	-0.28	-0.16	-0.14	-0.11
R		1.00	0.61	0.62	0.43	0.73	0.99	0.39	0.29	0.56	0.70	0.69	0.57	0.80	0.85
LE_w		0.61	1.00	0.28	0.87	0.31	0.57	0.20	0.23	0.27	0.31	0.28	0.24	0.39	0.60
N_w		0.62		1.00	0.16	0.84	0.66	0.45	0.32	0.81	0.86	0.88	0.81	0.30	0.38
R		0.43	[illegible]		1.00	0.25	0.40	0.36	0.41	0.25	0.14	0.16	0.22	0.20	0.48
C_m		0.73		[illegible]		1.00	0.79	0.67	0.58	0.93	0.81	0.94	0.96	0.35	0.53
PC		[illegible]	0.57	0.66	0.40	0.79	1.00	0.44	0.33	0.62	0.74	0.74	0.63	0.77	0.83
K_m		0.39		0.45	0.36	0.67	0.44	1.00	0.93	0.74	0.33	0.51	0.72	0.08	0.48
K_w					0.41	0.58		[illegible]	1.00	0.70	0.19	0.38	0.67	-0.01	0.43
N_{mw}		0.56		[illegible]		[illegible]	0.62	0.74	0.70	1.00	0.71	0.85	0.98	0.11	0.40
N_{uw}		0.70		[illegible]		[illegible]	0.74			0.71	1.00	0.94	0.72	0.46	0.44
W_f		0.69		[illegible]		[illegible]	0.74	0.51	0.38	[illegible]	[illegible]	1.00	0.89	0.36	0.44
W_{fmw}		0.57		[illegible]		[illegible]	0.63	0.72	0.67	[illegible]	0.72	[illegible]	1.00	0.13	0.39
LE_{mw}		[illegible]	0.39			0.35	0.77				0.46	0.36		1.00	0.84
LE_{uw}		[illegible]	0.60	0.38	0.48	0.53	[illegible]	0.48	0.43	0.40	0.44	0.44	0.39	[illegible]	1.00

A diagram of the mutual influence of indicators of furniture industry enterprises has been built based on the correlation analysis (figure 1).

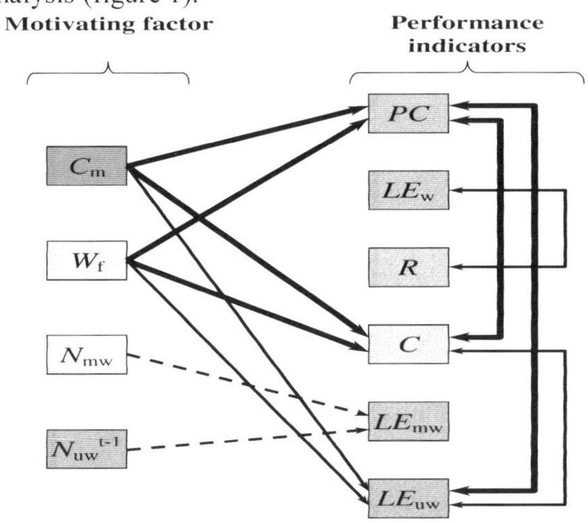

Figure 1. Diagram of the influence of motivating factors on the performance indicators of furniture industry enterprises

IOP Conf. Series: Materials Science and Engineering **483** (2019) 012025 doi:10.1088/1757-899X/483/1/012025

Correlations of indicators provide information about the nature of management of furniture industry enterprises [10], which may be the initial data for building an optimization model of the motivational process as the basis of management decisions.

Table 1 and table 2 have been used in scheme construction, summarizing information on the mutual influence of indicators for various enterprises.

Figure 1 shows thick and thin lines, which, respectively, show strongly and weakly expressed relationships. Solid and dashed lines show direct (with an increase in one indicator - the second also increases) and reverse (with an increase in one indicator - the second decreases) interrelations.

This scheme of interconnection can be taken as the basis for developing an optimization model of the motivational process as the basis of management decisions. With an increase in the share of motivated workers among all workers, an increase in the efficiency of the enterprise can be expected.

4. Conclusions and recommendations

With the increase in the number of motivated workers N_{mw} revenue R and labor efficiency LE of motivated workers LE_{mw} increase according to the square law.

Thus, the increase in the share of motivated workers leads to a significant square growth of the financial results of the furniture industry enterprise as an effective basis for taking right management decisions.

Matrix to assess motivation and incentives is built based on the developed optimization model, and then the following management solutions are offered (table 3).

Table 3. Management decisions depending on the degree of motivation and incentives

Management decision	Actions depending on the degree of effectiveness of the system of motivation and incentives
Management decision 1.	Ineffective motivation This combination of factors indicates critical state of the enterprise The following set of measures is recommended: 1) To improve the system of motivation and incentives 2) To introduce motivation criteria
Management decision 2.	Weak motivation This combination of factors indicates insufficiently developed system of motivation and incentives at the enterprise The following set of measures is recommended: 1) Maintenance and improvement of the current system of motivation and incentives 2) Product-line expansion
Management decision 3.	Effective motivation This combination of factors indicates the accuracy of enterprise performance The following set of measures is recommended: 1) Maintenance and improvement of the current system of motivation and incentives
Management decision 4.	Highly effective motivation This combination of factors indicates absolutely accurate and well-built system of motivation and incentives. No recommendations.

From this the following conclusions can be made:

– It is advisable to motivate the largest possible share of employees to improve the efficiency of the enterprise;

– If some of the workers are already motivated, it is advisable to use motivating measures for another share of the workers, since financial indicators increase with increasing N_{mw} according to a stronger law than linear one;

– If the number of motivated employees is small (less than 20%), the company's financial indicators are extremely low or negative.

Increase in the number of motivated workers N_{mw} linearly leads to an increase in the cost of production PC. A non-obvious pattern has been found: with an increase in the share of motivated workers, starting from 10-20%, the productivity of unmotivated LE_{uw} workers increases significantly according to a square law.

The influence of motivating factors on management decisions has been revealed using correlation analysis.

Correlation analysis has led to the following conclusions:

– Revenues and cost of production increase with an increase in the size of material incentives and wage fund;

– Increase in labor productivity of unmotivated workers is directly related to the growth of revenues and production cost.

References

[1] Bezrukova T L, Stepanova Yu N, Shanin I I, Busarina Yu V and Nesterov S Yu 2017 *European Research Studies Journal* **20 №3** 183-92
[2] Silin Y P, Animatsa Y G and Novikova N Y 2017 *Economy Region* **13(3)** 684-96
[3] Kauppila O -P 2018 *Journal Of Organizaitional Behavior* **38(8)** 941-55
[4] Lau C M, Scully G and Lee A 2018 *Journal of Business Research* **90** 247-59
[5] Graves L M and Sarkis J 2018 *Journal of Cleaner Production* **196** 576-87
[6] Duarte H and Lopes D 2018 *International Journal of Manpower* **39(5)** 746-63
[7] Batova T 2018 *Journal of Business and Technical Communication* **32(3)** 308-46
[8] Monnot M J 2018 *Thunderbird International Business Review* **60(4)** 675-89
[9] Boyd N, Nowell B, Yang Z and Hano M C 2018 *American Rewiew of Public Administration* **48(5)** 428-43
[10] Zinoveva I S, Yakovlev A V and Pecherskaya O A 2019 *Advances In Intelligent Systems And Computing* **726** 337-43

Enhancing foreign language communication skills in international business environment

I V Skrynnikova, E G Grigorieva

Department of Germanic and Romance Philology, Volgograd State University, 100 Prospekt Universitetskij, Volgograd, 400062, Russia

i.skrynnikova@volsu.ru, e.grigorieva@volsu.ru

Abstract. The article analyzes the peculiarities of enhancing written communication skills in the interethnic environment of transnational corporations. The authors explore the rhetorical and pragmalinguistic features as well as the linguistic realizations and strategies used by Russian business communicants in writing business letters in English and contrast them to the samples of business letters of native English speakers. The paper proposes recommendations for preparing managerial personnel for international business interaction in the workplace through integrating authentic models and contexts into training strategies.

1. Introduction

Modern globalization processes changing the world economy structure, close intertwining of national economies, geo-economy and its infrastructure emergence are accompanied by a rapid growth in the rates of capital and productive resources transfers, labor migration, and establishing a significant number of international or transnational corporations.

Business communication with current and potential partners, competitors, consumers and the public is the basis of any enterprise effective operation. In the context of a transnational corporation, communication in a multilingual setting becomes multi-polar where interpersonal professional communication among employees is of particular importance for the financial viability of the company.

The executives of a transnational corporation when creating their own corporate culture should take into account the national specificity of a country where it is located. In this case, of particular value is an expert in the field of intercultural communication who trains managers and employees of international companies to work in a multicultural environment as well as acquaints them with the values and behavioral norms of another culture and conducts training sessions on communication in an interethnic group and on keeping business documentation records in a foreign language. In the course of cross-cultural adaptation trainings, the current or potential employees of transnational corporations are inculcated with social and personal skills necessary for interacting with people from other cultures in a non-native language [1]. Thus, a foreign language is becoming an orientation tool in modern technological world and a link for multicultural and interdisciplinary thinking formation, its knowledge being a significant individual and professional characteristic of a modern person [2].

The issues of achieving communication environment stability in a company are set by international organizations, public and private educational institutions, and the business community. The Council of Europe, in particular, launched a number of projects, such as ELAN (Effects on the European Language Skills in Enterprise 2005-2008); Dylan (Language Dynamics and Management of Diversity 2006-2011), aimed at identifying problems in the field of foreign language professional communication of employees of small, medium and large businesses. According to the survey, “11% of respondents (195 SMEs) had lost a contract as a result of lack of language skills <…> and at least 50% of respondents believed that they would need additional language skills in the next there years. <…> As for large enterprises, English is more extensively used as an intermediary language than was the case with SMEs, possibly reflecting its use also as a corporate language in many multinationals" [3].

Other studies also confirm the paramount importance of English which has become the official language of the business community in the world today being used in 85% of international organizations [4]. Some scholars have gone as far as to propose the term "English as lingua corporative", i.e. English as a corporate language. According to them, communicating in English enables companies to integrate their branches in all regions of the world, successfully compete in world markets and attract the best specialists [5]. However, the level of proficiency in the common corporate language is not always adequate to the requirements of a communicative situation [6]. Researchers draw attention to the fact that "the issue of mastering special vocabulary often becomes critical in situations where a discussion in English takes place and one has to make important decisions for the company. Insufficient knowledge of professional vocabulary does not enable the discussion participants to defend their position in a foreign language, and as a result, no serious factors can be taken into account" [7].

Possible solutions to the communication problem in an interethnic environment of a transnational corporation include the Council of Europe analysts' recommendations to use such language management approaches as recruiting native speakers, adapting websites, use the services of translators and interpreters [8]. A more important aspect of settling this issue, in our opinion, is the enhancement of foreign language culture among "in-corporate" employees.

Multinationals in Russia have successfully applied a staff training scheme under the Corporate Training program. As it is obvious from business experience, modern labor market requires not merely translators and interpreters, but highly qualified specialists with near-native foreign-language proficiency. Corporate employees are offered educational programs, such as "Foreign Languages in Professional Activity", "English for Successful Business", "Business Translation Course" [9].

In higher educational institutions of the Russian Federation, senior and master students, as well as employees of international companies are offered "English for Special Purposes" (ESP) training programs in "Business Foreign Language", "Business Communication", "Business Correspondence and Documentation in English", "Business Language and Speech Culture", "Fundamentals of Business English", "Foreign Language of Business and Professional Communication", "National Mentality and Language of Business Communication", "Business Contacts in a Foreign Language" and some others.

Obviously, such a diversity of training programs for professionals speaking a business foreign language is predetermined by the complexity and versatility of business communication in international companies where communicators' interaction is required at several levels: external, internal and interpersonal. Communication can also be oral (e.g. negotiations, consultations, meetings) or written (e.g. contracts, agreements, all types of business letters). This paper dwells on identifying specific features of written business communication.

2. Intercultural business communication environment and its features

Intercultural business communication today can be viewed as a special interaction environment. It comprises a special language and conventions, its own social hierarchy arising as a result of business-to-business communication. Such an environment is characterized by much greater social ambiguity – both due to its dynamics and rich communication opportunities.

According to J. Stock [10], to succeed in modern business conditions one should manage complex communication processes in any of its fields. The socio-cultural environment of a globally operating business consists of a number of components: language, education level, religion, values, technologies, social organization, political techniques, infrastructure development level and regulation systems. Each component, both independently and in combination with others, can affect the activities of managers in different ways. For instance, language differences can cause difficulties in compiling documents, which describe techniques and procedures and formulate strategies for moving products both within and between individual countries and regions.

It is critical to keep in mind that on intercultural soil conflicts most frequently arise between top and middle management of Russian businesses, acquired or established by foreign companies. Such conflicts have major similarities and follow almost the same scenario. Expats are usually the heads of such businesses while the middle management has Russian roots. It is between these two groups of managers that intercultural misunderstanding arises most frequently, that is, communication and behavioral barriers are built.

In old days cultural differences were believed to be a barrier and interfere with communication and joint activities. However, today astute business leaders believe that by properly managing national cultural differences, they can be turned into sources of additional resources, rather than be potential problems. Many of them assume that cultural diversity increases the competitiveness of a multinational organization. However, failure to properly use it can result in intercultural conflicts and communication disruptions. Ignorance of the cultural characteristics of a host country translates into numerous problems, although it does not affect all the company departments. The manager does not have to master cultural knowledge in full, but is supposed to have a clear idea of the basic theories of culture and the major problems of this phenomenon. Cultural concepts in the form of complementary and mutually enriching tools for studying the national and cultural characteristics of individual employees should be considered as increasing the efficiency of their work and the firm's activities as a whole, ensuring the national-cultural environment within the latter [11]. A study of the fundamentals of cross-cultural written communication will contribute to improving the efficiency of the company and its competitiveness in the market.

3. Peculiarities of business English letter writing

Contacts with English-speaking countries form the basis for business communication in multinationals, with English being used as a means of international interaction. In this respect, Russia is of a great interest as it can boast the abundance of natural resources, strong agricultural and industrial potential, relative political stability and dynamic markets development.

Establishing ongoing mutually beneficial relations with foreign partners primarily assumes comprehensive training of highly qualified experts for local businesses [12]. Consequently, there is an ever-increasing demand for English-speaking professionals. It is worth mentioning that knowledge of business English has become the fundamental requirement for both white-collar and blue-collar positions.

One of the most critical fields of the English language use is international business correspondence. While establishing and maintaining their interaction with foreign companies, business partners tend to face challenges and pitfalls related to the following: intercultural differences, becoming obvious only after mastering various ways of writing business correspondence taking into consideration the national-cultural conventions of communicants [13], [14], [15] [16]; some cross linguistic differences in writing styles and strategies [17], [18], [19]; sufficient understanding of business writing culture across various languages [20], [21], [22].

Thus, particular value and topicality of studying business texts and correspondence in international business communication based on the primary use of English is hardly disputable. On the one hand, one should be aware of cultural differences and the strife to maximize the effectiveness of business communication. On the other hand, those compiling business documents in English seek certain variety in writing styles, emphasizing the culture-specific traditions of the communicants [23]. Therefore, one of the major purposes of this paper is to reveal the pragmalinguistic and rhetorical features and the linguistic strategies applied by Russian business communicants in writing business letters in English and to contrast them to those of native English speakers. The results obtained will enable to elaborate recommendations for preparing future managers for international business interaction in the professional settings through integrating authentic models and contexts into training effective writing strategies. This requires addressing the following: identifying rhetorical and pragmalinguistic peculiarities of business correspondence of Russian and English-speaking managers; explaining diverging business letters writing strategies between native and non-native speakers;

integrating authentic writing strategies into the course of English for Specific Purposes at Russian universities to train future business personnel.

4. Method, data and analysis procedure

The current paper dwells on the sales promotional letters written by university students taking the English for Business Purposes Course at Volgograd State University (Volgograd, Russia) as compared to authentic sales letters of the USA and the UK-based native English speakers. The cross-cultural variation in business letter writing presupposes applying contrastive text linguistic and pragmalinguistic procedures. According to Connor [24], contrastive rhetoric studies rest on contrastive linguistic studies of texts that examine, compare and contrast the ways of texts generation and interpretation across languages and cultures applying written discourse analysis methods. The contrastive text analysis in the present paper involves analyzing rhetorical structures and their linguistic implementation (e.g. linguistic features).

The pragmatic character of business correspondence manifests itself in politeness and persuasion strategies (e.g. formality, indirectness, the delayed introduction of purpose in requests, and excessive politeness, etc.) and some culture-specific discourse patterns. Its linguistic implementation in our corpus includes lexical, syntactic and semantic features viewed in relation to moves with authentic examples. In this study, moves and steps form are the units of the analysis. According to Upton and Connor [25], moves are semantic or functional units of texts, identifiable through their communicative purposes and certain linguistic boundaries typical of the moves. Bhatia [26] sees moves as 'rhetorical instruments that implement a sub-set of specific communicative purposes associated with a genre'. Swales stresses that recurrence and typicality of particular moves in genre-specific texts enable viewers to recognize these texts as exemplars of the same genre [27].

The research corpus comprises 116 sales letters, including 78 sales letters written by VolSU students (business and finance majors) and 38 authentic sales promotional letters of native English speakers, collected between November 2016 and March 2018. As for the students' English language writing proficiency, over 70% of them reported they could write in most cases, but occasionally needed some help, while 20% of the students did not experience any difficulty in writing.

Data were collected through the use of a scenario describing a business situation for students. The respondents specified what they would mention in a sales letter to promote the product or service of their company.

Scenario: *Imagine you work for an agricultural company in Russia. Write a sales letter to promote your company to your potential British customers, so that they could become your customers in future. You may want to consider the following points: the way you address your potential customers; the major advantages and benefits of your company; tools you will use to persuade your customers.*

The scenario subsequently asked students to provide additional information about the problems/challenges they faced while writing the sales letters in English, the ways to cope with them, and the strategies they use to make their letter more effective and clearer for the potential customers.

Numerous researchers apply Bhatia's move analysis [28] to reveal the specificity of cross linguistic business correspondence. This method has enabled to identify rhetorical features in business correspondence in English and Chinese [29], English and Thai [30], English and Taiwan [31] business correspondence. The analysis implies seven moves in each sales promotion letter including three compulsory moves and four optional ones [32] (see table 1):

Table 1. Bhatia's Move Analysis of Sales Letters

Move 1	Establishing credentials	obligatory
Move 2	Introducing the offer	obligatory
(i)	Offering the product or service	
(ii)	Essential detailing of the offer	
(iii)	Indicating value of the offer	
Move 3	Offering incentives	optional
Move 4	Enclosing documents	optional
Move 5	Soliciting response	obligatory
Move 6	Using pressure tactics	optional
Move 7	Ending politely	optional

In our view, Bhatia's move analysis proves to be the most reliable research method, capable of demonstrating peculiarities in the business correspondence of native and non-native speakers. Therefore, following Bhatia's procedure, the authors studied the texts applying the method of analyzing rhetorical structures (e.g. moves, structural representation). As for the experimental research, the analysis undertaken has revealed some peculiarities. The limited availability of the letters has resulted in insufficient number of sample letters. Such factors can interfere with summarizing our results about international sales promotional letters both for native speakers and non-native speakers. However, despite the fact that our research was limited only to sales letters, we assume similar research can also be done into other categories of business correspondence. The topicality and importance of the further research in this sphere is hardly disputable as it could also be carried out in the occupational environment to define whether there are similar cultural distinctions in particular letter-writing techniques.

5. Findings

This section of the paper presents the results of native and non-native speakers' sales letters analysis with the emphasis on rhetorical moves and their linguistic implementation. Following Bhatia's move analysis [33], cross-cultural variations are found from contrastive rhetorical perspectives to identify differences found between the promotional writing styles of Russian and native English speakers.

The authors selected two sample letters, most typical of native and non-native speakers' writing styles respectively.

Sample 1. Sales Letter by a Native Speaker of English

From the desk of Jane Watson
March 5, 2016
Ms. A. Prokopjeva
18 Novosibirskaya Str., apt.
1514
Moscow, Russia

Dear Ms Prokopjeva,	
Achieving the highest standards of security has always been top priority for Citibank, as we are among the first Banks in Sweden to introduce chip debit cards very soon.	**Move 1 – Establishing credentials**
Following the launch of MasterCard Chip cards, Citibank will cease issuing new and renewing existing traditional Cirrus, Cirrus Maestro and MasterCard bank cards. Also, renewed cards will be issued as MasterCard Chip cards, except if the Customer notifies Citibank in writing at least two months prior to the expiry of the existing bank card that they no longer need it, in which case the card service of the Bank will no longer be used together with additional services such as our CitiPhone Banking telephone customer service and the Citibank Online internet banking service that are available exclusively to card holder Customers. With the launch of MasterCard Chip cards, the visual design of the debit cards of Citibank will be modified to replace the current appearance of them.	**Move 2 – Introducing the offer** (i) Offering product or service (ii)Essential detailing of the offer

Please also note that in case of a Chip card, in addition to the magnetic stripe, the chip integrated into the front of the card also contains the data of the card and the card holder, which is a much safer method of storage of these data than the magnetic stripe, hence also makes it safer to use the card.	(iii) Indicating the value of the offer
We are pleased to inform you that if your company sponsors 6 or more staff for the MasterCard Chip cards, we will offer you a discount of US$100 per person.	**Move 3 – Offering incentives**
For your convenience, I enclose a reservation form which should be completed and returned directly to me.	**Move 4 – Enclosing documents**
If you have any questions concerning the use of smart cards, please visit our website at www.citibank.hu or call the Citiphone Banking phone customer service on 06-40-24-84-24.	**Move 5 – Soliciting response**
Your response is required by 3/30/01: To qualify for no fee balance transfers at a rate as low as 0% - plus the benefits shown on the reverse side – mail your Acceptance Form today.	**Move 6 – Using pressure tactics**
Thank you very much for your kind consideration. Sincerely, Jane Watson Managing Director	**Move7 – Ending politely**

As is clear from the sample above, this letter has all seven moves which replicates Bhatia's results [34], identifying moves typical for sales letters written by native English speakers.

The following letter is written by a student, non-native speaker of English.

Sample 2. Sales Letter by a Non-native Speaker of English

The Republican Product Corporation
Myasnitskaya str., 14/1
Tel: +7-7051198807
November 7, 2017

Dear Customers!

Our company is engaged in processing grain on the most modern mill equipment, we have a large number of the most modern agricultural machinery: "John Deere", "Amazone", "Catros" etc.	**Move 2 – Introducing the offer** (i) Offering the product or service
We give special attention to quality of grown-up wheat and barley. We professionally are engaged in seed farming, and our seeds of several grades of wheat buy many companies for improvement of own crops.	(ii) Essential detailing of the offer
For more details you may call us today for a Free Quote on +7-7051198807.	**Move 5 – Soliciting response**
Additionally you can view our quality furnishings on our website at www.agroholding.com. Our website also lists our current specials – there are even greater savings to be made!	**Move 3 – Offering incentives**
As a Free Gift to you we have enclosed the Agroholding Corporation pen containing our website address and contact information along with our complete catalogue for 2012.	**Move 4 – Enclosing documents**
If you have any further questions, please contact me directly on +77758268792. Mrs. Krivosheeva.	**MOVE 5 – Soliciting response**

As you can see, some differences from the previous sample are obvious. There obligatory Moves 1, 6 and 7 are missing. The writer finalizes the letter with Move 4 and Move 5 which are found in the same passage instead of a polite ending. We- orientation is also extensively used by the student.

The remaining letters' analysis focuses on each move in Bhatia's procedure [35]. The frequency of occurrence of each step was calculated on the basis of the total number of units analyzed, and presented as a percentage for both corpora. For example, Figure 1 below presents the results of move analysis of the both corpora. An interpretation of the results will follow, with each move being dealt with individually.

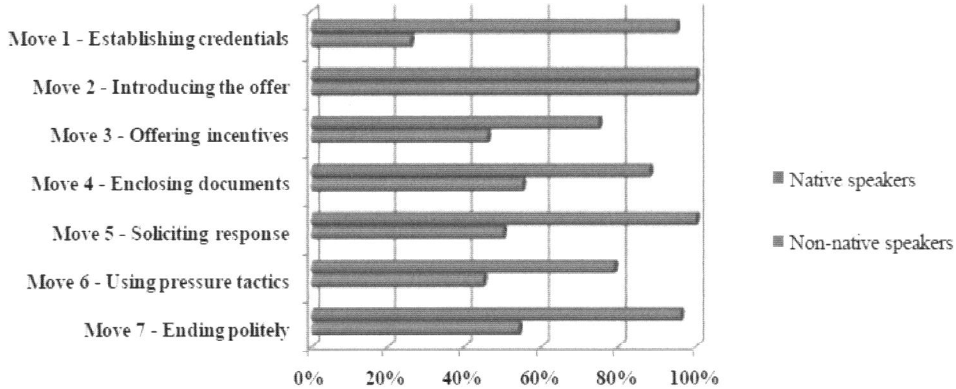

Figure 1. Analysis of rhetorical moves in the sales letters by native and non-native speakers.

It cannot escape our notice that Move 2 appears in every letter while Moves 4 and 6 occur occasionally while Moves 3 and 5 cannot be found in almost 50% of the letters. In accordance with Bhatia's findings for his Singaporean data, Offering Incentives is a culture-specific feature of South-Eastern Asia. The compulsory Move 5 is missing in 60% of Russian university students' letters. It may be attributed to the fact that native English speakers' business texts seek some "feedback", that is the aspiration of a sender to stress the importance of the addressee's response and, thereby to encourage them to respond, thus raising the probability of repeated contact. In contrast to other cultural backgrounds, the addressees and senders in the English business culture are normally individuals who account for a considerable share of sales letters and requests' recipients. In Russian business communication, the addressees are frequently legal entities and organizations or those who speak on their behalf. The most important parts of business texts include repeated contacts between communicants, and the sender does not commonly use additional tools to ask for "feedback". Meanwhile, 45% of the letters contained Move 6, which is pervasively typical of European cultures.

6. Discussion and implications

The results of contrasting business texts of English and Russian communicative cultures representatives provide some evidence that the collected sales letters reveal both similarities and differences in all the aspects and text levels in question. The differences are indicators of cultural specificity of sales letters texts. The possible diverging features of the contrasted business texts are inherent differences in terms of their graphical organization as well as orthography and punctuation formatting. These differences provide the information of both the genre-stylistic and national-cultural character. Thus, it is worth noting that in some collected sales and promotional letters university students used exclamation marks addressing their potential customers, for instance, "Dear Customer!" in accordance with Russian salutation conventions. For business English texts, the presence of salutation is obligatory, the forms of which may vary depending on a degree of acquaintance of addressees, the frequency of contacts between them, low or high social distance and their professional status. Thus, one should meet the requirements of greetings and a form of ending the letter. Business English letter etiquette encompassing the salutation "Dear Mrs. Jonson" requires the form "Yours sincerely" at the end, and a greeting to the addressee or to the group, whose names are unknown: "Dear Madam" or "Dear Sirs" require "Yours faithfully". One more etiquette rule is an address to a female business partner without referring to her marital status and preference of the form "Ms Hartford".

Another peculiarity lies in the fact that non-native speakers' greetings are more formal which provides some evidence in favour of foreign business partners' willingness to progress to a more friendly level resulting in some misunderstanding. However, Russian business professionals can appear to be interested in neither mutual cooperation nor use of a basic official style and a cold and unfriendly tone.

The above mentioned findings lead us to conclude that teaching linguistic and cultural peculiarities of sales genres will assist future business people gain deeper insights into purposeful business communication. In addition, business correspondence training should address specific linguistic forms and strategies used in each letter. Thus, integrating more authentic materials in teaching business writing within ESP courses might result in higher foreign language proficiency in naturally occurring professional setting and raise cultural competence of communicants in the context of global business and communication.

7. Conclusion

The results obtained confirm the crucial role of the English language in written business discourse in Russia. The analysis of business letters samples shows (a) significant differences in writing strategies applied Russian university students taking the course of English for Business Purposes and native speakers in terms of communicative, pragmatic, rhetorical, graphical and genre-stylistic aspects; (b) greater adherence to the conventional writing style in native English business correspondence as compared to non-native speakers; (c) non-native speakers' unawareness of their foreign partners expectations and the cultural context in which their letters would be received.

These conclusions support the findings of several studies (cf. [36]). Thus, sociocultural structures lie at the heart of language structures which means overcoming the language barrier is not sufficient to maintain effective business communication between representatives of different cultures. Such state of affairs calls for breaking a cultural barrier. The presented study provides several reasons for communication failures in business letters: (a) mixing different national and cultural standards in the structure and language of sales letters as well as features of different functional styles in one text; (b) combining several communicative tasks; (c) relatively low English language proficiency of business letters authors (spelling, grammar, lexical and stylistic errors).

The results substantiate the need for teaching business communication for non-native speakers aimed at developing skills in writing business letters for English language learners to enhance the efficiency of professional business correspondence. The findings can shed some light on the prospects for further research into the language of intercultural business communication. The research contributes to the theory of languages for specific purposes, teaching a language of business communication, as well as to raising business actors' awareness of intercultural competence for the development of business contacts on the global scale.

The major results of the paper may serve as recommendations for training business correspondence to students majoring in Management, Economics, Finance and related fields: to allow sufficient number of hours to explore the peculiarities of writing business letters in English; to pay special attention to the knowledge of intercultural business communication; to consider, in more detail, the structure of a letter; to emphasize pragmatic factors exemplified in the appeal to potential customers and polite ending of the letter.

8. References

[1] Bogatyreva M A 2010 Context approach to teaching the language of specialty in the corporation *Higher education today* Moscow: Logos 66

[2] Neuman S Yu, Burenkova S V, Gruenko S E, Dalke, S G, Dolgova T V, Kozhevnikova I Yu, Okhotnikova E P, Pevtsova V A and Sokhryakova E S 2015 *Prospects of continuous education system development in new socio-economic conditions* Omsk: Omsk State Institute of Service 74

[3] ELAN: Effects on the European Economy of Sportages of Foreign Language Skills in Enterprise 2005-2008 [Electronic resource] URL: http://ec.europa.eu/assets/eac/languages/policy/strategic-framework/documents/elan_en.pdf

[4] Hahaleva A Yu 2012 Discourse characteristics of the situation of business communication in the context of globalization *Bulletin of Moscow State Linguistic University: Humanitarian sciences* Moscow: Moscow State Linguistic University 157
[5] Borodina D S 2018 The English language in a specialized domain of Swedish business communication *Political linguistics* Yekaterinburg: Ural State Pedagogical University 91
[6] Rönnlöf H 2014 Language practices in the worlplace. Ethnographic insights from two multilingual companies in Sweden [Electronic resource] URL: http://www.diva-portal.org/smash/get/diva2:733464/FULLTEXT01
[7] Piekkari R, Oxelheim L, Randøy T 2013 *The Role of Language in Corporate Governance: The Case of Board Internalization* Stockholm: Research Institute of Industrial Economics 31
[8] ELAN: Effects on the European Economy of Sportages of Foreign Language Skills in Enterprise 2005-2008 [Electronic resource] URL: http://ec.europa.eu/assets/eac/languages/policy/strategic-framework/documents/elan_en.pdf
[9] Neuman S Yu, Burenkova S V, Gruenko S E, Dalke, S G, Dolgova T V, Kozhevnikova I Yu, Okhotnikova E P, Pevtsova V A, Sokhryakova E S 2015 *Prospects of continuous education system development in new socio-economic conditions* Omsk: Omsk State Institute of Service 169
[10] Stock J M 2005 *Strategic Logistics Management* Moscow: INFRA-M
[11] Persikova T N 2002 *Intercultural communication and corporate culture* Moscow: Logos 67
[12] Smith M 2012 The Kazakhstan opportunity *World Monitor-Industrial Development* **5**(36) 38-9
[13] Bell A H, Becker H and Dillon W T 1995 German memo and letter style *Journal of Business and Technical Communication* **5**(2) 219-27
[14] Cheng W 2003 *Intercultural Conversation* Amsterdam: Benjamins Publishing
[15] Kaplan R 1966 Cultural thought patterns in intercultural education *Language Learning* **16**(1,2) 1-20
[16] Sarangi S and Candlin C 2011 Professional and organisational practice: A discourse/communication perspective *Handbook of Professional Communication in Professions and Organisations* Berlin: De Gruyter 1-49
[17] Akar D 2002 The macro contextual factors shaping business discourse: The Turkish case *International Review of Applied Linguistics* **40** 305-22
[18] Hewings M 1999 The Academy Meets the Real World: Response to Audience in Academic Business Writing *Business English: Research into Practice* ed M Hewings and C Nickerson Harlow: Longman.
[19] Scollon R and Scollon S W 2001 *Intercultural Communication: A Discourse Approach* Malden, Mass.: Blackwell
[20] Du-Babcock 2013 English as Business Lingua Franca: A comparative analysis of communication behaviour and strategies in Asian and European contexts *Ibérica* **26** 99-131
[21] dos Santos and Pinto Valeria B M 2002 Genre analysis of business letters of negotiation *English for Specific Purposes* **21** 167-99
[22] Li, Wei, Zhu, Hua, and Li Yue 2001 Conversational Management and Involvement in Chinese-English Business Talk *Language and Intercultural Communication* **1**(2) 135-150
[23] Aimoldina A and Zharkynbekova S 2014 Incorporating authentic models into business letter writing within the ESP course at Kazakhstani universities *Hacettepe Üniversitesi Eğitim Fakültesi Dergisi [Hacettepe University Journal of Education]* **29**(2) 16-33
[24] Connor U 1996 *Contrastive Rhetoric: Cross-Cultural Aspects of Second Language Writing* Cambridge: Cambridge University Press
[25] Upton T and Connor U 2001 Using Computerized Corpus Analysis to Investigate the Textlinguistic Discourse Moves of a Genre *English for Specific Purposes: An International Journal* **20** 313–29.
[26] Bhatia V K 2001 Analysing Genre: Some Conceptual Issues *Academic Writing in Context: Implications and Applications* ed M Hewings University of Birmingham Press 84

[27] Swales J 1990 *Genre analysis: English in academic and research settings* Cambridge University Press
[28] Bhatia V K 1993 *Analysing Genre: Language Use in Professional Settings* London: Longman
[29] Zhu Y 2005 *Written Communication across Cultures: A Sociocognitive Perspective on Business Genres* ed J Benjamins Amsterdam
[30] Ora-Ong Chakorn 2000 Contrastive Rhetoric of English Persuasive Correspondence in the Thai Business Context: Cross-Cultural Sales Promotion, Request and Invitation *Centre for English Language Teacher Education* University of Warwick PhD thesis
[31] Hsiao-I Hou and Min-Yu Li 2011 A Contrastive Rhetoric Analysis of Internship Cover Letters Written by Taiwanese and Canadian Hospitality Majors *International Journal of Linguistics* **3**(1) E43
[32] Bhatia V K 1993 *Analysing Genre: Language Use in Professional Settings* London: Longman
[33] Bhatia V K 1993 *Analysing Genre: Language Use in Professional Settings* London: Longman
[34] Bhatia V K 1993 *Analysing Genre: Language Use in Professional Settings* London: Longman
[35] Bhatia V K 1993 *Analysing Genre: Language Use in Professional Settings* London: Longman
[36] Varner I 1988 A comparison of American and French business correspondence *The Journal of Business Communication* **25**(4) 55-65.

IOP Conf. Series: Materials Science and Engineering **483** (2019) 012027 doi:10.1088/1757-899X/483/1/012027

School - university - production: a chain of effective communications

O U Soloveva, N V Gretsova, M V Gretsov

Volgograd State Technical University, Volgograd , 400028, Russia

e-mail: soloveva8@yandex.ru

Abstract. The article describes various methods of increasing the effectiveness of communications in the chain "school - university - production" both at the tactical and at the strategic level. The main methods of solving the problem of shortage of production personnel in the sector of skilled labor are identified: participation of large enterprises in effective vocational guidance for adolescents, creation of the Center for Youth Innovative Technologies (TSMIT), implementation of a modular principle of training in cooperation with leading professional and higher education institutions, social networking resources, encouragement of employees for participation in career counseling communications.

1. Introduction

For all enterprises and especially for industries, communication as the art of providing clear, timely and relevant information has now become the key factor for survival in the current market situation. With proper management, internal production communication has a huge impact on the processes in the company, on its customer focus, on image, on customer loyalty and on the main factor is profit.

On the one hand, according to the results of the NRU HSE ISSEK survey of 90 large employing organizations (universities, research institutes, NGOs in 2016), a significant proportion of organizations (61%) currently experience a shortage of personnel and in none of them do not observe a surplus. Many researchers have noted a structural shortage of personnel: more than half of the respondents noted a shortage of personnel in certain specialties, and 36% declared insufficient skill level of workers.

A third of organizations-respondents declare insufficient inflow of young personnel and their weak training. On average, only half of young employees qualify for their positions.

On the other hand, annual sociological studies show that school graduates are poorly vocational. The absolute majority of graduates determine their future profession, relying on the "prestige" of work and not on the demands of the labor market and the interest of employers in it.

A graduate is guided by one or several motives in choosing a profession. Conventionally they can be divided as follows:

- ✓ material (the pursuit of high-paying jobs and benefits);
- ✓ moral (the desire to benefit humanity);
- ✓ social (the desire to occupy a worthy place in society);
- ✓ "prestige" (the desire to build a career);
- ✓ professional and cognitive (development in the profession);
- ✓ utilitarian (the desire to work in close proximity to home, in a clean office, etc.);
- ✓ creative (fantasy and creativity).

From the classification it is clear that in order to choose a profession that is related to industrial production, enterprises need to use certain methods of presenting information about themselves.

Vocational orientation of teenagers is, as a rule, a program that includes several techniques. Let's give some for an example.

"Map of interests of A. Golomshtok". [1] The purpose of this technique is to determine the structure of interests of the graduate. The full methodology consists of 144 questions and correlates the results in the following categories of knowledge: physics and mathematics; chemistry and biology; radio engineering and electronics; mechanics and design; geography and geology; literature and art; history and politics; pedagogy and medicine; entrepreneurship and home economics; sport and military affairs.

"Career anchors of E. Shane".[2] By responding to 41 statements, some career orientations are identified: professional competence; management; autonomy; stability; service; call; life style integration; entrepreneurship.

Differential diagnostic questionnaire of E. Klimov.[3], [4] The technique consists of 20 pairs of professions, of which a teenager chooses one in each pair. According to the test results, the scores correspond to five types of professions: "man-nature" - professions related to plant and animal husbandry, forestry; "Man-equipment" - technical specialties; "Person-to-person" professions built on communication and service for people; "Person-sign" - working with numbers and calculations; "Person-artistic image" – professions, which are based on creativity.

An objective picture of the vocational orientation of a teenager can be constructed by using several test methods, practical exercises and on the basis of an individual conversation with each subject.

However any testing gives little information if the senior student does not have a clue about what people of a particular profession do.

The purpose of this work is to analyze several interaction options in the "school-university-production" chain to increase the awareness of a graduate of any level when choosing a future profession.

2. State of the problem abroad.

Let us turn to the experience of solving personnel problems in some countries.

In Germany, the professional education system is based on the needs of business and government support. The training of working specialties in Germany is carried out by the Vocational Training and Skills Development Centers (ZAW Zentrum für Aus- und Weiterbildung), located in all major industrial regions of the country. Responsible for professional training is Chamber of Commerce (CCI). There are two main areas of industrial and technical training courses: training on the basis of industrial partnerships for enterprises and advanced training courses. Primary professional training is carried out in accordance with the protocol of professional training and production needs. In most cases, the student learns already from the specific enterprise that funds the training. Federal and regional authorities are actively stimulating enterprises. The formation of a set of training courses takes into account the orders of enterprises. Pupils are mainly students after the 10th grade (in Germany the thirteen-year study cycle). In Germany, the so-called dual training system operates. It involves two stages. At the first stage in an educational institution, a person will first master the theory, study the principles of work and labor law. The second stage will apply the knowledge gained in practice, and the focus is on practice. Training classes of such institutions are divided according to their functional characteristics and the level of equipment located in them. For example, from manual drawing of the simplest drawings to three-dimensional design on modern software. The development of working specialties by young people allows them to receive 1.500 Euros per month in a little over 20 years old and lead an independent life. Thus, the prestige of working professions is formed.

In the US, there is a Career Development course (career building), which has become a part of the educational system. It introduces the professions of children aged 5 to 14 years. Students learn their abilities and taking them into account will be able to plan a career. This program allows teenagers to try themselves in different professions during the summer holidays. It also pays great attention to the correct approach to the choice of profession and the ability to be ready to change their specialty. At the same time, a special monitoring of the student's activity is conducted. On its basis, recommendations on professional readiness are given and an individual self-realization program is proposed. In Japan children are required to "try" different professions. There career guidance begins with the seventh grade. Each year a student must test himself in 16 professions and in various areas: business, agriculture, and so on. As a result, during the school years, the graduate accumulates more than four dozen “samples” for all types of work.

3. Some ways to build communications in the system "school - university - production"

In the modern interpretation of the term communication has basically three interpretations: the medium; information transfer; exposure information. In all cases, the term communication is considered synonymous with the word connection.

Communicative event is a complex of communicative acts, united by a common task and situational conditions. Until recently, it was customary in the scientific literature to distinguish the domestic and industrial spheres of communication. However, in the conditions of wide spreading of professional communities on the Internet, household communications flow smoothly into professional ones. (Guskova, "Theory of Communications").

The sphere of industrial communication is in the field of view of the scientific interests of engineering psychology. The manager's style of communication with subordinates, employee interaction, office workflow and other aspects of production activity can both improve and ruin the work of the company.

From the point of view of the participation of industrial communications in the process of vocational guidance of schoolchildren, graduates, students, it is assumed that there is effective feedback.

All participants in the economic process are divided into large (FIG, concerns, consortia), medium, small enterprises and companies. Each of them has its own life in a particular sector, the path of development and degradation.

Large companies and companies with long-term development prospects can become participants in the school-university-production communication in the following way.

Firstly, to participate in the development of federal, regional, and regional sites of children's, youthful, and student scientific creativity (TSDOD, TSMIT, Kvantorium). On the basis of such sites it is effective to provide information about the enterprises of the high-tech manufacturing sector directly. Attracting talented and technically competent young professionals in this chain has a long-term perspective.

Secondly, to participate in the creation on the basis of specialized higher and secondary educational institutions of production departments and platforms for practices. In this case, the production division directly meets the alumni contingent and selects promising employees for its team.

Thirdly, to take an active part in job fairs, which are held on the basis of specialized educational institutions. The criterion of the effectiveness of an educational institution - the employment of graduates is a resource for production in communication with the university.

Fourth, use a promising social networking resource. Modern teenagers are the generation of gadgets, it is very important to teach them to properly use the possibilities of the Internet to choose their future profession.

Fifthly, to take an active part in the organization of dual training. Dual training is a type of vocational education, where the practical part of training takes place in the workplace, and the theoretical part is based on an educational organization. The system of dual training involves the joint financing of training programs for a specific workplace by commercial enterprises interested in qualified personnel and regional authorities interested in developing the economy and raising the standard of living in the region. Dual training is especially widely implemented in the system of secondary vocational education, focused on training workers and middle-level specialists. [5.].

In several regions of the country, a project was launched to train workers on the basis of dual education.

Large and medium-sized industries, on the one hand, and the state as a whole, on the other hand, should be interested in introducing production-oriented practices in schools with the encouragement of student labor.

IOP Conf. Series: Materials Science and Engineering **483** (2019) 012027 doi:10.1088/1757-899X/483/1/012027

Our country in the past also had a positive experience. There were specialized vocational schools (VS), factories, training programs at workplaces in enterprises, there was a system of mentoring and apprenticeship. The stimulus to work and career growth of a worker was the policy of admitting to the Communist Party of the Soviet Union (hereinafter referred to as the CPSU), workers and collective farmers had an advantage in joining the CPSU. They formed the country's elite, which gave a number of social benefits and advantages of professional growth. Planning a career for a young worker: trainee - apprentice - young worker - experienced worker - party experienced worker - mentor. In case of successful completion of the previous stages, subject to higher education, a CPSU member mentor, became a site master or foreman, then deputy shop manager or shop foreman, and positions of chief mechanics or chief engineers were available to the most successful and interested in professional growth. A retraining program for workers also existed in the form of industry refresher courses where the best workers were sent on the recommendation of several heads of a plant, a workshop or a site. The increase in the level of work was a direct consequence of the results of compulsory education. The entire career path of the Soviet worker was accompanied by continuous on-the-job training based on the strategic goals of the enterprise. The system of education and training in the USSR was created from the very beginning as a unified system for the improvement and development of workers. The collapse of the USSR and long-term economic transformations changed the status of workers. Formed groups working in municipal, state, joint-stock enterprises, as well as in private business. Such a variety of forms of ownership could not but lead to a total differentiation of wages depending on the sector of industrial production and, especially, to differences in working conditions. All these circumstances have caused a decline in the socio-professional status of workers and the attitude of society to work.

For companies of all sizes and most directions, such forms of career guidance communications as excursions to production are available. We should not forget about the factor of continuity of generations, when children walk in the footsteps of their parents. This is how the dynasties of professionals are formed. This is, of course, a small percentage of the renewal of personnel, but high professionalism.

In large companies with a long-term development perspective, a strategic approach to the planning of personnel renewal is necessary. In this sense, it is difficult to overestimate the participation in the organization of children's art sites for primary and secondary schoolchildren. These are external centers on the basis of children's clubs, their own "laboratories of young engineers", and the educational and entertainment sector (kinder malls, kinder - city, "cities of professions"). Business is ready to invest in career guidance for children. Companies help to create specialized playgrounds in the children's town so that children can try different professions. For example, in Tatarstan, for example, Tattelekom - partner of the Salon of Communications station, Unistroy - supervises the construction, Behetle created a mini-store, and Ak Bars Bank organized a children's bank and trains young visitors to financial literacy. In Volgograd, the City of Professions operates on the basis of the Kinder Mall shopping center. Here children can get acquainted with the professions of a doctor, photographer, banker, actor, hairdresser and fireman. It is too little. Russian President Vladimir Putin instructed to develop a national directory of professions in demand in the labor market, as well as a list of the most sought-after professions in the labor market and specialties requiring secondary vocational education, highlighting the 50 most promising and sought-after professions and specialties. This handbook includes information and communication technologies, construction, maintenance of transport and logistics, automation, radio engineering and electronics, services, engineering. As a region with a huge value of traffic intersection, the Volgograd region is interested in the influx of specialists in this industry and in the training of personnel for the future.

Small and medium-sized enterprises and companies interested in the influx of personnel should encourage employees to provide vocational guidance on the principle of mini-excursions of children to the workplace. To do this, a group of children (classmates of a son-daughter) is organized and, according to an agreed schedule, an event is held "find out what mom or dad are doing at work and try to replace parents in the workplace". This format is effective in that children in a playful way get

IOP Conf. Series: Materials Science and Engineering **483** (2019) 012027 doi:10.1088/1757-899X/483/1/012027

acquainted with the main professions; get a basic understanding of the diversity of manifestations in the profession.

4.Conclusion

In our country at the moment there is already a shortage of personnel in high-tech industries. If the policy of the state and large companies does not change, this situation will only get worse. Currently a well-thought-out strategy in the development of personnel policy is needed.

References

[1] Golomshtok A E 1979 *Career Choice and Personal Education of a Student* (Moscow: Prosveshcheniye) p 235

[2] Chicker V A 2006 *Psychological Diagnosis of the Organization and Staff* (Saint-Petersburg: Rech)

[3] Klimov E A 1993 *Person Developing in the World of Professions* (Obninsk: Printer, Moscow State University) p 57

[4] Klimov E A 1993 *Person Developing in the World of Professions* (Obninsk: Printer) p 340

[5] Kutumova A A and Yarkova G A 2016 Dual Education System as a Methodology for Baccalaureate of Professional Education in the Modern Realia of the Pedagogical University *Modern Science-Based Technologies* **4-1** 139-142

IOP Conf. Series: Materials Science and Engineering **483** (2019) 012028 doi:10.1088/1757-899X/483/1/012028

Effect of cross-cultural differences on industrial communication

Elena Eltanskaya, Yulia Kulichenko, Lyudmila Medvedeva

Volgograd State University, 100, Prospect Universitetsky, Volgograd, 400062, Russia

yeltanskaya@volsu.ru, kulichenko@volsu.ru, milamed@volsu.ru

Abstract. This research deals with the important circumstances influencing cross-cultural communication and disagreements arising within it. Current social changes affect the world we live in and as a result the relationships between countries and individuals. Globalisation, economic integration and labour migration have a significant impact on industrial communication. The successful operation and development of an enterprise are heavily dependent on effective communication between co-workers who often come from different cultural background nowadays. The specificity of this type of communication is described in the article. The special attention is paid to the possible problems in cross-cultural interaction. The chain of causes and effects leading to miscommunication is determined. The article also provides analysis of the barriers to effective communication and describes the problems caused by them. The importance of cross-cultural competence is emphasised and the principles of effective management as a way of solving cross-cultural conflicts in communication are characterized.

1. Introduction

The problems of intercultural communication and tolerance are very important nowadays due to the current social and economic situation characterized by the dynamics and complexity. Globalisation has made cross-cultural communication inevitable. Business, education, industry, trade and other spheres of social activity are characterised by communication with different cultures. Globalisation has changed the context in which cross-cultural communication takes place [1]. If earlier people had to communicate with individuals or even groups of other cultural origin mostly while travelling, nowadays cross-cultural communication is a part of our everyday activity at the university or at work. Many developed countries have become multi-cultural societies that leads to different cultural issues in the workplace. "Worldwide intercultural cooperation drives corporate growth and development across the globe resulting in a heightened demand for a qualified but diverse workforce" [2]. Due to the complex, diverse and mobile workforce, nowadays the need for cultural awareness and cross-cultural communication is greater than ever [3].

Economic integration and advances in transportation and telecommunication have broken down geographical isolation. The world is more global and mobile than ever before. Globalisation is linked by new technology, communication is, for example, intensified via electronic media facilitating trade contacts and international projects.

When companies attempt to expand their activity abroad there is always the risk associated with the work in unfamiliar business environment. If managers and employees see cultural diversity as something that can be understood and learnt, they try to develop and improve their cultural intelligence and competence as part of their communication skills, their companies succeed.

In the modern world cultural intelligence is a critical competence [4]. This concept was formulated in order to ease and understand multicultural interactions in business organizations. Cross-cultural intelligence assists adaptive cultural adjustment. Due to increased number of cultural contacts and clashes; it has attracted academic interest [5]. Cultural intelligence leads to an enhanced and successful experience while a lack of it causes challenges [6]. Intercultural competence implies knowledge of traditions, customs, social norms, specificity of communication and non-verbal behaviour as well as the relevant skills.

As a result of ongoing integration in the world and regional economy, many companies are increasingly becoming aware of the potential that a multinational market gives to businesses. They can operate worldwide and set up subsidiaries, branches and offices abroad that leads to new markets and employment of cheaper workforce. This provides opportunities for the expansion of the business and brings about a substantial increase of organizations. "However, there are many serious conflicts which may arise owing to the fact that there may be entirely different ways to run the companies and different opinions on how the employees should handle their work, especially the barriers between internal staff, hence cross-cultural communication conflicts" [7].

Countless immigrants and refugees leave their familiar surroundings and move to the new countries. They have to adapt to a new cultural environment, establish and maintain a stable working relationship with new colleagues.

All kinds of co-worker relations are associated with psychological well-being of employees and their job satisfaction. The development of positive cross-cultural communication is essential for host country national employees as well as immigrants [8].

2. Problems in cross-cultural communication

Effective business communication depends on a lot of factors that have to be considered.

2.1 Communication

Communication is a type of human activity that implies knowledge sharing, mutual influence and interaction between partners. It occurs when a sender has an idea, a feeling or an emotion that he or she intends to communicate to another person (a receiver), then he or she expresses it in the form of message (either verbal or non-verbal). When the receiver gets this message, he or she usually gives feedback consciously or unconsciously. It is a return message that is of no less significance than the previous one. Thus, communication is an ongoing process.

A person cooperating with other people to ensure one's life activity is the subject of communication. The system of human relationships is mediated by culture, which determines the nature and effectiveness of human communication. But in the course of various types of human activity, there is a need for more advanced and effective forms of communication, which gives rise to different roles and significance of certain forms of communication. In other words, depending on various reasons, various forms of communication acquire an appropriate purpose in the life activity of people, that is, a function. Functional analysis of communication clarifies the social role that it performs in society, and helps to understand its essence more accurately.

2.2 Cross-cultural communication

Cross-cultural communication is characterized by different features. First of all, the sender and the receiver of the message have to belong to different cultures. The participants of the communication should be aware of the cultural differences of each other. The realization and attitude towards these differences affect the type, form and result of the contact. Each participant of the cross-cultural communication has his or her own set of rules both verbal and non-verbal. Besides cultural differences there are other factors that affect the process of interpretation such as age, gender, profession, social status of an individual. That is why the act of communication depends on tolerance, character and personal experience of participants.

2.3 Non-verbal communication

Non-verbal communication plays a key role in face-to-face interactions [9]. Body position, gestures and eye contact are very important part of non-verbal communication affecting the business and working relationships. It is necessary to take into consideration not only social norms and cultural traditions but also the role of body language in intercultural communication. For instance, prolonged eye contact can symbolize honesty and confidence in Europe and the USA but at the same time it can be considered rude in some Asian countries. The differences in the interpretation of body position, gestures and handshakes are numerous and affect the business communication directly. The same is also true about the working relationship in the same company. The managers must train their employees in the proper way to be aware of these cultural differences in order to maintain effective communication with colleagues and clients.

2.4 Possible problems in industrial communication

The sequence of possible problems in cross-cultural communication can be presented in the following way: lack of knowledge → communication barrier → misunderstanding → impossibility to achieve the goal of communication → stress → generalisation of the problem → biased perception of communicative situation → conflicts or avoidance of further direct contacts.

General knowledge of language, culture, customs, traditions, social norms and typical nonverbal behaviour of people from other countries is essential for effective cross-cultural communication and lack of it causes a lot of problems.

There are different types of communication barriers such as physical, semantic or individual ones. They are caused by a variety of sources. Physical barriers like walls, doors, noise or distance are determined by external factors.

Language can cause semantic barriers even if it is native for both of speakers. The use of unknown terms, slang, jargon, ambiguous words can lead to misunderstanding. In cross-cultural communication this aspect is especially important.

Barriers created by the sender of the message or the receiver are usually called individual barriers. They include the differences in personality and perception. In the case of cross-cultural communication this type of barrier is often caused by cultural differences.

Misunderstanding is unavoidable element of communication [10]. Due to misunderstanding, i.e. inability to understand the message of the sender at all or incorrect interpretation of it, it gets impossible to achieve the goal of communication that causes a large amount of stress to interlocutors. As a result, the whole communicative situation is perceived in a wrong, biased way that negatively affects the attitude towards the speaker and the subject of the conversation. All these factors cause a lot of problems at work and can result in avoidance of any further direct contacts with people of different origin. If contacts are inevitable this type of miscommunication can lead to numerous conflicts.

3. Principles of effective cross-cultural communication in business environment

The interaction between the participants of interethnic communication is based on the certain principles. The empathic approach that implies the awareness of cultural differences is dominant. Empathy that seeks to understand the problems of cross-cultural communication and solutions to them increases the ability to understand the interlocutor and as the result the effectiveness of communication.

The main principles of cross-cultural communication are:
- to treat your colleagues with respect;
- to listen attentively;
- to attempt to understand the way your colleagues feel;
- to have sincere interest in cooperation with co-workers;
- to have ability to understand opposite point of view;
- to expand knowledge of different cultures and traditions;
- to avoid stereotyping;
- to enhance cross-cultural competence;
- to increase tolerance and empathy.

The knowledge of similarities and differences between speech acts in different languages contributes to successful contacts and prevents misinterpretation and misunderstanding [11].

4. The culture of inter-ethnic communication

It is clear that people of different cultural origins operate within such group and not all of its members are able to perceive the cultural values of different nations correctly (as they are sometimes incomprehensible and unaccustomed), but co-workers should be able to show tolerance and tact.

A person's attitude to a particular nationality largely depends on the personal experience of communicating with its representatives, with friends and acquaintances. Such experience can be not only positive, but also negative: narrowness of assessments, absolutization of individual traits depending on personal sympathy and antipathy, etc. are possible. This may result in an unfavorable atmosphere with a national tone in the team.

The moral climate in public organizations relating to organic culture is determined by the range of social and moral values shared by their members; public ones belonging to bureaucratic organizational culture are governed by officially adopted laws and codes for public servants to follow. In commercial organizations, goals are achieved by creating a highly sustainable motivation of employees - members of the organization, including moral motivation.

Manufacturing, energy, mining, construction, wholesale and retail trade, hotels, restaurants, social work and healthcare are characterized by the high level foreign workforce employment. Migrants are often employed in low-paid and hazardous jobs that do not require any special skills.

The increasing number of migrant workers in the labour market and the establishment of multinational or even global firms make employment one of the areas in which the importance and impact of cultural diversity increases considerably at both the national and the organization level.

As a result of economic globalisation, cross-cultural differences are becoming more and more relevant in businesses that expand into other countries and appoint managers of different nationalities.

The specific character of a culture of inter-ethnic communication is determined primarily by the value orientations of society, which are relatively widespread in the ethnic community. The implementation of national norms of behavior, traditionally determining the style, level and nature of relations with representatives of other nations is a part of the interpersonal culture of communication with representatives of various nationalities.

Behavioral characteristics of an individual in inter-ethnic communication include friendliness, the willingness and ability to make contacts with persons of other ethnic groups and maintain friendly relations with them, taking into account the ethnic specificity of people who come into contact, well-bred behavior towards people of other nationalities in any, even conflict situations.

The culture of inter-ethnic communication is a person's respectful attitude towards people of different nations, their culture, traditions, languages, history, and national dignity, which implies the fulfillment of legal, moral and ethical standards in a multinational environment by people. The culture of inter-ethnic communication includes moral and aesthetic literacy, which helps individuals to understand the psychological state of people involved in an ethno-conflict situation and to determine ways of solving the problem in a rational way. It also includes knowledge of human rights, nations and national relations, different cultures and their mutual influence, national, social, universal and interfaith conflicts and their causes.

Thus, the culture of inter-ethnic communication acts as a system of ideas and beliefs, ways, forms and types of specific cultural activities distinctive for an individual, nation and social groups or society as a whole carried out in order to deepen the process of interaction and interference of people of different nationalities.

The culture of inter-ethnic communication is specific interrelation and intercommunication in which people of different nationalities exchange experiences, values, thoughts, feelings, experiences. Not only personal qualities, but also features of the psychology, traditions, and culture of the peoples, whose representatives are communicating, show up in the process of inter-ethnic communication.

The culture of inter-ethnic communication is a part of intercultural communication, which should be considered as a combination of various forms of relations and communication between individuals and groups belonging to different cultures. The goal of intercultural communication is to enrich every national culture while preserving its unique character.

Within each sphere, intercultural communication takes place at different levels. Inter-ethnic communication is defined as intercultural communication at the micro level. It is communication between persons representing different nations (ethnic groups) as typically a society consists of various ethnic groups that create and share their subcultures.

Communicative influence as a result of the exchange of information is possible only when the sender (the person directing the information) and the receiver have a single or similar codification and decoding system.

5. Conclusion

Economic integration due to the development of technology, transportation, trade and telecommunication has led to many significant changes and broken down geographical isolation. Nowadays the world is more global and dynamic that is has ever been before and the importance of cross-cultural communication is vital.

Awareness of cultural peculiarities of co-workers and following the main principles of effective cross-cultural communication help people avoid problems arising in cooperation which may result in bad performance of any enterprise.

Effect of cross-cultural differences on effective industrial communication is immense as they have an impact on co-worker's relations, job satisfaction and even the result of the work. If the company wants to succeed, its managers and employees should broaden their minds to be ready for acquaintance with new cultures and traditions.

As a result of the presented study the following recommendations would be useful for businesses:

1. Enterprises and businesses should give opportunities for their employees of all levels to go abroad to be engaged in cross-cultural communication and to obtain the essential knowledge, skills and directions for managing the cultural diversity. Obtained language training and cross-cultural knowledge will result in improvement of intercultural competence.

2. A new enterprise culture should be established on the basis of mutual understanding of culture and the requirements of the enterprises' development strategy and mutual benefit policy should be enforced for the mutual benefit of both parties.

3. The idea of a cooperative team formation should be realized in enterprises and sponsor cultural programmes should be organized in their host countries periodically.

References

[1] Yoshida T, Yashiro K and Suzuki Y 2013 Intercultural communication skills: What Japanese business today need *International Journal of Intercultural Relations* **37** 72-85

[2] Congden S W, Matveev A V and Desplaces D E 2009 Cross-cultural Communication and Multicultural Team Performance: A German and American Comparison *Journal of Comparative International Management* **12** 73-89

[3] Bortnikova T G 2014 Student Mobility and Problems of Intercultural Communication Development *Bulletin of TSTU* **20** 407-413

[4] Iskhakova M 2018 Does Cross-cultural Competence Matter when Going Global: Cultural Intelligence and Its Impact on the Performance of International Students in Australia *Journal of Intercultural Communication Research* **47** 121-140

[5] Sharma N and Hussain D 2017 Current Status and Future Directions for Cultural Intelligence *Journal of Intercultural Communication Research* **46** 96-110

[6] Qi H, Smith K A and Yeoman I 2018 Cross-cultural event volunteering: Challenge and intelligence *Tourism Management* **69** 596-604

[7] Ike Nnia Mba Sr 2015 Conflicts Encountered by Multinational Corporations in Cross-Cultural Communication and its Solutions *Journal of International Business and Economics* **3** 86-92
[8] Bergbom B and Kinnunen U 2014 Immigrants and host nationals at work: Associations of co-worker relations with employee well-being *International Journal of Intercultural Relations* **43** 165-176
[9] Vargas-Urpi M 2013 Coping with Nonverbal Communication in Public Service Interpreting with Chinese Immigrants *Journal of Intercultural Communication Research* **42** 340-360
[10] Edwards R, Bybee B T, Frost J K, Harvey A J and Navarro M 2017 That's Not What I Meant: How Misunderstanding Is Related to Channel and Perspective-Taking *Journal of Language and Social Psychology* **36** 188-210
[11] Farsian M R and Fesanghari A 2017 Exploring the cultural variables in the communications between the Iranian and the French based on language patterns *Language Related Research* **7** 173-195

Best available technique in the environmental management system of ceramic enterprises

L M Generalova, E A Eltanskaya, L M Rebrina

Volgograd State University, 400062, Universitetskiy av., Volgograd, Russia

L.M.generalowa@volsu.ru, yeltanskaya@volsu.ru, LNrebrina@volsu.ru

Abstract. The article discusses the advantages of introducing the Best Available Technique (BAT) as a set of various processes, methods and other ways to protect the environment in the ceramic industry. The article touches upon the efficiency of referring tools and methods of system of ecological management to BAT as one of the most important directions of ensuring ecological safety. The work defines that ecological safety is extremely important for development and activity of any enterprise as for profitability of production and competitiveness of the enterprise it is necessary to prove not only profitability and efficiency of its work, but also ecological safety of manufactured production. The introduction of the Best Available Technique into production is a highly effective method of increasing the level of environmental safety. The research analyzes the theoretical foundations of the concept of the BAT available in domestic and foreign sources, existing technological processes, equipment, technical methods, as well as other methods of environmental protection referred to the Best Available Technique in the ceramic industry. The article states that introducing the Best Available Technique in production involves an overall system of environmental protection measures, which achieve a significant minimization of the negative impact on the environment and population.

1. Introduction

In the modern world, environmental protection is one of the most urgent and global problems. The adverse impact of human activities on the natural environment has been growing at an alarming rate every year and in some cases, it has reached global proportions and is comparable to large-scale technogenic disasters. One of the most important indicators of national security in modern society is environmental security, which provides a state of protection of vital human interests, society and the environment from the possible negative impact of economic and other anthropogenic activities.

Since modern man is not ready today to abandon the products of engineering and technical design, the scientific and technological progress should be civilized and different from all previous technogenic transformations. Due to the constantly deteriorating environmental situation it is necessary to improve the legal regulation in the field of environmental safety at the government level. Environmental control and environmental monitoring are important parts of the organizational and legal mechanism, the purpose of which is to verify compliance with the requirements of environmental legislation, compliance with standards and regulations and environmental safety.

2. Results and Discussion

2.1. Development of the concept of BAT in the ceramic industry

At the end of the last century there was a change in the dominant environmental values and relations associated with the global environmental crisis as well as with the awareness of society of the urgent need to limit the negative impact of large industrial and agricultural enterprises on the environment. The change in the environmental paradigm has resulted in the creation of the concept of BAT (Best Available Technique), which has contributed to limiting the consumption of natural resources in order to preserve them for future generations.

IOP Conf. Series: Materials Science and Engineering **483** (2019) 012029 doi:10.1088/1757-899X/483/1/012029

Under the term of Best Available Technique we mean the most effective latest developments for various activities, processes and modes of operation indicating the practical feasibility of using specific technologies as a basis for established permits for emissions/discharges (pollutants) into the environment in order to prevent pollution. Otherwise when prevention is impossible, it is necessary to minimize emissions/discharges into the environment as a whole" [8]. This concept was widely used in Europe, the USA and the USSR.

Legislative environmental regulation of enterprises based on the concept of BAT was defined in the European Union in 1996 by Directive 96/61/EC «Integrated pollution prevention and control». This Directive addressed the relationship between industrial technologies and their impact on the environment. It was later replaced by Directive 2008/1/EC on integrated pollution prevention and control and subsequently by Directive 2010/75/EC «On industrial emissions and or discharges (integrated pollution prevention and control)» [8, 7]. International experience in the implementation of BAT in industrial production, considered on the base of EU Directives, was subsequently the basis for the creation and development of the concept of BAT in Russia [6].

As part of the introduction of BAT in the industrial production of ceramic products, the European Union Handbook on the Best Available Technique «production of ceramic products» was created [9]. According to this Handbook the main issues in the implementation of Directive 2010/75/EC are the following: reduction of emissions/discharges of pollutants, energy efficiency, resource conservation, minimization, disposal and recycling of waste and effective management systems [6, 8].

In 2014 by the Act of Federal Agency on Technical Regulation and Metrology from December 3-d, 2014 №192 in Russia there was the Russian Bureau of BAT created, the main activity of which was the development of information and technical guides on the Best Available Technique. Development of handbooks is carried out by leading technical working groups based on European experience as well as on the base of the analysis of resource efficiency and environmental performance of enterprises. At the same time the first domestic handbook on BAT ETS 4 is being developed and implemented in production taking into account the peculiarities of producing ceramic products in the Russian Federation. The Handbook contains a description of the processes used in the production of ceramic products, equipment, technical methods, methods of reducing a negative impact on the environment and improving energy efficiency and methods allowing to ensure resource conservation in the ceramic industry [5]. Nowadays, there are more than 30 information technology handbooks on the Best Available Technique in the Russian Federation.

2.2. Environmental safety in the ceramic industry as an object of scientific research

Regarding the above statistics it is obvious that in recent years the environmental safety in the ceramic industry has been of genuine interest to domestic researchers. Close attention is paid to the disposal of waste products of ceramic production, especially ceramic bricks and tiles in the scientific research works of E. M. Averochkina, Ya.P. Molchanova, T.V. Guseva, M.V. Begak, A.V. Mironova and others [1,2,3]. Scientists consider in detail not only the problem of processing waste ceramic production, but also the prospects for the introduction of BAT reference books in the production of ceramic products as well as peculiarities of implementing the concept of the Best Available Technique in the enterprises of the Russian Federation. The problem of implementation of these technologies is discussed not only at the stage of manufacturing final products, but also at the stage of design, maintenance, creation and operation of the enterprise.

Scientists point out the fact that the technologies used must necessarily be developed on a scale that allows them to be implemented in the relevant industry in an economically and technically feasible way taking into account the relevant costs and benefits, in order to effectively achieve the necessary level of environmental protection in the whole.

The aim of these developments is to assist local authorities in implementing the best environmental technologies into the waste management system in order to minimize the negative impact on the environment.

Currently, such tools of environmental safety as best practices/means (Best Practical Means) and best practical environmental solutions/options (Best Practical Environmental Options) are widely used providing the search for methods to achieve a high level of environmental protection in the whole.

Russian scientists and specialists by the term «Best Available Technique» mean, first, best practices, methods and environmental regulations. To solve specific problems an extensive database including various technical proposals for environmental safety is formed, as well as practices for the management of processing production waste.

Besides, it was found out, that despite the active development of the process of modernization of the ceramic industry in the Russian Federation and the development and implementation of BAT in various sub-sectors of the production of ceramic products, this process is rather complicated. It is stipulated by the fact that a significant lack of criteria for categorization of objects having a negative impact on the environment, according to many environmentalists, is explained by the focus on the capacity of enterprises, and not on the negative impact in the latest research.

In addition, a clear drawback of the system of environmental safety at many enterprises of the ceramic industry is the lack of environmental services, environmental laboratories and the presence of an environmental engineer. Some of the enterprises are not willing to implement the Best Available Technique, as this only leads to improving environmental performance, but not economic, as these activities are often considered economically impracticable.

2.3. Environmental management system as part of the enterprise management system

In the Russian Federation approved general methods of environmental protection for all industries of ceramic products are the methods and means of environmental management system, which are necessary for the development and implementation of environmental policy and management of environmental aspects of the enterprise management system.

Environmental aspect is a key concept of EMS allowing to correlate the enterprise activities and its interaction with the environment, which includes the following priority environmental aspects:

- emissions of pollutants into the atmosphere;
- discharges of pollutants into water bodies;
- waste formation;
- consumption of energy, raw materials [4, 5].

The key principles of EMS are pollution prevention and consistent improvement of the environment, as well as the periodic process of improving the environmental management system in order to improve the overall environmental performance consistent with the environmental policy of the organization. This process is implemented by setting environmental goals and objectives, allocating resources and distributing liability for their achievement and implementation, in particular, the development and implementation of environmental management programs.

The development of environmental management programs implies improving the practice of industrial environmental monitoring and control, including the selection, justification and organization of measurements of key parameters, because for the formulation and verification of the tasks of the EMS it is necessary to provide a system for assessing performance indicators.

The effectiveness of EMS is ensured through the development, implementation and compliance with the basic procedures necessary for the management of environmental aspects, which may include working criteria for the normal implementation of the stage, action in case of deviation from the norm or criteria for the selection of subsequent stages.

The most important component of the economic efficiency of environmental activities, enhancing the implementation of EMS, is the principle of pollution prevention. This principle is the most effective, and sometimes the only possible, as it is the prevention of negative impact on the environment by affecting the processes (the root cause) of its cause.

The above methods of pollution prevention are often efficient and cost-effective. First, it concerns organizational approaches related to the selection of raw materials, production process control, recycling or processing materials, production logistics, etc.

2.4. Applying BAT in the ceramic industry

Let us consider in more detail the technological and technical solutions applicable as BAT for the industry of ceramic products in the whole.

In the field of energy consumption in the production of ceramic products improving the energy efficiency of production as a whole, as well as reducing the specific emissions of pollutants entering the air due to fuel combustion allows to reduce specific energy consumption [10]. Reduction of fuel consumption in the production of ceramic products is achieved through the introduction of BAT, which is a reduction in energy consumption using a combination of the following technological solutions and techniques:

- modernization of furnaces and dryers;
- recovery of excessive heat of furnaces, especially from the cooling zone. In this case, the excess heat of the furnace in the form of hot air can be directed to the heating of dryers for drying raw materials or semi-finished products;
- optimization of the shape, size, composition and structure of the blanks significantly increases the energy efficiency of drying and firing with the use of appropriate dryers and furnaces.

In the field of reducing emissions of pollutants into the air, reducing dust emissions in the production of ceramic products is achieved through the introduction of BAT-reduction of unorganized dust emissions using a set of technical solutions [5]:

- carrying out technological operations, accompanied by the formation of dust in a closed volume;
- equipment of mixers, crushers and feeders with protective covers and exhaust installations;
- application of storage bins of the corresponding capacity, level sensors with cut-offs and filters for cleaning of the dusty air displaced at filling of the bunker;
- remove of the dusting material using closed conveyors;
- reduction of leakages and elimination of their sources, sealing of installations;
- for bulk storage areas of raw materials:
- fencing sites bulk storage of raw materials using screens, walls or large green spaces;
- equalization of the level of unloading of raw materials with the height of the receiving area;
- reduction of the speed of unloading raw materials;
- humidification with spray systems source dusting.

In the field of reducing emissions of pollutants with waste gases during the firing of ceramic products the technology of BAT is to reduce emissions of pollutants (carbon monoxide, nitrogen oxides, sulfur, volatile organic compounds, etc.), formed during the firing of ceramic products, through the use of a set of technical solutions [5]:

- optimization of the fuel combustion process to reduce emissions of carbon monoxide and nitrogen oxides, control of the excess air ratio for the completeness of fuel combustion;
- reduction of the maximum firing temperature and its duration;
- optimization of the firing mode by regulating the rate of temperature rise in intervals of up to 400 °C (possible reduction in the rate for complete binding of fluorine and sulfur) and after 400 °C (possible increase in the rate to reduce emissions);
- replacing technological ligaments and components of the raw materials that form during firing VOCs (volatile organic compounds) on the ligaments and the parts that make up fewer VOCs or do not form them at all;
- reducing the supply of sources of pollutants, namely: the use of raw materials with a low content of sulfur and nitrogen, fluorine, chlorine or reducing the amount of raw materials with their high content.

In the field of reducing the formation of wastewater the BAT of reducing the amount of industrial wastewater in the production of ceramic products is used [11]. In this case the BAT is the reduction of water consumption and the reduction of pollutants in wastewater through the use of a set of technical solutions [3]:

- reduction of water consumption through the application of measures to optimize the process, including:

- prevention of water leakage throughout the water supply system by installing automatic valves and meters;
- installation of systems for collecting glaze and waste water from various stages of the technological process in the places of their formation;
- reuse of water, in particular repeated use of water, after appropriate treatment;
- use of wastewater treatment systems (if necessary, depending on the characteristics of the sub-sector).

For industrial wastewater returning to the mass treatment process, treatment may not be required (except in cases of high content of soluble salts in water). For industrial wastewater used for washing, additional preparation is required.

In the field of re-use of sludge is used the BAT of re-using of sludge in the production of ceramic products. The BAT is the recycling (re-use) of sludge through the application of one or more technological solutions [5]:

- re-use of sludge in the production process by adding to the molding mixture, mass or slicker with pre-treatment (averaging, separation, aging, drying, etc.) in accordance with the technological capabilities of production and formulation [12].

In the field of minimization of production waste, the following BAT is used – minimization of production waste and technological losses. The BAT is the reduction of solid waste generation through the use of a combination of the following technological solutions [5]:

- return of technological wastes (dust, scraps, battle, baked products) in production;
- use of solid waste gypsum molds and ammunition for other industries (gypsum-for the cement industry) or their own production (combat ammunition-for its production);
- replacement of plaster molds with plastic with greater turnover;
- decrease of the ratio of one pipe/fired product by applying nonsagger firing, fast firing in roller kilns and optimization of cages products;
- optimization of firing conditions, leading to the decline of marriage goods and the time of operation of kiln furniture.

In the field of noise reduction the following BAT is applied – reducing the noise impact of the production of ceramic products. The BAT is the reduction of noise by applying one or more of the following technological solutions [5]:

- sealing and proper maintenance of equipment that is a source of increased noise;
- vibration compaction of equipment, especially equipment for grinding and pressing;
- use of sound insulation of fans and blowers;
- application of low-speed fans;
- sound insulation of windows and doors;
- use of soundproof building materials in the construction and repair of industrial premises;
- carrying out noisy work only in the daytime.

3. Conclusion

Thus, it can be concluded that the Russian Federation is actively developing the process of modernization of the ceramic industry, as well as the development and implementation of the Best Available Technique in various sub-sectors of the production of ceramic products [13]. At the same time, it should be noted that it is permissible to classify not only technological processes, equipment, technical methods, but also other methods of environmental protection as the Best Available Technique. In this regard, quite effective are such tools and methods of environmental management system as BAT. Nowadays various Best Available Technique have already been developed, applicable for the entire industry of ceramic products in General, and for individual sub-sectors.

The optimal implementation of the environmental management system as BAT in the ceramic industry will allow to control the causes and conditions of the impact of environmental aspects, which will lead to the solution of the problem of preventing and (or) reducing the negative impact of the production organization on the environment, that is, to ensure its protection.

Practice shows that the environmental management system plays the role of a starting mechanism in the production, in which the search and application of pollution prevention approaches takes a regular and systematic character, and organizational and management decisions are implemented most successfully.

References

[1] Averochkin E M 2014 Instruments of environmental regulation of enterprises for the production of ceramic goods (as exemplified by Best Available Technique: abstract (Dissertation of Candidate of Technical Sciences) 43 p

[2] Ergunova O T, Lizunkov V G, Malushko E Yu, Marchuk V I, Ignatenko A Yu 2017 Forming the system of strategic innovation management at the high-tech engineering enterprises *IOP Conference Series: Materials Science and Engineering* **177 (1)** 012046

[3] Ergunova O, Lizunkov V, Blaginin V, Politsinskaya E, Shaykina O I 2018 Formation of the price mechanism for energy resources in Russia and the countries of the European union *International Journal of Energy Economics and Policy* **8 (1)** 122-127

[4] Fedotov N G, Moiseev A V, Syemov A A, Lizunkov V G, Kindaev A Y 2017 New Methods of Three-Dimensional Images Recognition Based on Stochastic Geometry and Functional Analysis *IOP Conference Series: Materials Science and Engineering* **177 (1)** 012047

[5] Guseva T V, Molchanova Ya P, Averochkin E M, Vartanian M A 2014 Best Available Technique: prospects for proliferation in the chemical industry in Russia *VI international conference of Russian chemical society named after D. I. Mendeleyev* 21-22

[6] Guseva T V, Begak M V, Molchanova Ya P, Averochkin E M, Vartanian M A 2014 Prospects of implementation of Best Available Technique and transition to integrated environmental permissions in the production of glass and ceramics *Glass and ceramics* **7** 26-36

[7] Guseva T V, Molchanova Ya P, Mironov A V, A V Malkov 2015 BAT: a new ecological measure of quality in the industry of construction materials *Competence* **8 (129)** 4-7

[8] Information and technical guide to the Best Available Technique: production of ceramic goods 2015 (Moscow: Bureau of BAT) 235 p

[9] Begak M V 2010 Best Available Technique and integrated environmental solutions: prospects of application in Russia (M.: LLC «YurInfoR-Press») 220 p

[10] Directive 96/61/EC of 24 September 1996. Concerning integrated pollution prevention and control Official Journal of the European Union, 10.10.1996 p 26

[11] Directive 2010/75/EU of the European Parliament and of the Council of 24 November 2010 on industrial emissions (integrated pollution prevention and control) *Official Journal of the European Union*

[12] Lizunkov V, Politsinskaya E, Malushko E, Kindaev A, Minin M 2018 Population of the world and regions as the principal energy consumer *International journal of energy economics and policy* **8 (3)** 250-257

[13] Reference Document on Best Available Technique in Ceramic Manufacturing Industry. The European IPPC Bureau 2007

IOP Conf. Series: Materials Science and Engineering **483** (2019) 012030 doi:10.1088/1757-899X/483/1/012030

Determination of the actual zone of influence of an industrial enteprise on the basis of the quality assessment of the environmental components

A A Tihonova[1], V V Yanina[1], E A Eltanskaya[1]

[1]Volgograd State Technical University, 28 , Lenina avenue, the city of Volgograd, 400005, Russia

tihonova@volsu.ru, yanina230172@volsu.ru, : yeltanskaya@volsu.ru

Abstract. In this work, we determined the actual concentrations of heavy metals in the soil cover within the assumed zone of influence of a specific local source of anthropogenic impact. The joint-stock company Volgograd Metallurgical Plant (VMP) "Red October", located in close proximity to the residential zone, was selected as an object of environmental impact. Sampling was carried out at control points at a distance of 0-4.5 km from the enterprise. The study was conducted using the precision method of atomic absorption spectrometry with preliminary preparation of samples for analysis by the method of acid decomposition of solid soil samples in open vessels. The presented measurement results, for example, concentration of labile forms of Cu, Mn, Zn in the soil cover indicate the possibility of assessing the actual zone of influence of the industrial object under study, which is relevant in determining the quality of the urban environment.

1. Introduction

Currently, the problem of reliable assessment of the current state and quality of the urban environment, taking into account local man-made pollution of terrestrial ecosystems located in close proximity to industrial enterprises in the city, has become particularly relevant [9]. One of the criteria reflecting the quality of the environment and the level of ecological well-being of residents in the area is assessment of the actual state of the soil cover of the territory and, in particular, determination of degree of contamination of various kinds of toxic elements (including heavy metals).

The real connection between the systems of chemical elements of each component of the biosphere is provided by various groups of chemical compounds specific for each of the natural environments. Chemical pollution complicates this interaction, and this fact determines the increasing attention to the chemical composition of the soil.

Urban soils, in contrast to the soils of background areas, function, among other things, in altered water and temperature conditions inherent in urban ecosystems, as well as with an increased flow of dust, the composition of which is determined by the flow of vehicles, fuel and industrial production developed in the city. Pollutant fluxes that fall into the soils of urban areas along with precipitated atmospheric emissions accumulate, some of them can be kept in the soil for tens and hundreds of years, resulting in a significant impact on soil formation processes and the evolution of urban soils [15].

So, in the soils of cities, where enterprises of the metallurgical industry are located, there is an increased concentration and activity of heavy metals. Hg, Pb, Cd, Zn, As are among the priority pollutants, since the accumulation of these elements in the environment goes at the highest rates, which, in turn, causes an increase in their concentration in plants and a further transition through trophic chains [10, 13].

With long-term intake from stationary sources, concentration of heavy metals in soils is comparable to their number in natural geochemical anomalies or even exceeds it and, thus, so-called "zones of influence" with a high concentration of metals in soil are formed around large enterprises.

Pollution can be traced at a distance of up to 10-12 km from the source of metal-containing emissions. At the same time, the degree of contamination of individual areas of the zone of influence varies considerably: the soils in the immediate vicinity of the plant contain more metals than the soils removed, and the areas that are under the emission plume are the most polluted. As the distance to the source of pollution increases, the migration capacity of metals also increases due to the dominance of vapor-gas and finely dispersed forms and greater solubility of compounds [3].

Heavy metals, entering the atmosphere with emissions from a metallurgical plant, are deposited and fall on the soil cover, where they are firmly retained due to biogeochemical barriers. In this case, soil is the main reservoir which receives and fixes microelements, and their source for other components of the natural environment, biota, for soil and plant microorganisms in particular.

Migrating through food chains, trace elements can accumulate in toxic concentrations in the organs and tissues of plant and animal organisms. Taking into account the circulation of substances and energy, it should be noted that, entering the plant component of biogeocenosis, heavy metals are redistributed along other rows - consumer goods (animals and humans) and decomposers, and are characterized by their durability and practical non-withdrawal from the system: soil - plants - animals - human [5, 16].

This must be taken into account, since the final link of the food chain in this case is a man. Accordingly, soil contamination with metals, in addition to the negative effects on ecosystems, can also affect the health of the local population indirectly through plants, worsening the quality of the urban environment [11, 12, 17]. This is due to the fact that, agricultural products for personal consumption are grown in households and in summer cottages both within the city and in the zones of influence of industrial enterprises [14].

2. Materials and research methods

For the northern part of Volgograd, one of the main sources of heavy metals in the environment are emissions from enterprises of the northern industrial hub, **which take a significant part in** formation of the quality of the environment, its individual components (soil in particular) and the general ecological well-being of the urban environment.

The joint-stock company VMP "Red October" [7, 8] was selected as the investigated local object of environmental impact. This enterprise is currently one of the largest producers of high-quality metal-roll special grades of steel for the automotive industry and aviation industry, chemical, oil and power engineering, oil and gas industry in the Russian Federation, and, therefore, is a major source of pollutants (including heavy metals) into the environment

In accordance with the data on the composition of the company's emissions, three metals were selected for the study - typical representative emissions of this enterprise - zinc, copper, and manganese belonging to hazard classes I, II, and III, respectively, resulting in the study covering all hazard categories of substances in the soil. In addition, for the selected elements, the maximum allowable concentration (MAC) standards have been developed, that is, substances that are rationed by their concentration in soil and are subject to control as part of an environmental monitoring system [2].

Control points for soil sampling were laid in accordance with State Standard of the USSR 17.4.4.02-84 [1] within the territory of the proposed zone of influence of the enterprise. The total distance from "Red October" was 0-4.5 km to the north-west from the enterprise border, which corresponds to 3 times the size of the sanitary protection zone (3 × 500 m) and slightly exceeds it in order to clarify the results. When selecting control points, such factor as prevailing wind direction was also taken into account (based on the analysis of the wind rose of the northern part of Volgograd).

The study of soils was based on quantification of the concentration of labile metal forms and subsequent analytical studies and interpretations of the data obtained. Determining the concentrations of the investigated elements at each of the sampling point was carried out using the precision atomic absorption spectrometry method [18] with preliminary preparation of samples for the analysis by acid decomposition of solid soil samples in open vessels (based on the sample preparation method used for analytical equipment [6]).

The atomic absorption method is most developed for determining the elemental composition of liquid samples, Accordingly, the method of implementation of the analytical process included two stages. At the first stage, in oder to transfer the studied elements into the solution and prepare the working solution for analysis, a sample of a known mass was taken from a solid sample and then dissolved. Then the analyzed solution is directly entered into the atomizer, creating an absorbing layer of atomic vapor and measuring the analytical signal.

The analysis was carried out using an atomic absorption spectrometer "KVANT.Z", the main function of which is to determine the concentration of lots of elements in liquid samples of different origin and composition. The principle of operation of the spectrometer is based on the analysis of the degree of selective absorption of resonant spectral lines by the atomic pair of the element being determined. In addition, each element has its own analytical resonance line, providing maximum absorption of light, the source of which is a lamp with a full cathode for each element being determined (linear radiation).

The process of atomization (conversion of the analyzed sample into atomic vapor) is performed in the analytical cell of the electrothermal atomizer. The graphite furnace is heated to the temperature required for evaporation of the sample and atomization of the element by an electric current.

Due to the presence in the analytical cell, in addition to the atoms of the element being determined, gaseous components capable of absorbing the incident light, nonatomic or background absorption takes place, which, in turn, is a source of systematic measurement error. For the automatic correction of this type of absorption in the spectrometer, the reverse Zeeman effect is used, provided by the placement of a graphite furnace in a longitudinal alternating magnetic field.

The data obtained as a result of measurements make it possible to judge the concentration of the elements being analyzed in the studied environment in a specific area, in this case, in the zone of influence of a local source of environmental pollution (JSC VMP "Red October").

3. Results and discussion

An isolinear map of the obtained values of the concentration of labile forms of copper in the soil cover in the study area is shown in Figure 1. As a result, soil ranges with different levels of copper, that is, with varying degrees of contamination, were identified. On the presented map, the ranges are displayed as a color gradient, reflecting the levels of copper concentration, correlated with the established MAC.

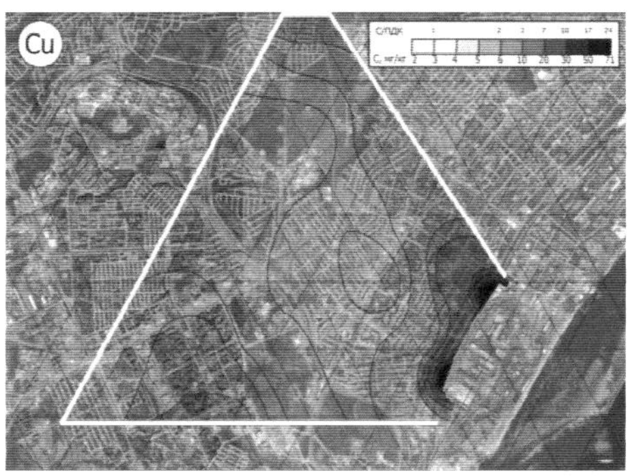

Figure 1. The ratio of the actual and regulatory concentration of labile forms of copper in the soil cover in the assumed zone of influence of JSC VMP "Red October"

According to hygienic standards 6229-91, the maximum allowable concentration of labile forms of copper in soils is 3.0 mg / kg [2]. The foci of maximum pollution are associated with the activity of steel-smelting shops No. 2 and No. 3 and are located in the immediate vicinity of the border of the Red October production zone; copper concentrations at control points exceed the MAC by a factor of ten (from 2.4 to 20 time). At the border of the sanitary protection zone (500 m from the enterprise), the values fluctuate within 1.5-2.2 MAC. Further, in the north-west direction within the study area, the concentration of copper in the soils in most cases does not exceed 2 MAC, but only in 5 points the copper concentration is within the norm. To the west and north-west of the enterprise, in the extreme Foci of soil contamination with copper with metal concentrations of 5–6 mg / kg are also located in the areas of the study region. These foci are largely limited from the zone of influence of JSC VMP Red October to areas with low copper values in the soil, which indicates the presence of other local sources of exposure that fell into the studied area [5].

The isolinear map of the obtained values of the concentration of labile forms of manganese in the soil cover in the area under study is presented in Figure 2.

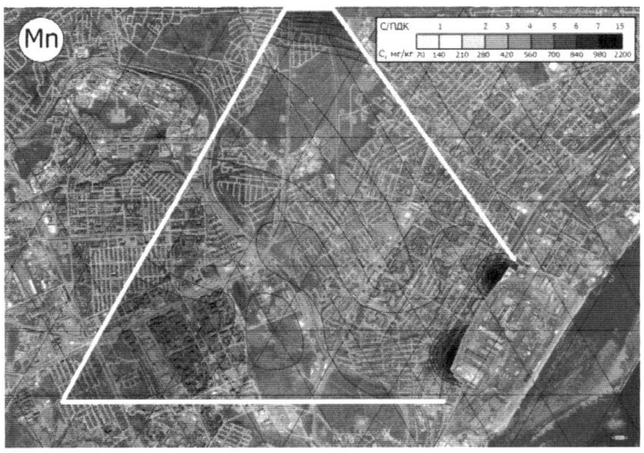

Figure 2. The ratio of the actual and regulatory concentration of labile forms of manganese in the soil cover in the potential zone of influence of the JSC VMP "Red October"

The maximum allowable concentration of labile forms of manganese in soils is 140.0 mg / kg [2]. The foci of maximum pollution are also associated with the activities of steel-smelting shops No. 2 and

No. 3 and are similarly localized in space. Within the sanitary protection zone, the concentration of manganese at control points exceeds the MAC by an order of magnitude and more: from 1.2 to 6-15 times. From the border of the sanitary protection zone and to a distance of 3.5 km to the north-west of the enterprise, it does not exceed 2 MACs anywhere. At a distance of 3.5-4.5 km, there are 3 cases of excess concentrations above 2 MAC, which is connected with the influence of other local sources of pollution. In general, concentration of manganese in most points does not exceed 2 MAC, but only in 3 points concentration is within the standard [5].

An isolinear map of the obtained values of the concentration of labile forms of manganese in the soil cover in the area under study is presented in Figure 3.

IOP Conf. Series: Materials Science and Engineering **483** (2019) 012030 doi:10.1088/1757-899X/483/1/012030

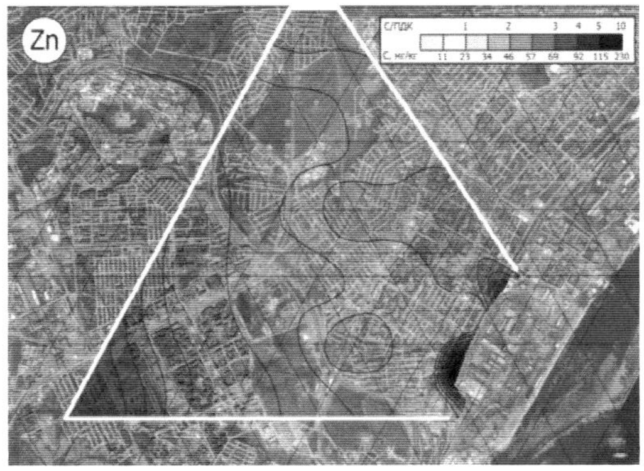

Figure 3. The ratio of the actual and regulatory concentration of labile forms of zinc in the soil cover in the potential zone of influence of JSC VMP "Red October"

The maximum allowable concentration of labile forms of zinc in soils is 23.0 mg / kg [2]. The centers of maximum zinc contamination, as well as with copper and manganese, are related to the activities of steel-smelting shops of the enterprise No 2 and No 3. In the immediate vicinity of the enterprise's border, zinc concentration varies greatly from values within the limits of the standard to maximum allowable concentrations exceeding 9.4 times. On the border of the sanitary protection zone, the zinc concentration in the soil is already set within the standard or with a slight excess. Further, in the north-west direction within the study area, the concentration of zinc in soils in some cases does not exceed the MAC standard, at 5 points the concentration is below 1.5 MAC. At a distance of 4.5 km, at one point only, an increase in concentration up to 2.6 MAC is observed, which is connected, as in the cases with copper and manganese, with the influence of other local sources of pollution [5].

4. Conclusion

Thus, on the basis of the obtained results, it can be concluded that the actual zone of influence of JSC VMP "Red October", taking into account the background metal concentrations for the northern part of Volgograd, and also taking into account the prevailing winds, is limited by a distance of 4.0–4.5 km to northwest of the enterprise. The zone of influence, taking into consideration the MAC of metals, instead of their background concentrations, is somewhat smaller and extends 3.5-3.7 km. The first value reflects the absolute zone of influence of the enterprise, where any changes in environmental parameters are observed. The second value determines the zone of influence, where there are noticeable exceedances of standards for the metals content in the soil and negative trends in the state of natural communities, especially vegetation, can be observed and indirectly affect the health and well-being of residents in the area [5].

Thus, the actual state and level of soil contamination with heavy metals are most important for residents of the individual development zone, which occupies 30% of the territory of the Krasnooktyabrsky district (in the zone of influence of the Red October complex) within the boundaries of the region. At the same time, for the entire private sector, elevated concentrations are observed either by one of the studied elements or by their combination, which must be taken into account when growing agricultural products in private farms, since the accumulation of elements to toxic levels in edible parts of cultivated plants is potentially possible.

In addition, several foci of pollution were found within the study area, presumably related to the activities of other sources of anthropogenic impact or unfavorable ecological condition of natural objects, and this confirms once more the relevance of applying soil monitoring data in the areas of actual influence of local industrial facilities in assessing the current quality of the urban environment.

References

[1] GOST 17.4.4.02-84. State Standard of the USSR. Protection of Nature. The soil. Methods of selection and preparation of samples for chemical, bacteriological, helminthological analysis (Standard of 12/12/1984 N 4100) Consultant Plus: inform. system 2018

[2] The List of Maximum Permissible Concentrations (MPC) and Approximately Permissible Amounts (ODC) of Chemicals in the Soil" Consultant Plus: inform. system 2018

[3] Mandzhieva S S 2014 Ecological state of soils and plants in the natural and man-made sphere: monograph *Southern Federal University* p. 264

[4] Warm G A 2013 Heavy metals as a factor of environmental pollution *Astrakhan Bulletin of Environmental Education* **1** 182-192

[5] Tikhonov A.A, Bolgov I A 2017 The significance of data on the actual zone of influence of a local industrial facility in assessing the quality of the urban environment (using the example of the Red October metro station) *Modern Ecology: education, science, practice* **2** 108-114

[6] M-MVI-80-2008 2018 Methods for measuring the mass fraction of elements in soil samples, soils and bottom sediments by atomic emission and atomic absorption spectrometry (approved. And put into effect on 06/02/2008 Monitoring LLC) *Information system MEGANORM* Access mode: http://meganorm.ru/Index2/1/4293824/4293824289.htm.

[7] Object 18-0134-000164-P JSC "VMK KrasnyOktyabr" VET UONVOS: From -open registry of objects NVOS from 05.14.2018. - Access mode: https://onv.fsrpn.ru/#/public/registry/18-0134-000164-%D0%9F

[8] Information about the enterprise VMK "KrasnyOktyabr" VMK "Kra October October 14.05.2018 - Access mode: http://www.vmzko.ru/spravka.html

[9] C von Brömssen, J Fölster, M Futter, K McEwan 2018 Statistical models for evaluating suspected artefacts in long-term environmental monitoring data *Environmental Monitoring and Assessment* URL: https://doi.org/10.1007/s10661-018-6900-3

[10] G Tóth, T Hermann, G Szatmári, L Pásztor 2016 Maps of heavy metals in the soils of the European Union and proposed priority areas for detailed assessment *Science of The Total Environment* **565** 1054-1062

[11] G Tepanosyan, L Sahakyan, O Belyaeva, S Asmaryan, A Saghatelyan 2018 Continuous impact of mining activities on soil heavy metals levels and human health *Science of The Total Environment* **639** 900-909

[12] I Sá, M Semedo, M E Cunha 2016 Kidney cancer. Heavy metals as a risk factor *Porto Biomedical Journal* **1** **(1)** 25-28

[13] M Woszczyk, W Spychalski, L Boluspaeva 2018 Trace metal (Cd, Cu, Pb, Zn) fractionation in urban-industrial soils of Ust-Kamenogorsk (Oskemen), Kazakhstan—implications for the assessment of environmental quality *Environmental Monitoring and Assessment* **190(6)** 362

[14] N V Stepanova, S F Fomina, E R Valeeva, A I Ziyatdinova 2018 Heavy metals as criteria of health and ecological well-being of the urban environment *Journal of Trace Elements in Medicine and Biology* **50** 646-651

[15] N E Kosheleva, D V Vlasov, I D Korlyakov, N S Kasimov 2018 Contamination of urban soils with heavy metals in Moscow as affected by building development *Science of The Total Environment* **636** 854-860

[16] Qi Wang, Jianfeng Liub, Zhao Chen, Fangbai Li, Huanyun Yu 2018 A causation-based method developed for an integrated risk assessment of heavy metals in soil *Science of The Total Environment* **642** 1396-1405

[17] S Türtscher, P Berger, L Lindebner, T W 2017 Berger Declining atmospheric deposition of heavy metals over the last three decades is reflected in soil and foliage of 97 beech (Fagus sylvatica) stands in the Vienna Woods *Environmental Pollutio* **230** 561-573

[18] Shi-Hua Chen, Yi-Xiang Li, Pei-Hua Li, Xiang-Yu Xiao, Min Jiang, Shan-Shan Li, Wen-Yi Zhou, Meng Yang, Xing-Jiu Huang, Wen-Qing Liu 2018 Electrochemical spectral methods for trace detection of heavy metals: *TrAC Trends in Analytical Chemistry* **106** 139-150

Timber industry terminology translation: case study

N G Sivtseva[1], N A Paskova[2], N Yu Berezhnykh[3] and I S Shilnikova[2]

[1]Irkutsk National Research Technical University, 83 Lermontov Street, Irkutsk, 664074, Russia
[2]Irkutsk State University, 1 Karl Marx Street, Irkutsk, 664003, Russia
[3]Irkutsk State Transport University, 15 Chernyshevski Street, Irkutsk, 664074, Russia

s_nataly77@mail.ru

Abstract. This paper is an attempt to reveal and analyse potential challenges a translator can face when translating technical texts connected with the timber industry. In modern linguistics there are several approaches to the definition of technical translation based on characteristic features of translated texts. As there is some difference between technical and scientific types of translation there are various requirements put to these types of translation. However, the terminology is an integral part of both types of texts and sometimes it can be difficult to translate it. Though there is no one generally accepted definition of the notion "term" linguists mention its unambiguous nature, lack of synonyms, precise meaning. The analysis of timber industry texts showed that these qualities of the terms can be lost when it involves translation. It is conditioned by different meanings of the term in different spheres of knowledge, its belonging both to terminology and spoken language, different scopes of the concept in different languages, influence of linguistic and situational contexts, lack of equivalents and usage of the shortened forms.

Introduction

As a new discipline and a separate field, technical translation has been identified, examined, and developed since the 1960s [1], [2]. Currently, technical translation takes an important place in the world's total translation output. It accounts for more than 90% of all professionally translated work [3]. It is not surprising given the variety and abundance of technical information written in different languages, the companies' international cooperation in scientific, technological and industrial activities, their focus on exporting manufacturing equipment, the laws and regulations of international trade organizations requiring accurate and comprehensive technical documentation in the languages of collaborating countries and the necessity of the effective use of translated information. The listed facts emphasize the significance of the field of technical translation, the importance of highly qualified professional technical translators and well-done translations of technical texts.

The trend of the increased necessity for technical translation is furthered by the issues of legal requirements. In the European context, for example, Annex I section 1.7.4 of the Directive 2006/42/EC regulating the health and safety requirements relating to the use of machinery contains the requirement that "all the machinery must be accompanied by instructions in the official Community language or languages of the Member State in which it is placed on the market and/or put into service" [4].

Thus, the practical approach to this process outlines the serious problems the Russian technical translators face. The matter is that during the Soviet period of time the translation of technical texts was done by specially educated state translators, particularly from all-USSR Translation Center. Besides, special methodical recommendations for technical translators were worked out [5]. A new trend in the technical translation theory appeared after the breakup of the Soviet Union when private agencies and freelancers were involved into the translation process. Only several editions of Recommendations to technical translators have been issued since then [6]. The main aim of these works is to unify demands towards the translation on the basis of Soviet Union translators' experience [6]. The absence of the legal requirements could not affect the quality of technical translation.

One of the most difficult issues in the field of technical translation is translation of terminology. It may seem that terms translation has indisputable character as they have exact meanings, non-complicated nature, etc. However, the research and translation practice show that a translator meets certain difficulties while translating terms; as they are of multi-aspect character and can have various meanings and define various objects in different industries. Besides, some linguists believe that terms are ordinary words having some special function, namely, they are means of designating a special concept. Thus, the translator's task becomes even more complicated; he/she should find out the function of a word, recognize it as a term and interpret it within the context of a definite industry. Translation of the timber industry terminology sets similar problems. These terms can define different objects, moreover, the timber industry loans terms from chemistry, mechanic engineering, machine tools usage, etc. Consequently, in the given research the multi-aspect nature of timber terms and issues of their translation are revealed and analyzed.

1. Theoretical framework

Nowadays, there are various factors that lead to an unprecedented demand for scientific and technical translation: increased information exchange volume, intensified knowledge transfer due to increasingly international character of science and technology, globalization and diversification of commercial activities and the ever greater industrial products sophistication. Not to have any misconceptions it is essential to cast light on the term 'technical translation' and to distinguish among scientific, technical and scientific and technical translations. In the works of Russian and non-Russian academics there are different approaches to defining this term. J. Byrne, the British author of a number of translation books, stresses the difference between these types of translation. Moreover, he believes that one of the greatest fallacies when discussing technical translation is to somehow lump it together with scientific translation, or worse still, to use the two terms interchangeably. Byrne explains that technical translation deals with technological texts and more precisely with texts on the subjects based on applied knowledge from natural sciences, scientific translation relates to pure science in all of its theoretical, esoteric glory and scientific and technical translation is a generic term which is used to refer to pure sciences, applied scientific research and technology [7].

Two American translation scholars Sue Ellen and Leland Wright define technical translation as translation of special language texts, i.e. texts written using Languages for Specific Purposes. In their point of view technical translation includes not only translation of texts in engineering or medicine, but also such disciplines as economics, psychology and law [8]. Maeve Olohan, a contemporary British scientist, states that the term "technical translation" is often used to refer to translation of texts outside the fields of science and technology and that some scholars see technical translation as a synonym for specialized translation [9].

The analysis of Russian linguists' works shows the interchangeable use of the above-mentioned terms. The explanatory translation dictionary points out that the frequently used term 'scientific translation', as a rule, coincides in its meaning with the term 'scientific and technical translation' and the phrase 'translation of technical texts' has approximately the same meaning. The definition of technical translation given in this dictionary says that technical translation is a translation used to exchange special scientific and technical information between people speaking different languages. Scientific and technical translation is understood as one of the types of special translation serving the areas of communication on scientific and technical topics or translation of special texts (documents) of scientific and technical nature, carried out by specific methods different from fiction translation in compliance with such requirements as accuracy, conciseness, clarity, etc. [10].

Similarly, V.A. Usachev, a Russian contemporary scholar, uses the terms 'scientific and technical translation' and 'technical translation' interchangeably. Nevertheless, his definition of technical translation deserves attention and can be considered as accurate and complete. He states that technical translation is one of the most difficult types and it means translation of texts on technical subjects, in particular, documents of different specialties, all kinds of reference books, various dictionaries, certificates of conformance, operating instructions, engineering plans, scientific and technical articles, business contracts and other commercial technical proposals [11].

One of the integral parts of translated technical texts is terminology usage. Currently, modern linguistics does not have a single definition of the word "term". According to L.M. Alekseeva, the most difficult task is to determine the nature of the term, since it is a multidimensional and internally contradictory object of study [12]. Thus, in the modern linguistics there are several approaches to the definition of the notion of term. Traditionally, some scientists use a logical definition of the term, while another ones describe it within the context of its characteristic features. The others define the term by contrasting it with any other concept. Such a variety of definitions of the notion "term" is explained by the fact that it is an object of several sciences. The same term can be often included in different terminological systems of the given language. This fact leads to inter-scientific terminological homonymy [13]. Thus, V.M. Leichik believes that it is fundamentally and logically impossible to create a universal definition of the "term" because of its multi-aspect nature [14].

Consequently, there are several approaches to the definition of the term. O.S. Akhmanova defines the term as "a word or a word combination of some special (scientific, technical, etc.) language created (accepted, borrowed, etc.) for the exact expression of special concepts and designation of special objects" [15]. Z.I. Komarova emphasizes the nominative nature of the term: "a term is an invariant (a word or a phrase) that designates a special object or a scientific concept limited to its definition and its place in the certain system of terms" [16].

According to the definition of B.N. Golovin, "a term is a single word or a subordinate noun phrase, denoting a professional concept and intended to meet the specific needs of communication in a particular profession (scientific, technical, industrial, managerial)" [17]. V.M. Leychik gives the following definition of the term: "a term is a lexical unit of a language for specific purposes, denoting a common (concrete or abstract) concept of the theory of a specific special field of knowledge or activity" [14]. This definition emphasizes the highly specialized nature of the term and represents it as an integral part of the natural language. According to A.N. Baranov, "terms can be defined as words (phrases) of the metalanguage of science and languages of scientific disciplines, as well as words denoting specific realities of the areas of practical human activity" [18]. S.V. Latysheva researches the terms in view of their belonging to different types of discourse: Discourse of Consensus and Discourse of Expert Community [19].

Therefore, it can be concluded that the term is a word or a phrase having a definition and denoting any special concept or subject of a certain area of knowledge or human activity. The problem of the relationship between a term and a word has long been one of the most important in terminology. S.V. Grinev-Grinevich writes that the term primarily refers to the general class of lexical units and its belonging to special vocabulary is secondary [20]. The author determines a number of term characteristics: specificity of use (the term is related to the special area of usage as it is used for naming concepts); substantial accuracy – clarity, limited meaning of the term (a special concept has precise boundaries established by a scientific definition); context independence; stylistic neutrality (the term should not give rise to any additional associations); conventionality (necessity for naming new concepts requires creation or careful selection of lexical meanings of existing terms); nominative character (terms are mostly nouns) [20].

Thus, within the context of the above-mentioned definitions it can be concluded, that a technical or scientific term is a symbolic unit and it must meet a number of criteria, namely, the requirements to the term meaning and form as well as pragmatic requirements (usage specificity, definition, unambiguity, full meaning, absence of synonyms, conformity to literary norms, stylistic neutrality, general acceptance, motivation).

V.M. Leichik expresses the idea that terms are not special words but words with a special function. Analyzing the roles that the term performs when defining a special concept, the author determines the following basic functions of the term: nominative, significative, communicative, learning, pragmatic, cognitive and heuristic [14]. A.A. Reformatski also writes about the reflective function of the term, as terms are special words limited to their specific purpose, they tend to be unambiguous as exact lexical representations of concepts [13].

2. Research methods

The issues of timber terminology translation were researched and analyzed on the basis of technical texts (as understood by J. Byrne). When researching the material, the following methods and approaches were used: definitional analysis approach (it allows determining conceptually conditioned relations between English and Russian terms); contextual analysis method (it allows getting the idea about the complete notion of the term in case of the definition lack); continuous sampling method; contrastive analysis approach (target and original texts are compared to define the necessity and validity of translation transformations).

3. Results and discussion

When translating a term, a translator should perform several acts: to recognize the given original word as a term, to understand it correctly, to find the appropriate equivalent or analogue in the target language. There are some reasons making it difficult for a translator to perform one or all of the above mentioned acts: interference of the native language, intralingual homonymy, heterogeneous form of a term, complicated structure of a term.

As it was already mentioned when discussing the theoretical framework of the research there are some requirements to the notion of term; among them there is unambiguity and absence of synonyms. These requirements are sometimes not met when it involves translation.

The most obvious reason of this inconsistency is different meanings of the same term in the different spheres of knowledge. For example, the word *pocket* can have the following special meanings: «воздушная яма» (in the aviation), «окружение» (in the military business), «мертвая зона» (in the radio-electronics), «гнездо месторождения» (in the geology), «коробка» (in the timber industry). Moreover, in this case a translator should decide if this word is used in its terminological or ordinary meaning («карман»). Sometimes this task is rather difficult, a translator should have some extra-linguistic knowledge to understand the terminological nature of the meaning. For example, in view of the timber terminology the word combination *secondary colour* that consists of non-terminological units can be translated only as «смешанный цвет»; this aspect often leads to the mistakes in translation and misunderstanding.

One more reason of this deviation involves different scopes of the definite concept in the source and target languages, besides, the concept scope is always wider than the range of lexical means of its representing. Thus, more detailed description of the same object in the target language can cause the variety of equivalents. For instance, the scope of the Russian concept «лес» is rather wide and it includes the aspects that can be represented in English with the following lexemes: *timber, lumber, wood, forest.* Among these words only the word *forest* is not used in the meaning «древесина», it always means «лес» (forestry, forestland, woodland). The word *lumber* has the meanings «лес (материал), пиломатериал, древесина», *timber* has the meanings «лесоматериал, пиломатериал, доски, древесина», *wood* can also be used in these meanings but in special word combinations, such as *phenol-impregnated modified wood* «дельта-древесина». However, the word *tree* having the meaning «дерево» is almost never used in technical literature to define the material used for building or wooden works.

Another case of different scopes of the concept and its asymmetry can be exemplified by English *pulp* and Russian «древесина». In English there are many terms derived from the word *pulp* the translation and meaning of which depend on the attributes. Two groups can be classified among them: *pulp* having the meaning or translated as «целлюлоза» (*bleached pulp* «беленая целлюлоза», *pulp lap* «лист влажной целлюлозы», *semichemical pulp* «полуцеллюлоза», *sulfite pulp* «сульфит-целлюлоза», *undercooked pulp* «непровар», *pulpability* «непровариваемость», *pulpmaker* «целлюлозник (рабочий)»); *pulp* having the meaning or translated as «древесина» (*pulp density meter* «пульпомер», *pulp handsheets* «отливка», *pulp line* «пульповод», *pulp machine* «пресспат», *pulp pump* «пульпонасос», *pulp saver* «массоловушка», *pulp trap* «мезголовушка», *pulp washer* «массомойка», *beet pulp* «жом», *raw pulp* «полумасса», *pulpwood* «баланс, балансы»).

The second reason of arising difficulties is caused by the existence of two types of contexts: linguistic and situational. Linguistic context is made up by source language units in a source text while situational context includes temporal, spatial and other circumstances under which a source text was produced as well as all the facts that a receptor is expected to know to interpret the message adequately. Having assessed the meanings of the source language units against linguistic and situational contexts a translator can discover what they mean in the particular case and what equivalents should be chosen as their substitutes.

Within the framework of the given conception the following examples can be analyzed. One of the characteristic features of technical texts is the usage of special words (regarded here as terms) for the objects that are named in the spoken language in a different way. Timber industry terminology contains a lot of names of different species of trees. The concepts represented by these names can belong to two pictures of the world simultaneously: to the scientific one and to the naïve one. And the translator's task here is to define the situational context, to recognize the type of the world picture and to find the adequate variant of translation. When the Russian lexeme «ель» belongs to the situational context of Christmas it is translated in English as *fir(-tree)*. However, if it is used in the linguistic and situational context of the timber industry it must be translated terminologically as *spruce*. The similar situation is with the Russian word «кедр»; it can be translated as *cedar* and *pine*. However, the lead in explaining the translation adequacy in this case is taken by the linguistic context. The translation depends on the attributes (*cedar of Lebanon* – «ливанский кедр»; *Cembran (Arolla) pine* – «европейский кедр»); if the trees of this species grow in Siberia the word *pine* is used to represent the notion of «сибирский кедр».

The translation of timber terms in view of two context types may also pose another difficulty. For example, the term *larch* has two meanings «лиственница» (the tree itself) and «древесина лиственницы» (the wood of this tree), so its translation fully depends on the context:

The larch in this forest sector is especially strong and conditioned – «Лиственница в этом лесном секторе является особенно крепкой и кондиционной» (the tree itself);

The larch is used in the building production extensively – «Древесина лиственницы широко используется в строительной промышленности» (the wood of this tree).

The same problem arises with the term *pocket* that has already been discussed. Despite the fact that it has different meanings in different technological descriptions and sciences it may be translated differently within the context of the timber industry: «паз», «смоляной кармашек», «приемное устройство, коробка». In this case the analysis of the linguistic and situational contexts may be helpful as well.

The fact that a source language unit corresponds to a number of regular equivalents does not necessarily mean that one of them will be used in each particular translation. In the analyzed translations the translator's skill to make a good choice among such equivalents is often well demonstrated. But sometimes the situational context does not allow a translator to choose any of the regular equivalents. In timber industry technical texts, it would be a mistake to translate the terms *degree* and *uppers* not as «сорт древесины» and «высшие сорта древесины» consequently. Besides, in a particular context even singular and plural numbers of the same noun can have semantically distinguishing meaning: the term *chip* means «щепа» in the timber industry, whereas its plural form *chips* means «варочная щепа».

Another reason that cause problematic issues in translation is the lack of equivalents or non-equivalent words. It is obvious, that in manufacturing processes of the timber industry the number of equivalent-lacking terms is not so great. However, there are some "ordinary" words for which the target language may have no equivalent lexical units: *fluid, conservationist* and so on. Such words appear in the language due to the emergence of new advanced technologies or techniques. Lack of regular equivalents does not imply that the meaning of an equivalent-lacking source language unit cannot be rendered in translation or that its translation may be less accurate.

Coming across an equivalent-lacking word a translator often uses different types of occasional equivalents, which can be later used as terms if these contextual substitutes turned out to be successful and adopted by the members of the target language community. To that end, different ways of translation can be used:

1) loan-words imitating the source language form (*feeder* «фидер» – machine attachment transferring the wood; *trawler-drifter* «дрифтер-траулер»);

2) approximate substitutes, that is target language words with similar meaning which is logically developed to convey additional information (*thermoresponsive switch* «термовыключатель; термо-предохранитель»);

3) descriptive translation explaining the meaning of the source language unit; in the timber industry this method turns out to be rather productive: *combustion furnace* «печь для органического анализа»; *"green" chain* «сортировочная цепь на лесозаводе», *craft specialties* «специальные сорта бумаги из жесткой сульфатной целлюлозы», *post colour number* «число изменений цвета после отбелки», *test digester "now down"* «анализ варочного цикла в период снижения давления в конце варки», *digester blow test* «конечный анализ варочного щелока».This method is sometimes used in conjunction with the first one when the introduction of a loan-word is followed by a foot-note explaining the meaning of an equivalent-lacking word in the source text.

The problem of equivalents lack is often associated with the usage of shortened widely-spread terms (abbreviations or acronyms). It is obvious that adequate translation presupposes decoding of shortened forms and their replacement with the Russian corresponding ones. But the problem is that there are often no corresponding analogues in the target language and creation of new terminological abbreviations is not encouraged. In such cases translators often decode the abbreviation meaning and loan it directly in its native form, after that they are free to use this newly-created substitute. It may really be helpful when it involves the inscriptions on machine-tools, mechanisms, chemical substances containers: *ALK (alkali)* «щелочь», *dft (draft)* «набросок, схема, чертеж», *ga. (gauge)* «калибр, масштаб», *pbd (paperboard)* «картон», *P.P. (paper pulp)* «целлюлоза», *b.o. (back order)* «в обратном порядке», *dw (dead weight)* «вес конструкции, мертвый вес», etc.

4. Conclusion

The goal of the given research was to find out ways of translating timber terminology in technical texts that allow creating adequate and equivalent translation.

Even if a source language unit has a regular equivalent in the target language, this equivalent cannot be used in the target text whenever this unit is met in the source text. An equivalent is but a potential substitute, as the translator's choice is, to a large extent, dependent on the context in which the source language unit is met in the source text. Having assessed linguistic and situational contexts and used all the background knowledge and available information on the timber industry a translator can make the adequate choice of one of the substitutes, that can be regular equivalents, contextual substitutes or analogues.

As terminology and special lexis pose a lot of difficulties in translation, the method of explication or descriptive translation is frequently applied; it is often accompanied by the appearance of a new term created on the basis of transcription, transliteration or loan translation. It is conditioned by the fact, that mechanisms of categorization of the world and creation of new terms are not alike in different countries, the progress is going forward, so newly created terms, abbreviations and special words need to be explained properly before putting them into practice.

The study and analysis of mechanisms of recognizing, understanding and translating the timber industry terms conditioned by different types of the contexts and their semantic nature will make the work of industry-specific translators more efficient and competent.

References

[1] Pumpyanski A L 1965 *Introduction to the Practice of Translation of Scientific and Technical Literature into English* (Moscow: Nauka) 303

[2] Finch C 1969 *An Approach to Technical Translation: An Introductory Guide for Scientific Readers* (New York: Pergamon Press)
[3] Kinscott G 2002 *Perspectives: Studies in Translatology* **10:4** 247
[4] Krüger R 2015 *The Interface between Scientific and Technical Translation Studies and Cognitive Linguistics. With Particular Emphasis on Explication and Implication as Indicators of Translational Text-Context Interaction* (Berlin: Frank & Timme GmbH. Verlag für Wissentshaftliche Literatur) 423
[5] Lyutkin I D 1991 *Scientific-Technical Translation from Russian into English* (Moscow: The All-Union Centre of Translation of Scientific-Technical Literature and Documentation) 125
[6] Duplenski N 2016 *Written Translation – Recommendations to the Translator, Employer and Editor* ed E. Maslovski (Moscow: The Russian Association of the Translators) 123
[7] Byrne J 2006 *Technical Translation: Usability Strategies for Translating Technical Documentation* (Dordrecht: Springer) 3–8.
[8] Wright S and Wright L 1993 *American Translators Association Scholarly Monograph Series* **VI** (Amsterdam/Philadelphia: John Benjamins Publishing Company) 1
[9] Olohan M 2011 *Special Issue. The Translator. Science in Translation* **17:2** 246
[10] *Explanatory Translation Dictionary* 2003 ed L L Nelyubin (Moscow: Flinta: Nauka) 320
[11] Usachev V A *Bulletin of Donetsk National University of Economics and Trade* **2** 58
[12] Alekseeva L M 1998 *Issues of Terms and Term Formation* (Perm: GSSRU) 120
[13] Reformatski A A 1996 *Introduction into Lingustics* (Moscow: ASPENT PRESS) 536
[14] Leichik V M 2007 *Terminology Studies: Object, Methods, Structure* (Moscow: Librocom) 256
[15] Akhmanova O S 1956 *The Dictionary of the Lingustic Terms* (Moscow: Sovetskaya Entsiklopediya) 607
[16] Komarova Z I 1991 *Semantic Structure of the Special Word and Its Lexicographic Description* Sverdlovsk: USU) 156
[17] Golovin B N and Kobrin R Yu 1987 *Lingustic Studies of Terms* (Moscow: Vysshaya Shkola) 103
[18] Baranov A N 2001 *Introduction into Applied Lingustics* (Moscow: Editorial URSS) 360
[19] Latysheva S V 2018 *Discourse-Competence Approach to Design of ESP Courses* SHS Web of Conferences **50** 01095
[20] Grinev-Grinevich S V 2008 *Terminology Studies* (Moscow: Academia) 304

IOP Conf. Series: Materials Science and Engineering **483** (2019) 012032 doi:10.1088/1757-899X/483/1/012032

Features of training and retraining specialists in the technical sphere in higher educational institutions

O I Vaganova[1], N Yu Tuaeva[2], Zh V Smirnova[1], L K Parsieva[3], E A Aleshugina[4]

[1]Minin Nizhny Novgorod State Pedagogical University, ul. Ulyanova,1, Nizhny Novgorod, (603004), Russia
[2] Vladikavkaz Institute of Management Vladikavkaz, Russia, 14 Borodinskaja Street, 362025
[3] North Ossetian State University, 44-46, Vatutina street, Vladikavkaz, 362025.
[4] Nizhny Novgorod State University of Architecture and Civil Engineering, Ilyinskaya str., 65, Nizhny Novgorod, (653950), Russia

z.v.smirnova@mininuniver.ru

Abstract. This article describes the process of teaching students in a technical sphere in higher educational institutions. Characteristic features of their training are revealed. It is necessary to take into account current conditions of the educational process in higher education institutions where the tendency to success is increasing. Therefore, it was demanded to search for innovative ways of organizing the training process at a high level. Thus, the aim of the article is to identify features of this process in order to provide competent training. The authors point out the need to develop special teaching techniques for each discipline that will help students to learn better. Students are expected to study the subject profoundly and consolidate the content. This article describes the process of teaching students in a technical sphere in higher educational institutions. Characteristic features of their training are revealed. Current development of educational process in higher education institutions tends to reduce classroom time allocated to collaborate with teachers. Thus, special attention is paid to students' independent work and the necessity to provide them with fundamental education in the technical field.

1 Introduction

Constantly increasing amounts of information that students need to master studying technical subjects at the university, reduction in classroom pressure and increase in the share of independent work encourage teachers to search for effective training approaches and suitable ways to work with information. Training of students in a technical sphere is based on fundamental sciences. It is the basis of all training as it is impossible not to provide students with this knowledge despite the changed goals of education, its forms, methods and the role of the teacher in this process due to the changing meaning of educational paradigm. Today, students should be able to solve problems independently and creatively, be able to apply their knowledge in practice. It is aimed at the competence approach which is an integral part of modern higher education. The implementation of the competency-based approach makes it possible to form a competent, well-trained graduate. However, in this process, there are many features that need to be identified and taken into account when teaching students of technical universities.

As we have already said, the share of independent work of students is increasing, however, students who are considered by us need to be supervised by the teacher (this is especially important during the first years of study). Therefore, e-courses including open online courses are being actively introduced into the training process. Today, they are incorporated into education to implement the requirements of the Federal State Educational Standard which, in turn, provides students with academic mobility, freedom to exercise their creativity and develops their independence.

Consequently, we cannot but take these factors into account when training and retraining technical students.

2 Methodology

In the article we consider the process of teaching students in a technical sphere with the help of electronic tools in higher education institutions. We worked with students whose majors are «Construction», «Land transport and technological complexes», «Operation of railways». We designed e-courses for each subject to provide high-quality basic training for full-time students and retraining of students who study online.

As practice shows, students master the content deeper and more confidently using electronic technologies. The results of the final tests arranged with the use of electronic courses revealed higher results and showed further interest in studies compared to traditional approaches of students' training.

3 Results and discussions

To carry out the experiment we chose students whose majors are «Construction», «Land transport and technological complexes» and «Operation of railways». It is worth noting that graphic training is extremely important for these students [1]. It should take place throughout the entire learning process. The acquisition of basic graphic competencies occurs in the first and second years when students study technical and mathematical courses [2]. Graphic competence is the basis for acquiring professional competence of future engineers, technicians, designers and technologists [3]. Graphic disciplines are necessary for students to acquire professional knowledge and skills to perform, design and deal with design documentation. The block of graphic disciplines includes:

- descriptive geometry [4];
- engineering graphics;
- graphic means of PC;
- geometric modeling [5].

They are studied from the first until the fourth semesters and there are connections between them. Each stage involves the solution of certain pedagogical problems.

The first stage is «Descriptive Geometry». Here, students learn basic methods of obtaining images, theoretical rules, form presentation and transformation skills [6]. The course of descriptive geometry was created in order to provide students with theoretical foundations of drafting. This course is designed to develop creative thinking:

- figurative [7];
- spatial;
- logical;
- algorithmic [8].

It is important to say that traditional learning can be complemented by e-courses and blended learning. The use of such approaches in high school showed good results and proved to be efficient [9].

After completing the study of descriptive geometry, a student must learn how to:

- read the drawings [10];
- perform geometric constructions;
- be able to imagine the spatial image of an object according to its projections [11];
- draw up drawings according to established standards [12].

Here the student must learn to recreate three-dimensional images of an object from its flat images [13]. At this stage visualization is important. For better understanding of tasks the content must be presented on the screen [14]. These manuals require constant access so that students can refer to them at any time and recall something if there is a necessity. E-courses provided by educational platforms will help to solve this problem. It can act as an auxiliary tool for full-time study and the main one to retrain correspondent students.

The second stage is «Engineering Graphics». Its goal is to form students' competencies related to the development and execution of documentation. Engineering graphics provides the basis for studying other general engineering disciplines: materials resistance, the theory of mechanisms and machines and machine parts [15]. If students study an e-course in the classroom environment, both the teacher and students can refer to it. The course provides necessary tools to analyze learning tasks, illustrates the application of theoretical rules to solving practical problems and provides content to master skills. These tools include:

- video lectures [16];
- simulators for practicing skills with the help of theoretical material [17];
- test tasks to check the degree of gained knowledge [18];
- laboratory workshops in computer graphics.

The third stage is «Graphic means of a personal computer» and «Geometric modeling». Students master a lot of specialized software systems for engineering modeling, design and engineering [19].

Students become familiar with industry standards in drawing design and the basics of spatial modeling.

The aim of the course is to teach students to develop models of information systems, models of databases and human-electronic computing interfaces, components of hardware-software systems and databases using modern tools and programming technologies.

When studying these courses, visualization is important. For better understanding of tasks, materials must be presented in the monitor screen [20]. These manuals require constant access so that students can refer to them at any time and recall necessary elements.

We believe that in order to master basic courses (graphics courses) better it is necessary to develop additional electronic courses that would serve as a supplementary element when teaching in the classroom [21].

Each topic should be provided with lecture content, presentations and analysis of tasks.

Some topics should include laboratory work, test questions and assignments that allow students to be certified. Each topic must be accompanied by a video [22].

In the training of engineers, as we have already mentioned, visualization is important. Therefore, 3D models, electronic posters and presentations can provide electronic platforms and should be used by students alongside courses developed by teachers [23]. The advantage of computer models is the ability to reproduce not only geometric shapes but also physical and mechanical characteristics (hardness, plasticity, bending, reflectivity, torsion), as well as animation of movement and interaction. As a result, students acquire a correct visual image of spatial forms.

In addition, 3D-visualization allows you to contribute to students' desire to self-master the technology of three-dimensional modeling. In order to achieve it, students have to study the objects independently and thoroughly being able to understand their properties and to work through the content of the topics more deeply. Nowadays students need the support of electronic courses which are available and accessible at any convenient time. Open access allows the learner to allocate time to do extracurricular work most effectively. Some of the materials can be placed locally, in computer classrooms for example, and some tasks can be found on the Internet.

Electronic technologies allow students to conduct experiments independently or with minimal help from a teacher. For example, when studying the topic «Surface formation», a student can conduct an experiment in which the impact of the surface shape on the shape of determinants in the kinematic method of surface formation is revealed and established.

In our opinion, in addition to e-courses developed by teachers for university use, massive open online courses should be included. However, teachers must, first of all, pass such courses themselves and evaluate their usefulness and effectiveness before involving students into working with it.

4 Conclusions

We have reviewed the process of teaching students in a technical sphere in higher educational institutions, the study by future engineers of the disciplines of the graphic cycle.

We found out that control and support from teachers are extremely important for such students because they have to deal with huge amount of tasks independently. To solve the problem we suggested creating e-courses for each discipline that will support both students and teachers. As practice shows, this type of training is of great interest among students. They can independently carry out experiments, have self-testing and they are engaged into extracurricular activities in the electronic environment, since the training content is publicly accessible. The results of the final testing at the end of the course are higher than without the use of electronic courses. Another reason that proves the importance of developing such courses is the need for clarity of examples to visualize information. Electronic platforms provide enough tools to complete this task. Future engineers acquire a correct visual image of spatial forms. As a result, high educational institutions contribute to teaching competent and independent specialists in the technical sphere.

References

[1] Barber M, Donnelly K, Rizvi S, Summers L 2013 An avalanche is coming. *Higher Education and the revolution ahead* 73

[2] Bulaeva M N, Vaganova O I, Koldina M I, Lapshova A V, Khizhnyi A V 2017 Preparation of Bachelors of Professional Training Using MOODLE. *In International conference on Humans as an Object of Study by Modern Science* 406-411

[3] Bicheva I B, Filatova O M 2017 Formation of the teacher-leader in the educational process of the university *Vestnik of Minin University* **3 (20)** 5

[4] Vaganova O I, Koldina M I, Trutanova A V 2017 Development content of professional pedagogical education in the conditions of realization of competence approach *Baltic Humanitarian Journal* 6 **2 (19)** 97-99

[5] Vaganova O I, Ilyashenko L K The main directions of implementation technologies of student-centered education in high school *Vesknik of Minin University* 2018. vol. 6 **3** 2

[6] Vaganova O I, Smirnova ZH V, Trutanova A V 2017 Organizatsiya nauchno-issledovatel'skoy deyatel'nosti bakalavra professional'nogo obucheniya v elektronnoy srede *Azimuth of Scientific Research: Pedagogy and Psychology* **6(3)**, 239-241

[7] Iltaldinova E Yu, Filchenkova I F, Frolova S V 2017 Peculiarities of the organization of postgraduate support of graduates of the targeted training program in the context of supporting the life cycle of the teacher's profession *Vesknik of Minin University* **3 (20)** 2

[8] Ilyashenko L K, Vaganova O I, Smirnova Z V, Prokhorova M P, Gladkova M N 2018 Forming the competence of future engineers in the conditions of context training *International Journal of Mechanical Engineering and Technology* Vol 9,Issue **4**

[9] A Z Ibatova, N V Ippolitova 2018 Structuring the Content of Disciplines in Higher School Using a Block-Modular Rating System for Future Oil and Gas Engineers *International Journal of Civil Engineering and Technology* **9(3)** 394-399

[10] Arkhipova M V, Belova E E, Gavrikova Y A, Lyulyaeva N A, Shapiro E D 2018 Blended learning in teaching EFL to different age groups *Advances in Intelligent Systems and Computing* **622** 380-386

[11] Abramova N S, Gladkova M N, Vaganova O I 2017 Features of the development of evaluation materials in the context of the implementation of the competence approach *Problems of Modern Pedagogical Education* **57-1** 3-9

[12] Ajeenkya D, Patil Y, Dr Gagandeep Nagra, Dr. Gopal R 2014 A Study on Total Quality Management in Higher Education *International Journal of Management* 5(5) 1-6

[13] Braine G 2013 Non-native educators in English language teaching. Routledge

[14] A Manikandan, M Muthumeenakshi 2018 Role of Engineering Education in Sustaining the Economic Development of India *International Journal of Mechanical Engineering and Technology* **9(3)** 706-710

[15] Ilyashenko L K, Smirnova Z V, Vaganova O I, Prokhorova M P, Abramova N S 2018 The role of network interaction in the professional training of future engineers *International Journal*

of Mechanical Engineering and Technology. T. 9 **4** 1097-1105
[16] Kutepov M M, Vaganova O I, Trutanova, A V 2017 Possibilities of health-saving technologies in the formation of a healthy lifestyle. *Baltic Humanitarian Journal* **6(3)** 210-213
[17] L K Ilyashenko 2018 Pedagogical Conditions of Formation of Communicative Competence of Future Engineers in the Process of Studying Humanitarian Disciplines *International Journal of Civil Engineering and Technology* **9(3)** 607–616
[18] Markova S M, Sedykh E P, Tsyplakova S A, Polunin V Y 2018 Perspective trends of development of professional pedagogics as a science *Advances in Intelligent Systems and Computing* **622** 129-135
[19] Markova S M, Sedykh E P, Tsyplakova S A, Polunin V Y 2017 Perspective Trends of Development of Professional Pedagogics as a Science *In International conference on Humans as an Object of Study by Modern Science* 129-135
[20] Smirnova ZH. V, Gruzdeva M L, Krasikova O G 2017 Otkrytyye elektronnyye kursy v obrazovatel'noy deyatel'nosti vuza.*Vestnik of Minin University* **(4)** 3
[21] Smirnova ZH V, Vaganova O I, Trutanova A V 2017 Itogovaya gosudarstvennaya attestatsiya kak sposob kompleksnoy otsenki kompetentsiy. *Karelian Scientific Journal* **6(3)** 74-77
[22] Smirnova Zh V, Mukhina M V, Kutepova L I, Kutepov M M, Vaganova O I 2018 Organization of the research activities of service majors trainees *Advances in Intelligent Systems and Computing* **622** 187-193
[23] Tsyplakova S A, Grishanova M N, Korovina E A, Somova N M 2016 Theoretical bases of designing of educational systems *Azimuth of Scientific Research: Pedagogy and Psychology* Vol 5 **1 (14)** 131-133

IOP Conf. Series: Materials Science and Engineering **483** (2019) 012033 doi:10.1088/1757-899X/483/1/012033

Teaching Package Development for Engineering Training Programs

Zh V Smirnova[1], O I Vaganova[1], L B Gatsalova[2], O V Golubeva[1], E.A.Chelnokova[1]

[1]Minin Nizhny Novgorod State Pedagogical University, ul. Ulyanova,1, Nizhny Novgorod, 603004, Russia
[2] North Ossetian State University, 362025 44-46, Vatutina street, Vladikavkaz, Republic North Ossetia-Alania, Russia.

z.v.smirnova@mininuniver.ru

Abstract. Nowadays society has a great need for efficient engineers. The article deals with relevant problems of engineering training programs productivity. The programs differ both in content and in the nature of educational process organization. Therefore, there is a need for appropriate teaching package development which provides educational and methodological support. Educational and methodological support is aimed at improving the quality of vocational education providing a specific combination of training tools and technologies. This process is designed by the teacher. The article focuses not only on teaching package development of the educational process but also on the content, monitoring and evaluation tools. Effectiveness research of vocational training program and their teaching package is carried out. The practical value of the research lies in the fact that the ways of teaching package development provide more efficient organization of the educational process. It leads to efficiency and high quality of education. We have conducted a study that resulted in a unified system concerning organizational and pedagogical conditions of academic and methodological support. We identified relevant requirements for teaching package development for students whose major is "Construction".

1 Introduction

The reform and modernization of education in Russia is aimed at improving its quality, preserving fundamentality, increasing variability and strengthening humanistic orientation with the view to make the whole teaching process technological. As practice shows, there is a constant need for competent engineers. Thus, the demand for such specialists is increasing. It entails the demand for highly qualified specialists who train engineering students at universities.

At the same time, the level of graduates' qualifications and skills differ both in content and in the way the educational process is organized. It requires the need for appropriate academic and methodological support presented in a teaching package[1].

Once again we define the importance of the engineering sector for society as engineers are required in many spheres of life. Therefore, training specialists in this field is extremely important and necessary. At the same time, we will focus on the quality of education received by students. Paying attention to improving the quality of education in this area, we pay attention to the development of the country's economy.

Problems of organizing professional training of future engineers were highlighted in the works of A.V. Brushlinsky, S.D. Smirnova, V.A. Bigeeva. They devoted their work to a methodology of vocational education, the development of the concept of the vocational education system, oral communication teaching of future engineers. Besides, some research works are devoted to the institutional structure of engineering field.

In this regard, teaching package development for the vocational training system will be effective if modern requirements are observed. It is necessary:

- to have data for analysis and comparison;
- to summarize pedagogical experience, taking into account new achievements of modern pedagogical science;
- to ensure communication of the studied material with production and future field of activity;
- to be specific and brief; meeting the requirements of the Federal State Educational Standards [24].

It is significant to intensify scientific-pedagogical search and development of teaching effectiveness as well as methodological support while training specialists [9]. It is vital to increase education knowledge-intensiveness and culture-intensiveness, taking into account the ongoing reassessment of human values, professional opportunities, expansion of specialists' field of activity, the emergence of integrative educational institutions [8].

Thus, the study of the relevant problem allows determining main aspects in educational activities - teaching package development for engineering specialists training - in pedagogical theory and practice.

It is important to develop appropriate teaching package for engineering training programs as there is an increasing need for high-quality competitive specialists and improving the level of professional and personal competencies of engineering graduates. However, it is associated with insufficient quality of academic and methodological support presented in a teaching package [14].

2 Methodology

At the first stage, we carried out analysis of future graduates training programs, identifying features and problems in their organization. We also studied the research topic and its development in pedagogical and psychological literature. The bibliography and research project [18] completed the stage.

At the second stage, ways and possible solutions to the problem were selected, management process of the training system of future specialists was analyzed, effectiveness of teaching package development as well as its organizational and pedagogical conditions was proved [19]. At this stage, an experiment was conducted where the motive of choosing a field of study, students' satisfaction with the educational process was analyzed. There was also another experiment, where the development, analysis and verification of organizational and pedagogical conditions of the teaching package development system took place [7].

The third stage is marked by the analysis and generalization of the obtained data, identifying the prospects for further research in the framework of our problems. Conditions for the creation of educational programs, curricula and other types of teaching package for engineering education were tested [13].

The key point is the opportunity for students to be engaged in their own education at any time, at any place and at any time in their lives [10].

The greatest effect is achieved in improvement of educational process teaching package through the acquisition of computer equipment and programs [4].

In this regard, introduction of teaching support is the provision of a system for engineers training as a process and result of management structures at various levels in the field of education. Its aim is to provide methodological support in operating and developing the educational process [5]. Its main purpose is to improve the quality of educational process by providing faculty and students with materials that are necessary for the development of educational programs, taking into account the observance of the following principles: fundamentality, universality, integrity, integrity, professionalism, variability, multi-level preparation. [11].

Innovative pedagogical technologies contribute to a successful solution of professionally significant tasks. They should be based on a solid scientific foundation that guarantees the optimal combination of fundamental and practical knowledge, the use of interactive learning technologies, the development of strategies and technologies for applying knowledge, the relationship of the studied material with everyday life problems [3].

To confirm the effectiveness of the teaching package development for vocational training programs an experimental study was conducted.

The quality of future specialists training reflects: the quality of the educational process; the level of academic performance and level of training; quality of methodical work; quality of curricula and programs; quality of educational work; quality of material support; quality of teaching staff (teacher qualifications, certification results, involvement in innovative activities) [12]; quality of learning outcomes - the level of graduates skills, their level of demand, admission to higher professional educational institutions, the level of satisfaction of personal inquiries, teachers achievements [6].

The manual within the vocational teaching package should contain the following essential elements: annotation, table of contents, introduction, main part, conclusion and bibliography.

Additional elements in the textbook include preface, illustrations, glossary, a list of conditional abbreviations.

3 Results and discussions

The research results contribute to the systematic updating of the content of the main educational engineering program. Organizational and pedagogical conditions of teaching and methodological support, substantiated in work theoretically and verified in practice are aimed at improving example programs of the professional courses.

In pedagogical practice preparation of a competent specialist, the use of input, current, boundary, final testing is used.

Testing tools are both paper (test questions, jobs, tests, crossword puzzles, test tasks and coursework tasks, examination questions) and technical ones (computer monitoring programs should be highlighted) [22].

Developing methodological content to check students' knowledge we should pay attention to:

- variety of forms, teaching methods and methods of testing;
- variability of test tasks, both in the classroom and for homework (taking into account the level of preparedness of students), options for distant learning [21];
- educational and methodological support of testing procedures, focusing on the need and desirability of using specific visual (illustrative) materials including those made by students;
- evaluation criteria for all kinds of independent work.

Studies show that the coefficient of mastering the material in the course "Engineering and computer graphics" must be at least 0.7. Only in this case we can talk about students' achievements.

During the test, we calculated the coefficient within three intervals in the range from 0.7 to 1.0, and their ratio at a 5-point scale (table 1).

Table 1. Criteria for assessing students' knowledge at a 5-point scale.

Coefficient	Evaluation
0.7-0.8	Satisfactorily
0.8-.09	Good
0.9-1	Fine

Students had to do a certain number of tasks that corresponded to a specific assessment.

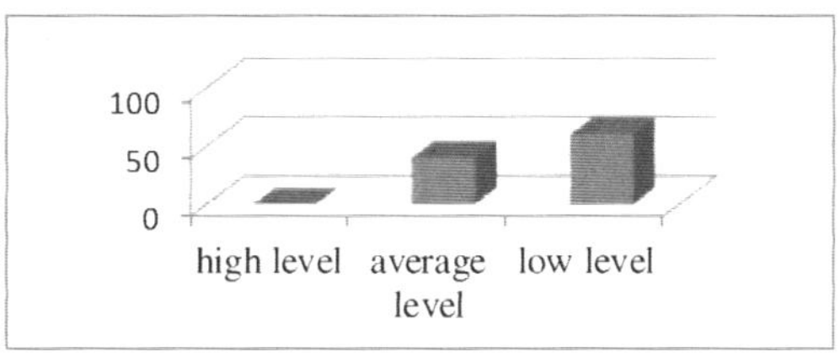

Figure 1. The results of the ascertaining phase of the experiment.

Thus, the results of the test indicate a low level of effectiveness of the used teaching package development in the course "Engineering and Computer Graphics."

In the process of didactic data creation insufficiently developed educational and methodological support was revealed. Thus, it was necessary to implement computer graphics during the study of the course "Engineering and Computer Graphics."

To develop this project, a thorough analysis of the content of the course program should be carried out.

The experiment was attended by students of Nizhny Novgorod State Pedagogical University named after Kozma Minin whose major is "Vocational education (by industry)" training profile "Construction" (7 people) and "Service in housing and communal services" (9 people), In total, 16 students took part in the experiment.

The procedure of monitoring the training course "Engineering and Computer Graphics" can consist of several stages:

1) monitoring the development of theoretical material on each topic of the course;

2) an interview on the implementation of any tasks on each topic of the course, writing essays, listening practice [16].

The first stage is recommended to be carried out at the end of the course study since students are not allowed preparation time to pass them - testing reveals general knowledge of the course (concepts, classification of concepts, laws, patterns, principles, methods, etc.) and basic mental skills [20].

Students who do not have positive assessment cannot be admitted to the main stage of the final test, approved by the department.

The overall assessment is the subject of intermediate tests at each stage.

The content of the testing stages:

1. The results of the assessment of theoretical knowledge are evaluated at a four-point scale ("excellent", "good", "satisfactory", "unsatisfactory").

2. The interview is the main stage of the final test which is conducted on the proposed situations, containing several tasks (or one complex), the implementation of which confirms that the student has the skills defined by the objectives of studying the discipline [23].

Then comes the questions of self-control as well as a list of questions for the exam in the course "Engineering and Computer Graphics", which will be reflected in the annex to the final qualifying work.

The purpose of teaching package development was to test the effectiveness of its use in a vocational training course (using the example of testing students in the "Engineering and Computer Graphics" discipline) for the "Construction" training profile in Nizhny Novgorod State Pedagogical University named after Kozma Minin.

The use of teaching package development for the course "Engineering and Computer Graphics" showed that the use of this tool allows you to improve the effectiveness of students' vocational training to the standards of required professional skills.

The study of educational material through the use of methodological support allows the teacher to pay more attention to the consolidation of knowledge and the implementation of exercises by students, which guarantees the acquisition by students more durable skills [15].

Conducting the ascertaining stage at the university shows that the development of educational information without the use of properly developed educational and methodological documentation can be difficult.

The need to learn a large amount of information distracts from its reflection. This negatively affects the future specialists' training, prevents the transition of motivation of their professional activities to a higher level [2]. The use of educational and methodological documentation in the process of mastering the discipline is intended to eliminate this drawback.

In the course of the study, as part of a pedagogical experiment, two groups of students whose major is "Construction" and "Service in the Housing and Public Utilities" were involved. As well as at the ascertaining stage, the verification of the effectiveness of the implemented educational-methodical documentation was carried out by conducting testing in the topics of the course "Engineering and Computer Graphics" at the university.

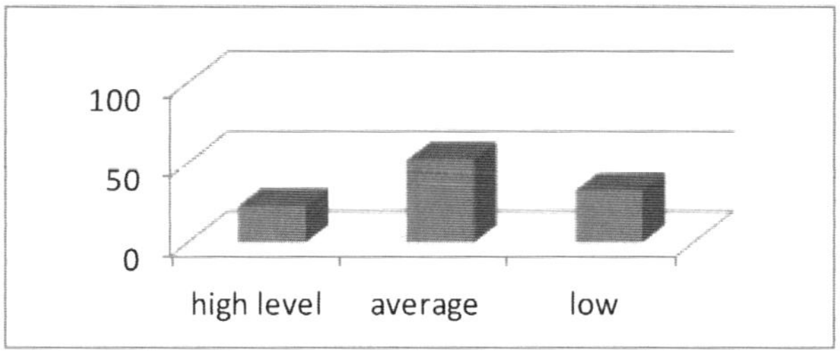

Figure 2. Results of the final stage

Evaluation of students' knowledge was carried out with the use of test control and reflected the learning of knowledge on the topics of the discipline in accordance with educational and methodical instructions.

The success of the teaching package development for the "Engineering and Computer Graphics" course is seen in the improvement of the list of recommended literature.

Thus, the results of testing in the discipline "Engineering and Computer Graphics" showed a higher level of material learning by students, the results confirmed the effectiveness of the implemented educational and methodological support.

4 Conclusions

We have identified specific features of modern requirements for teaching which affected students whose major is "Construction". Developing a teaching package we attracted employers, an expert assessment.

We have established the importance and the need to improve education in the engineering field since in modern society the services of engineers are extremely in demand. The better the training of future graduates in this area, the more these needs are met, which generally affects the country's economy. Engineering is one of the most promising areas, so higher schools should develop it further using the latest teaching tools, including technological ones which have become an integral element in both training and professional and everyday life.

Summing up the research, it must be said that for effective vocational training of students in modern conditions, teachers need to develop and use new teaching methods.

In the process of research the theoretical and methodological aspects of vocational course teaching package development are considered.

Teaching package development is aimed at improving the quality of vocational education using a combination of training tools and technologies for their use. This process is designed by the teacher [17].

The principles of the formation of a teaching package are: the principles of compliance of the content of education with its goals, which are determined by the needs of the development of society, science, culture and personality; target orientation, integrated mastering of knowledge and skills; feedback; functionality; integrity of the formation of competence and a number of others.

The intermediate stage reflected the low quality of students' knowledge obtained in the course of the educational process using the current teaching package.

This factor contributed to the need for a new educational program documentation development in accordance with modern trends in educational process and civil law relations. Introduction of methodological documentation developed in the study and verification of its success by conducting testing revealed the increase in the quality of knowledge gained by students.

As a result, the interdependence of the teaching package efficiency and the quality of the learning process is revealed. The restructuring of the content of methodological activity requires a turn from

IOP Conf. Series: Materials Science and Engineering **483** (2019) 012033 doi:10.1088/1757-899X/483/1/012033

normative regulation of teaching activity to stimulating their creative activity, active introduction of innovative approaches in the system of education.

References

[1] Ajeenkya D, Patil Y, Gagandeep Nagra, Gopal R 2014 A study on total quality management in higher education *International Journal of Management* **(5)** 1-6

[2] Brusakova I A, Kaznacheeva S N, Perova T V 2016 The development of the cognitive activity of students through training in professional self-determination. In Strategic Partnership of Universities and Enterprises of Hi-Tech Branches *Science. Education. Innovations IEEE V Forum* 113-114

[3] Gorkaeva E Yu 2014 Features of teaching and methodological support in college *Young scientist* **18** 538-539

[4] Gusev P P 2013 Methodical readiness of teachers to create an integrated educational and methodological support of the educational process *Secondary professional education* **3** 22-25

[5] Garina E P, Kuznetsov V P, Lapaev D N, Romanovskaya E V, Yashin S N 2017 Formation of the production system elements and R&D product development processes in the early stages of the project *Journal of Applied Economic Sciences* **12 (2)** 538-542

[6] Garina E P, Garin A P, Kuznetsov V P, Popkova E G, Potashnik Y S 2018 Comparison of approaches to development of industrial production in the context of the development of a complex product *Advances in Intelligent Systems and Computing* **622** 422-431

[7] Grishina A V, Volkova E N 2017 Features of the subjectivity structure of teenagers with different level of computer game addiction *Journal of Pharmaceutical Sciences and Research,* **9 (7)** 1108-1116

[8] Gruzdeva M L, Smirnova Zh V 2017 Results of the introduction of the model of management of independent work of students in the educational process of the university *Vestnik of Minin University* **(1)** 12

[9] Egorova A, Yaschkova E, Sineva N, Schkunova A, Semenov S, Klyueva Y 2016 Mapping of losses within organization of service activity for effective use of equipment *International Journal of Environmental and Science Education,* **11 (18)**11819-11830

[10] Zulkharnaeva A, Vinokurova N, Krivdina I, Martilova N, Badin M 2017 Training of a geography teacher in the process of getting master's degree to form the experience of making decisions on environmental problems by students of 10 grades: *Theory and methodological experience Man in India* **97 (15)** 559-571

[11] Ibatova A Z, Ippolitova N V, Akhmetov L G 2018 Professional training of future specialists in higher school: Intercultural component *International Journal of Civil Engineering and Technology* **9 (3)** 508-514

[12] Bulaeva M N, Vaganova O I, Gladkova M N 2018 Activity technologies in a professional educational institution *Baltic Humanitarian Journal* T 7 **3 (24)** 167-170

[13] Ilyashenko L K, Vaganova O I, Smirnova Z V, Gruzdeva M L, Chanchina A V 2018 Structure and content of the electronic school-methodical complex on the discipline "mechanics of soils, foundations and foundations" *International Journal of Mechanical Engineering and Technology* **9 (4)** 1088-1096

[14] Kaznacheeva S N, Lazutina A L, Perova T V, Smirnova J V, Chelnokova E A 2018 Research trends of HR management in tourism *Advances in Intelligent Systems and Computing,* **622** 448-455

[15] Kashtanova S N, Kudryavtsev V A 2017 Career Routing as a Strategic Line for the Professional and Personal Development of Persons with Disabilities and Disabilities *Vesknik of Minin University* **6 (2)** 3

[16] Markova S M, Sedykh E P, Tsyplakova S A, Polunin V Y 2018 Perspective trends of development of professional pedagogics as a science *Advances in Intelligent Systems and Computing* **622** 129-135

[17] Prokhorova M P, Semchenko A A 2018 Involving of trainees-future teachers of professional training in project activities in the discipline *Vestnik of Minin University* vol 6 **2** 6
[18] Smirnova ZH V, Gruzdeva M L, Krasikova O G 2017 Open Electronic Courses in The Educational Activity of The University *Vestnik of Minin University* **(4)** 3
[19] Smirnova Z V, Mukhina M V, Kutepova L I, Kutepov M M, Vaganova O I 2018 Organization of the research activities of service majors trainees *Advances in Intelligent Systems and Computing* **622** 187-193
[20] N S Selezneva, N G Alekseeva, S V Zotov 2015 With. Didactic quality assurance of secondary vocational education in the context of the formation of a multi-level educational space: Sat. articles /; MOIN Chelyab. region, ChelIRPO. - Chelyabinsk 121 p
[21] Fedorov A A, Paputkova G A, Ilaltdinova E Y, Filchenkova I F, Solovev M Y 2017 Model for employer-sponsored education of teachers: Opportunities and challenges *Man in India* **97 (11)** 101-102
[22] Kutepov M M, Vaganova O I, Trutanova A V 2017 Possibilities of health-saving technologies in the formation of a healthy lifestyle *Baltic Humanitarian Journal* 6(3) 210-213
[23] Vaganova O I, Smirnova ZH V, Trutanova A V 2017 Organization of research activities of bachelor of professional education in electronic form *Azimuth of Scientific Research: Pedagogy and Psychology* **6(3)** 239-241
[24] Tsyplakova S A, Grishanova M N, Korovina E A, Somova N M 2016 Theoretical bases of designing of educational systems *Azimuth of Scientific Research: Pedagogy and Psychology* Vol 5 **1 (14)** 131-133

IOP Conf. Series: Materials Science and Engineering **483** (2019) 012034 doi:10.1088/1757-899X/483/1/012034

The specifics of academic discourse while training translators as professional staff for work in the chemical industry

N B Egorchenkova[1], O V Korobova[2], Zh V Smirnova[3]

[1]Volgograd, 400082, Fadeeva street, 51, apt. 70
[2]Volgograd, 400081, Tvardovskogo street, 11, apt. 7
[3]Minin Nizhny Novgorod State Pedagogical University, ul. Ulyanova,1, Nizhny Novgorod, (603004), Russia

natalia.egorchenkova@volsu.ru

Abstract. In the article, the authors turn to the question of the implementation of academic discourse as a part of training translators as professional personnel for work in modern plants of the chemical industry. The article describes the main approaches to understanding discourse as a complex and multifaceted object of study. Besides, academic discourse is analyzed from the standpoint of its institutionalization and constitutive features. The article discusses the relationship between the participants of academic discourse and their strategies, and also the main values forming in the framework of institutional discourse. Special attention is paid to the formation of the professional competence of the translator, in particular, the discursive competence that is part of the communicative competence. The article substantiates the possibility of expanding the understanding of discursive competence from the standpoint of the linguistic component of perception and generation of texts, on the one hand, and mastering a set of communicative and interactive skills which are necessary for working in chemical production, on the other hand. The article describes the basic principles of the organization of industrial production in general and discusses the feature of the work of a translator at a chemical plant, in particular. The authors conclude that in a situation of interpreting an object is perceived by a translator at different modal levels, reducing its multimodality to a language code. In this connection, the article shows the need to introduce new methods for the formation of the discursive competences of the translator, allowing to improve the work of modern plants of the chemical industry.

1. Introduction

According to the Concept of the Long-Term Social and Economic Development of the Russian Federation, approved by the Government of the Russian Federation on November 17, 2008, one of the priority directions of the innovation development of the economic policy of Russia is the formation of competitive advantages directly due to the professional training of highly qualified personnel.

In accordance with this concept, the national innovation system should be restored, contributing to improving the competitiveness of the country based on non-raw material factors, and improving the quality of professional training and qualification of staff in the industrial sector of the economy.

Despite the political and economic situation, the critical conditions and the process of globalization, the chemical industry is in its heyday, since it is chemical products that can solve many of the pressing problems of mankind. The world market of products is growing steadily. Specialties related to work in the chemical industry are becoming more and more popular, especially in our region, in which there are several chemical plants operating internationally and increasing their participation in the global chemical industry.

Therefore, training highly qualified translators with deep knowledge in such professional field as the chemical industry, as well as training aimed at developing a number of competencies required in the practice of translators in various fields of professional activity, fostering a critical attitude to the results of the translation process and its editing is important for state economic policy strategically and for the economic policies of the Volgograd region.

In the light of the new education paradigm focused on the principles of the Bologna process, the global goal of translator training is the formation of professional competence of a specialist in the process of developing his potential and disclosing personal resources, which is carried out primarily in the framework of academic discourse as a result of interactive interaction between the professor and student.

2. Institutional and constitutive signs of academic discourse

Discourse category is a complex and multifaceted object of study. This complexity is caused by a variety of existing industry concepts of discourse: philosophical, literary, linguistic. However, there is not still universally accepted definition of «discourse» that would cover all cases of its use. Our analysis of interpretations and definitions of the phenomenon allows us to conclude that in modern linguistic science there are two main positions in understanding the discourse nature.

According to the first of them, discourse is a difficult complex of interrelated texts which function within the same communicative sphere, and certain areas of human cognition and communication are special discourses, for example, medical, legal, etc. According to the second position, discourse is understood as «a specific communicative event, carried out in a certain cognitively and typologically determined communicative space, which has a probabilistic nature and is often characterized by the absence of clear boundaries» [Dake van T., 1989, p. 11-14].

In our work, following K. F. Sedov, by «discourse» we mean the process of social interaction of people imprinted in semiotically different signs [Sedov, 2011, p. 13].

From the standpoint of sociolinguistics, it is customary to select two main types of discourse: personal (person-oriented) and institutional (socially-oriented) [Atman, 2012, p. 78-82]. As for personal discourse, according to V.I. Karasik, it exists in two main varieties: household and being communication [Karasik, 2012, p. 240]. Household communication is communication between well-known people, which is necessary to maintain contact and solve everyday problems. In the being discourse, the inner world of communicants is revealed in all its richness. Communication has a developed, extremely meaningful character and all forms of speech are used on a base of the literary language.

Institutional discourse is communication within the framework of status-role relationships. In modern society there are such institutional discourses as political, diplomatic, administrative, law, military, pedagogical, religious, mystical, medical, business, advertising, sports, scientific, stage and mass information.

In identifying the institutional sphere of discourse manifestations, there are communication pairs. Communication of these couples should be carried out according to the norms and standards of communication inside one of the following five basic social institutions: institution of family and marriage (husband / wife, father / son-daughter, mother / daughter-son); institution of state (officer / soldier, chief / subordinate); economic institution (seller / buyer, taxi driver / passenger, manager / client); institution of education (teacher / pupil, professor / student) and institution of religion (preacher / parishioner) [Golubeva, 2008, p. 13-18]. When selecting typical communicative pairs inside each of the institutions, it is taken into account that each member of the couple plays a certain social role that «reflects the interactional conditions between the subject of the communicative action and its object» [Krysin, p. 57].

There are three types of communicative-social relations within any of the mentioned institutions: equal status of subjects (K1 = K2), subordination (K1 <K2) and dominant (K1> K2), which determine the status of each of the communicants. For example, at the institution of education, the statuses of the social roles of the teacher and the pupil do not allow the child to dictate his terms of conducting a dialogue through his speech actions since his status is defined as « dependence». At the institution of the army, the statuses of an officer and a soldier are related as an officer> soldier, therefore the officer feels his right to conflict openly, while the soldier feels addicted. The status inside the communicative couple suggests the right or absence of the right to certain speech acts in a communication situation. Indicators of status in interactions are markers that indicate both the social status of the recipient and his unequal position in the discourse.

Systems of roles, statuses and sanctions are created in the form of social institutions, which are the most complex and important types of social connections for society.

The institutionalization of activity and social relations suggests their well-known standardization and consolidation in certain forms, «including linguistic ones» [Penkov, 1972, p. 51]. The speech behavior of people in a socially-oriented society has a number of features. In social interaction, «speech is aimed at the organization of joint activities of people». This feature predetermines a much stricter regulation of speech behavior, and the observance of «norms of speech behavior in the sphere of socially-oriented communication is accompanied by much more strict control» [Romanov, 2009, p. 87].

Relationships between partners, due to the social role status, are marked by certain linguistic means, signaling the interlocutor about the status of his partner. Moreover, the choice of linguistic means in a particular type of interaction when implementing the same communicative intention (intention) depends on the relationship between partners and their social role status [Romanov, 1992, p. 45-46].

All participants of the institutional discourse are divided into agents (representatives of the institute) and clients (people who apply to the representatives of the institute). The first are those who play an active role in institutional communication, the second is those who are forced to turn to agents and act as representatives of society as a whole in relation to representatives of the institute. For example, a teacher and a pupil, a doctor and a patient, a politician and an elector, a priest and a parishioner, an officer and a soldier.

In this way, institutional discourse is a specialized kind of communication between people, viewed from the perspective of their belonging to a particular social group, to a typical speech-behavioral situation. At the same time, the status-role speech interaction is carried out inside of established public institutions, the number and nature of which are determined by the needs of society at a particular phase of its development.

The institutionalization of academic discourse is determined by the purpose of communication and the status-role communication of social action agents. The purpose of academic discourse should be considered professional pedagogical communication in higher education, aimed at training highly qualified specialists. The participants of the academic discourse are agents - professors and clients - students. Communication between professors and students has unequal, bidirectional tone.

In the process of learning a student receives knowledge and besides is attached to cultural values, learns the norms and rules of behavior in society. In this case, the professor as an agent of academic discourse assesses the level of socialization of the student as a client of academic discourse.

Participants of academic discourse – professors and students of higher educational institutions - are characterized by certain distinctive features that are essential for this type of discourse. Professors need to have academic degrees, qualification categories, conducting research to work in higher education. Students are academic discourse clients who finished school. Besides, it is a conscious choice of a professional direction, specialized financing of training (payment in the case of contractual training, a system of scholarships and grants).

The chronotope of academic discourse is a combination of typical time and place of this discourse. In a broad sense - the period of study of a student in an appropriate institution and educational institution as a special territory for such activity. In the narrow sense - lectures, seminars, exams and tests, independent work in the library and an academic audience as a place at the university for conducting classes.

The values of academic discourse are difficult to separate from universal, general cultural and pedagogical ones. The basis of personal development is universal values. Then, in the process of socialization, a person learns cultural values. Besides, each individual forms his or her value scale, depending on personal taste and preferences. Based on the fact**,** university education is a voluntary choice, getting a profession is a value for clients, suggesting the opportunity to get a certain job and occupy a high social position**.**

The institutionalization of academic discourse suggests two types of strategies: professor strategies and student strategies. Professor strategies derive from the functional characteristics of the pedagogical profession and are determined by the targets of modern education. The strategies of the teacher are aimed at the development of the educational process and perform the function of regulating this process. The most important of these are strategies for learning, nurturing, monitoring and evaluating. The strategy of nurturing is aimed at the development of life and personality, as well as social and professional qualities of students. This strategy is realised through the organization of a particular subject activity.

Learning strategy consists of explanation, assistance, transfer of experience. The control strategy is realised at different stages of training. Speaking in general about the educational process, it should be noted that this strategy is used constantly: at the beginning of the training period to identify the general level of theoretical training of students, at the stage of mastering knowledge to establish feedback when explaining a new material, during the preparation for the final certification in the form of intermediate surveys or testing before the exam.

The social role of a professor determines the following strategy of academic discourse - an assessment strategy. First of all, we are talking about the assessment of students' knowledge, which is expressed in the rating or traditional points system for the semester. This assessment is given officially in the statement for each discipline. Besides, there is an informal assessment, but no less significant for the student's personality: praise, censure, criticism. The assessment strategy is verbally reflected in the professors' speech.

Student strategies are related to the needs of the individual to create their own system of knowledge for further self-realization. We distinguish two leading strategies of students: the strategy of knowledge and the strategy of self-realization. The first one is realised in the process of learning and communicating with professors and classmates. The second one provides the realization of a common life strategy at a certain life stage and it is a stepping stone for the realization of an individual's plans for studying, getting a profession and socialization.

The system-forming features of academic discourse include specific discursive formulas expressing the fragmented experience of communicative behaviour [Motorina, 2015, p. 23]. There are scripts, a special type of such mental formations with an emphasis on their value component. We consider linguocultural scripts as a set of instructions for language and behavioural code used for small articulate value situations.

3. Formation of the competence of translators as professional staff for work in the chemical industry within the framework of academic discourse

The main task of academic discourse is still the formation of the professional competence of the translator, which is interpreted as a hierarchical system of interacting and complementary competencies. Among the key professional competencies of the translator in the new Federal State Educational Standard is the discursive competence, which, as a subject of formation, was included in the communicative competence in the 90s of the last century. For a long time, the composition of skills included in the discursive competence has changed. An analysis of existing research has shown that most authors define discursive competence as the ability to create connected oral or written texts of different genres and functional styles in different communication situations [Common European competencies of foreign language proficiency: study, training, evaluation, 2003, p. 76; Bachmann, 1990, p. 227; Canale, 1980, p. 32]. At the same time, in modern communicative linguistics, discourse is understood not only as linguistic and verbal but also as a social phenomenon (M. L. Makarov, V. I.

Karasik, Yu. S. Stepanov, E. S. Kubryakova, T. V. Milevskaya and others). Concerning, it is possible to expand the understanding of the term «discursive competence». Two skills groups can be distinguished in relation to the teaching of oral speech and oral sequential translation: 1) skills related to the linguistic component of the utterance, that is, the perception and generation of coherent and logical utterances in oral speech with an adequate choice of tone, register and style of utterance (in the discourse of chemical production - the correct choice of terms); 2) skills associated with communicative and interactive components of discourse (in the discourse of chemical production - recognition of the intention and dominant strategies of interaction between negotiators, knowledge of situational and cultural conventions typical for the functioning of a chemical plant, a set of speech acts typically for discourse of chemical production and the ability to use this knowledge in an adequate situation of communication.

Therefore, it is necessary to consider the object of interpretation «a text in which not only linguistic factors are actualized - rules for the compatibility of words and the sequence of statements, their intonation design, forms of questioning, interrupting a partner, types of reaction to a question, etc., but also non-linguistic extralinguistic factors - cognitive, ethnographic, sociocultural, psychological» [Konetskaya, 1997, p. 106]. Undoubtedly, such a text is a complex object of perception for the translator. The result of perception, comprehension and understanding becomes a subjective image that arises in the minds of the recipient and includes images of the content of the speech work, the image of the communicative situation, the image of the specifics of the interaction of partners, as well as the image of each of the participants of communication. So, the understanding of oral speech is connected with making decisions by the recipient at the linguistic, cognitive, situational and pragmatic levels.

In the situation of oral translation, the object of translation is presented in the form of a multimodal (creative), unfolding in time and space of text that has a quantum character. This text may be generated by one speaker as well as by two or more participants (dialogue or polylogue) belonging to different cultural communities, with the direct participation of the translator as an intermediary of communication.

Therefore, in the situation of interpretation, the cultural component of discourse is actualized at all levels: linguistic, non-verbal, extralinguistic, mental. The source of information is not only sounding speech with all the richness of its rhetoric, intonation, and emotional coloring, but also facial expressions, gestures, and behavior of the interlocutors, and often their environment (interior of the room, landscape, people around, etc.).

The holistic impression of perceived information arises only as a result of the synthesis of such different semiotic systems because with perception of sounding speech, the addressee sees mimicry, gestures, the posture of the addressee, etc., involved in creating the communicative meaning and guiding of its decoding. [Schmitt, 2004, p. 32]. The interactive behavior of communicants is characterized by a multi-modality parameter, which is based on the simultaneous «community» of heterogeneous components that implement one communicative task. At the same time, the features of «multimodal interaction» allow defining «as key properties polycodularity as conjugation in a single space of works of different semiotic nature and multimodality as the use of different sensory modalities of an individual's perception» [Sonin, 2006, p. 23].

Western linguists (for example, A. Deppermann, L. Mondada, R. Schmitt), and also some Russian scientists (for example, A. A. Kibrik, A. G. Sonin, A. A. Petrova), who study communication as a multimodal event , consider that the «genuine» communication between the author and the addressee is carried out only in oral spontaneous discourse, that is, only when the recipient of the message receives information in its multimodality [Deppermann, 2013, p. 95; Mondada, 2008, 2; Schmitt, 2004, p. 32; Cybric, 2008, p. 135; Sonin, 2006, p. 23; Petrova, 2008, p. 108]. In oral spontaneous discourse, there is that multi-signaling that provides optimal decoding of the meaning of the message, and the «text in the head» of the addresser of the message is as close as possible to the «text in the head» of its addressee. In a certain way, a fixed prepared discourse is a forced form of communication, reducing multimodality to a (predominantly) language code [Petrova, 2008, p. 105].

In the discourse of industrial production, the situation of translation has its own characteristics due to the principles of the organization of the production process.

The production process is a combination of all the actions of people and labor tools which are necessary in the plant for the manufacture of products.

The main principles of the organization of the production process include, among others, the principle of proportionality (proportional productivity per unit of time of all production units of a plant), the principle of differentiation (separation of the production process of manufacturing products of the same name between separate units of a plant), the principle of combination (the combination of all or part of different processes for the manufacture of a particular product within one unit). At a chemical plant, for example, the production process is carried out using the methods of chemical industry technology in the production of products.

The chances for success of international plants depend on the competence of the translator, because translation is designed to ensure compliance with both the above-mentioned principles of differentiation and combination, and the principles of standardization (development, establishment and application of uniform conditions ensuring the best course of the production process), parallelism (simultaneous implementation of technological process on all or some of its operations) [Nepomniaschy, email. resource]. Therefore, the specifics of the work of a translator in a chemical plant consist, firstly, in the need to own a constantly growing corpus of chemical terms, abbreviations and phrases, which often have several translation options. However, the most often used in the certain production may be only one of them. For example, a chemical product such as PVC resin (polyvinyl chloride) is produced at the chemical plant JSC «Kaustik». This term has several translation options: Polyvinylchlorid or PVC-Harz. However, at the plant this chemical product is simply called «resin», which can confuse an unprepared translator, who in the output text will give an incorrect translation. Therefore, secondly, when preparing an interpreter for work in an industry, it is necessary to take into account the situational conditionality of his activity.

Obviously, it is the discursive competence to help the translator to do his tasks within the framework of the realization of production activities because this competence allows you to link knowledge, skills and ability to act (create texts) in any situation.

4. Conclusion

So, in the situation of interpreting an object is presented in the form of a discourse and is perceived by the translator in its multimodality. So, the ability to perceive and understand information at various modal levels, as well as take into account perceived information when formulating the meaning of a message, is an integral part of the future translator's discursive competence, which ensures the success of the translation process in oral communication. In this way, to provide the success of the translation process in oral communication, for more efficient training of translators in the professional field, within the framework of academic discourse, it is advisable to introduce new methods of developing professional competence of translators who could improve the functioning of their training system for work in production and the work of modern industrial plant in general

References

[1] Atman OU 2012 Discourse of the Saturday Radio Address as a genre of political communication Modern problems of linguistics and linguodidactics: concepts and prospects: materials of the second international correspondence scientific-methodical conference (Volgograd: VolSU Publishing House) 78-82.

[2] Golubeva TM 2008 Gender and the strategy of discrediting in the pre-election discourse Bulletin of the Nizhny Novgorod University. N.I. Lobachevsky. *Series: Philology. Art Criticism* **5** 13-18

[3] Dake van T. 1989 Cognition. Communication (Moscow: Progress) 312 p.

[4] Karasik V and 2012 The Language Culture Matrix (Volgograd: Paradigm) 448 p.

[5] Karasik V and 2002 The language circle: personality, concepts, discourse (Volgograd: Change)

477 p.

[6] Kibrik AA 2008 Multimodal linguistics: research directions V.N. Polyakov (Ed.) Text Processing and Cognitive Technologies (Cognitive Modeling in Linguistics Proceedings of the X International Conference) (Kazan: Kazan University Press) p. 132-145.

[7] Konetskaya VP 1997 The Sociology of Communications: a textbook (Moscow: International University of Business and Management) 304 p.

[8] The concept of long-term socio-economic development of the Russian Federation until 2020. Approved by the Order of the Chairman of the Government of the Russian Federation No. 1662-p dated 11/17/2008.

[9] Krysin LP 1976 Speech communication and social roles of speakers Socio-linguistic studies (Moscow: Science) p. 57.

[10] *Common European Framework of Reference for Foreign Language Proficiency: Study, Training, Assessment 2003 (Moscow: MSLU)* 256 p

[11] Motorina N. In 2015 Linguocultural scripts of communicative behavior: monograph (Volgograd: VolgGTU) 112 p.

[12] Nepomnyashchiy E G Economics and Enterprise Management: lecture notes [Electronic resource] (Taganrog: Publishing house of TSU) Available at: http://www.aup.ru/books/m83/7.htm (accessed 1997)

[13] Penkov E M 1972 Social Norms-Regulators of Personal Behavior: Some Questions of Methodology and Theory (Moscow: Thought) 51 p.

[14] Petrova A A 2008 The Multimodal aspects of the Interaction Study Vestnik Volgograd. state un-that. Ser. 2, Linguistics number 2 (8). (Volgograd: Volgograd Publishing House) p. 105-111

[15] Romanov A A 1992 Communicative initiative of the speaker in the dialogue Text as a Structure (Moscow: Institute of Nuclear Sciences of the USSR Academy of Sciences) p. 45-46.

[16] Romanov A A The macrostructure of discourse. Institutional and linguistic and cultural characteristics of communicators A A Romanov N A 2009 Belous Fundamentals of Discourse (Ulyanovsk: UlSU) p. 87

[17] Sedov K F 2011 Discourse as suggestion: Irrational influence in interpersonal communication (Moscow: Labyrinth)

[18] Sonin A G 2006 Modeling the mechanisms for understanding polycode codes: Diss. for competition scholar. degree doctor filol. Sciences: 10.02.19 (Moscow) 310 p.

[19] Tsukhlo S In Russian industry in October 2012. *Scientific Herald of IEP. Gaidar* **56**

[20] Bachmann L F 1990 *Fundamental Concepts in Language Testing* (Oxford:Oxford University Press) p 408

[21] Canale M Swain M 1980 *Theoretical bases of communicative approaches to second language teaching and testing (Applied Linguistics 1)* pp 1–47

[22] Deppermann A 2013 Conversation Analytic Studies of Multimodal Interactio .*Journal of Pragmatics* **46, 1** pp. 1-7, 91-121

[23] Mondada L 2008 *Doing video for a sequential and multimodal analysis of social interaction: Videotaping institutional telephone calls.* Internet: www.qualitativeresearch.net

[24] Schmitt R 2004 *Bericht über das 1. Arbeitstreffen «Multimodale Kommunikation* (Sprachreport 1) 31–34

IOP Conf. Series: Materials Science and Engineering **483** (2019) 012035 doi:10.1088/1757-899X/483/1/012035

Information and analysis support for pharmaceutical business management

A V Evstratov[1], I A Ezangina[1] and E V Novozhenina[1]

[1]Volgograd State Technical University, 28 , Lenina avenue, the city of Volgograd, 400005, Russia

E-mail: evstratov.mail@gmail.com, ezangirina@rambler.ru, elena.nvzh@mail.ru

Abstract. The present study emphasizes growing importance of information support for pharmaceutical companies. Information provides not only decision-making and its transfer between firm's departments but also increasing the company's competitiveness by obtaining and collecting some additional data on plan numbers deviations. It also ensures adding new scientific laws, inventions and novelties to the total body and capacity of knowledge and technical solutions.

1. Introduction

The topicality of the pharmaceutical market research is, first and foremost, related to its increasing social significance, which is revealed in immanent population need for recovery and a limited ability of manufactures and suppliers to meet this need. A high level of social significance in a pharmaceutical market is due to a low level of demand elasticity for the price of medicines. For example, in the mid-2000s, the Government of the Russian Federation began to pay more attention to the health care sector including development of the pharmaceutical market. In 2009 the strategy of the pharmaceutical industry development for the period up to 2020 was approved. According to this programme, some strategic actions are supposed to be done in due sequence. They will enable to come up to a totally new level of the pharmaceutical market development and to provide the population of the Russian Federation with medicines and health care products. Achieving a totally new level of the pharmaceutical market development is associated with intensification of information and analytical activity of pharmaceutical companies both in the sphere of drug design and production and in the sphere of circulation and realization of medicines by wholesale and retail suppliers.

Information analytics (hereinafter referred to as IA) of a company involves obtaining, analysing, collecting and accumulating information with further foresight to the work and development of a pharmaceutical company. Using foresight in the strategy of transnational business provides an in-depth analysis of alternatives and proves to be a foundation for understanding strategic corporation challenges and opportunities [1]. In future, new and coming into service technologies in the field of information support and analysis will objectively lead to the transformation of a labor market [2].

According to many researchers, information management and analysis can greatly increase labor efficiency of enterprises [3]. Its enhancement will give a necessary boost for the development of the economy of the whole country [4] taking into account the existing macroeconomic situation in the long run [5]. But in any case, the possibility of accelerating economic development is connected with changing the growth model [6] and also with pursuing a well-balanced monetary policy [7].

2. Results and Discussion

The need to develop and bring new medicines to a market is connected with the search for competitive advantages of pharmaceutical companies which result from cost savings. The formation of the competitive environment of a pharmaceutical market takes place under the influence of a large number of factors. They can be divided into internal factors such as: internal resources of the company, pricing policy, product policy, personnel policy, strategic and financial management, and external ones like development of the world economy, government regulation, level of technological development, level of cultural development, availability of natural resources.

Globalization of the world economy and implementation of the concept of sustainable development result in increased competition in a pharmaceutical market. Cost savings are linked not only with production activities of a pharmaceutical company but also with savings on advertising and marketing communications. Apart from that, mergers and acquisitions in a pharmaceutical market contribute much to cost savings thanks to joint research efforts. Development of health care as a sector of the economy is determined by the growth of innovation activity [8], investment attractiveness of the pharmaceutical industry [9] and a pharmaceutical market as a whole. Health care costs continue to increase in the world every year. This growth is usually associated with GDP level, and it is most noticeable in mature markets, which form the bulk of pharmaceutical industry's revenues. As a rule, the higher GDP level of a country, the greater is its cost of the health care system.

An important factor in the smooth running of a holding pharmaceutical company is transfer of reliable and timely information. The IA structure of a pharmaceutical company is given in Figure 1.

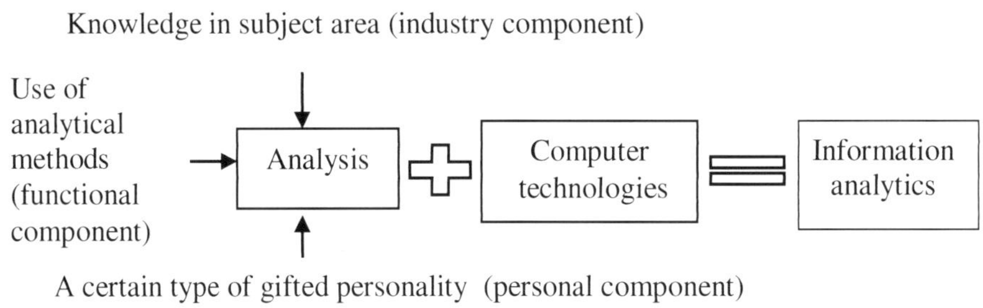

Figure 1. IA structure of a pharmaceutical company (authors').

To transfer reliable and updated information it is necessary to use various data processing systems capable of solving problems in various areas (trade and procurement, accounting, warehousing), and in the field of accounting automation, organization of the technological process of medicines production and medical products. Improving efficiency of pharmaceutical businesses is one of the priorities for the development of the Russian economy in accordance with the state programme of the development of the pharmaceutical and medical industry for 2013-2020 [10]. First of all, this refers to supplying pharmaceutical companies with commodities from external and internal warehouses via intra-company channels, because information is data that helps to reduce costs and increase efficiency of a pharmaceutical company (Table 1).

Table 1. IA tasks within a pharmaceutical company.

Task	IA types	Necessary data	Information sources
Increasing company's competitiveness	Analysis of the market situation. Defining the company's market share	Market structure and development trends data	Statistics and analytics of market conditions
	Analysis of market segments in accordance with the selected criteria	Market characteristics	Questionnaire of professionals and consumers

Task	IA types	Necessary data	Information sources
Development of goods production program	Identification of consumer preferences and the degree of their satisfaction	Main characteristics of potential customers	Data on the criteria and characteristics of market segmentation
	Process of evaluation of goods by buyers	Attitude of buyers to goods	Testing product name and packaging quality using group discussions and focus groups
	Evaluation of the program of goods production in terms of the economic goal of the company	Market prices on goods and quantity of sold products	Primary and secondary information
	Selection of promising areas of product range development	Main market trends and conditions	Primary and secondary information
Defining price policy	Selection of pricing methods in the context of actual market prices and forecast data	Data on the minimum price and competitors' prices	Statistical price level data and audit data for companies' pricing policy
Selection of distribution channels	Selection of distribution channel participants and their motivation	Transportation costs and sales revenue data.	Corporate reports and surveys of intermediaries and buyers
	Selection of stocking location, methods of storage and the inventory management system	Data on purchasing power of the area and storage costs evaluation	Social and demographic statistics
	Evaluation of efficiency of wholesale and retail companies	Data on potential sellers and buyers	Special inquiry, monitoring, survey
Advertising campaign development	Evaluation of efficiency of promotional activities	Data on the market saturation degree in terms of product life cycle	Corporate reports and advertising costs analysis
	Evaluation and selection of advertising media	Data on the availability of advertising media	Cost of standard advertising media
	Monitoring results of advertising activity	Data on economic advertising efficiency	Corporate reports and sales volumes

The pharmaceutical industry is recognized to be socially significant nowadays. That is why receiving timely information aimed at further processing and using obtained results in the process of production activities is crucial. A key step towards increase of pharmaceutical company's competitiveness and its internationalization is a fully developed management system. Mergers and acquisitions occurring in a pharmaceutical market lead to reorganization of a newly formed company [11]. There is also a need to optimize a range of products, but first of all, it is necessary to change an organizational structure. The organizational structure of a pharmaceutical company is a structure consisting of several branches that have a management company as a centralized monitoring and management body. Automation of such management system allows reducing the risk of making management mistakes.

The topicality of this study lies in the fact that leading pharmaceutical companies spend significant financial resources on information technologies which make it possible to reduce the costs of planning and managing all information flows concerning raw materials, products, services, and processing transactions. They also serve to retain and strengthen their positions in the global market in view of the internationalization of the global economy.

According to the authors, in future, business-entities that are not, for any reason, responding to the changing conjuncture of a pharmaceutical market may face the problem of reducing their competitiveness and attractiveness of their shares. Virtually, implementation of modern information technologies in pharmaceutical companies will contribute much to business development. They bring the following benefits:

1. electronic document management, which allows to automate the stages of the life cycle of documents from design to archiving. Advantages of electronic document management should be recognized as follows:

- consumables expenses decrease, such as: payment for postal and courier services, photocopying materials, minimizing labor costs;
- work of all departments and divisions of a company in one information space;
- safety and security of documents thanks to use of data encryption and electronic signatures which minimizes the risk of information leakage.

When staff members deal with document flow their productivity increases and mistakes in documents processing due to the human factor are reduced.

2. production technology management which contributes to increasing labor productivity due to automatic and semi-automatic production lines, product quality control, reduction of technological errors associated with human factors;

3. automation of accounting and planning and decision-making system. To make management more effective it is necessary for the manager to change some parameters of the visual assessment of the events happening. The introduction of automated accounting systems mediates:

- acceleration of obtaining operational statistics;
- elimination of errors in the calculations, entailing additional time and money costs;
- improving the quality of labor of accountants due to the minimization of routine work.

In the context of this study, it should be noted, that over the past few years, the pharmaceutical industry has undergone important changes related to updating molecular approaches. In the future, these approaches will allow diagnosing diseases more accurately and proposing solutions on patient's health advocacy.

In accordance with the choice of the company's development strategy, information technologies, given below, can be an important key to this transformation:

1. computing systems with petaflop-level performance (1015 operations per second). The generation of machines with such productivity in the pharmaceutical industry will create conditions for the mass use of biomolecular simulation;

2. Grid technology which is a rapidly developing tool uniting resources of thousands (millions) of personal computers, a huge «virtual» system with powerful computational capabilities. The implementation of this technology means solving tasks in the field of screening for the coincidence of DNA sequences, conducting analytical actions on sales volumes and promotion devices in real time mode;

3. bio-simulation which is the process of creating a virtual or real model of the human body based on the information of a CT scanner, NMR scanner or ultrasound scan results. It is used for improving quality in diagnosis, modeling planned operations, producing individual prostheses or implants, analysing treatment results. Predictive bio-simulation allows pharmaceutical companies to reduce the number of laboratory tests devoted to finding out potential drugs, their health effect in terms of efficiency and safety;

4. intelligent markers and radio frequency identifiers designed to enhance product control at any production stage, as well as at the distribution stage. The main advantage of RFID is the replacement

of slow and inefficient traditional production processes and the switch of pharmaceutical manufacturers to the production of a range of complex drugs in smaller batches;

5. modern servers which are virtual networks of information storage with transparent integrated file and archiving management systems focused on storing vast amount of information;

6. Production Process Analysis (PAT) technology which continuously and automatically monitors production processes in real time mode. The introduction of PAT technology largely improves the quality of production processes, resource conservation, since it is more profitable to carry out the correction online than to reject products that are beyond the limits of tolerance from the standard.

3. Conclusion

The future development of a pharmaceutical market is associated with consumers' need to be healthy and to be satisfied with the need for medicines. Based on business competition, a pharmaceutical market being an emerging economic system of interaction between its main subjects of production and consumption meets the population needs, and namely, it provides people with socially significant products in maintaining their health.

The existence of a modern pharmaceutical market is thus predetermined by its vital priority. It is due to the fact that consumers demonstrate their inherent need for recovery but meeting this demand is associated with manufacturers' limited capacities based on some technological and legal factors which affect manufacturers. In modern economies, only healthy people, with other things being equal, will be more competitive in a labor market. Therefore, they will be able to create greater added value from the production factors at their disposal and thus improve their well-being.

This study allowed the authors of the article to draw the conclusion that pharmaceutical industry has a significant investment potential. While developing it should be supported by the permanent implementation of information technology in production and financial cycles. In this regard, the advantages of using the corporate portal of the organization, electronic document management, systems of client, supplier and partner interaction management have been updated. However, it has been found by the authors that automation of information technology processes is accompanied by many problems such as staff competence, difficulty of setting up and adapting to individual business characteristics, lack of resources and poor development of the financing mechanism of training companies' staff to information technology skills.

A pharmaceutical market is one of the markets with developed competition, associated with a large volume and variety of medicines, a large number of suppliers (including importers), significant interchangeability of drugs within groups, development of pharmacy chains. Increased market competition in the pharmaceutical industry makes companies look for new ways of development. Therefore, leadership in a pharmaceutical market will belong to those companies that will be able to recognize the emerging perspectives and opportunities for their development before their main competitors can do that.

References

[1] Bereznoy A 2017 *Foresight and STI Governance* **11(1)** (Moscow: National Research University Higher School of Economics) 9-22

[2] Sorgner A 2017 *Foresight and STI Governance* **11(3)** (Moscow: National Research University Higher School of Economics) 37-48

[3] Bessonova, E V 2018 *Voprosy Ekonomiki* **7** (Moscow: Non-profit Partnership "Redaktsiya zhurnala 'Voprosy Ekonomiki'") 96-118

[4] Oreshkin M S 2018 *Ekonomicheskaya politika* **13(3)** (Moscow: Autonomous Non-profit Organization Editorial Board of the Journal "Economic Policy") 8-27

[5] Manevich V E 2018 *Voprosy Ekonomiki* **1** (Moscow: Non-profit Partnership "Redaktsiya zhurnala 'Voprosy Ekonomiki'") 40-63

[6] Akindinova N V, Nazarova A G and Chepel A A 2018 *Journal of the New Economic Association* **3(39)** (Moscow: Autonomous non-profit organization Journal of the New

IOP Conf. Series: Materials Science and Engineering **483** (2019) 012035 doi:10.1088/1757-899X/483/1/012035

Economic Association) 128-137

[7] Ezangina I A, Evstratov A V and Jovanovic T G 2016 *International Journal of Economics and Financial Issues* **6(2)** (Mersin: Econjournals) 58-64

[8] Manturov D V 2018 *Ekonomicheskaya politika* **13(2)** (Moscow: Autonomous Non-profit Organization Editorial Board of the Journal "Economic Policy") 64-77

[9] Petrov A P and Petrov P P 2011 *Economy of region* **2** (Yekaterinburg: Institute of Economics, Ural Branch of the Russian Academy of Sciences) 213-220

[10] Sapir E V, Karachev I A and Chzhan M 2016 *Economy of region* **12(4)** (Yekaterinburg: Institute of Economics, Ural Branch of the Russian Academy of Sciences) 1194-1204

[11] Evstratov A V 2014 *World Applied Sciences Journal* **32(7)** (Dubai: International Digital Organization for Scientific Information) 1400-03

Conflict communications in the production management

A S Sukhova, I G Selezneva, I N Naumov

Volgograd State Technical University, 28 Lenin Avenue, Volgograd, 400005, Russia

E-mail: belousov34@yandex.ru

Abstract. In this article, the nature of conflict communications between the executive and the subordinates in the sphere of production management is investigated, as well as possible strategies and patterns of behavior for both parties are determined. Also, the causes, mechanisms, and solutions for conflict communications are analyzed. This article presents the results of the practical research carried out at Volgograd's two enterprises. In this research, the way the executive's style of behavior in a conflict situation influenced the relationship with subordinates was studied.

1. Introduction

Conflict communications are found in all spheres of human activity and are often considered to be the significant and thrilling events of human life. Conflict communications – as Darwin, Marx, and Freud have shown – are a necessary and important element of life. However, their consequences may have negative effects sometimes, and these effects can be mitigated provided one possesses the skills of constructive conflict resolution. Revealing the deep foundations of conflict behavior, the psychological research is a key element in the process of cognition and management of conflict communications.

Conflict communications are a necessary component of the development of one's personality, interpersonal relations. They also play an important role in the development of any organization and society as a whole. There are potential resources for conflict communications in any organization, and these resources usually bear the imprint of the social system and the environment in which the organization operates.

Conflict communications do not always contribute to the development of the organization. For instance, conflict communications can be regarded as beneficial if they take place between the parts of the organization, thus raising the effectiveness and helping achieve the main goals[1].

The executive of any level should strive to achieve the best results so it is necessary to ensure proper interaction both between groups and between individual members, including both competition and conflict communication. Competition between groups in the organization, as well as between individual members, can be beneficial forcing them to work better, rallying the team [2].

Conflict communications are a strong radical means and its use by an opponent (a leader or a group) is justified only in extreme situations when other measures are not successful. Conflict communications violate the normal course of events and the team's well-being and are considered socially undesirable, and in acute and abrupt forms – and as an unacceptable phenomenon. To solve such situations, it is necessary to eliminate the causes which gave rise to conflict and to restore normal relations.

2. Discussion

The study of the methods of interaction between the executive and the subordinate is associated with the definition of tactics or styles of interaction. The number of behavioral tactics, highlighted by different researchers, varies from two to five; on the other hand, there is no unity of ideas about their content characteristics [3]. However, the most productive way to study the interaction between the executive and the subordinate is to diagnose behavior styles using methods based on the ideas of Thomas E Phipps JR [4].

Various combinations of quality and degree of attitudes intensity of interacting parties determine the way in which their interaction will unfold. On the one hand, understanding the basic styles of behavior helps predict the tactics of interaction, and on the other hand, to assess the effectiveness of one's own behavior.

The following basic interaction styles are described:

1) The use of power – subordination. There is a high level of interest in achieving results combined with a lack of concern for the interests of the other side.

2) Smoothing of contradictions based on flexibility and tolerance. This interaction style is characterized by a high focus on the interests of the subordinate, and a low focus on the result. Interaction is often achieved by parties abandoning their original positions.

3) Neutrality maintenance no matter what it takes (the lowest values for both measurements).

4) The average position on both coordinates determines the strategy when the positions come to the agreement. The agreement is the result of negotiations.

5) A Paternalistic approach to interaction is possible with high support of the interests of another person and a high interest in finding a solution to the problem. With this orientation, partners are able to reason sensibly and to find out who is right by comparing the facts, logical analysis, and evaluation.

As the authors highlight, none of these styles characterize one person compared to another, though each style can be dominant. In addition, the dominance of one style can be temporary; a person can abandon it and change it to another if it is inefficient. Further conceptual development of the ideas of Blake, Mouton, and Thomas was carried out in the study of N V Grishina, in which the explanatory models mentioned above are directly applied to the subordinate relations.

In order to identify the dominant factors in conflict communications between the head of the enterprise and the subordinates, in 2017, the practical studies were conducted at two enterprises in Volgograd, examining the influence of the executive's behavior style in a conflict situation on his relationship with the subordinates. The main hypothesis of the study was based on the assumption that the relationship between the head and the subordinates is determined by the strategies of behavior used by the head in a conflict situation. Thus, the objects of the study were 10 chiefs and 2 directors of the enterprises. The sample size was 280 people. The survey was chosen as the method of data collection. The data obtained were encoded; automatic processing of the obtained sociological information was carried out with the help of Microsoft Excel editor.

The data collection procedure was carried out in two stages. In the first stage, each of the 12 executives was tested by the method of R Kilmann, K Thomas (in the adaptation of N V Grishina) aimed at identifying the strategies of behavior used in conflict situations. In the second stage, each executive was evaluated by his subordinates according to the following parameters: 1) the quality of the executive's work; 2) the level of special knowledge possessed; 3) independence; 4) a sense of responsibility; 5) perseverance; 6) the team management; 7) the attitude to subordinates; 8) the relationship with the superior managers; 9) self-control; 10) overall assessment. Each parameter was estimated on a five-point scale. Various combinations of quality and the degree of attitudes intensity of interacting parties determine how the process of interaction between the executive and the subordinate will unfold [5]. In this study, particular attention is paid to the problem of conflict communication between the executive and the subordinate, and possible strategies of the behavior of both parties are identified. As a tool for the analysis and settlement of conflict communications in the office, a package of diagnostic techniques and practical recommendations for the analysis, management, prevention, and resolution of conflicts is given to the executive. The results show that even in the groups with good relationships and in the organizations with effective management some conflict communication is not only possible, but even desirable. Undoubtedly, they are not always positive. In some cases, they may interfere with the satisfaction of the needs and desires of the subordinates and the achievement of the goals of the organization. Based on the analysis of each executive's performance, the following conclusions can be drawn.

3. Results

In cases when the executive's activity is mainly connected with working with people (human resources, administrative service), the subordinates demand from the executive such parameters as self-control, the appropriate relationship with subordinates. Such executives get a higher assessment if the prevailing strategies of behavior in the conflict are a strategy of cooperation.

The attitude of subordinates to the chiefs, whose direct duty in addition to the management of subordinates is the performance of production activities, involving the presence of deep professional knowledge (in the field of construction, etc.), is determined by a number of factors. These factors are the quality of work, expertise, and leadership; and to a lesser extent - self-control and the relationship with the subordinates [6].

Such executives are regarded by the subordinates as, first of all, the experts and their errors and mistakes in the management are not always noticed. In this case, the attitude of subordinates to their executive is less determined by the strategies of behavior in the conflict they prefer. It should also be noted that the overall assessment of the head of the personnel or administrative service is higher than that of the "production manager", though insignificant (at the level of statistically insignificant differences).

Almost every participant is looking for supporters, which means that the original conflict is replenished with new contradictions that reflect the interests of those who joined it. And since the conflict can bring less benefit, compared to material, moral, and psychological damage to the team, and can endanger physical and mental condition of the participants, the executive has to take measures. The executive's inaction can be regarded as indifference, cowardice, and incompetence.

Analyzing the behavioral tactics of managers in interaction with the subordinates, we can distinguish two main dimensions:

1) The aim to satisfy the interests of the subordinate involving cooperation and associated with an integrative style of behavior;

2) The aim to satisfy the personal interests of the executive and the achievement of his own goals in the first place.

The combination of these basic styles with the account of the degree of their severity create five tactics of interaction:

– A persistent and unable to unite (non-cooperative) leader is likely to choose the tactics of perseverance, competition, competition;
– A persistent and cooperative leader is likely to choose cooperation tactics;
– A non-persistent and non-cooperative leader is likely to choose avoidance tactics;
– A non-persistent and cooperative leader is likely to choose adaptation tactics;
– The average degree of expression of each of these styles is likely to lead to compromise tactics.

Analyzing the answers of respondents which characterize the influence of the behavioral style of the executive in a conflict situation on his relationship with subordinates, we can state the following. As can be seen (Table 1), if the activity of the executive implies the possession of serious professional knowledge, then his relations with the subordinates are determined not only by the strategy of behavior of this executive in the conflict communication but also by his assessment as a specialist. The assessment of the executive, whose work is mainly connected with interaction with people, is largely determined by the strategies of his behavior and such quality as self-control.

Table 1. Ways of resolving conflict communications in the workplace

Ways of resolving conflict communications	Executives, HR, and administrative services	Manufacturing managers
Conflict resolution according to the nature and content of the conflict		№ 3 № 10
Conflict resolution according to its aims		№ 4 № 5

Conflict resolution according to its functional sides	№ 2	
Conflict resolution based on emotional states	№ 7	
Conflict resolution according to its causal-motivational part	№ 11	
Conflict resolution according to the peculiarities of its participants	№ 6	
Conflict resolution according to its dynamic side		№ 8
Conflict resolution according to the tactics of influence on the conflicting parties		№ 9 № 12

Conflict communications between the head and the subordinate are inevitably associated with the aggravation of the emotional tone of the relationship and their progressive destabilization. However, the open contradiction arising in the course of industrial relations performs at the same time a kind of cognitive function when the development and dynamics of this situation lead to a systemic understanding of the conflict situation.

One reason for the conflict between an executive and a subordinate may lead to conflict communication of different psychological types with different development models, different consequences for the participants. It depends on such psychological factors as participants' perception of the situation, their attitude to it, the strategy of their behavior. The study conducted by the authors helps identify the following main models of the development of conflict communications in the workplace [7].

The "Industrial dispute" model. In this type of situations, there are contradictions on production questions between the executive and the subordinate and these questions are usually connected with their shared activity. Both the executive and the subordinate believe they can reach the agreement and look for the ways to achieve it. Their communication becomes intense, they discuss the situation, advantages and disadvantages of both points of view, use arguments, try to justify their position. The relations between the executive and the subordinate are characterized by goodwill, concern, and companionship. If conflict communication in the workplace develops according to this scheme, the subject of the proceedings will be the specific circumstance that caused the contradiction, and these differences will not be transferred to other aspects of the work. In this model, there is a high probability of making a really constructive decision that will satisfy both sides of the argument.

The "Formalization of communications" model. In the case of expanding the zone of contradiction in the workplace, arising on a wider range of issues, there are conflict situations developing under the scheme of the "production dispute". Recognizing a wide range of existing contradictions, the executive and the subordinate usually doubt that it is possible to reach the agreement and sometimes simply do not want to discuss controversial issues. If the decision is made, the parties turn to external, formal, and official methods of decision ("You are the boss, you decide", "Let the team members decide"). If one of the opponents is superior, he uses his right to make a final decision. The interaction of the participants in a conflict situation can be characterized as the interaction in the workplace. In such situations, the executive and the subordinate refuse to make a mutual decision but state their disagreements and claims, show the inability to agree, which causes serious damage to their future relations. Communications become strictly official, the communication becomes limited, all personal leaves the relationship, i.e. the relationship becomes strictly formal.

The "Psychological resistance" model. In this type of conflict communication, the real zone of disagreements in the workplace is not defined and tends to expand since the executive and the subordinate find it difficult to clearly determine the disagreements and tend to exaggerate them ("We are completely different people"). Communication in such a situation is almost impossible, opponents have a negative attitude towards each other and do not hide it either from themselves or from the partner, or from others. Any word can become a source for insults and disagreements. Both parties try to affect each other's feelings through openly hostile actions. In this conflict situation, it is not the fact of disagreements that is decisive but the development of this situation, the nature of communication, and relations between its participants.

Thus, in order to minimize conflict communications in the workplace between the executive and the subordinate the following factors must be taken into account:

1) Constructiveness depends on the friendly tone of the executive. To talk in a friendly way sounds simple and natural. It makes sense to try to speak in a friendly way when it comes even to the exchange of information in the workplace. Talking to the subordinate in a friendly tone means showing respect, concern, and eagerness to act in his interests;

2) The research conducted by psychologists concluded that those who report information and those who accept it, 80% act under the influence of feelings and only 20% – under the influence of the mind. Therefore, it is necessary to keep in mind the emotional side of communication when dealing with subordinates. The executive should perceive subordinates as a complex and should not be afraid to show his feelings in the conversation. Such an executive can win the subordinates' trust. The emotional part of the information is more difficult to decipher than its business content;

3) In the executive-subordinate conversation on production questions, it is important not only what the executive wants to report but also how much of the information the subordinate understands and how he processes what has been said;

4) The main resource for minimizing the conflicts in the workplace is the executive's patience. He should be able to listen to the subordinate and find enough time to do this;

5) Favorable interaction in the workplace helps maintain the atmosphere of frankness. Thus, the executive helps the subordinate take the right position;

6) If the executive wants the conversation to be effective, he should not resort to double-meaning and vague formulations. The subordinates will rather understand unambiguous orders, short, and clear phrases;

7) The executive should not avoid responsibility. He also has the right to make mistakes, but the employees can ascribe to him only the mistakes that he has really made;

8) The executive should be able to defend his projects or position. These qualities subordinates will appreciate.

9) If the subordinate is not ready for a serious conversation or is not able to conduct a constructive conversation, it is better to postpone the conversation. The proposal to postpone the conversation can come from both the subordinate and the executive. If they are not honest with each other, the conversation will not be constructive and each will justify the behavior of the other participant in different ways.

Each individual has his own personality that requires a respectful attitude. A positive attitude creates a good mood, inspires a person to new labor feats [8]. The executive's unfair assessment discourages the subordinates, narrows the consciousness and creativity, the subordinates quickly get tired under the "burden of work", which in itself feeds the emergence of conflict communications. The most frequently mentioned causes of conflict communication are the following:

– The errors in the application of sanctions and incentives, especially when this system is poorly thought out and the executive encourages the employee on the basis of personal sympathy, not an objective assessment of his work;

– The flaws in the distribution of work between the employees according to their knowledge, experience, position, and salary;

– Every conscientious worker expects promotion which is associated with his financial position, prestige in the team and, finally, self-esteem but the executive invites someone "from the outside" when there is a candidate inside the team;

– The salary determination which disrupts the balance "contribution-salary";

– The executive is sensitive to the standing of the subordinate, neglects his standing, and tries to harm him, to undermine the credibility;

– The executive accuses the employee, undermines his reputation in front of other people;
– The vague formulation of the assigned tasks with a deliberately negative assessment of the result;
– Suspicion and secret supervision of the documents, workplace, etc;
– The executive's underestimation of the importance of tasks explanation, information relevance cause mistrust, suspicion, and intrigues [9].

In this study, the subordinates noted that these mistakes of the executive have a negative impact on the moral and psychological climate in the team. Thus, this helps identify the dynamics of the conflict impulses in the everyday space of the relations of production.

According to N V Grishina, conflicts can be seen as people's reaction to obstacles they face when trying to achieve various goals in a shared activity, the reaction to others' behavior that does not meet their expectations, as well as a reaction based on the incompatibility of characters, dissimilarity of cultural foundations and needs [10].

4. Conclusion

The results of the practical study showed that the responsibility of the head of the organization is to choose an appropriate strategy which can serve a basis for the most productive ideas as well as attract people who can turn these ideas into reality. In order to make the right choice, the executive should have great knowledge and skills in the field of management psychology. In particular, the ability to behave in the conflict situations is important for an executive of any level. The results of the study and their analysis make it reasonable to put forward the thesis that the head of any Department, regardless of the content of his work, should possess certain skills in the field of conflict communications and the methodology for their settlement [11].

The stage of conflict behavior is inevitably associated with the aggravation of the emotional tone of the relationship between the executive and the subordinate, and their progressive destabilization. However, the executive's actions perform at the same time a kind of cognitive function, when the escalation and development of the conflict lead to a deeper, though not always a more accurate understanding of the situation.

In terms of the socio-psychological analysis, the possible conflict communications in the field of management can be of the following types: aspect (trait or set of traits) of personality – aspect of the personality, person - person, person – group, and group – group. In the first case, we usually deal with the conflict impulse of individual characteristics of a person and people's behavior; in the second, with their open confrontation about their needs, motives, values, and attitudes; in the third and fourth, an open clash of an individual with a group or a group with a group [12].

Almost all of these variations of conflict communications take place in the situation of business interaction since contradictions in it can be associated with the peculiarities of people's behavior, for example, their lack of compatibility (the first case), as well as with an open confrontation of the executive and subordinate, or the whole team [13].

The management in any conditions remains organizational and communicative. The styles of the interaction between the executive and subordinates we have examined are open to being filled with the psychological content.

References

[1] Horita M 2000 Group Decision and Negotiation **9(1)** 63–83
[2] Schelling T 2007 *The Strategy of Conflict* (Moscow: IRISAN) p 373
[3] Kilmann R H and Thomas R W 1977 Educational and Psychological Measurement **37(2)** 309-25
[4] Thomas E Phipps JR 1961 Journal of Conflict Resolution **5(3)** 274–8
[5] Tremblay R Ch, Nikolenyi C and Otmar L 2003 Journal of Comparative Policy Analysis **5(2–3)** 125–48
[6] Kriesberg L 1999 Contemporary Sociology **28(4)** 470–1
[7] Sukhova A S, Selezneva I G and Lyasina I Yu 2018 Primo Aspectu 34(**2)** 46–52

IOP Conf. Series: Materials Science and Engineering **483** (2019) 012036 doi:10.1088/1757-899X/483/1/012036

[8] Chernenilov V I 1981 Psychology Issues: Actual Problems of Psychology and Management **4** 38–50
[9] Kozlov S A 2007 Scientific Bulletin of the Moscow State Technical University of Civil Aviation **113** 129–34
[10] Grishina N V 2010 *Psychology of Conflict* (St. Petersburg: Peter) p 544
[11] Ashmore R, Jussim L and Wilder D 2002 Contemporary Sociology **31(6)** 711–2
[12] Aubert V 1963 Journal of Conflict Resolution **7(1)** 26–42
[13] Groshev I V 2007 Sociological research **6** 122–30

IOP Conf. Series: Materials Science and Engineering **483** (2019) 012037 doi:10.1088/1757-899X/483/1/012037

Strategies applied by technical interpreters in Khanty-Mansiysk and Murmansk based industries

O V Afanaseva[1], N S Gilmanova[2]

[1]Murmansk State Technical University, Egorov str. 15, Murmansk, 183038, Russia
[2]Ugra State University, Chekhov str. 16, Khanty-Mansiysk, 628012, Russia

afanaseva.olesya@gmail.com

Abstract. This article identifies and classifies some of the communication strategies used by technical interpreters in Khanty-Mansiysk and Murmansk based industries. In these regions interpreters are widely involved into the production process at either international enterprises co-founded by legal entities of different countries or at domestic enterprises having partners worldwide. The communication strategies most frequently applied by the local interpreters are viewed with account of such notions as production process, consecutive technical interpretation and interpreter competence.

1. Introduction

Oil has been the largest source of energy throughout the world for years regardless of the globalisation process, imposed quotas and sanctions, fall in oil prices or rising electricity consumption rate. Therefore international cooperation in oil and gas production and processing industries is highly welcome. Russian market capacity, abundant natural resources, skilled labour force attract foreign investment into the Russian economy.

Well-qualified technical interpreters are highly demanded in this country, in the Khanty-Mansiysk Autonomous Okrug and the Murmansk Region in particular. Due to the business and industrial environment in these regions interpreters are often required to enable professional communication during oil and gas production and processing, ship surveys, shipyard maintenance, spent nuclear fuel management and within many other industry-wise activities. The International hydraulic fracturing company the Calfrac which operates in Canada, the United States, Argentina and Russia has a subsidiary, the CWS (Calfrac Well Services) International LLC, in Ugra, the Khanty-Mansiysk Autonomous Okrug. The largest independent natural gas producer Novatek is currently realising its grand Arctic project in the Murmansk Region.

2. Production process characteristics to comply with

To characterise a production process, researchers single out 13 different principles, including proportionality, differentiation, combination etc. [1]. Although all of them effect interpreters' work and are to be taken into account, the most vitally important principle to be ensured by interpreters is the production process continuity, i.e. preventing interruptions in the production process.

Other principles are also to be supported by high-quality interpretation. These are the principles of differentiation, combination, standardization, streamlined, etc. However, it is the principle of continuity that is the most relevant for the implementation.

The principle of production continuity is provided mostly by the atmosphere of trust. It stands to reason that interpreters with an interpreted text or a discourse they generate play a leading role in creating such atmosphere.

3. Competencies of Interpreters

For obvious reasons, the above-mentioned companies frequently employ local residents. Furthermore, it is strictly required to select personnel from the local population in the CWS International LLC, the Khanty-Mansiysk Autonomous Okrug.

The interpreter in such a company is not a production employee of oil and gas industry and does not have the authority of one. However, they are familiar with both the specifics of the production process of well treatment and maintenance, repair of equipment, as well as peculiarities of national characters of both local and foreign colleagues.

The competencies of the interpreter are identified as follows:

- highly developed memory and skills for its development;
- ability to work with terminology, readiness to constantly perceive new terms, evaluate their valency, applicability, ability to work with them in context [2];
- ability to quickly acquire missing experience in working with terminology, in particular, deduction and induction methods: clarification on practical experience, deduction of meaning from a common understanding;
- the so called "falsification" skill, i.e. translation of terminology without qualification in oil and gas field. It can be said that often a technical interpreter has to "imitate" work experience, studying topics, compiling glossaries, reading special literature, linguistically guessing, as well as the psychological element, which is communicative competence. The interpreter necessarily and at the beginning of his career frequently addresses both Russian and English-speaking production employees to verify directly generated oral discourse;
- high motivation and life-long learning. The interpreter is given a very short time for adaptation in a stressful situation of misunderstanding that is constantly present at the beginning of work;
- highly developed psychological mobility. Being involved into production might impose much stress, as it implies fly in/fly out work on rotational basis, urgent phone calls, frequently at night.

4. Characteristic of interpreting during the production process

Researchers of interpreting and translation [3-9]. note that there are a great many definitions of translation/interpretation, due to a high and long-standing demand for this activity.

The production process in international industries implies consecutive interpreting in up to 70% of interpreters' work. It is usually required when the number of participants is relatively small, since consecutive interpretation is effective for the "mobile" communication, which is typical of the production process. This kind of communication is carried out when working at industrial facilities, negotiations, performing repairs directly on oil bushes, etc. The author of generated discourse specifically makes periodic pauses in speech and these pauses are usually very short. A professional interpreter usually generates the text while the speech is being heard and pronounces it during the pause. Consecutive interpretation requires good operational memory.

Obviously, the type of interpreting demanded in the industries can be defined as a technical one. Technical interpretation can be defined as one used for the exchange of special scientific and technical information between people speaking different languages. This is a formal logical (collective) style, characterised by accuracy, non-emotionality and impersonality. It is also the use of a variety of terms, often of Latin or Greek origin. Researchers comprehensively describe its specific grammar with special norms and rules to be used. These rules are as follows: 1) the syntax structures are indefinitely personal and impersonal; 2) passive structures are widely used; 3) verbs are most often used in non-personal forms; 4) inversion in the sentences of technical texts serves as a way to implement a certain logical selection.

Technical interpreting during the production process implies special communicative strategies.

5. Linguistic studies of communication strategies

As any social action is always accompanied by a communication strategy produced or reproduced by this action, supporting the old or creating a new way of communication, "which can then be consolidated as a meaningful basis of culture in general. In this sense, we are not talking about communication, but about a communication strategy, thereby introducing the content of the choice, the content of the perspective into this tool, and pointing to the need for constant efforts to maintain this choice and that perspective" [10].

Study of communication strategies and tactics originates from T. A. Van Dyke, who was one of the first to explore speech tactics and communication strategies. He says that the choice and use of these tactics depend on a variety of factors, such as purpose of communication, knowledge, cultural background, situation and sphere of communication, as well as text genre.

N. Papina notes that utterance is characterised as holistic, consistent and coherent. This is achieved by observing certain norms of a formal and semantic organization of the text. The organization of discourse directly depends on a communicative situation, an author's attitude, as well as on the attitude of a person receiving the utterance. In general, communication is defined as an interaction of two active parties, the speaker and the listener. The first implements the communicative intention, while the latter decrypts it. The most active position belongs to the sender of the message, i.e. the speaker. In the act of translation this is the sender of the original message. An interpreter translates a speaker's communicative attitude or intention [11].

Researchers study both a speaker and a listener. Communicative functions are highlighted for both. Communicative functions of a speaker are as follows: 1) messaging, nominative, denotative, reference; 2) cognitive, gnostic; 3) appeal (impact, "imposing" his own picture of the world); 4) expressive (expressive form of the mental state of a speaker); 5) emotive (creation of certain feelings in the recipient); 6) poetic (installation of the author on the sign itself, appeal to aesthetic feeling of the addressee); 7) contact-setting (to prepare for perception). Communicative functions of a listener are as follows: 1) communication (this is readiness for perception, understanding and communication); 2) perception (understanding of speech, social, language situations, decoding of semantic, lexical, phraseological, grammatical signs and symbols); 3) actual contact, with the maintenance of verbal communication, attention to the message); 4) aesthetic (textual satisfaction of needs of the recipient, his tastes, adequate perception, commonality of "pictures of the world".

L. Vvedenskaya suggests that there are various ways of communicating ideas, as well as different communication codes and systems [12].

In her work E. Dyakova highlights that communication of people is carried out for the purpose of rapprochement or distancing. Hence, rapprochement and distancing can be understood as global communication strategies or, according to the author's terminology, macrostrategies. Communicative macrostrategy includes the following concepts : 1) strategy - a set of theoretical speaker's moves aimed at achieving a communicative goal; 2) intention - a certain principle according to which the above moves are combined into a certain strategy; 3) purpose - an idea of the result of a certain communicative act; 4) tactics - direct practical moves that the speaker realises in speech; 5) intention - a move, a part of tactics, that is, something like a unit of the communication process; 6) experience - a set of conclusions speaking about those tactics, moves and intentions that were implemented earlier and marked as successful / unsuccessful. Based on all the selected units, the speaker implements a certain communicative strategy based on experience and engaging communicative competence [13].

O. Issers analysing the concept of communication strategy notes that the strategy is based on interpretation, which, to a certain extent, makes the concept of speech strategy similar to the concept of translation. It is basis of speech influence. Interpretation is made on the knowledge of a partner: their value categories, emotions and will. Communication strategies are actions that are precisely aimed at interpretation. Strategy always presupposes a motive: the interpretation of desire, the will of the subject, these are typical motives of human behaviour — to preserve good, to achieve better. Situations in which a message does not presuppose a subject (notifications, informational messages) are not purely speech strategies, in which the strategy may exist as some element of the situation. Issers gives a typology of communication strategies as specific ways of speech behaviour, carried out at the control of a "global intention" [10].

Communication strategy is a semantic, stylistic and pragmatic choice of a speaker. According to speech acts classification by J. Searle, A. Wezhbicka and others, strategies can be general and private. From the functional point of view, the main (semantic, cognitive) and auxiliary strategies are distinguished, which, in turn, are subdivided into pragmatic, interactive, rhetorical strategies, depending on the goals of communication.

The theory of translation defines the term "strategy" as follows. This is the operational activity of the interpreter, a set of actions and techniques used by the interpreter to achieve the goal. In studies of Russian scientists engaged in translation studies, a single approach to the definition of a "translation strategy" has not been developed. That is the very reason for a great variety of definitions of the given concept. The "strategy of the interpreter", "translation thinking" are also worth mentioning here. N. Georgieva suggests the following definitions. Translation strategy is consciously developed by an interpreter in the course of expert communicative activity aloof of his actions aimed at creating the project - a text of the translation, with the obligatory account of the professional ethics of the interpreter. The translation strategy in the framework of the translation process is a general plan of action of the interpreter aimed at translating a specific text (in our case, statements), determined by a specific purpose of the translation taking into account the linguistic and extralinguistic characteristics of a specific translation situation, relevant at a particular time period [5].

All the above provisions lead us to the conclusion that they are components of the translation situation in which a technical interpreter has to act.

According to Barkhudarov the translation situation is 1) a communication situation, that is, an environment in which a communicative act takes place; 2) the subject of a message, that is, the setting (set of facts) described in a text/discourse; 3) participants of communication, that is, the speaker (the writer) and the listener (the reader).

6. Communication strategies used by the local technical interpreters

6.1 Types of the interpreting/translation situations

Technical interpreters of the regions find themselves in the following interpreting/translation situations:

- interpreting communication between Russian and English-speaking experts during the maintenance and repair of well servicing equipment in the workshop;
- interpreting of communication between Russian and English-speaking experts during service maintenance and repair of well servicing equipment directly at the well;
- opening and closing work orders, entering information into the database;
- translation of correspondence for managers and supervisors.

Interpreting/translation situations are characterized, firstly, by a number of extralinguistic components, such as the above mentioned production process principles, an interpreter's competences, etc.; secondly, by linguistic components, such as technical discourse; various terms, including unknown ones.

6.2 Strategies applied by the local technical interpreters

The following strategies have been identified.

1) Preserving production process continuity.

An interpreter chooses the most accurate and concise equivalent, ignoring any extra or irrelevant information.

Table 1. Examples of preserving production process continuity

English discourse	Russian Interpretation
We ordered CV-joints and axle shafts. Expecting them in.	ШРУСы и *валы моста* заказаны, скоро будут.
Driving is the most dangerous thing you will do on the job, or off the job.	Вождение - самый опасный вид деятельности как в рабочее, так и в свободное время.

2) Filling in the gaps.

Paraphrasing unknown words, word combinations in case of unawareness of details.

Table 2. Examples of filling in the gaps.

English discourse	Russian Interpretation
crossover	передача вахты

3) Politeness, keeping the contact, ignoring excessively emotional elements of discourse that do not influence the meaning.

Table 3. Examples of keeping the contact.

English discourse	Russian Interpretation
I would like to ask you a question which is a task in itself.	Я бы хотел задать вам вопрос, который, возможно, трансформируется в просьбу.

4) Control over the topic, i.e. use of applicable combinations and equivalents within given interpreting situation.

Table 4. Examples of controlling the topic.

English discourse	Russian Interpretation
This is not to be repaired, drive shaft spline part is broken, do not try to…Drive shaft is to be replaced, it is beyond repairs.	Кардан неисправен, сломана шлицевая часть, требуется замена.

5) Explanation.

When Russian production employees do not understand the term or concept, the interpreter provides an interpretation similar to the translator's note; the transformation caused by the prepositional attributive noun, while in Russian it requires the genitive case.

Table 5. Examples of explanation.

English Discourse	Russian Interpretation
points	контакты прерывателя
grease gun	смазывающий инструмент
forklift mast load roller	ролик мачты вилочного погрузчика

6) Localization, which means depending on the production site and workshop.

Table 6. Examples of localization.

English Discourse	Russian Interpretation
bulbs	автолампы
coupling	переходник
spine	шлицевая часть кардана

7) Transference.

The text of the target language, or, more precisely, parts of the target language, will have to have the meanings established in the source language, in other words, have the meanings of the source language. The transfer can occur at all language levels, starting with the morphological one, and the elements of the target language at the same time acquire formal and contextual meaning from the systems and structures of the source language [8].

Table 7. Examples of transference.

English Discourse	Russian Interpretation
shutdown box	блок шатдауна

7. Conclusion

The discourse analysis of specific components of interpreting/translation situations for over 700 oral Russian statements generated by technical interpreters makes it possible to identify the following communication strategies used in the industries: preserving the continuity of the process, politeness, keeping contact, controlling the topic and others. The list of the strategies is not complete and can be continued. The communication strategies applied by the local interpreters are identified with account of such notions as production process, consecutive technical interpretation, and interpreter's competence.

References

[1] Nepomnyashny E *Economics and Enterprize Management* 1997 (Taganrog: TRTU)
[2] Robinson D *How to Become a Translator* 2005 (Moscow: Kudic-Obraz)
[3] Barhudarov L *Language and Translation* 1975 (Moscow: International Affairs)
[4] Volkova T 2011 Discourse-Communicative Translation Model in Institutional Discourse Study *Chel. Univ. Her.* **55** 37-42
[5] Georgieva N 2016 Adaptive Strategies of Temporal Meanings Translation in Cross-cultural Business Discourse *VSU Her.* **2** 119-24
[6] Gilmanova N 2016 On Well-Servicing Units Maintenance Terms Translation *YuSU Her.* **1** 65-72
[7] Gureeva A 2013 Interpreters Communication Actions in Communal Interpretation Situation *VSU Her.* **2** 159-162
[8] Catford J *Translation Linguistic theory* 2004 (Moscow: URSS)
[9] Sdobnikov V 2011 Discourse-Communicative Translation Model in Institutional Discourse Study *VGLU Her.* **1** 165-72
[10] Issers O *Russian Speech Strategies and Tactics* 2008 (Moscow: URSS)
[11] Papina A *Text, Units and Global Categories* 2002 (Moscow: Editorial URSS)
[12] Vvedenskaya L and Chervinskij P *Russian Speech Theory and Practice* 2005 (Spb: Piter)
[13] Dyakova E 2006 Discourse Strategies of Institutional Discourse *VSU Her.* 149-55

Teaching resources in professionally oriented foreign language learning

I V Aleshchanova[1], N A Frolova[1], M R Zheltukhina[2]

[1]Kamyshin Technological Institute (branch) of Federal State Budgetary Educational Establishment of Higher Education "Volgograd State Technical University", Lenin Street 6A, Kamyshin, 403874, Russia
[2] Volgograd State Socio-Pedagogical University, English Philology Department, Lenin Avenue 27, Volgograd, 400066, Russia

E-mail: dfl@kti.ru; zzmr@mail.ru

Abstract. This problem is important because it is necessary to improve the methodical and didactic perspective of specialists' professional language learning process at an organization. The purpose of the article is to consider the role of a study aid as one of the methodical and didactic tools in the system of skills development. Research methods of abstract analysis, methodical and didactic sampling, modeling, forecasting were used. Significant criteria for a study aid structure have been identified to develop the personnel's dominant foreign language communicative skills at an organization. The object of the research is study guide materials providing the process of professionally oriented language learning. The subject of the research is study aids forming the professional foreign language competence of economists and technical specialists (on the material of the German language). The study has revealed that we should abandon the idea of existence of universal, effective methods of teaching a foreign language and attempts to develop an "ideal" study aid solving all methodical and didactic problems. This applies not only to educational materials on teaching German as a foreign language, developed in Russia, but to original ones published in Germany.

1. Introduction

Professional competence of a specialist is a complex of general scientific and special knowledge. In this system an important role belongs to a foreign language and a professional and business component of foreign language communication, which forms skills for professional intercultural communication in a foreign language.

The right choice of methodical and didactic tools system is one of the significant factors influencing the process and result of professionally oriented foreign language teaching. Didactic tools cover all elements of the learning environment consciously used for better organization of the educational process, and, consequently, they increase the effectiveness of learning. The most important of them include textbooks, study aids, study guides.

Efficient educational materials development and use form conditions for successful implementation of the above approach. Availability of numerous foreign language textbooks developed in Russia and abroad in the system of postgraduate education sets the strategic task of choosing proper study guides to organize educational process in the framework of development programs. In this connection it is necessary to take into account characteristics of an educational institution, age and individual characteristics of students, methodical training specificity, the target audience national and cultural mentality.

On the one hand, the textbook is a foundation for language information and practice obtained by students in the classroom; on the other hand, it is an auxiliary means of teaching. In any case, none of the methods of foreign language teaching used nowadays can be applied without the extensive use of textbooks (study aids). Thus, the study of how textbooks should be chosen, organized, used and adapted is an important part of the teacher's professional skills.

2. Methodology

The question of textbooks role in educational process is actively discussed in modern studies. According to G. Neuer, the textbook is considered the main tool of foreign language teaching. It occupies a leading position in the modern methodology of teaching foreign languages [1]. I.L. Bim classifies study guides (textbooks) in the system of modern education as an information model, which is determined by goals and objectives, reflects theory, methodology and content of education [2]. The role of a textbook in the material support of students' classroom, extracurricular and independent work and increasing interest in learning a foreign language is discussed in the paper of S.Yu. Tyurina [3]. V.I. Blinov defines a textbook as a tool for university teachers to create their own methodological system [4]. Criteria for a teacher's creative potential development with the help of a textbook are set out in H. Acosta and D. Cajas's work [5]. P. Farrokh explores prospects of study guides use in terms of their ability to guarantee students' autonomy in foreign languages learning [6]. S.S. Bykova draws researchers' attention to the need to contrast the textbooks compliance with a communicative approach to foreign language teaching requirements [7]. L. Li emphasizes textbook value in integration of thinking skills when learning a foreign language [8]. N. Košuta and S.V. Ivanović review textbooks in the context of European language portfolio learning styles [9]. H.C. Liao and Y.U.H. Chen identify factors contributing to professionally oriented foreign language learning [10]. K. Umeda highlights significance of teaching materials development using multimedia [11]. Y. Tobe defines the role of audio-visual accompanying material for foreign language textbooks [12]. Y.P. Heinrich indicates the importance of the problem of general language policy in the textbook [13]. N. Enriquez pays special attention to supplementary electronic educational materials in textbooks [14]. T.L. Ballman emphasizes the role of the textbook subject content specified by context [15]. J. Jr. and J.C. Bookout discuss the idea of creating intensive foreign language courses in order to increase learning process efficiency [16].

Development of problems of professional communication teaching, particular issues of organizing textual material in a textbook, the optimal algorithm for teaching various types of reading, listening comprehension, a system of exercises, and rationalizing formation of terminological vocabulary are considered by the authors according to theories proposed by E.I. Passov [17, 20] and E.I. Kuznetsova [18, 20]. I.A. Zymnya's works help to take into account the psychological aspects of study guides, the need to develop motivation using a textbook [19].

Thus, in the present work, it is advisable to consider study aids for professionally oriented foreign language teaching from the standpoint of participants in educational process. On the one hand, this is the position of the teacher, expressed, according to Bimmel Peter and Rampillon Ute [21], in accordance with pedagogical activity to the goals of teaching the specified subject. In addition, A.A. Verbitsky and N.A. Bakshaeva [22] indicate the need to take into account students' psychological characteristics. On the other hand, A.N. Leontiev [23] emphasizes that from the position of students, it is important to establish the degree of their involvement in learning process. The authors' point of view on the question under consideration is reflected in N.A. Frolova's work [24] on motivation increasing factors and I.V. Aleschanova's work [25] on ways of communicative competence formation.

2.1. Materials and methods

The study was conducted at the educational Linguistic Center at Kamyshin Technological Institute (branch) of Volgograd State Technical University. Theoretical analysis of psychological, pedagogical and methodological literature on the question under consideration, the study and synthesis of teaching experience of the Department of Foreign Languages and Humanities, diagnostic methods (observation, testing, survey), analysis of educational outcomes of participants at the Linguistic Center, statistical methods of the study results processing were used to achieve the stated goals and objectives.

Methodological basis of the study is classical didactic requirements for a foreign language textbook. In this regard, a foreign language textbook should pursue the following common goals: introduce the lifestyle, culture, including professional one of the country where the learned language is spoken; contribute to communication skills development with native speakers; promote preparation for a foreign language use in professional activities.

Study guides of leading Russian and European publishers were sampled and subjected to analytical, structural and informative analysis to find out an optimal textbook for a professionally oriented foreign language course.

Educational and methodological materials aimed at ensuring the process of postgraduate professionally oriented language learning, forming professional foreign language competence of economic and technical specialists were analyzed. A survey among foreign language teachers at higher education establishments was conducted to identify subjective perception of study guides qualitative component (30 respondents).

3. Results and Discussion

A survey among foreign language teachers of technical universities allowed finding out a number of criteria for subjective perception of the qualitative component of the textbooks used for teaching a foreign language (German). Thirty teachers were interviewed. The outcomes of the survey are presented in tables 1-5 (table 1, table 2, table 3, table 4 and table 5).

Table 1. Are you satisfied with the textbook used for teaching a foreign language?

Yes, it corresponds to the Federal State Educational Standards and does not require additional materials.	33% (10 people)
Yes, but some additions are required.	50% (15 people)
No, I use it in the educational process, because it is listed as a textbook recommended by the Ministry of Education.	17% (5 people)

Table 2. Did you have to make adjustments to the structure of the textbook, depending on conditions of a specific target audience?

No, it is not necessary, although sometimes you have to skip some sections of topics. The textbook material is easily adaptable to any target audience.	30% (9 people)
Yes, but some additions are required.	20% (6 people)
No, I work according to a given structure recommended by the ministry.	50% (15 people)

Table 3. What supplementary materials should be included in your basic foreign language textbook, in your opinion?

There is no need to include supplementary materials	27% (8 people)
More exercises to learn vocabulary and grammar	13% (4 people)
Authentic audio and video materials	60% (18 people)

Table 4. Do your students experience dissatisfaction with their basic textbook?

Yes, very often	24% (7 people)

IOP Conf. Series: Materials Science and Engineering **483** (2019) 012038 doi:10.1088/1757-899X/483/1/012038

Sometimes, but only when they cannot do their tasks well	26% (8 people)
No, on the whole, they are satisfied with the textbook	50% (15 people)

Table 5. Do you think that a textbook solving all existing problems can be created in future?

Yes	30% (9 people)
No	70% (21 people)

According to the survey results, the majority of teachers adjust the available educational materials, adapt them taking into account the target audience needs. Many teachers have to supplement textbook materials with up-to-date additional information, especially authentic audio and video materials. But despite this, almost all teachers believe that textbooks are a key component in most foreign language syllabi.

It was experimentally proved that all material components of the reviewed textbooks content can have a methodological significance only if they are related to the actions that should be based on them.

Exercises become the main direct material means of organizing the teacher and students' activities, determining the quality and value of a textbook, study aid or study guide [32].

In this article an exercise is interpreted as a small scenario of a communication act, in which the program of actions of the speaker and listener, as well as the reader, is set. In all the reviewed study guides, three main groups of exercises were found: exercises and tasks preparing oral communication; exercises and tasks building and structuring the communication process; exercises and tasks simulating communication.

All the exercises, in turn, are divided into two large groups: exercises for skills formation and exercises for speech skills development. These are respectively simulating communicative exercises and actual communicative ones. Each type is adequate to the purpose served. This means that the quality of the exercises in each category is correlated with the objectives defined in the study aids.

In other words, the system of exercises in all textbooks is a combination of the necessary types and varieties of exercises. Sequencing of these exercises performance and their quantity is aimed at teaching patterns in formation of communicative skills. Depending on the goals of doing the exercises, the following hierarchical sequence has been built.

The study of the exercises has shown that professional vocabulary development forms the centerpiece of knowledge of word selection rules and their compatibility patterns in various contexts in the process of specialists' professional communicative competence development. The observation of special terminology acquisition was carried out on the basis of the most frequent terms of the German syllabi, which were selected from branch terminological dictionaries and textbooks in German.

In many cases professional terminology teaching includes reading special texts that provides new vocabulary on the studied field of science. The need to optimize the process of learning terminological vocabulary becomes apparent in such a situation. From a methodological point of view, professional texts perception is complicated by their substantial aspect characteristics as they contain significant percentage of unfamiliar or completely unknown special vocabulary to students. A way to solve this educational problem is suggested in the work. It is close interaction with further education students, discussion of the developed teaching materials with relevant disciplines teachers, rational organization of educational process according to the principle "from simple (study of general scientific, adapted texts) - to complex (more complex special authentic texts) ".

In methodological literature, vocabulary learning is divided into linguistic and methodological aspects [2]. Linguistic aspect covers formation of a vocabulary corpus for solving communicative tasks at a concrete educational stage. Constant repetition of the studied vocabulary and its maximum rotation is one of the key factors, which ensures vocabulary development. Hence professionally oriented vocabulary learning includes various types of exercises: 1) non-communicative exercises aimed at conscious learning the specialized subjects vocabulary; 2) conditional communicative exercises; 3) real communicative exercises.

Methodological component of vocabulary learning includes the necessary explanations and instructions on use of dictionaries, the form of maintaining individual dictionaries and cards with new vocabulary, ways to systematize the vocabulary learned. This is the knowledge and skills that allow the student to work independently on the vocabulary. A large role in this is also played by independent extracurricular foreign language learning in the form of fulfilling a semester assignment to study authentic professionally oriented texts and their special vocabulary.

Thus, foreign language teaching should be correlated with the specialists' profession features. Professional language needs make it necessary to acquire terminological vocabulary that may be the best applicable in specific situations of professional communication. Consequently, a systematic and integrated approach to the choice of teaching methods and means is reflected in all the considered study aids. It is to solve the problem of professional linguistic skills formation.

However, Russian and foreign educational materials have their own significant differences, determining their advantages and disadvantages.

Russian textbooks positive aspects include:

- availability of translations into Russian or analogues for German words and constructions. The need to associate foreign words and constructions with a world perception coded in their native language is deliberate for many students. Even if classes are given only in a foreign language, students will inevitably ask questions about the translation of unknown information into their native language or refer to auxiliary information sources. In this, a word translation in some unit of the study guide may not coincide with the one given in the dictionary;
- optimal set of translation exercises. There is the dominant idea of developing students' skill to speak a foreign language "immediately", without double translation. Nevertheless, a good system of translation exercises can eliminate interference errors and facilitate foreign language information decoding. In the first stages of foreign language learning, translation exercises, in our opinion, will help correctly link a foreign language system to a native language, automate the code switching system. Over time, translation exercises can be minimized or even temporarily excluded, and then used again at a later stage, when it is necessary to develop translation skills;
- grammatical and other explanations in the native language. Grammatical explanations in Russian help avoid the difficulties of perceiving grammatical rules in a foreign language;
- despite their attractive design and illustrations, foreign textbooks cause stress for students due to their "incomprehensibility". Russian textbooks, on the contrary, are perceived as a reference book, which facilitates search for the necessary information;
- cost factor. The cost of educational materials set published abroad is much more expensive, moreover they can be used for no more than one semester.

Russian textbooks have some disadvantages. The vision of the world, mentality, cultural contents in domestic textbooks are Russian. In this regard, there are a lot of cultural, lexical and grammatical errors in them, provided that textbooks are not verified by native speakers. Monologue and dialogic texts are unnatural to native speakers. There is confusion in the use of areal options. There is lack of audio materials or they are insufficient. In the process of learning, it is important that the "input" (what the students hear and read) will increase manifold what they produce.

A number of factors that have a significant impact on the textbooks development and finalization, for example, social order, basic goals and teaching methods, specific technologies and techniques, the dominant linguistic theories, etc., have been found out. All these factors are systematized into four main groups: socio-cultural factors; general pedagogical and general didactic factors; subject specific factors; subject learning factors. A separate group of authors and publishers can be identified as additional, but also significant one for the teaching aids aspects.

It was found that Russian study aids are characterized by the grammar-translational method of presenting material. It means they have been developed in the traditional method framework of foreign language learning, based on systematic study of grammar, phonetics, formation of reading and translation skills. The compilers emphasize that professional foreign language knowledge implies different types of communicative activity: speaking, listening comprehension, writing, reading and reading comprehension [33; 34]. Each of these types requires specific working methods. Therefore, from the point of view of different tasks, "traditional" study guides have their own strong and weak points.

The specificity of material presenting according to traditional method is to master a certain amount of all types of communicative activity. This implies a thorough integrated study of all language aspects: grammar, phonetics, vocabulary. In this connection teachers' skills to work with study guides are determined by the ability to "link" all the presented aspects and components into a single, effectively working complex. Russian professionally oriented study guides require systematic learning for quite a long time (at least a year), although the timing may be different depending on the available tasks. If the course of professional communication lasts for a year and contains about sixty-four academic hours of practical learning, it hardly makes sense to take on all types of communicative activity. In this case, it is better to focus on one or two aspects, paying the least attention to the rest.

An insufficient level of text materials development for professional foreign language education is Russian textbooks significant shortcoming. This can be explained by the lack of a system of professionally oriented texts presentation, uninteresting text subjects; differences in the degree of text material understanding and comprehensibility; dominance of monologic speech forms in textbooks, tasks aimed at teaching clichéd speech, absence of creative tasks for modeling potentially possible situations of professional communication.

In modern conditions, reading of professional texts is not the main purpose of teaching foreign languages at a technical college. A foreign language is considered as a means of oral and written communication in the professional field, which predetermines development of fundamentally new textbooks for professional purposes.

Leading German publishers' textbooks are based on principles of a communicative approach in foreign language learning, and, as the name implies, are aimed at developing communication skills. Special attention is paid to speaking and listening comprehension of all components of language learning. Study aids do not contain complex syntactic constructions or active vocabulary. However, this does not mean that communicative principle of construction is intended only for teaching easy small talk. This method is designed, first of all, to remove the fear of communication. All grammatical rules in the reviewed study guides are illustrated with examples taken from real oral and written contexts of various functional styles. The examples show how a particular language phenomenon is actualized in various contexts, how it is used by native speakers in appropriate communicative situations. In this, the contexts are chosen in such a way that, having studied several examples, the learners could themselves come to a conclusion about how and when this form is used. In the field of vocabulary, the main focus is on those word combinations that are used in real communication. The considered study guides have two types of structural organization: a) sequence of topics focuses directly on a phased, step-by-step course of study, the so-called "closed concept of the material presentation"; b) educational material has a flexible system of construction, the so-called "open concept of the material presentation" [1].

The first type includes Russian study aids. At first glance it takes a teacher a minimum of time to prepare for the lesson using them. But detailed study shows that their use in teaching causes difficulties in adapting to each specific educational situation. German study aides with their openness and non-strict sequence of educational material presentation allow the teacher to be more mobile, and what is more important, creative in their approach to the organization of practical learning.

An analysis of Russian and European educational literature has revealed that well-known methodologists and authors' textbooks are, as a rule, of a general nature and are represented by general scientific themes, without taking into account the professional specifics of a particular training profile. The above textbooks can be used in educational process only as additional materials for solving particular problems: for example, teaching professionally oriented reading and translation.

The analysis of the above study aides allowed us to highlight the list of criteria, which will ensure educational process with optimal and rational study guides (table 6).

Table 6. List of criteria

clear structure	clear grammatical progression and interesting content related exercises
focus on the European criteria for assessing knowledge of a foreign language (Der gemeinsame Europäische Referenzrahmen)	The optimal set of exercises on pronunciation and phonetics
suitability of methodological development for a specific target group	four-skill training (speaking, listening, reading, writing)
clear sketch, readable font	impetus for a teacher to further develop and create their own pedagogical strategies and methods of work
proportionate and relevant illustrative material	a teacher's methodological and didactic support, including Online
Interesting topics, assignments and exercises	facilitating preparation of a low-cost lesson
various topics and exercises appropriate for different types of learners	CD and / or DVD availability
clear instructions for assignments	rational additional material, including online materials
systematic training of professionally oriented and specialized vocabulary	cost
extensive authentic material	

Educational materials development for teaching students foreign language communication should be based on the following principles: unity of educational and developing technologies; personal orientation in foreign language education; integration in mastering all types of communicative activity; focus on formation of communicative skills for understanding and generating foreign-language discourse; educational materials authenticity and novelty; use of various types of professionally oriented discourse; reflection, self-assessment and self-development; interdisciplinary links integration in the content components selection for foreign language professional discourse learning.

Conclusion

The conducted study revealed that textbooks will continue to play an important role in foreign language teaching and be an effective auxiliary tool for both teachers and students despite the influence of new pedagogical technologies. Good textbooks are guides to curriculum and syllabus development for the discipline learnt. They are considered as one of the numerous means for a teacher to conduct an effective practical lesson. But the positive motivation formation in professionally oriented foreign language study largely depends on the teacher's personal contribution to lesson preparation, educational materials adaptation and modification, as well as the copyright educational material creation. Availability of high-quality educational materials creates a favorable pedagogical situation, marked by students' positive reactions and educational activities activation. Teaching students a foreign language is considered rather complex and multidimensional methodical problem. The results of the educational work under consideration are given as an illustrative means of learning process efficiency increasing, further development of educational and methodological support forms and methods.

Thus, a modern study aid (textbook) as the main means of students' foreign language teaching should reflect its competence content. All materials must meet the main learning objectives and allow acquiring a certain level of foreign language skills.

References

[1] Neuer G, Hunfeld H 1993 *Methoden des fremdsprachlichen Deutschunterrichts Fernstudieneinheit 4.* (Berlin-München: Langenscheidt) p 184
[2] Bim I L 1977 *Methodology of foreign language teaching as a science and school textbook problems* (Moscow: Education) p 268
[3] Tyurina S Yu 2014 *Modern problems of science and education* **4** 13848
[4] Blinov V I 2018 *Methodology of higher school teaching: educational and practical manual* (Moscow: Yurait) p 315
[5] Acosta H, Cajas D 2018 *Indonesian Journal of Applied Linguistics* **8(1)** 100
[6] Farrokh P 2018 *Chinese Journal of Applied Linguistics* **41(1)** 101

[7] Bykova S S, Kuvaldina E A and Mashkovtseva L M, Malova T V 2018 *European Journal of Contemporary Education* **7(2)** 275
[8] Li L 2016 *Thinking Skills and Creativity* **22** 273
[9] Košuta N, Ivanović S V 2015 *Croatian Journal of Education* **17** 187
[10] Liao H C, Chen Y U H 2012 *English Language Teaching* **5(6)** 89
[11] Umeda K, Yokota M, Uchida T and Ejima T, Nozaki H 2007 *15th International Conference on Computers in Education: Supporting Learning Flow through Integrative Technologies* 49
[12] Tobe Y., Fujita S, Hosaka T 2006 *7th International Conference on Information Technology Based Higher Education and Training ITHET* 9
[13] Heinrich P 2005 *International Journal of the Sociology of Language* **175-176** 213-232
[14] Enriquez N, Rodriguez M I S, Letelier R Q 2004 *An experimental study ReCALL* **16(2)** 539
[15] Ballman T L 1997 *Foreign Language Annals* **30(2)** 173
[16] Jr J J D, Bookout J C 1976 *Foreign Language Annals* **9(1)** 58
[17] Passov E I, Kuznetsova E S 2002 *Teaching foreign language communication: training manual* (Voronezh: Interlingua) 40
[18] Passov E I, Kuznetsova E S 2002 *The problem of skills in foreign language teaching: training manual* (Voronezh: Interlingua) p 40
[19] Zimniaya I A 1991 *Psychology of school foreign language teaching* (Moscow: Education) p 223
[20] Passov E I, Kuznetsova E S 2002 *Teacher's methodological excellence: training manual* (Voronezh: Interlingua) 40
[21] Bimmel P, Rampillon U 2000 *Lernerautonomie und Lernstrategien: Fernstudieneinheit 23* (Berlin-München: Langenscheidt) 20
[22] Verbitskiy A A, Bakshaeva N A 1997 *Issues of psychology* **3** 12-22
[23] Leont'ev A N 1975 *Performance. Conscience. Personality* (Moscow: Science) 181
[24] Frolova N A, Aleshchanova I V 2007 *News of VSTU Series: New educational techniques at a higher educational establishment* **8(23)** 81-83
[25] Aleshchanova I V, Frolova N A 2010 *Modern problems of science and education* **4** 87-90
[26] Basova N V 2013 *German for economists: study guide* (Moscow: KnoRus) 384
[27] Basova N V 2013 *German for technical universities: textbook* (Moscow: KnoRus) 512
[28] Vasil'eva M M 2013 *German for economists: textbook* (Moscow: Al'fa-M, SIC INFRA-M) 350
[29] Guschina T I 2003 *German for textile and light industry specialities: textbook* (Moscow: Higher school) 215
[30] Zinov'eva A F 2016 *German language: a textbook and practical work for bachelor's and master's degree programms: textbook* (Lyubertsy: Yuright) 348
[31] Kravchenko A P 2013 *German language for bachelors: study guide* (Rostov/Don: Feniks) 413
[32] Frolova N A 2012 *Innovations in German teaching and learning: materials of the international science practical conference Ul'yanovsk, 29-30 March 2012* (Ul'yanovsk: Ul'yanovsk state pedagogical university) 185-190
[33] Aleshchanova I V, Frolova N A, Morozova E V, Zheltukhina M R 2017 Psychological and Acmeological Aspect of Educational Cognitive Competence Development *Proceedings of the 7th International Scientific and Practical Conference Current Issues of Linguistics and Didactics: The Interdisciplinary Approach in Humanities (CILDIAH 2017) Advances in Social Science, Education and Humanities Research (ASSEHR)* **97** 19-24
[34] Aleshchanova I V, Frolova N A, Zheltukhina M R 2018 Creative approach to development of competence "writing" in foreign language classes. *SHS Web Conf.* **50** 01025

Protecting surfaces of parts with wear-resistant vibration-damping coatings

E S Ankuda[1], V V Kalmykov[1], M V Musokhranov[1], I K Ustinov[1]

[1]The Kaluga branch of Bauman Moscow Student Technical University, 2 Bazhenovastreet, Kaluga, 248000, Russia

e.ankuda@mail.ru

Abstract. The article is a review describing and justifying the necessity creation and using wear-resistant anti-vibration coatings in tool-making and machine-building industry. The basic methods of applying wear-resistant and anti-vibration coatings were presented and analysed, a search and study of the closest analogues of the studied coatings were carried out, and appropriate conclusions were drawn on the acceptability of using certain methods of applying such coatings.

1. The need for applying wear-resistant vibration damping coatings

Tear and wear of tribosystems is one of the most pressing problems in the vast majority of manufacturing industries. Development and production of new equipment is necessary to increase the power, productivity and bearing capacity of machines, which is inevitably accompanied by an increase in loads and internal stresses leading to fatigue damage, cracking and deformation of the structure of materials. These factors entail instability of the processes occurring during operation of the system and the probability of failures, which can cause equipment failure or adverse effects on the health of workers.

Vibrations, in turn, contribute to reducing fatigue strength of machines and structures, which leads to premature failures, and reduces the service life of the system. The harmful effects of vibrations followed by an increase in the rate of wear must be suppressed, and this can be done by using coatings with both wear-resistant and anti-vibration properties.

In production structures of various purposes, there is often a joint effect of fretting and tribological wear, that is, wear under conditions of small vibrational relative displacements of the contacting bodies, while the effect of vibrations inevitably leads to tribological wear. As a result, there is intensive wear of working surfaces, even in the presence of special wear-resistant lubricants.

The presence of surface energy preconditions interaction of the surface with substances in the environment surrounding the working surface (including lubrication) [1]. This phenomenon is called adsorption. The modern lubricants used contain surface-active substances producing the wedging effect of molecules, which inevitably leads to the growth of microcracks and their subsequent transformation into cracks. This phenomenon is described by the Rehbinder effect (Figure 1), which consists in formation of a thin plastic layer in the process of friction due to the adhesion activity of surfaces [2]. Adhesion is the cause of the formation of boundary layers, it leads to the mandatory appearance of elements of neighbouring phases on the surface. A thin plastic layer has low resistance to shear deformation causing adhesive joints to break.

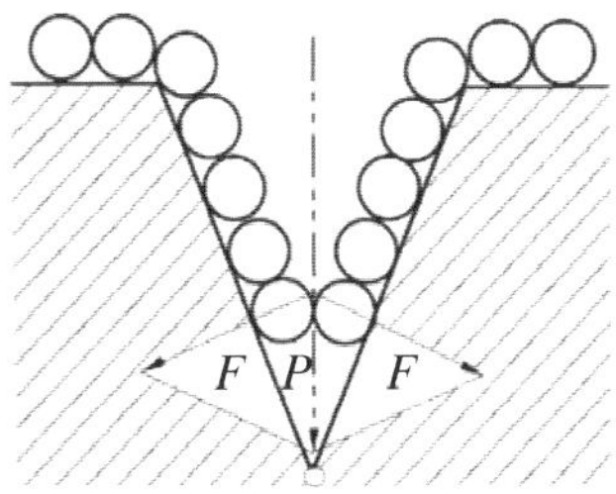

Figure 1. Rehbinder effect: P is the load caused by the adsorption layer; F is the wedging force.

The adsorption activity leads to the formation of a thin boundary layer on them, differing from the main material of the part body in structure and properties. Upon contact interaction of the deformed bodies during their relative movement, the state of such a boundary layer changes. The unlubricated tribological contact of working surfaces 1 and 2 for new parts (Figure 2, a) is replaced by lubricated contact in the process of running in. This process is characterized by formation of a boundary layer with a structure different from the base material (Figure 2, b). The phenomenon described above enables an adhesion breakdown which leads to wear.

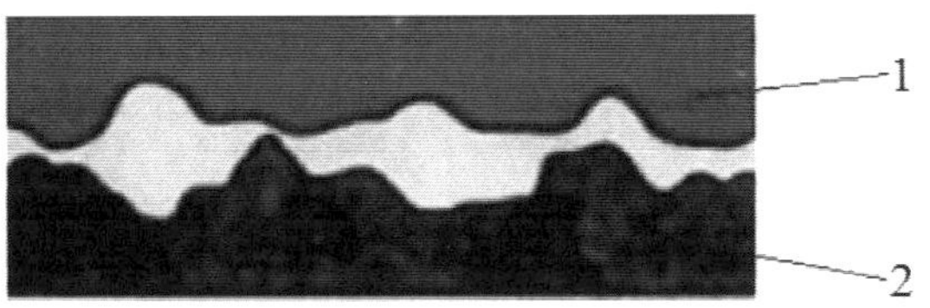

a)Unlubricated tribological contact.

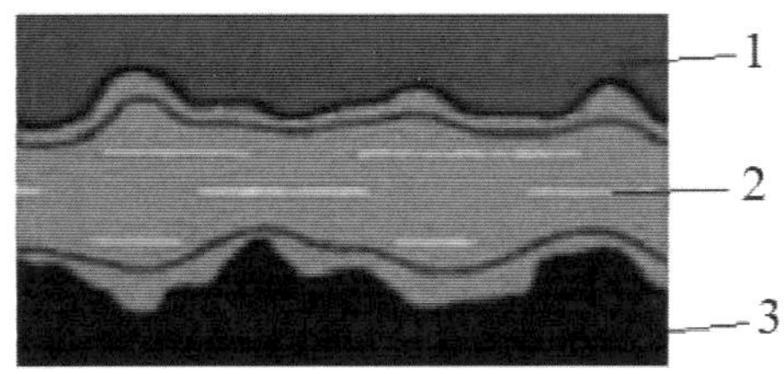

b) Lubricated tribological contact.

Figure 2. Tribological contact of the working surfaces of parts.

2. Major and most common methods of applying wear-resistant coatings

2.1. Gas-thermal coating

A gas or plasma stream heats, crushes and transfers the condensed particles to the substrate. This forms a layer of the required material with the specified properties [3]. It is possible to apply any meltable material and several layers with different properties. The substrate may have almost any shape, and its dimensions may vary in a wide range. During the process, there is no mixing of the sprayed material and the substrate material. Besides, there are no errors, defects and beads. The method features relatively low heating of the treated surface (about 150°C) and relatively high environmental friendliness. Due to the high feed rate, strong adhesion of the sprayed material to the surface occurs [4, 5].

2.2.Chemical deposition of coatings

Deposition occurs when chemical reactions take place on the substrate surface. Continuous layers of material on the substrate are formed as a result of deposition of substances from the gas phase, plasma or colloidal solution. The structure of the coatings obtained can be different: it can be single-crystal, epitaxial, amorphous, polycrystalline, etc. The method allows for applying coatings on parts of various sizes and shapes, however, the accuracy of the coating is achieved by a high accuracy control of the flow of the substances deposited. It is possible to obtain high-density homogeneous clean coatings with good adhesion. The main drawback of the method is the probability of chemical and fire hazard, caused in some cases by toxicity, corrosion activity, flammability and explosiveness of the precursor. Relative difficulty is posed by deposition of multicomponent materials, which is associated with the difference in the evaporation rates of the precursors used. The method is also limited by the ability to conduct separate chemical reactions taking into account deposition of the required components on a given substrate. The method is widely used in the tool-making industry [6,7,8].

2.3. Physical deposition of coatings from the gas phase

Deposition takes place in vacuum from the vapor (gas) phase by direct condensation of the vapor of the material applied on the substrate [9]. In the process of deposition, the sprayed material is heated to high temperatures to transfer enough energy to atoms or molecules to break chemical bonds and leave the substance, thereby the liquid phase evaporates and the solids sublimate. The process takes place in a vacuum environment, since the pressure of the gas medium is relatively low, but it reduces the pollution of the gas phase to a rather low level and produces a clean film on the substrate surface, but at the same time it complicates the method and increases its cost. Just like with chemical deposition, the synthesis of multicomponent surfaces is hindered, since many materials, when in the form of chemical compounds, partially dissociate during thermal evaporation, which leads to the formation of non-stoichiometric precipitation [10]. It is difficult to obtain coatings on surfaces of complex shape, since the method is a version of one-sided directional deposition [11].

2.4. Electrospark doping with electrodes from EDM powders
They resist abrasive and corrosive wear and consist in powders obtained by recycling wastes from solid alloys [12]. They found application in mechanical engineering, agricultural, food processing and many other industries. During operation of an element with this coating, there is no direct wear of the material of the element body; however, such a coating must be periodically reapplied. When applying such coating for a cutting tool, the resistance of the latter increases by 3.8-4.8 times [13].

2.5. Electrochemical deposition of electroplated coatings
It allows you to restore the performance of worn parts, as well as to prevent further loss of material during their operation. The method allows you to adjust the thickness of the coating at the stage of its formation and to form it on surfaces of complex geometric shape [14]. The major disadvantages of the method include non-ecological vapors of the electrolyte and the duration of the process [15].

2.6. Cathode ion bombardment
The surface is generated in vapors of the evaporating electrode using an arc discharge. Condensation of applied coatings of carbides or nitrides is carried out by means of chemical reactions in the plasma under the effect of ion bombardment in vacuum. The method is highly efficient because the ion flux can be accelerated by setting the sprayed product to a negative potential or by increasing the density of the ion flux [16]. This process is environmentally friendly, efficient, it provides the possibility of automation and high adhesion of the sprayed surface [17].

2.7. Magnetron deposition method
The method consists in vacuum application of thin films by magnetron spraying of the target and allows you to adjust the thickness of the films applied [18]. The main advantages of the method are ease of use, high quality of the coatings applied and a large supply of the material sprayed [19]. The films feature high quality and record-breaking physical characteristics: thickness, porosity, adhesion, etc., and are created at the level of atomic planes. However, the method allows you to apply coatings on very small areas, which is a significant disadvantage for the engineering industry. The method is mainly used in the instrument-making industry, namely in microelectronics [20].

3. Anti-vibration coatings
Vibrations entail fatigue damage to working surfaces; therefore, anti-vibration coatings must protect the part from the appearance and development of fatigue stresses. Two types of anti-vibration coatings can be used to protect parts of machine-building and instrument-making products. There are coatings with a high coefficient of elasticity, and coatings with a low coefficient of friction. In the first case, the reduction of the vibration load occurs due to the absorption of vibrations, and in the second case, the vibration waves are damped in the direction of the tribocoupling movement.

4. Coatings with wear-resistant and vibration-damping properties
Thus, the use of standard wear-resistant lubricants, coatings and vibration protection means does not fully meet the needs of modern machine building and instrument making. This raises the issue whether

it is relevant to create and use competitive coatings. Such coatings must feature acceptable adhesive resistance, low impact on the surface of the material and combined properties. Anti -vibration and wear-resistant properties must be combined. Such coatings will help reduce the harmful effects of vibrations on the operation of the mechanism and increase its service life without reducing the operational accuracy and without running in the working surfaces of parts which would lead to a change in configuration, and, consequently, to an even greater increase in vibration exposure and wear.

The life cycle of equipment must be increased by coating the loaded parts with anti-vibration and wear-resistant material due to the high adhesion of the material to the metal, and the adhesion strength is achieved by controlling the surface energy of the substrate. There are no complete analogues of the products offered, and partial ones do not fully meet the needs of the engineering industry, or feature a narrow focus.

In order to choose the right wear-resistant coating, it is necessary to take into account a number of factors, including the hardness ratio of the materials of the response parts, the temperature of the working contact, the rate and angle of attack of the abrasive material, anti-friction properties, brittleness, adhesion and cohesion of the surface and many others. The correct choice of the coating can significantly extend the service life of the product.

References

[1] Geguzin Ya E and Ovcharenko N N 1962 *Sov. Phys. Usp.***76**283-328
[2] Myshkin N K, Petrokovets M I 2007 *Friction, Lubrication, Wear. Physical Foundations and Technical Applications of Tribology*(Moscow: FIZMATLIT) p 368
[3] Khan A N, Lu J 2007 *Surface and Coatings Technology*vol 201**8**4653-8
[4] Lyudagovsky A V 2006 *Gas-Thermal Dusting of Coatings* (Moscow: Russian State Open Technical University of Communication Ways) p 43
[5] Puzryakov A F 2008 *Theoretical Foundations of Plasma Spraying Technology* (Moscow: BMSTU) p 360
[6] Syrkin V G 2000 *CVD-Method. Chemical Vapor Deposition* (Moscow: Science) p 482
[7] Danilin B S 1989 *The Use of Low-Tempetarure Plasma for Deposition of Thin Films*(Moscow: Energoatomizdat) p 328
[8] Ivanovsky G F and Petrov V I 1986 *Ion-Plasma Processing of Materials* (Moscow: Radio and Communication) p 232
[9] Oura K, Lifshits V G, Saranin A A et al 2006 *Introduction to Surface Physics* ed V I Sergienko (Moscow: Science) p 490
[10] Gurin V N, Korsukova M M, Nikanorov S P, Nechitaylov A P, Derkachenko L I, Uspenskaya Z I and Zimkin I N 2003 *Letters to the Journal of Technical Physics* **29** 51-5
[11] Mattox D M 1998 *Handbook of Physical Vapor Deposition (PVD) Processing: Film Formation, Adhesion, Surface Preparation and Contamination Control* (Westwood, N.J.: Noyes Publications) p 917
[12] Sidorov A I 1987 *Restoration of Machine Parts by Spraying and Surfacing* (Moscow: Mashinostroenie) p 192
[13] Ageev E V, Davydov A A, Ageeva E V, Bondarev A S and Novikov E P 2013 *News of South-West State University. Series: Technique and Technology* **1** 32-8
[14] Panteleenko F I, Sarantsev V, Panteleenko EF, Borovik DI and Khina BB 2009 *Strengthening Technologies and Coatings* **4 (52)** 27-33
[15] Vityaz P A 2004 *Friction and Wear* vol 25**6** 593–601
[16] Shaparev A V, Aymurzina A D and Savina AI 2017*Scientific Herald***1 (11)** 165-179
[17] Savin I A 2012 *Procurement Production in Mechanical Engineering***9** 41-4
[18] Richter F, Welzel T, Kleinhempel R, Dunger T, Knoth T, Dimer M and Milde F 2009 *Surface and Coatings Technology*vol 204**6-7** 845-9
[19] Sagatelyan G R, Shishlov A V 2014 *Science and education: a scientific publication MSTU. N.E.*

Bauman **11** 458-81

[20] Berlin E V, Seidman L A 2010 *Ion-plasma processes in thin-film technology*(Moscow: Tekhnosfera Publishing House) p 457

Melodic features of emphatic apology speech formulae in business communication

E Eltanskaya, A Arzhanovskaya, Y Linkova

Prospect Universitetsky 100, 400062 Volgograd, Russia

arjanovskaya@volsu.ru

Abstract. The article presents the main results of the research on the melodic features of apology speech formulae realization in the modern English language. Special attention is paid to the study and discussion of melodic features of emphatic apology speech formulae realization in business communication. Emphatic speech acts are used when the speaker realizes the seriousness of his fault, takes responsibility for the damage and demonstrates the emotional attitude to the situation. Speech acts of regret, speech acts of sympathy and speech acts of condolence are classified as emphatic speech acts. The materials of the study are formulae with the semantics of emphatic apology in modern English. The study material was based on the following films: "As good as it gets", "The Devil wears Prada" and "Sweet November". The steps and methodology of an integrated perceptional and acoustic experiment are described. In the audience analysis the increase / decrease of rate of pronouncing formulae with the semantics of apology are examined. The quantitative data for each type of emphatic apology are analyzed, the general and specific terminal tones typical of emphatic apology are determined, and the linguistic interpretation of the obtained results is represented.

1. Introduction

Communication is an essential element in the functioning of any organization. Communication flows connect the organization as a system into a single whole, help to solve production and social problems. The implementation of communicative competence is associated with the parameters of typical communicative situations that structure any sphere of communication: everyday, educational, industrial, etc [1].

Negotiating partners will be able to come to an agreement if they strive to establish a favourable psychological and communicative microclimate of the dialogue and maintain a friendly tone of conversation. Establishing and maintaining business contacts at a high level of culture, adhering to the rules of business etiquette improves the efficiency of solving substantive tasks. Establishing and maintaining business contacts includes an appeal, greeting, acquaintance, order, offer, request, gratitude, praise, compliment, apology, farewell and so on [2].

Below we will consider situations related to the functioning of the etiquette-speech formulae in the communication.

Speech etiquette is an important element of culture, the product of human cultural activity and an instrument of such activity. English speech etiquette is a set of special words and expressions that give a polite form of English speech, as well as the rules according to which these words and expressions are used in practice in various situations of communication. One of the most important and integral components of speech etiquette of any language is an apology [3, 4].

An apology is a culturally and socially determined communicative phenomenon that has significant differences in different sociocultural contexts, both in terms of the conditions for its implementation and in terms of its language implementation strategies [5].

Speech formulae of apology are widely used in business communication in many sociocultural communities, being a means of realizing an extremely important social function that ensures the harmony of interpersonal relations support between the participants of communicative interaction [6].

Despite the considerable amount of works devoted to the study of an apology, a number of problems related to the peculiarities of the apology formulae functioning in speech and especially in business communication, including the intonational features of this speech act, have not been studied yet. Relevance of such analysis determines the importance and urgency of the proposed work.

The study object of this research is English language formulae with the semantics of an apology.

The study subject is the melodic features of formulae with the semantics of apology in modern English and especially in business communication.

The study material was based on the following films: "As good as it gets", "The Devil wears Prada" and "Sweet November".

Depending on the conditions of the communication situation, the speech act "apology" is realized in speech in one of two variants: emphatic apology and phatic apology [7].

In this article we will dwell in more detail on the peculiarities of the emphatic apology implementation in the English language.

Emphatic speech acts are used when the speaker realizes the seriousness of his fault, takes responsibility for the damage, demonstrates the emotional attitude to the situation, which manifests itself in special forms of verbal behavior associated with the expression of regret, the desire to make amends for the addressee, sympathy, the desire to share someone's pain.

Speech acts of regret, speech acts of sympathy and speech acts of condolence are classified as emphatic speech acts.

2. Procedure of phonetic analysis

2.1. Audience analysis

The audience analysis of the experimental material was focused on identifying the melodic peculiarities of emphatic apology in the English language.

The main goal of the considered audience analysis is to determine the type of nuclear tone (NT), its level and range characteristics using the following perceptually discrete gradation of the tonal characteristic:

Falling NT (Low Fall, Mid Fall, High Fall): … ╲ …; … ╲ …; … ╲ …;

Rising NT (Low Rise, Mid Rise, High Rise): … ╱ …; … ╱ …; … ╱ …;

Mid-Level NT: … – …;

Fall-Rise NT: … ᵛ …;

Rise-Fall NT: … ^ …;

3. Perceptual-acoustic features of emphatic apology

The purpose of this stage of the research was to identify the melodic features of emphatic apology. In total 90 examples of emphatic apology were analyzed. Among them: 76% - speech acts of regret; 20% - speech acts of sympathy and 4% - speech acts of condolence.

3.1. The results of the audience analysis. Melodic features of emphatic apology.

3.1.1. Speech acts of regret. Melodic features of this type of apology were revealed on the basis of the performed analysis. In most cases, the Speaker uses High Fall (47%) or Low Fall (32%). Mid Fall is used less frequently (21%).

Using one of the falling tones, the Speaker regrets the circumstances, apologizes for being late or for the unperformed work:

- I don't understand why it's so difficult to confirm an appointment.

-I know. I'm so ╲sorry, Miranda. I actually did confirm last night

- Details of your incompetence do not interest me. (The Devil wears Prada)

Miranda complained about her senior assistant Emily because she did not confirm the appointment. Emily, in turn, tries to excuse herself and claims that she approved it last night. Emily uses Low Fall when she pronounced an excuse remark, as this tone is often used to express sincere regret and apology.

Let us give an example of the use of High Fall in Speech acts of regret:

No I can't , I'm ＼sorry but I have to go

Here is an intonogram of apology from this sentence (figure 3).

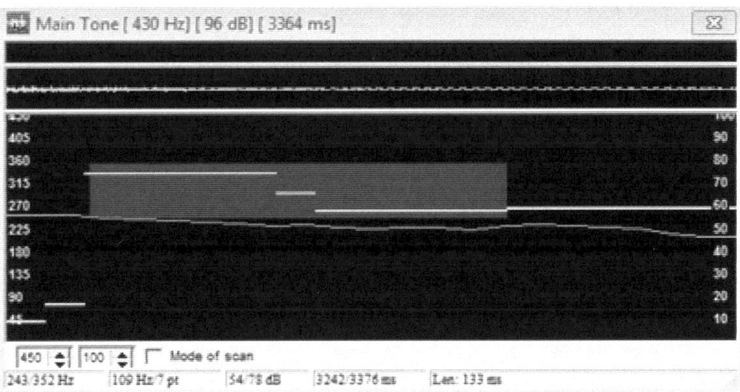

Figure 1. intonogram of apology from

In this example, Andy has to refuse the offer of her friend Christian to visit a party where he can introduce Andy to a man who can offer her a very good job that interests her. But Andy can not accept this offer, as Christian is actively caring for her, knowing that she is in a relationship. When expressing rejection, Andy uses High Fall tone, as it conveys certain emotions, such as completeness, finality, confidence. Lowering the tone sounds distinctly, confidently and convincingly.

In some cases (21%) the speaker uses a High Fall tone, (more often it is going on in an informal setting) this can be seen in the following examples:

Slow down. We need to talk.

-Talk about what?

-Us.

-Us. Right. Hey, isn't there some kind a limit on 'us' talks?

-You forgot about my parents, didn't you?

- I can't make it. ***- I'm ＼sorry.***

-Nelson, this is their third trip to the city. This is important to me.

- And this account is very important to me. (Sweet November)

The conversation takes place between the couple-Nelson and Angelica. Nelson hurries to work and he is very concerned about the new advertising project of the company he works for but Angelica believes that they need to have it out with each other. Her parents have to come to the city where Angelica lives with Nelson, but Nelson is completely absorbed in work, he forgot about arrival of her parents. He does not want to sort things out, because at the moment he is only concerned about the work. He apologizes for this, but apologizes are High Fall tone, because he is completely indifferent to the situation.

3.2. Speech act of sympathy

To implement this Speech act, Low Fall tone (60%), High Fall tone (20%) and Mid Fall tone (20%) are used. A specialized tone of sympathy expression is Low Fall tone, which is observed in the following examples:

*1.- **I'm so** ˎ**sorry** Miranda. (The Devil wears Prada)*

*2.- I heard about this terrible event. **ˎI'm sorry**. (The Devil wears Prada)*

*3.-Oh, ˎ**I'm so sorry**, dear. I can't believe in it! (The Devil wears Prada)*

Using this tone, the speaker tries to express sincere and deep regret about what happened. For example, Andy regrets Miranda's divorce, dismissal of her creative Director Nigel and her friend's problems. This tone is used to express the speaker's involvement in the problems of the interlocutor and to express sympathy.

However, sometimes the speaker uses High Fall tone:

- ***I'm** ˋ**sorry** about your job. (The Devil wears Prada)*

Andy apologizes in Mid Fall tone, as her interlocutor is not her close friend and the purpose of this apology is to pay attention to what happened.

In addition, when expressing sympathy, there is the use of the High Fall tone by the speaker:

- *It's non- Hodkin's lymphoma. It's a type of cancer. You shouldn't find out like this. ˋ**I'm sorry**, Nelson. She's gonna be asleep for hours. You want to go for a walk, have a chat or something?*
- *Yeah.*
- *Yeah, ok. (Sweet November)*

Sarah had a sudden attack, she was taken to the hospital. Nelson is desperate, he doesn't know what's going on. Sarah's best friend explains that she is sick: Sarah has cancer. He expresses sympathy to Nelson through the use of High Fall tone, as it is very emphatic and expresses the sincere involvement of the speaker in the problems of his partner [8].

3.2.1. Speech act of condolences

To express condolences, the speaker uses High Fall tone in all analyzed examples, as through this tone the speaker demonstrates sincere empathy and involvement in the problems of the communication partner [9].

- Hey, by the way, when do I get to meet your parents?

- My parents are dead.

- ***I'm** ˋ**sorry**. (Sweet November)*

Sarah and Nelson have saddled up with and entered into relationship. Sarah wonders when she will meet with the parents of Nelson. Nelson replies that his parents are died. Sara apologizes in High Fall tone, as she genuinely sympathizes with Nelson, and this tone is often used to express the speaker's involvement in the problems of communication partner.[10].

Thus, in the course of the analysis of various types of emphatic apology, the following tonal features were revealed: in most cases, the construction containing an apology is pronounced in High

Fall tone (44%), for example, when the Speaker seeks to provide moral support to the Addressee in a difficult situation, when he expresses his sadness and regret and demonstrates involvement in the conversation.

There are cases where we can see the use of the Mid Fall (20%) and Low Fall (36%) tone. It happens when the speaker is in a frustrated state, when he sets a goal his speech to be restrained and calm. For clarity, we present the results of the study of melodic features of the emphatic apology in the diagram:

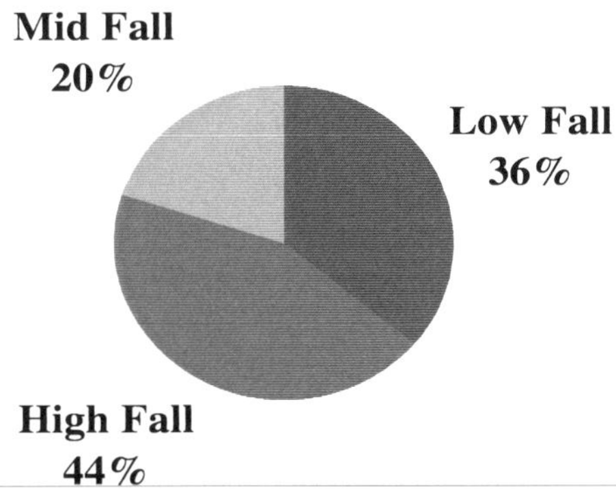

Figure 2. Diagram melodic features of emphatic apology

Without special communication skills and abilities, even a great specialist in his field will not be able to support a business conversation, hold a business meeting, take part in the discussion, and defend his point of view. It means that businessman except his professional competencies must master the communicative competence, i.e. knowledge of the psychological, intonational, substantive (content) and language-specific components needed for the understanding negotiating partner.

References

[1] Eltanskaya E, Popova O, Linkova Yu Arzhanovskaya A and Kulichenko Yu 2017 Types of Interactive Methods in Teaching English to Students of Economics *ASSEHR* **97** pp 100-102
[2] Blum-Kulka S, House J and Kasper G. 1989 *Cross-cultural pragmatics: Requests and apologies* (Norwood NJ : Ablex Publishing Corporation) p 290
[3] Leont'ev O 1997 *English Speech Etiquette Course* (Moscow) p 398
[4] Formanovskaya N 1990 *Speech Etiquette. Russian-English Correlations* (Moscow) p 90
[5] Pletneva E 2007 Functional filling of the apology formulae in the English and Russian languages (Voronezh) p 120
[6] Coulmas F 1981 *Conversational Routine: Explorations in Standardized Communication Situations and Prepatterned Speech* (New York : Mouton Press) pp 69 - 91
[7] Bolinger D 1993 *Intonation in American English* (Cambridge) p 487
[8] Crystal D 1969 *Prosodic systems and intonation in English* (Cambridge: Cambridge University Press) p 382
[9] SHahbagova D 1982 *Phonetic Features of the Pronunciation in the English Language* (Moscow: Vysshaya Shkola Press) p 128
[10] Matusevich M 1976 *Speech Melody* (Moscow: Vysshaya Shkola Press) p 151

IOP Conf. Series: Materials Science and Engineering **483** (2019) 012041 doi:10.1088/1757-899X/483/1/012041

Improvement of design and calculation method for contact air humidifiers

A G Averkin, A I Yeremkin, T I Korolyova, K O Chichirov and S O Kiselyov

Penza State University of Architecture and Construction, 28 German Titov Street, Penza, 440028, Russia

E-mail:algraw@mail.ru

Abstract. Described are designs, principle of operation and calculation technique for contact-type water-cooled air humidifiers in air conditioning systems. Presented is a range of rotor-type and film-type contact apparatuses working in adiabatic air humidification conditions. The said equipment has been devised as a result of experimental investigation in the University and is characterized with reduced specific consumption of materials, power and cost. Positive results are achieved by increasing the time of working media contact in the process of heat-mass trans fer and increase in heat-mass transfer area. A procedure has been devised to calculate the heat-mass transfer area, based on application of humid air I-d-diagram and number of transfer units which is determined by graphic integration method.

1. Introduction.

An artificial air humidification mainly in the cold period of year is a prerequisite for creating comfortable air conditions in rooms, as well as providing micro climatic parameters meeting specific requirements of modern technological processes practically in all regions of Russia [1].

To humidify air in air-conditioning systems (ACS), contact heat-and-mass exchangers are used. In the exchangers air flow contacts water directly.

Variety of types of contact apparatuses produced for hygrothermal air conditioning is accounted for by various ways for creating a phase interface in working fluids.

To humidify and cool air adiabatically, water is used whose temperature should equal wet bulb air temperature. For this purpose, water is to be circulated in the contact apparatus (Figure 1) [1]- [4].

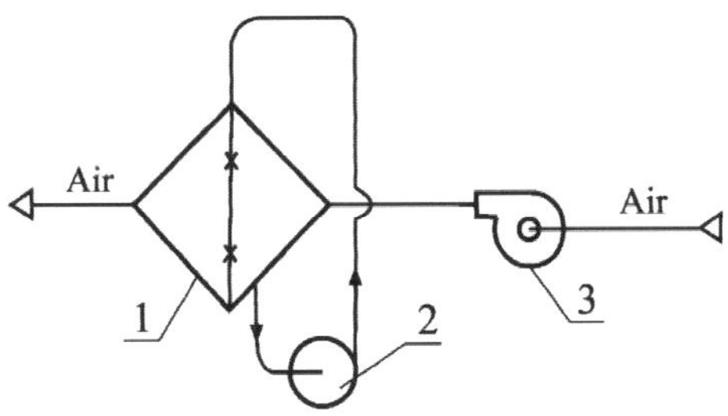

Figure 1. Diagram of contact apparatus with water Recirculation 1 – contact apparatus; 2 – water pump; 3 – fan

2. Basic part

To allow engineering modification aimed at reducing consumption of materials, power and cost of contact apparatuses for adiabatic humidification of air, a range of rotor-type and film-type contact apparatuses (Figures 2, 3, 4) has been devised at the University department "Heat and Gas Supply and Ventilation".

The contact apparatus of rotor-type (Figure 2, RF Patent 2270958) is composed of an air washer whose lower part has a water-collecting sump. Inside the air washer, there is a cruciform turbine with its horizontal shaft.

The lower part of the blades is immersed in water by some millimeters. At the chamber inlet, there are guide plates that allow changing air flow direction with regard to the turbine blades and controlling the turbine speed in a unit time, whereas at the outlet there is a drop catcher that prevents entrainment of liquid particles with air flow. A centrifugal fan is attached to the chamber end face through a flexible duct connector.

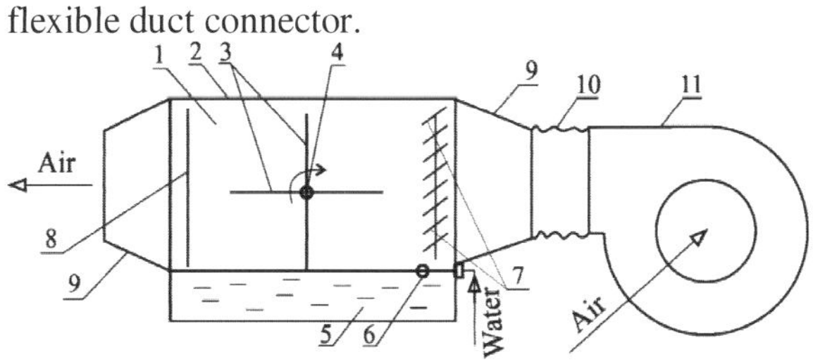

Figure 2. Diagram of spinning disk humidifier
1 – air channel; 2 – chamber; 3 – turbine blades; 4 – axle; 5 – water sump; 6 – float; 7 – guide plates; 8 – drop catcher; 9 – branch pipes; 10 – flexible duct connector; 11 – fan.

The device works as follows. The centrifugal fan delivers air to the chamber. At that time, the cruciform turbine starts rotating under the dynamic pressure of air flow and friction forces. During rotation the lower part of the turbine blades throws out part of liquid onto air flow and neighboring blades, i.e. water is sprayed in air and at the same time the blade hygroscopic material (gauze) is becoming moist. Air becomes wet due to evaporation of some sprayed water particles and film-type contact of air flow with the wet hygroscopic material of the cruciform turbine blades. Non-evaporated water droplets fall down back to the water sump, i.e. aqueous phase recirculates in the chamber. Its temperature stabilizes at a level of wet bulb air temperature.

Hygrothermal air conditioning takes place when air flows horizontally along the chamber thanks to the centrifugal fan operation.

This contact unit is notable for its compactness and less power consumption because it has neither turbine drive mechanism nor water recirculation pump. The unit is the most adaptable to operate in ducted ventilation and air conditioning systems.

Another modification of the spinning disc humidifier is represented by the design with air flow recirculation (Figure 3, RF Patent 2294490).

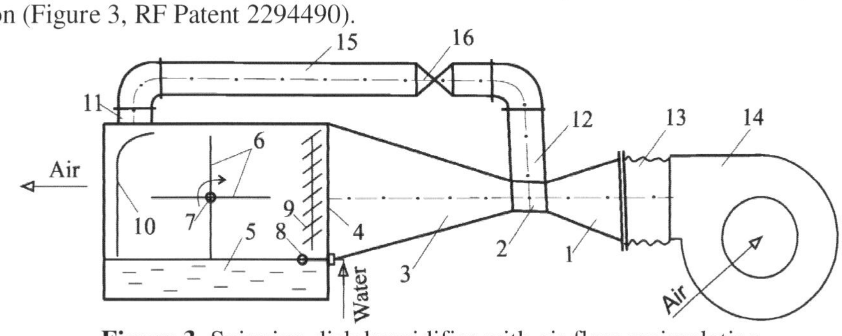

Figure 3. Spinning disk humidifier with air flow recirculation
1 – confuser; 2 – mouth; 3 – diffuser; 4 – chamber casing; 5 – water sump; 6 – turbine; 7 – turbine axis; 8 – float device; 9 – guide plates; 10 – drop catcher; 11, 12 – pipe branches; 13 – flexible duct connector; 14 – fan; 15 – recirculation air duct; 16 – valve

With the help of pipe branches 11, 12 of by-pass air duct 15 the chamber outlet channel is connected to the chamber inlet channel where Venturi tube is located. The Venturi tube consists of confuser 1, mouth 2 and diffuser 3. The Venturi tube is connected to centrifugal fan 14 through flexible duct connector 13. Air flow through by-pass 15 is controlled with the help of valve 16.

When the unit is running, air coming out of chamber 4 is ejected through by-pass air duct 15 into mouth 2 of Venturi tube, i.e. the process of hygrothermal air conditioning goes on with application of humidified air recirculation principle.

Ejection takes place due to reducing air static pressure, increasing air dynamic pressure in the mouth in full conformity with the equation of air flow total pressure:

$$dP_{\text{ñëí}} = dP_s + dP_d = dP_s + \frac{\rho v^2}{2}, \tag{1}$$

where $dP_{\text{ñëí}}$, dP_s, dP_d – are, respectively, total pressure, static pressure, dynamic pressure of air flow in the mouth section, Pa; ρ – air density, kg/m^3; v – air-flow rate, m/sec.

The principle of recirculating air through a contact apparatus allows increasing the durability of air-aqueous phase contact, which intensifies heat-mass exchange, i.e. enhances the effectiveness of the adiabatic air humidification process. This comes from the convective heat exchange equation (2) and convective mass transfer equation (3) [5]-[7]:

$$d\left(\frac{Q}{F_{\delta}}\right) = dq = \alpha(t_1 - t_2)d\tau, \tag{2}$$

$$d\left(\frac{M}{F_{\text{ì}}}\right) = dj = \beta(d_1 - d_2)d\tau, \tag{3}$$

where Q – is quantity of heat transferred from air flow to water, J; M – is quantity of water vapor coming from water to air flow, kg; $F_{\text{т}}$ – is heat exchange area, m^2; $F_{\text{м}}$ – is mass exchange area, m^2; dq – is heat flow density differential, J/m^2; α – coefficient of heat transfer from air flow to water, W/(m^2·°C); t_1, t_2 – are, respectively, air flow temperature and water temperature in the contact apparatus, °C; dj – is differential in specific water vapor mass discharge, kg/m^2; β – is coefficient of heat transfer from water to air flow, m/sec; d_1, d_2 – are, respectively, air moisture content in the boundary layer near the water surface and in the air flow, kg/kg; $d\tau$ – is moveout time (duration of heat-mass exchange), sec.

In addition, a contact apparatus with a vibrated plate packing has been devised, manufactured and installed to humidify air (Figure 4, RF Patent 2581982).

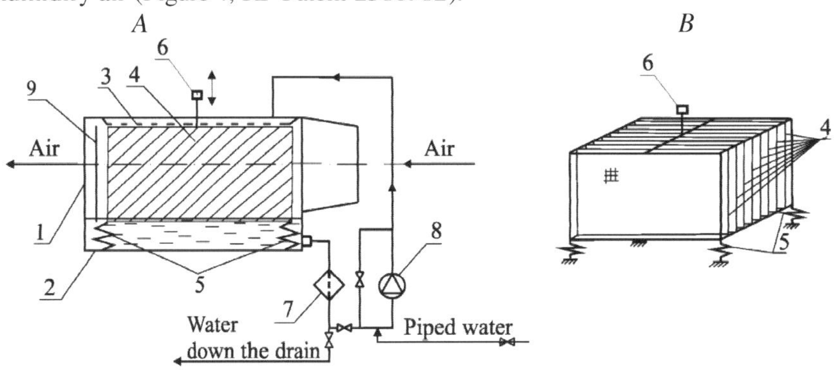

Figure 4. Contact apparatus with vibrated plate packing A – diagram; B – contact assembly with plate packing 1 – casing; 2 – lower tray; 3 – upper perforated tray; 4 – plate packing unit; 5 – spring shock absorber; 6 – mechanical vibrator; 7 – water filter; 8 – pump; 9 – drop catcher

The said apparatus is intended for hygrothermal air conditioning and works as follows.

With the help of a fan (not shown in figure 4) air comes into the humidifying unit of the vibrated plate packing 4. The vibrated plate packing is kept under mechanical vibration due to operation of mechanical vibrator 6 consisting of an eccentric flywheel secured on the motor shaft (not shown in the figure). The plate packing is continuously showered with recirculated water coming from the lower water sump 5 into the upper perforated tray 3 with the help of pump 8.

Water temperature stabilizes at a level of wet bulb air thermometer, i.e. hygrothermal air conditioning meets the adiabatic humidification and cooling mode. Air is moistened thanks to a film-like contact of air flow with packing wet plates made of hygroscopic material (viscose) 4 and in addition due to evaporation of the halo of water droplets and splashes caused by mechanical vibration of the packing wet plates.

The process of heat-mass exchange is intensified by increasing the heat-mass transfer area.

To calculate air-conditioning contact apparatuses, a method using number of transfer units has been prepared [1].

The thermotechnical calculation of contact apparatuses for air-conditioning systems is suggested to be as follows.

1. Set initial data for adiabatic humidification (cooling) of air - t_1, I_1, t_2;
2. Construct the ray of hygrothermal air conditioning process on *I-d* - diagram in accordance with figure 5.

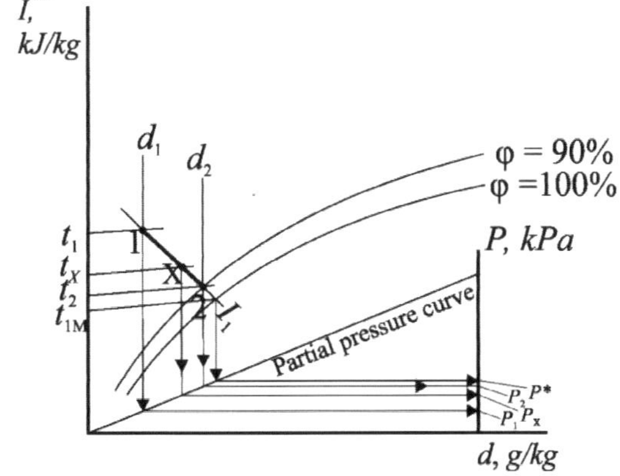

Figure 5. Determination of current and equilibrium values of water temperature as well as current and equilibrium values of water vapor partial pressure under adiabatic air humidification and cooling

3. Determine the number of transfer units n_t, for implementation of heat-exchange processes and the number of transfer units n_p – for mass-exchange processes.

Based on figure 5, plot characteristic curves for adiabatic air humidification (cooling) (Figure 6).

$$\frac{1}{t-t^*} = f(t); \quad (4)$$

$$\frac{1}{P^* - P} = f(P); \quad (5)$$

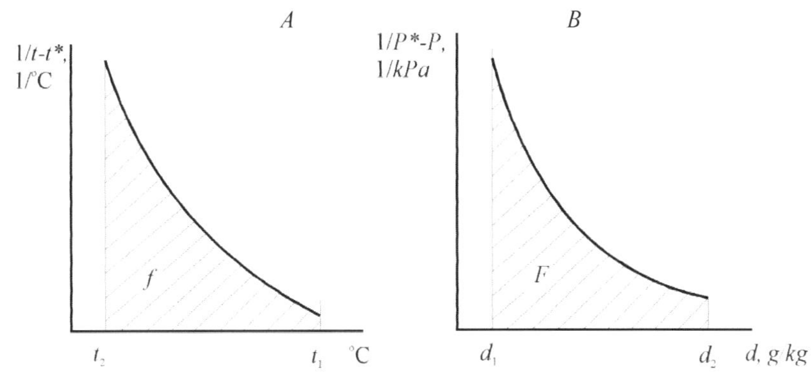

Figure 6. Characteristic curves for determining the number of transfer units

The number of transfer units is calculated from below formulas:

$$n_t = \int_{t_1}^{t_2} \frac{dt}{t-t^*} = f \cdot m_1 \cdot m_2 \; ; \quad (6)$$

$$n_p = \int_{P_1}^{P_2} \frac{d(d)}{P^* - P} = F \cdot M_1 \cdot M_2 \; ; \quad (7)$$

here *t*, *t** – are, respectively, current (working) and equilibrium values of air temperature under hygrothermal conditioning,°C; *f*,*F* - is curvilinear trapezoid area in units of the selected scale (figure

6*A*);
m_1, m_2, M_1, M_2 – is the scale of values to be plotted along abscissa axis and ordinate axis; P, P^* – are, respectively, current (working) and equilibrium values of partial pressure in the air humidification mode, *kPa* (figure 6*B*)
4. Determine the kinetic coefficients of heat-mass transfer on the basis of thermal and diffusion similarity criteria.

The heat-transfer coefficient is determined from the below formula:

$$\alpha = \mathrm{Nu}\frac{\lambda}{l}; \tag{8}$$

The mass transfer coefficient is calculated from the below equation:

$$\beta = \mathrm{Nu}'\frac{D}{l}; \tag{9}$$

here Nu, Nu^l – are, respectively, Nusselt numbers for heating process and diffusion process. These are determined by criteria equation (see the table) [1], [8] – [10]; λ – is air thermal conductivity, W/(m·°C); l – is determinant geometrical dimension, m; D – is diffusion coefficient of water vapor in air, m^2/sec.
5. Determine the required surface of heat-exchange F^t and mass-exchange F^p in the contact apparatus for air humidification in accordance with the initial data (item 1):

$$F^t = \frac{n_t \cdot G}{\alpha}; \tag{10}$$

$$F^p = \frac{n_p \cdot G}{\beta}; \tag{11}$$

Table. Criteria equations for calculating heat-mass exchange in the process of water evaporation from body surfaces into air flow

Serial number	Author, year of publication	Criteria equation form
1	Yu.Krisher	$\mathrm{Nu} = 0{,}8 \cdot \mathrm{Re}^{0,5}$
2	D.Vyrubov, 1969	$\mathrm{Nu} = 2 + 0{,}03 \cdot \mathrm{Re}^{0,54}$
3	A.Sokolsky, F.Timofeyeva	$\mathrm{Nu} = 2 + 0{,}16 \cdot \mathrm{Re}^{0,67}$
4	Polhausen	$\mathrm{Nu} = 0{,}664 \cdot \mathrm{Re}^{0,5} \cdot \mathrm{Pr}^{0,33}$
5	E.Jillyland, T.Sherwood, 1934; with correction by K.Bennett, 1966	$\mathrm{Nu}^l = 0{,}023 \cdot \mathrm{Re}^{0,8} \cdot (\mathrm{Pr}^l)^{0,33}$
6	S.Ilyina, 2009	$\mathrm{Nu} = C^l \cdot \mathrm{Re}^{0,5} \cdot \mathrm{Pr}^{0,44}$
7	N.Fresling,1938; R.Dreik, 1961	$\mathrm{Nu} = 2 + 0{,}45 \cdot \mathrm{Re}^{0,5} \cdot \mathrm{Pr}^{0,33}$
8	B.Katsnelson, F.Timofeyeva,1948	$\mathrm{Nu} = 2 + 0{,}03 \cdot \mathrm{Re}^{0,54} \cdot \mathrm{Pr}^{0,33} + 0{,}35 \cdot \mathrm{Re}^{0,8} \cdot \mathrm{Pr}^{0,356}$
9	S.Kutateladze, 1957	$\mathrm{Nu} = 2 + 0{,}03 \cdot \mathrm{Re}^{0,54} \cdot \mathrm{Pr}^{0,33} + 0{,}35 \cdot \mathrm{Re}^{0,58} \cdot \mathrm{Pr}^{0,56}$
10	V.Isachenko et al., 1961	$\mathrm{Nu} = 4{,}55 \cdot 10^{-3} \cdot \mathrm{Re}^{0,8} \cdot K^{0,4}$
11	A.Fokin, V.Mushtayev, 1969	$\mathrm{Nu} = A + B \cdot (\mathrm{Re} \cdot \mathrm{Pr}^{0,33})$

12	A.Nesterenko, 1954	$Nu = 2 + 1{,}05 \cdot Re^{0,5} \cdot Pr^{0,33} \cdot Gu^{0,175}$ $Nu^{1} = 2 + 0{,}85 Re^{0,52} (Pr^{1})^{0,33} Gu^{0,135}$

3. Findings.

To humidify air in air-conditioning systems, contact apparatuses have been designed. These are notable for low consumption of power and materials. Modern technique has been proposed to calculate contact apparatuses. The technique makes it possible to determine the required heat-mass transfer surface under adiabatic humidification of air.

References

[1] Averkin A G and Eremkin A I 2015 *Improvement of Devices for Heat and Humidity Treatment of Air and Methods for Calculating Climate Equipment* (Penza: PGUAS) p 204
[2] Lewis W K 1933 ASME Vol 55 **9** 567-8
[3] Bogoslovsky V N and Gvozdkov A N Proceeding HB·94 Budapest Vol **1** 381-4
[4] Merkel F 1925 *Evaporation Cooling. Research Work, From the Field of Engineer* (Being Issue **275** V D I) p 48
[5] Ramachandran S et al 2016 *Thermal Engineering* II (Airwalk Publications) p 766
[6] Rash R 1970 *Theory and practice of Luftwashers in the technique of aeration* Air conditioning B 12 **9** pp 12-31
[7] Averkin A G and Kiselev S O 2017 *Slovenia Journal of scientific thought* Vol 1 **7** 80-3
[8] Hong S W 1974 *Laminar Flow Heat Transfer in Ordinary and Augmented Tubes* (Ames: Iowa State University) p 291
[9] Wibulswas P 1966 *Laminar-Flow Heat-Transfer in Non-Circular Ducts* (London: University College London) p 150
[10] Raju K S N 2016 *Fluid Mechanics, Heat Transfer, and Mass Transfer* (Chemical Engineering Practice Published by John Wiley Sons Inc Hoboken New Jersey) p 736

IOP Conf. Series: Materials Science and Engineering **483** (2019) 012042 doi:10.1088/1757-899X/483/1/012042

Minority languages of Brittany in regional economy and regional enterprises

A S Bukhonkina, V P Sviridonova, U S Dzubenko

Volgograd state University, 100 Universitetskiy Ave., Volgograd, 400062, Russia

E-mail: bukanna@volsu.ru, v.p.tim@volsu.ru, dz_julie@volsu.ru

Abstract. The article deals with the analysis of the situation of Brittany minority languages in the economy of the region and its enterprises. Minority languages of the region include the Breton language and the Gallo language. Until the mid-20th century, these languages were on the periphery, their functional relevance was limited. By the mid-20th century language's policy vector has gradually changed. Over the past few decades the Breton language and the Gallo language received renewed attention of the society and the authorities of different levels. Over the past decade, the Breton language and the Gallo language had strengthened its position in the education, media and economic spheres. A number of enterprises in the region started using the Breton language for creating websites, the number of jobs related to minority languages significantly increased.

1. Introduction

French Brittany region stands out among other French regions, standing out on the background of the high level of local identity. The region is not characterized by only cultural, but also linguistic identity, since three languages with varying degrees of vitality and functional integrity here coexist. It's the dominant French, the Breton language spoken in the western part of the Peninsula, and the Gallo language spoken in Eastern (or Romanic) zone of the region.

The policy of linguistic and cultural homogenization of regions, which peaked at the end of XIX-beginning of XX centuries, reversed any manifestation of otherness and deviations from the French ideological maxims "one nation, one language". As a result of the policy of cultural and linguistic unification, that was undertaken sometimes by the very rigid ways, the number of Breton and Gallo native speakers has fallen strongly. The functional relevance of minority languages was limited because of the cliff of their generational transmission.

By the middle of the XX century the French language became the dominant means of communication in all areas, including the production and the labour market. In the first half of the 20th century a number of non-linguistic factors determined the change of functional status and relevance of the Breton language. In the beginning of XX century 75 per cent of the population in Lower Brittany spoke the Breton language. Most of these people were monolinguals and didn't know the French language but in the next decades the situation changed. The improvement of the means of communication, the introduction of school education in French and the participation of Bretons in two world wars affected the minority languages. The socio-economic and psychological factors adjusted communicative rating Breton language and contributed to its weakening.

By the mid-20th century vector language policy has gradually changed. With regard to the new linguistic policy, laws had been passed and institutions established to guarantee the proper functioning of minority languages in Brittany. The law Dekson in 1951 and the amendment to the Constitution in 2008 allow regional languages to teach and recognize them as part of the cultural-historical heritage of France. Thanks to all these efforts the attitudes towards minority languages changed at the level of both national authorities and at the regional level.

In Brittany, since the 90s, a vast education campaign has been undertaken to promote the Breton language. Bilingual schools have been opened (like "Divan"), mass media began to actively use the

Breton language, the bilingual French-Breton signage and signposting had become standard. Finally, the Breton language went back into production. Launched in 2001 the company "Ya d'ar brezhoneg" ("Yes" to the Breton language), has brought tangible results in many spheres of life. However, in Brittany, in addition to the Breton language there is another minority language, the Gallo language. This language little is known, not only outside of France, but even in Brittany.

The Breton language being the basis of the Breton identity has always been considered the only true language of historical Brittany and outmatched Gallo with this region being positioned as a successor of Celtic culture tradition. Breton has always had the status of the language even if it was supplanted from the local communicative environment. According to H.Walter, "the Breton language, apparently being different from French, seems more Breton as Romance forms of speech in Upper Brittany poorly contribute to image creating of Breton, the non-romance region. In this regard, residents of Upper Brittany are not considered to be Bretons, unlike residents of Lower Brittany" [1].

Gallo, that distribution area is the upper Brittany, largely shared the fate of the Breton language. Being close to the French language, he has long perceived as his "wrong" option. For a long time the status of the Gallo language was contentious. They did not recognize it as a language, even its native speakers (mainly inhabitants of rural settlements) called it a dialect or "patois". Only in 2004, the Regional Council of Brittany acknowledged the Gallo language and the Breton language as "historical languages of Brittany along with the French language". Although the Breton language and the Gallo language are officially recognized as minority languages of the region, their functional status is not equivalent. The Breton language is spoken everywhere in different social spheres. It fills an important symbolic function and it is perceived not only by residents of the region, but his guests as more obvious symbol of Brittany.

According to some researchers [2, 3], Breton and Gallo did not have identical status in the eyes of the inhabitants of Brittany. As noted by R. Le Squère "Breton and Gallo do not occupy the same position related to the functions to be performed. Languages of Brittany is much more than just a means of communication. They are both vectors of identity and cultural symbols, therefore in the collective consciousness of the two languages do not have the same value and do not appear to be one and the same "[4]. Moreover for the majority of people, Gallo is still considered to be 'patois'. It is indeed difficult for a foreigner to hear people spontaneously speaking Gallo as this form of speech is used within families [5].

Contacts between Gallo and Breton speakers has not always proved to be harmonious and would appear to confirm Nelde's statement that "there is—in the field of European languages—no contact situation which cannot be described as a linguistic conflict at the same time" [6]. Brittany's cultural specificity is frequently located in its Celtic past, as expressed by Humphreys: "Brittany's present distinctiveness can be attributed to British immigration originally organized as a defensive measure in a threatened Roman empire"[7]. Gallo, as a result of its close linguistic relationship with French, is contested in popular opinion and frequently referred to as a patois or dialect of French. This contestation is further aggravated when Gallo is contrasted with Breton, which in popular consciousness is a more firmly established regional language [8]. In addition, as a result of its linguistic proximity to French, spoken vernacular French and Gallo share the same sociolinguistic landscape. According to Blanchet (9), many Bretons in Upper Brittany use a mixed linguistic system, oscillating between two polarities that can be called gallo francisé ("Frenchified Gallo") and français gallésé ("Gallicized French"), which is also called regional French when this interference is relatively stabilized.

The advantage of the Breton language over Gallo in the first place becomes evident when analyzing the economic sphere of the region, in particular the labor market. The fact is that over the past decade, one of the main components of regional policy has been the widespread improvement of the image of Brittany, its positioning as one of the most distinctive French regions in order to improve investment attractiveness. Sociolinguistic features of Brittany - its Celtic roots, the presence of two minority languages at once, belonging to different linguistic groups - came in handy. The use of the Breton language and the Breton culture to improve the cultural and economic climate has coincided with a growing public demand for a regional (Breton) identity, for a surge of interest in everything Breton, primarily in the language.

2. The Breton economy: a brief overview of the main sectors

Agriculture and fishing are still important in the Breton economy. At the same time, strong industrial activity has developed in some key sectors of activity (agrifood, telecoms, automobile and shipbuilding) and the services industry has experienced significant growth. In Brittany, one of the few French regions to have continued to create jobs over the past decade, the unemployment rate is still lower than the national average by one or two percentage points. The total labour force consists of 1.24 million people: agriculture and fishing still accounted for 6% of these workers (despite a sharp decline), industry for 25% and the tertiary sector (trade and services) for 70%. While Brittany is home to major French and foreign groups, the regional economy is characterised by a dense network of small and medium-sized enterprises. Brittany is one of the only regions in France where industrial employment has continued to increase since 1980. It is centred around four main sectors of activity.

The main sector (with 43% of the companies and a third of the industrial jobs), the agrifood industry has experienced significant growth in Brittany over the past twenty-five years: processing of animal protein (meat, fish, milk), vegetable production, animal feed. The automobile industry centred around PSA. The establishment of Citroën in Rennes in the 1960s led to the development of the Brittany's automobile industry around a major network of parts manufacturers and subcontractors. Brittany's vehicle industry (Véhicules Bretagne) is distinguished by its experience and expertise in the field of specific and dedicated vehicles; it is developing expertise and skills in the small series and/or low production rate domains across the entire supply chain: design, industrialisation, tools, manufacturing, logistics, marketing and commercialisation, etc.

The shipbuilding tradition is the fourth leading industrial sector in Brittany, brings together building sites and military ship repair (despite the recent restructuring in Brest and Lorient), as well as those with jobs in fishing, sailing, oceanography, passenger transport, etc. Brittany is the second leading national centre for telecommunications and the fifth leading French region when it comes to electronics, and the Breton industry has built a global reputation in this field. The establishment of research centres, French and foreign industrial groups and many SMEs has encouraged its construction around three hubs: Rennes, Lannion and Brest.

3. The regional economy: the Breton vs. the Gallo

Unlike the Breton language, the Gallo language is poorly represented in regional economic life: there are no professions that would require knowledge of the Gallo language, i.e. owning them is not an advantage during the job interview. There is no demand for knowledge of the Gallo language neither in the trade sphere, nor in the sphere of social services. The Breton language, on the contrary, is in demand in different areas: in education (schools, preschool institutions, language centers, libraries), in medical and social institutions (nursing homes), in trade. Some Breton enterprises and commercial associations still make timid attempts to support the gallo and visualize it in the names of their products or on their websites. For example, "Clic ta bérouette" (association of local producers from the region of Plemerl), or "Le dret en Goule" (producer of cakes from Lambal). The following manufacturers also use words and phrases from Gallo like «Drao» (cafes in Meless) or "Lait Berbis du Broutay" (cheese maker), "Miron Mirette", etc. According to the National Institute of Intellectual Property, nine brands of manufacturers and businesses use the term "gallo" ou "gallèse" in their names. Among these names there are an amusement park "La mézon du câroujmaison des jeux gallo-bretons", a hotel "Le clos gallo", an enterprise for the production of sausage products "Charcuterie artisanale du pays gallo", a transport company "Gallo'bus", a dairy enterprise "Les prées natur'les produits laitiers du pays gallo", an yogurt and dairy desserts enterprise "Le p'tit gallo", etc. If the Gallo language is poorly represented in the industrial sector of the region, then its position in the field of education and culture is much stronger. In general, there are more than 60 associations directly related to the Gallo language and gallo culture. However, there are very few jobs requiring knowledge of gallo. All of these associations use the Gallo language either in the name or in the communication, but they do not create additional places in employment. This may be due to the fact that for a long time, the majority of people considered using Gallo, especially at school, to be a speech mistake and that is confirmed by the expression “se remettre à parler gallo" (resume speaking gallo), which has a meaning “to make a mistake”. The attitude to Gallo is ambivalent. On the one hand, people are ashamed of it and perceive it a speech flaw. On the other hand, people are very attached to it as to a traditional way of communicating within a family. For the most part, Gallo remains to be the language used in rural areas. Moreover, representatives of different age categories who live both in cities and in rural areas point out that Gallo is a language for “relatives”, exclusively for a narrow circle of family members.

As for the Breton language, its position in the regional economy is represented more actively. Of course, we are not talking about the fact that Breton can replace the French language in the industrial or commercial areas at the local level. However, many companies in Brittany are trying to use the Breton language as a marker of loyalty to the favorable linguistic policy pursued by the regional authorities in relation to the Breton language. Various production companies, farmer enterprises try to use Breton words, phrases or even Breton interjections as a marketing ploy capturing the public demand for authenticity and Breton identity.

Any indication that a product or service belongs directly to a region, to Celtic roots serves as a kind of consumer lure. It may be Breton words or easily recognizable symbols of Brittany. According to a number of polls regional consumers have a positive perception of using of their native language in regional economy because many of them identify themselves as Bretons. Now in Brittany you will not be surprised by the bright packaging of many products or the memorable slogan in the Breton language. However, some companies they have gone further to made the site in the Breton language. For example, Askorn Medical, a company specializing in the production of medical prostheses, also offers a Breton version of its website, in addition to French and English versions. According to its director, D. Pishon, “this not only does not repel customers, but, on the contrary, allows them to attract more attention. For the realization of our products and the successful work of the entire production, using of the Breton language is not necessary. However, this is our choice and effective market tool. Our goal is to make our company different from other competitors, and Brittany and the Breton language attract a great interest and sympathy, especially on the regional market”.

The reaction of people to the using of the Breton language is always positive, for many it looks unusual, but it evokes respect and interest in Brittany and its language. According to D. Pichon, "even people living in Brittany and not speaking Breton, perceive positively its using in the producing sphere". As for the personnel of the enterprise, at first its attitude towards the choice of language was neutral. Then, the company attracted attention of many associations to promote the Breton language, it was cited as an example when it came to real steps to support the region’s languages at work. After that the attitude of staff has changed and the employees began to take it in a positive way. Moreover, in 2012 the company was awarded the "Prix régional de l'avenir du Breton" (Regional prize for the future of the Breton language) in the category "Breton language in the economic sphere". Its founder – the Public Council for the support of the Breton language (Office public de la langue bretonne) - with the support of the regional authorities annually presents this award to private persons, associations, companies and public organizations to promote the Breton language.

8 people work at D. Pishon’s enterprise, two of them speak Breton and the two other understand this language, the company's accountant is able to draw up checks in the Breton language. “Employees have a choice, no one imposes the using of the Breton language on them. - emphasizes the director. - When hiring an employee, first of all I pay attention to his qualifications and professional skills. But with equal competencies, I prefer the one who speaks Breton". In addition to the production of orthopedic prostheses, one of the main goals of the enterprise, stated on the website, is "strengthening respect for the cultural environment in which production is localized".

In addition to the using of the Breton language on the site and directly in production, the company carry out a number of activities aimed at supporting a minority language. In particular, it accepts interns who speak the Breton language, thereby making knowledge of the Breton language an additional advantage when applying for a job, and encouraging its study among young people. Along with other Breton enterprises, the company participates in the program "Ya d'ar brezhoneg" ("Yes to the Breton language"), aimed at promoting the Breton language in the social and socio-economic life of the region.

Apart from that, Askorn Medical participates in various forums of the Diwan private school network, which is taught in the Breton language. Furthermore, Askorn Medical supports the Breton language learning centers (Skol an Emsav, Stumdi), and also participates in exhibitions of products and goods under the auspices of associations of regional producers "Produit en Bretagne", "Bretagne Commerce International", "CCCOB "(Communauté de commune de Center Ouest de Bretagne). Employees of the company take part in the annual Redadeg Marathon, the route of which passes through most of the communes in which they speak Breton. The publishing is another area of economic activity in which the Breton language is used. In Brittany, there are more than 300 publishing houses, of which dozens publish books and magazines in the Breton language, and 6 publishing houses publish products exclusively in the Breton language. With the introduction of new technologies, new professions have emerged that require knowledge of the Breton language as well. This includes mockup artists, software developers in Breton, creators of the Breton version of websites, etc. In addition, we should not forget about the traditional professions for this field related to the Breton language.

As an example of the involvement of the region's minority languages in the publishing business, it's enough to recall the "Ti-Embann ar Skolioù" publishing house that develops learning resources for schools and language centers where the Breton language is taught. Since 1994, this structure has been publishing and delivering printed and digital products to private and public schools that teach in Breton. Publishing products are focused on a variety of age and language levels, from primary school age to the senior class. The regional authorities (Le Conseil régional de Bretagne) provide comprehensive, including financial, support to the publishing house, allocating subsidies to educational organizations for the purchase of its products. In addition, the Ministry of National Education of France pays for the work of the publishing house's employees.

All employees speak Breton and the company's working language is Breton. Not all employees of the publishing house are Breton-speaking from birth, among them are those who have mastered it as a foreign language. For example, one of the mockup artist Anna, is a native of Haute Savoie. She began to study the Breton language on her own and, by her own admission, it was precisely because of this that she was hired by a publishing house. For another employee, infographics artist M. Le Maine, Breton was the language of interfamily communication, he spoke it from his childhood. He also got a job at a publishing house thanks to his knowledge of the Breton language.

Breton is also in demand in the field of support staff in social centers working with older people. In this regard, many vocational schools in Brittany made it necessary to study the Breton language for those specialties that are focused on working with older people, such as social workers or carers. "For older people who are lost in some everyday situations, it is pleasant to hear the Breton speech. This calms them down and inspires confidence", - explains Jan-Pierre Dérödel, a teacher of the Breton language at a vocational school.

4. The studies about the involvement of the Breton language in the regional economy

In 2006 and 2012 thanks to the efforts of the State Agency for the Support of the Breton Language (Office Publique de la Langue Bretonne), an analysis was made of the involvement of the Breton language in the regional economy, primarily the presence of workers in the Breton language. Thanks to a full-scale survey of 500 regional structures (associations, manufacturing enterprises, public organizations, educational institutions), three important socio-economic indicators were analysed.

Firstly, the current number of jobs occupied by people speaking the Breton language, for whom Breton is necessary to perform their professional tasks. Secondly, the potential jobs that could be occupied by brythonophones, and thirdly, the prospects for the development of the labour market for the Breton language. According to the survey, in the region there are 1,300 jobs, which requires knowledge of the Breton language, and which are occupied by people who speak it. Most of these jobs, more than 1000 associated with the field of education.

78% of them are teachers of elementary and secondary schools, 14% of them are assistant teachers in pre-primary education and facilitators of extracurricular activities in high school, 4.5% of them are teachers of Breton adult audience in the language centers and less than 1% of them are teachers and researchers of higher education. In addition to the educational sphere, Breton is in demand in the management area, especially in non-governmental organizations. These are the leaders of various associations, employees of the secretariat, accounting, etc. The media space over the past decade has also been replenished with new jobs that require knowledge of the Breton language.

In addition, according to the same survey, the number of workplaces with the Breton language at work, in trade, in health care, and among artisans has increased. At the same time, there is an annual increase in jobs with the Breton language by 5-8%, and not only in familiar areas, but also in new ones.

It is important to note that most of these workers belong to the category of permanent (contrat à durée indéterminée). In addition to jobs employed directly by native speakers of the Breton language, the survey confirmed the presence of additional places in the labor market for which knowledge of Breton would be desirable.

5. Conclusion

Thus, the using of minority languages in the regional economy means an attempt to label them as part of the cultural heritage. However the Breton language in comparison with the Gallo language has a great identification potential due to a number of factors (greater number of speakers, the well-established written tradition, the status of the language entrenched in the collective consciousness). This becomes especially obvious when it comes to the development of the labor market of the region. A weaker position of Gallo in sociolinguistic Breton hierarchy shows itself in high attention to the Breton language on the part of regional authorities and the bigger number of education programs in Breton, the amount of TV and radio broadcasts, printed publication and literature, unequal representation of Breton and Gallo on the Internet, etc.

The Breton language performs the function of one of the symbols of the region such as the sea, Celtic roots, flora and fauna of the region, festivals, etc. Unlike the Breton language, the Gallo language is poorly represented in regional economic life: there are no professions that would require knowledge of Gallo i.e. the Gallo speaking skills is not an advantage during the job interview. There is no demand for knowledge of the Gallo language neither in the trade sphere, nor in the sphere of social services. The Breton language on the contrary is in demand in different areas: in education (schools, preschool institutions, language centers, libraries), in medical and social institutions (nursing homes), in commerce and industry. Some Breton enterprises and commercial associations still make timid attempts to support the gallo and visualize it in the names of their products or on their websites.

In other words Breton acts as a kind of "trick", a well-recognized brand, which can contribute to the promotion of the region, Breton communes or city individual landmarks. As noted by Le Squère "Breton promotes what could be called "tourism values" (tourisme des valeurs)" [10]. Representation of minority languages in the regional economy is associated with their perception of the collective consciousness as cultural symbols of Brittany, performing an identification function. A more favorable attitude towards the use of the Breton language is explained by the fact that it is associated with a return to the traditions and cultural sources of the region. In this regard, the using of minority languages in the regional economy in the context of the modern sociolinguistic situation of Brittany is a way to legitimize them.

References

[1] Walter H 2012 *Adventures and misfortunes of the languages of France* (Paris: Champion classics) p 177

IOP Conf. Series: Materials Science and Engineering **483** (2019) 012042 doi:10.1088/1757-899X/483/1/012042

[2] Bulot T 2006 *French Language Studies* **16** 305-33
[3] Chevalier G 2007 *Sociolinguistics Books* **12** 75-09
[4] Le Squère R, Ar Rouz D 2005 *Linguistic Margins* **10**-7
[5] Blanchet Ph, Le Coq A 2007 *Sociolinguistics Books* **12** 11-9
[6] Nelde P 2007 *Maintaining Minority Languages in Transnational Contexts* (Basingstoke: Palgrave Macmillan) pp 59-77
[7] Humphreys L 1993 *The Celtic Languages* (London: Routledge) pp 606-643
[8] Nolan J 2011 Handbook of Language and Ethnic Identity **2** 310-23
[9] Blanchet Ph 2005 Linguistic Margins **10** 23-9
[10] Le Squère R 2007 An analysis sociolinguistique marking of the territory in Brittany: toponymy, bilingual display, cultural identites and developpement regional (Rennes:Linguistics. Rennes University) p 123

The role of English deverbative terms in industrial communication

N E Gorskaya[1], V E Glyzina[2] and AV Fedoryuk[3]

[1]Irkutsk National Research Technical University, Institute of Linguistics and Cross-Cultural Communication , 83 Lermontova Street, Irkutsk, 664074, Russia
[2]Baikla State University, 11 Lenina Street, Irkutsk, 664003, Russia
[3]Irkutsk State University, Pedagogical Institute, 1 Karla Marxa Street, Irkutsk, 664003, Russia

Fedoryuk@rambler.ru

Abstract. The paper aims at analyzing deverbative terms containing component "time" in their semantics. Being the means of conceptual representation of "time" in the language and particularly in terminology, English deverbative terms are of apparent interest for industrial communication. Semantics of time is analyzed irrespective of grammatical aspect, through the human's perception and understanding time manifested in communication. Semantic feature of "time", being a linguistic category, allows to reveal reflection of some real time outline in a language, determining accuracy of terminological semantics in technical translation and in industrial communication. It is concluded that there is a whole layer of deverbative terms that have a temporal meaning due to a genetic connection with verbs denoting an event that takes place in time. The use of deverbative terms in industrial communication allows to specify the relation of events and processes denoted in their roots to time.

1. Introduction

The subject of dissertation researches on philology during the last decades testifies to a significant increase of interest in the study of the semantics of English deverbative terms integrating the time component. This is due, first of all, to the fact that terms perform the communicative function in industrial communication, which consists in providing mutual understanding between representatives of various professions. The effectiveness of such communication depends on the availability of a theoretical basis in each particular field of knowledge; degree of specialists' proficiency in the stock of common, learned and special words; in terminological apparatus of the sublanguage of the system of scientific knowledge used by the specialists and professionals.

2. Terminology and terms

2.1 Terminology

Language as a tool of creation, transfer and storage of information is connected with the practical activities of humans. The professional language, or terminology, is one of its manifestation. Fundamental approaches to investigations in terminology in Russia are based on researches of D.S. Lotte and other Soviet linguists, as well as on the achievements of modern terminological theory and practice in general. As for the terminology, D.S. Lotte considers it as a set of terms that corresponds to the system of concepts of a particular field of science or technology. It can be assumed that the system of terms represents a system of concepts of a certain area of knowledge [1]. There are different definitions of terminology in linguistics. O.S.Akhmanova defines terminology as the complex of the terms of a particular manufacturing sector, activity, knowledge, forming a special layer of the vocabulary, which is most easily susceptible to conscious regulation and

IOP Conf. Series: Materials Science and Engineering **483** (2019) 012043 doi:10.1088/1757-899X/483/1/012043

arrangement [2]. V.M. Leitchik considers terminology as a language formation of a paradigmatic type, a spontaneously created set of lexical units having semantic generality and similarity in formal structure. These lexical units function in a language for specific purposes, denoting general concepts which are peculiar to a certain field of knowledge or activity [3]. Yu.V. Slozhenkina believes that terminology is the only one of the actualized parts of the whole creative process in the field of science, where there are objective and subjective beginnings, conservatism and novelty of discovery [4]. According to A.V.Superanskaya, N.V. Podolskaya and N.V.Vasilieva terminology as a system of terms constitutes the part of a special lexicon [5].

Terminology tends to be changed, the emergence of new terms and rethinking the old ones is the result of its dynamic character. Special words when appeared in the language become an integral part of a person's daily life and a part of a specialized picture of the world of professionals. Communicative activity of a person is directly intertwined in his or her production activity, being an integral part of the latter it determines features of nomination processes. Undoubtedly, terminology is the most dynamic and mobile lexical system of the language, therefore researches in this field often have a historical orientation, showing how terminology arises, develops and changes in time, depending on the development of relevant sciences and the general style of thinking of the era. Formation of terminology in this or that branch of knowledge depends more on conscious efforts of people developing a certain branch of knowledge.

2.2 Terms

Problems relating to terms' functions and properties, their role in technical translation and in industrial communication are considered in the works of some linguists [6], [7], [8], [9], [10], [11], [12], [13], [14], [15], [16]. Terms of any branch of knowledge are the most dynamic and integral part of any vocabulary of a language so as through the emergence of new terms a language vocabulary is enriched and developed. M.N. Minakova notes that earlier the terms were referred to the periphery of the language, nowadays there is no doubt about their functional significance, since they are a part of the general lexical system of the language, but only through a specific terminological system [17].

According to the traditional point of view on terms accepted in the Soviet linguistic science of the 1960s and 1970s, terms are special words or phrases that due to their semantic and grammatical structure are opposed to commonly used words or free phrases. That is why the linguists initially studied an ideal term, that is, a monosymantic term, a noun, accurately denoting a scientific and technical concept, the content of which is revealed in production process. As a result of this approach, the terms were attributed properties that many of them actually lacked.

At the beginning of the 1980s, linguists faced the task of investigating what was really not an ideal or desirable term, but a real term. As a result the point of view on the "term" as a word or phrase of special (scientific, technical, etc.) language, created (accepted, borrowed, etc.) for precise expression of special concepts and for designating special items was seriously criticized.

In contemporary Russian terminology, there is no apparent disagreement in understanding the "term" as a word or phrase associated with a concept belonging to any area of knowledge or activity. However, this does not necessarily imply the recognition of once and for all this and final connection of a particular term with a specific scientific (technical) concept. The lexical unit receives the signs of the term insofar as it appears in the terminological function, and loses them as soon as it ceases to be a term.

Terms essentially differ from other lexico-semantic groups of the language both as functional language units, and as an object of linguistic analysis. The peculiarities of the term include: 1) systemic character; 2) the availability of definitions for most terms; 3) the tendency towards monosemy within its terminological field; 4) lack of expression; 5) stylistic neutrality. Strict accuracy of terminological semantics refers to the characteristics of the term that are sustainably reproduced in linguistic literature. Nevertheless, the term retains the ability to transform its meaning depending on the conditions of communication. The accuracy of the terminological semantics is not provided by unity of a linguistic sign with the only possible conceptual content, it is determined by the ability of the author of a scientific statement to relate the term to the variants of its definitions. The change in the communicative task and the functional-stylistic affiliation of the text leads toe determinologization of the lexeme. Performing the most important communicative function - providing mutual understanding between representatives of various fields of knowledge and scientific disciplines – terms, as we noted above, are a representative layer of the lexicon of modern languages.

Terms have a special role in industrial communication, but it should be noted that terms are not created uniformly in each language in the terminologization of a part of its vocabulary. Some languages have a greater impact on terminological word building. Consequently, some languages contribute more to the development of terminological systems than other languages.

An undisputable factor of influence on the quality and peculiarities of professional speech is formal and informal communication. Besides this quality is changing dramatically, even if it is an industrial communication, but not in a strict formal setting. In order to perceive terms that relate to a certain sphere adequately, it is necessary to create basic rules intended for understanding existing terms and creating new ones. However, it should be noted that the language, which absorbed many changes, reflects cultural and anthropological states of language communities. An increase in the amount of violation of the˘ norms of the codified literary language is accompanied by fixation of vernacular language forms in speech. The factor of formality of spontaneous speech makes speakers use only those means of the language that are on the surface of linguistic consciousness.

3. Deverbative terms containing component "time" in their semantics

3.1 Time as a linguistic category

The investigation of lexical units with the seme "time" from the point of view of cognitive linguistics represents a model expressing definite ideas of a human with regard to some phenomena. The human idea of "time" is considered to be a certain real time reflection in a language, the real time being closely connected with extra linguistic activities of a person through the acquisition of, first of all, the spatial nature (objects, places). Time is one of the most important constituents of human self-existence. Many scientists believe that the daily experience is enough for absorbing the theory of the time nature. Nevertheless, it is evident that neither real physical time (which is a subject matter in natural sciences), nor psychological phenomenon of "the time being lived" i.e. "time of life being lived through" does not appear to be an object of a linguistic study, although they are directly relevant to the language. As for the linguistic aspect of time, usually the notion of "time" is considered to be a reflection of some real time outline in the language.

The language is known to form a continuum with other forms of a human behaviour. Therefore, the category of time in the language can be referred to such notions which can be used to determine the correlation of linguistic activities with an extra linguistic one of a human being and which contribute to transforming the real world into a "projected world" of a human consciousness. So, if event specifications of quantity are found as basic ones in building physical time outlines, in building linguistic time outlines specifications of quality are in predominance to deal with temporal relations. It should be noted that the problem of a language representation of time in the linguistics is most frequently connected with the temporality and aspect of a verb, as the verb essentially means an activity in the broad sense of the word, an activity always running in the time.

3.2 Deverbative terms with the seme "time" in industrial communication

The relationship between language and society manifests itself in a different form: the use of linguistic means are determined by the social characteristics of native speakers (age, professional, educational, etc.), by roles of communicants in society, and communicative situations. One of the aims of our research was to study the temporal semantics in such parts of speech as deverbative terms that are widely used in industrial communication, as well as in technical translation. Time as a linguistic category allows to see a certain reflection of real time in a language, it is closely associated with extralinguistic activity of a person through perception.

The scientific works of O. Ducrot and T.Todorov are of special interest for our study, in these works the semantics of time is analyzed irrespective of the grammatical aspect, through person's perception and understanding of time manifested in speech activity [18]. The analysis allowed the scientists to conclude that temporality can be contained in all parts of speech, including deverbative terms. In this connection one may state that the semantic includes a temporal component, denoting the subject and object of action, which is related to time. We adhere to T.V.Bulygina's point of view concerning the predicates/verbs' availability of such attributes as: absence/presence of temporal connection, static/dynamic, duration/ non-duration, temporal perspectives [19]. For us, these attributes are of fundamental importance, since they represent some dynamic, temporal or temporally-conditioned signs belonging to the verbal lexeme as such.

In the English language H. Wendler investigated the semantic features of time that the verbal predicates have, he believed that the time is inherent to any verbal word, with all the verbs on this basis corresponding to different time schemata, and through them are associated with the tense forms of finite verbs [20]. From the very beginning, the reference to time for verbal lexemes is not homogeneous. Temporal attributes of verbal names are practically not marked in the dictionary, because they accompany each word, denoting event in a broad sense. Another thing is the term, which being used requires precision in both translation and in industrial communication. In this regard, the fact that the majority of verbal names under investigation turned out to be terms is not accidental. The data from contemporary British, American electronic and traditional explanatory dictionaries as well as from electronic encyclopedias were the material for our research (about 3000 examples). Our investigation has revealed a layer of deverbative terms that denote:

- specific time periods:
  gestation-*the period of* development in the uterus from conception until birth; pregnancy;
  after ripening - *a period during which certain changes must occur* in some dormant but fully developed seeds before germination;
  inning - *one of the nine divisions or of the periods of a regular game* in which each team has a turn at the expense of three outs;
  chronaxie - the minimum *interval of time* necessary to electrically stimulate a muscle or nerve;
- specific points in time:
  cusp - a *transitional point or time*, as between two astrological signs;
  dimension - one of the least number of independent coordinates required to specify uniquely a *point in time* and space;
  singularity - a *point in time* at which gravitation forces cause of matter to have infinite density;
  sailing - departure *or time* of departure from a port;
- processes in which time plays a special role:
  aquatint - a *process* of etching capable of producing several tones by varying the etching *time* of different areas of a copper plate;
  macroevolution - large-scale evolution occurring *over geological time;*
  coalification - compression and hardening *over a period of time;*
  melioration - the linguistic *process* by which a word *over a period of time* grows more elevated in meaning;
- the completely nominalized verbal names that denote objects:
  printing - all the copies of a publication, such as a book, that are printed *at one time;*
  mailing - a batch of mail dispatched *at one time* by a sender;
  pressing - a number of recordings *at the same time;*
  washing - articles washed or meant to be washed *at one time.*

4. Conclusion

Thus, we can conclude that deverbative terms have the seme of time due to a genetic connection with the verb denoting an event that takes place over time. The use of deverbative terms in the industrial communication allows deverbative terms to clarify the relations between the events and processes indicated in the root and the time. The results obtained are of great theoretical and practical importance not only for the production process, but also for the general theory of the word and its semantic structure. In particular, they indicate the presence of a greater number of semantic features in the peripheral sphere of the deverbative terms and also about the prototypical nature of these features themselves, which is of undoubted interest for research in the field of modern terminology.

References

[1] Lotte D S 1971 *Brief Forms of Scientific and Technical Terms* (Moscow: Nauka) p 84
[2] Akhmanova O S 1966 *Dictionary of Linguistic Terms* (Moscow: Soviet encyclopedia) p 608
[3] Leitchik V M 2009 *Terminology : Subject, Methods, Structure* (Moscow: Librocom) p 256
[4] Slozhenkina Yu 2006 *Vestnik Samara S.U.* **1 (41)** 126
[5] Supernanskaya A V, Podolskaya NV and Vasilyeva N V 2009 *General Terminology: Questions of Theory* (Moscow: Librocom) p 248
[6] Reformatsky A A 2005 *Introduction to Linguistics* (Moscow: Aspect Press) p 536
[7] Komarova Yu A 2007 *Philology and Man* **4** 77-87
[8] Yunusova I R 2009 *Vestnik Baskir.Univ* **14(3)** 857-60
[9] Cabré M T, Sage J C and De Cesaris J A 1999 *Terminology: Theory, Methods and Applications* (John Benjamins Publishing Co) p 247
[10] Justeson J S and Katz S M 1995 *Natural Language Engineering* **1(1)** 9-27
[11] Cabré M T 1995 *Terminology* **2:1** 1-17
[12] Felber H 1983 *Standartization of Technical Terminology: Principles and Practicies* (ASTM Special Technical Publication) pp 3-14
[13] Sager J C 1990 *A Practical Course in Terminology Processing* (John Benjamins Publishing Co) p 252
[14] Ahmad K and Rogers M 2001 *Handbook of Terminology Management* vol 2 (John Benjamins Publishing Company) 725 - 60
[15] Pearson J 1998 *Terms in Context* (Amsterdam :John Benjamins Publishing Co.) p 242
[16] Kageura K 2002 *The Dynamics of Terminology: a Descriptive Theory of Term Formation and Terminological Growth* (John Benjamins Publishing Company) p 322
[17] Minakova N A 1985 *Features of the Formation and Structure of the Construction Terminology of the Russian Language* (Moscow) p 177
[18] Ducrot O and Todorov T 1987 *Encyclopedic Dictionary of Sciences of Language* (Oxford : Blackwell) p 380.
[19] Bulygina T V and Shmelev A D 1997 *Pragmatics. Semantics. Lexicography* (M .: Publishing House School of Languages of Russian Culture) p 574
[20] Vendler Z 1957 *Philosophical Review* vol 66 **2** 143-60

IOP Conf. Series: Materials Science and Engineering **483** (2019) 012044 doi:10.1088/1757-899X/483/1/012044

Sociolinguistic aspects of effective communication of a foreign specialist at a production department

N L Shamne, M V Milovanova, E V Terentyeva

Volgograd State University, 100 Universitetskij Ave, Volgograd, 400062, Russia

E-mail: nikolay.shamne@volsu.ru

Abstract. The problem of integrating migrants into the host society is particularly relevant in relation to the production team since it occurs within the framework of a professional productive activity. A non-native speaker of the official state language faces various difficulties which can largely affect the results of his/her work and the effectiveness of production communication in general. The aim of the study was to establish sociolinguistic parameters, ensuring the efficient communication of a foreign specialist at a production department. The main methods used were surveys and interviews organized at the same time. The "qualitative method" of an in-depth interview was used as one of the approaches practiced within the framework of the so-called "grounded theory". The authors suggest that effective communication can be provided for the foreign people by creating (on the basis of single universities in the region) linguoeconomic schools for migrants (non-natives), linking their future with the region, organization of specialized training. These workshops are proposed to include different types of classes, taking into account the future specialty and a field of study, for example, business meeting is operational, organizational, planning, controlling, and it is recommended to use a variety of educational technologies. Thus, the efficiency of migrants' adaptation (non-natives) in the host society and the production team is largely dependent on joint systematic sequential activities of the state regional agencies, relevant professionals, and media.

1. Introduction

The beginning of the twenty-first century was marked by forming a new worldview in the minds of people, which reflected major changes in all spheres of human life, including the process of globalization, accompanied by population migration, largely a labor one. Since the beginning of 2000-ies various regions of Russia began to take on more and more migration burden. Thus, according to sociological studies, the main influx of migrants was recorded in Central and North-Western Federal districts, a slightly smaller flow of migrants was observed in the Volga Federal district, and "the constituent entities of the Russian Federation which are a part of the Southern Federal district accounted for up to 15% of the all-Russian migration growth in recent years. In particular, the Volgograd region is the one experiencing active migration pressure" [1].

In connection with this global problem, the leading European States, including Russia, are carrying out consistent work to adapt migrants to the host society.

The analysis of foreign and domestic experience proves that successful adaptation of migrants requires the development of the so-called adaptation infrastructure, that is, a network of educational institutions implementing language and socio-cultural adaptation programs.

To maintain security and stability of the situation within the host country, including production, today we need a solid, purposeful, comprehensive, linguistic, and ethno-cultural policy concerning ethnic and national issues, which should be built both at the state and regional level to take into account both the fundamental issues of national security and local specifics [2].

Today it becomes obvious that the implementation of the proposed Concept of the state migration policy of Russia for the period up to 2025 requires not only the organization of language courses and testing centers, but also the creation of migrant schools. Work in these schools should be carried out under special programs, including the modular system [3]. The programs may contain, for example, a language module that includes Russian language courses by level of proficiency, and a special course on the basics of business Russian language; a legal module that provides studying the basics of legal behavior and the status of foreigners in the Russian Federation, the basics of labor and social legislation, as well as a module of pedagogical support and psychological assistance [4]. Fairness requires us to note that there are training programs for migrants and migrant children, developed by the UNESCO chair at the Moscow Institute of open education. However, these programs are designed for Moscow and the Moscow region and are not focused on the specifics of a particular region [5]. The problem of integrating migrants into the host society is particularly relevant in relation to the production team since it occurs within the framework of a professional productive activity. A non-native speaker of the official state language (here Russian) faces various difficulties which can largely affect the results of his/her work and the effectiveness of production communication in general.

2. Methodology.

To identify obstacles faced by non-native Russian language speakers when implementing production communication, we have carried out a series of interviews with graduate migrants (non-natives) at separate Volgograd universities: Volgograd State University, Volgograd State Medical University, Volgograd State Technical University. In total, 100 people graduated in 2010 – 2017 were interviewed.

The main methods used were surveys and interviews organized at the same time. The "qualitative method" of an in-depth interview was used as one of the approaches practiced within the framework of the so-called "grounded theory". The peculiarity of the methodological approach within a "grounded theory" is that in the study of certain phenomena of social life it is impossible to apply measuring procedures that would fit the framework of traditional variables. This is due to the complexity and individualization of social life, sociological methodology resulted in more attention to the so-called "soft" categories (in the terminology by *Piotr Sztompka*) [6], such as values, friendship, love, trust, etc. This "soft" categories, of course, can also include language and its use in different situations. Qualitative research can be called interpretative, it allows you to get information about the motives of behavior, attitudes of people.

So, we used the method of deep unstructured (unformalized focused) qualitative interview.

Interview, as you know, is a kind of conversation conducted according to a certain plan, involving direct contact between the interviewer and the respondent, and the answers of the latter are either recorded by the interviewer (the assistant), or mechanically (through the recording device).

For interviews we purposively identified informants for whom Russian is not their native language (the people from the national republics of the Russian Federation, other non-natives). In sociology, as a rule, it is considered that questions in the framework of qualitative in-depth interview should be formulated based on the following requirements: they should be monosyllabic, be clear to the informant, not contain a hidden enforcement of the answer.

The interview guide was developed on the basis of aspect dichotomies formulated taking into account the objectives of the study:

1. In the aspect of production communication special interest is given to establish the relationship of " efficiency of production communication :: language quality" in relation to how the use of the language affects communication in the production process.
2. In the aspect of institutional communication the interest is given to establish the relationship of "personal speech settings :: speech expectations about the partner" in relation to the institutional status of the respondent.

In the aspect of tolerance, and hence the efficiency of production communication, it is of interest to establish the relationship "willingness to forgive mistakes in the speech of the communication partner/elements of language hybridization :: the desire to condescend to one's mistakes in speech" in relation to the interlocutor's nationality [7].

2.1. Production communication. Interview. Step one.

Let's focus on the first block of the interview. The questions, asked within the framework of this block, and the answers to them will allow in the future to identify possible options for the use of the language in different situations of production communication, to operationalize concepts and to create a questionnaire for a wider quantitative survey. The first block included the following questions: 1) How often do you use your native language in the process of production communication? 2) In what situations is it necessary to use Russian correctly?

Since the situations of production communication are considered, the status of the speaker is of an official nature (doctor-colleague, head of the Department, University teacher, etc.), therefore, the requirements for the language quality increases. Here is a fragment of the interview.

Informant A: «Interviewer: When your work colleagues make mistakes, for example, in speech, despite they are native Russian speakers or representatives of other nationalities, do you notice these errors?

Informant: Yes.

Interviewer: How do you treat that?

Informant: Not good. It's terrible.

Interviewer: And why it is in one case...

Informant: The emphasis звонИт-звОнит makes me crazy! Just want to correct. Well, that's if the Russians say that. But, if the non-Russians, it is forgivable.

Interviewer: It turns out that you are condescending to errors in the speech of non-native speakers of the Russian language, and not condescending if they are native speakers. Explain to me why.

Informant: Well, because they are Russians, they are at work and must speak their language competently.

The following fragment shows the great role of the purposes of production interaction. So, if we are talking about informal communication in the process of production communication, the requirements for the language quality are reduced:

«Interviewer: Well, in what conditions do you think a person can be wrong?

Respondent: ... Do you mean when communicating at work?

Interviewer: Yes.

Respondent: Well, how to say, if you communicate on some not quite work-related topics and with colleagues close to you, then you can be wrong. Why not? And if with someone unfamiliar, it may be better not to be mistaken".

2.2. Production communication. Interview. Step two.

The second block included the following questions: Does the use of the language depend on the status of the interlocutor and how?

In different institutional conditions of production communication, respondents usually have different attitudes-expectations regarding the partner's speech. The interviews have clearly demonstrated the dependence of language use on moral norms realized when communicating with elders or people with a higher status at a production department. In particular, this is found in the following extracts from the interview:

«Interviewer: Does the use of the language depend on the status of the interlocutor and how? On the age of the interlocutor? Do you change the language and then the style of speech?

Informant: It happens. With fellow Azeris in Azerbaijan, we sometimes start speaking Azerbaijanian at first, attract a couple of claims, a couple of arguments and then change to Russian.

Interviewer: Why does this happen?

Informant: But I can not name the reason as it happens by itself. It happens sometimes that you can not find the right words in Azerbaijanian. And then work starts and, of course, you speak in Russian as the Russians do not like when non-Russian language is used.

Interviewer: But is the style of communication changing there depending on age?

Informant: Yes, certainly. I always try if there is an older person. I communicate respectfully to him even if he's not my boss: "Good day," or "Dear Mr/Ms" is usually used.

Interviewer: I see. Does the use of the language depend on the professional status of the interlocutor?

Informant: Well, it depends. If I talk to my boss, I try to think more so as not to make mistakes. And I try not to use any spoken expressions there.

Furthermore, in relation to the issue of integrating migrants (non-natives) in the host society we had to consider how we can engage representatives of different nationalities in the single production team, i.e. how effective is the communication process. It was necessary to find out the level of tolerance to mistakes made within production communication by the interlocutors and the respondent himself. The range of issues included: 1) Do you allow mistakes in the partner's speech in the process of production communication when communicating with representatives of other nationalities? How do you feel about such errors (if the language is native/national or if the language is not native)? Do you correct the interlocutor? 2) Do you make mistakes in your own speech in a foreign language and do they complicate the process of communication at work?

Conducted interviews show that non-Russian informants as a whole show a high degree of tolerance to interlocutors' errors in the framework of production communication (we speak about Russian language proficiency). Especially high level of tolerance towards those for whom the Russian language is not native (in this case, often no-one often corrects the interlocutor, even with serious errors). Exceptions are cases when the interlocutor himself asks to correct him.

As for the cases when the informants expressed a willingness to correct the interlocutor and consider it important, two obvious approaches to this issue were highlighted in the interviews. The first approach points to the need to correct mainly oral speech, while the second approach focuses mainly on the written language.

"It is necessary to correct (speaking)»:

«Interviewer: And do you think that errors in the speech are permissible? How do you treat that?

Informant: Mistakes, of course, made by non-Russians are permissible, but I still think they need to be corrected. If you are corrected in stress, for example, or other difficult language aspects, I think it is necessary to remember it, or you can look it up in the dictionary if you are in doubt. And you need to fix this error, you should not go on speaking "катАлог" or "directOry", otherwise, colleagues will think you don't want to achieve language proficiency.

Interviewer: Well. You said you have an international team in your production department. And how do you feel when your Russian-speaking and non-Russian-speaking interlocutors both make language mistakes?

Informant: Sometimes I notice these mistakes, sometimes I try to correct them more or less delicately. Well, it depends on how closely I communicate with colleagues. If closely, I can try to correct both a non-Russian and Russian speakers. But I actually try to correct only non-Russians.

Interviewer: And how do they respond to these corrections? Is someone irritated or on the contrary, understands?

Informant: Some are actually trying to correct their own mistakes and others just ignore them. Well, Russian, of course, may not like that they are corrected by a foreigner.

"It is necessary to correct (writing)»:

«Interviewer: Is it important for you to speak correct Russian? And why, if so?

Informant: I do not like, for example, when I receive emails and there are many errors. I can immediately understand that a person is non-Russian. And if in writing someone makes a lot of mistakes, you do not want to work with such a person. For me it is a very important indicator. So I try to always check everything in my speech before I send a letter.

Regarding their own mistakes in speech the informants have shown a fairly high level of demands on themselves. It should be noted that this circumstance is probably due to the fact that they all received higher education and seek to prove themselves as specialists in a production team.

2.3. Production communication. Interview. Step three.
And finally, the last block. Within the framework of this block, it was necessary to establish a self-assessment of the respondent's level of language proficiency (his satisfaction with the quality) first, and then to find out whether and how the respondent is ready to work (or is already working) on the quality of his speech to improve the efficiency of production communication. The following range of issues was proposed: 1) How do you work and do you actually work on improving the literacy in a non-native language? 2) Do you think you speak a foreign language well? If so, are you ready to proceed to improve its quality anyway? If not, and it prevents you from effective production communication, how are you going to solve such difficulties?

All interviews are recorded on tape, followed by transcription and further categorial analysis. A categorial analysis is carried out on the principle of 'bottom-up', that is, when categories are derived from the text and are not "superimposed" on it as a matrix. Such a technique in the English sociological literature is also called "discursive analysis" or "qualitative content analysis". In the Russian sociological tradition, where content analysis is considered to be more quantitative, the term "discursive analysis" is more often used. As already mentioned above, "discourse analysis" reveals the meaning of a phenomenon and does not reveal the frequency or nature of its manifestation. This method is closely related to the methodology of "comprehensive sociology" by *Maximilian Weber* [8]. When using this technique, all interviews are "broken" into certain semantic pieces corresponding to the categories derived from the interviews themselves.

This method is not representative in the sense that we cannot transfer its data to the whole entity due to representative generalization. This is not the goal. The task of this method is to form a "map of opinions" and characterize the most typical logic of informants' reasoning, which subsequently will help in completing a detailed questionnaire or formalized (structured) interview. In this sense, generalization is possible at a category level.

It should be noted that almost none of the interviewed informants assesses his speech as ideal and most are ready to work or are already improving it. The interviews clearly revealed two main approaches to work on improving one's speech. The first approach is connected with readiness to change the speech, being guided by competent native Russian-speakers in a production team. The second approach is related to the willingness to work on the quality of speech through the use of books, dictionaries and reference books. As a result, we can say that regardless of the informants' nationality, all of them are ready to solve the difficulties arising in production communication and to improve the quality of the language used.

All informants noted that the level of knowledge of the official language of the production team (in our case – the Russian language) largely affects the effectiveness of professional activities, so the non-natives are ready to improve their knowledge of the Russian language to achieve efficient production activity. Most of the informants also noted that their own efforts were not always enough. When asked by the interviewer how they would like to improve their level of Russian, the majority of informants said that they would like to attend any courses that would focus on their interests related to work in the production team..

3. Results and Discussion
The conducted focused (in-depth) interviews allowed us to trace the specifics of the relationship between the level of language proficiency in the host language society and the effectiveness of production communication (for example, in the Volgograd region).

Interviews illustrate that a language is an essential channel for shaping social practices. This is manifested, in particular, in compliance with certain language requirements that are implemented in communication within the framework of production communication.

As a medium of communication, a language acts, in terms of sociology by *Niklas Luhmann* [8], as a means of creating a generalized (social, group) trust. In particular, errors in oral speech, as the interview showed, can be perceived as marks (symbols, indicators) of intra-group trust. At this level, language (oral speech) can become a tool to facilitate, simplify (in the categories by Luhmann) life in a society or in a separate production team. In this context, at the same time, there is a link between literacy in written language and social status, which is of an institutional nature. In terms by *Talcott Parsons* [9] the Institute is a set of roles and expectations and related status relations that form the context of the corresponding social interaction. Language in this sense, as revealed by in-depth interviews, acts as a medium through which these social roles and expectations are formed. In general, the interviews confirmed the constructivist hypothesis about language as one of the tools for forming the social structure. There appropriate discourses are set through language and manifest themselves, in particular, in authority.

Based on the results obtained in the course of the interviews, the problem of integration of non-Russian migrants (we are talking about foreign students studying in Russian universities) in the host society, in particular in the production team, we offer the following activities:

- conducting specialized workshops on the correction of language behavior in the system of higher education;
– organizing (on the basis of separate universities in the region) linguoeconomic schools for migrants (non-natives).

First trend includes: - '*conducting specialized workshops on the correction of language behavior in the system of higher education*' involves the organization of work with students whose Russian language is not native.

Specialized workshops can be organized at the departments working within the Federal target program "Russian language". These workshops are proposed to include the following types of classes taking into account future specialization and fields of study: 1) preparation and conduct of business and role-playing games (for example, a business meeting demonstrates operational, organizational, planning, controlling functions); preparation and conduct of a linguistic business game "Negotiations"; analysis of specific situations (for example, negotiations have reached an impasse; business letters have to be responded to); 2) conduct of various psychological trainings (for example, competence in "intra-corporate business communication"; "competent contact with a colleague"; "increasing stress resistance in business contacts"), etc.

To achieve the planned results in these specialized workshops it is recommended to use a variety of educational technologies: 1) information-communication and developing technologies aimed at the formation of knowledge systems, memorization and free operation (the use of new information technologies for independent replenishment of knowledge, including the use of technical and electronic means of information); 2) activity- and practice-oriented technologies aimed at forming a system of professional practical skills that provide the ability to perform professional activities (analysis of speech and behavioral errors, role-play games, creation of sample documents, search for optimal solutions to a specific production problem by brainstorming); 3) developing problem-oriented technologies aimed at formation and development of problem thinking, mental activity, an ability to see and formulate problems, choose ways and means to solve them.

The second activity includes the organization (on the basis of individual universities in the region) of linguoeconomic schools for migrants (non-natives) and facilitates the integration of migrants linking their future with the region and the socio-cultural space of the region. Within these schools non-natives working in different production teams in the region can improve their level of Russian language proficiency to achieve more rapid adaptation in a team.

Within the framework of these schools it is advisable:

- to develop a program of migrants' adaptation, which includes modular themes and the programme of activities and training aimed at cultural and linguistic adaptation of migrants in a particular region; educational and methodical complexes for social and cultural-linguistic adaptation of migrants based on inclusion of the course "Russian as a foreign language"; training programmes for specialists in different fields with the mandatory inclusion of a section "Culture of interethnic communication»;

- to carry out explanatory work on main provisions of the Constitution of the Russian Federation, legislation, "Code of interethnic communication", containing guidelines for forming high-level culture of interethnic communication;
- to expand the network of cultural and information national centers, to conduct reviews and competitions among them for the best staging;
- to establish systematic coverage of the problems and the best examples of interethnic communication, to stimulate the activities of journalists specializing in interethnic relations, to provide them with the necessary information, to hold press conferences, competitions for the best publication about the lives of people of different nationalities, about the experience of interethnic communication;
- to organize permanent exhibitions in museums, palaces of culture, cinemas, cultural centers with a display of history, traditions, cultural achievements of the peoples living in the region.

The work of the proposed linguoecological schools should be systematic, consistent and widely covered in the media of the region. For effective operation of schools it is necessary to:

- develop urban, district and regional programs of integration and cultural and language adaptation of migrants, which reflect the entire system of activities with agreed upon goals, objectives, time, performers, resources, taking into account local characteristics in interethnic relations;
- conduct monitoring of interethnic relations, oral and written surveys, statistical and scientific surveys in order to timely identify problems in interethnic communication and find ways to solve them;
- create regional, city and district coordination councils, designed to ensure joint, coordinated work of all structures in the field of interethnic relations;
- study, generalize and disseminate the most effective models of interethnic interaction;
- hold conferences, "round tables", dialogue platforms, discussions, exchanges of views on the most important and controversial issues of interethnic cooperation and thereby prevent the emergence of ethnic conflicts in the workplace.

4. Conclusion

The interviews showed that the level of knowledge of the official language of the production team (in our case – the Russian language) largely affects the effectiveness of professional activities. The non-natives are ready to improve their knowledge of the Russian language to achieve efficient production activity, but their own efforts often not enough to achieve a positive result. Non-Russian respondents stressed that they would like to attend any courses that would focus on their interests related to work in the production team.

The solution of the mentioned problems is possible with organizing and holding specialized workshops for the correction of linguistic behavior in the system of higher education, establishing linguoeconomic schools for non-Russian migrants on the basis of separate universities in the region.

Thus, the efficiency of migrants' adaptation (non-natives) in the host society and the production team is largely dependent on joint systematic sequential activities of the state regional agencies, relevant professionals, and media.

References

[1] Ignatova YU E, Nikolenko N A 2011 The impact of regional conditions on the adaptation processes of migrant workers (on the example of the Volgograd region) *Vestnik VolGU. Series 7. Philosophy.* **3 (15)**

[2] Shamne N L 2012 Development of short-term forecasts of the development of the Russian literary language in the Lower Volga region. *Vlast'.* **7**

[3] Milovanova M V, Terent'eva E V 2011 The language space of a multi-ethnic region:

IOP Conf. Series: Materials Science and Engineering **483** (2019) 012044 doi:10.1088/1757-899X/483/1/012044

environmental problems (Volgograd: Izdatel'stvo VolGU) 292
[4] Omel'chenko E A From the experience of teaching Russian and socio-cultural adaptation of migrant workers: problems and prospects.: Available at: http:// refugee.ru/Events/2013/omelchenko_ehtnosfera.pdf.
[5] Kalenkova O N 2013 Lessons of Russian language. CMD for children of preschool and primary school age. CMD for children of preschool and primary school age. Moscow: Russkij yazyk, Kursy
[6] Sztompka P 1991 Society in Action: The Theory of Social Becoming (Cambridge: Polity Press and University of Chicago Press)
[7] Sztompka P 1993 The Sociology of Social Change (Oxford and Cambridge: Blackwell)
[8] Shamne N, Milovanova M V, Terentyeva E V, Velibekova I M 2016 The polyethnolect phenomenon in the current state of language in Germany and Russia *XLinguae*. T. 9. **4**
[9] Veber M 2016 Economy and society: essays understanding sociology. Sociology. Moscow: dom Vysshej shkoly ehkonomiki
[10] Luhmann N 1995 Social Systems (Stanford: Stanford University Press)
[11] Parsons T 2018 Social system. Moscow: Akademicheskij proekt

Dependence of physical and mechanical properties of metal surfaces on microgeometric parameters

V V Kalmykov[1], M V Mousokhranov[1], E V Logutenkova[1]

[1]Kaluga Branch of Federal State Budgetary Educational Institution of Higher Education "Moscow State Technical University named after N.E. Bauman (National Research University)," 2, Bazhenova Street, Kaluga 248600,Russia

E-mail: kalmykovvv@bmstu.ru

Annotation. This study examines the interrelation between surface microgeometry parameters after external influence by line-by-line microdeformation and the level of surface energy. The measurements were performed on experimental samples of C20, C45, X10Cr13. The values of surface energy were determined by the static capacitor method. It was established that the highest and lowest levels of surface energy correspond to the microgeometric parameters of the substrate. The results of the study can be used for industrial provision of the required physical and mechanical properties.

1. Introduction

Substrates are characterized by microgeometric parameters (Fig. 1) and physical and mechanical properties. They are formed by production operations accompanied by external mechanical influence. Traces from the cutting tool can be seen on the surface of the substrates, the shape and direction of which depend on the pattern and type of influence. External influence is carried out line by line and the choice of the line interval depends on the requirements for accuracy and surface cleanliness [1-3]. To ensure high values of these parameters, additional tool movements are used to compensate for errors caused by the elastic properties of structural materials. Besides, there is an informed judgement that additional movements of the tool on the surface of the substrate change its energy state [4–9].

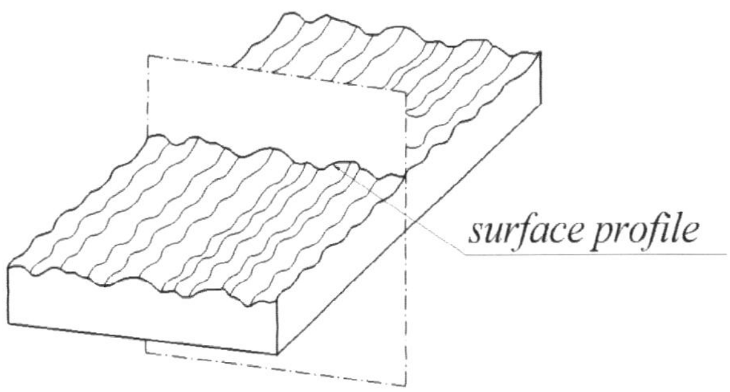

Figure 1. Schematic representation of microgeometric parameters of the surface

External impact leads to microdeformation of the surface layer, local compression and distortion of the crystal lattice, local breakdown of lattice parts and individual atoms due to frictional forces, a change in the surface microgeometry, which leads to a change in the physical and mechanical properties of the structural material's boundary layer [10, 11].

This study aims to examine the relationship between surface microgeometry parameters after line-by-line microdeformation and the level of surface energy (SE).

2. Materials and methods

The changes in the SE of materials and microgeometry of surfaces under external mechanical influence with line-by-line microdeformation were studied on experimental samples made of C20,

C45, X10Cr13. The material of the influencing tool was hard alloy GC1040. The blunting radius of the edges of the tool was 0.02 mm. Microdeforming was performed on a vertical milling machining centre with a MiniMill 450 CNC machine. The SE value was determined using the static capacitor method [12–15]; the surface microgeometry parameters were determined using an ABRIS–PM7 profilograph-profilometer. The hardness of the material under study was determined on the Brinell scale using a hardness meter.

Samples of rigid structures ($h_y \approx 0$ µm) and structures prone to elastic recovery after surface shaping ($h_y \approx 2$ µm and $h_y \approx 4$ µm) were studied.

3. Results and discussion

A comparative analysis of the mean absolute error of the substrate profile from the ideal plane within the evaluation length was performed. The profilogram of the surface after preliminary shaping by removing a layer of material ($h_y = 300$ µm) and microdeforming with external influence traces line interval $f_z = 40$ µm is shown in Fig. 2.

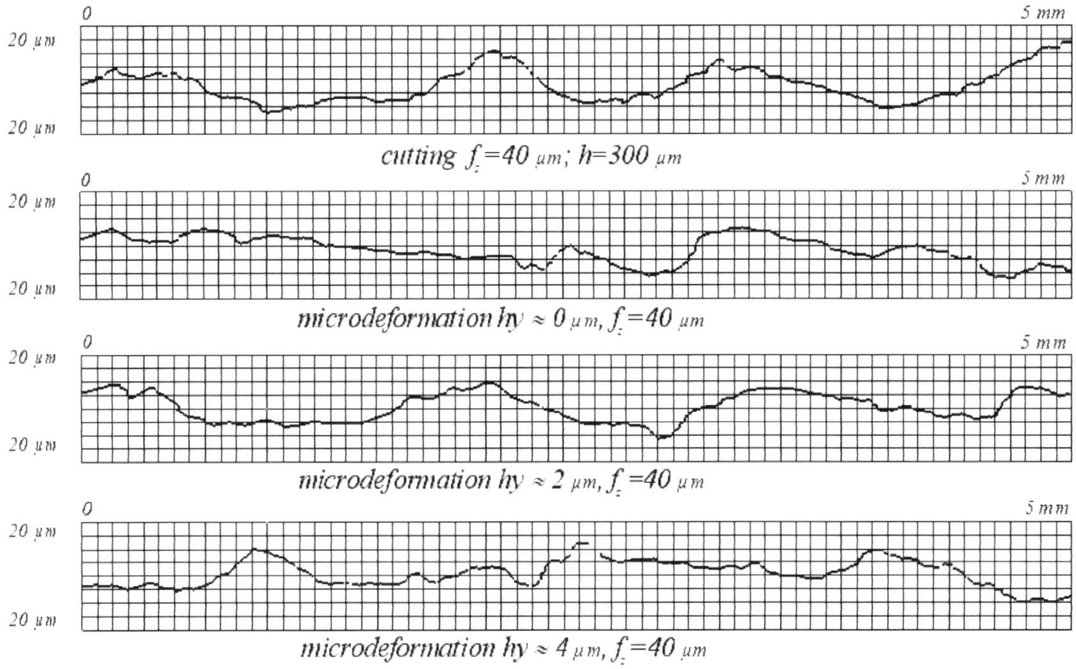

Fig. 2. The profilogram of the surface after preliminary shaping by removing a layer of material and microdeforming (C45)

The profilogram (Fig. 2, microdeformation $h_y \approx 0$ µm) shows peaks of the surface irregularities jammed into depressions, which corresponds to the highest SE values. The profilogram (Fig. 2, microdeformation $h_y \approx 2$ µm) shows slightly jammed peaks of the irregularities without the material moving into depressions. The profilogram (Fig. 2, microdeformation $h_y \approx 0$ µm) shows jammed peaks of the irregularities with the material moving into depressions, which corresponds to the lowest SE values. The deformation of the height combs of the irregularities is mostly observed during microdeformation of the surface of rigid structures ($h_y \approx 0$ µm).

As a result of the analysis of the microgeometry of the surfaces exposed to external mechanical influence, it can be concluded that the greatest manifestation of ruptured zones caused by the distortion of crystal lattices occurs in the substrate of rigid structures. This corresponds to the highest SE values.

Also, a comparative analysis of the mean absolute error of the profile was performed within the evaluation length after preliminary surface shaping and microdeformation of steel 20. The surface profilogram with line interval f_z=40 μm is presented in Fig.3.

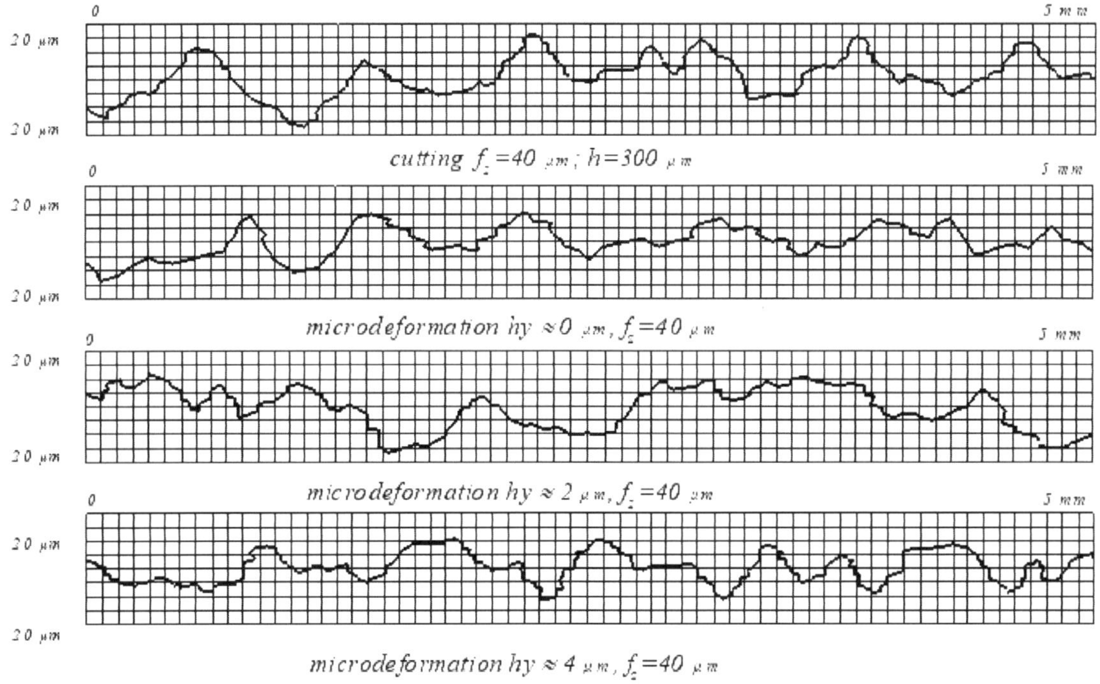

Fig. 3. The profilogram of the surface after preliminary shaping by removing a layer of material and microdeforming (C20)

The analysis of the results showed that deformation of the height combs of the irregularities is mostly observed during microdeformation of the surface of rigid structures ($h_y \approx 0\ \mu m$).

Also, the mean absolute error of the substrate profile was analysed within the evaluation length after preliminary surface shaping and microdeformation of X10Cr13. The surface profilogram with line interval f_z = 40 μm is presented in Fig.4.

The deformation of the height combs of the irregularities is also mostly observed during microdeformation of the surface of rigid structures ($h_y \approx 0\ \mu m$).

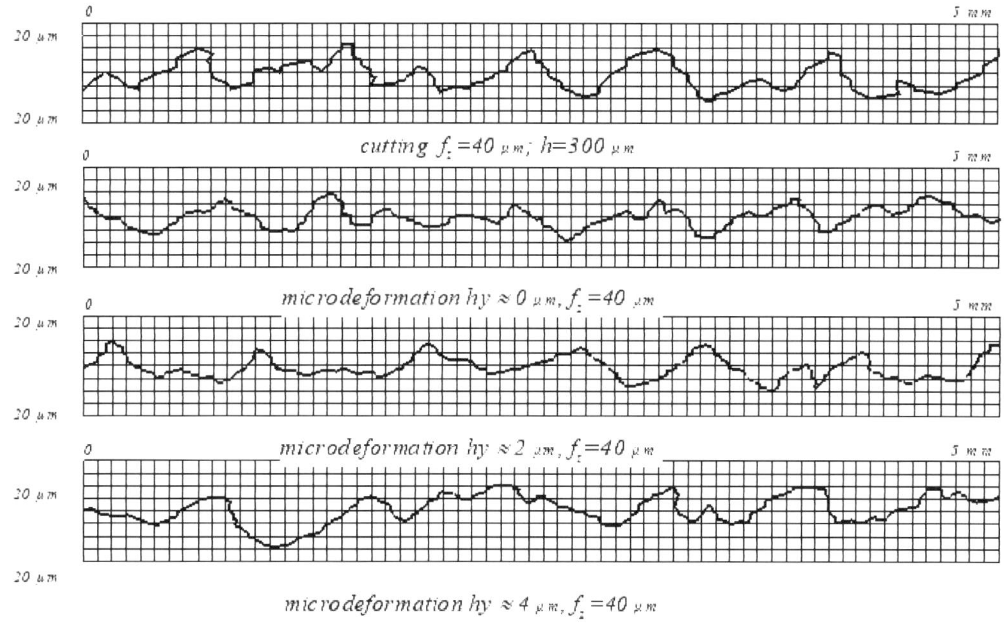

Fig. 4. The profilogram of the surface after preliminary shaping by removing a layer of material and microdeforming (X10Cr13)

4. Conclusions

As a result of the analysis of the change in microgeometry of the surfaces exposed to external mechanical influence by line-by-line microdeformation, it can be concluded that the greatest manifestation of ruptured zones caused by the distortion of crystal lattices occurs in the substrate of rigid structures. This corresponds to the highest SE values.

References

[1] Musokhranov M V 2005 Surface energy as an indicator of the quality of the surface layer *Directory. Engineering journal with the app* **12 (105)** 62-64

[2] Musokhranov M V, Kalmykov V V, Avramenko M U 2016 Technological prerequisites for improving the performance of the guide elements *Fundamental research* **8-1** 55-58

[3] Mousokhranov M V 2005 The role of surface energy in the formation of parts in precision engineering *Assembly in mechanical engineering, instrumentation* **6** 9-11

[4] Kachejev A Z, Kumykov V K, Manukyanc A R, Sergeev I N, Sozaev V A 2008 Pressure Influence on the surface energy of metals *In the book: Physics of low-dimensional systems* 289-292

[5] Malyshev E N, Musokhranov M V, Kalmykov V V, Antanuk F I 2015 Distribution of energy technological impact in the formation of working surfaces of machine parts made of metal materials *Modern problems of science and education* **1-1** 58

[6] Mousokhranov M V, Kalmykov V V, Logutenkova E V, Sorokin S P 2016 Energy state of the surface layer of machine parts *Modern high technology* **5-2** 276-279

[7] Kalmykov V V, Musokhranov M V, Logutenkova E V, Malyshev E N, Gorbunov A K 2017 Generation of surface energy in metals using row-by-row microdeformation *International Journal of Applied Engineering Research* T. 12. **24** 15621-15626

[8] Musokhranov M V, Kalmykov V V, Malyshev E N 2017 Experimental research of variability of surface energy value of Fe37-3FN, C45 and 41Cr4 steels *International Journal of Applied Engineering Research* T 12 **17** 6428-6433

[9] Musokhranov M V, Kalmykov V V, Malyshev E N, Zenkin N V 2015 Energy of surface layer

of metals as a tool of impact on the value of friction coefficient *Fundamental research.* **2-2** 251-254

[10] Chertovskikh S V, Semenov V I., Shuster L Sh 2015 The Influence of surface energies of the contacting materials on the tribological characteristics of Actual directions of scientific researches of the XXI century: theory and practice Vol.**3** 4-1 (15-1) 134-141

[11] Patent Nikolin S V, Timashev M Yu, Volkov S S, Suvorov G V, The method of measuring the contact potential difference *FindPatent.ru -patent search*

[12] Kushnir F. V. 1983 electrical measurements (textbook for universities Leningr. otd-nie) 320 p

[13] GOST R 8.736-2011 state system for ensuring the unity of measurements (GSI). The direct measurement repeated. Methods of processing of measurement results. Fundamentals

[14] Frolenkova L Yu, Shorkin V S 2013 Method of calculation of surface energy and adhesion of elastic bodies *Bulletin of Perm national research Polytechnic University. Mechanics* **1** 235-259

IOP Conf. Series: Materials Science and Engineering **483** (2019) 012046 doi:10.1088/1757-899X/483/1/012046

Oil and gas industry lexis translation: discourse-cognitive approach (notions versus terms)

S V Latysheva

Irkutsk National Research Technical University, 83 Lermontov Street, Irkutsk, 664074, Russia

E-mail: lasveta1@yandex.ru

Abstract. This work reveals a correlation of factors determining the adequacy of oil and gas lexis translation from English into Russian. They govern decision-making processes when translators choose between the two basic strategies (domestication and foreignization) limited by discursive practices and cognitive structures of source texts. When used within a specific type of discourse, specialized lexical units obtain influential characteristics affecting the choice of translation strategies, but in many cases these discursive parameters are not taken into account, as foreignization strategy predominates in specialized vocabulary translation. However, domestication strategy appears to be more effective in particular cases. We argue that discursive and cognitive analysis resulting in conscious strategy choice will allow translators to make adequate decisions about oil and gas lexis translation methods, as they also correspond to translation strategies: the strategy of foreignization leads to translation techniques that preserve formal characteristics of original lexical units, while the strategy of domestication imposes semantic and functional methods of translation. Consequently, the study presents a translation algorithm for specialized lexis in oil and gas source texts to bolster the decision-making effectiveness and to increase translation adequacy in target tests. The algorithm is based on discourse-cognitive approach drawing the connections between alternative discursive parameters representing different types of knowledge by means of oil and gas lexical units and translation strategies with correlating translation techniques.

Introduction

The research focuses on lexical aspects of specialised texts translation. Living in the world of multilingual communication, experts often face problems of specialized lexis translation and systematization of industry-specific lexical units. The fact that the European Union has the annual budget of one billion Euro to optimize communication in education, science, and technology proves this issue to be urgent. The biggest part of this allocation funds terminology management projects to control this problematic and controversial sphere of speech activity.

On the one hand, terminology has a long tradition of studying that can be expressed with the words of Antoine Laurent Lavoisier, a French naturalist of the 18th century who wrote, “It is impossible to separate neither science from terminology nor terminology from science because every science has its three essential elements: a number of facts that constitute its scientific content, the representations or notions of these facts, and the terms expressing these notions”.

On the other hand, the main characteristic of terms, i.e. their relation to scientific concepts and technical objects, is a problematic field for terminology management, since the development of these concepts, as well as the nomination of technological innovations are uncontrollable in most cases. Terminology starts chaotically, so terms often are not systematically arranged even within one language and one science. Thus, the concept “term” comes from the Latin root “to limit”. In other words, it is necessary to take control of something that is almost impossible to control.

The EU efforts to implement the policy of multilingualism, aimed both at efficiency improvement of cross-language communication and at preservation of national and cultural specifics of languages, led to creation of various management tools applied to terminological sphere. These tools allow scholars to address a number of practical issues, for example, to permit or not permit the use of a term, variations of the same term, or variations of the same term as different parts of speech. Consequently, the most important aspects of terms creation are fixed in industry-specific standards, glossaries, patents and specifications, compiled by both national and international teams of experts. These sources have become the foundation to create terminological databases.

For example, in 1975, the EU experts created a terminology database known as Eurodicautom. Its aim was to save time and efforts made by translators and other staff of the European Commission. In 2007, Eurodicautom was replaced with another tool called IATE (Inter-Active Terminology for Europe) that is a multilingual database of terms used in documents of the European Union. The purpose of the project, launched in 1999, was to give access to all terminology resources of the EU providing information and supporting the standards among the EU bodies. The International Organization for Standardization (ISO) has also created conceptual and terminological database, using original standards and their official translations.

However, it is not always possible to apply all these achievements of terminology management to translation of specialized texts from English into Russian. Firstly, most terminology platforms and databases are not free, so it increases the costs of translation services. In addition, highly demanded technical translation in strategically important sectors of Russian economy, such as oil and gas industry, causes a rise in terminological variability. Numerous unconfirmed technical translations are available on the Internet; therefore, the analysis of terminological meaning is time consuming, but it takes much more time to choose between alternative techniques resulting in translation variants. Industry-specific translators need to be careful in their choices in order to maintain the basic feature of specialized lexis that is semantic and functional unambiguity.

For the last decades, this mission has become urgent, as the prevailing strategy of foreignization prevents translators from applying meaning-based translation techniques. This means that, when translating specialized lexis, the major methods of translation are transliteration, transcription, loan translation, or even preservation of original forms of lexical units. Total rejection of domestication strategy often leads to negative consequences such as breaking of customary usage rules, semantic deformation of original meaning, terminological doublets preventing addressees from perceiving the information. These communicative fallouts ultimately result in a decrease of effectiveness in cross-language interaction.

The case described by Evgeny Bartov in his blog for industry-specific translators illustrates this problem. A Russian defence plant has contacted his company to get a presentation for a foreign exhibition translated from Russian into English. To complete this task the company hired an American translator. There were two specialized lexical units in the text of this presentation: загоризонтный радар (zagorizontnyi radar) and надгоризонтный радар (nadgorizontnyi radar). The translator used loan translation to transfer these terms into English – beyond-the-horizon radar station and over-the-horizon radar station, respectively. Translation editors disapproved of using the foreignization strategy and appealed to Pentagon documents. The analysis showed that the first unit represents the kind of radars that are installed along the coast of the United States. They spread a wave on water surface; therefore, such radars are called high frequency surface wave radar stations. As for the second unit, the American air defence system does not have any analogue radars. Therefore, American military experts called them Daryal-type radars, nominating the Russian city, where this equipment was first installed. This case proves that foreignization strategy does not always provide adequate translation. For these lexical units, a method of functional replacement corresponding to domestication strategy appears to be more effective. It corresponds to domestic terms making American experts refer to appropriate types of radars. As for the method of loan translation within foreignization strategy, proposed by the American translator, it would have led to difficulties in perception of the presentation content. Thus, it appears to be a less effective method of translation in this case. Consequently, the choice of translation methods determined by two translation strategies needs to be explained to achieve the adequacy of specialized lexis translation and to optimize industry-specific translation services.

1. The theoretical underpinnings of the study

The study of terminology problems involves various approaches of different academic schools, including translatology research. Scientific approaches vary considerably, but there is a limited number of mainstreams in studying specialized vocabulary and its functioning. A work by M. Teresa Cabré presents the three main perspectives of terms examining: linguistic, cognitive and communicative or discursive [1]. Linguistics focuses on lexical units of specialized value that demonstrate precise meanings shared by all the representatives of expert communities in each scientific area. Cognitive research describes conceptual units representing nodes of specific knowledge arranged as a system and conceived as a conceptual network that represents some scientific field. Discourse analysis studies discourse units that mark people as members of a professional group allowing them to be successful and effective in their field of expertise.

All the perspectives contributed to the idea that terminology is complicated enough to be viewed as an independent discipline [2]. However, the excessive variability of specialized lexis entails a broader sphere of categorization, so recent studies consider terminology to be an interdisciplinary and transdisciplinary field associated with lexicology, semantics, cognitive linguistics, sociolinguistics, communication theory, philosophy, language planning, corpus-based studies, translation theory, etc. [3].

From the practical point of view, specialized vocabulary studies are aimed at solving such real life problems as better access to electronic texts or the development of user-friendly tools for vocabulary analysis of corpora. This has led to increasing number of works studying the lexis in specialized discourse domains, both in universal spheres like general academic vocabulary [4], and in more specific areas such as agricultural research articles [5] and pharmaceutical pamphlets [6].

There is an alternative trend of terminology examining based on the idea that understanding the semantic nuances and pragmatic uses of specialized lexis builds expertise of a particular field enhancing professional skills [7], [8]. Both approaches have similar outcomes providing the standards for expert communities and institutions to use, to coin, and to arrange terminology creating the theoretical basis for terminology management. This aspect is relevant for translation studies as professional translators are supposed to use various translation tools including terminology management software to meet the demands of translation market [9].

All the research areas discussed above prove the necessity of a holistic approach to specialized vocabulary studies to integrate previous objectives and achievements. The example of such an attempt is a cognitive shift that demonstrates a greater emphasis on meaning as well as conceptual structures underlying texts and language in general. For example, the complex analysis of terminology and specialised translation performed by Pamela Faber Benítez shows that terminology theory is evolving from prescriptive to descriptive with a growing focus on the study of specialized language units from a social, linguistic and cognitive perspective [10].

This work is also an integration of the tree prospects involving the traditional designation of a specialized conceptual entity representing, preserving and transferring specialized knowledge. The functional aspect of terminology traditionally considered to be the most important in translation studies. For example, Eugene Nida wrote that we should use the terms according to "their practical usefulness and their explanatory power" [11, p. 489]. Therefore, to complete translation tasks effectively, researchers need to employ a holistic approach to transfer the basic features and functions of specialized vocabulary. Such an approach was proposed by Alexander Kaplunenko, who described the functioning of alternative semiotic entities representing different structures of knowledge. They are concepts, notions, and terms examined in three types of discourse – Discourse of Differences, Discourse of Consensus, and Discourse of Expert Community, respectively [12]. The study of Tatyana Tyurneva contributed to further development of this approach [13] that is also applicable to the sphere of ESP training [14].

The problem of knowledge evolution is directly connected with specialized lexis, and it is clearly seen through the diachronic analysis of specialized vocabulary [15]. It shows that two concepts are not always strictly identical, even if their names appear to be equivalent or translated from one language into another. Concepts that may be considered as well-known and easy to understand sometimes appear to be more complex than what they seem to be. It proves that translators and terminologists would highly benefit from a more regular use of all the information. We argue that the integration of discourse-cognitive approach with the opposition of notions and terms representing different types of knowledge will contribute to establishing the effective holistic terminology translation analysis.

Within the Discourse of Differences, language signs represent concepts. They have poor content and almost unlimited scope, and they represent phenomenological knowledge. This leads to individual and sometimes opposing interpretations. The conflicting positions of discourse arise from different perception and understanding of concepts fueled by personal experience.

Within the Discourse of Consensus, language signs represent notions. Unlike concepts, they have limited scope and content, so the empirical part of interpretation is not of primary importance. In this discursive practice, unstable characteristics of concepts become more sustainable, and concepts turn into notions.

The language signs, functioning in the Discourse of Expert Community, are terms. Only a limited group of experts are familiar with them. With the most narrow scope and specific contents, terms need to be monosemantic, because representatives of any expert community are interested in it.

2. Research methods

Considering all the above, translation of terminological vocabulary should be based on its cognitive and discursive characteristics. First, it is necessary to identify a discursive practice to which translated lexical units belong. It is typical of specialized texts to belong to Discourse of Consensus and Discourse of Expert Community. In real texts, these discourse practices blend, and up-leveling text specialization corresponds to a higher level of pansophy. However, this does not mean that all lexical units are terms if they function in Discourse of Consensus or Discourse of Expert Community. Therefore, the next step is to determine what type of knowledge a lexical unit represents: a concept, a notion or a term. This will influence the choice of translation methodology, aimed at (1) clarifying the personal meaning and transferring individual experience (for a concept), (2) identifying and transferring the entire scope and content (for a notion), (3) selecting a matching terminological unit (for a term). The choice of translation strategies (foreignization or domestication) results from cognitive and discursive characteristics described in the previous paragraphs.

Thus, Discourse of Expert Community activates highly specialized terms representing universal knowledge, so it is more efficient to apply foreignization strategy for their translation. This strategy is implemented through translation methods preserving formal characteristics of original lexis. It represents current expertise of a specific field employing terms that are scientific universals, so they are comprehensible to experts regardless of language systems they use. Such universals include, for example, Latin terms in medicine, acronyms of computer sphere, symbols of physical quantities and chemical elements, symbols denoting patented technical innovations, inventions and brands that have become so popular that they went beyond their original culture. The implementation of this strategy involves methods of transcription, transliteration, loan translation and form preservation of original lexical units.

Discourse of Consensus demonstrates alternative parameters of specialized lexis, as in this case experts have not reached the universal level of knowledge representation. Consequently, they operate not with terms, but with notions. Despite being limited in their scope and content, notions can have various nominations in different languages. For example, in Russian, the nominations for different kinds of radars indicate their position relative to the horizon, while their English nominations indicate location, function, and origin. As the missile defence systems of Russia and the USA are competing industries, their experts do not have a high degree of coordination and consensus. Accordingly, there is little agreement in nominating similar devices. In this case, the strategy of domestication is more preferable, as it applies alternative techniques preserving meanings and functions of transferred units, such as functional replacement, specification, generalization, explication and differentiation of meanings.

3. Results and discussion

The language materials used for the research includes: (1) Source Text that is a patent for the invention of the method of partially upgrading heavy oil at well-site published in 2017 in the English language; (2) Target Text that is an authentic translation of this document published by the Eurasian Patent Office. The original text is characterized by a high degree of specialization within oil and gas industry, as it describes a specific technological process innovative for oil and gas production. It is a topical issue for Russian conditions of HC extraction and transportation. The translation performed by professional industry-specific translators is the official description of this invention in the Russian language.

In the first phase of the research, we performed comparative analysis of ST and TT specialized lexis in order to identify and describe its translation methods. Then we divided these methods into two groups: (1) methods conserving formal characteristics of specialized vocabulary, and (2) methods transferring semantic and functional features. This allowed us to detect the cases of domestication and foreignization strategies. Table 1 selectively shows the results of the first phase:

Table 1. Differentiation of translation methods within foreignization and domestication strategies

Foreignization	Domestication
1) the preservation of original form	1) explication
an **API** gravity of 20	a pressure of 0 to 0.1 MPa**G**
2) transliteration	2) functional replacement[1]
cracking	3) generalization[2]
3) loan translation	soaking[1] drum[2]
liquid phase	4) specification
4) combined methods	injecting **stripping** steam
to have an S-value	5) differentiation
(the preservation of original form + loan translation)	at a thermal cracking **extent** of fractions

In the second phase of the research, we analyzed translation decisions made by professional industry-specific translators to transfer TT specialized lexical units: we identified their discursive-cognitive characteristics and their correlation influencing the choice of translation strategies.

The analysis shows that foreignization strategy was applicable to ST specialized lexical units constituting Discourse of Expert Community and representing terms. It proves a high degree of consistency in usage of analyzed ST and TT units among oil and gas industry experts belonging not only to a specific culture, but also to different language communities. In addition, specialized literature uses these terminological matches corresponding to similar referents.

For example, ST abbreviation API (American Petroleum Institute), as a part of terminological combination "API gravity" retains its original form in TT (градус API). The convention was adopted by oil and gas expert community, since this sign is a measure of oil density determining this characteristic relative to the density of water at the same temperature. This unit of measurement was developed by the American Petroleum Institute. In Russia the corresponding term is implemented by official technical documentation (GOST P 51069-97) "The method of determining density, relative density and density in API degrees by an aerometer". This document presents an authentic translation of the US National Standard ASTM D 1298 "Standard Test Method for Density, Relative Density, or

API Gravity of Crude Petroleum and Liquid Petroleum Products by Hydrometer Method", and it is recommended to be used in Russia.

Another example of implementing foreignization strategy due to cognitive and discursive characteristics of ST units is transliteration technique transferring the term "cracking". The matching TT term (крекинг) indicates high temperature processing of oil and its fractions aimed at obtaining the products of lower molecular weight. It historically derives from Discourse of Expert Community because the first industrial complex for continuous thermal cracking of oil was created and patented by a Russian engineer V.G. Shukhov and his assistant S. P. Gavrilov in 1891 (patent of the Russian Empire No. 12926, issued on November 27, 1891). The scientific and engineering solutions of V. G. Shukhov were repeated by U. Barton who constructed the first industrial plants in the United States in 1915-1918. Therefore, despite the fact that this lexical unit comes from English "cracking" as the technology makes oil molecules crack, within Discourse of Expert Community at the international level it comes from Russian specialized literature. The first domestic industrial complex for cracking was built by V. G. Shukhov in 1934; the plant called "Советский крекинг" was situated in Baku.

The analysis of domestication strategy implementation shows that ST specialized lexis functions in Discourse of Consensus and represents notions. This becomes clearly seen through the varying degree of inconsistency in the use of these terminological units demonstrated by experts from different language communities. The structures of knowledge assigned to them have not yet reached a universal level, so their semantics is not limited as it is typical of terms in Discourse of Expert Community. In this regard, there is a need for semantic transformations performed by translators.

For example, ST combination "a pressure of 0 to 0.1 MPaG" has a component (G) transferred by explication method, e.g.: давлении от 0 до 0.1 МПа (манометрических). G denotes a measurement unit for pressure and stands for "gauge". This term indicates a certain type and method of pressure measurement, when it is necessary to calculate the positive difference between the pressure in the pipe measured with a specific device (gauge) and the atmospheric pressure. The technical literature published in Russian refers to this type of pressure using the expression "избыточное давление". In English reference literature gauge is a polysemantic lexical unit functioning in many scientific fields. If we limit the scope of its use and take into account only measurements, it corresponds to the combination "measuring device", and it can measure not only pressure. The Russian word "манометр" corresponds to several English expressions: manometer; air-gauge; steam gauge; monoscope. Such variability in linguistic representations of the concept suggests that the structure of knowledge is still evolving and require clarity and coherence that is achieved by means of explication. It is reasonable to use it in the description of the technology presented in the patent, as it is important to point out the specific kind of pressure.

4. Conclusion.

Based on the foregoing, we propose an algorithm for differentiating translation strategies of specialized lexical units functioning in oil and gas industry:

Step1. Determine cognitive and discursive characteristics of ST specified lexical units (notions for Discourse of Consensus or terms for Discourse of Expert Community). To determine this it is necessary to specify:

- How narrow is the field of knowledge specialization, represented by ST lexical unit?
- How consistent is the expertise about a phenomenon represented by ST lexical unit?
- How consistent are the ways of representation of this knowledge in the framework of the universal scientific discourse?
- How limited is the scope and content of a notion, represented by a lexical unit in ST and TT?
- How universal is the specialized literature, which contains the matching ST and TT lexical unit?

Step 2. Choose a discourse-cognitive translation strategy according to the suggested pattern:

When detecting a high degree of characteristics described in Step 1, a lexical unit should be attributed to Discourse of Expert Community, so foreignization strategy is to be used.

When detecting a low degree of characteristics described in Step 1 a unit should be attributed to Discourse of Consensus, domestication strategy is to be used.

Step 3. Choose translation methods correlating the strategies:

Methods transferring formal constituents of ST lexical units (original form preservation, transcription, transliteration, loan translation) are in accordance with foreignization strategy.

Methods transferring semantic and functional features of ST lexical units (functional replacement, generalization, specification, modulation, explication, meaning differentiation) are in accordance with domestication strategy.

The proposed translation decision algorithm for the choice of discourse and cognitive related translation strategy can be applied to specialized lexis of other disciplines and other language pairs. A theoretically grounded mechanism of applying strategies and methods will allow industry-specific translators to optimize the translation services and to accelerate the process of knowledge universalization increasing the effectiveness of interaction in engineering and technical sphere.

References

[1] Teresa Cabré Castellví M 2010 *Terminology and translation* Handbook of Translation Studies: Volume 1 ed by Gambier Y and Doorslaer Lvan (Benjamins) 356-65
[2] Felber H 1984 *Terminology Manual* (Paris: Infoterm) p 426
[3] Thelen M 2015 *The Interaction between Terminology and Translation Or Where Terminology and Translation Meet* Trans-kom **8** 347-81
[4] Gardner D and Davies M 2014 *A new academic vocabulary list* Applied Linguistics **35(3)** 305-27
[5] Martinez I and Beck S 2009 *Academic vocabulary in agriculture research articles: A corpus-based study* English for Specific Purposes **28(3)** 183-98
[6] Grabowski L 2015 *Keywords and lexical bundles within English pharmaceutical discourse: A corpus-driven description* English for Specific Purposes **38(2)** 23-33
[7] Coxhead A and Demecheleer M 2018 *Investigating the technical vocabulary of Plumbing* English for Specific Purposes **51** 84-97
[8] Tongpoon-Patanasorn A 2018 *Developing a frequent technical words list for finance: A hybrid approach* English for Specific Purposes **51** 45-54
[9] Chen J and Tian Ch 2016 *Terminology management and translation teaching in Mainland China* IJRRE **3** 1-12
[10] Faber Benítez P 2009 *The cognitive shift in terminology and specialized translation* MonTI **1** 107-34
[11] Nida E Taber Ch 1969 *The Theory and Practice of Translation* (Leiden: E. J. Brill) 197
[12] Kaplunenko A 2007 *Concept – Notion – Term: Evolution of Semiotic Entities in the Context of Discursive practice* Int. Scien. Conf. Pr. (Irkutsk: ISTU) 115-20
[13] Tyurneva T 2012 *Experience in Analyzing the Lingvosemiotic Context of EDUCATION: Concept > Notion > Term (in the English Language)* Thesis for a Candidate Degree in Philology (Irkutsk) 203
[14] Latysheva S 2018 *Discourse-Competence Approach to Design of ESP Courses* SHS Web of Conferences **50** 01095
[15] Dury P 2005 *Terminology and Specialized Translation: the Relevance of the Diachronic Approach* LSP & Professional Communication **5** 31-41

IOP Conf. Series: Materials Science and Engineering **483** (2019) 012047 doi:10.1088/1757-899X/483/1/012047

Formation of adhesive properties of surfaces of multicomponent materials

E V Logutenkova[1], V V Kalmykov[1], M V Mousokhranov[1]

[1]Kaluga Branch of Federal State Budgetary Educational Institution of Higher Education "Moscow State Technical University named after N.E. Bauman (National Research University)," 2, Bazhenova Street, Kaluga 248600,Russia

kalmykovvv@bmstu.ru

Annotation. An experiment was conducted to determine the change in the adhesive strength of PF-115 enamel depending on the energy properties of the surfaces of various metals to assess the adhesive strength of paint-and-lacquer coatings. The adhesive strength was evaluated using the method of lattice cuts. Samples of brass CUZN39PBL, aluminium alloy 2024T35, C45, X10CrNiTi18-10 were taken as the test object.

1. Introduction

Along with electrochemical and chemical ways to protect metal products and structures from aggressive environmental impact, paint-and-lacquer coatings are widely used. This method is most popular due to its low cost and ease of recovery, it does not require expensive equipment, and also allows for achieving the required appearance and colour.

Paint-and-lacquer coatings are multicomponent mixtures. Such mixtures are applied to a product and then dried, thus forming an attractive film which protects the surface from corrosion. The quality of paint-and-lacquer coatings is determined by their mechanical, chemical and adhesive properties and the painting process itself. The adhesive strength of the coating is its most important operational characteristic [1-4]. Enamels, primers, putties and other paint-and-lacquer coatings are most frequently used for painting.

The adhesive strength depends on the conditions of paint coating application, its thickness, as well as on the area of the surface to be coated.

2. Materials and methods

To assess the adhesive strength of paint-and-lacquer coatings, an experiment was conducted to determine the change in the adhesive strength of PF-115 enamel depending on the energy properties of the surfaces of various metals. The adhesive strength was evaluated using the method of lattice cuts. The essence of the method is to evaluate the degree of adhesion of a paint film to the substrate judging by the number of cells that fall away from the substrate when the film is cut. To determine the stability, experiments were carried out in the laboratory using additional equipment: Mitutoyo micator, measurement accuracy up to 0.001 mm, white spirit, green PF-115 enamel, paint sprayer (Fig.1).

PF-115 enamels are intended for painting metal, wood and other surfaces exposed to weather impact, and for painting indoors. The advantages of using PF-115 are: the enamel coating is weather-resistant, resistant to water, cleaning solutions, industrial oils, resistant to temperature changes from minus 50 to plus 60°C.

Figure 1. Micator, white spirit, PF-115 enamels

Samples of brass CUZN39PBL, aluminium alloy 2024T35, C45, X10CrNiTi18-10 were taken as the test object. The surface of some samples of each material was subjected to grinding, and the other part of the samples was diamond smoothed. This allowed us to obtain different levels of surface energy of the samples. Before enamel was applied, each sample was treated with a cloth moistened in white spirit, followed by wiping. According to the method selected, the thickness of the enamel applied should not exceed 60 μm. It was controlled by measuring the height of the samples using the micator, before and after application. In this experiment, enamel was applied to the treated surfaces using the method of pneumatic spraying with a paint sprayer. The paint is applied on the samples for several seconds. According to standards, enamel was dried at a temperature of (20±2)°C for 48 hours (Fig.2).

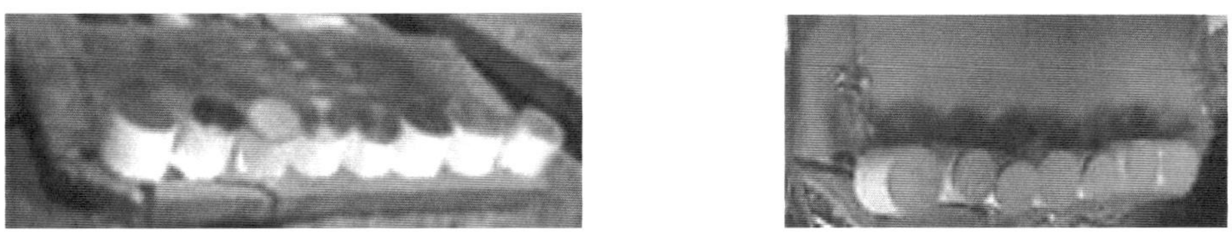

Figure 2. Sample

After the enamel dried up completely, a special tool was used to make 5 parallel cuts on the surface of the samples to the substrate separated by 1 mm from each other. The same number of similar cuts were made perpendicularly to the first ones. Adhesion was assessed by the state of the cuts in the coating expressed in points (according to a four-point system) in accordance with the table below:

3. Results and discussion

The results obtained during the experiment are added to Table 1, 2.

Table 1. The value of points depending on the percentage of the number of dead cells.

The number of cells that fell off, %	0	Less than 35	More than 35	70-100
Points	1	2	3	4

Table 2. Experimental results.

Sample	The number of cells that fell off	The number of cells that fell off, %	Point
ground CUZN39PBL	1	4	2

smoothed CUZN39PBL	0	0	0
ground 2024T35	14	56	3
smoothed 2024T35	0	0	0
ground C45	6	24	2
smoothed C45	1	4	2
ground X10CrNiTi18-10	20	80	4
smoothed X10CrNiTi18-10	12	48	3

The results of adhesive strength depending on the level of surface energy are presented in the graphs (Fig.3, Fig.4, Fig.5, Fig.6):

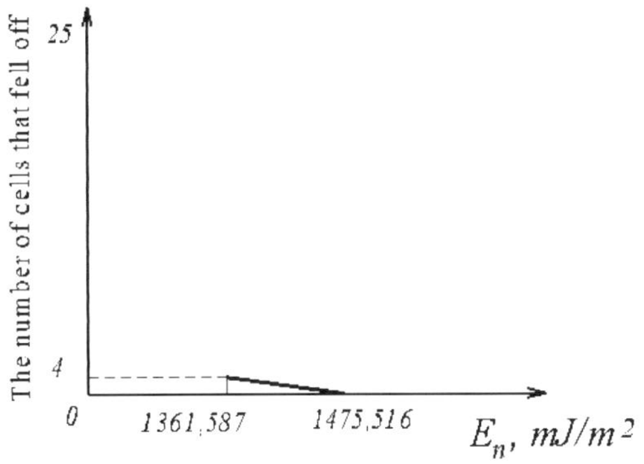

Figure 3. The results of adhesive strength depending on the level of surface energy (CUZN39PBL)

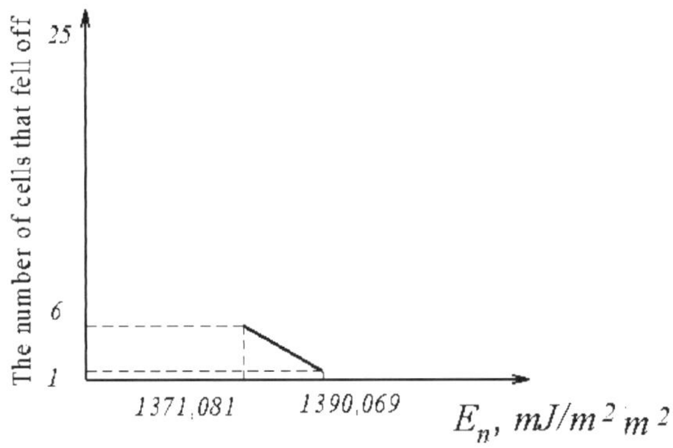

Figure 4. The results of adhesive strength depending on the level of surface energy (C45)

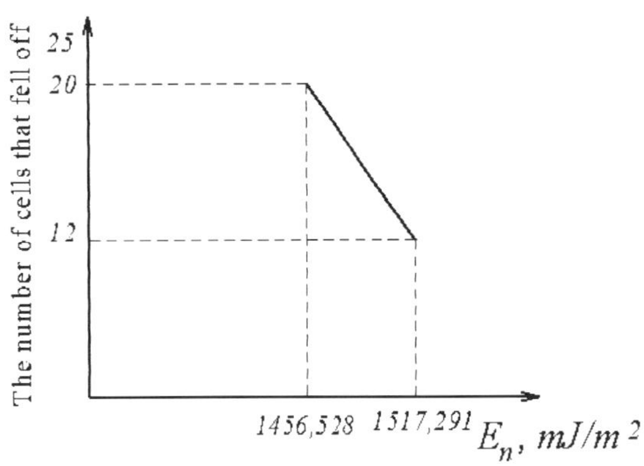

Figure 5. The results of adhesive strength depending on the level of surface energy (X10CrNiTi18-10)

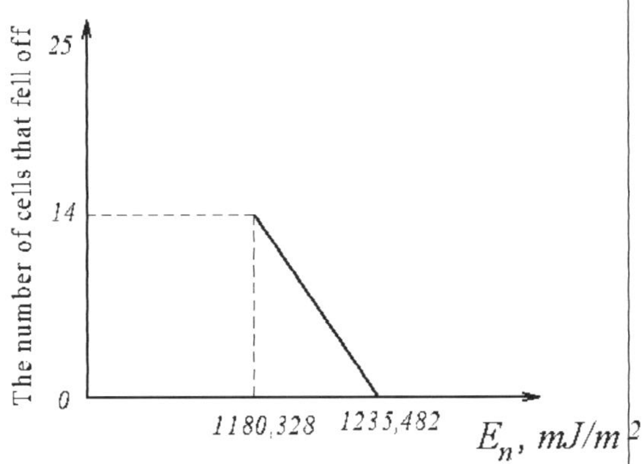

Figure 6. The results of adhesive strength depending on the level of surface energy (2024T35)

The graphs show that the samples subjected to diamond smoothing had the highest energy value, unlike the samples with a ground surface. Thus, the higher the level of surface energy, the higher the adhesion strength.

4. Conclusions

According to the Young model, the larger the surface energy of the sample to be coated, the smaller the wetting angle of the surface. In other words, liquid enamel spreads better on a surface with a large value of surface energy. And the improved spreadability of enamel contributes to filling the surface irregularities, which in turn provides adhesive strength. Due to surface energy, surface tension appears. According to studies [5-13], the method of processing the surface of metals has an impact on the amount of surface energy. It may be recommended choosing surface treatment methods providing a higher level of surface energy to ensure adhesion strength of coatings.

References

[1] Pashley M D, Pethica J B, Tabor D 1984 Adhesion and micromechanical properties of metal surfaces *Wear* T 100 **1-3** 7-31
[2] Gottenbos B, Van der Mei H C, Busscher H J 1999 Models for studying initial adhesion and surface growth in biofilm formation on surfaces *Methods in Enzymology* 523-533
[3] Topolyansky P 2005 Investigation of adhesion properties and the mechanism of formation of coatings by plasma hardening *Materials* 316-333
[4] Shautsukov A G 2006 The modern idea of the possible mechanisms of adhesion of metal films to various substrates *Journal of applied physics* **5** 16-22
[5] Kachejev A Z, Kumykov V K, Manukyanc A R, Sergeev I N, Sozaev V A 2008 Pressure Influence on the surface energy of metals *In the book: Physics of low-dimensional systems* 289-292
[6] Kalmykov V V, Musokhranov M V, Logutenkova E V, Malyshev E N, Gorbunov A K 2017 Generation of surface energy in metals using row-by-row microdeformation *International Journal of Applied Engineering Research* T 12 **24** 15621-15626
[7] Mousokhranov M V, Kalmykov V V, Logutenkova E V, Sorokin S P 2016 Energy state of the surface layer of machine parts *Modern high technology* **5-2** 276-279
[8] Musokhranov M V 2005 Surface energy as an indicator of the quality of the surface layer *Directory. Engineering journal with the app* **12 (105)** 62-64
[9] Chertovskikh S V, Semenov V I., Shuster L Sh 2015 The Influence of surface energies of the contacting materials on the tribological characteristics of Actual directions of scientific researches of the XXI century: theory and practice Vol 3 **4-1 (15-1)** 134-141
[10] Frolenkova L Yu, Shorkin V S 2013 Method of calculation of surface energy and adhesion of elastic bodies *Bulletin of Perm national research Polytechnic University. Mechanics* **1** 235-259
[11] Musokhranov M V, Kalmykov V V, Avramenko M U 2016 Technological prerequisites for improving the performance of the guide elements *Fundamental research* **8-1** 55-58
[12] Mousokhranov M V 2005 The role of surface energy in the formation of parts in precision engineering *Assembly in mechanical engineering, instrumentation* **6** 9-11
[13] Malyshev E N, Musokhranov M V, Kalmykov V V, Antonuk F I 2015 Distribution of energy technological impact in the formation of working surfaces of machine parts made of metal materials *Modern problems of science and education* **1-1** 58

Perseptron for assessing the students' task performance in learning a foreign language according to the competencies using bi data

N A Prom, E A Litvinova, A V Shokhnekh, T G Yovanovich, N I Lomakin

Russia, 400005, Volgograd, Prospect V I Lenin, 28

natalyprom77@mail.ru, srdjan@live.ru, shokhnekh@yandex.ru, elen-gordienko@yandex.ru, tel9033176642@yahoo.com

Abstract. The article presents the Perceptron neural network program developed by the authors, which enables assessing in points the quality of the task performed by students in a foreign language. The theoretical basis for the artificial intelligence systems applied in the learning process has been considered. The studies have shown that the online education market has grown in the world, and the trend marked persists. The trends in use of neural networks in learning have been summarized. Analysts noted that the use of artificial intelligence systems has been becoming more widely used in the learning process. It has been hypothesized and proved that a neural network makes it possible to give a grade for the quality of a control text read by a student according to the requirements of competencies using the BI data resulted from processing with an artificial intelligence system. The algorithm proposed by the authors uses the capabilities of the PyPhox mobile application and the Deductor platform. The algorithm is based on the following scheme. The teacher records a piece of a text into a file on the PhyPhox mobile application; its digitized version is then used as a reference to compare the quality of the assignment done by a Master student. The resulting reference file is exported to XL and then imported into the Deductor for further processing. The students do the same; their control files are paired with the reference, combined in XL and imported into the Deductor for processing. In the final, based on the comparison of the coefficients of correlation between the control and reference files, the Perceptron program assigns a score corresponding to the "excellent," "good," "satisfactory" and "not satisfactory" grades. The neural network formed is trained on data sampling, where they add semester points and competence points. The relevance lies in the fact that the program enables automating the learning process and replacing the teacher's work at the stage of the assignment control according to the competencies, which is of practical importance.

1. Introduction

The study presented was conducted due to the fact that the market of online education is constantly growing in the world, and this trend persists. In order to automate the work of a foreign language teacher (English), an attempt was made to computerize the process of assessing the quality of students' reading a text according to the requirements of competencies at the end of the control week.

The relevance is determined by the fact that the algorithm proposed automates the learning process and replaces the teacher's work at the stage of assignment quality control according to the competencies, which is of practical importance.

It has been hypothesized and proved that a neural network applied makes it possible to assess the quality of student's reading the control according to the requirements of competencies, using the BI data obtained as a result of processing by the artificial intelligence system.

2. Results and discussion

To develop the AI-system program for digital control and assessment of the compliance of a student's oral speech according to the requirements of competencies, initial data were taken, i.e., two digitized sound files, one of them was a reference file recorded by the teacher.

2.1. Results

The algorithm proposed by the authors used the capabilities of the PyPhox mobile application and the Deductor platform [1]. The algorithm was based on the following scheme. The teacher recorded a piece of a text to a file on the PhyPhox application, which was then used as a reference for comparing the quality of the test done by a Master student. The resulting reference file was exported to XL, then into the Deductor for further processing. The students came in the same way; their control files were combined in pairs in XL and imported into the Deductor for processing.

In the final, based on the comparison of the correlation coefficients between the control and reference files, the Perceptron program, previously trained on data sampling, assigned a score corresponding to the marks "excellent," "good," "satisfactory" and "not satisfactory."

The sound track was digitized according to the parameters: time (s), Period (s) and Frequency (Hz) with the help of the free-of-charge mobile application Phyphox (Fig. 1).

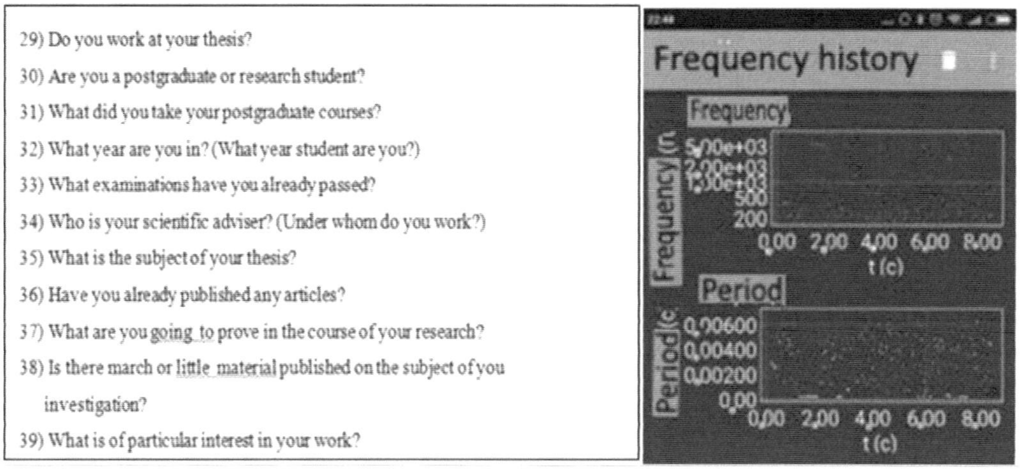
29) Do you work at your thesis?
30) Are you a postgraduate or research student?
31) What did you take your postgraduate courses?
32) What year are you in? (What year student are you?)
33) What examinations have you already passed?
34) Who is your scientific adviser? (Under whom do you work?)
35) What is the subject of your thesis?
36) Have you already published any articles?
37) What are you going to prove in the course of your research?
38) Is there march or little material published on the subject of you investigation?
39) What is of particular interest in your work?

Figure 1. A text fragment, the quality of reading the text was assessed; and a mobile phone screen with the file recorded

A text fragment, the quality of reading the text was assessed; and a mobile phone screen with the file recorded.

The voice file obtained in a digital format for control was exported to an XL file and then into the Deductor (Table 1).

Table 1. A fragment of the initial data, i.e., two digitized files in XL

	A	B	C	D	E	F	G	H	I	J
1	Time (s)	Period (s)	Frequenc	Time (s)	Period (s)	Frequency (Hz)				
2	0.179	0.00382	261.7486	0.148	0.000209	4790				
3	0.228	0.0038	263.1868	0.187	0.006367	157.0492				
4	0.248	0.0038	263.1868	0.228	0.001169	855.3571				
5	0.284	0.003779	264.6409	0.268	0.003946	253.4392				
6	0.331	0.003841	260.3261	0.308	0.002067	483.8384				
7	0.365	0.003862	258.9189	0.349	0.000209	4790				
8	0.406	0.004008	249.4792	0.389	0.000209	4790				
9	0.444	0.005219	191.6	0.826	0.000209	4790				
10	0.486	0.004551	219.7248	0.865	0.000146	6842.857				
11	0.528	0.003841	260.3261	0.905	0.000146	6842.857				
12	0.565	0.00309	323.6486	0.945	0.000188	5322.222				

The next step of the algorithm was processing of the data imported into the Deductor. In particular, the digitized files were aligned in length (by filling in gaps), processed by quantization for subsequent visualization in a multidimensional diagram (by grouping) and used to calculate multiple correlation coefficients for subsequent comparison of their identity in the Perceptron in order to assign an appropriate estimate (Table 2).

Table 2. A fragment of initial files containing 996 lines in the Deductor

Table

36 / 996

Time (s)	Period (s)	Frequency (Hz)	Time (s)	Period (s)	Frequency (Hz)
0,99916402180639	0,00386138534705818	252,754754958199	1,03126175013616	0,00301102309707738	92,7924622271235
1,03712881164659	0,0036053629175733	312,744244007155	1,03096570672781	0,00184576402103594	1579,19518956835
-0,105401065106211	0,00378409342813115	277,698538155389	-0,174184421502822	0,000716394916348666	3473,09538154637
-0,0654389060985672	0,00446012097117731	190,065727352319	-0,0097872401955037	6,10322960477166E-5	4976,49036875255
0,791662557845484	0,00487407884387598	185,749257517371	1,09048203666588	-3,83696294765012E-6	5870,35590086156
0,831738174766297	0,00453334825187051	262,330278942602	1,25943922739421	0,000256531567058598	6154,1925027694
0,256304484034295	0,00386659927052014	287,023110000461	0,648704942473127	0,000343443927331362	6120,92741043902
0,296235953583206	0,00337003068031256	255,012981942226	0,641665143239891	8,2454518284912E-5	6112,45199503166
0,823627830676021	0,00297092561239289	301,490834065047	1,19863020046112	-8,32333777128296E-5	5844,52062675216
0,862323756794709	0,00275098762867791	413,207097059126	1,27736946578395	0,000290743615958458	4820,88851695008
0,501624933951014	0,00328551748750061	402,678785423877	0,911957734816117	0,000904960856022112	3621,65302699765

2.2 Discussion

The files loaded in the Deductor were aligned in size using the "Fill-in gap" function and processed by the quantization method for later visualization in a multidimensional diagram.

The quantization in signal processing is known to be a partition of the range of signal reference value into a finite number of levels and rounding of these values to one of the two levels closest to them. It should be noted that in this case, the signal value might be rounded to either the nearest level or to a lower or higher level of the nearest ones, depending on the encoding method. In practice, this provided an effect of "compressing" the data (Table 3).

Table 3. A fragment of data after quantization

Table

1 / 996

Time (s)	Period (s)	Frequency (Hz)	Time (s)	Period (s)	Frequency (Hz)
before 3,885	from 0,003778706	from 248,1865285 до 264,640884	before 4,385	from 0,000208768 before 0,000688935	from 4790
before 3,885	from 0,003778706	from 248,1865285 до 264,640884	before 4,385	from 0,005093946	before 197,1193416
before 3,885	from 0,003778706	from 248,1865285 before 264,640884	before 4,385	from 0,000688935 before 0,001941545	from 526,3736264 before 1451,515152
before 3,885	from 0,003778706	from 264,640884 before 290,3030303	before 4,385	from 0,003820459 before 0,004154489	from 240,7035176 before 261,7486339
before 3,885	from 0,003778706	from 248,1865285 before 264,640884	before 4,385	from 0,001941545 before 0,00256785	from 389 4308943 before 526,3736264
before 3,885	from 0,003778706	from 248,1865285 before 264,640884	before 4,385	from 0,000208768 before 0,000688935	from 4790
before 3,885	from 0,003778706	from 248,1865285 before 264,640884	before 4,385	from 0,000208768 before 0,000688935	from 4790
before 3,885	from 0,004780793	before 209,1703057	before 4,385	from 0,000208768 before 0,000688935	from 4790
before 3,885	from 0,004300626	from 209,1703057 before 233,6585366	before 4,385	before 0,000208768	from 4790
before 3,885	from 0,003778706	from 248,1865285 before 264,640884	before 4,385	before 0,000208768	from 4790
before 3,885	from 0,002776618	from 319,3333333 before 362,8787879	before 4,385	before 0,000208768	from 4790
before 3,885	from 0,003131524	from 290,3030303 before 319,3333333	before 4,385	before 0,000208768	from 4790
before 3,885	from 0,003131524	from 290,3030303 before 319,3333333	before 4,385	before 0,000208768	from 4790
before 3,885	from 0,001461378	from 460,5769231 before 684,2857143	before 4,385	before 0,000208768	from 4790
before 3,885	from 0,004780793	before 209,1703057	before 4,385	from 0,001941545 before 0,00256785	from 389,4308943 before 526,3736264

The authors found it appropriate to visually compare the resulting sound files digitized that were quite different in parameters, i.e., time (s), period (s) and frequency (Hz) (Fig. 2).

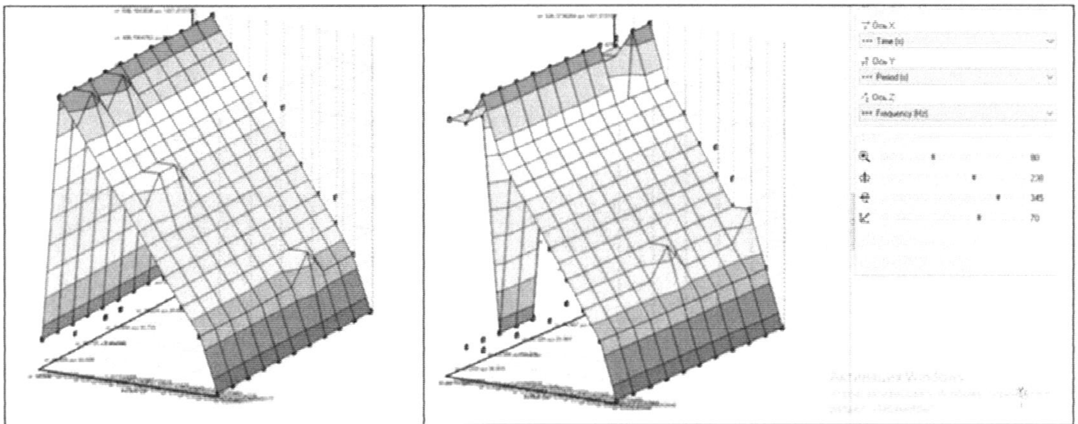

"Standard" - the file of the teacher	"Control" - student file

Figure 2. Visual comparison of files by parameters: time (s), period (s) and frequency (Hz)

In the algorithm presented, the initial data was transformed and used in the BI format.

So, the voice files in digital format received from students to be assessed were alternately exported into an XL file that already contained a reference file, then imported into the Deductor, processed there by quantization and compared. Thus, the information was converted to the BI, later used in the algorithm developed.

Business intelligence (abbreviated as BI) is the designation of computer methods and tools for organizations that translate transactional business information into a human-readable form suitable for business analysis and means for mass work with such information processed [2].

In order to compare the quality of a control file with the reference one, a neural network in the Deductor was formed to conduct the correlation analysis (Fig. 3).

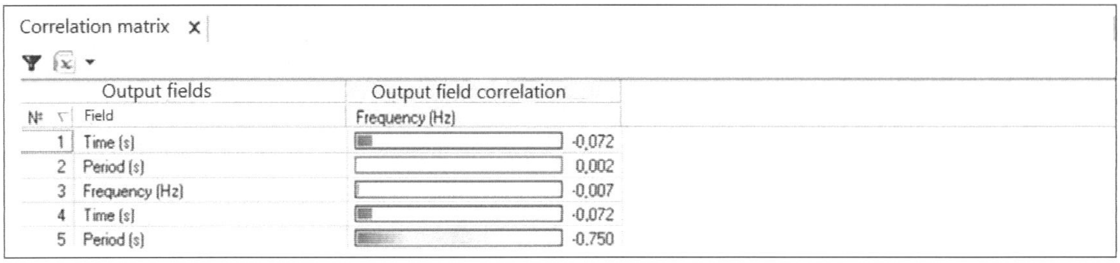

Correlation matrix

Output fields		Output field correlation
№	Field	Frequency (Hz)
1	Time (s)	-0,072
2	Period (s)	0,002
3	Frequency (Hz)	-0,007
4	Time (s)	-0,072
5	Period (s)	-0,750

Figure 3. Values of the correlation matrix

The values obtained were accumulated in a table. The correspondence of the correlation coefficients values to the teacher's estimates obtained in the control is presented in Table 4.

Table 4. The correspondence scale of the assessment level to the correlation parameters in the Perceptron training sample

Period (s)	Frequency (Hz)	Competence Score	Semester Score	Evaluation	Evaluation_OUT	Evaluation_ERR
-0,750	1	35	40	90	89,9613659600236	5,97035617960575E-7
0,0142	0,0541	25	30	75	75,0001189915279	5,66359348464927E-12
0,068	0,014	15	20	60	59,9996327848067	5,39387992699675E-11
0,892	0,0025	5	10	40	40,0306079829408	3,74739447881388E-7

The Perceptron formed during the research study has five neurons in the input layer, i.e., Period (s) _0, Period (s) _1, Frequency (Hz), competence score and semester score; one hidden layer with two neurons; in the output layer there is one output parameter-the estimate (Fig. 4).

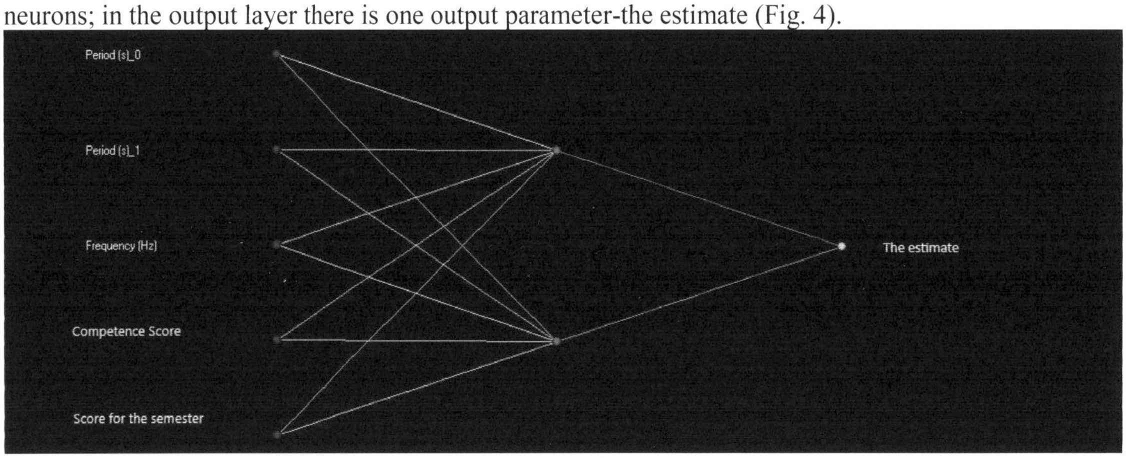

Figure 4. The Perceptron graph

With reference to the field of education, the term “competence” means a set of legal rights and obligations (duties) of state and/or authorized non-state bodies, institutions or organizations for making legal decisions in the course of legislative or law enforcement activity in education [3].

The score according to the competencies of a graduate student was put down by the teacher according to the “Fund of appraisal funds” of the Steering document for the discipline “Business Foreign Language (English)” in the categories “know,” “know how” and “have skills” [4].

The federal and state authorities’ competence established in Articles 28 and 29 of the Education Law of the Russian Federation in federal subjects of the Russian Federation is exhaustive and can be changed only by law.

The semester points were given to the graduate by the teacher according to the control list in the dean's office.

IOP Conf. Series: Materials Science and Engineering **483** (2019) 012048 doi:10.1088/1757-899X/483/1/012048

Thus, in the algorithm proposed for scoring the test, a trained neural network, the Perceptron [5] was used. When using the "what-if" function based on the parameters entered the neural network, we finally got an estimate (Fig. 5).

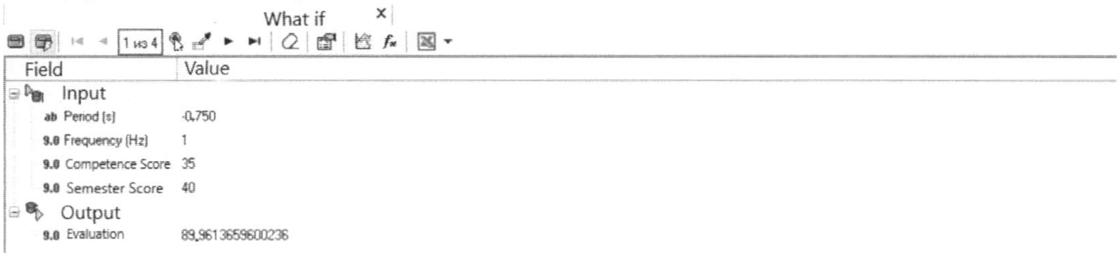

Figure 5. Scoring using the "what-if" function

Thus, the Perceptron program for assessing the task performance by students in a foreign language according to competences based on the BI-data according to the results of the neural network has been successfully formed.

It should be noted that the program presented differs from the "Neural network for assessing the students' competence in the discipline of Organization of entrepreneurial activity" previously proposed because it is more advanced regarding the automatic generation of a sound file and effective operation of the data transmission system [6].

The studies showed that in modern conditions the use of artificial intelligence systems in the learning is increasing.

For example, Parla startup raised € 2 million investment for the development of a virtual English teacher. The developers threatened to replace teachers and tutors with artificial intelligence soon. The Parla developers predicted the death of traditional teaching methods. Having replaced the English teacher, they plan to release applications for the study of physics, mathematics, biology and other subjects.

The analysts at J'son & Partners Consulting found out that the online education market has grown by 23% in the world and according to Global Market Insights, makes $ 165 billion or about 3% of the global education market for the past five years.

The study of the Russian market of online education and educational technologies established that the market will grow annually in the country by at least 5%. In Russia, in 2016, the language learning segment using games was worth $ 315.7 million. According to Skyeng, English accounts for 76% of the Russian market for learning foreign languages, German 10% and French 7%. According to Technavio, the volume of the global digital English language learning market in 2016 was $ 4.83 billion and is projected to rise to $ 13.8 billion by 2021 [7].

The main purpose of the Parla program was to create learning techniques that are comparable with speaking of native speakers. So, you can understand the properties of the language, synonyms, word-play and collocations [8].

The experts' opinions regarding the future of such technologies vary. We observed several trajectories of the development of artificial intelligence applied in education:

- machine learning and neural networks for the compilation of educational content for the users' interests;
- selection of teacher-student pairs using computer vision systems and machine intelligence; and
- cognitive computations [9].

Cognitive computations based on neural networks and deep learning use knowledge of cognitive science to create systems that simulate human thought processes. However, instead of focusing on a unique set of technologies, cognitive computing encompassed several disciplines, including machine learning, natural language processing and human-computer vision and interaction [10, 11].

A promising direction is the use of innovative IBM open source projects [12]. Deep machine learning is in the trend, for example, Tensor Flow that is an open source software library for numerical computation using data flow graphs, in other words, the best way to create deep learning models [13, 14].

3. Conclusions

Based on the study conducted, we can draw the following conclusions.

Firstly, the use of artificial intelligence systems has broad prospects; therefore, research of automating the control of students' knowledge in a foreign language based on a neural network is important [15, 16].

Secondly, it was hypothesized and proven that based on the neural network, the quality of the control text read by a student can be assessed, according to the requirements of the competencies, using the BI data obtained as a result of processing by the artificial intelligence system.

References

[1] Basegroup Advanced analytics without programming 2018 URL: https://basegroup.ru/deductor/description
[2] Business Intelligence 2018 URL: https://ru.wikipedia.org/wiki/Business_Intelligence
[3] Federal Center for Educational Legislation 2018 URL: http://www.lexed.ru/obrazovatelnoe-pravo/knigi/yagofarov2005/422.php
[4] Steering document "Business Foreign Language (English)." Field of education 09.04.01 Informatics and computer engineering. Programs "Information and software of automated systems," "Methods of analysis and synthesis of design solutions" and "Information and software of production automated systems." Master's level of training. Full-time form of education. The term of study is normative. Volgograd 2017 32 p.
[5] Serebryakova S V, Milostivaya A I 2017 Semantic emergence as a translation problem *Vestnik Volgogradskogo Gosudarstvennogo Universiteta-seriya 2-yazykoznanie* **16 (3)** 48-57
[6] Ergunova O T, Lizunkov V G, Malushko E Yu, Marchuk V I, Ignatenko A Yu 2017 Forming the system of strategic innovation management at the high-tech engineering enterprises *IOP Conference Series: Materials Science and Engineering* **177 (1)** 012046
[7] Perseptron 2018 URL: https://ru.wikipedia.org/wiki/
[8] Lomakin N I, Lukyanov G I, Plaksunova T A, Litvinova A V, Loginova E V, Kozlova E A, Skobor E A, Odintsov A E, Zavyalova M N, Korochinskaya S V 2017 Certificate of registration of a computer program **2017618928** Russian Federation. Neural network for assessing the students' competence in the discipline Organization of entrepreneurial activity in the field of education 38.04.01 Economics according to the program Economics of a firm and industry markets *VSTU*
[9] Shamne N L, Shishkina E V 2017 German popular scientific medical online media: structural and functional aspects *Vestnik Volgogradskogo Gosudarstvennogo Universiteta-seriya 2-yazykoznanie* **16 (2)** 143-151
[10] Malushko E, Maletina O, Lizunkov V, Tsybaneva V 2016 Use of virtual learning system for educating students with disabilities and special needs *International Multidisciplinary Scientific Conferences on Social Sciences and Arts* 481-487
[11] Fedotov N G, Moiseev A V, Syemov A A, Lizunkov V G, Kindaev A Y 2017 New Methods of Three-Dimensional Images Recognition Based on Stochastic Geometry and Functional Analysis *IOP Conference Series: Materials Science and Engineering* **177 (1)** 012047
[12] How artificial intelligence replaces living teachers 2018 URL: https://www.rbc.ru/newspaper/2017/09/21/59c25e659a7947f26210ac80
[13] Tim Jones M 2018 A guide for beginners in the fields of artificial intelligence, machine learning and cognitive computing URL: https://developer.ibm.com/articles/cc-beginner-guide-

IOP Conf. Series: Materials Science and Engineering **483** (2019) 012048 doi:10.1088/1757-899X/483/1/012048

machine-learning-ai-cognitive/
[14] IBM open source projects 2018 URL: https://developer.ibm.com/open/code/
[15] Lizunkov V, Politsinskaya E, Malushko E, Kindaev A, Minin M 2018 Population of the world and regions as the principal energy consumer *International journal of energy economics and policy* **8 (3)** 250-257
[16] Awesome Tensor Flow 2018 URL: https://github.com/jtoy/awesome-tensorflow

IOP Conf. Series: Materials Science and Engineering **483** (2019) 012049 doi:10.1088/1757-899X/483/1/012049

Research of layered composites by using subminiature eddy-current probes

A A Bagaev[2], A V Ishkov[2] and V N Malikov[1]

[1]Altai State University, Barnaul
[2]Altai State Agricultural University, Barnaul

E-mail: bagaev710@mail.ru, buvarton@mail.ru, Osys11@gmail.com

Abstract. On the basis of subminiature eddy-current probes, a measuring system is designed that allows scanning of composite materials for detection of small defects. The results of the study of discontinuity defects simulated in metal-polymer layered composites of aluminum-polyethylene-aluminum systems up to 10-layer composite are described. Composites were made by alternating layers of aluminum or copper foil, 20...100 μm thick, with layers of a film, 20 μm thick. Defects in the structure of the material were modeled by skipping or increasing the number of separate layers, and defects in continuity and bridges were modeled by cutting a round or rectangular hole in the foil or dielectric layer. Visual images of model defects are obtained by Fourier-transformer of the IENM-5FA device. The dependence of the eddy current probe signal on the presence of a defect in the layered structure is established.

1. Introduction

The eddy-current method (ECM) together with ultrasonic scanning and fluoroscopy is one of the main methods to inspect defects for the purpose to control various materials and products in modern technology. The method is based on the analysis of an interaction of the electromagnetic field of a special sensor — the eddy-current probe (ECP) — and the object of control. The properties of the investigated object, the nature and topology of the defects detected by ECM, design of such probes and the principles taken as a basis for defects examination change depending on the goals.

At present, metal-polymer laminated composites (MPLC) hold a specific place among objects to non-destructive testing.

Defects inspection of MPLC is performed to determine the following standard defects: defect of the metallic and (or) polymer layer uniformity, the number and thickness of the layers, conducting and non-conducting strips between layers, deformation of the metal layer surface, change in the state of boundary between the metal and polymer layers of the MPLC. One of the parameters sensitive to all the listed defects of MPLC is the local electrical conductivity of the material and its distribution over the surface, associated with the topology of the electromagnetic field in the material interacting with these defects.

According to the literature review, there are many types of inspections to evaluate composites and each of the proposed methods can be implemented in many ways [1-7]. Electromagnetic Testing (ET) methods use magnetism and electricity to detect and evaluate fractures, faults, corrosion or other conditions of materials. ET induces electric currents, magnetic fields, or both inside a test object and observes the electromagnetic response. Electromagnetic (EM) methods include Eddy Current Testing (EC) [8], Remote Field Testing, Magnetic Flux Leakage and Alternating Current Field Measurement. The physics behind each of these methods is different because various electromagnetic fields can be described by different equations. [9].

Nondestructive testing (NDT) measurements of realistic composite specimens have shown that material defects at the object surfaces and inside the material can be detected, e.g. [10]. Inner defects, which are not visible at the surface, can be found by time domain spectroscopy methods [11]. For technical composites, like laminated glass fiber reinforced plastics (GFRP), inner defects of the fiber rovings can be only in the dimensions of below 1 mm [12] but can also reduce the stability of the entire specimen and may decrease its quality and the desired performance and safety targets [13]. Detecting these small material defects inside the woven fiber roving structure itself gets more difficult if the composite is built out of multiple fabric and matrix resin layers.

Earlier, we reported on the successful development and testing of a virtualized measuring instrument - a conductivity meter for non-ferromagnetic materials — CMNFM-5FA.

The subminiature ECP [14, 15] of the original design is used as a sensor in this device, it is made according to a differential scheme of switching on of the coils of a transformer ECT and allowing to localize the control area up to 50 μm^2. Measurement of local electrical conductivity on the surface of a sample under investigation signal of the measuring coil of SMECT in real-time mode makes it possible to use this device as the defectoscope, comparing the data obtained at the object under investigation with the image of the defect previously obtained on the material model.

Timeliness is subject to the need to assess and predict safe operation life of products manufactured using metal-polymer laminated composites.

The purpose of this work was to investigate the possibility of detecting defects such as discontinuity of MPLC based on aluminum and LDPE type A1-(LDPE-Me)p-A1, as well as to visualize such defects by means of CMNFM-5FA.

Materials and methods

Developed a measuring system based on subminiaturized eddy-current probes aimed at examining locally the defects of metal-polymer laminated composites. A subminiature surface eddy-current transducer of the transformer type developed, and the optimum shape and size of the core and number of inductance coils and turns in them were defined, this provides efficient localization of electromagnetic field. On the basis of eddy-current transducer, a software and hardware complex was developed to control the eddy-current transducer (to generate alternating current of various frequencies, to provide its supply to the eddy-current transducer, to receive a useful signal from transducers to ensure its convenient visualization). To conduct the research, the CMNFM-5FA was used, equipped with a wireless hand-held SMECT with a locality of 0.1-0.5 mm^2.

The excitation parameters of ECP: the amperage is 30-35 mA, the voltage is 3.5-4.5 V, the operating frequency is 0.3-300 kHz. Absolute calibration of the device and the communication line with the sensor was carried out on the standard copper sample with the certified conductivity of 415±2 MS/m.

A set of measurements was performed on metal-polymer laminated composites.

MPLC and models of their defects were made by interchanging layers of aluminum foil with thickness of 100 μm and with layers of polyethylene film with thickness of 20 μm. Samples were cut out of these materials with a size of 20x50 mm, a package of the given structure was formed, a foil layer or dielectric with a defect was inserted. Defects in the MPLC structure were simulated by skipping or increasing the number of separate layers.

A superminiature ECP (SMECP) is designed to control the parameters of aluminum alloy products in small areas [16-18]. The defined parameter is the electrical conductivity of the material and its distribution on the surface and thickness of the object of control. The ECP is connected to the Arduino board, controlled by a personal computer with special software developed in C++ for the Windows operating system. The software controls the supply voltage to the exciting coil of the ECP and reads the voltage values (in conventional units) from the measuring coil of the ECP, and then, considering the pre-calibration converts the data into electrical conductivity values.

Information exchange between Arduino and the computer is performed via virtual COM-port. The developed software controls the frequency of the signal on the Arduino board, which generates a sinusoidal signal of a given frequency. The signal passes through a gain-controlled amplifier, thereby achieving the desired signal amplitude. After amplification, the signal enters the exciting coil of the ECP, inducing eddy currents in the object of control. The composite field of the exciting coil and eddy currents induces EMF in the measuring coil, carrying information about the object of control. This EMF is the output signal of the ECP. The output signal is amplified and filtered using a modified Delyann filter combined with a selective signal amplifier, and then is received by the Arduino ADC, where it is converted into a digital signal, which is transmitted to a personal computer and displayed in a convenient form for processing.

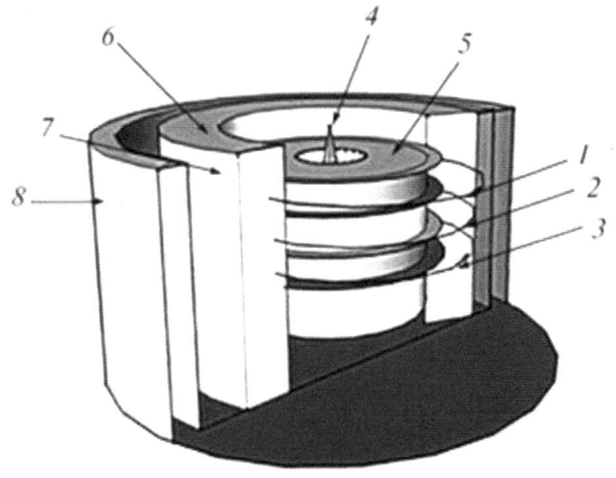

Figure 1. Eddy-current probe

ECP winding coils consist of a copper wire with the thickness of 5 μm. The core is made of ferrite 2000NM3 with an initial magnetic permittivity value of 2000 and has a pyramidic shape. Characteristics of the developed transducer make it possible to achieve high localization of the control. The characteristics of these ECPs allow for the localization of the magnetic field within the areas of 2500 μm^2 and, when using sufficiently low (100-700 Hz) frequencies on the exciting coil, allow to achieve a significant depth of its penetration into the object of control. The software coded in C++ for Windows allows controlling the signal on the energizing winding and receiving the signal from the measuring winding. With the help of the software it is possible to effectively control the signal, which is applied directly to the energizing winding. This software makes it also possible to receive a signal directly from the measuring winding.

2. Experimental results and discussion

The main characteristic of MPLC sensitive to investigated defects is the electrical conductivity of the material of the metal layer.

As it is shown in [18], the contribution to the value of the voltage introduced into the measuring winding of the SMECP with high locality and the ratio of the radii of the measuring and transmitting coils is not less than 0.2-0.4 from each new layer will be from 10 to 25% which is enough to record it by any measuring instrument with an absolute permissible error not exceeding 3.5%.

The measurements carried out by us for model defects of the type of violation of the number and order of layers for multilayer MPLC systems aluminium-polyethylene showed a change in the reference electrical conductivity up to 3-7 mV in the defect area, with an absolute error of 0,1-0.1-0.3 conventional units, i.e. 20-30-time excess of the defect signal over the noise. The appearance of any of these defects in the study area causes a sudden change in the field topology and relative device readings.

The defect may also be represented by amplitude-time dependence of the readings, that has been already implemented in CMNFM-5FA, with manual or automatic scanning of the object's surface. Diagnostics of MPLC continuity defects is possible with direct contact scanning of the sample surface by a sensor. Therein, time is converted into the position of the sensor relative to the starting point of the movement, taking into account the speed of the sensor. Before the measurements were started, the sensor was been calibrated. The measured characteristic was the voltage that is induced by the field of eddy currents that arise in the tested object. In this experiment a sector for calibration was selected on an intentionally flaw-free plate that was manufactured from an identical MPLC. In this case the dependence of the amplitude of the sensor signal on the sensor position in real time, displayed by CMNFM-5FA, can serve as a defect image.

In this case, the main frequency and amplitude of the signal are displayed on the left screen, and the probe coordinate — along the X axis.

The results of the experiment are presented in Fig. 2,3.

Figure 2 shows typical images of some MPLC model defects of layer discontinuity type, obtained with a device, when a sample of MPLC was placed on a dielectric base.

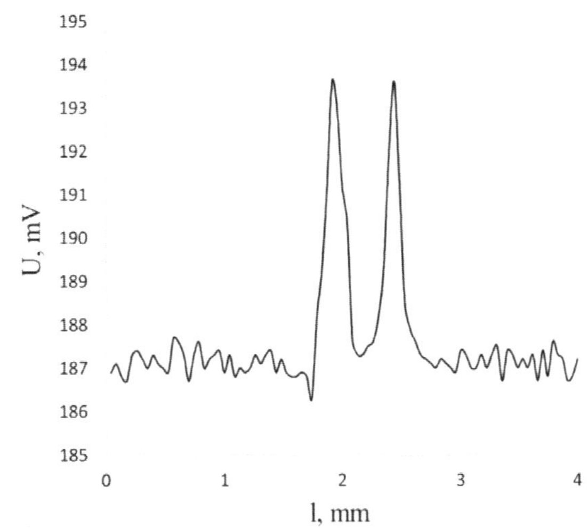

Figure 2. Signal amplitude obtained when scanning a defect of layer discontinuity type, hidden in a 2-layer MPLC

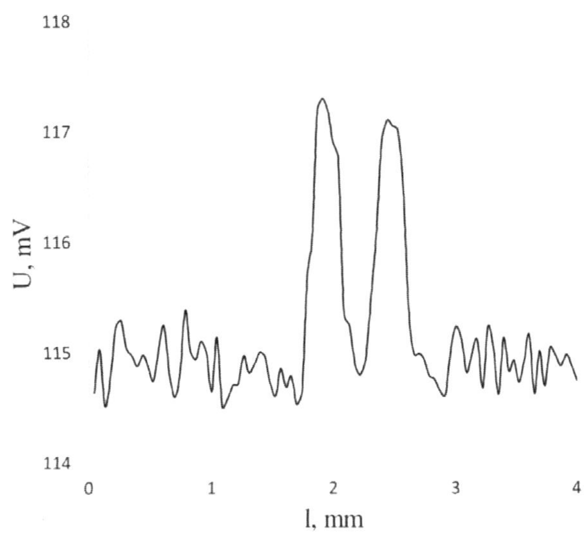

Figure 3. Signal amplitude obtained from the visible defect

It can be seen from Fig. 3 that the signal amplitude obtained when scanning a defect of layer discontinuity type, hidden in a 2-layer MPLC, as per selected control mode identical to that to the signal amplitude obtained from the visible defect (Fig. 2).

When obtaining a signal from such a defect, the device displays the dependence of the amplitude of its signal on the probe position. This allows to perform initial adjustment of the device and provides the operator's training in the diagnosis of defects directly from their visual models, hidden between the layers of MPLC.

3. Conclusion

1. The readings of CMNFM-5FA when measuring the electrical conductivity of MPLC type aluminium-polyethylene using SMECP with locality up to 50 μm^2 do not depend on the gap in the range from 100 to 250 μm.
2. High locality of the sensor and the determined dependence of electrical conductivity of MPLC on model defects make it possible to carry out defects inspection of materials directly as per CMNFM-5FA readings.
3. Visual images of model defects are obtained by Fourier-transformer of the IENM-5FA device. The dependence of the eddy current probe signal on the presence of a defect in the layered structure is established.

References

[1] Katunin A 2015 *Arch. of Civ. Mech. Eng.* **15** 436
[2] StaszewskiW and KostkaP 2009 *Comp. Sci. Tech.* **69**1 678
[3] Katunin A, Dragan K 2015*Comp. structur.* **127**1
[4] Maierhofer C, MyrachP and Steinfurth H 2014 *Comp. Part B: Eng.***57** 35
[5] Loutas T H 2012 *Exp. Sys. with Applic.***39** 8412
[6] Rizos D D and Fassois S 2008 *Mech. Sys.Signal Proc.***22** 315
[7] Loutas T H 2012 *Exp. Sys. with Applic.***39** 8412
[8] Koyama K and Hoshikawa H 2013 *Jour. Press. Vess. Tech.***135**
[9] Yang S H, Kim K B and Oh HG 2013 *NDT & E Int.***57** 45
[10] StoikC 2008 Proc 33rd International Conference on Infrared, Millimeter and Terahertz Waves (Pasadena), vol 1 (Pasadena: IEEE Xplore) p 574
[11] Nezadal M and Schur J 2013 Proc 5th International Symposium on NDT in Aerospace (Singapore) vol 1 (Saarbrücken: Saarland University) p 2145
[12] Vaara Pand LeinonenJ 2012*Technology Survey on NDT of Carbon-fiber Composites* (Kemi:Tornio University of Applied Sciences) p 44
[13] Schultz R 1996*Nichtlineare Struktur- und Versagensanalyse von Faserverbundsystemen mit Mehrschichten-Schalenelementen, Dissertatio* (Bochum: Institut fur konstruktiven Ingenieurbau, Ruhr-Universitat Bochum) p 257
[14] DmitrievS F, KatasonovA O and SagalakovA M 2015 *IOP Conf., Ser.: Mater. Sci. Eng.* **71**
[15] Dmitriev S F, Katasonov A O, Sagalakov A M andShevtsova L I 2015 *Journal of Physics: Conference Series* **643**
[16] Malikov V N and Ishkov A V2016 *AIP Conference Proceedings* **1785**1
[17] Dmitriev S F and Malikov V N 2017*Advances in Intelligent Systems and Computing***692**
[18] Dmitriev S F, Malikov V N, Sagalakov A M 2017 *Welding International* **31** 608

IOP Conf. Series: Materials Science and Engineering **483** (2019) 012050 doi:10.1088/1757-899X/483/1/012050

Museum dialogue as an important component of marketing communication of a brand

Tatiana Melnik[1], Oksana Maletina [2]

1 Volgograd State University, Marketing Department, Volgograd, Russia,
2 Volgograd State University, German and Romanic Philology Department, Volgograd, Russia

melnikti@list.ru, o.maletina@volsu.ru

Abstract. Transformation of marketing communication is firstly determined by such tendencies as decrease of influence of traditional advertising, growth of awareness and consumer independence and creation of impression culture. The response of business to new tendencies is the implementation of new methods of marketing communication with target consumers. The article considers brand museum as a part of brand communication and therefore brand companies can meet consumers' requirements, communicate with consumers and create consumer value, implement brand into their lifestyle. The analysis is carried out with the help of Th. Gad's four-dimensional branding model. According to the analysis we identify typical characteristics which are necessary for successful brand communication: creation of brand space where consumers can get more information, discover and investigate; interactivity as an opportunity to get one's own impression from interaction with a brand (to play a role, to test a product, ask questions (to take a sensible part in museum events)) and share your opinion with others; dialogue relationships with consumers; brand enrichment due to impressions which are not connected with trade.

1. Introduction

Museums are different: some of them are devoted to outstanding personalities, others are devoted to famous brands. Nowadays more and more museums are interactive, they are created by companies which own brands in different fields and spheres of activity. Thus Volkswagen Group opened a museum and park complex Autostadt (Wolfsburg, Germany) where all brands of the concern are exhibited. Heineken created Heineken Experience in Amsterdam, this exhibition centre allows visitors to learn the history of the company and follow the process of brewing interactively and taste beer. We can also name the following museums: Intel (Santa Clara, California, USA), Corning glass museum (Corning, New York, USA), Swarowski (Wattens, Austria), Gucci (Florence, Italy), Christian Dior (Granville, France), virtual Valentino's museum (http://www.valentinogaravanimuseum.com), Burberry World Live (London, UK), Ramen noodle Museum (Yokohama, Japan), Luxottica (Agordo, Italy), La Maison des Fondateurs of Audemars Piguet.

The modern competitive situation on the market requires companies and their brands to create and save effective and mutually beneficial relationships with real and potential clients. As a result, entrepreneurs are constantly looking for ways of contacting with target audience or improving the existing ones. Contact ways are numerous and different situations, places and interfaces of clients' interaction with brand. In other words, every time when consumers contact with brand, real and potential clients take buying decisions which are critically important for business. If the company has more effective contact ways, it means that it expands and strengthens its brand in a consumer consciousness. There are universal contact features almost in all fields and spheres of business: commodity (its name, design and package, layout in a shop, user's guide, guarantee and etc.), service (its name, its presentation, consumers' comments, quality and conformity certificates), advertising, commercial offers, company's documentation, site, groups in social networks, corporate cars, trade personnel of the company (competence, appearance, uniform, smile, posture and speech patterns) and personnel of call-centres, office space decoration and etc. It is necessary to note that corporate culture is an important component of brand marketing communication because namely company employees, who contact with consumers, form the image of any brand [1].

However, the use of contact brand features is certainly an unsuccessful communication strategy because brand-owner company loses an opportunity to compete in a consumer consciousness, to occupy mental space more widely and also to interact with different target audiences. That is why each company tries to find a number of unique long-term contact ways in order to occupy a strong position in consumers' consciousness. Brand museums are unique long-term ways of contacting with consumers.

2. Analysis of brand communication with the help of Th. Gad's branding model

2.1. Four-dimensional brand space model by Th. Gad

Let's analyze practices of marketing brand communication with target audiences in the form of museum with the help of Th. Gad's four-dimensional model. According to P. Feldwick brand is a number of perceptions in a consumer consciousness [2]. Thus brand is a property of a consumer because it is formed in his consciousness. In its turn the place, which it occupies, is a brand space. Th. Gad thinks that brand space is the reflection of company's individuality and it includes the following dimensions in the memory of potential clients:

1. Functional dimension of space means concrete benefit of commodity use, its unique characteristics (for example, specific style, presentation's methods and specific knowledge).

2. Social dimension characterized by social activity determines company's attraction which forms loyalty, expresses how brand supports communication with its potential clients and identifies itself with a certain social group (for example, a number of contacts, clients' comments, creation of brand community).

3. Mental dimension means a brand capability to influence people; to help them to develop personally when they contact with a company; to reflect ideas supported by brand (client training, seminars, books and messages: as a rule, mental dimension is expressed in an advertising slogan).

4. Spiritual dimension is a brand capability to contribute to welfare of each person and to social development (for example, environmental responsibility for future generations). In many cases this dimension is reflected in the mission of organization (sponsorship and charity) [3, 4].

Surely, the division into functional, social, mental and spiritual dimensions is very relative. Their interconnection is explained biologically, psychologically and socially. For example, there is no mentality as a separate psychological process in practice (and thus there is no mental dimension of brand space separated from the rest): mentality needs to get direct impressions from objects of the world (their functional characteristics through different sense organs). In addition, to support its image brand needs to exist not only in one dimension, but it should constantly interact with consumers using as many dimensions as it can. Otherwise, the perception integrity will be broken and brand will disappear in consumers' consciousness, it can be forgotten or has a passive form in people's consciousness. Thus, brand concentrated only on functionality (quality, technical product innovation and etc.) is oriented to failure when its functionality is copied by competitors. Besides, on the overcrowded competitive market functional characteristics have the second position and they concede emotional features which are formed by other three dimensions. However, if brand concentrates on feelings (social and spiritual dimension) or consumer idea (mental dimension), it will be sold rather better than brand concentrated on functionality but during a limited period and it won't make a revolution in its field.

Thus, the division of dimensions is relative and its aim is to describe the conception and content of marketing communication process which reflects the most important features in a concrete situation. This division also allows to analyze the effectiveness of management decisions in the field of brand marketing communication.

2.2. Analysis of BMW's museum as an example of brand communication

We are going to analyze a brand museum with the help of Th. Gad's model. The creation of the museum is aimed at strengthening brand's positions in consumers' consciousness and searching new contacts with them according to four dimensions. For example, BMW Group opened BMW's museum in 1972 (Munich, Germany). Nowadays the museum consists of four thematic zones: BMW's world, BMW's plant, BMW Group Classic. The museum organisation reflects four brand space dimensions of BMW according to Th. Gad's model. BMW's museum includes exhibitions of modern and conceptual cars, motorcycles, aircraft engines, propellers and aviation equipment. Exhibits are aimed to fill the functional dimension of the brand in consumers' consciousness and to attract attention to definite functional features and advantages of the plant's production as a whole. There are several reasons to create the museum zone which is responsible for the brand space in consumers' consciousness:

- creating the controlled image of the brand for a target audience which is more informative than any other instrument of marketing communication (for example, site, brochure or social networks): the museum is the source of information which accumulates facts about activity, ordinary life and unique facts from the history of the company;
- giving opportunities of comparative competitive analysis, demonstration of functional competitive advantages of the production brand (consumers look for definite features assessed in the terms of relative value and choose that commodity from the diversification on the market because they think that these goods are very useful from their point of view (combination of necessary features));
- influencing different sense organs of potential clients (exhibits allow consumers to get information about a product with the help of all sense organs, other formats of brand communication rarely provide such opportunities: consumer can interact with the product actively (see, hear, touch and etc.));
- demonstrating status of a brand owner (experience, professional activity and stability).

As we have noted, the limitation of one functional component of the product is not reasonable for the future brand: functional characteristics of a product make clients compare market offers of different producers; moreover, the more options are offered to study, the more efforts and time are necessary to make a buying decision, and the probability of choosing a competitor's product increases. That is why BMW is looking for new ways to capture social, mental and spiritual dimensions of the brand in consumers' consciousness.

According to social dimension consumers are interested in other people's opinions, and this interest influences their buying decisions because they are eager to make desired impressions on their social environment. To use social space brand-owner companies attract famous persons because target audience can associate itself with these celebrities. By possessing BMW brand person associates himself with other people and cultures. This technique is mostly used and aimed at associating with ideals: companies attract celebrities who are well-known in society; this association provides consumers with a psychological feeling of uniting with their idol. There are different scenarios of celebrities' participation: for example, he can recommend (praise benefits of a product thanks to his personal experience); give permission to use his name or image for advertising a product without being an expert in this field, he can participate as an actor in the advertising and etc.

Filling mental space of BMW brand in consumers' consciousness means to address to his intelligence, and it is aimed at learning, solving problems, analyzing and generalizing, it makes consumers surprise and get interested in novelty, originality, comparative effectiveness, thus promoted product and brand become a part of the field of cognitive and evaluative activity of consumers. As a result, consumers can have intellectual feelings which reflect their attitudes, thoughts, reasons, process and results of their intellectual activity: ability of being surprised is a stimulus of cognitive activity; consumers feel doubt when discrepancy of hypotheses and fact assumptions causes a thorough check of data; work results in feeling of confidence and satisfaction or disappointment and dissatisfaction and etc. In other words, the main task of marketing communication, which is aimed at filling mental space of BMW brand in consumers' consciousness, is provision of information about a promoted product and activation of thinking process, creation of cognitive feedback. In brand communication focused on mental dimension the following features of mental processes can be used:

- reference to experience in memory, knowledge of previous generations (establishing connections and relations between things and phenomena a human relies on influence of these things and relations, his previous experience from his memory, and uses knowledge and experience of previous generations);
- tight connection of thinking and speech (external and internal);
- human's thinking with the help of abstract concepts and symbols.

Communication campaigns of this type have different graduation and are usually followed by offscreen comments to the screen text, and always use such thinking operations as association, comparison, abstraction and concretization, analysis, synthesis, generalization, analogy, reasoning and etc.

The example of activation of a cognitive need is teaser advertising of BMW which intrigues consumers and causes curiosity. Creating an incomplete image in consumers' consciousness teaser advertisement generates a cognitive need which is satisfied by information about benefits of advertised product contained in its main part. The effectiveness of this advertising is determined by data of psychological researches which show that 80 % of all people feel a strong discomfort in cases of incomplete images in their consciousness [5].

Spiritual dimension of brand space means leaving pure commerce and basic consumption. In the example this dimension manifests in the union of BMW with art. Thus, BMW creates art guide (private collection guide), restaurants on the territory of BMW's World, and also slogan 'Fruede am Fahren' which attracts attention of potential consumers not to functional benefits of a product but to pleasure of driving the definite car. BMW museum sends a message to external audience thus referring to spiritual space of the brand. For example, the use of modern materials and technologies in the exposition shows that all modern technologies are used in production process of the plant. Art object becomes a metaphor of technology showing how it works under unusual conditions. Charity is one more way to use spiritual component of brand space [6]. Examples of BMW advertising charity in Russia are different social projects aimed at improving life of children (Easter party for children from orphan asylums and etc.).

Thus BMW uses four dimensions of Th. Gad's model which allows this brand to emphasize its creative character and provide their clients with new impressions that are not connected with such definitions as good product or good service. Moreover, initiatives in the sphere of art help BMW to create an emotional connection and have a dialogue with the whole audience discussing interesting themes. BMW museum provides facts from the history and helps to explain and prove the importance of the enterprise activity for a country and for a person. The attitude to the history allows to analyze projects and developments of the company over time.

3. Marketing space in brand museums

The tendency of creating special zones of communication with brands in the form of the museum is being used more actively nowadays. Their main task is to create brand communication between a brand-owner company and its potential consumers. This museum is a completely special type of marketing communication which means dialogue with a client. Visiting a museum opens the channel of communication which is responsible for process of learning information and its memorization. This communicative technique reflects observer and there is an effect of mirror where each visitor of a museum can reflex his own problems and can get a new internal experience creating a dialogue model of communication.

Brand is becoming a social and cultural phenomenon which interacts with society interactively. It gets features of culture, thus spiritual dimension of brand help consumers to find their role and identify their mission in society, take responsibility [7]. Undoubtedly, the formation of this model of communication demands efforts to create conditions which provide similar personal discoveries. Such efforts include offering another type of spending time, different from visiting traditional museums.

Brand museums offer breathtaking stories about past and future of their brands thanks to logic of their existence and development. The logic of museum exposition can be based on owner's autobiography which can be interesting for audience. Brand museum tells real and potential clients its legend in form of emotional dialogue which is the basis of other kinds of communication. This legend contains important message of brand: how and when it appeared.

As a rule, museum collections include exhibits connected with its legend, and these exhibits can be seen, tasted, touched, heard or smelled. Emotions and empirical feelings are additional non-material characteristics of brand museum. Services of such museums are specific: they can be perceived, felt, created and clients can participate because educational and entertaining elements are an integral part of brand. Exhibits in such museums are the following: packages, labels, advertising brochures, costumes from advertisements and etc. Brand museums offer creative communication and consumer experience interacting with brand. If we compare brand museums with traditional museums, we can notice the greatest difference between them, because in brand museums you can touch everything: children and adults push buttons, pull the rope, look into windows, smell, taste and etc. As a result, impressions from a museum visit are greater than from watching a commercial [8].

Brand museum is like a huge children entreating room where you can shout, be awkward, laugh, meet your favourite characters and overeat sweets. Thus consumers in the atmosphere of holiday and fun become less critical and more trustful easily. Consumers are moving towards the choice of pleasure and game and there is even a term 'kidult' when adults strive to become children. Brand museums allow to focus marketing communication on the internal world of a consumer, to use children's perception of the world and thinking, to add fantasy and illusion to the space of reality. In this atmosphere brand penetrates easily into the consumers' consciousness. Brand-owner companies with the help of museum space create a material world turning their brands from a number of associations and images into the objective reality which has a structure, rules of organization understood by every visitor of a museum. All similar museums are united by the theme of a definite brand and connected with a definite product category: chocolate, children's constructor, car and etc. Each museum is divided into thematic zones showing stories: each zone presents a definite product (chocolate or car). Visitors can feel the atmosphere of a brand thanks to designers and artists who create a reality of the history of a company. Consumers' emotions involve them into the atmosphere of pleasure and entertainment and subsequently they are associated with brand provoking emotional connections [9]. Marketing communication with potential consumers is subjective, thanks to positive emotions visitors feel satisfied and happy and they are grateful to brand museum where they manage to get such feelings. Brand strives to make consumers' dreams come true (for example, Christmas magic world of Coca-Cola), to create unforgettable impressions and emotions in the process of communication and thus to contribute to stable emotional connection of brand with the target audience. Brand offers their guests to satisfy their needs (dreams, emotions) through goods which can subsequently become elements of consumers' lifestyles. In a museum brand uses not only material but psychological, emotional and social needs of people. Consumers strive to satisfy their emotional and aesthetic needs in the process of brand communication. As a result brand in consumers' consciousness begins to be a combination of important values and emotions, and it is able to create and strengthen specific relationships with clients. Loyalty to brand is created gradually through definite emotional feelings and emotional experience from interaction with brand [10].

That is why brand museums try to increase the probability of rich and various museum feelings by offering the diversity of interaction opportunities, different events for museum visitors of different ages.

4. Conclusion

Thus, the result of this interaction is the following: consumers change their attitude to brand and to company, they begin to perceive brand as a living creature with its character, individual and unique features. Associations with a personality help brands to provide a desired differentiation on the competitive market. Brand becomes a part of people's life offering definite emotional benefits from possessing it: bright and impressive moment of life. Afterwards the created emotional connection between a brand and a consumer results in making decisions about purchases due to emotional feelings and experience, and it decreases influence of rational thoughts based on logic. Hence, the form of museum communication allows brands:

- to provide recognition;
- to guarantee emotional connection with a consumer;
- to create psychological benefits from contact with him;
- to provide great involvement of consumers;
- to form and increase loyalty;
- to distance from competitors.

Brand museums have all prerequisites to become leading in modern marketing communication processes of company. The object environment of these museums with their specific atmosphere of brand worship is simultaneously space, condition and reason for brand communication. Here visitors-consumers have time to think about brand, to talk about it, to speak with their family members and museum visitors about brand. Museum exposition provokes communication including virtual talks because it contains special photo zones that make consumers share their impressions with their friends, colleagues in social networks in the Internet.

Finally, brand museum helps to improve communication, to inform consumers about product and its producer. Brand museums are aimed to create relationships and contacts with governments, local community and target audiences. Ideally, in brand museums visitors begin to share brand values and beliefs, and they lead to further promotion of brand.

References

[1] Meethan K 2001*Tourism in Global Society: place, culture, consumption* (NY, Palgrave) p 117
[2] Feldwick P *Defining a Brand* URL:http:// paulfeldwick.com/selected-publications/ bwilson (accessed on 4th November 2018)
[3] Gad Th 2002 *4-D Branding: Cracking the Corporate Code of the Network Economy* (London, Financial Times-Prentice Hall) 230 p
[4] Gad Th 2001 *4D Branding: Cracking the Corporate Code of the Network Economy* (Saint-Petersburg, Stockholm school of Economy in Saint-Petersburg) 230 p
[5] Webster K *Branding the Non-Profit* URL:http:// www.marketingprofs.com/2/kwebster1.asp. (accessed on 14th September 2018)
[6] Levinson J 1985 *Guerilla Marketing: Secrets for Making Big Profits from Your Small Business* (Boston, Houghton Mifflin Harcourt) 226 p
[7] Schmitt B 1999 *Experiential marketing. How to Get Customers to sense, feel, think, act, and relate to Your Company and Brands* (London, Financial Times-Prentice Hall) 250 p
[8] Pine J and Gilmore H 1999 *Experience economy* (Boston, Massachusetts, Harvard Business school press) p 169
[9] Schmitt B and Rogers D 2004 *There's No Business that's Not Show Business: Marketing in an Experience Culture* (London, Financial Times-Prentice Hall) 271 p
[10] Kotler N and Kotler Ph 1998 *Museum Strategy and Marketing* (San Francisco, Jossey-Bass Publishers) p 39

IOP Conf. Series: Materials Science and Engineering **483** (2019) 012051 doi:10.1088/1757-899X/483/1/012051

Foreign language competence in manager professional structure and training

E V Meshcheryakova, E A Loktyshina, J V Meshcheryakova

Volgograd State Social Pedagogical University, 27, Lenina Av., 400013, Russia

E-mail: elenam7@yandex.ru

Abstract. The article is devoted to the problems of professional and foreign language training for manager staff. The authors study the importance of communicative component as an element of professional readiness, identify inefficiency causes of the existing system of language training for future managers, reveal the nature of professional and language experience and prove the need for their joint development in the process of quasi-professional activity as a condition of manager professional development. The approach to the content construction of teaching foreign languages in the higher professional education structure on the basis of language experience integration in professional activity is being revealed.

1. Introduction

Russian economy progressive movement and our country integration into the world education space serve as an international cooperation and partnership incentive in the field of personnel management, which is reflected in the growing need for highly qualified specialists who speak a foreign language at a productive level, that is, capable of conducting business foreign language communication and having socio-cultural knowledge.

The current economic situation in Russia allows managers to cooperate directly with foreign partners at home and abroad, to make deals, to expand markets, to participate in international exhibitions, presentations, conferences and congresses.

Meanwhile, studies in the field of professional linguistic didactics show that Russian managers are generally not ready for professional communication with foreign partners. Modern understanding of the culture of a leader is limited today only by emphasizing the importance of professional and General humanitarian competence, breadth of horizons and knowledge of basic ethical norms, leaving without due attention the importance of communicative training, foreign language skills, which is especially important for specialists of management profile, for which foreign language competence is an integral part of professional competence.

The discussed problem is also related to the fact that the globalization process taking place around the world has led to multi-national enterprises, where successful management decisions depends on the ability of managers to operate effectively in a multilingual environment. *The activity of large multinational companies in our country* is characterized by an intensive exchange of knowledge, ideas, human resources and products between the parent and subsidiary enterprises. An important feature of modern business is various European languages involving in the economic activity structure of the enterprise at the level of not only management decisions, but also over their execution control.

Professional foreign language competence of management personnel affects both its position in a company and regional labor market as a whole, and the language policy pursued by the company, communicative climate, system of language training, and language policy of the region play a decisive role in these globalization processes.

2. Methodology

The issues of specialist training for successful functioning in multipurpose professional activities designed in modern society are to be engaged in higher professional education. Researchers in the

IOP Conf. Series: Materials Science and Engineering **483** (2019) 012051 doi:10.1088/1757-899X/483/1/012051

field of vocational education note that the main purpose of vocational education is not only to acquire skills and knowledge, but also to find an answer to the question of how knowledge and skills are combined with personal qualities when it comes to choosing a particular job. Being "suitable" for a job is determined by social and cultural factors, and in this case vocational education – from employment to free entry into the market – plays a powerful role in this process, being an intermediary between the class-gender component and the requirements of the labor market [1].

However, if the authors agree on the role of vocational education, they have different views on its place in the training system. According to some authors, vocational education and training can be seen as the development and application of knowledge and skills for middle-level jobs needed by society from time to time. This definition is based on the concept of applied knowledge, as in the definition of Aristotle's art, or applied science, which requires clarification. This definition also places vocational education and training ***at the professional level in the middle rank***, thus translating it into an applied hierarchy and economic structure [2].

In addition, educational policies should focus on understanding vocational education and training, young people needs and their success, and not only on responding to current and future labour market needs. In this regard, we need to study more young people trajectories who receive or have already received professional education [3].

Problems of training specialists in non-linguistic areas to solve professional problems on the basis of foreign language proficiency, as one of the goals of professional education, are in the sphere of interests of theorists and practitioners in the field of language education. A theoretical model of foreign language communicative competence, revealing the psychological and pedagogical mechanisms of its development are presented in works of L. Bachmann, I. L. Bim, N. D. Galscova, T. Dudley-Evans, M. Channels, J. Manby, R. P. Milrud, A. M. Mitina, M. Swaine, D. Himse, etc.). Attempts of practical implementation of these models are reflected in the concept of "Language for special purposes" and its variant "Subject and language integrated learning", as well as in the writings of A. P. Minyar-Beloruchev, P. Strevens, P. V. Sysoev, S. G. Ter-Minasova, A. Waters, E. Frendo, T. Hutchinson, and others.

The purpose of vocational education today is in the specialist training for whom a foreign language is one of the main tools of professional activity and means of personal development.

3. Professional foreign language competence as structure component of specialist's readiness for professional activity

3.1 Approaches to the organization of language training in the higher professional education system

Approaches to the implementation of the language component in the higher education systems of developed countries in the management personnel training are guided by international requirements and standards and, at the same time, bear the imprint of national traditions, take into account the factors of the language environment, political, economic, geographical, ethno-cultural and other conditions of the country. The greatest effect, as can be assumed, is achieved where language learning, multilingualism become an integral part of a person's lifestyle from early childhood, and education expands and conceptualizes this life situation. Language experience is organically included in the educational, educational, project, search and research, organizational and communication activities. Language learning acts as a necessary component of project activities, serves as a prerequisite for achieving its goals. The demand for language by the quasi-professional process changes the motivation and psychological mechanism of language acquisition, reduces the barrier to its study. In this regard, the training of a specialist who speaks a foreign language is considered as a new type of professional education, dictating changes in the strategies of language education and training strategies.

3.2 Language education structure in vocational training

The need of manager personnel in language training in modern conditions of globalization is considered as an aspect of the specialist personality formation. Researchers of pedagogy and linguistics problems proceed from the fact that language education is education in all modern (native and non-native) languages and cultures [4], that language education should be considered from the position of a holistic approach as: 1) value, 2) process, 3) result, 4) system [5].

3.2.1 Language education as a value. Taking as a fundamental point of view of S. Guershunsky, we can conclude that the value of language education at the higher professional school, which has increased over the past decade, is determined by the demand for foreign languages as a result of globalization of the integration processes of society, the introduction of higher professional education and the development of joint educational programs at the interstate level, the opened up opportunities of familiarization with the world experience and knowledge through the use of new information technologies.

In addition to the state and public importance of language education is also undeniable personal value for a future manager, which is due to: a conscious need for a foreign language as a means of communication in the modern multicultural world; the requirements of a competitive professional environment to the level of its language training, and the degree of compliance with these requirements; the desire for continuous self-improvement, using the advantages of foreign language proficiency, which determines for a particular person the prestige of language and language education.

That is, the need for language education as a value is determined by socio-economic, socio-cultural, psychological and personal factors. *Socio-economic factors* include the prestige of language skills; social need for language as a means of communication; competitiveness in the labor market; the possibility of real use of language in management; mobility; traditions in the study and choice of language. Under the *socio-cultural factors* are proximity and community cultures; attitude to people, culture and society; the system of cultural and social values, accepted in society; an interest of a different ethnic culture; the desire to find compromises in dealing with controversial issues, and others. *The psychological factors* reflect the need of the individual in the communication; ownership of the mechanisms of speech-thinking activity; the psychology and ethics of business communication. T*he personal factors* refer to the individual abilities of the individual; its subjective position; desires, intentions, interests, plans, level of development, motivation in the study of languages, etc.

3.2.2 Language education as a process. Language education as a process of mastering a foreign language culture is accompanied by the formation of the ability to conduct intercultural communication in the field of management decision-making, which involves adequate interaction in everyday and professional situations with speakers of different languages and cultures with the mismatch of national stereotypes and communicative behavior. The guarantee of the effectiveness of such interaction is educational technologies, which provide the construction of the subject-subject activity of the teacher and students, aimed at strengthening the motivation in the study of languages and mastering the basics of the future profession, awareness of their ability to real communication and responsibility for the results of this process.

A necessary condition for the effectiveness of this process is to rely on the real experience of students, which they acquire in the process of mastering special disciplines. This experience is refracted to the situations of intercultural communication, where there is its inclusion in the language experience and thus maintained an interest in the continuous improvement of their foreign language training. That is, language education as a process is designed to create such pedagogical conditions that each student has the opportunity to receive high-quality linguistic and cultural training, which would allow him to improve his previously acquired foreign language and cultural experience throughout his life.

3.2.3 Language education as a result. The result of language education is the achievement of a certain level of language proficiency, which is understood as a set of competencies that allow a specialist to speak, write, read and understand the foreign language. That is, knowledge of a foreign language implies the possibility of self-realization of the individual through new forms of expression, which are mastered in the process of mastering the language. Moreover, knowledge of a foreign language is, first of all, introduction to a specific discourse marked by a certain professional environment and national and cultural characteristics of the speakers of this linguistic culture. And finally, mastering the language, a person learns the complex of some extra-linguistic socio-cultural knowledge, skills and abilities that characterize a particular foreign student as representative of a certain linguistic and ethnic society [5].

3.2.4 Language education as a system. Language education as a system is aimed at creating psychological and pedagogical conditions for the development and realization of the creative potential of all its subjects through mastering a different linguistic culture in the process of intercultural interaction and knowledge.

To achieve a positive result, it is very important that language learning is not just in-depth, but has acquired a professional status.

The system of professional training should provide an opportunity to integrate professional experience into the content of foreign language education through the creation of situations where success in the professional and language fields are organically interrelated, and this requires a long period of time sufficient for the development of both language and professional experience. The need to find new approaches to the formation of foreign language competence of management personnel is also due to the fact that the use of translation work is economically inefficient, since the translator is usually not a specialist in a particular field.

The system of language education in higher education is a relatively autonomous subsystem of a single language macro system, in the center of which there are people who determine its purposeful functioning and development. The whole system works to achieve specific goals, where the result is a system-forming factor that ensures the ordering of all components of the system. Thus, **language education in higher education can be considered as a *process* of forming the ability to intercultural communication in a valuable professional context as a result of the integration of language experience in the structure of professional activity through the system of future manager professional training.**

4. Professional activity and language experience model

Speaking about the system of higher professional education, it should be recognized the need for a different approach to the design of foreign language education in the field of personnel management, which involves a revision of the goals and content of language education with a focus on building a **multidimensional model of professional activity with included language experience**. Professional activity is understood as a set of four components: functional (professional activity as a system of tasks solved by a specialist); competence (professional activity as a set of competencies); cognitive (professional activity as an integral part of knowledge, as a result of functioning of two factors: internal (i.e. activity of consciousness) and external (professional culture); personal (value-semantic, professional activity as a system of motives and meanings of activity).

4.1 Functional component in the professional activity structure

The system of tasks solved by a specialist as a **functional** component in the structure of professional activity reflects the nature of this activity and should be adequate to it. The psychological condition for the implementation of this system is the reliance on the relevant type of experience [6]. The inclusion of language experience in the process of solving professional problems should ensure their more effective implementation, the achievement of goals in situations of intercultural communication.

Training of specialists in the field of management is to form the experience of communicative actions in the system of professional tasks that would allow professional contacts with the use of a foreign language in various fields and situations, to function as a powerful linguistic personality in professionally significant speech events of different types.

4.2 Competence component in the professional activity structure

Competence component essence in the structure of professional activity with included language experience is that language experience is manifested not in speaking about their profession in a foreign language, and to manage this activity on the basis of a set of competencies to ensure its successful implementation.

Analysis of existing research within the competence approach allows identifying three key competencies of a specialist with experience in foreign language activities: linguistic, discursive and intercultural.

Linguistic competence involves the acquisition of a certain amount of formal knowledge and relevant skills related to various aspects of the language, for the construction of grammatical forms and syntactic constructions that will allow carrying out their communicative intentions in situations of professional communication in accordance with the level of language proficiency.

If the presence of linguistic competence indicates the formation of ideas about the language and tends to focus on the language outside the context, *discursive* competence reflects the willingness to use the language, taking into account the professional specifics of the management environment. The key point in determining the degree of formation of discursive competence is the readiness for oral and written interaction within the professional context using relevant strategies in each specific discourse in the appropriate genres, styles and registers.

Intercultural competence is considered to be an integral part of foreign language communicative competence. A modern manager should be ready to carry out professional activities in accordance with the requirements of intercultural communication, which is interpreted as a set of specific processes of interaction of communication partners belonging to different linguistic and cultural communities [4]. Intercultural competence is determined by a variety of factors, including the environment, behaviors, values, age, ethical standards. Forms of interaction between people are constantly changing, and the manager must be ready for an adequate response to these changes. National culture, national identity, organizational, corporate, professional culture, stereotypes – all these concepts are used to categorize different types of cultures and people in them. Knowledge of their characteristics is necessary to identify and analyze the problem in order to adapt to a new situation or form of behavior.

Intercultural competence therefore includes the ability to interact with a foreign-language ethnic environment and a constant willingness to change their behavior in accordance with the parameters of this environment through the recognition and use of cultural differences to create a favorable business environment.

4.3 Cognitive component in the structure of professional activity

The **cognitive** component of professional activity is related to such categories as knowledge, thinking and understanding processes involved in the performance of professional duties. The essence of this component is that language experience as a component of readiness for professional activity reflects the interaction between psychological, communicative, functional and cultural factors. This is expressed in the ability of flexible use of various types of information, the ability to effectively construct mental representations in the process of activity, based on its foreign-language potential. There is an introduction to the world culture, to the culture of the peoples of the partner countries in business, an understanding of their lifestyle. A specialist who speaks a foreign language and is able to use it in the performance of functional duties, has the opportunity to learn the spiritual heritage of countries and peoples, their historical and cultural memory, to find more effective ways to achieve cultural understanding.

The synthesis of knowledge about the specifics of native culture and foreign culture, common knowledge about cultures and communication allows you to learn rational techniques of communicative activities, creates the desire and willingness of a specialist to engage in interpersonal interaction, determines his attitude to the partner in communication, which ultimately determines the quality and results of the joint activities are an indicator of the level of formation of professional culture.

4.4 Personal component in the professional activity structure

Personal (value-semantic) component of professional activity, as a system of motives and meanings of activity, reflects the degree of transfer of the individual to the level of self-development, self-realization, when realizing his subjectivity, a person seeks to bring new to the culture of professional activity, transforming it and him. The effectiveness of this activity at the same time increases immeasurably due to the fact that its content now acquires a qualitatively new personal meaning, acts as a meaning and value for the individual, becomes a source of personal experience.

5. The purpose of language education in vocational training

Therefore, the purpose of language education in the system of management reserve training is the formation of foreign language professional competence of the future manager in the unity of all its components through the development of professional and language experience in situations of intercultural interaction, and the integration of language experience in the structure of professional activity, which implies:

- assignment of experience of foreign language activities, which is implemented through a system of tasks aimed at providing future managers with the opportunity to master a foreign language at a level that allows them to solve professional issues in the chosen field, to successfully operate in their environment, taking into account the specifics of intercultural communication (functional aspect);
- formation of a system of competencies that involves knowledge and knowledge of the main aspects of modern professional communication, practical knowledge of language tools, the ability to present the content side of their professional field with these tools, analysis and assessment of real business communication, decision-making in relation to their speech behavior, finding more effective ways to achieve cultural understanding (competence aspect);
- ability and readiness to cognize the surrounding reality through the use of language means, introduction to the world culture, knowledge of the spiritual heritage of countries and peoples, possession of the culture of professional activity (cognitive aspect);
- perception of language as a value that has a personal meaning, and the specialist's desire to subjective analysis of their communicative potential, control their speech and actions of their business partners and to change their speech behavior if necessary (personal aspect).

6. Language education in vocational training

Implementation of the goals is carried out, first of all, through the **content** of education, which is designed to provide: an appropriate world level of general and professional culture; integration of the individual into the national and world culture.

The content of any education is culture; the content of foreign language education is foreign language culture. The foreign language speaking implies communication in a foreign language in the space of a foreign linguistic culture, as well as the use of language to achieve specific communicative purposes [6]. Culture is understood today as a generalized civilized space, i.e. as a product of human thought and activity. This is an experience and norms that define and regulate human life, people's attitude to new and different ideas, ideological systems and social forms. Speaking about foreign language education in higher education, culture should be understood as a holistic professional and speech experience, which should be mastered by a specialist, and which can be represented as a holistic system of professional tasks, for which it is necessary to know a foreign language. The criterion of the integrity of such experience is the degree of professional activities effectiveness, the ability to solve these problems, the achievement of goals.

Considering the structure of professional activity and the goals of foreign language education in higher education, it can be assumed that the content of higher professional language education consists of the gradual assimilation of **professional** experience, which occurs in the process of mastering the subject-activity side of special disciplines, and **language** experience, acquired on the basis of inter-subject interaction in situations of intercultural communication in the form of reflection, creativity and self-development, and, moreover, the personality of a specialist acts as a product of integration of professional and language experience.

If we interpret the content of education as a certain set of experience types that students should master, the development of this content can be represented as a continuous complication of the composition and structure of the types of experience included in it, which is due to the continuous development of all spheres of social culture [7].

Underestimation of the role of professional experience in language education is the cause of inefficiency of the existing system of education at the University. Professional experience is the experience of performing professional functions, the content of which can be presented in different ways, depending on the professional environment in which the implementation of future activities is planned. The specifics of professional experience as a component in the content of language education are that it simultaneously has a purely subject (a certain scope of application of this experience) and non-subject (universal strategies and types of communication, regardless of profession) aspects. Language experience, therefore, cannot be considered autonomous in relation to the subject matter of academic disciplines. It is characterized by specific ways of development, involving the entry of the subject in a professionally-oriented educational situation, and an auxiliary role in relation to the professional component in the content of education.

Mastering this qualitatively new competence cannot be achieved within the traditional discipline "Foreign Language". It requires the creation of qualitatively new practical training modules that ensure the formation of professional experience in solving communication problems on the basis of orientation in professional and language situations [8].

Problems also arise in the clash of interests of professional and educational bodies in organizing out-of-class training which makes educational process practically oriented. In general, the idea of the representatives of the labor market according to the forms of cooperation, venue, duration, and the criteria for assessing the effectiveness of this training remain unchanged [9].

Nevertheless, all the scholars agree on the need to coordinate efforts to develop common approaches and requirements to vocational training. As it is pointed out, universities, governmental organizations and employers' associations need to get together to create a uniform set of short, straightforward and easily memorable definitions of key skills in order to facilitate shared understanding. Consensus in this regard on the parts of universities, government organizations and employers can only be to the benefit of future graduates [10].

7. Conclusion

Therefore, professional-foreign language competence as the purpose of Manager training, assumes its ability to adoption of administrative decisions on the basis of implementation of foreign language communicative competence in situations when the solution of professional tasks at the level of the world standards of quality of professional activity is impossible without the foreign language opening access to the world information bases.

This competence is a new type of education, including the development of integrated professional and language experience, which is a system of subject knowledge, knowledge of ways of functioning in a professional environment, awareness of their own value as a subject of this environment on the basis of mastering a foreign language as a tool of professional activity, an integral part of the culture of modern management.

References

[1] Colley H, James D, Diment K and Tedder M 2003 *Journal of Vocational Education & Training* **4(55)** 471-98
[2] Moodie G 2002 *Journal of Vocational Education & Training* **2(54)** 249-66
[3] Molgat M, Deschenaux F and LeBlanc P 2011 *Journal of Vocational Education & Training* **4(63)** 505-24

[4] Galskova N D, Gez N A 2004 *Theory of foreign language teaching: Didactics and methodology: Textbook for linguistic students* (Moscow: Publishing Center Academia)
[5] Guershunsky B S 1997 *Philosophy of education for the XXI century (in search of practice-oriented educational concepts)* p 38
[6] Meshcheryakova E V, Loktyushina E A, Meshcheryakova Ju V 2016 *Russian Linguistic Bulletin* **2(6)** 5
[7] Serikov V V 2008 *Competence approach in higher professional education: theory, methodology, technology* (Moscow: Publishing house NOU VTO "SVGA") p 30
[8] Serikov V V, Loktyushina E A and Pichugina V K 2014 *International Education Studies* **7(12)**
[9] Akomaning E, Voogt J M and Pieters J M 2011 *Journal of Vocational Education & Training* **4(63)** 575-92
[10] Bennett R 2002 *Journal of Vocational Education & Training* **4(54)** 457-76

The means of cognitive information conveyance in technical translation

A Milostivaya[1], I Makhova[2], O Tchudnova[2], A Sidelnikova1 and A Simonyan[3]

[1]Department of Translation Studies, North-Caucasus Federal University, 1 Pushkin Street, 355009 Stavropol, Russian Federation
[2]Department of Foreign Languages, Stavropol State Agrarian University, 12 Zootechnicheskiy Ln, 355017 Stavropol, Russian Federation
[3]Department of Theory of Language and Cross-Cultural Communication, Russian-Armenian University, 123 Hovsep Emin Street, 0051 Yerevan, Armenia

xyscha@mail.ru

Abstract. The purpose of this article is to review various approaches to translation of cognitive information in English technical text into Russian. The focus of this study is to specify the equivalent translation strategies used to convey English technical terms and abbreviations being the source of cognitive information. The analysis of the empirical material, i.e., the original and translation of the English photographic equipment manual, revealed that, possible equivalents, equivalent counterparts and such translational transformations as omission of redundant elements, lexeme addition, generalization, semantic modulation and complete transformation are predominant choices. We have described the following two basic strategies for translating abbreviations. The first one is applied to common, frequently used technical terms and actualizes the reference to the dictionary equivalents; the second strategy is applied to terms specific for the analyzed photo printer manual and results in the use of the source text words.

1. Introduction

Translation plays an important role in the cultural development of mankind. Through translation, people in one country learn about life, daily routines, history, literature and scientific achievements of people in other countries. The expansion of economic and trade relations between nations as well as the development of scientific and technological progress contributes to the increase of technical texts ratio among total volume of translated materials. This results in greater attention from the modern translation studies community to the problems of technical translation [1; 2; 3; 4]. This fact also confirms the relevance of this article.

In addition, modern linguistic concepts recognize the cognitive aspect as basic and essential in the scientific text structure and semantics. This fact fits perfectly into modern linguistics and translation studies.

2. Material and Methods

The research material is the original and translation of the English manual for photographic equipment (Konica Minolta Photo Imaging, inc. No. 1 Sakura-machi, Hino-shi, Tokyo 191-8511, Japan). The conclusions material is examples card file, extracted from the text and containing 175 terms and 67 abbreviations.

The primary research method is comparative analysis of the source and target texts aimed to establish equivalent counterparts. Also, we analyse dictionary definitions to evaluate correctness of terminology and abbreviations translation. In addition to it, we apply contextual analysis to study lexical transformations in terms of achieving equivalence when dealing with translation of the English terminological word-groups. Finally, we employ general scientific analysis, synthesis and description methods.

3. Results and Discussion

The dominant stylistic feature of any technical text is an accurate and clear material presentation. Expressive means with emotional coloring are nearly absent, whereas the main emphasis is on the logical rather than the emotional side of the subject [5; 6; 7]. Hence the importance of cognitive information for a scientific text [8; 9; 10]. Characteristic features of the technical style include meaningfulness, logic (strict consequentiality, clear coherence between the basic idea and the details), accuracy, objectivity, clarity and apprehensibility. Various technical texts may possess these features to a greater or lesser extent, yet they all contain linguistic means aimed at providing professional communication between the members of scientific and engineering communities. In the lexical aspect, it involves the use of technical-scientific terminology and special lexis.

As a result of the technical text study, we define its main stylistic feature as accurate and precise material presentation without any expressive elements that add emotional coloring to the utterance. Metaphors, metonymic transpositions and other figures of speech, widely used in fictional works, are seldom found in scholarly literature. Being stylistically antipathic to common language, technical texts still may contain a number of "colour-free" technical phraseological combinations. The scientific-and-technological translation should meet the following requirements: accuracy (correct representation of all thesis in the source text); conciseness (concise and succinct conveying); clarity (compactness of the target text should not hinder its comprehension); literariness (translated text must comply with conventional norms of the target language).

In this research, we found upon modern translation theorist G D Voskoboynik and his scientific work "Linguophilosophical Foundations of the General Cognitive Translation Theory". Applying cognitive approach, he described how a distinctive text type influences the equivalence / adequacy ratio in translation. Thus, we regard equivalence as conveying cognitive information by contrast with adequacy, which conveys emotional and evaluative information: "The equivalence of translation is typical for texts with the communicative purpose of changing the addressee's practical activities. The scientific and technical texts, device manuals etc. are exemplary for the "Actions" realm. The concept of adequacy, on the other hand, exists in an axiological context, therefore translators of poetry and fiction in general refer to adequacy" [11]. Translated technical text should correctly convey the meaning in a form that as much as possible close to the initial one. Any deviations have to be justified by peculiarities of the Russian language as well as by style requirements. The translation should not be neither word-for-word nor a loose adaptation, although elements of both are inevitably present. To make the Russian technical text equivalent to its English original, it is essential to convey the original cognitive information, presented in the following parameters:

- translation of terms;
- translation of abbreviations.

As the basic definition of a term, we adopt the one suggested by L M Alexeyeva: "A term is a word or a phrase that denotes a concept (object, phenomenon, property, relation, process) and is specific to a given branch of knowledge, technology, art or public life" [12]. The material analysis revealed that the preponderance of the technical text vocabulary are terms with all their typical attributes: unambiguity, neutrality and context insensitivity.

According to our research, possible, i.e., optional equivalents predominate in term translation. Thus, within each particular word meaning, one of the regular term translation techniques is applicable:

*Press the yes **key**. - Nazhmite **klavishu** "yes".*

*Are the **rollers** stained or damaged? – Est 'li zagryazneniya i povrezhdeniya **valikov**.*

*Be sure to clean the magnetic **head** after completing the daily printing work of 240 **film**. – Obyazatel'no ochistite magnitnuyu **golovku** posle ezhednevnoj obrabotki 240 **plenok**.*

*Remove any **play**. – Dvigajte sektsiyu vpravo i vlevo dlya ustraneniya **lyufta**.*

To convey cognitive information represented by terms in English manuals, we often use calquing, i.e., morpheme-for-morpheme or word-for-word translation. Calquing is sometimes accompanied by lexical extension, which is decomposition of the original complex word into constituents, i.e., individual words in the target language:

Set the 135 ANS on the ***turntable****. – Ustanovite 135 ANS na* ***povorotnyj stol****.*

In most cases, the number of lexemes involved in calquing remains identical in both the source and the target language. However, parts of speech alter in the target language. Thus, English attributive nouns are replaced with attributive adjectives in Russian:

Raise the mirror cover and draw out the ***mirror tunnel****. – Podnimite kryshku zerkala i vytashhite* ***zerkal'nyj tunnel '****.*

Set the ***calibration plate*** *on the densitometer. – Ustanovite v densitometr* ***kalibrovochnuyu plastinu****.*

In translation of terminological phrase components, the cases of the part-of-speech altering are frequent. As a rule, when parts of speech remain unaltered, the change of word order occurs:

Cleaning the ***Exposure System*** *- Ochistka* ***sistemy ehksponirovaniya***

Sometimes, terminological phrase translation is accompanied by the use of translation transformations. One of them is the addition of lexical units specifying the denotative properties of the described object (in the following example, the word "ob"ektiva" is added), another one is the omission of repeating lexemes (the words "poverkhnosti"):

Do not touch the ***mirror surface*** *and* ***lens surface*** *with you bare hands. – Ne prikasajtes' golymi rukami k* ***poverkhnosti zerkala*** *i* ***linz ob"ektiva****.*

The terms are also transmitted into the Russian language by means of transliteration:

Check for any stains and scratches on the ***diffuser****. - Prover'te* ***diffuzer*** *na predmet otsutstviya zagryaznenij i tsarapin.*

This display shows the Auto Setup method by using the optional build-in Konica ***Densitometer****. – Na ehkrane otobrazhaetsya rezhim avtomaticheskoj nastrojki vstroennogo* ***densitometra*** *Konica.*

As mentioned, in some instances, the cognitive information in English technical terms is translated by means of various transformations. To describe them, we use the classification by V N Komissarov [13], given that this classification is the most common in translation studies and reflects well the logic of our study object. In the following example, the omission (film-film) and the addition (the word "podachi") are utilized simultaneously:

Remove the ***negative film carrier****. - Snimite* ***ustrojstvo podachi negativov****.*

The omission technique is quite frequent (in this example, the word "size" is omitted):

In case to operate the moveable ***mirror for 120 size****...– Dlya podvizhnogo* ***zerkala 120 ispol'zujte****...*

Sometimes, semantic modulation occurs. In the following example, we translated the word "values" as "znachenie", although the dictionary only contains the following meanings: "tsennost '", "stoimost'", "vazhnost '":

Enter adjusted ***values****. – Vvod kalibrovochnykh* ***znachenij****.*

In the examples below, we applied such type of transformation as generalization, using a more general concept in the target language than the one in the source language:

Use the ***ten-keypad*** *to enter the Y, M and C values in order. – S pomoshh'yu* ***tsifrovoj klaviatury*** *posledovatel'no vvedite znachenie dlya Y, M i C.*

Open the upper cover of the ***processing section****. – Otkrojte verkhnyuyu panel'* ***protsessora****.*

In the following example, descriptive translation was employed in order to maintain the conventional structure of a technical manual used in the target language:

Check items. – Na chto obratit 'vnimanie.

In this text fragment, the conventional norms of the target language require a holistic form transformation.

Next, we review abbreviations as ways of explicating the cognitive information in a technical text. In our opinion, the most accurate definition of abbreviation is provided in the scientific work by E N

Ozhogin: "Abbreviation is the process of creating the secondary naming units, which entails reducing the linear length of the original name and, after applying a number of formal operations, resulting in a reduced structural version of this name" [14].

In modern linguistics, the fundamental classification of abbreviations, i.e., according to their structure, is traditionally represented as follows:

1. syllabic – part or parts of a single word;
2. multi-syllabic, comprising words with reduced linear length as well as full words;
3. initial, comprising the initial letters (sounds) of words in the abbreviated phrase [15].

The abbreviations of the three types reviewed are found in English manuals and function as cognitive information carriers. An abbreviation poses a problem for translator only when the meaning of its denotation is unknown. Otherwise, its cognitive elaboration during translation does not differ from elaboration of any other terminological lexeme.

Defining the complete form of abbreviations itself does not present a particular difficulty for the translator. Many of the so-called morphological abbreviations are intuitively comprehensible (*ERR, Func*), and most abbreviations in general, including initial (*ml, ANC*), are listed in established dictionaries, as a rule at the end, in a typical appendix, as well as in special abbreviation dictionaries, containing more specific scientific abbreviations.

Consequently, the challenges in translation of such lexis are not of semantical nature: even with a clear meaning defined, finding an equivalent-counterpart, or an interlingual synonym, is usually not an easy task. Being a word, according to expert opinion, the abbreviation should on an equal basis comply with the lexis translation rules. These rules are based on semantical correspondences between the correlative units of a given language pair. Similarity to terms, i.e., the absence of connotations, further highlights the importance of conveying the meaning of abbreviations. Strictly schematically, source language word

1. is translated – its meaning is transmitted by appropriate means of the target language (equivalents, counterparts) or
2. is transferred to the target text as original, partly or fully preserving its form (loan word, transcription, transliteration).

However, the abbreviation is a peculiar word, different from other words. In the same way that a stenograph functions as a representative, a substitute of an ordinary word, the abbreviation represents a different unit, potentially contained in it and the original one, from the "fragments" of which it is composed.

In the course of the technical manual original and translation analysis, we encountered two main types of cognitive information transmission using abbreviations. We believe that translator should convey the cognitive information to the fullest extent possible in a given communicative situation. This fact determines the strategy for transmitting this type of information using abbreviations. Those containing standard and frequent for any technical text data, are translated into the target language by means of equivalent abbreviations. On the other hand, the abbreviations, determined by the specifics of this particular photo printer manual, are included in the Russian text as the source text words in their original form (foreignisms). The latter group comprises device model names and the photo printer part names displayed during device performance as well as the comments in the source language, provided in the manual. These English comments are copied as original into the manual translation so as to make device user feel confident while using the translated text.

Let us take a look at some examples of the abbreviation translation.

As we have mentioned, conventional abbreviations of length and volume measures are transmitted into the target language by means of equivalent abbreviations, i.e., fixed equivalents consisting of initial letters in a morpheme or a word:

Use the monitor at the distance of 80 ***cm*** *in a bright room. – Ispol'zujte monitor v osveshhennoj komnate na rasstoyanii 80* ***sm****.*

Do not position any lighting fixtures within 2 ***m*** *from the operation table. – Ne razmeshhajte nikakikh osvetitel'nykh priborov v predelakh 2* ***m*** *ot rabochego stola.*

The temperature parameters of the corresponding processes during equipment performance are translated into Russian, as well as into English, by means of conventional Latin alphabetic acronyms:

*Operate this system under an ambient temperature of 18º-28º **C**. – Trebovaniya k okruzhayushhej srede, v kotoroj budet ehkspluatirovat'sya sistema, komnatnaya temperatura 18º-28º **C**.*

Volume measures are usually transmitted into the target language by means of fixed equivalents, i.e., the initial morpheme equivalents:

*Should tablets or processing solution be swallowed, rinse you mouth out with water or drink one or two cups (180 **ml**) or more of water. – V sluchae popadaniya tabletok ili obrabatyvayushhikh rastvorov v organizm propoloskajte rot ili vypejte odin-dva stakana vody (bolee 180 **ml**).*

Some abbreviations in English manuals are transmitted into Russian as full-word counterparts. In particular, this applies to a common Russian lexeme "naprimer":

*In case of any problems in the cutter section (**e. g.,** blade sharpness). – Neispravnost' v sektsii rezaka (**naprimer**, lezviya zatupilis').*

Sometimes, as shown in the example below, English conventional non-technical abbreviations are transferred into the target language as foreignisms:

*When judgment is "Almost **OK**" or "Once again", press the yes key to make another test print and enter it until the judgment becomes **OK**. – Esli rezul'tat sravneniya "Almost **OK**" ili "Once again", nazhmite klavishu yes dlya polucheniya drugogo testovogo otpechatka i vvoda ego v ustrojstvo, poka rezul'tat ne stanet "**OK**".*

A photographic equipment manual often contains abbreviations of the parameter names seen on the computer display during device operation. Such abbreviations are transmitted in translation as original, i.e., in Latin letters, so that device user can respond to the messages adequately:

*Should "**WARN-0121**" be displayed when "Light Compensation" is selected... – Esli v rezhime "Kompensatsiya sveta" vydano preduprezhdayushhee soobshhenie "**WARN-0121**"...*

In other cases, a full lexeme in the source language is translated as a syllable abbreviation. In the analyzed text, such is the case for the word "page", which is abbreviated in the Russian version, perhaps because the Russian word "stranitsa" is much longer than its English counterpart:

*"Cleaning Inside the Printer" (**page** 3. 32) – "Ochistka vnutrennej chasti printera" (**str**. 3.32)*

Many abbreviations in the technical text refer to the device model and series. In the analyzed text, they constitute a majority (65%) and are always transmitted into the target language as foreignisms:

*Maintenance and Inspection of the **ECOJET** component. – Osmotr i ochistka zheloba **ECOJET**.*

*Thank you for purchasing the **NPS-838 Super-SQA**. – Blagodarim Vas za priobretenie sistemy **NPS-838 Super-SQA**.*

In manuals, there is always a group of abbreviations denoting technical characteristics of a certain ongoing process. They are often transmitted into the target language as full equivalents, with the original abbreviation, i.e., a foreignism, given next to it:

*For **ERR-0212** [Paper Jam Processor Exit] Display. – **Oshibka ERR-0212** (Zamyatie bumagi na vykhode iz protsessora).*

In some cases, the abbreviations of this type keep their original form in the translated version, and their complete form is provided further in the text:

*Writing file data (**PP→FD**) – Zapis' dannykh v fajl (**PP→FD**)*

In the example above, the lexemes "Printer Processor" and "Floppy disc" are further translated as full-form words "printer-protsessor" and "disketa".

*Remove the **Auto Negative Carrier (ANC)** from the main unit. – Otsoedinite **avtomaticheskoe ustrojstvo podachi negativa (ANC)**.*

Thus, it is possible to transfer the complete term form and its bracketed abbreviation in both the source and the target languages. Moreover, the full version is translated by the use of calquing, whereas the contraction is conveyed in its original form. In such cases, further in the text the abbreviation functions without its full form.

4. Conclusion

The analyzed text evidently belongs to the category of technical texts and therefore provides an accurate material presentation with the main focus on the logical rather than the emotive side of the message. In this article, we have specified most distinctive features of the scientific-technical style of writing. They are: informativity (thoroughness), logic (strict consequentiality, clear coherence between the basic concept and details), accuracy, objectivity, and, subsequently, clarity and comprehensibility.

We recognize terms and abbreviations as primary markers of cognitive information contained in English technical texts. As our studies have shown, possible equivalents constitute the predominant means of term translation (55%), i.e., to convey the precise meaning into Russian, we apply one of the regular translation techniques for each particular word meaning as opposed to a fixed equivalent, which is a ready-to-use solution and is easily found in a dictionary. Calquing of terms (12%) is also a common method for conveying cognitive information in technical manuals. This technique comprises morpheme-for-morpheme or word-for-word translation, which is sometimes accompanied by lexical extension, i.e., we decompose the original compound into words-constituents of the target language. However, applying this method, translator should pay special attention to maintaining the authentic meaning of the initial compound and correctly define the semantic core (i.e., the central word) of the Russian phrase. The terms, which fall under the category of fixed equivalents (counterparts) are transmitted into Russian by means of transliteration (15%).

In some cases, we applied translation transformations. To describe them, we adopted the classification by renowned Russian translatology theorist V N Komissarov, given the fact that his arrangement is considered most common and respected by the scholars in the field of translation studies. This classification reflects well the logic of our study object. The analysis of the empirical data showed that most frequently applied transformations are: omission of redundant lexical units (5%), addition of lexical units (5%), generalization (4%), semantic modulation (2%) and holistic transformation (2%).

Abbreviations are found in English technical instructions and, in conjunction with technical terms, serve as carriers of cognitive information. Translating an abbreviation is not a challenging task for translator, except when the meaning is unknown. Otherwise, its cognitive representation does not differ from the procedure towards any other technical term. We believe that translator should be able to convey cognitive information to the fullest extent possible in a certain communicative situation. Given the fact that a typical technical text is heavily infused not only with terminological lexemes, but also with abbreviations, studying the specifics of their transmission into Russian is essential. In our article, we have listed, according to the established classification, three types of abbreviations, which are: syllabic, multi-syllabic and initial.

Whether dealing with full-word terms or their abbreviations, after the analysis of the original manual and its translation, two main types of conveying cognitive information have proved to be efficient. Thus, the abbreviations denoting a typical and commonly used in any technical text terms are translated through the use of equivalent abbreviations. On the other side, the context-specific abbreviations (in our research, those determined by the specifics of this particular photo printer manual) are transmitted into the Russian text as source text words (foreignisms). This group includes the abbreviations with the device information (e.g., model name, series etc.), the photo printer part names or command signals displayed during device performance and finally, the abbreviations denoting the comments in the source language, provided in the manual. These English comments are copied as original into the manual translation so as to make the device user feel confident while relying on the translated text.

References

[1] Aixelá F 2009 An overview of interference in scientific and technical translation *Journal of Specialised Translation* **11** 75–85

[2] Kingscott G 2002 Technical Translation and Related Disciplines *Perspectives: Studies in Translatology* **10 (4)** 247–55

[3] Byrne J 2012 *Scientific and technical translation explained* (Manchester: St. Jerome Publishing)
[4] Stolze R 2009 Dealing with cultural elements in technical texts for translation *Journal of Specialised Translation* **11** 124–42
[5] Zehtsen K K 1999 The Dogmas of Technical Translation – Are They Still Valid? *Hermes* **23** 65–75
[6] Krein-Kühle M 2011 Register Shifts in Scientific and Technical Translation *The Translator* **17(2)** 391–413
[7] Archer J 2002 Internationalisation, technology and translation *Perspectives: Studies in Translatology* **10(2)** 87–117
[8] Montgomery S L 2009 English and Science: realities and issues for translation in the age of an expanding lingua franca *Journal of Specialised Translation* **11** 6–16
[9] Göpferich S 1995 A pragmatic classification of LSP texts in science and Technology *Target* **7(2)** 305–26
[10] Schubert K 2009 Positioning Translation in Technical Communication Studies *Journal of Specialised Translation* **11** 17–30
[11] Voskoboynik G D 2004 *Linguophilosophical foundations of a general cognitive translation theory* Doctor of Science diss. Moscow State Linguistic University, Moscow
[12] Alexeyeva L M 2006 A Cognitive Approach to Terminology *Modern Approaches to Terminology Theories and Applications* ed H Picht (Bern: Peter Lang) pp 25–34
[13] Komissarov V N 2001 *Modern Translation Studies* (Moscow: ETS)
[14] Ozhogin E N 1999 *Abbreviations in military sublanguage* (Moscow: Military University Publishing House)
[15] Alexeyev D I 2018 *Abbreviations in Russian* (Moscow: URSS)

Strategic marketing planning in the petrochemical market

N S Mushketova, S V Fedorova

Volgograd State University Institute of Economics and Regional Management
Marketing Department
Volgograd, Russia

nmushketova@volsu.ru, fedorovasv@volsu.ru

Abstract. The market of petrochemical products is a key segment of the Russian economic system, and its subjects play a key role in the formation of the revenue part of the state budget and the active balance of payments of the Russian Federation. Hydrocarbon raw materials are indispensable and having no alternative base of technological functioning of petrochemical enterprises. Oil and gas are irreplaceable strategic resources, so their rational use, deepening of processing, is a task of national scale. It is the transition from direct export of hydrocarbons to their deep processing, to the strategic development of oil and gas and petrochemical complexes that will minimise the dependence of the Russian economy on adverse fluctuations in the world market. In the current geopolitical conditions, the subjects of the market will have to move in line with the ongoing transformation of the industry, making maximum use of strategic marketing planning tools to strengthen their own market positions, adequate to the market of increasing production volumes, winning new markets, increasing sales of products with high added value.

1. Introduction

Strategic marketing planning is a sufficient process of developing adaptive measures to changes in the external environment, which allows to increase the effectiveness of the use of internal factors of production for sustainable and successful development. In the process of marketing strategic planning, optimal goals and strategies of market participants' behaviour are developed, the implementation of which ensures their viability in the long term.

Strategic goals of any organisation are generated primarily in the external market environment, in other words - in the area of marketing interests. There is a whole range of proven marketing tools to keep the company in a leading position in a highly competitive environment. At the same time, marketing in theory and practice is considered as a strategically oriented concept, which is the foundation of management activities and is based, in turn, on legal, financial, information, intellectual and expert-analytical support.

The whole process of strategic marketing planning is a logical sequence of the following stages:

1. Strategic audit.
2. Development of the company's mission.
3. Goal setting (building a tree of goals).
4. Development of marketing strategy.
5. Development of a mechanism to monitor the implementation of the strategy. [1, p. 68]

Without the implemented mechanism of strategic marketing planning, no enterprise of the petrochemical industry is able to survive.

2. Results and Discussion

The strategic vector of modern development of world petrochemistry in developed countries is an intensive movement towards the creation of powerful industrial clusters based on the integration of oil and gas production, oil and gas processing, oil and gas chemistry. At the same time, such consolidation is accompanied by state support in the form of participation in the creation of modern infrastructure, as well as the use of financial and tax instruments to stimulate investment. The periphery of the cluster core of petrochemical industries is the centers of engineering, consulting, scientific, educational and other services. To increase the investment attractiveness of the region, developed and developing countries create special economic zones (SEZs) around such clusters, attracting international companies. Economic support of cluster production by the state is aimed at improving the environmental friendliness and energy efficiency of production [2, 3].

The positive foreign experience of strategic development of petrochemical clusters confirmed by practice should be taken into account by all subjects of the domestic market in the development of their own long-term strategic marketing plans.

Since the world and Russian markets of oil and petrochemical products are characterised by high dynamics and intense competition, the most important stage in the development of the marketing strategy is a strategic audit of the external environment of the organisation, as well as the subsequent monitoring of the company's environment in the process of implementing the strategy.

Market research itself in the petrochemical industry has its own characteristics. This process involves the analysis of a large array of data from which you want to select only the relevant information. The audit of such elements as the structure and dynamics of development of oil and petrochemical, as well as related industries, the configuration of the world market and its impact on the domestic petrochemical sector, the key factors of success in the industry, the analysis of consumers and competitors are extremely important.

The authors' analysis of the Russian market of petrochemical products revealed the following specific features of this sector:

- Russia has significant reserves of hydrocarbons, however, the petrochemical complex, having a high potential of monetisation of strategic reserves of oil and gas, in the Russian economy takes about 1.5%. The existing production base of the domestic petrochemical industry was formed in the second half of the last century;
- offer in the domestic market of the petrochemical industry does not provide the domestic demand, significant part of which is met through imports;
- petrochemical companies are more likely to operate in the B2B market, where the main consumers are manufacturing companies that produce fuel products, polymers, fertilizers, hence the demand for petrochemicals is derived from the demand in related markets (e.g. insulation materials, building materials, packaging products, synthetic rubbers, etc.).);
- the Russian petrochemical industry does not have its own specialised mechanisms to stimulate this sector, all available instruments are derived from the system of stimulation of the adjacent oil and gas industry (its state subsidies);
- the studied sector is characterised by a significant dependence on the state of fuel and raw materials markets and the conjunctural configuration of the world economy as a whole;
- the instability of commodity and product markets, as well as the need to improve production efficiency, are leading to the concentration of the industry.

In the process of analysis, the determinants that hinder the successful development of the petrochemical complex were identified.

The main internal barriers include high depreciation (up to 85%) of fixed assets of the industry with almost their maximum load, outdated technologies, lack of investment in the industry due to the low return on investment capital.

External factors impeding the development of petrochemistry are regional disproportion in the provision of territories with raw materials and energy resources, as well as in the asymmetry of their consumption; the growth of geopolitical tension, the aggravation of Russia's confrontation with a number of countries, a sharp restriction of access of Russian petrochemical companies to international credit resources, as well as to cutting-edge foreign technologies (Fig. 1).

Internal Factors	External Factors
• High depreciation of production assets (85 %); • Maximum capacity utilisation; • Outdated technologies and innovative backwardness in comparison with foreign complexes, high dependence on imported technologies and equipment; • Lack of investment in the petrochemical industry due to lower return on investment than investment in oil production and refining; • Insufficient development of the system of technical regulation, industry standards and quality control system of the chemical complex; • Low levels of reproduction of highly qualified personnel.	• Change in the structure of proven reserves of FER (Fuel and Energy Resources) in favour of hard-to-extract and multi-component ones; lagging of growth of explored hydrocarbons from their production; • Regional asymmetry in the supply of raw materials and energy resources and their consumption; • Growth of capital intensity of scientific and technical developments; • Lack of transport and logistics infrastructure, the complexity of its expansion and access to it; • High tariffs for electricity and railway transportation; • Increased geopolitical tensions and deteriorating relations between the Russian Federation and a number of leading Western countries limit Russian energy companies' access to credit resources of major world banks, as well as to advanced technologies.

Figure 1. External and internal barriers to the development of the petrochemical industry [4]

External marketing audit is a tool for prompt and symmetrical response to global trends.

The internal audit of the company, which allows to assess its strategic potential is no less important. The main directions of internal marketing analysis are shown in fig. 2.

1. Business-process marketing in the enterprise
   - SWOT-analysis and identification of key success factors (KSF);
   - Strategy, mission and corporate culture;
   - Methods of monitoring the implementation of the adopted strategy;
   - Place and principles of organisation, as well as goals and objectives of the marketing service in the company;
   - Coordination of the units' activities in the implementation of the strategy;
   - Regulation of business processes.
2. Marketing tools for creating a balanced product portfolio
   - Unique consumer value of the product;
   - Reproducible service quality and product value;
   - Assessment of the degree of importance of the types of products produced by the enterprise (ABC, XYZ, etc.);
   - Monitoring of competitors' prices: business process regulations;
   - Effective participation of the company in exhibitions: business process regulations.
3. Management of commodity circulation channels
   - Participants in the purchase process and typical stages of the transaction;
   - Structure of the sales organisation units and their interaction with the departments of marketing;
   - Role and tasks of marketing departments in preparation for participation in tenders: business process regulations;
   - Marketing scenario approach to motivation;
   - Use of Internet technologies in promoting products to the market.

Figure 2. Key aspects of the marketing audit of business processes of the enterprises of the petrochemical sector [5, 6]

When analysing the product portfolio, it is necessary to take into account such specific features of petrochemistry as the requirements of quality standards. In this industry, the quality of the product is determined primarily by natural indicators of the raw materials used, and secondly - the quality of production processes. It should be understood that compliance with technical and regulatory standards is extremely important for petrochemical products, while aesthetic and ergonomic parameters do not matter at all.

Therefore, maintaining a stable position in the market and ensuring high competitiveness for the petrochemical industry is possible only through the transition to environmental standards, innovative and resource-saving production technologies.

Marketing strategy of production companies operating in the market of petrochemical products is a comprehensive long-term plans for 3-5 years, designed to ensure the effective marketing of petrochemical products, the implementation of the company's mission and achievement of its long-term development goals.

In general, the algorithm for the development strategy of a petrochemical company is as follows (Fig. 3):

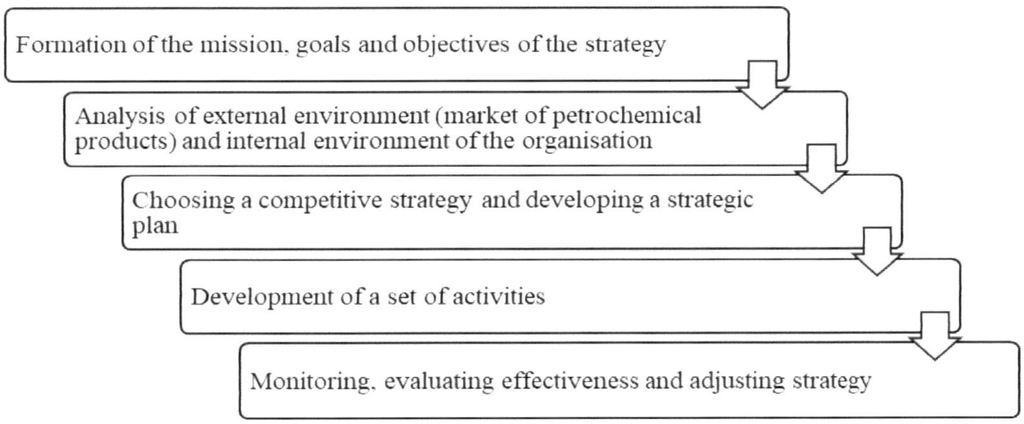

Figure 3. Stages of development of the strategy of the petrochemical complex

Thus, the consistent implementation of all stages of strategic business management in the market of petrochemical products will allow to take into account a wide range of market situations.

In the process of developing a strategy in the market of petrochemical products, enterprises should set the following targets:

- sustainable business growth;
- high quality products according to international standards;
- reliability of product supply;
- preservation of the environment and rational use of natural resources;
- high reputation in the business community;
- business transparency and social responsibility, etc.

Besides, the key trends of strategic development in this market should be taken into consideration.

Thus, according to experts, the cluster approach declared in the development plan of the gas and petrochemical complex of Russia for the period up to 2030 [7] is the most effective and expedient. It is the clusters that will be able to become the key points of growth of the petrochemical industry and increase its share in GDP, and the innovative infrastructure of cluster production should become a powerful catalyst for positive structural reforms in the economy of the country and its regions.

In accordance with the development Strategy until 2030, six petrochemical cluster production facilities should be established in Western Siberia, the Volga region, the Caspian region, Eastern Siberia, the North-West and the Far East. To date, only the Volga region production cluster has been fully formed, which has been loaded to full capacity. The main barriers to the effective launch of the remaining clusters are infrastructure constraints: weak capacity of railway and motor transport, the lack of pipe and product pipelines. Investment needs for the formation of a full-fledged infrastructure are estimated at 1 trillion. dollars [8, p. 67].

The main driver for the development of the petrochemical industry should be the increase in value added from processing of primary raw materials of the oil and gas sector, while the development of petrochemicals in the reference scenario and in the long term should be faster than oil and gas production. To do this, it is necessary to actively increase the production of high value added products in the petrochemical sector, to create common raw materials and technological chains, to involve foreign partner countries [9].

If large integrated companies are the centers of creation of petrochemical clusters, medium and small firms in the industry will have to use the full potential of the concept of strategic marketing management to enter into the periphery of the clusters and search for their own niche in it.

The efforts of small petrochemical companies should be focused on the implementation of the following areas of strategic development:

– expansion and deepening of the range of products (due to the production of new innovative products, such as polymers with desired properties, bioplastics, membranes, etc.));
– expansion and improvement of the raw material base through the involvement of new types of raw materials (e.g., bio-raw materials);
– search for new innovative technologies of deep processing (bio-and nanotechnologies);
– search for new market niches and consumer segments;
– the combination of broad cooperation with narrow specialisation in order to reduce the cost of production, etc.

Product marketing strategies will play a special role in the petrochemical segment of the market. Foreign manufacturing is now reoriented to the production of more high-tech products, petrochemical complexes, for example, are engaged in the development of new catalysts and technologies [10, 11]. In the domestic market, the most promising are product areas with significant potential for import substitution, that is the production of synthetic rubbers, polyamides and polystyrene.

In addition, according to experts, it is imperative to stimulate the final markets for the consumption of petrochemical products. Thus, construction is the main recognised driver of consumption of petrochemical products (especially polymers, insulation materials, etc.) all over the world. The second driver is the automotive industry. Consolidation of efforts of economic entities of the market and the state on territorial localisation of production in the territory of Russia, of both accessories and raw materials for their production, is the strategically important direction of development [12, p. 166].

3. Conclusion

In modern conditions, companies in the petrochemical industry should pay more attention to marketing strategies - in a highly competitive environment and the strong impact of negative factors on this sector, only effective strategic marketing will allow business to succeed in this market. The work of marketing services allows companies to identify the most promising market niches, new target segments, develop new products demanded by the market, form a recognisable brand and build a "road map" to achieve strategic goals.

The strategic target for the development of the domestic petrochemical complex is to increase the competitiveness of products, in particular, by the deepening of processing, import substitution, the introduction of product and technological innovations.

The implementation of the strategy involves the fulfillment of such tasks as stimulation of petrochemical industries, clustering of the petrochemical industry, further development of import substitution in the industry, large-scale investments in infrastructure, introduction of innovative technologies in petrochemical production, development of human and scientific potential, active promotion of the "swarm of small enterprises" in the structure of the cluster periphery of the industry.

References

[1] Pyanov R L 2015 Basic marketing strategies in the system of strategic planning of companies

Economic environment **4 (14)** 68-71
[2] Mariadoss B J , Tansuhaj P S , Mouri N 2011 Marketing capabilities and innovation-based strategies for environmental sustainability: An exploratory investigation of B2B firms *Industrial Marketing Management* Vol 40 **8** 1305-18
[3] Chaney D, Carrillat F A, Zouari A 2018 Uncovering institutional orientation as a new strategic orientation in industrial marketing *Industrial Marketing Management*
[4] Kutuzova A O 2017 Analysis of the specifics of the Russian oil market and assessment of the impact of world oil prices on the state of the economy of the Russian Federation economic relations Vol 7 **1** 65-86
[5] Ksenophontova J V, Etova E V 2015 Marketing as a tool of regulation of economic activities and reduction of risks for enterprises of the oil and gas industry *Scientific-methodical electronic journal "Concept"* **33** 136-140
[6] Pomirleanu N, Mariadoss J, Chennamaneni P R 2016; 2015 Managing service quality in high customer contact B2B services across domestic and international markets Indus*trial Marketing Management* **55** 131-143
[7] The order of the Ministry of Industry and Trade of the Russian Federation and the Ministry of Energy of the Russian Federation from 14.01.2016, No. 33/11 "On amendments to the Development Strategy of Chemical and Petrochemical complex for the period till 2030", approved by order of Minpromtorg of Russia and the RF Ministry of Energy 8.04.2014, No. 651/172" URL.: http://www.garant.ru/products/ipo/prime/doc/71224670/#ixzz5WqJYntgY.
[8] Khorokhoryn A E 2014 The Strategy of the Development of Modern Petrochemical Complexes, International Experience and Possibilities for Russia: thesis ... candidate of Economic Sciences: 08.00.14. Place of defense: Russian State University of Oil and Gas named after I. M. Gubkin Moscow 193 p
[9] Chennamaneni P R , Desiraju R 2011 Comarketing alliances: Should you contract on actions or outcomes? *Management Science* Volume 57 **4** 752-62.
[10] Hanna D, Parent M, Leyland P, Berthon P 2014 It's a secret: Marketing value and the denial of availability *Business Horizons* Vol 57 Issue **1** 49-59.
[11] Cortez R M, Johnston W J 2017 Needed b2b marketing capabilities: Insights from the USA and emerging Latin America *International Business Review* Vol 27 Issue **3** 594-609
[12] Volkova N A, Naboychenko A S 2016. Relevant aspects of the implementation of industrial policy in the oil and gas complex *The Bulletin of University (State University of Management)* **9** 8-15

The influence of technological parameters on physical and mechanical properties of surfaces

M V Mousokhranov[1], V V Kalmykov[1], E V Logutenkova[1]

[1]Kaluga Branch of Federal State Budgetary Educational Institution of Higher Education "Moscow State Technical University named after N.E. Bauman (National Research University)," 2, Bazhenova Street, Kaluga 248600,Russia

E-mail: marls77@ya.ru

Annotation. In this study, we examined how the parameters of external mechanical impact by microdeforming the surface influence the energy state of a solid body at the interphase boundary. We studied the dependence of the surface energy on microdeformation modes with different values of the line interval and the elastic recovery of multicomponent iron-based materials. The measurements were performed on experimental samples of C20, C45, X10Cr13. The results of the study can be used to ensure the required performance properties of products.

1. Introduction

To concur with the never-ceasing growth of technical requirements to engineering products and instrument-making, it becomes necessary to improve the existing technologies and to develop new ones. Several studies [1-8] show a stable relationship between the performance properties of products and the energy parameters of surfaces.

Examination of the features of energy parameters [9–13] of a substrate of structural materials with metallic properties helps get an insight into the change in the properties of near-surface areas. It is known that the energy state of surfaces of a solid body at the phase boundary is determined, among other things, by the structure and state of the crystal lattice [14–16]. Therefore, it is important to know how to form a certain level of energy state on a surface by means of external mechanical impact. The aim of this study was to examine the effect of the parameters of external impact by microdeforming the surface on the surface energy (SE) of multicomponent iron-based materials.

2. Materials and methods

To conduct an experimental study of the change in SE of materials when microdeforming line-by-line, impact on the surface of the samples was simulated. The materials of the experimental samples are C20, C45, and X10Cr13. The material of the tool (cutting edge) is GC1040. The blunting radius of the cutting edge of the tool ρ= 0.02 mm. Microdeforming was performed on a vertical milling machining centre with a MiniMill 450 CNC machine. In the process, the values of the surface energy were determined by the method described in [17]. The hardness of the material under study was determined on the Brinell scale using a hardness meter.

The influence of microdeformation modes on the surface energy at different values of elastic recovery hy was examined with regard to:

– materials of rigid structures – the hy value less than 1 μm;

– materials of the structures prone to elastic recovery after shaping by removing part of the surface – hy ≈ 2 μm and hy ≈ 4 μm.
The assessment was performed according to the methods described in [18].

3. Results and discussion

To determine the effect of the line interval (longitudinal tool feeding) on the SE change, its fz value was decreased and increased by 30% during microdeformation with the help of a milling tool. The graph of SE dependence on the line interval during microdeformation after removing part of the metal at depths h = 500 μm, h = 300 μm and h = 100 μm is shown in Fig. 1.

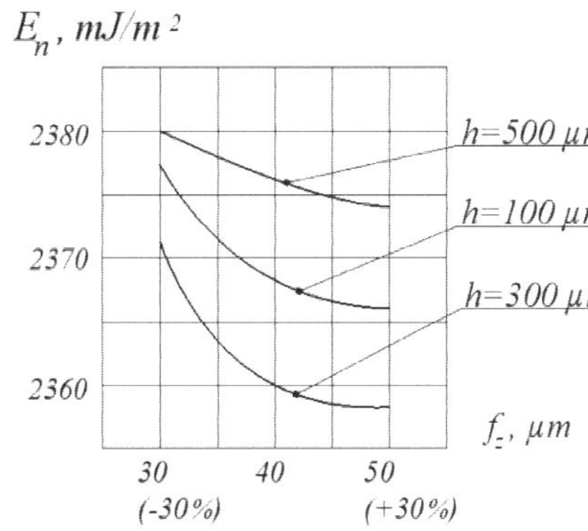

Figure 1. Dependence of surface energy on the line interval

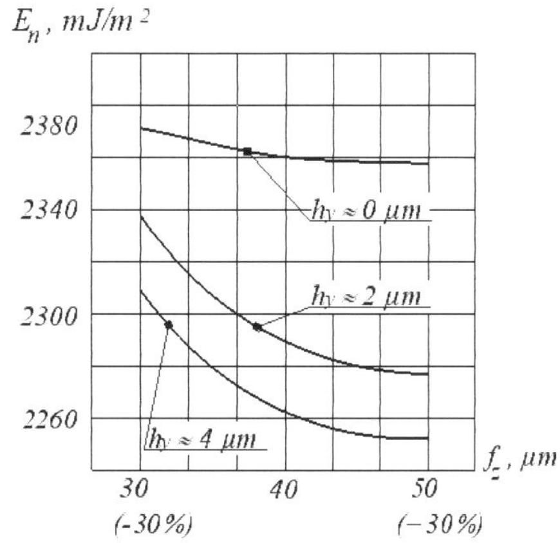

Figure 2. Dependence of the surface energy on line interval f_z and the depth of line-by-line microdeformation h_y after preliminary shaping of the surface by removing material to depth h = 300 μm

Analysing the graphs, we can conclude that as line interval fz is increased during microdeformation, the SE value decreases. The lowest value of SE is accumulated during microdeformation after removing a layer of depth h = 300 μm, and the highest value of SE – after removing material to the depth of h = 500 μm.

Then, the influence of microdeformation parameters on the SE value was determined. The surface energy resulting from the change in the value of elastic recovery after shaping (hy ≈ 0 μm, hy ≈ 2 μm and hy ≈ 4 μm) was measured. The results of the experiments with microdeformation of the surface after removing material to depth h=300 μm are shown in Fig. 2.

The analysis of the results revealed that as the depth of microdeformation h_y increases, the value of surface energy decreases. This is due to the smoothing of the microrelief vertices of the surface accompanied by unloading of the energy concentrator zones.

The effect of microdeformation parameters on the SE of other materials was studied in a similar way. The material of the samples is C20. The surface energy resulting from the change in the value of elastic recovery after shaping by cutting (h_y ≈ 0 μm, h_y ≈ 2 μm and h_y ≈ 4 μm) was measured. The results of the experiments with microdeformation of the surface after removing material to depth h=300 μm are shown in Fig. 3.

The analysis of the results revealed that as the depth of microdeformation h_y (due to the impact of elastic recovery) increases, the SE value decreases.

When examining samples from X10Cr13, the surface energy resulting from the change in the value of elastic recovery after shaping by removing a layer of material (h_y ≈ 0 μm, h_y ≈ 2 μm and h_y ≈ 4 μm) was measured. The results of the experiments with microdeformation of the surface after cutting to depth h=300 μm are shown in Fig. 4.

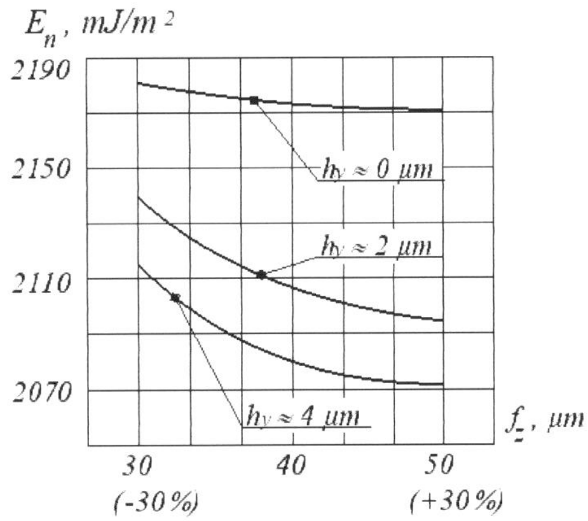

Figure 3. Dependence of the surface energy on line interval f_z and the depth of microdeformation h_y after shaping by removing material to depth h = 300 µm (C20)

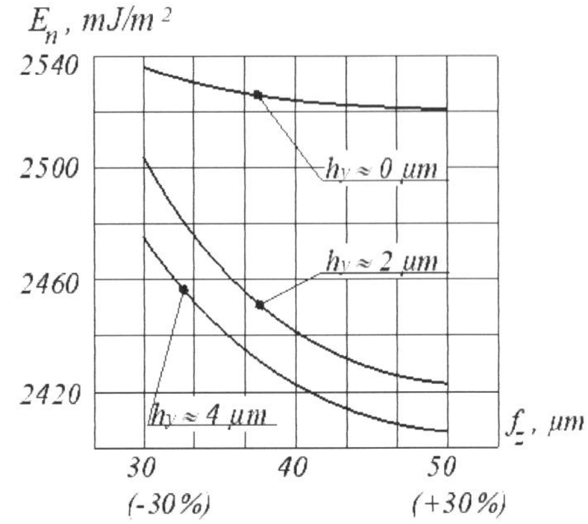

Figure 4. Dependence of the surface energy on line interval f_z and the depth of microdeformation h_y after shaping by removing material to depth h = 300 µm (X10Cr13)

The analysis of the results revealed that as the depth of microdeformation hy increases, the value of surface energy decreases.

4. Conclusions

For materials C20, C45 and steel X10Cr13, the SE measurement results revealed:

– in order to obtain the maximum level of SE, after removing a layer of material to depth h = 300 µm, microdeformation should be carried out;

– as line interval fz increases during microdeformation of materials, the SE value decreases;

– as the depth of microdeformation increases, the value of surface energy decreases.

References

[1] Musokhranov M V, Kalmykov V V, Avramenko M U 2016 Technological prerequisites for improving the performance of the guide elements *Fundamental research* **8-1** 55-58

[2] Musokhranov M V 2005 Surface energy as an indicator of the quality of the surface layer Directory. Engineering journal with the app **12 (105)** 62-64

[3] Malyshev E N, Musokhranov M V, Kalmykov V V, Antanuk F I 2015 Distribution of energy technological impact in the formation of working surfaces of machine parts made of metal materials *Modern problems of science and education* **1-1** 58.

[4] Oleshko V S, Pigovkin I S 2017 The influence of machining on the electron output of metal parts of aircraft *Defense complex-scientific and technical progress of Russia* **3 (135)** 24-28

[5] Goncharenko V I, Oleshko V S 2017 Determining the surface energy of tools in the aviation industry *Russian Engineering Research* T 37 **7** 628-630

[6] Shvetcov A N, Skuratov D L 2017 Evolution of the residual stresses formation from fec0.15cr12ni2 steel in the part surface during the diamond smoothing *Proceedings of the 3rd International Conference on Dynamics and Vibroacoustics of Machines, DVM 2016* 355-362

[7] Izmailov V V, Novoselova M V, Chaplygin S.A 2017 On the experimental techniques for assessing the specific friction force and its parameters *Journal of Friction and Wear* T 38 **5** 369-376

[8] Nosenko V A, Mitrofanov A P, Butov G M Impregnation of abrasive tools with foaming agents *Russian Engineering Research* 2011 T 31 **11** 1160-1163

[9] Mousokhranov M V, Kalmykov V V, Logutenkova E V, Sorokin S P 2016 Energy state of the surface layer of machine parts *Modern high technology* **5-2** 276-279.

[10] Mousokhranov M V 2005The role of surface energy in the formation of parts in precision engineering *Assembly in mechanical engineering, instrumentation* **6** 9-11

IOP Conf. Series: Materials Science and Engineering **483** (2019) 012054 doi:10.1088/1757-899X/483/1/012054

[11] Mousokhranov M V, Kalmykov V V, Logutenkova E V, Gerasimova N S 2016 Energy surfaces in the formation of mates in the Assembly of products *Modern high technology* **5-2** 249-252.
[12] Mitrophanov A P, Parsheva A K 2016 The electron Work function as a parameter to assess the mechanical state of the surface *the collection of Scientific-practical conference of students of VPI (branch) VSTU "youth Science: ideas, results, prospects"* 63-64.
[13] Arefeva L P, Shebzukhova I G 2016 Electron work function and surface energy of body-centered and face-centered cubic modifications of 4d- and 5d-metals *Physics of the Solid State* T 58 **7** 1289-1294.
[14] Aref'eva L P, Shebzukhova I G 2018 Interfacial energy on the boundary of the metal crystal-melt *Physics of the solid state* Vol 60 **7** 1270-1276.
[15] Aref'eva L P, Shebzukhova I G 2017 Surface energy thin films of niobium and tantalum under uniaxial deformation *Physical-chemical aspects of the study of clusters, nanostructures and nanomaterials* **9** 32-37.
[16] Shebzukhova I G, Aref'eva L P 2015 The anisotropy of the electron work function of 3d metal crystals *Bulletin of the Russian Academy of Sciences. Series physical* Vol 79 **6** 896
[17] Goncharenko V I, Oleshko V S 2017 Determination of the surface energy of the instrument in aircraft engineering *STIN* **2** 24-27.
[18] Kalmykov V V, Barkov A V 2016 Analytical and statistical estimation of elastic deformations during milling the contour of *Fundamental research* **8-1** 34-38

Influence of electromagnetic radiation of the Earth on ethnogenesis processes: to the problem of refinement of sources of «passionarity»

S A Yurchenko[1], N Yu Nikolaev[2]

[1]24/8a Naberezhnaya Street, Volzhsky, 404130, Russia
[2]Volzhsky Polytechnic Institute (branch) of Federal State Budget Educational Institution of Higher Education Volgograd State Technical University
42a Engelsa Street, Volzhsky, 404121, Russia

E-mail: nikcam@mail.ru

Abstract. In this paper new factors, that have influence on passionary impacts and ethnogenesis processes are considered. According to authors, along with traditional sources of energy (space energy, an internal radioactive background and energy of the Sun) Earth`s core energy, that comes out to a surface, should be taken into consideration.

1. Introduction

The widely known passionary theory of L N Gumilev produces many different estimates and opinions from enthusiastic ones to blames of quasi-historicism and scientific amateurishness [1]. Let's note that the greatest interest (as well as the majority of critical remarks) is caused by reasons for «passionarity». It must be admitted, that Gumilev formulated them without any scientific specificity. As the result, efforts of scientists` supporters are directed to detailing and further development of his ideas, first of all in their natural-science aspects [2]. In this paper authors made an attempt to specify passionary theory, having suggested to consider earth`s core energy, that comes out to a surface, as an energy source, influencing ethnogenesis explosions (passionary impacts).

2. Results and Discussion

««Passionarity» is a characterological dominant, – Gumilev wrote, it is an irresistible internal aspiration (conscious or, more often, unconscious) for activity, directed to realization of goals (often illusory)» [3]. He was considering «passionarity» as the major «ethnogenesis impulse», defining historical fate of ethnos, its birth, development and decline. The scientist defined ethos as an equally biological and geographical (geobiocoenosis) phenomenon [4]. Within an anthropogenic landscape, in his opinion, there is an energy impact on ethnogenesis processes [5]. So what are the sources of this energy impact?

Gumilev thought that there are three sources from which the biosphere, (and certainly a person, as its part), takes energy. Firstly, it is a «radiant energy of the Sun»; secondly, it is an energy, connected with radioactive decay inside the Earth; thirdly, the biosphere absorbs space energy of «scattered elements, coming from our galaxy» [6]. Energy of the Sun is accumulated in plants by photosynthesis. Through them energy «passes into flesh and blood of all living beings". Disintegration energy of radioactive elements negatively influences biological processes. The mankind consumes energy bunches from space in «small portions». Gumilev noted that this energy type influence is uneven, both in terms of scales, and relatively to periodicity [7].

The scientist considered that the inception point of any ethnogenesis is «a specific mutation of a small number of individuals in a geographical area» [8]. The mutation, in turn, is caused by energy impact. "Each living organism, – Gumilev wrote, – possesses an energy field, now we can already compare it with a description of ethnos features and, therefore, call it an ethnic field, that is created by biochemical energy of living substance... Accepting this energy model, model of a field of force and applying it to ethnos problem, ethnos can be imagined as a system of a certain ethnic field fluctuations» [9].

The scientist assigned a special role to space energy. It takes the biosphere out of the equilibrium state, generating an entropy state [10]. It is expressed in an unmotivated movement (lemming or locust migration) among animals. Gumilev considered that in a human community the surplus of energy leads to passionary impacts or ethnogenesis explosions [11]. Zones of passionary impacts represent comparatively narrow bands about 300 km wide in a latitude direction, and slightly more in a meridian direction, approximately 0.5 of the planet`s circumference. The researcher compared these zones to geodetic lines. According to him, passionary impacts occur with a frequency of two, three times in thousand years and almost never pass across the same territory. The same impact is able to create several centers with increase of «passionarity», depending on physiographic features of regions [12].

Recently, there have been made attempts to keep the passionary theory from Gumilev`s natural science mistakes and delusions, to make some kind of academic upgrade, taking into account new data of genetics and social anthropology. It is necessary to agree with the opinion, expressed by Gumilevs` supporters, that «passionarity» is not constantly arising new social and biological phenomenon, but «rather accurate and certain characteristic, that has similar manifestations among different ethnoses in different historical times» [13].

However, even convinced adherents of passionary theory acknowledge that examples of energy impact on the person, given by the scientist, are not always compelling. For example, dendrochronological analysis data shows that dates of passionary impacts, given by Gumilev, do not really match observed maximums of education (14C), which is a universal marker of external radiation intensity. Some reasonably interesting attempts are being made to link the impact of solar energy on passionary impacts with the violation of periodicity of Pluto`s passage through perihelion (point of its orbit, closest to the Sun) [14]. But this theory also requires additional scientific verification.

According to authors, it is worth considering earth`s core energy as the main source of energy impact on a person. It is authentically known that our planet has complicated structure, which features are not completely studied even nowadays. However, attempts to consider problems of geoenergy impact on a person, linking them to common problems of geophysics, have been taken repeatedly. For example, the problem of geoenergy impact on mankind was raised within the theory of the crystal-like structure of the Earth. Idea of our planet being a crystal was put forward by ancient Greek thinkers (Plato, Pythagoras, Archimedes) [15]. In the 20th century this hypothesis was developed by Soviet geophysics – N F Goncharov, V L Makarov, V S Morozov. They considered Earth`s structure as icosahedral-dodecahedral (further – IDEST). Basic concepts of this theory state that nodal points of a dodecahedron and a triangle`s edge are energy conductors, transmitting planet`s core energy to a surface. Authors of IDEST state that, according to their knowledge, the configuration of dodecahedron`s nodal points coincides with the location of ancient civilizations centers, which was associated with energy impact on people, living there [16].

It is not a secret that over time magnetic poles of our planet change their location relatively to the coordinate grid, and even inversions of poles are recorded, that is proved by a paleomagnetic research method. From this it follows that earth`s core changes its internal location over a certain period of time (figure 1). Planet`s core change leads to a location change of nodes, through which Earth's core energy comes out to a surface in the form of electromagnetic waves. Schematically these processes can be represented by A1 node location change, moving in the direction of arrow`s rotation to node A (figure 1).

IOP Conf. Series: Materials Science and Engineering **483** (2019) 012055 doi:10.1088/1757-899X/483/1/012055

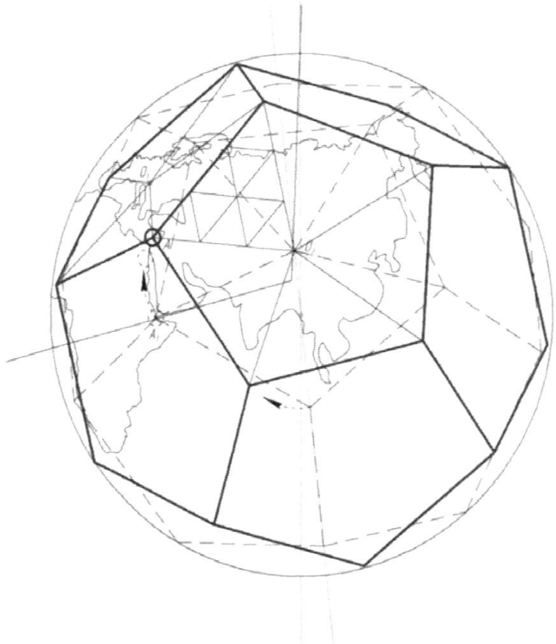

Figure 1. Location change of a dodecahedron inside the planet (dotted line is the previous location)

In figure 2 lines 1,2,4,6,7,9 are places, where earth`s core energy comes to its surface out of triangle`s edges, of which dodecahedron edges consist. This leads, in turn, to energy change of nature of living organisms in area of energy output (release).

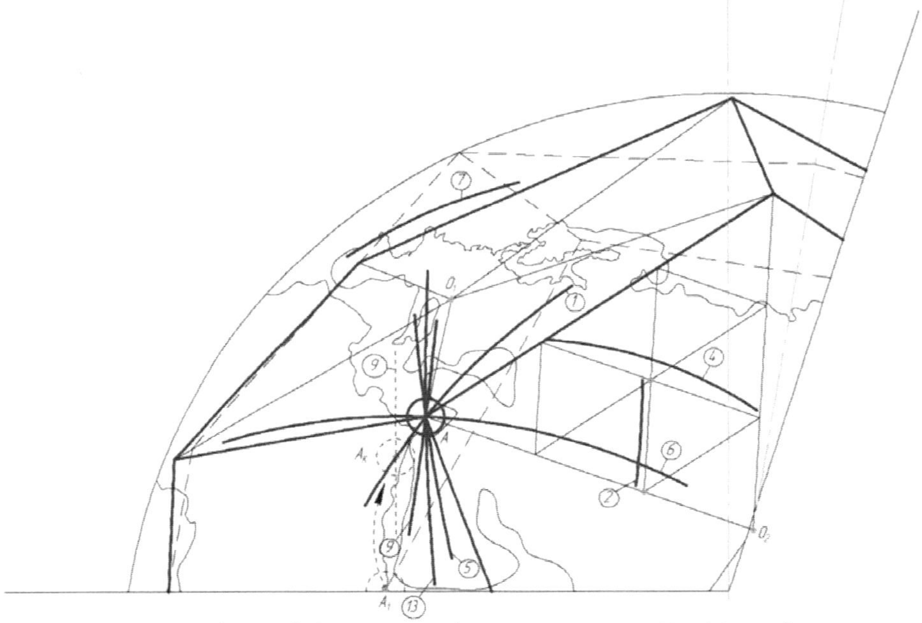

Figure 2. Places of electromagnetic energy output to Earth`s surface.

Let's specify that these lines on earth's surface are not constant, and depend on fluctuation or shift of the internal core, and consequently, parallel to each other lines 9 and 9` can be formed, and it is visible in nodal point A, lines 5, 9, 13 can turn near some center of node A.

As follows from the schemes, provided on figure 2 and figure 3, lines of increased Earth's core energy impact on the planet's surface (1 and I, 2 and II, 4 and IV, 5 and V, 6 and VI, 7 and VII, 9 and IX) are generally correlated with zones of passionary impacts, given by Gumilev in his work «Ethnogenesis and the Biosphere of the Earth».

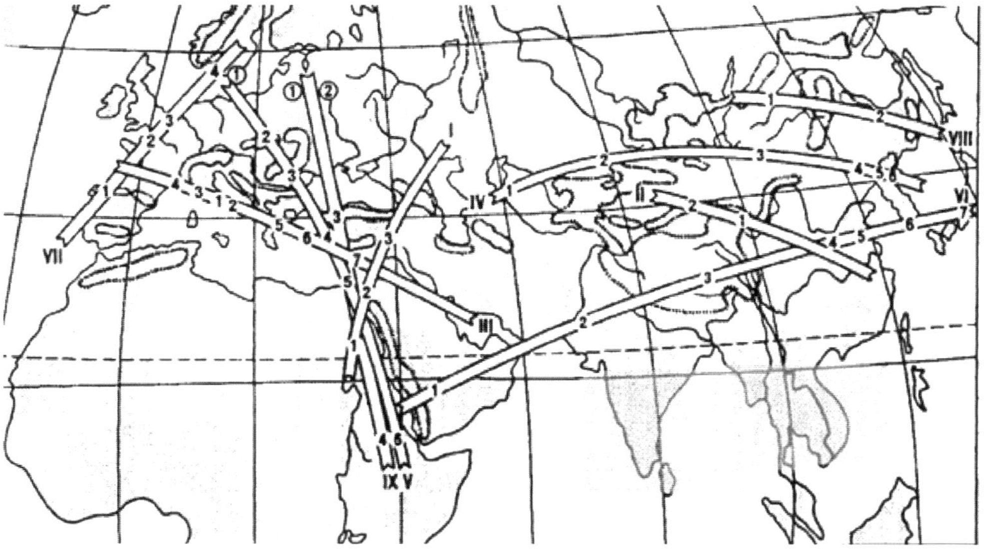

Figure 3. Passionary impacts, described by L N Gumilev in his work «Ethnogenesis and the Biosphere of the Earth».

According to A G Bakirov: – «The possibility is not excluded that the entire array … of nodes of a system or considerable part them, relating to a triangular-pentagonal grid, represents a projection to earth's surface of a space framework of some radially directed movable wave guides, through which a pulsing energy flow processes from mantle and deeper parts of the globe, giving rise to various geological processes and creating favorable conditions for emergence, development and evolution of the biosphere, plants, animal migration and the emergence of civilization centers» [17].

In other words, a person living in a certain place of the Earth is influenced by electromagnetic field of the Earth with a frequency of fluctuation, which is character to this area. Frequency of fluctuation has an impact on processes, taken place in brain, that leads to unique perception of the world around – an ethnocultural isolation. The group of people, living for a long time in the same place, generally experiences quite homogeneous impact of electromagnetic radiation on brain activity, and consequently the same perception of the world, identical requests and aspirations are formed. It is necessary to consider that the frequency of fluctuation of the planet's electromagnetic is identical and the distinction can be felt only in places with additional radiation from movable waveguides, located in earth's mantle. At the same time, from time to time waveguides move, thereby, creating favorable conditions for new center of ethnic «passionarity». This process takes place continuously. In geological time scale it is estimated as insignificantly short time and in historical scale as several hundreds of years.

It is worth to note that creators of IDEST emphasized the relation of crystal structure of the Earth and hydrography [18]. It is possible to assume that, as time goes by, movable waveguides change their location in earth's core space, and consequently, areas with «excessive» energy in comparison with other areas of land surface, are formed. Waterways can serve as natural directing channel of energy flow, especially because a part of them is localized in regions of edges and corners of earth's crystal structure. According to us, human settlements and especially objects of religious purpose (further – O.R.P.) were purposefully built in surrounding hydrocultural space, at the same time river bends acted as key locations. Localization of O.R.P. in meanders of big and small rivers is illustrated by authors on examples of both the most ancient Neolithic cultures of the Volga-Ural interfluve and the well-known religious buildings of Europe, Asia and America [19]. Therefore, it is reasonable to consider the energy impact theory on ethnogenesis processes, while taking into account the revealed patterns of O.R.P. location in river bends.

The revealed patterns can be illustrated with topographical schemes of Neolithic and Eneolithic O.R.P., located in Volga-Ural interfluve (figure 4).

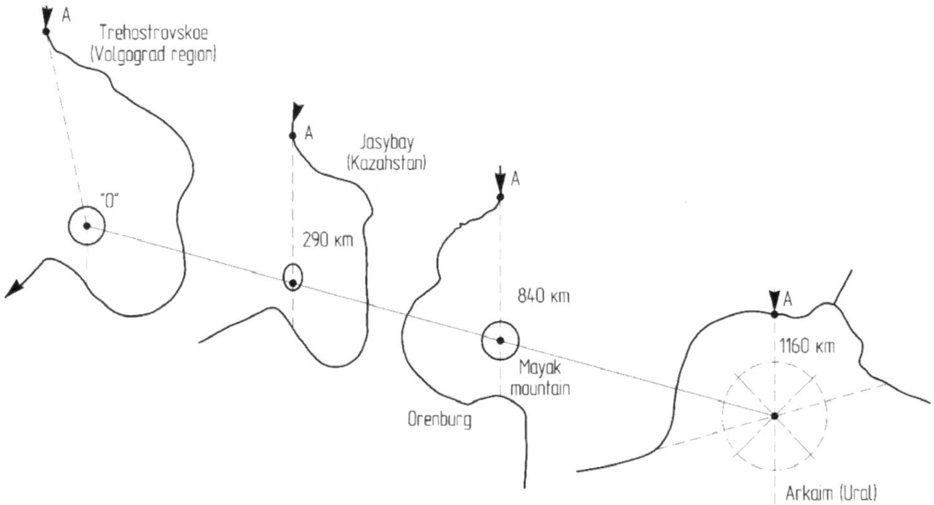

Figure 4. Area of passionary impacts in the Volga-Ural interfluve.

In our opinion the scheme, showed in figure 4, proves the existing correlation of IDEST with the passionary impacts theory. The line, drawn through O.R.Ps., located in river meanders from the Lower Volga to the South Ural (approximately 1166 km in length), through quite equal intervals connects centers of nomadic civilizations of the Volga-Ural interfluve throughout II-I thousand years B.C.

In our opinion hydrography and IDEST interrelation is able to «slightly open» mysteries of death of some civilization, found in droughty or desert area with arid climate. There are examples of disappeared civilizations, whose traces found on the territory, today deprived of considerable sources of fresh water. They are Tiwanaku on the Bolivian plateau, Garamantida in the Libyan Desert, Nazca civilization in the Southern Peru, objects of religious purpose of Nabta in the Nubian Desert and so forth. Assuming that there is direct link between the movement of waveguides in crystal structure of the Earth and change of hydrography on the surface of the planet, then rather logical theory appears, explaining environmental disasters at early stage of human history. Shift of dodecahedron`s edges in the planet led not only to passionary impacts, but also to change in hydrography, the emergence of new water flows and drying-up of the old ones. The consequence of it was natural migration of people from low-level water areas and gradual desertification of the surrounding area.

3. Conclusion

Thus, IDEST is not only accompanied with the passionary theory of L N Gumilev, but also significantly corrects it. Authors suggest that electromagnetic influence of the Earth`s core is one of the sources of energy impact on ethnogenesis processes (by passionary impacts). Nodal points and dodecahedron`s edges are not only the sources of energy, but also are locations of waterways. River bends, in turn, acted as the source of additional energy influence, that explained by location of objects of religious purpose in river meanders.

IOP Conf. Series: Materials Science and Engineering **483** (2019) 012055 doi:10.1088/1757-899X/483/1/012055

References

[1] Korenyako V 2006 On criticism of the concept of L N Gumilyov *Ethnographic Review* **5** pp 22-35
Shnirel'man V A, Panarin S A 2000 Lev Nikolaevich Gumilev: the founder of ethnology? *Herald of Eurasiai* **3(10)** pp 5-37
[2] Timashev S F On the physico-chemical nature of passionary impulses [Electronic resource]. – Access mode: http://gumilevica.kulichki.net/debate/Article08.htm (access date: 10.10.2018)
[3] Gumilev L N 2002 *The end and the beginning: popular lectures on ethnology* (Moscow: Rol'f) p 48
[4] Gumilev L N 2002 *Ethnogenesis and biosphere of the earth* (Moscow: Rol'f) p 28
[5] Gumilev L N 2002 *Ethnogenesis and biosphere of the earth* (Moscow: Rol'f) p 282
[6] Gumilev L N 2002 *Ethnogenesis and biosphere of the earth* (Moscow: Rol'f) p 331-3
[7] Gumilev L N 2002 *The end and the beginning: popular lectures on ethnology* (Moscow: Rol'f) pp 37-8
[8] Gumilev L N 2002 *The end and the beginning: popular lectures on ethnology* (Moscow: Rol'f) p 75
[9] Gumilev L N 2002 *The end and the beginning: popular lectures on ethnology* (Moscow: Rol'f) p 71
[10] Gumilev L N 2002 *Ethnogenesis and biosphere of the earth* (Moscow: Rol'f) p 523
[11] Gumilev L N 2002 *Ethnogenesis and biosphere of the earth* (Moscow: Rol'f) pp 333-4
[12] Gumilev L N 2002 *Ethnogenesis and biosphere of the earth* (Moscow: Rol'f) p 362
[13] Question: How can one characterize the «factor X» (passionary) from the point of view of modern biology [Electronic resource]. – Access mode: http://gumilevica.kulichki.net/faqs/faqs05.htm#Q1 (access date: 08.10.2018)
[14] Ivanov V V, Gorshkov E S 2012 On cosmic conditionality of passionary impulses [Electronic resource]. – Access mode: http://www.biophys.ru/archive/congress2012/proc-p155-d.pdf (access date: 08.10.2018)
[15] Lachugin K 2005 *Is the earth a big crystal?* (Moscow: Zaharov) p 9
[16] Lachugin K 2005 *Is the earth a big crystal?* (Moscow: Zaharov) pp 25-32
[17] Makarov V A 2010 The structure of the earth's crust, as a result of the operation of geocrystal power frames *Russian thought* **1-12** p 50
[18] Lachugin K 2005 *Is the earth a big crystal?* (Moscow: Zaharov) p 36
[19] Ramazanov S P, Nikolaev N Yu, Yurchenko S A 2017 River landscape and monuments topography in archaeological cultures of the Volga and the South Ural regions in the late eneolithic period of and the bronze age *Historical, Philosophical, Political and Law Sciences, Culturology and Study of Art. Issues of Theory and Practice* **2(76)** pp 166-70
Nikolaev N Yu, Yurchenko S A 2017 Cultural landscape and meanders: historical parallels, peculiarities of topography, hidden semantics *Historical, Philosophical, Political and Law Sciences, Culturology and Study of Art. Issues of Theory and Practice* **9(83)** pp 131-5

IOP Conf. Series: Materials Science and Engineering **483** (2019) 012056 doi:10.1088/1757-899X/483/1/012056

Industrial discourse in the context of court interpreting in Sweden

A M Polikarpov, A D Frantsuzova

Northern (Arctic) Federal University named after M.V. Lomonosov,
Translation and Applied Linguistics department,
Severnaya Dvina Emb. 17, Arkhangelsk, Russia; 163002

E-mail: polikarpov.ling@yandex.ru, info@alinafrants.se

Abstract. The article examines industrial discourse as part of spoken legal language in a courtroom setting in Sweden and its impact on the interpreting process and results. The paper draws up a brief overview of the discourse key features, the concept of communicational failure (a misfire) and presents analyses of authentic audio recordings from court trials interpreted from Swedish to Russian. The study is based on the linguistic semiotics approach with focus on the subfield of semantics. A special attention is paid to legal vocabulary and technical terminology. The eventual outcome of the article might be seen as an attempt to unravel the nature of mistakes and interactional misfires in connection with spoken industrial discourse.

1. Introduction

The sphere of legal language represents a "recognizable linguistic variety, differing from other kinds of language use" [1, p. 3]. At the same time it is not a single, unified concept, but rather "a cluster of uses" and the term itself is one of many other designations (e.g. *law and language, the language of the law, language in law, forensic linguistics*) used to describe a general field of study. Details and nuances between these phrases and what they denote have been much discussed in the literature. Nevertheless, this "fractured" field [2, p.11 as cited in 1, p.3] is relatively well-researched to conclude that legal language is a kind of discourse, in other words "a variably classifiable set of utterances with enough regularities to contrast fairly consistently with other kinds" [1, p.3]. At its broadest legal language can include both written and spoken legal discourse, linguistic evidence, legal education, translation and interpreting [1, p.3].

When put in one specific context – the courtroom – it is inevitable for the legal discourse, as it reflects events of life, to encompass other types of texts. Considering rapid technological development around the globe it is becoming difficult to ignore the occurrence of scientific, technical and business terminology in court discourse more and more often. In this paper, the term "industrial discourse" will be used to loosely describe a scope of special knowledge on and terminology associated with the development, manufacturing and merchandizing an industrial product. For example, when dealing with a theft, industrial discourse might be present when identifying stolen goods or equipment used for stealing, when recounting the circumstances of the event to establish available means and ways of payment and so on. And, of course, there is often technical evidence presented in many types of criminal cases – whether it is represented as an expert's statement or a part of a witness' or defendant's testimony, industrial discourse becomes a constituent of the legal language during the courtroom hearing because of its' direct connection to the merits of a case.

In line with the European Convention on Human Rights [3, art. 6], courts in Sweden are required to provide the right to a fair trial, including the right to free assistance of an interpreter. This obligation is incorporated into Swedish legislation, which clearly states that a so called "authorized interpreter" is to be engaged in the first place [4, kap.5 §6]. The national authorization is in place to ensure that a distinction is made between amateurs and qualified interpreters, who have passed an advanced professional examination and is subject to the supervision of The Swedish Legal, Financial and Administrative Services Agency (Kammarkollegiet). It is also possible to obtain a special authorization as a court interpreter in some languages, which is to indicate a higher quality of interpreting. Becoming a court interpreter in Sweden means that in addition to the basic authorization examination an interpreter must also complete a specialized test that focuses on legal interpretation. However, this test does not include a proficiency exam on either technical or business terminology.

Since both the institutional and the lay members of legal conversations expect high quality of interpreting throughout the interaction without regard to the type of text encountered in the courtroom, it is interesting to investigate if the terminology of industrial discourse presents any particular challenges for an interpreter and leads to communicative failures or just the opposite – creates a perfectly firm basis due to its semantical unambiguity when given a specific context.

This study sets out to assess the functioning of industrial terminology in interpreted court discourse on the basis of audio records of court interpreting in Sweden. These analytical procedures and the results obtained from them are described in the following sections.

2. Methods

The present study was conducted using sample examination of more than 35 hours of court interpreting between Russian and Swedish languages. The data was gathered during 2018 with significant geographical spread. The examined material was collected in form of audio records that include 22 hearings held in 9 different district courts in Sweden [Attunda District Court, Gothenburg District Court, Hässleholm District Court, Kalmar District Court, Lund District Court, Nacka District Court, Södertörn District Court, Stockholm District Court, Östersund District Court].

The records represent one specific stage of the hearing process which is examination of the Russian-speaking plaintiff, defendant or witness. At first the speaker is asked to present his or her version of the events and afterwards the parties and the judges can put questions to the speaker. It also means that the interpreter is working with both Russian and Swedish as alternating source and target languages, although in a few cases there are two one-way interpreters working in tandem. In most cases interpreting is consecutive, sometimes simultaneous. To the extent that there is no video image available, the only non-verbal indicators to allude to a communicative failure are hesitation pauses, coughs, sounds of hesitation and sighs.

With the purpose of facilitating analysis, transcriptions of the records were made in compliance with the authors' analytic perspective and needs. According to the used system only speech and pauses are reflected, where short pauses are conveyed with "/", long pauses –with "//".

3. Results

Analyses of data have demonstrated a variety of possible outcomes. They have shown that sometimes actualization of industrial discourse does not hinder the interpreter's performance and is covered by his or her thesaurus on the subject as illustrated in examples 1-4. It has also been revealed that terminology of industrial discourse might go beyond the scope of interpreter's active vocabulary, and thus might or might not cause communicative misfire as shown in examples 5-8 respectively.

Interestingly, the correspondence between communicative misfire and industrial discourse was observed not as a result of interpreter's failure of perception, understanding or delivery of an adequate and equivalent translation, but rather because of interlocutor's low information awareness and knowledge on a subject (example 9).

There is also a peculiar complication of the issue of occasional loanwords (taken from Swedish and uttered by Russian-speaking side during his or her speech) that in most cases is shown to cause a shorter pause or a slight delay in interpreting rather than a communicative misfire as illustrated in examples 4 and 8.

Overall, the findings were not unexpected; they correlate with the research hypothesis concerning the industrial discourse representing both a source of misunderstandings and translation mistakes, and also constituting a solid ground for meaning-making as one of court interpreting elements.

4. Discussion and Conclusion

This section is divided into four parts. The firstpart gives a brief overview of discourse key features. The main issues addressed at the beginning of the section are therefore: a) type of discourse from the perspective of interactional sociolinguistics (person-oriented /institutional), b) type and characteristics of discourse from the perspective of semiotic linguistics (legal / industrial), c) text form(written / spoken).

The second part deals with defining a concept of communicative misfire. The third part of the section presents the findings of the research, focusing on examination of translation strategies used to prevent or overcome failure in interpreted court discourse. The remaining part of the paper sums up the study and provides conclusions and suggested implementation of the research results in the field of oral legal translation.

4.1. Discourse key features

4.1.1. Type of discourse from the perspective of interactional sociolinguistics. From a sociolinguistic point of view legal language used in a courtroom setting, being opposed to informal communication, is a typical example of institutional discourse [5, p.89], which means "a discourse that takes place within a professional or work-based setting" [6, p.7]. In other words, it has prescriptive nature and is provided with a certain temporal set of reference. A further definition of institutional discourse is given by V. Karasik who describes it as "a specialized clichéd type of communication among people, who might not know each other in person, but should communicate in accordance with the regulations of this community" [7, p. 292]. Characteristic for institutional interactions as a set of typical communicative events is asymmetrical relations between the participants (with power, control and status unequally distributed between a member of an institution on one hand and laymen, defendant or witnesses on the other hand) [see 6, p.15].

Among other prominent attributes are occurrence of indirect speech acts (e.g. when "questions are intended as commands requiring compliance and therefore expect no reply at all" [6, p.15]), different rules of politeness [6, p.15], predetermined topic of conversation, highly conventionalized speech features. Judicial discourse is also characterized by more stringent structuring, maximum speech restrictions, fixed exchange of communicative roles, less conditional context, organization in accordance with the norms of a justice institution etc. [5, p.88-89].

4.1.2. Type and characteristics of discourse from the perspective of structural linguistics. In order to examine industrial discourse in the context of court interpreting, it is reasonable to proceed with considering how it looks like and what distinguishes it from legal discourse from perspective of linguistics, for example, stylistics and discourse analyses. Crystal describes the language of legislation as follows: "[it] depends a great deal on a fairly small set of grammatical and lexical features. For example, modal verbs (e.g. *must, shall, may*) distinguish between obligation and discretion. Pronouns (e.g. *all, whoever*) and generic nouns (*hypernyms, e.g. vehicle, person*) help foster a law's general applicability" [8, p. 374 as cited in 1, p.3]. Further descriptions are offered by T.A. Volkova, who notes: "multi-part closed syntactic constructions having a partially autonomous status within a sentence are used. The objectivity of information presentation is achieved through the use of present tense of a verb, the universal nature of information is transmitted, in particular, by pronouns with generalizing semantics (*each, none, everyone*)" [5, p. 91].

Both legal and industrial discourse have common stylistic features: legal texts [5, p.90], as well as scientific texts [9, p.3] are distinguished by objectivity, generalization, high informativity, logical organization of presentation (strict sequence, clear connection between the main idea and details), and thus by semantic clarity and distinctness.

It is significant that both types of investigated discourse are of systematical character in terms of vocabulary. As far as industrial discourse is concerned, special attention should be paid to terminology since continuous development of science, technology and business involves the emergence of a huge variety of terms, professionalisms, and slang. Moreover, terminology "is the most 'mobile' layer of vocabulary, subject to constant and active enrichment and change" [9, p.3]. It is against this background that systematic character of industrial discourse is formed.

Lexicon of industrial discourse is primarily built on the use of technical terminology and so-called special vocabulary. A term is usually understood to be a word (or a group of words), having a specific and unique meaning within a given industry or specialization. In addition, any possibility of understanding or interpretation, different or divergent from author's, should be excluded. A technical term must comply with following qualities: 1) designate only one referent; 2) be unambiguous and in this sense context independent, in other words, its exact meaning must be indicated by its definition in all cases of its use in any text; 3) only one term must correspond to each concept so that there are no synonyms with matching meanings.

The systematical nature of legal language, on the contrary, has limitative effect. According to Golev, legal language is a "special system that reveals its own significance, formed by opposition of special concepts and determination by current legislation" [10]. Thus, it is fair to state that interpretation of a text depends on the whole system of concepts and laws and its internal patterns. As Golev notes: "The decoding of this code requires special hermeneutic presumptions, special culture of thinking, formed by many generations of people with special legal thinking". Volkova observes that: "specification of a legal text, its understanding in connection with a certain situation of assignment create a closed context and completeness of understanding; in this case, interpretation is limited to context and in fact comes down to activities for understanding prescriptions. <...> Open context of judicial practice, the balance of legal dogma and cultural tradition form certain philological and legal openness and some incompleteness of a text, which confirms the need of its interpretation" [5, p.90].

In other words, legal terminology often functions as a framework that needs to be filled with context to form an exact meaning, it is abstract to a certain degree and needs to be defined more accurately within a context, whereas technical terms are designed to be context independent when identified within a broader context as a specific technical field.

Another interesting observation to support this conclusion was made by A. Tcherdantsev: "Qualified lawyers easily "grasp" the abstract meaning of the rule of law and in relation to specific situations without much effort. But <...> such <...> understanding of the rule of law is based on the experience of the past, individual interpretation and interpretation of other subjects" [11, p.14].

In line with a linguosemiotic approach the issue of terminology lies within a subfield of semantics which is therefore considered to be a primary object of the present study.

4.1.3. Text form. Since the undertaken research is being conducted on interpreting in the sphere of legal discourse, which throughout this paper is used to refer to legal language used in courtroom settings in Sweden, our main focus lays on spoken texts. The literature on spoken discourse has highlighted several key features: cooperation between participants, politeness and rules of turn-making: turn design, allocation, distribution and function [6, p.15]. People usually presume that a written text can be defined as highly structural, whereas oral speech is unstructured [6, p.15]. However, spoken texts are both rule-governed and have their own "norms and conventions, [which] contribute to creating distinctive and recognizable features of a genre" [6, p.15]. If we unexpectedly closed a brief meeting with *The End* or*In conclusion,* it would be recognized as rule-breaking and "the interlocutor or the audience would draw inferences" [6, p.15] from this marking or foregrounding manner of production.

4.2. On the concept of communicative misfire

Before moving on to examining examples of interpreted legal texts, it is necessary to discuss the concept of communication failure. The term 'misfire' was introduced by Austin in his speech act theory and initially referred to a putative act of speech that fails to perform because it either fails to meet felicity conditions or the addressee fails to respond with an appropriate uptake [12, p.16, 25-27]. While a variety of definitions of the term 'communication failure' has been suggested, in this paper we will use the definition suggested by E. Zemskaya who explains it as "complete or partial misunderstanding of a partner's statement of communication, i.e. non-fulfillment or incomplete realization of communicative intention of a speaker" [13, p.602].

4.3. Research findings

In order to consider the issues of industrial discourse in the context of legal interpreting let us now move on to examining excerpts from audio recordings of the court hearings:

(1) Prosec: Du behövde en skiftnyckel?
Interp:Vam nuzhen byl razvodnoy klyuch?

(2) Interp: Kakaya u vas spetsial'nost'?
Def-nt: Elektrik
Interp: Elektriker
Prosec: Svag ström stark ström eller // vilken typ av elektriker?
Interp: Kakogo tipa u vas // dopusk yest' na vysokoye napryazheniye na nizkoye napryazheniye?
Def-nt: Nu dopuskov u menya netu / no ya mogu rabotat' do 380 vol't
Interp: Jag har ingen särskilt behörighet men jag kan jobba med upp till 380 volt

(3) Def-nt: Skazal / chtovotzdes' on imennozabil v navigatsii tot magazin...
Interp: Det var Viktor som programmerade i sin GPS just den här butiken...

(4) Def-nt: U menya yest' svoya firma /my / "snikeristy"
Interp: Jag är snickare och vi har ett byggföretag

Examples 1 to 4 illustrate that although not having been prepared for dealing with industrial discourse during oral legal translation, the interpreter did not stumble, and no communication failure occurred. It could be assumed that these positive results were partly due to widespread use of technical terminology in our daily life (examples 1 and 3) which made us accustomed to some more or less frequently used technical terms, and partlydue to theinterpreter's broad vocabulary and knowledge of professional realia (as in example 2). It can also be argued that mono-semantically features of the term *"skiftnyckel"* (adjustable spanner) was of greater importance since the interpreter happened to know exactly what it denoted and had knowledge of an equivalent term in Russian. Another helpful feature of technical terminology in the present language pair is international unification of some units of measurement (e.g. *"volt"* from example 2). The latter example (4) is particularly interesting where the defendant uses a so-called barbarism – the Swedish word *"snickare"* (carpenter) with Russian morphology (agent suffix "*–ist*" and inflection outlining plural number *"–y"*) in order to name his occupation. Since *"snikeristy"* is not a real technical term the interpreter had to do the logical procedure of ruling out all other possible meanings until he or she made a rapid conclusion that it must be an adopted version of the Swedish term for "carpenter" due to the language interference.

Regarding difficulties in comprehending or translating industrial discourse by the interpreter, the following examples illustrate cases where communication between participants of court discourse was not disrupted despite some distortion of sense:

(5) Def-nt: Yasmotrel to shomnenado / nu raznyyeprofilya / raznyye tam // yazanimayus' yeshchedomaremontami
Interp: Jag tittade på det som var intressant för mig / ja tittade bland annat på olika verktyg som jag behöver för att jag sysslar med renoveringar hemma

(6) Prosec: När bestämde du dig för att ta<...>[unhearable] *utan att betala?*
Interp: Kogda vy prinyali resheniye vzyat' i uyti / ne zaplativ?

The strategy of generalization might be used to prevent a misfire and provide good communication flow. When as shown in example 5, the term *"profil"* (drywall profile) from construction industry appeared challenging for the interpreter, she chose to replace it with a hypernym *"verktyg"* (tool). It was wrong decision that created an incongruity between original and translated texts, since drywall profile doesn't share the same semantic field as the generic term "tool" and cannot therefore be put into hyper- and hyponimic relation. However, this translation mistake did not lead to a communication failure between the interlocutors, presumably because this technical term was not of interest for merits of the case and its precise translation was considered to be excessive. Another translation technique illustrated in example 6 is omission of challenging element, relying on the concept of "shared knowledge" – because of indistinct pronunciation the interpreter had trouble in hearing what the item was, so she relied on the fact that the information about stolen item was already known and accepted by the parties, and thus it was also established what item was being discussed, therefore it didn't need to be named explicitly and could be omitted.

Let us now turn to the examples of industrial discourse (7 and 8) that led to a misfire in the course of court interpreting:

(7) *Witness: Eto bylo 10 /Nebol'shoy zhiloy kompleks iz 10 domov*
Interp: Det var ett mindre villaområde / 10 villor
Witness: 175 kvartir
Interp: Ursäkta kankse inte villor // 10 fastigheter / totalt 175 lägenheter

As far as the word "*dom*" (house) is concerned, it is a part of common vocabulary and can refer to both a private villa and a high-rise. In the example above nothing alludes to what type of building is meant by the witness and so the interpreter makes a choice based on her presumption about real estate market in the region, which would suggest that "*zhiloy kompleks iz 10 domov*" is most likely to denote a small residential area with single-family housing. However, when the witness specifies "*175 kvartir*" (175 apartments) it is evident that her initial assumption has been wrong, and some corrections should be made on the preceding rendition. In this case, it is rather expectations of a technical term and its factual absence in the context of construction industry than polysemy of the word "*dom*" (house) that challenges the interpreter and leads to an unsuccessful communication.

Another example of a communication failure due to industrial discourse use illustrates what is probably the most common type of the source of misfire in that respect:

(8) *Def-nt: I / tak / poluchayetsya / my potom podoshli k samorezam // gde samorezy byli kryuki byliy eshche eti vesha-veshalki yest' / kotoryye veshayutsya / i my ikh nachali nabirat' / i oni u nas v kraske provalivalis'*
Interp: Och det är / det är // när vi gick runt såhär, som [tolken måste fråga]
Interp: Chto takoye samorezy? [asking the defendant]
Def-nt: "Skrover" "spiker" gvozdi shurupy
Interp: Aha aha och // skruvar och // bultar och liknande

Difficulties in interpreting illustrated in this example are caused by a designator of a specific type of fastener (a hardware device). Since the speech originator is not an expert witness, but a common citizen, the term "*samorezy*" (something that is "self-cutting" – in literal translation) might from the interpreter's perspective be perceived as either a colloquial name for a tool that the interpreter is familiar with or as an unknown technical term. In that way, a misfire here is caused by the act of reference and the issue of ascertainment of the signified. Beginning of the interpreter's line gives us a clue about her hesitation on the matter, which was materialized in form of repetitions and interpreting with somewhat vague sense, presumably to win some time to think. Without making any success on decoding the word by herself, the interpreter attempts to clarify the meaning of "*samorezy*" by asking the originator of the message for explanation. However, the defendant proceeds with enumeration of terms that could either be a series of synonyms or continuation on the items collected in the shop. This situation is similar to the one described in example 4 – by ruling out a possibility of "*skrover*" and "*spiker*" being real technical terms, the interpreter established that it was a result of language interference which represented the defendant's trying to pronounce these terms in Swedish. When having decoded these barbarisms in addition to other synonyms in Russian, the interpreter proceeds with her task, producing generalized translation into Russian (because "*samorezy*" is not a screw in general, but a *"self-tapping screw"*). This, however, doesn't have any negative impact on continued communication.

(9) *Def-nt: Net / v Belorussii yest' takoye ponyatiye kak CHUP // eto chastnoye unitarnoye predpriyatiye /eto kogda odin chelovek rukovodit 4-5 lyud'mi // i ya vot rabotal v takoy kompanii / my delali domain bani iz brusa / iz dereva*
Interp:Nej, asså jag har jobbat inom något som i Vitryssland motsvarar en enskild firma där man som ensam ägare kontrollerar ... ett arbetslag som består av 4-5 personer och där jag hållit på och satt ihop bastuhus
Prosec:Aa, så du, du är liksom arbetsledare, chef för arbetslagen, styr arbetet?
Interp:Det här var egentligen inte min firma, utan jag var en av arbetarna på en sån här firma.
<...>
Prosec:Så tolken har uppfattat dig fel när jag sa att du hade drivit en enskild firma och styrt arbetslag?

In the preceding turns of the defendant's talk he states to have been a construction worker, neither self-employed, nor working for some other company. Naturally prosecutor wanted a more precise answer about the defendant's state of working and in the reply to his question encountered a previously unknown to him term *"CHUP"* (Private Unitary Enterprise) that denotes a form of enterprise existing in Belarus. Notwithstanding rather clear information from the defendant and its correct interpreting to Swedish, the prosecutor still could not get the grip of the defendant's role in the company. At last, despite subconsciously admitting his own misunderstanding of the defendant's explanation, the prosecutor concluded that responsibility for a communication failure lies on the interpreter.

4.4. Conclusions and suggested implications of the research results

Having discussed key features of legal and industrial discourse, the concept of communicational failure (a misfire) and analyzed some real conversation practice, the final part of this section addresses results of the study and ways of its suggested implication.

The paper has raised important questions about the nature of industrial terminology in court interpreting and the scope of technical terminology in contrast to legal vocabulary. The research has shown that degree of accuracy on interpreting technical and business terminology as well as interpreting techniques used to deal with potential communication failures might differ.

Taken together, these results suggest that industrial discourse can contribute to occurrence of communication failures, and thus incongruities in interpreting due to interpreter's poor knowledge of technical terminology within the relevant technical field. This, however, can be helped by advance preparations on the topic or by using feedback-feature of a dialogue setting of an oral translation. The findings will be of interest to active court interpreters, members of institutional settings and other participants of court discourse.

References

[1] Durant A and Leung J H C 1969 *Language and Law* (New York: Routledge) p 241
[2] Tiersma P *Loyola-LA Legal Studies Paper* **2009-11**
[3] The European Convention on Human Rights (ECHR) URL:https://www.echr.coe.int/Documents/Convention_ENG.pdf
[4] Rättegångsbalken URL: https://www.riksdagen.se/sv/dokument-lagar/dokument/svensk-forfattningssamling/rattegangsbalk-1942740_sfs-1942-740
[5] Volkova T A 2010 *Diskursivno-kommunikativnaya Model' Perevoda* (Moscow: Flinta; Nauka) p 125
[6] Coulthard M, Johnson A and Wright D 2007 *An Introduction to Forensic Linguistics Language in Evidence* (New York: Routledge) p 235

[7] Karasik V I 2010 *Yazykovaya Kristallizatsiya Smysla* (Volgograd: Paradigma) p 428
[8] Crystal D 2010 *The Cambridge Encyclopedia of Language* 2nd Edition (Bangor: University of Wales) p. 572
[9] Gredina I V 2010 *Perevodv Nauchno-tekhnicheskoy Deyatel'nosti: Uchebnoye Posobiye* (Tomsk: Izd-vo Tomskogo politekhnicheskogo universiteta) p 121
[10] Golev N D Vzaimodeystviye yestestvennogo i yuridicheskogo yazyka kak bazovaya problema yurislingvistiki URL: http://lingvo.asu.ru/golev/articles/z09.html
[11] TcherdantsevAF1979 Tolkovaniye Sovetskogo Prava (Moscow: Yuridicheskaya literatura) p 166
[12] Austin J L 1975 How to Do Things with Words (Cambridge: Harvard University Press) p 168
[13] Zemskaya E A 2004 Yazyk kak Deyatel'nost' // Yazyki slavyanskoy kul'tury (Moscow: Flinta) p 685

Innovative technologies of personnel management of industrial enterprises in terms of introducing professional standards

Y Y Savchenko[1], O G Goleva[1], I A Korchagina[1], A V Ovsyannikova[1], Y.S. Lobanova[1]

[1]Altai State University, 61 Lenina Street, 656049, Barnaul, Russia

E-mail: kolmakova@rb.asu.ru

Abstract. The article is devoted to the problems of developing and adapting innovative personnel management technologies in the process of introducing professional standards. The authors dwell on innovative personnel technologies from the point of view of applying new methods and tools to the work of personnel, as well as the way these methods affect the increase in work efficiency. The article points out the main difficulties in introducing personnel innovations. The authors determine levels of adaptating personnel technologies at Russian industrial enterprises to the system of national qualifications. Increase in work efficiency is to become one of the final indicators of successful introducing the system of national qualifications. Taking into consideration the results of the rating published in the analytical periodical Market Watch, the auhors make an attempt to make a comparative analysis of Russia and other countries of the world in terms of work eficiency indicator. The article analyses the attitude of chief executives and HR-managers at industrial enterprises to the idea of introducing professional standards. The authors distinguish the main problems of introducing professional standards into the practical activity of Russian industrial enterprises. The authors come to the conclusion that the search of various innovative personnel technologies at industrial enterprises is necessary for successful functioning of the created system of professional standards.

1. Introduction

The topicality of the problem under analysis is determined by a number of factors and circumstances. Firstly, working out professional standads has been actively introduced into practice in Russia since 2012. Though the first attempts of transition to the already developed world system of professional standards were made in 2007 at the initiative of employers. Consequently, since 2012 the development of personnel technologies has been connected with reorganization of personnel management policy at enterprises, as well as search and use of personnel technologies which require additional research in terms of adapting to the newly introduced system of professional standards.

Secondly, according to the estimates given by chief executives and HR-managers, personnel innovations can be effective and adapted to the system of professional standards only partially, which determines the topicality of the problem under analysis.

Thirdly, introducing professional standards calls for the increase in efficiency of personnel management processes connected with understanding employees' needs and their motivation to continuous development.

Thus, research, development and adaptation of personnel innovations to the practice of introducing professional standards predetermines the topicality of our analysis.

2. Research methodology

Personnel technologies (HR-technologies, human resources technologies) is the term integrating various business functions connected with management of human resources in any institution. They include selection and hiring of personnel, motivation of employees, social protection of personnel, its development, analytical activity, training and competence management.

The methodology of human resource management distinguishes between two types of personnel technologies: traditional and innovative. Innovative personnel technologies assume application of new methods and tools in personnel management and are aimed at increasing labour efficiency.

Innovative personnel technologies are personnel management innovations connected with practical use of unconventional solutions and risk and aimed at creating and realizing competitive advantage (benefit), which will predetermine their further field and cross-field diffusion in the market.

Introducing professional standards into practical activity of organizations is impossible without innovations in personnel management. However, innovations, including personnel innovations, are always connected with something new and unconventional. Therefore, their introduction is accompanied by certain difficulties.

1. Innovations in personnel management system always lead to gaps in conventional behavior, criteria and management structure, which leads to more tangible manifestation of the phenomenon of resistance to changes, than it happens during the introduction of material innovations.
2. It is more difficult to present economic justification of personnel innovations, as it is necessary to compare costs of their introduction with estimated results, and the result of personnel innovations often has a more multidimensional and hardly changeable character.
3. On the one hand, personnel innovations require fewer resource expenses in terms of time consuming and financing. They are usually carried out with fewer one-time costs, but on the other hand, they can significantly affect the financial performance of the organization and thus can have a negative effect.
4. Uniqueness of human capital of the institution in the process of diffusion of personnel innovations requires their adaptation to its specific character.

Today, traditional procedures of hiring, selection and performance assessment of the organization can be complemented with innovative techniques of distinguishing motivatonal factors and leadership skills. Vocational training programs can be supplemented by innovative techniques of identifying the leading source of perception in the course of training. Under the conditions of introducing professional standards, innovative technologies of personnel management are to be focused on the development of intellectual component on the basis of parametrical and structural transformation of inner competences and motivations of employees.

Therefore, it is important to manage employees, their knowledge and competences in such a way that the organization can build a long-term competitive advantage [1].

Consequently, it is necessary to model such intraorganizational developing environment in which readiness for changes is perceived as an attribute of development.

Major employing organizations, in the person of their chief executives and HR-managers were used by us as respondents in our research. The sampled population was costituted by 196 industrial Russian enterprises which appeared in the rating of RAEX-600 (based on the volume of output) in 2017 [2].

Defining the level of adaptating personnel technologies to the system of national qualifications (SNQ) was carried out on a five-point scale: 1 –not adapted; 2 - adapted to low; 3- not fully adapted; 4 – implemented at a sufficient level, but requiress improvement; 5 – fully adapted.

Table 1. Summary table of assessing the adaptation level of personnel technologies to SNQ

Name of personnel technology	Total score
Selection and admission of new employees	3.7
Staff motivation	2.3
Development of staff	2.4
Personnel assessment	3.1

Formation of personnel reserve	2.4
Career development	2.9

The results of the conducted survey showed that two of the six personnel technologies offered to assessment "are adapted to low " and three of them "are not fully adapted". Also, only selective and enrolling technologies are at a stage of adaptation to the existing professional standards at the moment, though they require improvement as well. None of personnel technologies is not fully implemented in terms of adaptation to the system of national qualifications.

A consolidated assessment of the adaptation of personnel technologies at Russian industrial enterprises to the SNQ is presented in chart 1.

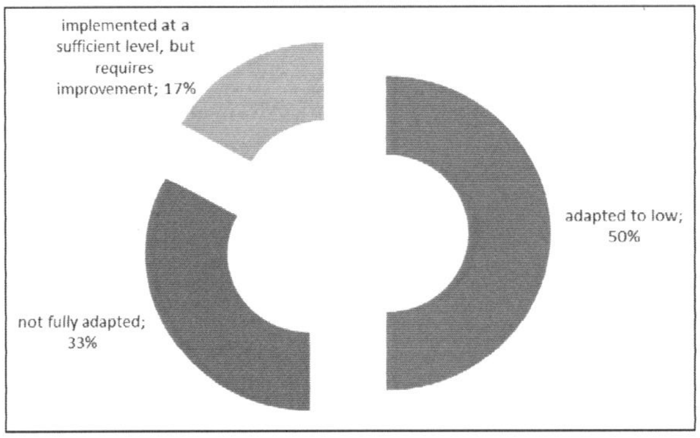

Figure 1. Summary assessment of the adaptation of personnel technologies at Russian industrial enterprises to the system of national qualifications

The received results showed that personnel technologies at Russian industrial enterprises are not implemented to the right degree, taking into account a large number of currently existing approved professional standards. However, it is the adaptation of personnel technologies of organizations to the system of national qualifications that constitutes one of fundamental elements of effective functioning of any system.

The increase in labor productivity is to become one of the resulting indicators of introducing the system of national qualifications. In Russia, as well as in many countries of Europe, there exists a complicated demographic situation which is characterized by decrease in number of able-bodied population. But unlike the countries of Europe, in Russia it is accompanied by low labor productivity. Developing and introducing the national system of qualifications is to promote increase in labor productivity by vocational training of labor force. However, in Russia we still have the lowest labor productivity in Europe.

The analytical periodical Market Watch made the rating of labor productivity. In 2016 Russia took the 32nd place [3]. But in 2017, Russian economy became even less productive and Russia dropped to the 34th place in the ranking. Experts predict a continuation of the downward trend in 2018-2019 (table 2).

At the same time the majority of the countries have much shorter working weeks than in Russia. So, in France, the Netherlands and Germany we can see the shortest week - 26 hours. Russia works 1914 hours a year, this indicator is higher only in Costa Rica, South Korea, Greece and Mexico.

Table 2. Labor productivity rating according to the analytical publication Market Watch.

2016 Rank	2017 Rank	Country	Annual GDP per capita	Annual Hours Worked	Productivity Per Person Per Hour	Productivity Per Person Per Day	Producti vity Per Person Per

							Week
1	1	Luxembourg	£78.320.90	1512	£51.80	£301.23	1506.17
2	2	Norway	£56.556.80	1424	£39.72	£217.53	1087.63
4	3	Switzerland	£60.249.00	1590	£37.89	£231.73	1158.63
7	4	Denmark	£40.710.51	1410	£28.87	£156.68	782.89
15	5	Iceland	£52.028.94	1883	£27.63	£200.11	1000.56
8	6	United States	£45.898.98	1783	£25.74	£176.53	882.67
3	7	Australia	£42.515.76	1669	£25.47	£163.52	817.61
9	8	Ireland	£47.805.77	1879	£25.44	£183.87	919.34
5	9	Netherlands	£34.383.76	1430	£24.04	£132.25	661.23
10	10	Sweden	£38.364.73	1621	£23.67	£147.56	737.78
6	11	Germany	£31.757.78	1363	£23.30	£122.15	610.73
11	12	Austria	£33.715.30	1601	£21.06	£129.67	648.37
14	13	Finland	£32.811.11	1653	£19.85	£126.20	630.98
13	14	Canada	£33.580.67	1703	£19.72	£129.16	645.78
12	15	France	£28.716.93	1472	£19.51	£110.45	552.25
20	16	New Zealand	£31.652.78	1752	£18.07	£121.74	608.71
16	17	United Kingdom	£29.115.63	1676	£17.37	£111.98	559.92
17	18	Japan	£29.476.82	1713	£17.21	£113.37	566.86
22	19	Israel	£30.126.80	1889	£15.95	£115.87	579.36
19	20	Italy	£22.905.29	1730	£13.24	£88.10	440.49
18	21	Spain	£20.515.38	1695	£12.10	£78.91	394.53
23	22	Korea	£22.418.33	2113	£10.61	£86.22	431.12
21	23	Slovenia	£16.217.40	1682	£9.64	£62.37	311.87
28	24	Portugal	£15.174.77	1842	£8.24	£58.36	291.82
24	25	Czech Republic	£14.271.44	1770	£8.06	£54.89	274.45
27	26	Estonia	£13.776.13	1855	£7.43	£52.99	264.93
25	27	Slovak Republic	£12.637.00	1740	£7.26	£48.60	243.02
31	28	Greece	£13.710.38	2035	£6.74	£52.73	263.66
26	29	Lithuania	£11.619.46	1885	£6.16	£44.69	223.45
33	30	Latvia	£10.924.45	1910	£5.72	£42.02	210.09
29	31	Hungary	£9.830.61	1761	£5.58	£37.81	189.05
34	32	Chile	£10.520.44	1974	£5.33	£40.46	202.32
30	33	Poland	£9.795.75	1928	£5.08	£37.68	188.38
32	**34**	**Russia**	**£8.381.82**	**1974**	**£4.25**	**£32.24**	**161.19**
36	35	Costa Rica	£9.264.74	2212	£4.19	35.63	178.17
35	36	Mexico	£6.154.74	2255	£2.73	23.67	118.36

3. Results of research

It remains an open question, whether introducing professional standards will result in increase of labor productivity through professional development of personnel. Unfortunately, currrently there are too many unresolved problems in the economy of Russia.

Capacity of Russian industrial enterprises to increase their productivity calls for effective organization of productionprocesses, modernization and technical reorganization. Heads of the industrial enterprises consider these tasks their top priority.

Hence, we observe the minimum attention to the problem of introducing professional standards to the practice. "It is possible to bring a horse to a watering place, but it is impossible to force it to drink". Today, the competitiveness of Russian industrial enterprises is based not on qualifications, but on totally different parameters.

There is an opinion that introducing professional standards to the practical activity of enterprises s can also produce a negative effect. Today the state remains the main customer and the main regulator of the system of national qualifications. Yet, implementing professional standards did not become a priority task for industrial enterprises and consequently did not lead to the emergence of innovative personnel technologies.

The state, certainly, has to act as an attractor in creating and developing similar systems. In Australia there is a signed agreement between the government and employers on the need of reducing the quantity of laborforce without proper qualifications from 47.1% to 23.6% by 2020. In Vietnam there exists a planned target to increase the share of workers possessing technical certificates and diplomas up to 23% by 2020.

The priority purpose of creating the system of national qualifications is achieving the maximum possible compliance of a product of the system of professional education with the needs of employers. Today in the course of professional staff training of estimation of frmed competences according to the accepted federal state educational and professional standards is carried out. But here again we are to face the question of reliability and objectivity of the used means of estimation, including those made by employers. Partially this question will be solved after developing the network of independent certification centers.

There is a question of obligatory application of professional standards. So, in the countries of the European Union, about 20% of jobs require certification, accreditation or licensing of the employee. In Great Britain this indicator is slightly higher than the average European one and constitutes 30%.

In Russia processes of introducing professional standards into practical activity began quite a long time ago, but according to the results of our research, employers are not active enough in their implementation.The employers cannot cope with the role of the intermediary between the system of professional education and labor market so far. Many employers in different sectors of industry consider the model of "corporative" standards their first priority. The replies to the question "Do you support the idea of Introduction of professional standards?" were distributed as follows (chart 2).

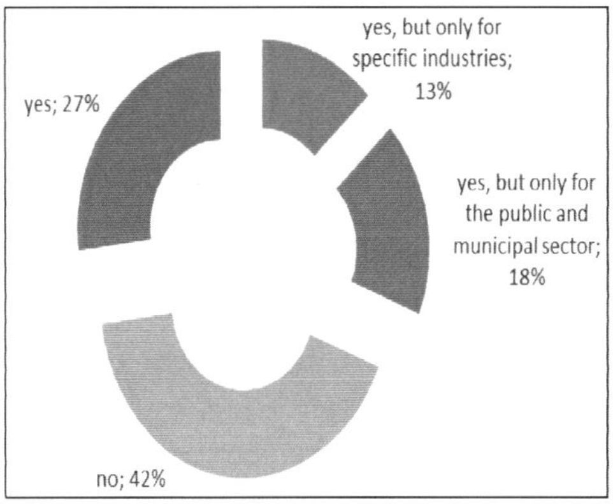

Figure 2. Attitude of heads and specialists of HR departments of the industrial enterprises towards the idea of introduction of professional standards

Nevertheless, despite the contradictory attitude of employers to introducing professional standards, they became an objective reality that, of course, demands a totally different approach to personnel technologies. It refers to the formation of innovative personnel management on the basis of so called "knowledge-relevant" approach which assumes operational aspect of mastering knowledge and skills.

A large number of articles is devoted to this problem. The article « Strategic HRM Practices & their impact on company performance in Chinese enterprises» describes the trends in the development of personnel management in Chinese enterprises and their impact on the effectiveness of their activities (Akhtar S R Ding D Z & Ge G L 2008) [4]. Becker B E and Huselid M A also emphasize the importance of differentiated HR architecture not only between firms but also within firms [6] . The influence of the social environment on the functioning of high-tech systems is presented in the work «High-performance work systems and organizational performance: The mediating role of internal social structure» (Evans W R & Davis W D 2005) [7]. The development of human resource management systems and their impact on organizational effectiveness is reflected in many other scientific studies [8, 9, 10, 11, 12].

The object of innovative personnel management is presented by human resources of the organization, and its subject consists in developing innovative personnel technologies for proper HR-management management

Traditional personnel technologies will be replaced by new techniques based on comparatively new innovative principles (table 3).

Table 3. Characteristics of main types of personnel technologies.

Elements of personnel technologies	Types of personnel technologies	
	Traditional	Innovative
Control object	Professional development in a planned order with orientation to preparing highly specialized experts	Programs of continuous training of personnel with orientation to requirements of professional standards and broader preparation
Career planning	The minimum functions, movement on hierarchy levels	Planning of career is based on competence-based approach taking into account a work position in innovations. Networking with educational institutions and the certified centers
Motivation	Appraisal and awarding on the basis of the standardized work indicators	Individual approach taking into account needs of the individual and encouragement of innovative enterprise approach: creative ideas, large-scale projects, possibilities of professional development, creative and career development
Selection	Standard techniques of assessment of professional characteristics. Screening and recruiting.	Process approach to assessment of candidates, abilities of the applicant, including innovative potential and leadership skills. Headhunting. Competence-based model of assessment.
Assessment	Compliance of the carried-out duties of duty regulations and to professional standards	Assessment of potential opportunities and creative initiatives
Training	Ensuring compliance of knowledge, skills of workers to requirements of professional standards	Advancing development of personnel

Unconditional is the fact that in dynamically changing economic conditions one of the most important conditions of growth of competitiveness of enterprises and industries is increase in potential of human resources. For its achievement the system of national qualifications providing elimination of a set of the problems existing in this system can be used rather successfully. In addition, employers are to have a clear idea of the need for adaptation of the used personnel technologies.

The professional standard will become "working" for improving competitiveness of the industrial enterprise only taking into account its organizational and production specifics.

From the point of view of the atate, it is optimal to use the instrument of compensation of a part of costs for developing competences of employees at industrial enterprises by involvement of experts, development of personnel, including training, teaching technical personnel and managers. Compensation of costs for the improvement of technological processes provides payment of the expenses that appear in the process of performing a technological audit, an energy audit, consultations in the field of information technologies, preparation of engineering decisions, development of industrial design. Along with it is offered to turn on the instrument of support of development of highly professional employees in the program.

Search of new innovative personnel technologies at industrial enterprises is a necessary condition for successful functioning of the created system of professional standards today. Innovative personnel technologies in human resource management at industrial enterprises represent instruments of more effective management of personnel on the basis of competence-based approach. Introducing professional standards into practical activity of industrial enterprises is impossible without the solution of this task.

4. Conclusions

The results of the conducted research allowed us to come to the following conclusions.

The level of introducing professional standards into practical activity of industrial enterprises in Russia is rather low. Personnel technologies are not adapted to the introduced system.

By far, the development of the system of professional standards is at the first stage of development. Stage similar to a phase of origin of an innovative product. At this stage there is a mass search and selection of innovative personnel technologies for ensuring effective functioning of the created system.

Potential of development of the system of national qualifications is rather high, but at this stage it does not allow us to achieve the main strategic objective, that is coordination of interests of professional education market and labor market.

There are possibilities of further increase and development of the staff.

Effective formation and development of the system of professional standards is impossible without development of personnel technologies at industrial enterprises in the innovative way on the basis of competence-based approach.

Increase in efficiency of implementation of personnel technologies will allow to create prerequisites for fuller realization of the revealed intra production reserves of increase in efficiency of functioning of each division and and the company in general.

References

[1] Amaeshi F U 2013 Human Resource Management and Technology Transition for Firm Competitiveness in Nigeria's telecommunication industry *International Journal of Advancements in Research & Technology* **2 (1)** 83-111

[2] Sectoral structure of RAEX-600 rating (by sales volume) *Expert RA Rating Agency* Homepage

[3] The Most Productive Countries in the World: 2017 *Expert Market* Homepage

[4] Akhtar S R, Ding D Z, Ge G L 2008 Strategic HRM Practices & their impact on company performance in Chinese enterprises *Human Resource Management* **47(1)** 15-32

[5] Becker B E & Huselid M A 2006 Strategic human resource management: Where do we go from here? *Journal of Management* **32(6)** 898–925

[6] Evans W R, Davis W D 2005 High-performance work systems and organizational performance: The mediating role of internal social structure *Journal of Management* **31** 758–775

[7] Katou A A & Budhwar P S 2006 Human resource management systems and organizational performance: A test of a mediating model in the Greek manufacturing context *International Journal of Human Resource Management* 17 1223-1253

[8] Beulen E 2009 The contribution of a global service provider's Human Resources Information Systems (HRIS) to staff retention in emerging markets Comparing issues& implications in

six developing countries *Information Technology and People* **22(3)** 270-288
[9] Rodriguez J M, Ventura J 2003 Human resource management systems & organizational performance: an analysis of the Spanish manufacturing industry *International Journal of Human Resource Managemen* **14(7)** 1206-1226
[10] Ulrich D, Smallwood N 2005 HR's new ROI: Return on intangibles *Human ResourceManagement* **44(2)** 137–142
[11] Shrivastava S, Shaw J B 2004 Liberating HR through technology *Human Resource Management* **42** 201-222
[12] Sadri J, Chatterjee V 2003 Building organizational character through HRIS *International Journal of Human Resources Development and Management* vol **3(1)** 84-98

IOP Conf. Series: Materials Science and Engineering **483** (2019) 012058 doi:10.1088/1757-899X/483/1/012058

The optimal placement of information resources on the nodes of a distributed information processing system based on a two-tier and three-tier client-server architecture

Scoba A N, Mikhaylov V K, Ayesh Ahmed Nafea Ayesh, Skorik N S
Platov South Russia State Polytechnic University (NPI), Novocherassk, Russia

Sec-kret@mail.ru

Abstract. This paper presents the results of numerical experiments. The models the optimal allocation of information resources on the nodes of the distributed information processing system (IPS) were compared. They are implemented on the basis of a two-tier and three-tier client-server architecture. The article presents a conceptual model of the functioning of a distributed IPS based on a three-tier client-server architecture. It also shows a mathematical model for determining the stationary probabilities of network states and the average system response time to user requests.

1. Introduction

The problem the optimum location of information resources on LAN nodes is one of the most important tasks. Development and implementation of a distributed information processing system (IPS) are impossible without its solution. This was noted earlier in the article [1]. The task includes the selection of a network topology, an efficiency criterion, the construction of a mathematical model, the development of an optimization algorithm and its software implementation. Publications [2-4] are devoted to this issue. Analysis publications have shown that at present only certain aspects of this problem can be considered. These works have drawbacks: the developed models are separated from the real network structure; the question determining the elements of the matrix, transition probabilities and service rates at network nodes remains open when developing analytical models of data transmission networks; the results of experimental studies and optimization algorithms are missing.

This article presents a conceptual model of the functioning of a distributed information processing system based on a three-tier client-server architecture. There had been obtained the mathematical model for this model. It determines the stationary probabilities of network states using the apparatus of exponential queuing networks. The problem of optimal placement of information resources on the nodes of the computer network is solved using a heuristic algorithm. This placement is made according to the criterion of the minimum average system response time to user requests. The results of numerical experiments are given in comparison with the model of a distributed information processing system. The system is based on a two-tier client-server architecture. Based on the analysis of the results, it is possible to more rationally organize the computational process in the system.

2. Conceptual model

This architecture differs from the two-tier client-server architecture in the physical separation of programs. The database server (responsible for data storage) is separated from the application server (processes data) [5]. This allows optimizing the load on the network and computing equipment. The architecture is implemented: the database server (MySQL server); application server (web server); any browser serves as a client

The article [5] describes the work of distributed IPS on the basis of a LAN three-tier client-server architecture. It is built as follows: DBMS and DB as a set of files are placed on the hard disk of dedicated computers (database server), business analysis software (business applications) is hosted on application servers; There are many client computers, each of which has a thin client installed — a client application that implements the user interface. The user has the ability to run an application on each of client computers. This application accesses the business analysis software through the user interface. The software corresponds to a business application hosted on an application server. This business application analyzes user requirements and generates SQL queries to the database located on database servers. Information on SQL queries is copied to the application server. Thereafter, it is returned to the user's client application using business analysis software. This software uses the user interface to display the result of the query on the computer screen.

The conceptual model of the functioning a distributed IPS on the basis of a three-tier client-server architecture differs from a two-tier client-server architecture (a conceptual model is presented in the article [6]). It is modified: devices simulating the operation of database servers are added – $DB_1,...,DB_s,...,DB_n$; buffer memory of database servers $-BDB_1,...,BDB_s,...,BDB_n$; devices simulating the work of application servers $-SP_1,...,SP_s,...,SP_n$; buffer memory application servers – $BSP_1,...,BSP_s,...,BSP_n$. A conceptual model of the functioning of a distributed IPS based on a three-tier client-server architecture is presented in figure 1.

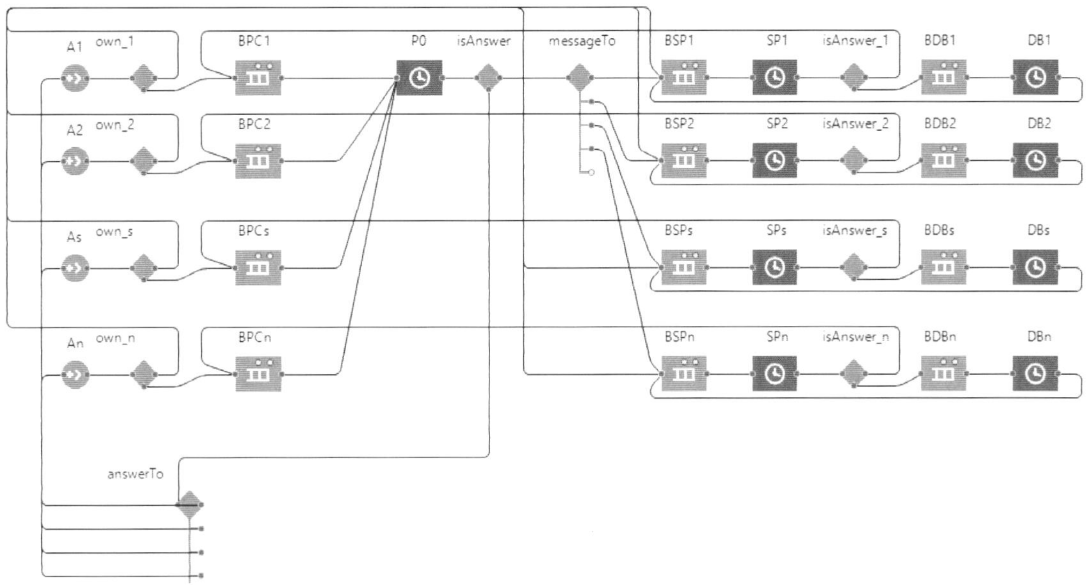

Figure 1. Conceptual functioning model of a distributed IPS based on a three-tier client-server architecture

Assumptions were made to formalize the formulation of the problem [7]: the network topology is a star; network architecture — three-tier, client-server; request type — one-time; subject area — information service; process type — parallel; information service mode — reading; the way to ensure data integrity is blocking at the level of the entire database. The following model assumptions should be considered:

1) The number of application servers coincides with the number of database servers, i.e. $|SP| = |DB|$.

2) The case is considered when initialization of a business application $r_j(j=\overline{1,s})$ requires relationships $R_k\,(k=\overline{1,d})$ such that $r_j \in SP_i \Leftrightarrow R_k \in DB_i, (i=\overline{1,n})$. It is required to find such a DDB location across database servers so that the average response time of a distributed IPS to requests from thin clients would be minimal.

3. Mathematical model

The characteristics of the distributed IPS that are of interest are determined by the stationary probabilities of the network states. Let $P(\bar{i})$ be the stationary probability that the network is in a state

$\bar{i}$, where $\bar{i} = i_{11},\ldots,i_{1r},\ldots,i_{1n};\ldots;i_{N,1},\ldots,i_{N,r},\ldots,i_{N,n}$. For a three-tier architecture - $N = 3n+1$, and for a two-tier architecture - $N = 2n+1$. The process of changing the states of such a network is described by a homogeneous regular Markov process. The equation of the global balance for a stationary mode, the functioning of such a network, according to [1,6], will have the form:

$$\sum_{k=1}^{N}\sum_{r=1}^{n} P(\bar{i})\mu_{kr} = \sum_{l=1}^{N}\sum_{k=1}^{N}\sum_{r=1}^{n} P(\bar{i} + \bar{1}_{lr} - \bar{1}_{kr})\mu_{lr} P_{lk}(r),$$

where $\mu_{kr}(r=\overline{1,n}, k=\overline{1,N})$ is service intensity in the k-th message center of the r-th user; $\mu_{lr}(r=\overline{1,n}, l=\overline{1,N})$ is service intensity in the l-th message center of the r-th user; $P_{lk}(r)$ is the probability that the message of the r-th user after servicing in the l-th center falls into the k-th center; $\overline{1_{kr}}, \overline{1_{lr}}$ – vector, in the k-th and l-th coordinate of which, respectively, on the r-th place is 1, and all other values are zero.

Calculation of the average response time of the system to the requests of "thin clients" is made by the formula:

$$\bar{T} = \left(\frac{1}{\sum_{s=1}^{n}\lambda_s}\right)\sum_{s=1}^{n}\lambda_s \bar{T}_s, \tag{1}$$

where $\lambda_s\ (s=\overline{1,n})$ – the intensity of query formation by the s-th «thin client»; $\bar{T}_s(s=\overline{1,n})$ – the average response time of the system to the request of the s-th «thin client». The quantity $\bar{T}_s$ is defined as $\bar{T}_s = \frac{\bar{N}_s}{\bar{\lambda}_s}$, where $\bar{N}_s,(s=\overline{1,n})$ is average number of requests of the s-th «thin client»; $\bar{\lambda}_s,(s=\overline{1,n})$ is average intensity of query formation by the s-th «thin client». The quantities $\bar{N}_s$ and $\bar{\lambda}_s$ define, as: $\bar{N}_s = 1 - P_s(d)$, $\bar{\lambda}_s = \lambda_s P_s(d)$, where $P_s(d)$ – the probability that the s-th «thin client» is in the active state (forms a query). It was shown in [6] that the calculation of the quantity $\bar{T}$ by formula (1) reduces to the calculation of the normalizing constant $G(N_1,\ldots,N_n)$, for which Buzen's recursive method can be used [8,9].

4. Results of numerical experiments

The paper [10] presents a heuristic algorithm for solving the problem of optimal placement of information resources on the nodes of a distributed information processing system by the criterion of the minimum average response time of the system to user requests. The algorithm is developed by the authors of this paper [11].

Present algorithm and the calculation of the value $\bar{T}$ are programmed in *C#*, on the platform *NET Framework 4.6.1*, and a computer based on *Intel* processor, with a clock speed of 1,8 GHz, the amount of RAM 8 GB, with numerical values of the source data: read speed in the s-th node $\{VSP_s, VDB_s\} \in [60000;100000]$ KB/s, $s=\overline{1,n}$; speed of writing to the memory of the s-th node $\{DSP_s, DDB_s\} \in [1\cdot 10^7; 3\cdot 10^7]$ KB/s, $s=\overline{1,n}$; processor performance of the s-th node $\{PSP_s, PDB_s\} \in [2.5\cdot 10^9; 3.5\cdot 10^9]$ operations/s, $s=\overline{1,n}$; the speed of data transmission over the communication channel $\theta \in [1000;10000]$ KB/s; constant transmission delay on the channel

$\theta_0 = 3 \cdot 10^{-4}$ c; constant processing delay in the node $\alpha_1 = \alpha_2 = 3 \cdot 10^{-6}$; volume j-th relations $V_j \in [50000;150000]$KB, $j = \overline{1,d}$; volume of information read $b_{lj} \in [1;1000]$KB $\left(l = \overline{1,q}, j = \overline{1,d}\right)$, $j = \overline{1,d}$ on the l-th request for reading from the j-th relation, on l-th request for reading from the j-th relation; $\bar{b}_{lj} \in [1;500]$KB $\left(l = \overline{1,q}, j = \overline{1,d}\right)$ – the amount of information received after processing by the l-th request for reading from the j-th relationship.

Using the developed software, a number of numerical experiments were carried out, the results of which are shown in figures 2-7 [12].

In figures 2-3 are graphs of the dependence of $\bar{T}$ on the dimension of the solved problems of the optimal allocation of the distributed databases to the distributed nodes IPS: $\bar{n}_i = (n \times q \times d), i = \overline{1,6}$, where $\bar{n}_1 = (5 \times 10 \times 10)$, $\bar{n}_2 = (7 \times 30 \times 20)$, $\bar{n}_3 = (9 \times 50 \times 30)$, $\bar{n}_4 = (11 \times 70 \times 40)$, $\bar{n}_5 = (13 \times 90 \times 50)$, $\bar{n}_6 = (15 \times 100 \times 60)$.

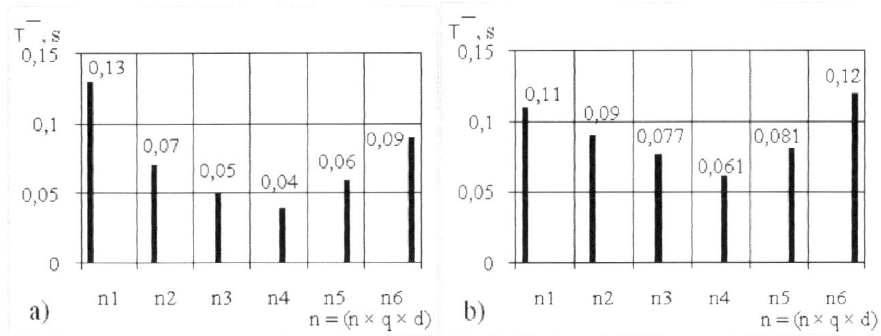

Figure 2. Dependence $\bar{T}$ on the size of the tasks to be solved, taking into account the locks on the basis of client-server: a - two-tier architecture; b - three-tier architecture.

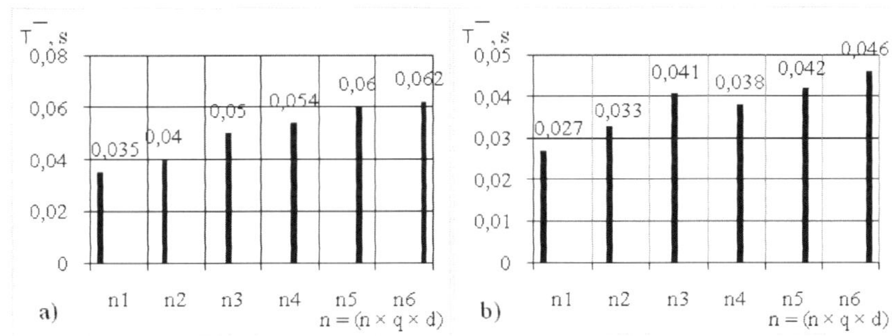

Figure 3. Dependence $\bar{T}$ on the size of the solved problems without taking into account the locks on the basis of client-server: a - two-tier architecture; b - three-tier architecture.

Figures 4-5 show graphs of the dependence of the value of $\bar{T}$ on the reading speed at network nodes (KB/s), for different task sizes $\bar{n}_i$ and for a fixed data rate on the channel $\theta = 5000$(KB/s) [13].

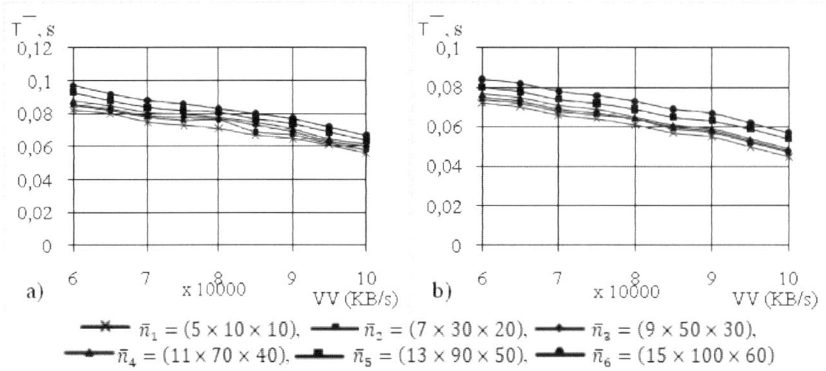

Figure 4. Dependence $\overline{T}$ on the read speed in the nodes, taking into account the locks on the basis of client-server: a - two-tier architecture; b - three-tier architecture.

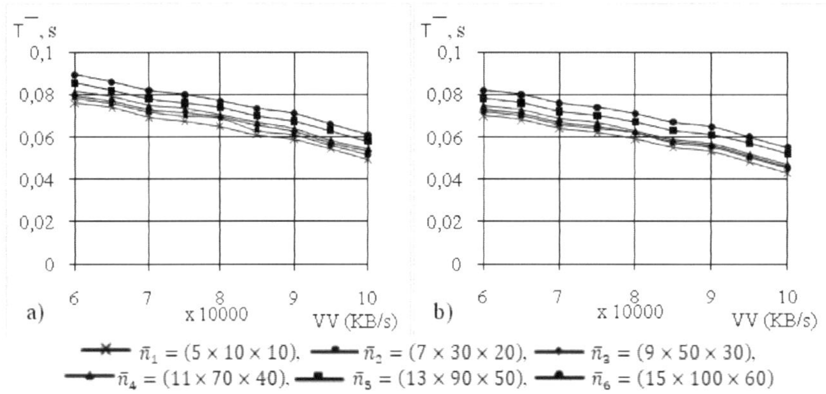

Figure 5. Dependence $\overline{T}$ on the speed of reading in network nodes without taking into account locks on the basis of client-server: a - two-tier architecture; b - three-tier architecture.

Figures 6-7 show graphs of the dependence of the value $\overline{T}$ on the transmission rate of data over a channel θ for different dimensions of tasks $\overline{n}_i$.

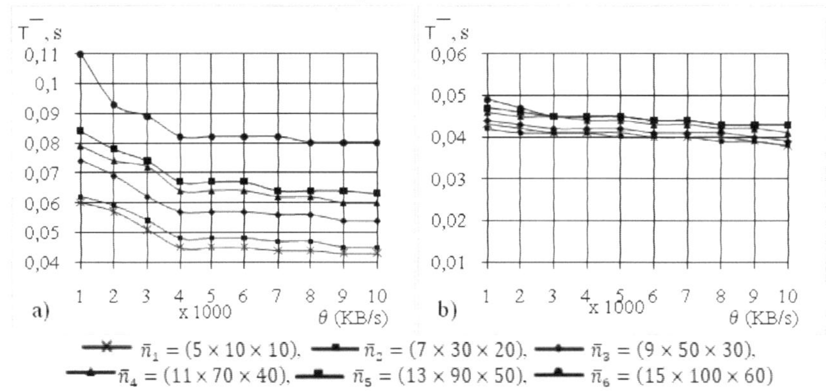

Figure 6. Dependence of $\overline{T}$ on the data transfer rate on the channel θ taking into account locks on the basis of client-server: a - two-tier architecture; b - three-tier architecture.

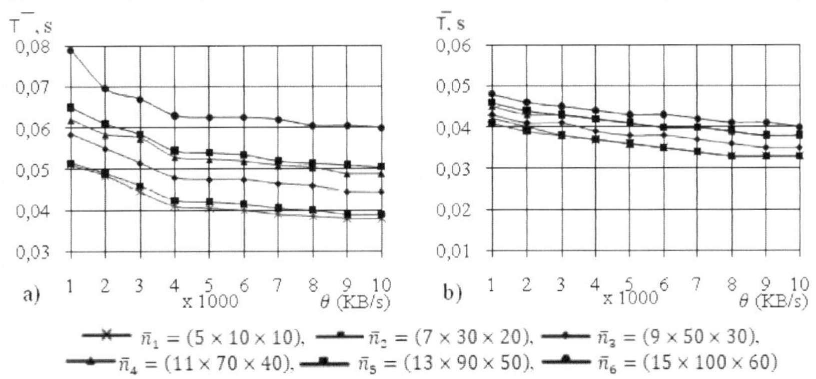

Figure 7. Dependence of $\overline{T}$ on the data rate on the channel θ without taking into account locks on the basis of client-server: a - two-tier architecture; b - three-tier architecture.

5. Conclusion

Obtained modeling results analysis showed that the presence in the model of blocking at the level of the whole database for small dimensions of allocating information resources over the distributed IPS nodes problem does not have a significant effect on the reactivity of the entire system as a whole; however, the influence of the blocking time on its reactivity becomes more significant and this value must be taken into account when constructing real models of computing systems [14].

The model of the optimal allocation of information resources by nodes of distributed IPS on the basis of the three-tier client-server architecture developed by the authors, taking into account the effect of locks, shows better results than the model based on a two-tier architecture and can be used to plan more rational organization of the computational process and to provide the required indicators of system reactivity.

References

[1] Scoba A N and Panfilov A N 2017 *Izv. Universities. Electromechanics* **2** 77-84

[2] Boyko V V and Savinkov V M 1989 *Designing Databases of Information Systems* (Moscow: Finance and Statistics) 351

[3] Kuznetsov N A, Kulba V V and Kosyachenko S A 2002 *Methods of Analysis and Synthesis of Modular Information-Control Systems* (Moscow: FIZMATLIT) 880

[4] Ahmad I, Karlapalem K and Kwook Y K 2002 *Distributed and Parallel Databases.* **1** 5-32

[5] Karpov A E 2007 *Architecture of Distributed Software Systems* (Moscow: MAKS Press) 130

[6] Lizunkov V, Politsinskaya E, Malushko E, Kindaev A, Minin M 2018 Population of the world and regions as the principal energy consumer *International journal of energy economics and policy* **8 (3)** 250-257

[7] Fedotov N G, Moiseev A V, Syemov A A, Lizunkov V G, Kindaev A Y 2017 New Methods of Three-Dimensional Images Recognition Based on Stochastic Geometry and Functional Analysis *IOP Conference Series: Materials Science and Engineering* **177 (1)** 012047

[8] Skoba A N and Sostina E V 2015 *Engineering Bulletin of the Don* **2** [Electronic resource] Access mode: http://ivdon.ru/en/magazine/archive/n2y2015/2882 (access date: 17.09.2018)

[9] Chernomorov G A, Vorobyov S P and Skoba A N 1997 *Engineering, Economics, Culture: Yubil. Sat. Sci. Tr. Prof.-Teacher. Composition of the NSTU.* (Rostov-on-Don: Hephaestus) 178-186

[10] Ergunova O T, Lizunkov V G, Malushko E Yu, Marchuk V I, Ignatenko A Yu 2017 Forming the system of strategic innovation management at the high-tech engineering enterprises *IOP Conference Series: Materials Science and Engineering* **177 (1)** 012046

[11] Zhozhikashvili V A and Vishnevsky V M 1988 *Network Queuing. Theory and Application to Computer Networks* (Moscow: Radio and Communication) 192

IOP Conf. Series: Materials Science and Engineering **483** (2019) 012058 doi:10.1088/1757-899X/483/1/012058

[12] Vishnevsky V M 2003 *Theoretical Bases of Designing Computer Networks* (Moscow: Technosphere) 512

[13] Ferreira R, Lizunkov V G, Politsinskaya E V 2017 Formation of entrepreneurial competencies of university graduates in conditions of transition to the universities of the third generation *Novosibirsk State Pedagogical University Bulletin* **7 (6)** 195-211

[14] Skoba A N, Ayesh Ahmed Nafey Ayesh and Mikhailov V K 2018 *Engineering Bulletin of the Don* **1** [Electronic resource] Access mode: http://ivdon.ru/en/magazine/archive/n1y2018/4774 (access date: 19.09.2018)

Organisational communications in the enterprise management system

Shindryaeva I V[1], Berezovaya A Y[2], Ovsyannikova T V[3], Shindryaev S O[4]

[1]Volgograd Institute of Management – Branch of the RANEPA under the RF President Volgograd, Russia
[2]Branch of the National Research University "MPEI" in Volzhsky Volgograd, Russia
[3]SEBI HE "The Volgograd State Institute of Arts and Culture" Volgograd, Russia
[4]Volgograd Institute of Management – Branch of the RANEPA under the RF President Volgograd

i.shindryaeva@yandex.ru, ovstvl@mail.ru, irina_sergei94@mail.ru

Abstract. In the context of dynamically developing market relations, a modern industrial enterprise should take into account economic, technical and political factors that affect its activities in close connection with the external environment. The problem of improving the efficiency of resource use by enterprises is inextricably linked to the use of timely, detailed, transparent and reliable information in real time. In this regard, communication of the enterprise becomes one of the most important elements of activity and significantly affects the competitiveness. The article analyses the key tasks of organisational communications.
According to the results of the analysis, the main problems of communication in enterprises are identified and recommendations for optimising the management of the system of organisational communications are proposed.

Introduction

Communications in modern conditions of globalisation of the world economy, the formation of post-industrial society are increasingly becoming the object of attention of scientists and specialists in all spheres of management. The reason for this is the generally recognised fact that communication is essential for the success of economic systems. Thus, 73% of American, 63% of English and 85% of Japanese chiefs consider communication to be the main obstacle to the effectiveness of their organisations. According to the survey, about 250 thousand employees from 2000 of various companies regard information exchange as one of the most difficult problems in organisations [1].

The increased role of communication interactions in the functioning and development of modern enterprises brings to the fore the problem of communication management both within the enterprise and between the enterprise and its environment, in order to conduct optimally favorable communication processes for the enterprise. Effective communication in practice is the main prerequisite for the successful achievement of the company's goals.

Communications in the organisational system are considered as a phenomenon and as a process. As a phenomenon, communications reflect the established norms (rules, instructions, regulations), as well as the principles and laws of relations between people in the organisation. As a process, communication is a special form of interaction between staff to exchange information within and outside the organisation, aimed at achieving the goals of the organisation, as well as staff development.

The development of the mechanism of communication system management at the industrial enterprises is a perspective direction of research.

1. Materials and Methods. Model

Subject of research-organisational communications in the production process.

Methods of content analysis of literary data, analytical and logical methods were used for achieving the goal of the study.

When considering the role of communications in the enterprise management system, a structural approach to the description of organisational communications was applied.

The streamlining of functions is done on the basis of a resource approach, which creates a better basis for the management of communications in the organisation.

2. Results and Discussion

Modern enterprises are an open social and economic system, which is in connection with both the system of public administration and other enterprises of this or related industries. And, of course, the most important function of the system is the management, by which we mean "a purposeful constant process of influence of the subject of management on the object of management, aimed at changing the state of the object and/or subjects (including yourself) according to a pre-considered plan" [1]. Each open system (which is the company) is constantly subject to changes in the organisational, managerial and social plan, which is reflected in the style of management, economic behaviour, the system of organisational communication links both external and internal.

Any enterprise is also a communication system built on the interaction between management and employees at different levels, structural units through the transfer of information both within the organisation and outside. Effective communication is the most important condition for achieving existing goals, solving management problems, ensuring the implementation of basic management functions (planning, organisation, management and control). Organisational communications have a great influence on the formation of the organisational culture of the enterprise, its social responsibility to society, staff adaptation [2].

Organisational communication is a process of coordination of the company's activities and formation of staff loyalty to management. We can also say that they are involved in promoting the image of the company, the accumulation of its symbolic capital.

Modern research in the field of organisational communication allows us to identify the following approaches to the definition of "organisational communication":

1) managemental (communication as a management function, as the transfer of orders and explanation of procedures and operations);

2) systematic (communication as a system of interacting elements providing functioning of the organisation);

3) cultural (communication as a way of expressing organisational culture);

3) humanistic (communication as a factor of human development).

Among the groups of tools that contribute to the increase of communicative efficiency, traditionally distinguish are the following: organisational (top management speeches, meetings, general meetings); informational (this includes corporate website, newsletter); analytical (sociological research to learn the views of staff) and communicative (corporate events, trainings). However, this research has shown that in reality the majority of companies (62%) simultaneously use on average no more than two, at least three (20%) tools of each group.

Almost everything that happens in the organisation has a direct or indirect connection with communication processes. They are an important means of ensuring its integrity and efficiency. Among the main forms of organisational communication are the following: external (communication with the external environment), internal (vertical and horizontal, formal and informal), as well as communication networks that cover all forms of communication and are based on information flows circulating in the structure of the enterprise and its management system [3].

At the same time, the researchers believe that only half of the transmitted information reaches the recipient and is clear to him. In other cases, the information is distorted and is not always perceived adequately. Therefore, the correct organisation of the information transfer process is very important, since poorly structured and redundant information generates distrust to its sources, and as a result, to the management system.

Various components can be an obstacle (or barriers) to communication: noise, distorting the process of information transmission, linguistic and semantic errors arising at any stage of information exchange, which cannot be avoided completely. Among the barriers to communication, the most significant ones are individual and organisational barriers. Individual barriers appear as a result of differences in the system of values of employees; discrepancy between the level of education and competence, income level, age and other social and demographic indicators. Organisational communication barriers are associated with a contradictory system of values and mission of the organisation, different needs of structural units, disproportionality of power distribution, etc. [4].

To prevent these organisational barriers to communication, it is necessary to create an atmosphere of trust and openness in the enterprise, special importance should be given to the effectiveness of feedback and the development of corporate culture. Effective use of communications is also associated with such factors as the professionalism of managers, the structure and characteristics of the organisation, the presence of stable interpersonal relations between its members, the culture of subordinates, the position of the organisation in the external environment, the competence of employees.

Our own study of the industry market on the example of enterprises of the Southern Federal District allowed us to identify the main factors of success and problems existing in the communicative sphere of these industries.

These activities are part of the food industry. The food industry is one of the largest and most important sectors of the economy of the Southern Federal District. In 2018, the authors conducted a survey of production managers, press services, departments of communication policy, personnel management, marketing structures and managers, specialists engaged in PR. In total, the study involved 42 people, representatives of 13 medium and 5 large enterprises.

The analysis of communicative activity of such enterprises shows that the communication system at the enterprises studied by us is quite effective. This was noted by 86% of respondents. It should be noted that the enterprises have structural units or specialists responsible for the development of communications and communication policy, as well as actively involved employees responsible for personnel management.

Each company has its own website, the information on which is updated at least twice a week, there is an electronic document management system, active interaction with the regional media is conducted (the example of Krasnodar is especially notable here), the distribution of powers at different levels of management.

At six enterprises there are corporate publications that reflect information about activities, news, achievements of employees, management decisions. The same information can be found on the website. Within companies, information stands, meetings and planning meetings, meetings with staff etc. are held. But paradoxically, the study pointed to the lack of information in the field. This problem is largely due to the fact that employees prefer oral transmission of information, and not always interested in such stands. The main channels of communication, as noted by 76% of respondents, are hierarchical communication on the principle of "management - subordination" and informal communication between employees.

In order to study the opinion of employees on the activities of the organisation, management decisions, regular sociological research is conducted at seven enterprises (questionnaires, interviews of employees, online surveys). All enterprises have a system of "feedback" between management and employees, which is expressed in the fact that any employee can ask questions to management during direct meetings or on the website. Also, the wishes of enterprise managers, communications specialists in terms of improving the existing system are always taken into account.

Specialists of five enterprises noted that previously all requests and wishes are discussed with their chiefs on the basis of which ascending vertical communications are organised. The chief arranges a monthly meeting where for every item of the reports instructions for further work are given, as well as manages and transmits information, received from the external environment. Thus, the company has established a vertical descending communication system.

Horisontal communication in all enterprises mainly occurs informally, directly through communication between employees. Informal communications are distributed to the celebration of birthdays, New Year, celebration of the Eighth of March, etc. Through such communication the chief tries to reduce the amount of rumours, leaks of information [5].

In the field of public relations, first of all, attention is paid to the creation of a certain image, the "image" of the enterprise on local television.

At the enterprises, despite the active construction and development of organisational communications, there are difficulties: late delivery of information to employees and management, which entails inconsistencies in the work, a complex social and psychological climate, difficulties in communication with managers. The next reason for difficulties in organisational communications is the overload of communication channels, which can be a consequence of both elementary failure of technical means of communication and information processing and redundancy of information flows.

To solve these problems, the following measures can be proposed: regular and timely informing employees about the real situation in the organisation at the moment, increasing the degree of involvement of employees in the company's activities, the introduction of the fundamental values of corporate culture; conducting collective and individual conversations with employees, general meetings, professional trainings, etc.

3. Conclusion

The need to manage organisational communications at the strategic level is now recognised by most scientists and practitioners, but the nature and content of strategic decisions regarding organisational communications have some uncertainty.

To solve the identified problems, the following measures can be proposed: regular and timely informing employees about the real situation in the organisation at the moment, increasing the degree of involvement of employees in the activities of the company, the introduction of the fundamental values of corporate culture; conducting collective and individual conversations with employees, general meetings, professional trainings, etc.

Optimisation of the management system of organisational communications will allow to organise the process of communication in a more constructive way, to increase the effectiveness of feedback, to minimise the barriers of communication, to strengthen organisational values, to form, maintain and develop a corporate culture, which will positively affect the effectiveness of the company's management, will allow to form a more positive image of the organisation for different categories of information consumers. Thus, the effectiveness of organisational communications affects the efficiency of the enterprise management system, so management should be interested in the development of communication policy, which affects the quality of management decisions.

References

[1] Pocheptsov G G Public relations for professionals M"Refl-book. Wackler" 2005 P 218

[2] Shiryaev V I, Baev I. A, Shiryaev E V Management of the enterprise: modeling, analysis, control Moscow: Librokom 2015 272 p

[3] Moskvichev Yu N, Badalova O V, Naida O A, Poleva Yu V, Pasharina E S, Rostovsky V Yu, Sarkisyan I F, Sgibneva O I, Cheremushnikova I K, Avhodeeva E A, Shindryaeva I V, Vodop'yanova N A, Berezovaya A Yu, Ovsyannikova T V, Boldina M Yu, Mironova Yu G, Ulanovskaya N S, Boldin S Yu Dynamics of socio - economic and socio-cultural processes: tradition and innovation Volgograd 2017 246 p.

[4] Shindryaeva I V 2016 Process of labour motivation and its relationship with organisational culture *Scientific Bulletin of the Volgograd branch of RANEPA. Series: Political Science and Sociology* **2** 71-75

[5] Ovsyannikova T V 2016 Social responsibility of the media in formation of civil society *Bulletin of the Eurasian Academy of Administrative Sciences* **1 (34)** 106-112

[6] Valtchanova V 2013 Implications of cloud culture for communication management *Public*

administration. Electronic Bulletin **37** 127 – 145
[7] Tertychko T V 2013 Theories of Communications in Managing Industrial Enterprises: Advantages and Disadvantages *Buisness Inform* **7** 368-372
[8] Zerfass A, Huck S 2007 Innovation, communication, and leadership: new developments in strategic communication *Internat. Journ. Of Strategic Communication* Vol 1 **2** 107–122

IOP Conf. Series: Materials Science and Engineering **483** (2019) 012060 doi:10.1088/1757-899X/483/1/012060

Innovative management in Russian production companies

V V Smirnov, V V Semenov· A N Zakharova, E N Kadyshev, G S Dulina

Chuvash State University, Cheboksary, Chuvash Republic, Russia

walera712006@mail.ru, zaharova_an@mail.ru

Abstract. The article exposes the necessity of effective management of innovations on the basis of the new scientific direction – innovative management. The growing attention of the Russian economists to the innovations management problems in the increasing economy complexity, the scientific and technological progress speeding up resulting in specific changes is presented. The research methodology is based on the typical changes caused by the wave fluctuations and business cycles in the public production. As the result of innovative activities growth rate evaluation, there was found out the tendency of the Russian production companies to the certain innovations. The innovative management quality evaluation in the Russian companies found out the lack of thereof and defined the factors negatively influencing the innovative development. The main means of innovative management able to increase its quality by influencing on the indices of commercialization, length of the works being completed in social, geographical, communicative, technological, science and technology, economic and financial spheres are explained.

1. Introduction

There is a gradual change of management benchmarks in the companies from prescriptive to innovative in the process of the Russian economy and production. The core of the management system is the same and concerns stability on the base of incentives. The innovations are these incentives in the new environments. It is important to take the changes caused by innovations as a new feature of the management system and an ability to reach the goal.

Under the current circumstances there is a necessity of effective management of innovations. As the result of this necessity a new scientific direction in the management system – the innovative management is being formed (*Innovation Management, IM*) (Robert Nowacki, Katarzyna Bachnik, 2016), (Chih-Wen Wu, Kun-Huang Huarng, 2015).

IM is a relatively new concept for the Russian scientific society and the business community. Nowadays the old forms and methods of management are being replaced with "innovative" ones. All the economic entities from state-level management companies to the limited liabilities companies in the small business sector and individual entrepreneurs have to undertake the innovative activities.

IM is the primary direction of the strategic management carried out at the top level management of the company (Viliam Lendel, Stefan Hittmar, Martin Latka, 2015). The goal of *IM* is finding out of the main directions of science and research and production activities of the company in the process of new products development and introduction, company's products modernization and improvement, the products in demand production development, ceasing the production of the outdated products.

In the market environment as a certain system of supply and demand economic regulation on the basis of which the market price is formed, the main components of *IM* are the improvements, investments and innovations. Improvements form the novation market, investments form the capital market, and innovations form the innovations - pure competitiveness market.

The scientific society, in general, defines the innovation as the final result of IM implemented as a new or improved product introduced in the market, a new or improved technological process used in practice or as a new way to meet the public needs. Meanwhile, the time period from the idea inception, novation creation and dissemination to its introduction and use is generally referred to as the innovation life cycle. Taking into account the works implementation stages the innovations life cycle is referred to as the innovative process.

On the modern stage of the economic science the rising attention of the Russian economists to the problems of *IM* is explained by the increasing complexity of the economy structure, speeding up of science and technology progress causing the contradictions between the decreasing time of the product market life and increasing the time of its potential possibilities, it is also explained by the limited possibilities for the economic growth. (Deniz Dilara Dereli, 2015).

2. Evolution of Innovation

The Russian economist Kondratiev N.D. published in 1925 the theory of the wave fluctuations in the public production thus paving the way for *IM*. Having investigated the vast statistic on the alternation of the changing stages cycling in industrial production he determined the existence of large economic cycles.

J. Schumpeter who worked in Harvard University, the USA developed Kondratiev's wave theory. He revealed in the wave theory the possibility to overcome the crises and declines in the industrial production through the innovative enhancement of the capital by the means of technical, organizational, economic and managerial innovations (Heinz D. Kurz, 2008). Three types of cycles are defined by J. Schumpeter in his work «Business cycles» (1939). The first cycle related to the Industrial Revolution triggered by the development of the industry in Great Britain. The second cycle coincided with the railways, mechanical engineering, steamships advent, it lasted till the 90-th of the XIX century. The third cycle was related to using the electric power, invention of the internal combustion engine, development of the chemical industry (Konstantinos N. Konstantakis, Panayotis G. Michaelides, 2017).

Such terms as the product life cycle and the production technology life cycle are determined according to the modern concept of *IM*.

The product life cycle consists of four stages. Investigation and design of the product novation are performed in the first stage. The stage is completed when the finalized technical documents are transmitted to the production units of the industrial organizations. Technological exploitation of the new products large-scale production is performed in the second stage. The first and especially the second stage are related to significant investment. The growth of the production is followed by the products cost lowering and profits increasing. It allows recouping the investment of the product life cycle at the first and second stages. Then the third stage starts, it is distinguished with the stabilization of the production volume. A gradual lowering of the production and sales takes place in the fourth stage.

The production technology life cycle also consists of four stages. The first stage is related to the birth of the process novation and carried out by a broad range of technological research. The second stage is connected with the process novation exploitation in the production facility. The proliferation and replication of the new technology with its multiple replicating in the other production facilities are carried out in the third stage. The fourth stage includes the using of novation on the routine basis i.e. the realization of the novation process in the stable full-time functioning units of the production facilities.

3. Solution Methodology

The "investment in novation" as the result of the new process, product or service practical application is defined as innovation. Novation is some novelty that was not applied before: a new phenomenon, discovery, invention, and a new way to meet public needs.

J. Schumpeter interprets the innovation as the new scientific and organizational combination of the production factors motivated by the entrepreneurial spirit. Schumpeter developed the theory of the entrepreneurship according to the technological innovations in his "Theory of Economic Development" (1912). He interprets the profit and the income of an entrepreneur as the result of the dynamic changes in the economy because of the innovative activities. The function of the entrepreneur amounts to the imbalance, creation of the imbalance state in the market as the result of the innovations which brings the profit. A person can be an entrepreneur only till he undertakes innovative activities and loses this status after business establishing, running it as a routine process without any further managerial, technological and other improvements (Nadide Sevil Tülüce, Asuman Koç Yurtkur, 2015).

Changes are the innovation specific content and the function of changing is the primary function of an innovative activity. J. Schumpeter defined the following typical changes: new equipment, new technological processes or new production marketing (purchase and sale) application; new featured products introduction; new raw materials application; changes in production management and its logistics; new sales markets formation (Kerk L. Phillips, Jeff Wrase, 2006).

Science and technic innovativeness and production applicability are the essential innovation properties. Tradability concerning innovation is an eventual property requiring some efforts to be reached. According to the commercial aspect the innovation is an economic necessity realized through the market needs. It is necessary to pay attention to the possibility of innovations "materialization" into new technically perfect kinds of industrial products, means and tools of labor, technologies and production management as well as commercialization converting the innovations into the source of income.

Innovations are an effective tool of competition in the market economy as they lead to the creation of new demands, cost price reduction, an influx of investments, increasing the reputation (rating) of the new goods producer, opening-up and capturing new markets including export ones (Massimiliano Bratti, Giulia Felice, 2018).

4. Result

IM includes scientific and technological, managerial, financial and marketing activity. *IM* changes creation and application in the field of new products, methods, their introduction, exploitation and replication.

Novelties, innovations and investments are the main components of *IM*. Novelties form novation market, the main product of which is intellectual property. It is covered by copyright, obtained according to the international, federal, corporate and other legislative and normative acts.

A novelty creation and its transformation into the innovation require material, time, finance and other types of investment. The volume of investment into the innovation, its recoup time, the profitability of the future innovation and failure risk are the basic criteria for evaluating the innovative changes undertaking perspectivity (Jin Chen, Ximing Yin, Liang Mei, 2018).

The results of the innovative activities growth rates evaluation (table 1) helped to define the weakest indicators of Russian companies' innovation development

Table 1. Innovative activities in Russia growth rate, per cent

Indicator	**Average, 2011 – 2017**
Companies undertaking ecological innovations ratio in the year under review, in the overall number of companies reviewed	-14.75
Companies undertaking marketing innovations ratio in the year under review, in the overall number of companies reviewed	-5.67
Companies undertaking managerial innovations ratio in the year under review, in the overall number of companies reviewed	-4.51
Investment in technologic innovations ratio in the delivered goods, completed works and services overall volume	-1.42
Companies undertaking technology innovations ratio in the year under review, in the overall number of companies reviewed	-0.49

Source: calculated by Federal Government Statistical Service data URL:http://www.gks.ru (Date of retrieval 02.10.2018).

The Russian companies tend to develop ecological and marketing innovations less and technologic innovations more. It reveals positive developments in technology for innovations introductions non-linearity.

IM can reduce positive developments in technology non-linearity; the quality (*Quality of Innovation Management, QIM*) of which is calculated by the following formula:

$$QIM = (I_p \times I_c)^{1/2}$$

Ip - Innovation potential, *Ic* - Innovation climate.

Ip – the level of commitment to the project or strategic innovative changes program realization, it is defined by the effectiveness of economic, financial, managerial and personnel resources. The main factors of *Ip* are:

-Scientific and technological potential: number of personnel with Academic Degree; quantity of rational proposals to one employer; quantity of patents;

- Commercialization indicators: new products share in the total amount of the produced goods; quantity of the license agreements;

- Duration of the works being carried out (rate of the innovative lag);

- Management system innovativeness: kinds of innovative activities stimulation in the company, top managers participation in innovative projects realization; innovative activity participants' degree of freedom.

Ic is a part of total team culture of the company contributing to the employee's efforts in new ideas generation and realization. To create a favorable *Ic* it is necessary to create the atmosphere of confidence, to detect and overcome the factors "blocking" creative efforts and a team searching work of the personnel, to broaden the powers of innovators in their job activities, to use managerial and psychological tools helping to generate the new ideas.

QIM rate is identic to *PMI* (*Purchasing Managers Index*), but expressed in shares, accordingly to *Ip* and *Ic* (table 2).

Table 2. QIM

	Base	**Calculated**
Ip	1.0	0.7
Scientific and technical potential (the number of Academic Degree Personnel, the number of rational proposals to one employer; quantity of patents)	0.3	0.3
Commercialization indicators (new products share in the total amount of the produced goods, the number of licensed agreements)	0.3	0.1
Duration of the works being carried out (rate of the innovative lag)	0.2	0.1
Management system innovativeness (kinds of innovative activities stimulation in the company, top managers participation in innovative project realization, innovative activity participants' degree of freedom)	0.2	0.2
Ic		0.7
Social, geographic and communication spheres (social strain, access to raw material, fuel, energy, material and technical resources, transport and information resources)	0.3	0.2
Technological, scientific and technical sphere (technologies, scientific and technical information market, availability of research institutions, consulting, engineering, venture and other companies).	0.3	0.2
Economic and financial sphere (taxes, benefits for R and D, federal investment climate, interested in innovations investors)	0.2	0.1
Political and legal sphere (federal and regional plans and programs, legal base regulating the R and D works).	0.2	0.2
QIM	1.0	0.49

Source: calculated by Federal Government Statistical Service data URL:http://www.gks.ru, Rating Agency RAEX ("Expert PA") URL https://raexpert.ru (Date of retrieval 02.10.2018).

The *QIM* evaluation results allowed revealing its low level (more than 0.5) and the factors negatively affecting on the Russian companies innovation development: commercialization; duration of the works being completed; social, geographical, communicative, technological, scientific and technical, economic and financial spheres.

5. Discussion

IM must start with the general goal formulation clear to each member of the personnel. Goal formulation is very important for the company business relations with the external environment, market and consumers. Company's general goal has to cover the main company's activity, working principles in the external environment (trading principles; treatment to the consumers; business relations); company's culture, its traditions and working climate.

The next stage of *IM* is innovative planning relying on company's external and internal environment analysis, during which changes happening in the planned period are evaluated, factors impending company's positions are revealed, factors favorable for the company's activities are investigated, innovations letting the company strengthen its market position are planned.

IM is a way affect agent management on object management including innovations, innovative process and business relations in the innovation introduction market. Innovations introduction always depends on the investment of the innovation. Therefore all the ways of innovative management should take into account money relations arising in the process of innovations introduction.

6. Management solutions

Let us consider the *IM* main ways capable to increase the *QIM* rate and improve Russian companies innovation activities lagged factors (commercialization/marketing index, duration of the works being carried out, social, geographic and communication sphere, technological, scientific and technical spheres, economic and financial sphere):

1) Innovations engineering is a scope of works on innovation project implementation including creation, realization, promotion and replication of some innovation. The goal of innovations engineering is to get the best economic effect from the investment into the new product and determination of the innovative activities perspective ways;

2) Reengineering is a variety of engineering method. Reengineering as the way of innovative management treats with the innovation process dedicated to the new products and operations production as well as their realization, promotion and replication. Reengineering in *IM* is connected with some certain goal of the innovation, i.e. the current necessity or strategic necessity in innovations;

3) Bench-marking is business entities', first of all, competitors' activities investigation to use their positive experience in your own activities. It includes a scope of means to find, evaluate all the positive sides of another's experience constantly and use them in your own work;

4) Brand-strategy (brand-innovations) is a system of the new product or operation distinctive properties that forms the consumer's consciousness and defines the innovation place in the market as well as its producer or seller. Brand makes a buyer to choose a certain product (operation) among all the available in the market.

5) Price control in *IM* represents a way to effect on the innovation sale through the price mechanism.

The main goal of all the methods is speeding-up the innovations sale at the greatest profit and effectiveness within the current period of time as well as the greatest advantage of this sale in the future.

7. Conclusion

The necessity of *IM* in the Russian companies is related to new requirements to the companies, their managerial activities content and methods in the raising competitiveness for the sales markets. *IM* in the Russian companies must represent itself a system of coordinated measures to increase *QIM* by all the factors optimal combination.

Innovative activities carrying out in a more perfect way providing *QIM* increase acknowledges the going to a higher level of production abilities and is an index of the innovative development of the company.

IM introduction process in the Russian companies includes the following stages: defining the innovation management goal; choosing the innovation management strategy; defining the innovation management ways; developing the program on innovation management, organizing works on the program carrying out; controlling the planned program carrying out; management ways effectiveness analyzing and evaluation; innovation management ways correction.

IM in the Russian companies must be budgeted at innovation creation and realization, i.e. in the process of innovation itself. Innovative process is the basic foundation and the effectiveness of *IM* and *QIM* ways and methods will depend on it.

References

[1] Bratti M, Felice G 2018 Product innovation by supplying domestic and foreign markets. *International Journal of Industrial Organization* **60** 126–178
[2] Chen J, Yin X, Mei L 2018 Holistic Innovation: An Emerging Innovation Paradigm. *International Journal of Innovation Studies* vol 2 iss **1** 1–13
[3] Dereli D D 2015 Innovation Management in Global Competition and Competitive Advantage *Procedia - Social and Behavioral Sciences* **195** 1365–1370
[4] Konstantakis K N, Michaelides P G 2017 Does technology cause business cycles in the USA? A Schumpeter-inspired approach *Structural Change and Economic Dynamics* **43** 15-26
[5] Kurz H D 2008 Innovations and profits: Schumpeter and the classical heritage. *Journal of Economic Behavior & Organization* vol 67 iss **1** 263–278
[6] Lendel V, Hittmar S, Latka M 2015 Application of Management of Innovation Processes in Enterprises: Management Approach, Problems and Recommendations *Procedia Economics and Finance* **34** 410–416
[7] Nowacki R, Bachnik K 2016 Innovations within knowledge management *Journal of Business Research* vol 69 iss **5** 1577–1581
[8] Phillips K L, Wrase J 2006 Is Schumpeterian 'creative destruction' a plausible source of endogenous real business cycle shocks? *Journal of Economic Dynamics and Control* vol 30 iss **11** 1885-1913
[9] Tuluce N S, Yurtkur A K 2015 Term of Strategic Entrepreneurship and Schumpeter's Creative Destruction Theory *Procedia - Social and Behavioral Sciences* **207** 720–728
[10] Wu C.-W, Huarng K-H 2015 Global entrepreneurship and innovation in management *Journal of Business Research* vol 68 iss **4** 743–747

IOP Conf. Series: Materials Science and Engineering **483** (2019) 012061 doi:10.1088/1757-899X/483/1/012061

Adjustment of induction high-temperature synthesis to in situ synchrotron study of SHS-mixtures on the example of Ti-Al system

A V Sobachkin[1], A Yu Myasnikov[1], A A Sitnikov[1] and M R Sharafutdinov[2]

[1] Polzunov Altai State Technical University, 46, Lenina ave., Barnaul, 656038, Russia
[2] Budker Institute of Nuclear Physics of Siberian Branch Russian Academy of Sciences, 11, Acad. Lavrentieva Pr., Novosibirsk, 630090, Russia

E-mail: anicpt@rambler.ru

Abstract. The paper discusses experimental complex, which allowing to carry out high-temperature synthesis of mechanically activated powder reagents by a method of induction heating in vacuum and inert environment. On the basis of this complex conducted in situ synchrotron research of the specific features of phase formation dynamics in mechanically activated 3Ti + Al powder composition. High temperature synthesis carried out under conditions of high volume inflammation by means of inductive heating. The developed complex is adapted to the method of dynamic diffractometry in synchrotron radiation beams to register the kinetics of phase transformations in the solid-phase reaction of the components of the powder mixture.

1. Introduction

These days big attention is paid to methods of nanostructured composition materials with targeted properties obtaining. An effective method of obtaining such materials is self-propagating high-temperature synthesis [1-3]. An effective way to influence the structure of initial reactionary furnace charge, thus regulating the kinetics of structural phase transition during the synthesis, is a method of mechanical pre-activation (MA) [4-7]. Background mechanical activation helps change the structure and phase composition of the final product, gives more way to chemical reactions, namely, extending concentration combustion limits, changing the temperature and speed of combustion, inflammation temperature, etc. [8-13].

To initiate the synthesis in a mechanically pre-activated system a heat impulse is required. It will fluctuate considerably depending on the type of powder composition. Another important factor for gaining a homogeneous structure and phase composition is speed of powdered composition heating. One mode to create an initiating temperature with high rate of heating is to apply extreme exposure to the powder system (blast wave, electric impulse), it can influence both macro kinetic parameters of self-propagating high-temperature synthesis and phase structure transfiguration [14-17]. However the most universal extreme treatment is inductive heating with the heat input of the 100 K/s order. The inductive heating method is noted for low energy consumption, high rate heating reaction, simple and reliable equipment, and pure final product [18-20]. But application of the inductive heating method, used for controlled synthesis of new materials with targeted properties, has not been thoroughly studied and it requires more research.

For better understanding of phase transformation during the mechanically activated systems synthesis by inductive heating one should know characteristics of synthesis under dynamic conditions. High speed of the self-propagating high-temperature synthesis implies application of the corresponding experimental research methods of the dynamics of initial composition transformation. Recently in situ research of structural change and chemical dynamics in the combustion zone in real time mode has become possible due to the application of synchrotron radiation. Today the synchrotron radiation method has no alternative in the research of high-rate processes which are made possible by means of:

- high speed of diffraction registration;
- high intensity facilitating operations with reverse beam;
- high special and time resolution of the detector, that altogether provide high accuracy [21, 22].

Many researchers apply synchrotron radiation of «diffraction movie», i.e. a consequential filming of diffractogram series enabling them to observe structural change of matter during its deformation, melting, crystallization, synthesis, etc. [23-25]. High density of synchrotron radiation of monochromatic beam and progress in detector rate resulted in getting an X-ray pattern per microsecond. In a classical registration scheme it is close to the limit, which is determined by the time of charge assembly and relaxation in the detector. Owing to the properties of synchrotron radiation it is possible to apply it in the process of phase and structural material analysis. Short wave length of synchrotron radiation is considered important. Its tolerance is commeasurable to interatomic distance in the points of the lattice.

In this paper a compound of Ti_3Al was chosen for the research. These α_2-phase alloys are widely used in aerospace engineering, automobile industry, shipbuilding and electric power engineering due to their high heat resistance, durability, resistance to aggressive environments, corrosion stability, low density, high resistance to fatigue failure and creep [26-30]. Reasoning from this fact there is a necessity to extend the research of structure and phase formation in Ti-Al system, and also to control the synthesis in order to obtain intermetallic compounds with targeted properties. The paper describes in situ experimental tests of the dynamics of phase formation of mechanically activated 3Ti + Al powder composition with different timing during high temperature synthesis in the mode of dynamic thermal explosion by means of heating the composition in quickly changing electromagnetic fields.

2. Materials and methods

The object of research in this paper is the installation of induction heating. In general, the induction plant consists of a generator and an induction coil – inductor. The most interesting of induction systems for purposes of present study are microwave induction units. The frequency of these induction units exceeds 100 kHz.

Thus, the microwave induction heater 6AV was used as a base for the creation of the experimental complex in this work. Frequency of inductor varies between 400-1100 kHz and the maximum output power is 6 kW. Induction heater is connected to the mains voltage of 220 V.

High-frequency setting automatically adjusts the inverting frequency from the specified range to any inductor.

Thus, in order to register the dynamics of phase formation in process of self-propagating high-temperature synthesis, experimental complex was constructed. It was based on high frequency electromagnetic heating unit 6AV that generates electromagnetic energy in a wide range of power (figure 1). It should be noted that advantage of induction method over classical heating methods (such as muffle furnace or an electric filament) is as follows: induction method allows flash heat of powder composition, and that can be crucial for mechanically activated systems with high content of non-equilibrium defects of structure [12].

Testing operation of the induction heater 6AV was performed on previously mechanically activated powders of titanium and aluminum.

The ratio of the powder used to make the composition was 16 wt.% Al + Ti. Afterwards, mechanically activated treatment of initial powder compound was carried out with planetary ball mill AGO-2.

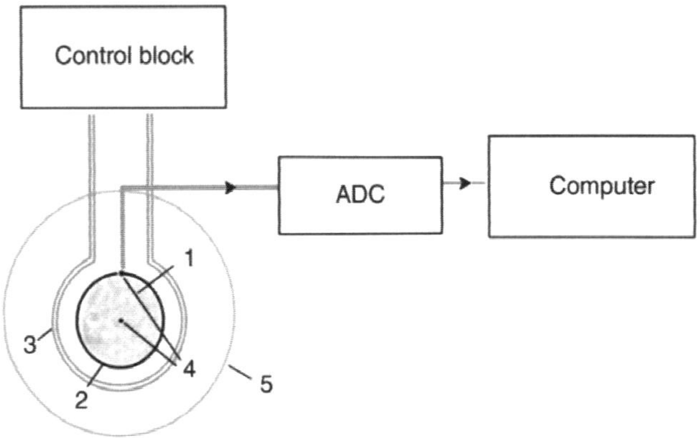

Figure 1. Experimental complex of high speed induction heating: 1 – powder composition; 2 – graphite melting pot; 3 – induction filament; 4 – wolframite thermocouple units; 5 – vacuum lid.

Next stage represented in situ research of dynamics of mechanically activated composition phase formation during high temperature synthesis with the method of dynamic diffraction analysis in synchrotron radiation beams. We used radiation of electron storage ring VEPP-3 of the 5b station («Diffraction movie») in Budker Institute of Nuclear Physics of Siberian Dept. of Russian Academy of Science (Russia, Novosibirsk). Nonstop filming of diffractograms was carried out with monochrome radiation, wave length being λ = 0.1505 nm, range of scanning angles being 30° – 70°, frame interval being 10 sec. System of electron storage ring is presented in figure 2. Method of diffraction analysis is based on radiation of electromagnetic waves with charged particles, particles moving from relativistic velocity within one magnetic field [31, 32].

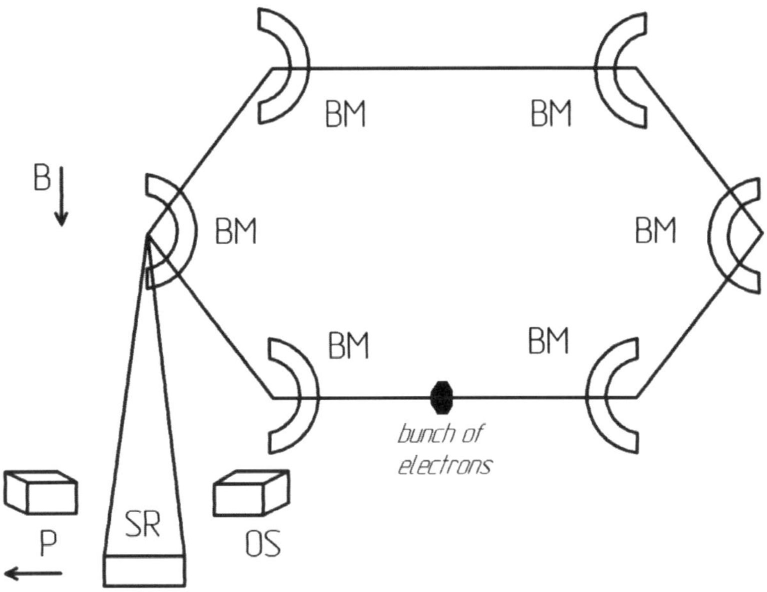

Figure 2. Electron storage ring structure: BM – bend magnets; B – magnetic field; P – photon polarization vector, radiated in electron orbit; OS – output slot, horizontally restricting synchrotron radiation beams.

From injector (linear accelerator) pre-accelerated electrons with relativistic velocity (energy limit equals approximately 300 MeV) get to circular orbit of accelerator and are held there by the magnetic field of bend magnets. Radiation is concentrated in a cone with an 1/Y angle and is tangential to trajectory of radiation point.

$$Y \approx \frac{m c^2}{E},$$

where Y – relativistic factor, m and E – mass and energy of particles, c – velocity of light in vacuum. There is a vacuum channel going out of cyclic orbit accelerator which is 10 – 15 m long and carries radiation to the experimental stations.

To conduct a phase analysis of a final synthesis product we used X-ray diffractometer DRON-6 with CuKα-radiation (λ = 0.15418 nm). Diffractograms were registered under similar conditions with scanning interval of 0.05 degrees and time-exposure of 3 seconds at each point. Processing and analysis of data were performed by means of PDWin program pack, designed for automatization of X-ray pattern processing [34].

3. Results and discussion

Preliminary tests have shown possibility of using an induction heater 6AV for study of dynamic synthesis diffractometry in Ti-Al system under extreme heating.

However, this heater must be adapted to conditions of the experiment. For this purpose, the installation was retrofitted with a mobile and productive water cooling system of inductor, as well as a specialized chamber that allows induction heating both in vacuum and in inert environment. Specialized chamber was also equipped with a cooling system. The scheme of the adapted complex is shown in figure 3.

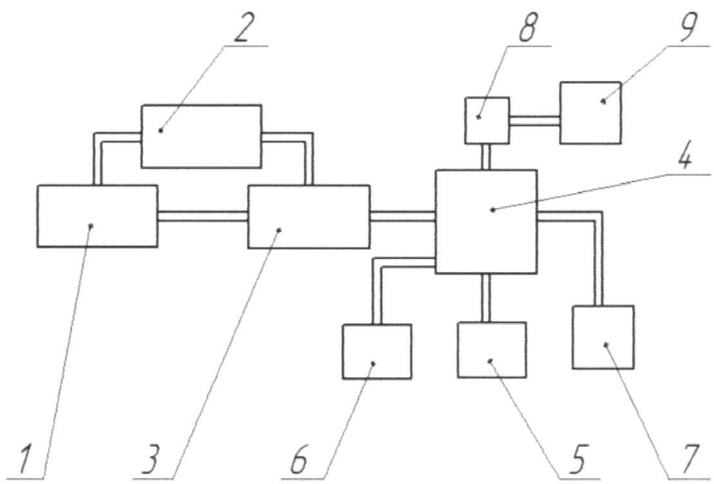

Figure 3. scheme of installation of the experimental complex of high-frequency induction heating:
1 – control unit, 2 – induction heater cooling system, 3 – induction heater, 4 – vacuum chamber, 5 – vacuum pump, 6 – inert gas supply system, 7 – vacuum chamber cooling system, 8 – multichannel LA-2usb board, 9 – personal computer.

One of important components of complex is a specialized vacuum chamber. Inside this chamber is placed a metal crucible with a mechanically-activated mixture of 3Ti + Al and a thermocouple necessary to control synthesis of substance. These temperature measurements were processed by multichannel board LA-2USB, connected to personal computer. Under the influence of magnetic field, powder mixture is heated to a certain temperature, as a result of which synthesis reaction of Ti_3Al compound is initiated.

Thus, mechanically activated composition was placed into graphite melting pot, which is open top cylinder, and then packed down. Then melting pot was isolated with asbestos layer. Signal from thermocouple unit was transmitted to analogue-digital converter (ADC).

System was covered with the vacuum lid, air being extracted and argon pumped down into it. Graphite melting pot was heated to high temperatures by high frequency electromagnetic field, and the system heated powder composition up to 1500 °C and more.

One peculiarity of method of dynamic diffraction analysis is the fact that in order to use it one needs composition materials with open surface to which synchrotron radiation beam fall. Width of synchrotron radiation beam is 2 mm. To carry out experiment a container was constructed, designed as a big hollow cylinder (figure 4).

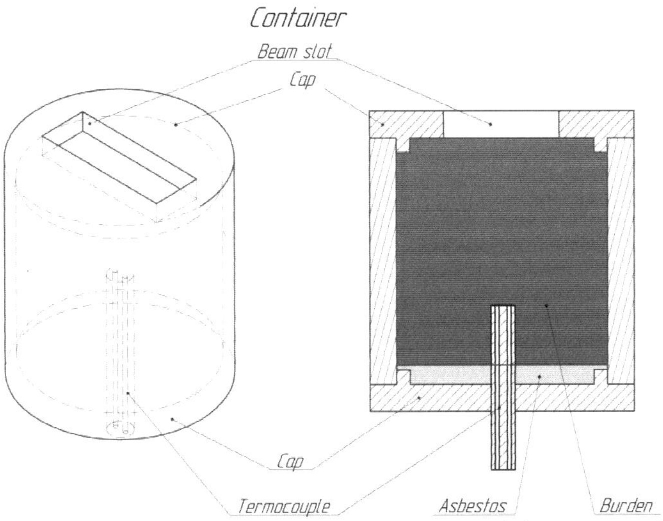

Figure 4. Container, adjusted for the research by means of synchrotron radiation

Specific feature of reactor is sub open surface of composition to provide penetration and reflection of synchrotron radiation beams. It was made possible due to thick plate made of metal with a slot 3 mm wide and 30 mm long. That is enough for a beam to fall to furnace charge surface and to be reflected at angle of 35-40 degrees. The use of metal plate provides additional fixation of composition surface at one level during heating and active chemical reactions.

Finally experimental complex was mounted to unit of 5b synchrotron radiation station of the VEPP-3 electron storage ring (figure 5). Filming of in situ dynamics of phase formation («diffraction movie») was started from the begging of system heating. The heating was switched off when reacting furnace charge reached its maximum temperature with consequent cooling to room temperature.

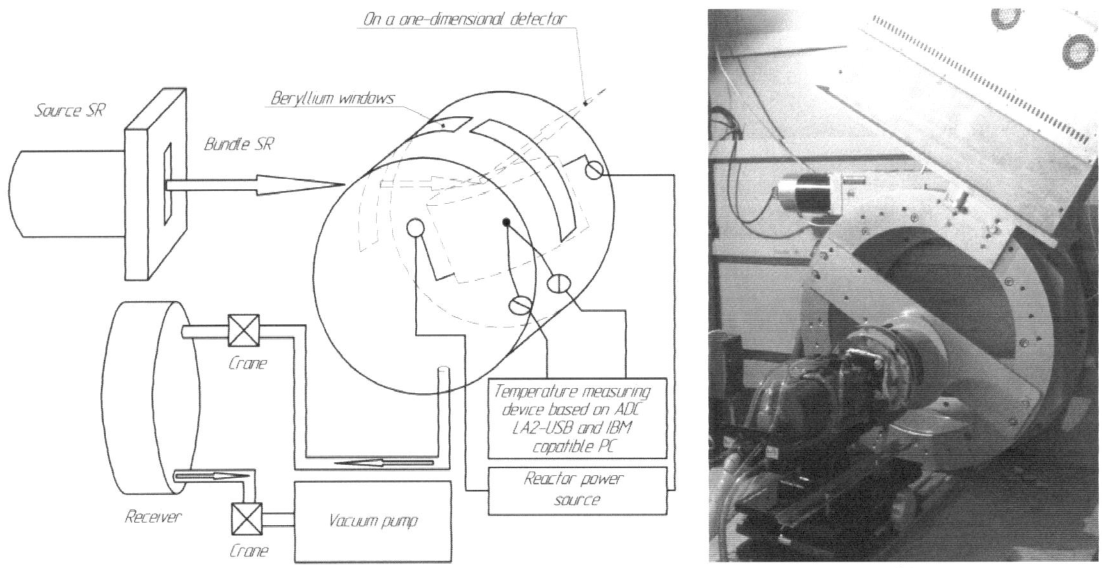

(a) (b)

Figure 5. «Diffraction movie» experimental station 5b of the electron storage ring VEPP-3: (a) Experimental complex for the registration of phase formation of titanium aluminide in the process of inductive heating in the mode of thermal explosion; (b) exterior view of the station with a mounting experimental device based on the inductive heating unit.

Figure 6 represents in situ dynamics of phase formation in conditions of inductive heating of mechanically activated 3Ti + Al composition with different periods of activation.

Analysis of dynamics of phase formation during high temperature synthesis for 1 min. activated 3Ti +Al powder composition proves that reflection on the first diffractogram (first frame of «diffraction movie») corresponds to initial titanium and aluminum components (figure 6).

After that comes the stage of initial structure formation, starting from 564 °C and on change of initial composition reflexes takes place as well as their gradual broadening with a shift to lower angles. That indicates average parameter increment of crystalline grids and phase changes. At that stage along with stable compounds Ti_3Al, $TiAl_3$, $TiAl_2$, some metastable virtual phases Ti_9Al_{23}, Ti_5Al_{11}, Ti_2Al_5 that had formed before the synthesis, were identified. After turning off heater compounds Ti_3Al, $TiAl_3$, Ti_5Al_{11}, Ti_9Al_{23} emerge as reaction products due to quick temperature decrease of furnace charge.

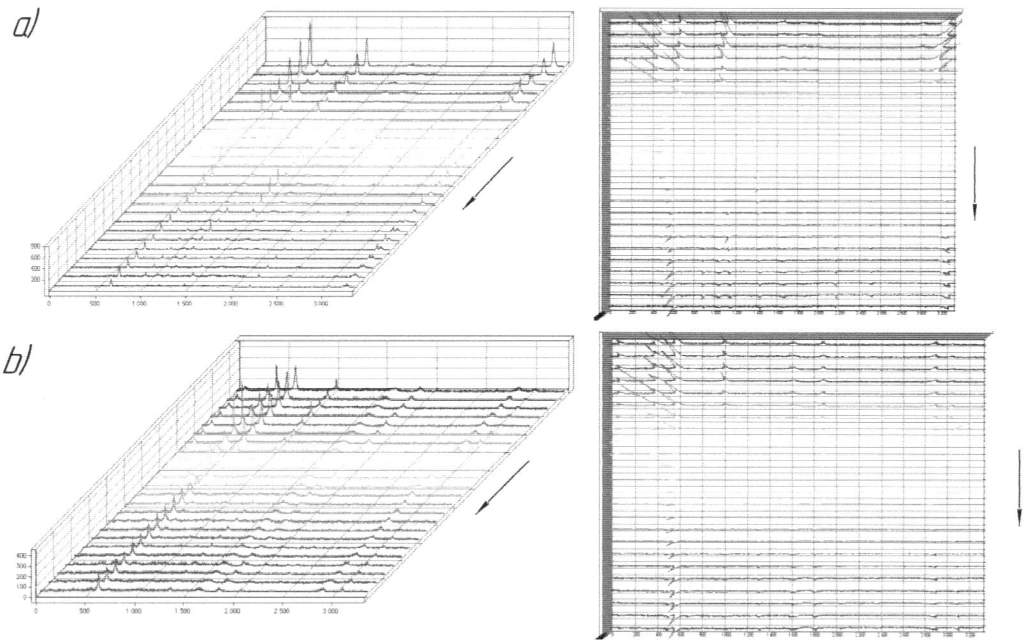

Figure 6. «Diffraction movie» synthesis of the mechanically activated 3Ti + Al composition. Horizontal axis – 2 Theta, degrees, axis of coordinates – intensity, a) MA – 1 min., b) MA – 3 min.

In order to figure out phase content of final product we made a diffractogram of the synthesis product upon cooling. As we can see dissipation of metastable phases Ti_9Al_{23}, Ti_5Al_{11} takes place in the process of furnace charge cooling. There are three phases present in final product – Ti_3Al, $TiAl_3$, $TiAl_2$, predominantly $TiAl_3$ phase of intermetallic compound.

4. Conclusion

The following conclusions were made:

1. Developed experimental system based on ultra-high frequency induction heater 6AV for research to study dynamics of phase formation processes in mechanically-activated systems, adapted to technique of dynamic diffraction of synchrotron radiation. Designed experimental complex is based on inductive heating and adjusted to self-propagating high-temperature synthesis by means of synchrotron radiation, it enabled us to carry out in situ synchrotron research of dynamics of phase formation during synthesis of mechanically activated 3Ti + Al composition.

2. The experimental complex developed and tested on the mechanically-activated Ti-Al system, adapted to the method of dynamic diffractometry in synchrotron radiation beams, which can be used to study dynamics of phase formation processes in induction heating mode, is of practical importance. Application of designed experimental complex is possible not only for registration of dynamics of phase formation of powder SHS Ti-Al systems in thermal explosion mode, but also for fuel air explosions in any SHS system due to fact that induction heated powder composition can be 1500 °C and more, what is enough for SHS formation in most systems.

3. As a result of pre-activation, reaction of high temperature synthesis of Ti-Al composition goes without liquid phase in mode of actual solid phase combustion what proves the emergence of new mechanisms of diffusion and mass-transfer in solid phase combustion of mechanically activated compositions [35, 36]. The studied dynamics of phase transformations of mechanically-activated system will allow to control high-temperature synthesis and to obtain ultrafine structures of synthesized materials-mechano-composites, providing increased properties of products, for example, in case of surfacing with similar composites [37].

5. Acknowledgments

This work was supported by Ministry of Science and Higher Education of Russian Federation (The task no. 11.1085.2017/4.6).

References

[1] Levashov E A, Rogachev A S, Yukhvid V I and Borovinskaya I P *Physico- chemical and technological bases of self-propagating high-temperature synthesis* (Moscow: Binom)
[2] Itin V I and Nyborodenko Yu S *High-temperature synthesis of intermetallic compounds* (Tomsk: Tomsk University press)
[3] Merzhanov A G *Combustion processes and synthesis of materials* (Chernogolovka: ISMAN publishing house)
[4] Bernard F and Gaffet E 2001 *International Journal of Self Propagating High Temperature Synthesis* **10 (2)** 109-131.
[5] Levashov E A, Kurbatkina V V, Rogachev A S and Kochetov N A 2007 *International Journal of Self Propagating High Temperature Synthesis* **16 (1)** 46-50.
[6] Philpot K A, Munir Z A and Holt J B 1987 *Journal of Materials Science* **22 (1)** 159-169
[7] Mukasyan A S, White J D E, Kovalev D Y, Kochetov N A, Ponomarev V I and Son S F 2010 *Physica B* **405 (2)** 778-784
[8] Grigor'eva T F, Korchagin M A, Barinova A P and Lyakhov N Z 1999 *Doklady AN* **369 (3)** 345-347.
[9] Loginova M V, Filimonov V Yu, Yakovlev V I, Sytnikov A A, Negodyaev A Z and Shreifer D V 2015 *Applied Mechanics and Materials* **788** 117-122
[10] Park Na-Ra, Lee Dong-Mok, Ko In-Yong, Yoon Jin-Kook and Shon In-Jin 2009 *Ceramics International* **35 (8)** 3147-3151
[11] Zhu X, Zhang T, Morris V and Marchant D 2010 *Intermetallics* **18 (6)** 1197-1204
[12] Mukasyan A S, Khina B B, Reeves R V and Son S F 2011 *Chemical Engineering Journal* **174 (2-3)** 677-686
[13] Loginova M V, Yakovlev V I, Sitnikov A A, Sobachkin A V, Ivanov S G, Negodyaev A Z and Gradoboev A V 2017 *Physics of Metals and Metallography* **118 (2)** 170-175
[14] Nazarenko V V, Pashinskaya E G, Ryabtsev A D and Pashinsky V V 2001 *Metal and casting of Ukraine* **7-9** 67-70
[15] Loginova M V, Yakovlev V I, Filimonov V Y, Sitnikov A A, Sobachkin A V, Ivanov S G and Gradoboev A V 2018 *Letters on Materials* **8 (2)** 129–134
[16] Medovar B I, Saenko V Ya, Medovar L B, Pomarin Yu M, Orlovsky V Yu, Tsykulenko A K, Fedorovsky B B, Landsman I A and Chernets A V 1996 *Advances in Special Electrometallurgy* **3** 3-8
[17] Grigorenko G M, Sheiko I V, Pomarin Yu M, Chadyuk E N and Orlovsky V Yu 2001 *Advances in Special Electrometallurgy* **1** 32-37
[18] Filimonov V Yu, Sitnikov A A, Afanas'ev A V, Loginova M V, Yakovlev V I, Negodyaev A Z, Schreifer D V and Solov'ev V A 2014 *International Journal of Self-Propagating High-Temperature Synthesis* **23 (1)** 18-25
[19] Filimonov VYu, Sitnikov A A, Loginova M V, Yakovlev V I, Negodyaev A Z and Schreifer D V 2015 *Basic Problems of Material Science* **12 (1)** 16-25

[20] Sobachkin A V, Loginova M V, Sitnikov A A, Yakovlev V I, Filimonov V Yu and Gradoboev A V 2018 *IOP Conference Series: Materials Science and Engineering* **327** 032051
[21] Gauthier V, Bernard F, Gaffet E, Josse C and Larpin J P 1999 *Materials Science and Engineering A* **272 (2)** 334-341
[22] Ternov I M and Mikhailin V V 1986 *Synchrotron radiation. Theory and experiment* (Moscow: Energoatomizdat)
[23] Lizunkov V, Politsinskaya E, Malushko E, Kindaev A, Minin M 2018 Population of the world and regions as the principal energy consumer *International journal of energy economics and policy* **8 (3)** 250-257
[24] Boldyrev V V, Lyakhov N Z and Tolochko B P 1989 *Diffraction using synchrotron radiation* (Novosibirsk: Nauka)
[25] Popova A A, Sobachkin A V, Nazarov I V, Yakovlev V I, Loginova M V, Sitnikov A A, Sharafutdinov M R and Lyakhov N Z 2013 *Bulletin of the Russian Academy of Sciences: Physics* **77 (2)** 120-122
[26] Ergunova O T, Lizunkov V G, Malushko E Yu, Marchuk V I, Ignatenko A Yu 2017 Forming the system of strategic innovation management at the high-tech engineering enterprises *IOP Conference Series: Materials Science and Engineering* **177 (1)** 012046
[27] Ternov I M 1999 *Synchrotron Radiation Theory and Its Development* (Singapore)
[28] Bartolotta P A and Krause D L 1999 *Proc. The Second Int. Symp. Gamma Titanium Aluminides* 3
[29] Shalin R E and Ilyenko V M 1995 *Titanium* **1-2** 23-29
[30] Anoshkin N F 1999 *Light alloy technology* **3** 39-43.
[31] Solonina O P and Glazunov S. G 1976 *Heat resistant titanium alloys* (Moscow: Metallurgy)
[32] Kim Y W and Froes F H 1990 *Physical metallurgy of titanium aluminides* (Indianapolis, TMS)
[33] Kunz K 1981 *Synchrotron radiation, properties and applications* (Moscow: Mir)
[34] Bordovitsyn V A 2002 *Radiation theory of relativistic particles* (Moscow: Fizmatlit)
[35] Loginova M V, Yakovlev V I, Sitnikov A A, Filimonov V Yu, Sobachkin A V and Gradoboev A V 2018 *Journal of Surface Investigation: X-ray, Synchrotron and Neutron Techniques* **12 (3)** 480-484
[36] Sobachkin A V, Sitnikov A A and Sviridov A P 2015 *Applied Mechanics and Materials* **698** 374-377
[37] Yakovlev V I, Sobachkin A V and Sitnikov A A 2013 *Applied Mechanics and Materials* **379** 173–177

IOP Conf. Series: Materials Science and Engineering **483** (2019) 012062 doi:10.1088/1757-899X/483/1/012062

Information technology in the digital document management in the tourism industry as a perspective tool in increasing effectiveness of a tourist enterprise

E V Stelnik, Y A Kiyashko and P I Lysikov

Department of Service and Tourism, Volgograd State University, Prosp. Universitetsky, 100, 400062 Volgograd, Russian Federation

kiyashko.azi@volsu.ru

Abstract. The article is dedicated to the problem of development of information technology in the digital document management in the tourism industry. The authors analyzed the information system "Electronic voucher" and assessed the level of electronic flow of documents in the tourism industry in the Russian Federation. The authors emphasize that the required implementation of this system, as well as its further improvement, will significantly increase the efficiency of management in tourism. It is supposed to lead to the work development of tourist enterprises. Increasing the capitalization of tourism enterprises is very important for development of regional and international tourism in the Russian Federation. So, implementation and development of information and communication management infrastructure promote the realization of two government programs such as the concept of tourism development in the Russian Federation and the programs for the country's transition to the digital economy by 2024

1. Introduction

The order of the government of the Russian Federation to confirm the concept of the government target program "Development of regional and international tourism in the Russian Federation (2019-2025)" [1] was accepted in May 2018.

This document is the continuation of the government target program «Development of regional and international tourism in the Russian Federation» (2011-2018) [2]. The government pays such a serious attention to tourism development due to the prospects of this industry in the context of the socio-economic development of the Russian Federation.

Special attention is paid to development of the regional tourism. The successful arrangement of some events helps to solve a number of problems such as an increase in the share of non-commodity exports in the total exports of the country and also the problem of the import substitution.

It is not possible to achieve the goal in the government program without creation, implementation and development of information and communication management infrastructure in tourism which is an integral part of the digital economy in the Russian Federation [3].

The constant information exchange can characterize mainstream tourism. According P. Evans and T. Wurster, tourism is turning more and more into an information business [4]. Nowadays there are only 30% of all technical opportunities which are used in tourist industry according to the experts [5]. It is apparently impossible to create the whole chain of value without professional management of IT field.

It is already incorrect to consider the electronic sphere in touristic business only as one of the parts the technical support. The permanent operation of electronic systems is the requirement of the daily work [6]. And the presence of the most modern technical equipment is perceived as a requirement of daily work and profitability multiplier. Tourism industry is more often and more successfully using new technical opportunities in its interests. The creation and practice of using of "an electronic voucher" fully refer to it.

2. Results and Discussion

The government law "About the basics of tourist activity in the Russian Federation" (then it is named as "Law") [7], which is the main source of the legal regulation of tourism in our country, provides a variety of mechanisms to ensure the financial security of the tourist during international tours. First of all, all tour operators must give the insurance contract or the bank guarantee to the tourists and these documents provide a proper execution of obligations to the tourist according to the contract. (art. 17.1). Besides, the tour operators, who carry out the international tours, are required to be members of the association of tour operators in the sphere of the international tourism (then –tour assistance) (art. 11.1 of the Law) and should pay the annual contributions to the reserve fund of the tourist assistance (art. 11.4), and also to the personal liability fund of the tour. The reserve fund is intended to be an emergency help for tourists who have travelled abroad, but the funds of the personal responsibility have to cover property damage caused to the consumers of the tourist product by the tour operator as a result of the failure or improper performance by the tour operator.

The obligation to form and supplement the funds of personal liability were given to the tour operators in the sphere of the outbound tourism in the framework of corrections to the Law which are written as the separate Federal Law [8]. They were adopted due to the spontaneous bankruptcy of a big number of tourist enterprises in summer and autumn of 2014 [9]. It was necessary to reform legislation in the sphere of tourism in the Russian Federation because paid money to the affected insurers according to the financial guarantees wasn't enough to satisfy even the small part of tourist's requirements.

It should be noted that the amount of tourist's taxes to the fund of personal responsibility depends on the total price of the tour product in the sphere of the outbound tourism which is a very important indicator of its work during last financial year. Annual contributions of the major tour operators to the funds of personal liability, constituting 1% of the total price of the tourism product for the last year, can reach several dozens of millions of rubles. Anyway, few tour operators, who work in the sphere of the outbound tourism, report to the Federal Agency for Tourism (then – Rosturizm) about the payment of similar amounts of contributions, using various tricks to hide the real numbers [10]. It is possible to verify if you look through the list of travel agencies included to the Unified Federal Register of Tour Operators (then - UFRT) [11].

In theory, unscrupulous tour operators who violate the procedure for determining the total price of a tourist product with purpose can be brought to responsibility on the basis of art. 14.51, part 2 of the Administrative Code of the Russian Federation. It provides an administrative fine in the total of 500 thousand to 1 million rubles for tour operators who make the amount of financial responsibility of tour operators less [12]. However, it makes the procedure of data recheck difficult in some extent which is given to Rosturizm by the tour operators for completing UFRT without serious audit of their activities [13]. At the same time Rosturizm does not have authorities to control tourist enterprises but only can initiate such check making a request to the controlling authority – the Ministry of Economic development of the Russian Federation [14]. It is going to be the same situation which happened in the sphere of the outbound tourism before the crisis of tour operators in 2014: Rosturizm has to deal with unverified data again and there is a danger that deceived tourists will not receive real compensation for the material damage caused by the wave of bankruptcy.

Anyway, the same Federal Law made tour operators form the funds of personal liability and amend the Law. It is about the developing of the system of electronic documents which was called informational system "an electronic voucher" (then IS EV). According to the art.1 of the Law, an electronic voucher is a document which is formed on the basis of a contract on the implementation of a tourist product by a tour operator or travel agent in the form of an electronic document, taking into account the special features defined by the Federal Law (according to the article 10.3). It is necessary to add that registration of the electronic voucher in IS EV carries out by putting it in the specified system by a tour operator (travel agent) [15].

According to the customer's website (Rosturizm), creation of IS EV pursues two main tasks:1) the state control of tourist activity and 2) the provision of state guarantees of its participants [16]. So, IP EV is an instrument of Public law regulation of the tourism industry, through which the state imposes conscientious execution of contractual obligations on market players, as well as the mechanism for improving the financial security of consumers of tourist products. Besides, an informational function of IS EV has a special meaning because it makes the conditions for the transparency of operations which are accomplished by the tour operator (separately or through the tour operators) for the service implementation written in the contract between the travel agency and the customer.

It is also necessary to note the next important aspect of a successful introduction of given informational system to the modern tourists industry in the Russian Federation: its tasks and potential don't have to violate the work of tour business. As it is said earlier, people, involved to the creation of IS EV, are supposed to provide an integration of the last one with the software products which are used in the process and the realization of tours. To clarify, not only completed systems of automation of tour operators and tour agencies (particularly, software which is designed by the companies "Megatek", "Samo-soft" and so on), and also own software are used by the major tour operators [17]. An integration of its software products to IP ES connected to the development of interaction of the subsystems (gateways) and by means of it the specified information system will be accessed to the database of the automated workplace (AWP) of the tour sales manager which is rather laborious and expensive process and also requiring a prior full transition of the tourist business to electronic document management [18].

Anyway, the introduction of IP EV as a compulsory for use by all participants of the tourist markets [19] (there is no doubt that this procedure should be carried out by the imperative method) will help to increase the effectiveness of management in tourism. In a perspective, it should lead to the improvement of work of a tourist enterprise. The head of the tour company, who doesn't use special software in their activity, will have a choice to give a position for the specialist for the registration of electronic vouchers in the interface of IS EV or to automate the working process, which is not only more profitable from an economic point of view, but also makes it possible to optimize the work of the enterprise in general. In addition, the automation of tour agent's activity will make the marketing research easier and also will help to forecast the demand for tourists' services and a strategic planning of the future development of the enterprise.

In conclusion, it is necessary to say that the elaboration of IS EV can be described as a rare example of the further going creation of standards in tourism from that point of view that its implementation will be one of the stimulating tools of tourist activity in the Russian Federation and probably will have a positive impact on the development of the entire industry.

3. Conclusion

At this stage, the main problem is the introduction of IS EV to a manufacturing operation due to the recent bankruptcy of its creator JSC "Asteros" [20] and the constant improvements and the changes which are inserted in the process of testing the given system. Taking into account the potential side of IS EV for the further development in the tourism industry in the Russian Federation, we can see reasonable justification for the introduction in a clause in the Law about the compulsory registration of sold tourist products in the system and, therefore, adding this procedure as a compulsory condition for the tourism business in our country.

The use of web-based architecture offers new opportunities for the tourist industry. Technological innovations significantly increase business potential. Electronic document management is a key concept in this high-tech sphere [21].

The tourism industry requires refusal of the traditional mechanistic view of electronic systems and the transition to the dynamic systems approach, which, should be based on the most modern information decisions.

Acknowledgement
The reported study was funded by RFBR and the government of Volgograd region according to the research project № 18-413-342003.

References
[1] Federal target program "Development of internal and inbound tourism in the Russian Federation (2019-2025)". [Electronic resource]: URL: http://government.ru/docs/32585
[2] Federal target program "Development of domestic and outbound tourism in the Russian Federation (2011-2018)". [Electronic resource]: URL: http://docs.cntd.ru/document/902295121
[3] The program "Digital Economy of the Russian Federation" [Electronic resource]: URL: http://ac.gov.ru/en/projects/014097
[4] Evans P E Wurster T. S. 1997 *Strategy and the New Economics of Information.* 71-82
[5] Papathanassis A. (2004) *Post-Merger Integration and the Management of Information and Communication Systems: An analytical framework and its applicationin tourism.* 302; Bender K. Ammann C. Meitner H. *Beherrschung von IT-Kosten in Post-Merger-Situationen.* 24-28
[6] Berensmann D. Spang S 1998 *IT Integration nach Unternehmenszusammenschliissen: Eine Herausforderung auch an das Top-management.* 35-42
[7] The federal law "About the basics of tourist activity in the Russian Federation". [Electronic resource]: URL: http://www.consultant.ru/document/cons_doc_LAW_194702
[8] The federal law "About the changes in separate legislative acts of the Russian Federation for the purpose of improving the legislation regulating tourist activities". [Electronic resource]: URL: http://www.consultant.ru/document/cons_doc_LAW_194702
[9] RBC, December 26 2014 *"The collapse of tourism: why 2014 turned out to be fatal for tour operators"* URL: https://www.rbc.ru/business/26/12/2014/549c196d9a79474ebffbf916
[10] RBC 2017*"Covering legal entity: how tour operators saved on payments in case of collapse"* URL:https://www.rbc.ru/business/14/09/2017/59a43ebe9a79474d350278ea?from=materials_on_subject
[11] The Unified Federal Register of Tour Operators 2018 the Federal Agency for Tourism, [Electronic resource]: URL: https://www.russiatourism.ru/operators
[12] Administrative Offenses Code of the Russian Federation of December 30 2001 [Electronic resource]: URL: http://www.consultant.ru/document/cons_doc_LAW_34661
[13] Turprom from November 2 (2016), "*Personal Responsibility Fund of Tour Operators: How, How Much and Where?",* [Electronic resource]: URL: https://www.tourprom.ru/news/34097
[14] Regulations on the Federal Agency for Tourism" dated December 31, **901** 2004 Federal Agency for Tourism. [Electronic resource]: URL: https://www.russiatourism.ru/contents/ob_agentstve/Regulation_of_the_Russiatourism
[15] Order of the Ministry of Culture of the Russian Federation of May 30, 2018. [Electronic resource]: URL: http://publication.pravo.gov.ru/Document/View/0001201805310034
[16] E-voucher 2018 "Federal Agency for Tourism", [Electronic resource]: URL:https://www.russiatourism.ru/operators/elektronnaya-putevka
[17] Profi.Travel of February 8 2017 "Electronic Voucher: how it will work", [Electronic resource]: URL: http://profi.travel/news/26292/details
[18] Hot Line. Tourism from February 9 2018 "Electronic ticket: is it a failure?", [Electronic resource]: URL: https://www.hotline.travel/obzory-i-analitika/elektronnaya-putevka-eto-proval
[19] Draft Federal Law 2018 "On Amendments to the Federal Law" On the Basics of Tourism Activities in the Russian Federation "in order to optimize the processes associated with the implementation of the tourist product", Official Internet portal of legal information,

IOP Conf. Series: Materials Science and Engineering **483** (2019) 012062 doi:10.1088/1757-899X/483/1/012062

[Electronic resource]: URL:http: // regulation .gov.ru / projects # npa = 82339
[20] Turprom from July 19 2018, *"The court declared bankrupt the IT company Asteros, which was involved in the creation of the"Electronic ticket*". [Electronic resource]: URL: https://www.tourprom.ru/news/39754
[21] Campos A, Rubiera E, Tejo C, Labra J E 2010 *The Web of Data and the Tourism Industry* 75-89

Gender-related specifics of communication in the work place

L I Stolyarchuk[1], I A Stolyarchuk[2], V I Chumakov[3]

[1]Volgograd State Social-Pedagogical University, 27, Lenin Avenue, Volgograd, 400066, Russia
[2]Volgograd State Academy of Postgraduate Education, 19 a, Novodvinskaya Street, Volgograd, 400012, Russia
[3]Volgograd State Medical University, 1, Pavshikh Bortsov Square, Volgograd, 400066, Russia

lisgender-vspu@mail.ru

Abstract. The paper discusses gender-related specifics of communication in the work place. The authors study the issue of gender-related stereotypes impeding business communication between men and women. The paper gives an insight into reasons for patriarchal self-consciousness of both genders, based on traditional views on sexes (physiological differences of men and women), which are wrongly applied to a concept of «gender» (social behavior, social attitudes) typical for communication in the work place. The authors reveal conditions, supporting efficient communication in the work place, which put no obstacles in relations between men and women, provide them with equal chances in various professional fields, and eliminate stereotype gender-related attitude and expectations existing in the society. The implementation of gender-related approach to communication in the work place is studied in educational institutions of Volgograd.

1. Introduction

Serious transformations taking place in all spheres of our life caused alteration in communication between men and women in the work place (changing from patriarchal, dominant to relations, where partners enjoy equal rights), that is why, it is necessary to carry out more detailed research into the specifics of gender-related communication in the work place, since investigation and application of gender-based theories and methodologies have already been a professional standard in the world's scientific community for a long time.

For collaboration with foreign partners, participation in international conferences, forums and exhibitions it is vital to study gender-based specifics of communication in the workplace, which is a priority task in the state policy of different countries, e.g. Germany, Canada, Luxembourg, Norway, the USA, Finland, France, Sweden etc., where important insights into gender-based peculiarities of communication in the workplace, potential gender-based conflicts in companies and organizations and their probable solutions have been accumulated.

Researchers, focusing on communication in the workplace, pursue an aim to improve business operation and strongly believe a style manager communicates with subordinates is a key factor in this case (equality of individuals, i.d. egalitarian style of communication); despite a common goal personal achievements of employees, their individual efforts, high mobility irrespectively to their gender specifics are important. On one hand, transformations taking place in the present day Russia should further the development of a tolerant society. There is a tendency to redress injustices relating to gender, promote young girls' rights in education, choice of occupation and professional development. On the other hand, traditional gender-related stereotypes existing in the society, being in conflict with a dynamically changing gender-related reality, still hinder successful development of communication in the workplace on the base of gender-related standards in the world educational space.

Furthermore, the issue under consideration is important because it is necessary to establish conditions of gender-related communication in the work place, which enable all communicants to become active subjects in the gender-related interaction. Under conditions of globalization equal gender-related interaction contributes to successful business operation, full scale evolvement of human resources, improvement of production processes and manufactured goods. The implementation of a gender-related approach to communication in the work place has a positive effect both on all employees and success of a company. Gender-related peculiarities of communication in the work place imply democratic relations in a company and take into account communicative gender-related competence of employees.

2. Ideological and theoretical fundamentals of gender-related specifics of communication in the work place

Production is considered to be direct creation of material and spiritual values with the purpose to meet human and social needs on the base of demand elasticity [1].

We view communication in the work place as a turnout of information between individuals in a form of oral or written messages, body language and speech and as a process of communication. It is a vital condition for human existence in the society, equal collaboration between men and women in an enterprise, in a sub-division (as a part of the company) and furthers their success. Therefore, American theoreticians of management (R. Blake and I. Mouton) emphasized importance of two managerial activities: «focus on production» and «care about employees» [2]. To stimulate the efficient work they called for encouragement of employees for creativity, interest in innovations, enthusiastic experimental activity and for training of employees to interact with each other.

A diversity of views on communication between men and women in the work place has been established over centuries (it was started as an idea of a destiny to carry out different jobs by men and women, so a gender-related labor division into male and female jobs).

First notions «gender», «gender-related approach» were used in psychology and sociology in the 1970s of the 20th century in connection with the concept proposed by Marxist feminists of a «dual system» (capitalism and patriarchate), oppression of women in family and society. However, before the opposition (sex-gender) was introduced, the issue under consideration had been discussed as «relations between genders» with respect to two approaches – biological and social ones.

The biological (traditional) approach (P. Abailard, Aristotle, Demokrit, D. Dominici, M.N. Katkov, N.S. Leskov, L.N. Tolstoy) was based on anatomical and physiological differences between men and women. In accordance with these differences a «natural» destiny for labor was established: men are to be engaged into the social sphere, whereas women – into housekeeping. «Useful mothers of families and good wives» were trained for housekeeping in classical educational institutions (gender-differentiated government-supported girls' gymnasiums) in Russia (second half of the 19th century – early 20th century). Émile Durkheim attributes public gender-related division of labor to the social development: «one of the genders took hold on emotional functions, the other one – on intellectual ones» [4], and reasons this «natural» labor division due to natural differences, which are automatically copied in the social sphere.

The social (radical) approach (Quintilianus, N.P. Ogarev, Platon, A.S. Simonovich, M. Wollstonecraft, N.G. Chernishevsky) originated as critics of traditional patriarchal views on «natural» splitting of labor into «male» and «female» one. It implied the idea of equality of men and women, offered women to leave «airless walls» of families and encouraged them for a social mission equally with men. Private girls' gymnasiums, special schools for young women and training courses in Russia in the second half of the 19th century – the early 20th century were inspired with the idea of emancipation of women, directed women towards the professional recognition (accountant, telegrapher, teacher, medical assistant etc.). Feminist sociologists West and Zimmerman [5] developed the fundamentals of social equality between men and women, required for their access to work equally with men, at least in several jobs.

The democratic approach relied on ideas of the social approach (like another offshoot of the radical one), but it was focused on search for the «golden mean» – elimination of gender-related discrimination in labor both for men and women. Its further development as a gender-related approach is research of Giddens [6] and Bem [7]. A strict norm-oriented approach in Bem's questionnaire on androgyny turned into the idea of variability of masculine and feminine properties in studies on gender-related characteristics viewed as variables in constantly changing communication in the work place.

To date, gender-related studies in Russia (T.V. Bendas, E.A. Zdravomyslova, I.S. Kon, I.S. Kletsina, O.I. Klyutchko, A. Temkina, L.V. Shtileva etc.) are focused mainly on the development of their theory and methodology, in the scope of the gender-related approach empirical studies are carried out (leadership development, specifics of present day competition between men and women in the field of production etc.). Russian (theoretical) discourse is thought to be open nowadays. It seems to be in the phase of learning, mastering, perception, absorption, and «digestion» of various theories with quite different origin. Its main feature implies co-existing and overlapping of theoretical models. Different paradigms are co-evolving – a gender-role approach coexists with its radical criticism, social-constructivist research discusses the idea of a «female experience», which hasn't become a well-established research domain in the Russian research discourse.

In recent days gender-related specifics of communication in the work place is thought to be a key factor for understanding of social transformations. Gender-related approach is accepted in a discourse space, which brings into question patriarchal traditions of communication men and women are involved in the work place, at the same time, it makes a contribution to methodology, subject area and methods or research, creating incentives for interdisciplinary studies.

3. Specifics of communication in the work place: gender-related dimension

3.1 Advances in communication in the work place – a factor of production growth

Prosperity of economic activity in an enterprise or a company is strongly dependent on advances in communication. That is assessment of a manufacturing process not only according to the physical output, but also with respect to advantageous research ideas of employees, their creative products, «godsends» with the aim to increase sales, consumer-friendly innovations to improve the quality of provided services and acknowledgment of successfully made decisions by employees and management – all these factors support favorable conditions of communication in the work place. For successful development of «production» and «communication» their efficient combination is needed in a real manufacturing process in companies and industries.

Gender-related peculiarities of communication in the work place are considered as interpretation of socially organized relations between genders within the professional framework.

3.2 Actual strategies in organization of gender-related communication in the work place

To date, a gender-related strategy in communication in the work place originating from the biological approach still exists (47%) and comes to the idea of biological male and female labor dichotomy. It refers to fundamentals of essentialism (sex differences between men and women), traditional expectations and stereotypes in the society. In the process of research (Volgograd State Social-Pedagogical University, Volgograd State Medical University, Volgograd State Academy of Physical Education, Volgograd State Academy of Postgraduate Education) typical gender-related attitudes – oppositions were disclosed: – the man initiates and creates, whereas the woman saves and implements; – the man is competition-centered, and the women is oriented towards partnership and establishment of relations; – the man is rational and prudent, whereas the woman – emotional and willing to empathy; – being in new conditions the man starts changing the space around him, the woman is more adaptive and conservative. Conditions of the gender-related mode thought as a state of play in communication in the work place can manifest themselves exclusively within encouraged manhood and womanhood. In spite of stable gender-related stereotypes this strategy has been losing its influence.

A social strategy, originated from ideas of the social approach and based on fundamentals of accidentalism implies equal rights of men and women in the work place (21 % educational institutions under consideration). The framework of the social strategy in the gender-related communication in the work place is not stable and supposed to study gender-related modes, gender contracts [9], breakdown of old patterns. Transformations might result from governmental actions (reforms etc.), due to changes in the work place (regulations, orders and other documents) and egalitarian position of the manager. The purpose of a manager (often a woman) is to eliminate gender-related stereotypes of employees. For instance, men have strong leadership skills, whereas women prefer to stay in the background, like to be advised and are willing to work selflessly. Naturally, men are more aggressive; being determined, therefore, to fight and compete, while women tend to find compromises without competing. Communication is supposed to be a means to achieve a business goal for the masculine type of behavior, and it is an end in itself, on the contrary, for the feminine type. Egalitarian managers assess critically gender stereotypes mentioned above and try to eliminate them in the process of communication in the work place. Attitudes towards them are reconsidered in companies, the initiative of women and their communicative skills are supported with respect to their professionalism, initiative, dynamism, and personal results.

A suggested integral strategy emphasizes the importance of eliminating any kinds of discrimination, equivalence of communication in the work place both for women and men. A unifying paradigm in the gender-related approach encourages new gender-related practices in conventional gender-related relations. A challenge initiated by a group or individual marginal is needed to change patriarchal gender-related stereotypes, creating a precedent for «new» gender behavior. For example, leaving the work place without completing tasks, a working mother is regarded a neglecting employee, but if she earns enough and can hire a babysitter to work without any stress, this person isn't thought to be a neglecting employee or a bad mother anymore. Initially, a single-parent man is also thought to be a marginal, since it obstacles business, career, apprehending reliability of this employee; but later it can become an acceptable combination of career advancement and parenthood. Although gender stereotypes still exist, there is an opportunity to "soften" gender relations in communication in the work place (32%). Social prescriptions and inequality between the sexes might be reflected critically and changed.

An integral strategy within the unifying paradigm can be «a useful methodology of analysis» to study power relations, based on cultural and symbolic sex determination, viewing gender as an achieved status. Gender-related approach, aimed at resolving the dilemma between biological and social approaches, admits and does not deny gender differences between men and women, arguing, however, these issues are beyond the scope of this research field. The methodology for analyzing gender relationships in the workplace aims to have insight into social relations, including gender-based ones; therefore, a biological strategy is thought to be unacceptable to study social relations. An integral strategy in organization of gender-related communication in the work place is supposed to be a new strategy aimed at reflecting gender relations in enterprises and companies, democratic changes and transformations. The focus of managers on goals of production implies free and equal competition of male and female employees, assessing fairly results of their work. Managers should have a non-prejudicial and equal attitude to individual characteristics and skills of men and women, free from gender-related narrow-mindedness and stereotypes.

4. An interdisciplinary model in organization of gender-related communication in the work place

For efficient organization of gender-related communication in the work place it is necessary to take into account global dynamic transformations, taking place in the society these days, and reconsider theoretical knowledge accumulated in different sciences: history and philosophy of education, sociology, psychology, theory of management in industry, theory of communication and theory of gender education with the purpose to develop an interdisciplinary model, providing an integrated system for resolving a multidimensional theoretical problem and its implementation in practice.

4.1 First component – «managerial»

The first component «managerial» in the interdisciplinary model of organization of gender-related communication in the work place relies on Blake and Mouton theory of management. Basic fundamentals of the theory imply search for the balance between «focus on production» and «care about people», so a combination of two core components. A «managerial net», oriented towards production, horizontally, and environment of cooperation, vertically, in their reasonable combination are still considered a base for the development of up-to-date theory of human resource management in industry, in different companies, and institutions. A list of statements suggested in the theory for assessment of a managerial style makes it possible to reveal «a small-minded style of management» because of neglecting production and mishandling employees. Focusing only on achieving high performance has an exhausting effect on creative energy of employees. Orientating exclusively on human relationships, it is possible to lose competitiveness of production, and company, resulting, in the worst case, in cessation of their economic activities. A management style oriented towards «focus on production» and «care about people» furthers achievement of better production results and keeps employees working. Both short-term results and long-term development of the organization are important for this managerial style, as well as trusting relationships of employees, achievement of company goals by people holding the same views, creating team spirit, mutual respect, and individual development of employees. The gender-related approach to the organization of communication in the work place is meant for contribution to the development of new gender rules. They are to meet requirements of social reality and internal needs of an individual, justice between genders, resulting as a consequence, in the increase of economic return of employees and their labor performance.

4.2. Second component – «communicational»

The second «communicational» component in the interdisciplinary model of organization of gender-related communication in the work place is based on social communication theories, which address communication as a necessary condition for a person's existence in a society. Communicability is supposed to be an ability to communicate, collaborate; it is an important factor promoting success in the modern society. Public communication is gaining in importance for any enterprise or organization alongside with external, internal, and interpersonal communication due to its importance for competition and relevance for public opinion and public image of a company. No doubt, public image of all employees, their position on a carrier ladder are important; they can be tracked in the Internet by managers, employers and colleagues. In gender-related communication, a so-called frame is manifested; i.d. a cognitive structure anticipating the development of a situation according to a social and cultural context. Today, it extends norms of masculinity and femininity, increasing the spectrum of individual behavioral manifestations. Framing of social reality gains in importance, as well as structuring of gender, "gender display" (I. Goffman, M. Sokolov, V. Vakhshtayn) as a theoretical and methodological basis for analysis of everyday communication at home and in the work place with reference to gender-related specifics of objective reality. Frame-based analysis of professional interaction has not been studied in details from the standpoint of gender dimension, so it can be considered a promising research domain of communication in the work place.

4.3. Third component – «Gender-related»

The third «gender-related» component in the interdisciplinary model of organization of gender-related communication in the work place is based on a gender-related approach to training and education of employees. In the present day world a lot of women make a career like men. In this case they just have to accept rules of the play in the business world: competition, leadership, strategy and orientation towards professionalism evolve such feminine characteristics as adaptability, ability to cooperate, collaborate, skills to be focused on details for a long time. The business world is thought to belong to men, that is why, both men and women have to meet its requirements to be successful.

A core means in implementation of a gender-related component in the interdisciplinary model of gender-related communication in the work place is business ethics (official relations are built like a hierarchy in the production process, activity of the upper member of a hierarchy and subordination of male and female employees), and an auxiliary means is an everyday ethics of man and woman relations (a man protects a women against inconveniences). Interrelations in the work place are to be based on equality and equivalence of labor achievements and personal characteristics of employees, both men and women. For instance, initiating a conversation a hand is given by a person with an upper status, even if it is a man, a woman shakes a hand first if communicants with equal right start conversation. Being asked to enter a manager's office, both male and female subordinates have to wait for an invitation to take a seat, although they may ask for a permission to sit down if it is not mentioned by a manager. A male manager has a right to enter a room first, as well as go upstairs and enter the lift, following a subordinate woman; however, gallantry towards women allows him to make a concession to a woman. Being in a car, a priority place is always given to a male manager, unless he offers it to his female subordinate. Due to subordination rules a male manager doesn't perform work in the office which does not agree with his status, e.g. carry chairs or heavy folders, however, if there are no other men in the office, he can help women, following rules of an everyday ethics. Business ethics, excludes sexual harassment at work, because it violates the principle of gender equality, undermines authority of a manager and credence of employees because of exceptional and unequal attitude towards them. To date, business environment and industrial sector are considered harsh, so men and women are advised to reflect their own gender stereotypes and prejudices to eliminate them. Instead of focusing on the gender and its features in gender-related communication in the work place it is more important to observe rules of business ethics and conditions of egalitarian interaction for successful adaptation to the world of business.

5. Conclusion

The approaches above (biological and social) have accumulated theoretical knowledge of different sciences: history and philosophy of education, sociology, psychology. They are heuristically important, although past and present day state of the art can't be compared on their base because of different social situations and contexts. Synthesizing scientific achievements available nowadays and relevant for understanding of gender-related communication in the work place in theory and practice of modern education, it is possible to identify origin of its present day strategies (biological, social, integral) in the context of gender-related transformations taking place nowadays.

The suggested integral strategy of organization of gender-related communication in the work place implies a unifying paradigm, a new methodology in the gender-related analysis. It expands the scope of gender ideas, making natural sex differences less significant in professional development of competitive specialists, and encourages for dialog and partnership. To date, gender-related labor division and standards of gender behavior tend to be multiple, variable and dynamic, therefore, organization of gender-related communication in the work place, oriented towards a new reality, is supposed to be prospective on the base of the interdisciplinary model. The interdisciplinary model, comprising managerial, communicational and gender-related components, establishes structural scientific combination between the theory of management in industry, theories of communication and frame-analysis (sociology), and gender-related approach (theory of gender education). It provides an integral system to solve a multidimensional problem of research; as a result, a new quality is formed – a professional with flexible gender-related attitude and skills of business ethics, open to global democratic transformations, which are strongly required in the present day Russian industry.

IOP Conf. Series: Materials Science and Engineering **483** (2019) 012063 doi:10.1088/1757-899X/483/1/012063

References

[1] Marshall A 1892 *Elements of economics of industry* (London: MacMillan and Co.) XIV 416
[2] Blake R R and Mouton I S 1981 *The versatile manager: a girl profile* (Irwin-Dorsey)
[3] Blake R R and Mouton J S 1985 *The managerial grid III* (Gulf Publishing)
[4] Durkheim E 1991 *On the division of social labor. Method of sociology* (Moscow: Nauka) 61
[5] West K and Zimmerman D 1997 *Gender Notebooks* (Proceedings of the St. Petersburg branch of the Institute of Sociology) **94 (24)**
[6] Giddens A 2009 *Sociology* (6th edition Polity Press) 1183
[7] Bem S L 1981 *Bem sex role inventory professional manual* (Consulting Psychologists Press)
[8] Zdramomyslova E and Temkina A 1999 *Journal of Social sciences and modernity* **177 (85)**
[9] Malin L and Jacob M *Journal of Vocational Education & Training*
[10] Colley H, James D, Diment K and Tedder M 2003 *Journal of Vocational Education & Training* **55 (4)**

IOP Conf. Series: Materials Science and Engineering **483** (2019) 012064 doi:10.1088/1757-899X/483/1/012064

Quasi-professional tasks in the foreign language education

Y Timkina, M Khlybova

Perm State Agro-Technological University named after Academician D.N. Pryanishnikov, Perm, Russia

timkinaj@mail.ru

Abstract. Foreign language education in a non-linguistic multidisciplinary university is a part of specialists' professional training. A foreign language is considered as a tool for solving professional problems in future work activities. It is possible to train students for solving problems by creating practically-oriented learning conditions. Quasi-professional activities in foreign language include content and activity aspects, that is, based on professionally-oriented materials, involve students in modeling certain practical activities. Quasi-professional tasks are based on the principles of communication, interdisciplinarity and problematicity. The basis of the quasi-professional problem in a foreign language environment is the communicative situation that ensures the interaction of participants in the process of solving the problem. The interdisciplinarity is defined as the integration of various sciences, namely linguistics and specialized professional training of students, as well as ways of mastering and expanding knowledge and skills. Problematicity defined as the presence of a task with unknowns, or contradictions, is necessary for activating the cognitive processes of a person in overcoming difficulties, which contributes to activating cognitive processes and producing foreign language speech. The aim of the presented work is to develop and justify the effectiveness of in-class work, designed to systematically solve gradually sophisticating quasi-professional tasks. The article provides examples of quasi-professional tasks developed for undergraduate students in the "Hunting, maintenance, breeding of wild animals and animals for commercial purposes" training course. The conducted experimental work revealed a number of benefits and disadvantages of this system. It is noted that the **apparent disadvantages are an incentive for development of teacher competence.**

Introduction

Foreign language education as a part of undergraduate and postgraduate students' training in higher educational institutions is aimed at preparing specialists to solve professional problems in a foreign language environment. The range of such tasks encompasses business communication, the study of world experience and innovations in the field of professional activity, presentation of their own developments for discussion within professional community and skills enhancement and professional development. Thus, the goal of foreign language education at all levels of training in higher educational institution is foreign language acquisition in the professional sphere [6, 12]. The modern methodology defines the foreign language acquisition by students of non-linguistic institutions as the formation of foreign language communicative competence with the following aspects: readiness and ability to carry out interpersonal foreign language communication [2], use of language means in speech activities, compensation for a limited number of language means, employment of different strategies and tactics of communication, use of knowledge of the national and cultural specificity of a foreign country in order to overcome cultural and linguistic barriers [14], solution of production tasks within cross-cultural environment [9].

Professionally-oriented foreign language teaching in recent decades has firmly strengthened its position in the methodology and practice in non-linguistic universities, focusing on the use of professional knowledge in the form of content of the foreign language teaching. It has now created the basis for integration of linguistic and professional education. The expansion of the area of "professional orientation" is realized through the inclusion in the educational process not only content that allows students to get acquainted with professional communication linguistic features, but, to a greater extent, the organization of the activity aspect that simulates professional activities in a cross-cultural communication [11, 13].

The integration of linguistic and professional training is based on the interdisciplinarity, the interdependence of disciplines in an integral professional training program and is aimed at developing the "abstract and professional thinking in the process of mastering and applying language material in situations of professional communication" [7, p. 33]. The orientation of the educational process toward the practice, the need to create conditions for mastering the ways of conducting professional communication and performing future labor functions determines the transformation vector of both in-class work and students' self-study. The in-class contact work should acquire a new status, during its conduct students should receive the necessary minimum knowledge that would allow them to confidently step forward in the self-study process [8]. The effectiveness of in-class work should be enhanced by active forms of education, namely, the systematic implementation of the quasi-professional tasks with gradually increasing sophistication.

Quasi-professional activity is considered as being "educational in form and professional in content" [13], based on authentic materials reflecting the real state of science and industry [15], based on motivation to learn and to obtain professional qualification [5]. The tasks of this type of activity are the situations that simulate the performance of labor functions of a specialist, containing specific problem or difficulty, which implies the search for solutions. Considering foreign language professional activity, it is necessary to outline such characteristics as the presence of a communicative situation for the implementation of various types of communicative activities (speaking (monologue, dialogue, polylogue), reading, listening, writing). The analysis of studies' results on organizing foreign language courses based on quasi-professional tasks indicates many advantages of this educational process, particularly the "close resemblance to professional activity, high levels of mental activity, creation of specific linguistic and professional environment that ensures the development of logical-communicative operations" [6], large educational value of foreign language as a means of obtaining professional knowledge [15].

However, systematic actualization of quasi-professional activities during foreign language courses implies solving certain problems, namely, defining tasks, applying task types in accordance with level of education and language knowledge, developing tasks aimed at improving cognitive skills and professional thinking, identifying benefits and disadvantages of the system being developed. The purpose of this work is development and methodological justification of the foreign language course encompassing quasi-professional activities during in-class work . The hypothesis of the study is the statement that systematic implementation of gradually sophisticating quasi-professional tasks will contribute to formation of foreign language communicative competence at a higher level.

1. Materials and Methods

The bases of organization of foreign language quasi-professional activities are provisions of the communicative approach with the presence of a communicative situation and professionally-oriented approach with interdisciplinarity and personal-activity approach. The communicative situation is a prerequisite for formation of receptive and productive speech and communicative skills. The development and consolidation of these skills is realized through the process of performing a speech act. Activation of speaking activities ensures obtaining practical experience by forming the ability and readiness for foreign language communication that constitute the foreign language communicative competence. The speaking skill activity is possible in presence of the interaction of interlocutors by which the developer of the communicative approach E. Passov assumes the interrelationships of communication participants, which have personal character and determine motivation to solve the verbal thinking task [10, p. 46]. Creating conditions for establishing relationships in a group of students during in-class work involves the following steps: identifying the subject of interaction, providing subjects of communication with the information being discussed and subject-linguistic context, establishing information communication, establishing the form and method of interaction, setting up communicative task [10, p. 56].

Interdisciplinarity of foreign language studies in the future professional activity is considered as the integration of two or more sciences such as linguistics as the language science and the scientific knowledge, which is fundamental for students of certain field of study [7, p. 90]. At the same time, not only the substantive part is in the focus of integration, but also the ways of interaction in professional communication, "the dynamics of students' personal and professional development", the involvement of future specialists in the "multicultural global world" of professional activity [7, p. 91]. An important factor ensuring interdisciplinarity is the cooperation of foreign language and special discipline teachers, as well as the involvement of employers and professional community in order to develop up-to-date teaching cases, monitoring and evaluating educational process from the point of view of modernity and adequacy of solved quasi-professional tasks, solutions and results to be achieved. Significant for our study is the fact that disciplines related to mastering a foreign language are studied in the first and second courses at the undergraduate and postgraduate courses, while core disciplines are studied at graduate courses, which causes gradual and consistent extension of professional knowledge in linguistic education. The prevalence of a foreign language at the beginning of a course provides preparation for further independent mastery of a foreign language and its use as a tool in acquiring professional competencies. A gradual increase in the amount of professional knowledge in a foreign language corresponds to development of future specialists' cognitive sphere in the training process, "realizes the transition from basic study with methodological support during the course to an independent creative professional form of activity" during self-study process at graduate courses [16].

Problematicity is considered in this article as the presence of certain tasks or contradictions that need to be solved by activating cognitive processes, such as analysis, comparison, use of deductive and inductive methods, etc. According to the personal-activity approach in vocational training, it is important to involve students in finding information and such ways to solve problems where knowledge accumulates, and their application mechanisms are improved by including a certain mental state of intellectual difficulty stimulating the cognitive processes and speech skills [1].

Quasi-professional tasks in foreign language education are designed to simulate professional interpersonal communication in a foreign language environment based on solving a certain problem that involves the activation of verbal and cognitive skills. The "educational in form, but professional in content" [13] tasks should reflect labor activities of graduates, which allow use of professional standards as guidelines in design. Studying and analyzing labor activities of specialists allow to identify the main areas of communication, determine the subject and object of communication, predict possible problem situations, determine the components of foreign language communicative competence necessary for the formation. The movement towards accepted professional standards determines the maximum likeness to future activities in accordance with the goals of higher education. In addition, it is possible to establish contact with employers for a substantive assessment of the training quasi-professional tasks, solution methods and obtained results.

It should be noted that possible diversity in levels of tasks' complexity varies with the level of students' foreign language skills and stage of professional training. The spectrum of tasks presented in the methodological literature includes different level types of tasks, such as situation - problem, evaluation, illustration, exercise [13], or according to the level of problematics depending on the unknown component, problem-presentative, linguistic-cognitive, communicative-active and professional-communicative tasks [1]. Designing quasi-professional tasks of varying complexity, from simple to complex, allows one to consider the stages of formation students' foreign language communicative competence. The structure of the quasi-professional task adapted by O.L. Felde and O.N. Igna includes the objective's formulating, context establishing for a specific situation, tasks leading to result, and assessment criteria [4]. Performing these step-by-step actions facilitates process of creating quasi-professional tasks and contributes to students' ability to solve tasks, and as a result, to achieving an educational result.

Preparation for solving problems during in-class work is carried out in the self-study of students which acquires a new status. Most of the time in the development of disciplines concerned with mastering a foreign language is spent on self-education process. This is the work that includes forming of foreign language competence in terms of mastering new language units (phonetics, morphology, etc.), performing speech exercises, mastering new grammatical structures, and terminological system of professional activity. In self-education process is possible to choose text, audio or video materials that meet individual needs and interests, in such a way that they stimulate and maintain motivation and develop independent learning skills [15].

Let us point out the main features of organization of educational process based on the quasi-professional activity: in-class work is based on solving gradually sophisticating quasi-professional tasks with the simple-to-complex principle; quasi-professional tasks contain linguistic or professional problem; quasi-professional task is connected with communicative act; self-education process takes over the majority of traditional methods of in-class work in regard to new language material introduction, communicative skills based on authentic materials development and application.

2. Results and Discussion

On the basis of theoretical principles of quasi-professional activity there was developed the course in English for undergraduate students of "Hunting, maintenance, breeding of wild and game animals" training course of the "Zootechnology" speciality of the Perm Agro-Technological University. The discipline includes 108 hours (3 *learning credits*), of which 42 hours is in-class work. The basis of the course are the quasi-professional tasks developed on the professional standard "Hunting" (2016) and the federal state educational standard "Zootechnology" (2016). 20 tasks for in-class work of three types – illustration, problem, assessment were developed. Let us demonstrate several sample tasks.
Example 1. The quasi-professional task-illustration.

Objective: making commentary on what hunting is, the importance of hunting in the modern world, traditions of hunting in Russia and abroad, difference between hunting and poaching, the main lines of hunting's specialty.

Subject content: the place of hunting in the modern world, the importance of hunting based on the historical development of society, history of hunting in Russia and in other countries, hunting traditions in different countries, the main components of hunting.

Communicative activity types: speaking (polylogue), listening, reading for gist, reading for specific information, and reading for detailed comprehension.

Implementation: searching for information, mastering new necessary vocabulary, drawing up a mental map of a particular part of the report (for example, hunting traditions in a particular country) or making a whole report during self-study process, justification of hunting's importance in the modern world, expression of agreement or disagreement between speakers.
Example 2. The quasi-professional task-problem.

Objective: determining the sustainability of natural resources' management, the measures taken for organization of environmental management and hunting.

Subject content: sustainable management of natural resources and hunting use, the concept, signs of rationality, irrational natural resources management, the examples, generally accepted principles, ways of improvement, examples of the world community.

Communicative activity types: reading, writing, speaking (presentation).

Implementation: small texts with a description of use of natural resources are offered to students in small groups, where it is necessary to assess the situation, determine the rational or irrational use of resources. On the basis of these texts, the principles of rational use should be drawn up (in writing).

Expected results: the creation of natural national parks with rich flora and fauna, protected landscapes by human activity, nature conservation combined with the organization of leisure activities and mass tourism, natural resources are used in a way that does not damage nature.

Designing a course based on similar tasks focuses the developers on a more thorough selection of lexical base for students' active vocabulary, functional grammatical structures based on statistical methods of corpus linguistics that meet the criteria of frequency, universality, breadth and utility. Particular attention is paid to phrases which help to organize communication and support the conversation.

Such an educational process forces a change in the teacher's role. The teacher performs the functions of a moderator and manages communication: puts tasks in context, asks questions, directs the conversation in the desired direction, sets the tone for communication and other things, but does not evaluate or make comments, corrections in the process of communication. The students' self-study process may consist of two parts: firstly, repetition of the problem being solved in the form of an essay, outline, mental map, etc. and secondly the preparation for the next task.

From the students' standpoint involvement in solving quasi-professional tasks allows them to realize their own preferences and interests, choosing the appropriate issues and means, creative abilities, and developing skills of academic communication. Freedom to choose the option of preparing for the task, use of authentic materials supports the motivation of learning and mastering a foreign language for professional purposes. The assessment of students' foreign language communicative competence is carried out on the basis of subcompetencies.

Analysis of the above characteristics of the educational process suggests that this organization of the process is quite effective. It forms relationship between teacher and students and is focused on developing future specialist' personality, professionally-motivated mastering a foreign language, using limited time for in-class work within the course framework for the revitalization of productive speech skills (speaking), the orientation of students' self-study process on meeting professional interests and needs.

3. Conclusion

The organization experience of quasi-professional tasks exhibits benefits and disadvantages of this system. Benefits include orientation of the educational process towards students' future professional activities, mastering a foreign language in practical activities by using it as a tool for the formation of professional competence, tightening up relations between teacher and students. The discussion and overcoming of emerging difficulties and problems contributes to immersion in educational activities, develops cognitive area of future specialists, supports motivation training, activates students' self-study, and, as a consequence, increases the level of foreign language knowledge. The main disadvantage is high labor-intensive development of tasks for teacher, which includes mastering new interdisciplinary content for the foreign language teacher himself, formulating tasks, developing teaching materials and methodological support for students' self-study, developing evaluation criteria and playing the role of moderator requires communication management skills that requires considerable preparation. Obviously, such organization of training stimulates teacher to enhance his own skills as well as communication skills and creates the prerequisites for a thorough study of the area of students' activity.

The presented system differs from traditional one by a deeper immersion into the area of future professional activity with the help of a foreign language on the basis of systematic solution of quasi-professional tasks during in-class work. Orientation of self-study on personal needs and interests' satisfying during preparing for solution quasi-professional tasks allows one to center the educational process on the development of future specialists' personality.

References

[1] Aitov V 2006 *Izvestia: Herzen University Journal of Humanities & Science* **17** 127
[2] Bezukladnikov K E, Kruze B A 2015 *Procedia - Social and Behavioral Sciences* **200** 393
[3] Chilingaryan K, Gorbatenko R 2015 *Procedia - Social and Behavioral Sciences* **214** 559
[4] Felde O L, Igna O N 2017 *TSPU Bulletin* **10 (187)** 155
[5] Gitman E K, Danilov A N, Stolbova I D 2017 *News of higher educational institutions. Technology of textile industry* **2** 321
[6] Khomiakova N P, Dubynina N V, Osetrova M G 2015 *Philology at MGIMO* **1** 132

[7] Krupchenko A K, Kuznetsov A N 2015 *Basics of professional linguodidactics* (Moscow: The Academy of Professional Development and Re-Training of Educators) p 232
[8] Leushin I, Leushina I 2017 *Higher Education in Russia* **6** 51
[9] Leushina I 2018 *Language and Culture* **42** 198
[10] Passov E 1989 *Basics of communicative methods of foreign language communication teaching* (Moscow: Russian language) p 276
[11] Pozdniakova S 2018 *SHS Web of Conferences* **50** 01141
[12] Serova T S, Krylov E G 2015 *Language and Culture* **3** 106
[13] Shliamova A A 2014 *European Social Science* **7-3** 204
[14] Sysoev P V 2014 *Language and Culture* **4** 184
[15] Vaičiūnienė V, Užpalienė D 2010 *Studies about Languages* **17** 94
[16] Verbitskiy A A 2018 *Moscow State Linguistic University Bulletin* **796** 126

IOP Conf. Series: Materials Science and Engineering **483** (2019) 012065 doi:10.1088/1757-899X/483/1/012065

Economic and mathematical model for calculation of preliminary production costs of multidimensional products for objects of protection having complex geometric shape

D A Tolstobrov[1,2], N A Tolstobrova[1], S A Fedoseev[1,3]

[1]Federal State-Funded Budgetary Educational Institution of Higher Education "Perm National Research Polytechnic University", 29 Komsomolsky prospekt, Perm, Perm krai, 614990, Russia
[2]LLC "NPO "PermNefteGaz", 15b Chernyshevsky Street, Perm, Perm krai, 614002, Russia
[3]CJSC "GELIKON PRO", 25b Lebedeva Street, Perm, Perm krai, 614107, Russia

2told@mail.ru

Annotation. The research presents a description of the operational and economic indicators of the protective multidimensional product. The structure of the protective product and the classification system of its model range depending on the type of protection object are described. The estimation of development dynamics of the protective products market is carried out. The process of formation of the industrial enterprise pricing policy on the basis of ABC method and the method of cutting by "Mueller and son" is presented, the mechanism of preliminary calculation of material consumption for the production of protective products on individual order is visualized

1. Introduction

In 2015, the industrial company "NPO "Permneftegaz" expanded its range of products, organizing the manufacture of protective products. The protective product limits the impact of negative and harmful factors on the equipment (hereinafter the object of protection) of the client or its maintenance personnel, creating a closed shell around the object of protection after installation.

Properties of the protective shell are defined by the customer requirements when ordering. The list of specified properties can be formed from the following list: conditions of constant or time-limited maintenance of the specified thermal regime for the object of protection, moisture resistance, vibration protection, noise insulation, dust protection, chemical insulation, visual isolation or hiding, indication of the occurrence of specified events, convenience and speed of installation and dismantling, constructive compliance with the specified form.

Implementation of safety products at the client's company allows to improve economic performance, developing the following tasks:

1. increase in working life of expensive equipment operated in a negative environment;
2. reduction of heat and energy losses of the production line;
3. exception, reducing the impact of harmful factors of the production environment on the staff, decreasing the cost of personal protective equipment (PPE) subject to compliance with labor protection standards.

Compliance with the requirements to the protective product set by the client is achieved by combining it with different properties of materials, components, selected design.

2. Multidimensional product

Protective products manufactured by the industrial enterprise LLC "NPO "Permneftegaz" are a multidimensional products, described with a set of consumer properties $X=(X_1, X_2,\ldots, X_N)$, amenable to the stage of coordination with the client adjustment, in accordance with the established external conditions while maintaining its overall structural scheme.

The proposed range of protective products consists of a standard model range and products manufactured individually in accordance with the project agreed with the client. Shut-off and adjustment valves were chosen as the basis for the standard model range of protective products, as well as other objects of protection, often repeated in the applications of customers. When agreeing on individual performance of a protective product, the client provides a detailed list of information about the object of protection: operating conditions, geometric dimensions, connector lines, viewing windows, doors, pockets, entry points, and the necessary components.

Compliance with the requirements set by the client to the protective product is achieved by the use of appropriate materials of the upper and lower coatings, filler, filler layer thickness, additional components.

3. Preliminary calculation of production costs

The market of manufacturers of protective products is a model of perfect competition. The client distributes an application for calculation with a limited amount of general information about the object of protection to potential performers in order to establish the range of prices offered on the market. Efficiency in the calculation of the application submitted by the client and not an inflated value of the set price among other competitors can be the key to the success of attracting the client. When carrying out such a preliminary calculation of the protective product cost, it is important for an industrial enterprise to immediately establish a price order taking into account the additional costs that may be incurred during the production process.

A general analysis of all the work carried out by the enterprise allowed to establish that about 10% of all processed applications from the client is confirmed by the order, and the rest pass either to competitors or were submitted by the client for an analytical assessment of the possible cost. Therefore, in order to ensure the stable development of the enterprise direction, it is necessary to increase the volume of processed applications and reduce the price offered to the client while maintaining the overall profitability of production. Reducing the cost of production while maintaining its quality indicators should be ensured by reducing the time of submission processing.

Objects of protection can be divided into several types: elements of the pipeline, shut-off and control valves, test equipment and objects of complex geometric shape. In this paper, we consider the mechanism of preliminary calculation of material consumption for objects of protection with complex geometric shapes.

The protective products produced by an industrial enterprise are manufactured using technical textile materials and fillers. Due to the fact that this area is rather new in comparison with the market of the clothing industry, it is characterized by the absence of uniform standards. At present, the level of development of the clothing industry has reached a stage where for each well-known cutting of a model of clothing or a textile product, depending on their size, the consumption of material for single or mass production is determined.

In the case when the cut is different from the generally accepted standards, the developers calculate its patterns and layout on the material. Carrying out such calculations using specialized software packages is a time-consuming process even for a highly qualified employee. Such an approach for the preliminary calculation of the consumption of material to perform non-standard protective products is not acceptable, and requires a different method of calculation.

The ABC method and the method of designing the cutting by “Muller and Son” were taken as the basis for the authors’ method of calculation. The ABC method is a functional cost calculation, where the calculated product is represented as a system of interrelated functions. The design technique by “Muller and Son” is based on the representation of a three-dimensional body into planar segments that can be measured in length and width.

To simplify the calculation system, we assume that a protection product can be represented as consisting of one or several shapes of a rectangular parallelepiped shape (RPS) for any object of protection of a complex geometric shape. RPS can be described by a combination of its overall dimensions (x, y, z) and the presence of each of its faces (A, B, C).

The amount of material needed to perform such a RPS product can be estimated by presenting it in the form of an unfolded surface consisting of all its faces. Figure 1 shows an example of an isometric product RPS (type a) with its unfolded surface (type b).

The amount of material needed to perform such a RPS product can be estimated by presenting it in the form of an unfolded surface consisting of all its faces. Figure 1 shows an example of an isometric product RPS (type a) with and its unfolded surface (type b).

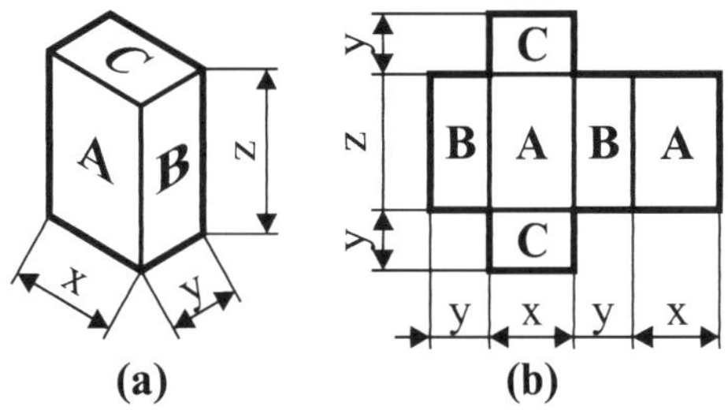

Figure 1. The product RPS and its surface

As can be seen in figure 1 (type b), each of the faces (A, B, C) has its own pair and has two dimensions from its constituent RPS.

The protective product is a combination of layers, consists of top, bottom coatings and filler. This means that such a protective product can be represented as three separate RPS products made of materials corresponding to each layer.

3.1. Estimation of the value occupied by the protective product surface on the material.

When determining the preliminary calculation of the material consumption for the product manufacture, it is necessary to take into account the quantity of the trimmed material obtained. Trimmed material is unused or rarely used in the future work of the enterprise, remains of this material require disposal or prolonged storage in the warehouse for further use.

The materials used by the enterprise are not subject to recovery on its territory, and their cost is significantly lost upon delivery to the processing points. Therefore, the cost of materials should be distributed in proportion to the total volume of their consumption for each protective product.

Each layer of a protective product can be represented as one or several details that can be described as rectangular ones with two overall dimensions - Xw and Lw, where the dimensions are across and along the roll, respectively. The designation of the dimensions of the part varies depending on the location of the part on the roll. Figure 2 shows graphical examples of placing a part with a size Xw smaller than the useful roll width (figure 2a) and exceeding it (figure 2b).

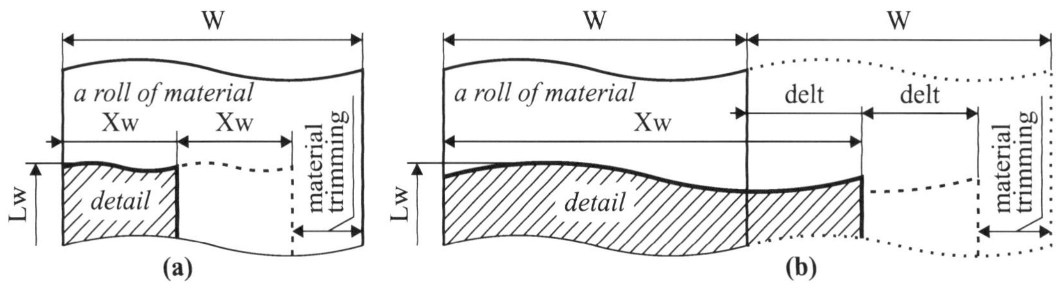

Figure 2. Graphic examples of placing the detail on a roll of material

3.1.1. KDR algorithm for calculating the filling of a roll detail. The input data for the algorithm are Xw (overall dimensions of the part along the roll), W (useful width of the roll), Rz (permission to cut the part along the length of the roll). Based on the accepted condition that each RPS can be represented in the form of one or more rectangular details, described by horizontal and vertical dimensions. Then for each part or combination of details you can set two coefficients (k) for filling the roll with them, the smallest of them will correspond to the optimal material consumption. Different values of the coefficient (k) are explained by the size of the material cut-off, depending on the multiplicity of dimensions of parts (Xw) to the useful width of the roll (W).

The following figure 3 shows such an algorithm, divided into the following blocks:

Steps 1 and 2. Data entering and analyzing it for completeness of information;

Step 3. Calculation and derivation of the coefficient for the detail, not exceeding the value of W;

Step 4. Check for permission to cut the detail;

Steps 5, 6, and 7. Setting the bandwidth (delt) of the part to be cut from the detail along the length of the roll, calculating the coefficient and deriving it.

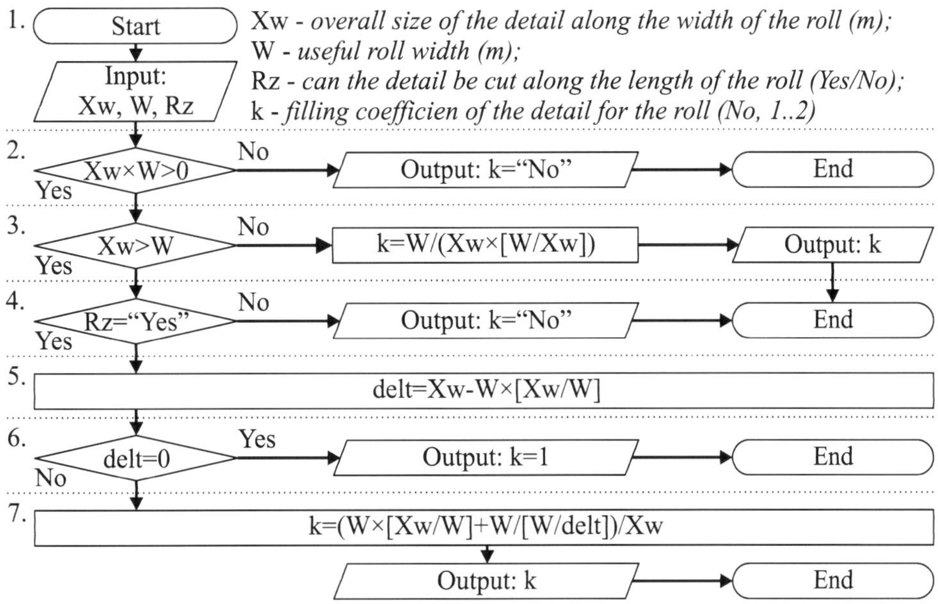

Figure 3. Algorithm KDR for calculating the filling coefficient of the detail on the roll

3.1.2. SRPSR algorithm for calculating the space occupied by the RPS on a roll. The input data for the algorithm are X, Y, Z (overall dimensions of RPS), nA, nB, nC (the required number of corresponding faces in the figure), W (useful roll width), Rz (permission to cut the edges along the length of the roll).

The calculation algorithm is based on the following statement: "On a roll of material, the layout of each face is considered separately, or its layout with the other in combination is one to two." Examples of the layout of faces A and B in various combinations are presented in figure 4. In types 4a and 4f, a combination of faces A and B is depicted in a one-to-one ratio. Views 4b, 4c, 4d, 4e, 4g, 4h, 4i, and 4j reflect different combinations of faces A and B in a ratio of one to two. Types 4b and 4g are not acceptable, since pieces of material equal to faces A and B, respectively, are lost on them in trimming.

From figure 4 it can be seen when calculating the space occupied by the RPS on a material roll, filling factors for a roll detail (k) are required for the following values: x, y, z, 2x, 2y, 2z, (x+y), (x+z) and (y+z), as well as the maximum values of the following pairs: (x, 2y), (x, 2z), (y, 2x), (y, 2z), (z, 2x), (z, 2y). These values will allow to determine the occupied area on a roll for various combinations of pairs of faces from A, B, C in ratios of one to one or one to two.

The input data for the algorithm are Xw (overall dimensions of the part along the roll), W (useful width of the roll), Rz (permission to cut the part along the length of the roll).

The SRPSR algorithm for calculating the space occupied by the RPS on a roll is shown in figure 5.

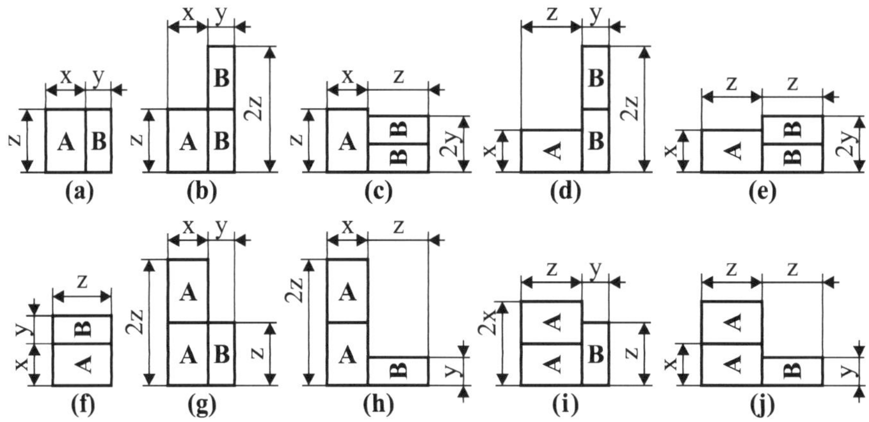

Figure 4. Variations of the layout of faces A and B

This algorithm consists of the following steps:

Stages 1 and 2. Data entering and analyzing it for completeness of information;

Stages 3 and 4. Calculation of intermediate data for determining the areas occupied on a roll of various combinations of faces;

Stages 5 and 6. Calculation of the spaces A, B and C occupied on a roll and their combinations AB, AC, CB, AB2, AC2, BA2, BC2, CA2, CB2. The space is defined by multiplication of its two dimensions to the minimum of the roll filling factor corresponding to these overall dimensions;

Stage 7. Calculation of ten different variations of the space occupied by the RPS based on the values established at stages 5 and 6.

Step 8. Determination of the minimum of ten variations which is equal to the least space occupied by the RPS and the removal of the spaces of the faces not included in the calculation according to the values of nA, nB and nC from it.

This calculation algorithm has the following limitation - only one or a pair of one type of faces is allowed to be removed from the calculation.

4. Conclusion

Usage of the algorithm presented in this work makes it possible to estimate quickly the required material consumption for each layer of a protective product without the involvement of a highly qualified specialist, to perform the optimal selection of the shape, size and materials of a protective product according to the preferences of the client.

Any additional viewing windows, doors and sleeves that are necessary for making a product can be calculated separately using the KDR roll filling algorithm and adding it to the total volume of materials.

The techniques used in modern programs for designing and calculating textiles are relatively complex and mostly applicable to confirmed orders. The proposed method of calculating textiles is much easier to implement. The efficiency of calculation provided by it allows one to select the properties of a protective product in the maximum accordance with the requirements of the consumer in the shortest possible time, namely, in the interactive mode with direct communication with the consumer.

The enterprise, which focuses on the responsiveness to consumer expectations and the completeness of compliance with them, implements a strategy of non-price competition. The portfolio of satisfied customersformed on this basis, allows the enterprise to provide the necessary balance between profit and workload. With such approach it receives the rights of an expert monopolist in a dedicated segment of the market, which guarantees a stable position and development.

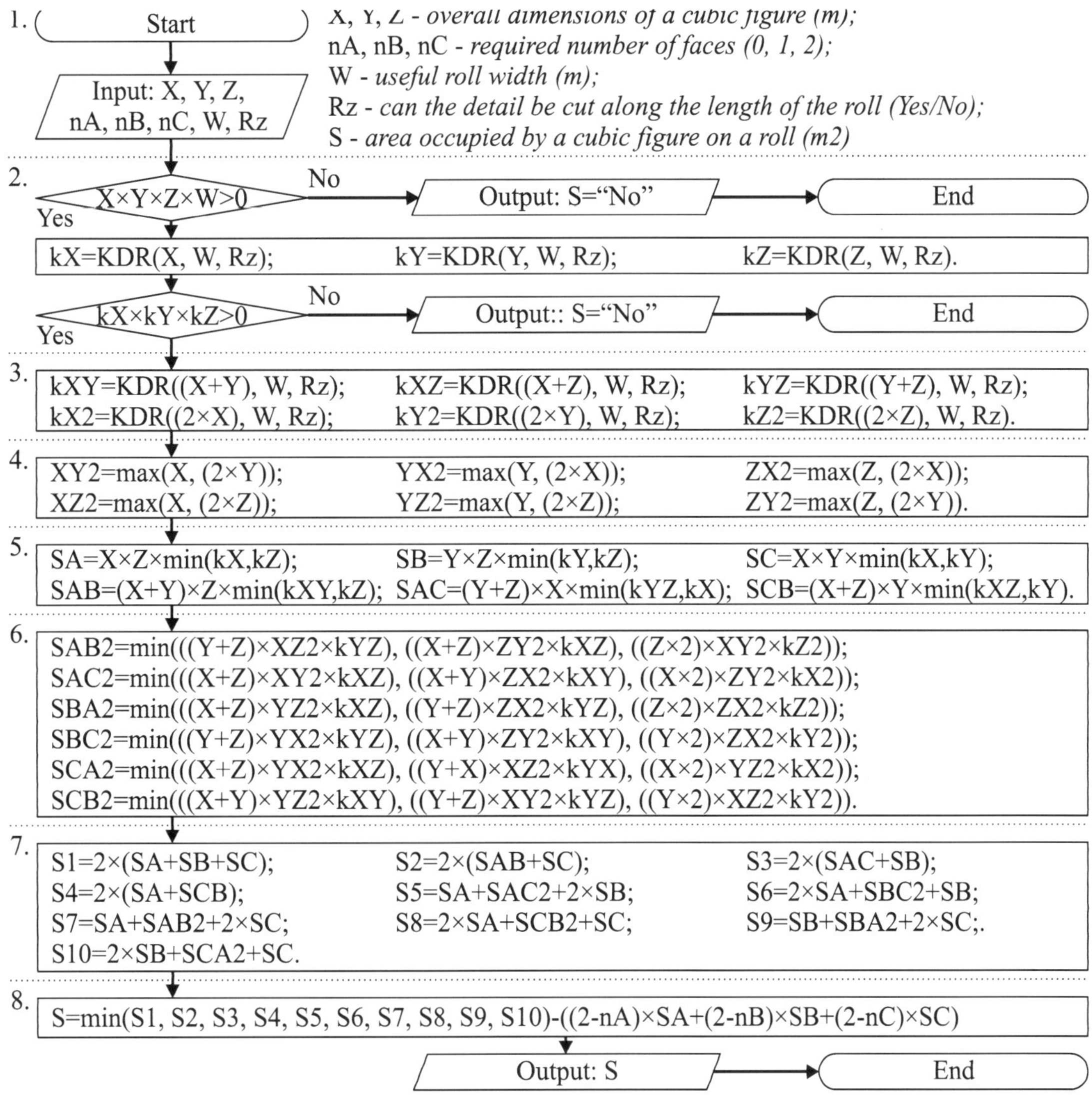

Figure 5. Algorithm SRPSR for calculating the area occupied by the RPS on a roll

References

[1] Tolstobrov D A 2017 *Schumpeterian readings: materials VII VII international scientific and practical conference* (Perm: PNRPU) pp 186–192

[2] Karaleu Y Y and Myshkovets Y A 2018 *Vestnik VSAWT* **55** 93–9

[3] Alhimenkova L V 2016 *Technological processes in the clothing industry. The complex process of preparing production for the transition to the release of new products* (Yekaterinburg: Architecton) p 134

[4] Golikov I A 2016 *Russian entrepreneurship* **17** 1251–1264

[5] Pivkin S A 2016 *International Accounting* **10** 21–34

[6] Prischenko E A and Nizovkina N G 2018 *World of Economics and Management* **18** 121–131

[7] Lugovoi O Y and Khasanov A S 2017 *ALLEY OF SCIENCE* **15** 218–223

Experience in the management of business processes with the use of digital technologies by Russian companies of a petrochemical complex

Deberdieva E M, Vechkasova M V, Golikava H S, Borisova A A, Lysenko A F

Tyumen Industrial University, Tyumen, Volodarskogo street 38, Russia
Polessky State University, Pinsk, street of Kirov24, Republic of Belarus
Novosibirsk State Technical University Novosibirsk, K. Marks Avenue 20, Russia
Don State Technical University, Rostov-na-Donu, Gagarin street 1, Russia

e-mail: vechkasovamv@tyuiu.ru

Abstract. This article discusses the prospects for the introduction of modern technologies in the business processes of industrial enterprises in the framework of the fourth industrial revolution. The factors determining the vector of development of modern technologies are revealed. The classification of tasks of the enterprises of non-resource sector of economy under which high technologies of "Industry 4.0" are introduced is given. The article also considers the impact of modern technologies on the business processes of industrial enterprises in the context of types of technologies. The analysis of the data of industrial enterprises, which are part of the largest petrochemical holding of the Russian Federation for compliance of the implemented technologies with the tasks of the considered enterprises is given. The study presents data on the possibilities of modern technologies and the results of the impact of technology on the business processes of enterprises.

1. Introduction

Currently, there is a growing interest in the technologies of "Industry 4.0", the digital economy is destroying the usual models of industry markets, which in turn increases the competitiveness of their participants. The positive impact of digitalisation is to determine the growth prospects of industrial enterprises and industries in general.

The demand for oil and gas chemical products is growing rapidly among the branches of the non-resource sector of the economy. The consumers of petrochemical products can be found in almost all industries: construction, engineering, energy, agriculture, medicine, electricity, space, etc. The demand for petrochemical products continues to grow, and, according to experts, will increase fivefold by 2030. Petrochemistry is a link between the oil and gas industry, processing and high-tech industries. At the same time, the petrochemical complex is the basis for further technological development. In this regard, the enterprises of the petrochemical complex must meet the requirements of the market, which in the era of technological breakthroughs is shifting towards customisation of the industry, so the transition to industry 4.0 and deep end-to-end automation of all activities of the enterprise requires fundamentally new technologies that change the usual business models. [7, 8]

In addition, the high demand for industry 4.0 is due to the interests of stakeholders. The lack of funds at high costs for the development and implementation of innovative and technological solutions with an incomprehensible economic effect does not allow to use high technologies to the full in the operating activities of the enterprise. Thus, the main task of the fourth industrial revolution is the evidence base of the efficiency of digital transformation of production. For the present study, petrochemical enterprises belonging to the largest petrochemical holding of the Russian Federation have been selected, those which have applied new technologies in their business processes. New technologies and their impact on the business processes of enterprises of the non-resource sector of the economy are the subject of the study.

The purpose of the study was to assess the impact of modern technologies on improving the efficiency of individual business processes of petrochemical enterprises. The achievement of the goal in the study required the formulation and solution of the following problems: to study the experience of industrial enterprises introducing modern technologies in the framework of the fourth industrial revolution; to identify the factors constraining the development and introduction of new technologies in the workplace and the factors contributing to the emergence of new technologies; to classify the tasks of enterprises of the non-resource sector of the economy, under which new technologies are introduced; to determine and classify the results obtained by the introduction of high technologies; to formulate the directions of the introduction of high technologies in the petrochemical industry.

2. Results and Discussion

As a rule, specific cases of using elements of "Industry 4.0." in Russian companies are discussed at specialised events (conferences, forums). One of the first Russian companies to embark on the path of introduction of "Industry 4.0" technologies is the group of companies of petrochemical holding - SIBUR. The level of technological equipment of the group's enterprises is currently one of the highest in Russia. The enterprises of the holding have already managed to reduce the cost of production due to automation, the level of which reached 84% last year. The holding implements such advanced solutions as advanced process management system (APC), production system (MES), laboratory system (LIMS), enterprise management system (SAP ERP), business process management system (BPMS). The holding was one of the first to introduce virtual reality (VR) technologies into the process of personnel training in repair, assembly/disassembly and maintenance of equipment. Employees of the group's enterprises begin to use mobile devices in their work, which contain information on the state of the equipment and current tasks. Data Science tools – online advisors and predictive Analytics - are also used in the holding to improve the reliability and efficiency of the equipment, identifying defects at an early stage and foreseeing the optimal set of parameters to optimise performance. Intellectual video surveillance, video analytics and technical vision system, allowing, respectively, to prevent emergency situations, analyse the environment and control the quality of products, are widely used in the company. Today, the holding company is studying the possibility of using wearable devices with telemetric modules – these can be "smart helmets" or bracelets that allow to track the location of the employee, his health status and much more up to the efficiency of activity [5.9].

On the other hand, according to experts, the impact of modern technologies on the improvement of business processes of industrial enterprises is different in terms of types of technologies and the period (at the stage of implementation and in the future), in addition, according to expert estimates, the larger the enterprise, the stronger the impact of digitalisation on business processes (table 1).

Table 1. The impact of digitalisation on the company's business process

Technology	The impact of technology on business process, %	The impact of technology on business process in future, %
Internet of things and automation of production	60	65
Digital design and modeling	58	55
Virtualisation technologies, remote access	57	67
Mobile technologies and cross-channel communications	55	60
Supercomputer system	42	50
Social networks	41	52

Big data and business analytics (BI)	38	34
Cloud technologies	36	48
Dialogue interfaces	34	41
Robotics	31	41
Additive technologies (3D printing)	31	43
Distributed registry technologies (Blockchain)	29	31
Artificial intelligence and machine learning	28	37
Virtual, mixed and augmented reality systems	22	38
Biological communication systems	22	31
Unmanned transport systems	21	31
Quantum technology	17	19

Source: built according to the data provided by NRU HSE (National Research University. Higher School of Economics)

The main technologies of "Industry 4.0" causing the greatest interest among the industrial enterprises of the chemical industry are the following (See table №2):

Table 2. Chemical Industry Trends in the use of "Industry 4.0" technologies

Name of the technology	Characteristics
Automation and robotics	Automated systems installed as software or hardware robots; features: work autonomously or collaborate with people
Sensors	Record physical conditions (machines, environment, etc.) and transmit data to make effective decisions
BIG DATA	Large amounts of structured and unstructured data
AI (Artificial Intelligence)	Intelligent devices replace people in solving problems, eliminating the human factor and make their own decisions
3D-printing	Additive technologies to enhance the technology and improve the condition of the equipment
IoT	A network of objects or devices equipped with sensors capable of collecting and exchanging data over the Internet.
Clouds	Data storage, access to data is possible to the set circle of persons from any location

Source: The Digital Transformation Initiative World Economic Forum 2017

In addition to the above given, there are technologies that are specific to petrochemical enterprises. Among them, the following are of particular interest [4]:

- Data Science - mathematical and algorithmic methods optimised for effective detection of complex patterns. This is a set of techniques and practices for solving advanced data analysis problems, including: preparation (cleaning), modeling with the use of machine learning methods, evaluation and verification, visualisation;

- counselor "Overfitting the model to the column K-410" as well as counselor "Product Quality LDPE" - Recommender systems the "Digital Guide" feedback. It allows you to get an accurate forecast of production, quality, information on equipment failures, as well as get recommendations for further actions. In turn, the presence of such a forecast allows you to make effective decisions. The basis of predictive analytics are algorithms for statistical modeling of specific devices, which can significantly increase the profitability of production. This system, based on the accumulated data such as: the technical parameters of the workpiece, its processing, consumption of tools, materials, provides recommendations on the quality of the final product, energy efficiency and productivity;

- SAP-based ERP systems are a designer of interrelated modules for production process control. SAP ERP enterprise resource management system covers all areas of financial and management accounting, personnel management, operational activities and services of the company. Provides full functionality required for the implementation of self-service information services and SAP ERP analytics;

- corporate information system based on ERP (enterprise resource planning) methodology and aimed at achieving optimal business process;

- predictive monitoring "Compressor M-1" - a method of management of maintenance and repair using analytical models that are based on the analysis of big data to monitor the condition of the equipment to predict the occurrence of an emergency. Allows you to choose the optimal mode of operation of the equipment, promotes the transition from planned repairs to support based on its actual condition, to the increase of operation safety and cost reduction;

- Big Data - these are various tools, approaches and methods of processing both structured and unstructured data in order to use them for specific tasks and purposes. Huge amounts of data are processed to ensure that a person can get specific and desired results for their further effective use. It is becoming impossible to process large volumes of the heterogeneous and rapidly incoming digital information by traditional tools. The data analysis itself allows you to see certain and invisible patterns that a human cannot see. This allows us to optimise all areas of our lives-from public administration to production and telecommunications;

- APC - Advanced Process Control - is a wide class of automation systems based on various technologies and methods that allow to optimise the technological process. APC manages the installation in automatic mode in "real time", interacting with the basic regulation (regulators APCS). APC is a sort of "autopilot" for the processing plant but with a much more advanced set of optimisation functions. The" heart " of the APC is a mathematical statistical model of the process, which allows to predict its behaviour in the near future. Properly configured APC allows you to: optimally and consistently manage the TP; adapt to changing economic conditions; proactively compensate for disturbances; stabilise the process better than the operator does;

- mobile M&R (Maintenance and Repair) is an application for mobile bypassing of equipment that will eliminate the pile of paper logs, the need to search for the information about the work of previous shifts, will allow to quickly collect data about the equipment in the vicinity of the installations, as well as will show safety instructions at any time. The application for obtaining a digital work order-admission already has a wide functionality: it offers to choose the parameters of the work, to fix the scheme of the object, to tighten the activities and possible risks, to choose the responsible ones, to remotely agree on the tolerances from the abundant device;

- virtual reality is an augmented reality glasses that give the opportunity, considering all agreements, to receive consultation of the expert for hours, accordingly, reducing the time of repair;

- 3D- printing - this is the construction of a real object based on a 3D-model created on a computer. Then the digital three-dimensional model is saved in the STL-file format, after which the 3D printer, which displays the file for printing, forms a real product;

- drones, UAVs-unmanned aerial vehicles, their configuration depends on the industry of the company and the work performed by it: drones to collect information should be small and maneuverable, as well as for warehouse accounting - they fly between the shelves, read the RFID-mark on the product or barcode from the box; monitoring of the assets of the enterprise: flying around the territory, it will record data on the number of certain objects, the area of sowing and the cultivated land.

These technologies are widely used in enterprises of non-resource sector of the economy and allow to solve problems related to improving the quality of products, document management, cost reduction, minimising various risks and losses, as well as to optimse the business processes of the enterprises under consideration (table 3).

Table 3. Compliance of implemented technologies with the tasks of business processes of enterprises

Enterprise	Implemented Technologies	Tasks covered by the Technology
OOO "SIBUR Tobolsk"	Automation of production processes, Projects Data Science; counselor "Overfitting the model to the column K-410"	improving product quality; reducing costs; decreasing the number of accidents; eliminating the human factor
JSC "SiburTyumenGaz"	SAP-based ERP systems	improving the quality of document management; minimising the burden on financial services; cost reduction; decreasing the human factor
LLC "Tomskneftekhim"	Data Science projects, Advisor "LDPE Product Quality"; predictive monitoring "Compressor M-1"; the project "Digital factory" (pilot): Big Data; Video analytics; Advanced Process Control (APC); mobile Maintenance and Repair (M&R); Virtual Reality (VR); 3D printing; medical gateway; drones	improvement of product quality; reduction of the cost of finding and training staff; decrease in the number of accidents; shortening the time of repairs and downtime due to them; elimination of the human factor; enhancement of security; increase in profit

Implemented technologies in industrial enterprises are an add-in to the business processes of the studied enterprises, covering a number of tasks, allow to eliminate their "bottlenecks". In turn, the main problem of "Industry 4.0" is the unproven economic effect of the introduction of technologies into production processes [5]. The degree of influence of technologies on business processes depends on the goals of the enterprise, on the complexity of the problem to be solved in the conditions of growing competition, on the solvency of the enterprise, on the interest of management in solving existing problems, on the physical condition of the fixed assets and innovative involvement of the enterprise, therefore, it is rather difficult to prove the economic effect of the introduction of modern technologies with the use of existing methods.

So, currently there are several methods for assessing the economic efficiency of IT projects in the enterprise: financial and economic; quality; probabilistic. The choice of specific methods for determining the effectiveness of IT projects depends on the specifics of the situation. There are no exact recommendations in this regard, as each project has its own individuality, the enterprise has its own specific goals. Therefore, what is so important for one company may not matter for another (table 4).

Table 4. Cost of introduction of new technologies, according to experts of the companies

The scale of the enterprise	Cost of implementation,

	mln. RUB.
Small market scale	1
A simple organisational structure, the small scale of the market	2
Medium enterprise	6
Large enterprise	10
Broad market coverage	150
Large-scale enterprise at the Federal level	1000

Source: built by the authors according to [6]

According to the table, the cost of implementing new digital solutions in the enterprise can vary significantly depending on the size of the enterprise and the complexity of its organisational structure. The analysis of the studied enterprises reveals the dependence of the improvement of business processes on the introduction of technologies.

3. Conclusion

According to experts, the projected results from the introduction of new technologies in the business processes of enterprises of the non-resource sector of the economy, depending on the scale and specifics of the enterprise tend to increase in the forecast period, as well as the growth of interest in technology (table №5).

Table 5. Potential results of the Industrial Revolution for non-resource sector enterprises

Application area	Opportunities	Result
Production Quality	digital quality management; advanced process control (APC); statistical process control (SPC)	10 – 20 % reduction in quality assurance costs %
Time to introduce the product to the market	fast modeling and experimentation; concurrent engineering; open innovation/collaboration with customers	reduction of time of introduction to the market by 20 – 50 %
Equipment operation modes	smart energy consumption; Informatisation of production; optimisation of equipment operation in real time	3 – 5 % increase in productivity
Loading of production equipment	the flexibility of the routing; flexibility in the use of equipment; remote monitoring and control; predictive maintenance; an augmented reality to maintain	30 - 50% reduction of equipment downtime
Efficiency and safety of labour	human-robot interaction; remote monitoring; digital performance management	45 - 55 % increase in productivity of technical functions due to automation of labour
Logistics	3D printing on site; real-time supply chain optimisation; optimisation of batch sizes	20 - 50 % cost reduction of storing inventory

Source: "Digital Russia: New Reality", July 2017, McKinsey, forecast to 2025.

The impact of new technologies on business processes can be grouped by classification criteria: reduction of logistics costs; reduction of document flow, optimisation of production processes. Among the main opportunities of digitalisation it should be noted: increase of labour productivity, decrease in risk of injuries of the personnel and probability of accidents, increase of efficiency of work of the equipment, increase of efficiency of decision-making, increase of satisfaction of clients and increase in sales volume, decrease in time of introduction of a new product to the market [1]. In turn, the implementation of these prospects requires the following measures: the development and implementation of long-term and comprehensive development programs in the field of digital technologies, activities to disseminate best practices and the formation of an innovative culture, monitoring of advanced technologies, knowledge accumulation, evaluation of economic efficiency, prototyping and piloting the most promising solutions, preparation of normative reference documentation for the implementation of the circulation of promising solutions [3].

Table 6. Factors impeding and conducive for the development of new technologies in the non-resource sector of the economy

Factors impeding the development	Factors conducive for the development
Resistance to changes in the company	State sponsorship
Lack of specialists	Reduced tax burden
High cost of solutions	Saving resources of the enterprise
The specificity of the climate (low temperature)	-

When implementing the above given measures, it is necessary to take into account the factors limiting the use of high technology in the non-resource sector of the economy (table 6).

The analysis of the studied objects shows the possibility of using new technologies in business processes, as they carry a high growth potential for business, but with due consideration of the conditions: the employment of highly qualified digital professionals, the integration of these technologies will require significant financial resources, due to the high cost of technology, as well as interaction with high-tech companies engaged in the development of the technologies under consideration.

References

[1] Order of the government of the Russian Federation of July 28 2017 1632 p

[2] Innovations in management of regional and industry development: collection of scientific works [Electronic resource] conference proceedings (TyumenTSOGU) 2015 254 p

[3] New technologies- to oil and gas region: Materials of the International Scientific-Practical Conference of Students, Postgraduates and Young Scientists. Improving the technology of construction of wells, drilling of oil and gas facilities in Western Siberia. Development and operation of oil, gas and gas condensate fields. Economics and management of enterprises, industries, complexes [Electronic resource] 2016 (Tyumen TSOGU) 398 p

[4] Official website of the company SIBUR [Electronic resource]. Mode of access: http://investors.sibur.com/events-and-presentations/investor-presentation.aspx?sc_lang=ru-RU

[5] Klochko S N 2017 Evaluation of the effectiveness of the introduction of information technology in the accounting system at the enterprise *Problems and prospects of economy and management: materials of the international. scientific. Conf.* 101-107

[6] Russian industry 4.0: how not to miss the train in the future [Electronic resource]: Access Mode: http://www.cnews.EN/reviews/it_v_promyshlennosti_2018/articles/rossijskaya_promyshlennost_40_kak_ne_opozdat_na_poezd_v_budushchee.

[7] Step into the future - RLC [Electronic resource]: Access Mode: http://t.rbcplus.ru/news/5ba8b6e97a8aa9587ebd995e?ruid=uUjlA1vlAbt3I0aOAwqBAg.

[8] Management of Development of Fuel and Energy Complex: Problems of Modernity and Foresight 2017 collective monograph (Tyumen TIU) 160 p.

[9] Vechkasova M V, Deberdieva E M 2017 Overview of world experience of development of the petrochemical industry *"Oil and gas of West Siberia" Proceedings of the International Scientific and Technical Conference* 67-70
[10] O V Lenkova, E M Deberdieva 2015 Innovative development of petrochemical enterprises Tyumen 124 p.
[11] Milani F, Luciano G B 2018 Blockchain and Principles of Business Process Re-Engineering for Process Innovation [Electronic resource] Mode of access: https://arxiv.org/pdf/1806.03054.pdf.

New approaches for exploring cult buildings: to the issue of introducing energy assessments into a comprehensive research method

S A Yurchenko

24/8a Naberezhnaya Street, Volzhsky, Volgograd region, 404130, Russian Federation

E-mail: stepan1974.yur@yandex.ru

Abstract. This paper deals with the subject of new methods in the study of objects of cult purposes. In author`s opinion data, collected from natural, humanitarian and technical sciences, makes possible introducing additional research parameters into historical research. Integrated approach used in the study of monuments of the past takes into account cartographic, topographical, landscape and other aspects, but ignores their energy impact on the environment. Author suggests using data on energy impact of buildings along with general criterions for some particular historical and cultural studies.

1. Introduction

Heinrich Schliemann was the first archeologist who used a comprehensive approach in archaeological researches. He introduced preliminary analysis of researched place, based on literary works, kept till our time. Thus, on the basis of the text of Homer`s "Illiad" he was able to define the location of Troy and the Minoan palace in Crete. Nowadays new receptions and methods, that sounded fantastical even a decade ago, are used by science. Satellite archeology can be an example of similar research innovations [1]. Recently, with its help a number of considerable discoveries were made, including those made on the territory, which was previously explored before with a help of traditional receptions and technologies. Using it for the last decade, researchers found a new unknown civilization on the border of modern Brazil and Bolivia, and to archeologists, working in Egypt, to find whole city buried under desert sands. In Central Asia aerial photography allowed to find ancient irrigation canals, fortresses and cities, covered by sands. Archeology keeps up to date and implements satellite archeology in an integrated approach for studying megalithic heritage, thereby confirms an aspiration to cross-disciplinary approach in historical researches [2]. However, offered aspects, which require research attention (cartographic, topographical, landscape, etc.), ignore energy parameters of a subject of scientific researches.

2. Results and Discussion

The objective of the research is to reveal patterns in location of cult buildings. According to author taking into account new achievements in natural and humanitarian sciences it is possible to use energy estimates (parameters) in their study. This research is intended to prove the necessity of inclusion of energy parameters in an integrated approach of the analysis of historical monuments, that will expand our ideas of their nature and functionality.

In his work "Symmetry of crystals", which became crystallography classics, academician E.S. Fedorov wrote: «Today the scientific knowledge does not represent separate independent studies, deprived of connection with each another. On the contrary, different areas of science are connected with each other by numerous and tangled threads, and success of one almost always significantly depends on achievements in others...» [3]. Thus, for combination of all known properties and

IOP Conf. Series: Materials Science and Engineering **483** (2019) 012067 doi:10.1088/1757-899X/483/1/012067

parameters of the object of the research, it is necessary to find the "thread", connecting each piece of the mosaic into unite, whole picture, satisfying all scientific directions.

On the basis of satellite data author revealed the known pattern in the structure of objects of cult purpose (further O.C.P.) across huge territory of the ancient cultural centers of Eurasia, the Middle East and South America [4]. Human settlements and primarily O.C.P. were purposefully embedded into surrounding hydrocultural space, at the same time river bends acted as key location. The choice of the place of construction of O.C.P. was not accidental as river beds acted as the natural channels of energy influence, transmitting energy of the flowing water in a meander space [5]. For purpose of additional energy storage of water flows new manufactured reservoirs were built, forming new focal points of natural energy storage, serving as a kind of "energy traps" like artifact lakes [6].

The revealed patterns of structure of O.C.P. and construction of special artifact buildings for natural energy focusing and accumulation allow to consider buildings, located in these places, as energy objects of cult purposes (further – E.O.C.P.). In author`s opinion confirmation of this pattern on purpose of E.O.C.P. was obtained by the study of pyramid of Cheops by scientists of the St. Petersburg University IMTO and the Laser center of Hanover. Result of conducted researches made it clear that the well-known pyramid can concentrate electromagnetic energy in itself due to its form, sizes and material. Calculations specify that in a resonant mode the pyramid is capable to concentrate electromagnetic energy in its internal chambers and in space under its base, where the third chamber is located. Computational modeling evaluated that radio waves within a range from 200 to 600 meters will be resonant for the pyramid [7].

Indeed, a detailed examination of pyramids` location (figure 1) shows that these buildings are located on vast sea cape (a bend of the Mediterranean Sea), and place of construction corresponds to the focal point of energy concentration of the Mediterranean Sea in Fig. 1 corresponds to sector 1. The location of the pyramids of Djoser, Userkaf and Teti in Fig. 1 sector 2, built before the pyramid of Giza, may correspond to the previous focal point of the lower level of the Mediterranean Sea, which nowadays is confirmed by archaeological researches [8].

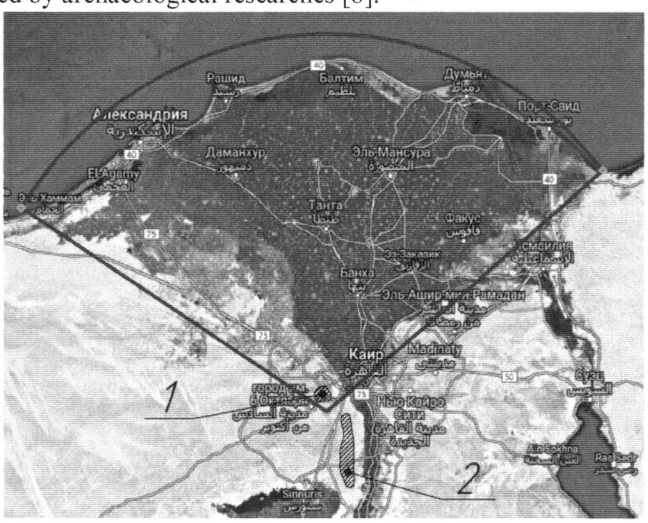

Figure 1. Cape-«bend» of the Mediterranean Sea. Area 1 corresponds to the pyramids of Giza. Area 2 corresponds to the pyramid of Djoser

Thus, it is possible to consider as a confirmed fact that the pyramid is E.O.C.P. as in terms of pattern of its position in the focus of collecting energy of the sea, and from a point of theoretical calculation of electromagnetic energy concentration.

Russian scientists repeatedly conducted researches on energy evaluation of cult buildings and their influence on the human beings [9]. It was found that some structures (Throne) normalize physiological state of a human body, while others (Tomb) cause an activity state. The Labyrinth structure caused the state of euphoria in most members of a research expedition. The general state of health of focus group improved in 20 days both in psychoemotional state and in physiological indicators of health [10].

At the same time, it was revealed that many dolmens of the Caucasus are located on a watershed. The flowing underground water created energy in rocks, that was confirmed by a difference of potentials between different sides of dolmens. A big charge accumulation in the form of difference of potentials leads to energy release, that is the discharging in the place of a mountain break, leading to dolmen charging [11]. Thus, dolmens can be also attributed to energy objects of cult purpose. Further studying of megaliths of the Southern Russia showed that the majority of ancient buildings are made of rocks with high content of quartz, which has piezoelectric effect, so influencing electromagnetic field`s voltage around this O.C.P. The result of experiments near megaliths was revelation of weak, but quite registered electromagnetic and acoustic radiation. Therefore, in our opinion, these monuments can also be attributed to the category of E.O.C.P.

The acoustic researches at underground megalithic sanctuary of Hypogeum (Malta), conducted by doctor Ian Cook from the University of California, showed that using a frequency of 110 Hz, which corresponds to resonance frequency in Chamber of the Oracle (the second level of the Hypogeum), results in a sharp shift of activity patterns of brains` prefrontal cortex, leading to temporary change of brain domination, from left-side to right-side. Respectively, using sound of a particular frequency and amplitude, lead to perception change of surrounding reality of the person, who was at a sanctuary. [12]. Therefore, the Hypogeum (Hal-Saflieni), as well as the Ancient Egyptian pyramids and dolmens are attributed to energy objects.

Let's note that in electromagnetic neurology works the effect of improving psychoemotional state and changes in the perception of consciousness under the influence of an electromagnetic field is revealed and proved with arguments. Special attention should be paid to researches of a brain state after impact of electromagnetic radiation (in the form of ultra-high and microwave frequencies) on it [13]. For better understanding of the effect of electromagnetic radiation impact on person the author suggests to use the concept of «insight», understanding it not only as traditional foresight (insight), but also as deep personal transformation, associated with general improvement of cognitive functions of the psyche. As the result a person acquires superconsciousness, that is «the ability to see a situation in multidimensional space, in space and time at once» [14]. Let's note that "insight" is used to characterize super abilities of orthodox monks [15].

3. Conclusion

Thus, in our opinion, the existing methodology of an integrated approach for cult buildings studying is not complete without an object`s energy component, its influence on electromagnetic fields and on the person`s consciousness. Therefore, introduction of an energy component when studying cult buildings will not only allow to expand research opportunities, but also provide new data on unknown abilities of human`s body and a brain.

References

[1] Menze B H., Ur J A 2012 Mapping patterns of long-term settlement in Northern Mesopotamia at a large scale *Proceedings of the National Academy of Sciences of the United States of America* PNAS(2012)109(14)

[2] Marsandalov L S, Parnina G N, Grigor'ev A A 2013 An integrated approach to the study of the megalithic heritage *Bulletin of Tomsk State University. History* **2(22)** pp 72-4

[3] Fedorov E S 1949 *Symmetry of crystals The main work* (Moscow: Nauka) p 9

[4] Ramazanov S P, Nikolaev N Yu, Yurchenko S A 2017 River landscape and topography of monuments in archaeological cultures of the Volga region and the Southern Urals in the era of the Late Aeneolithic and Bronze *Historical, philosophical, political and legal sciences, cultural studies and art history. Theory and practice* **2(76)** pp 166-70

Nikolaev N Yu, Yurchenko S A 2017 Cultural landscape and meander: historical parallels,

topography features, hidden semantics *Historical, philosophical, political and legal sciences, cultural studies and art history. Theory and practice* **9(83)** pp 131-5
[5] Yurchenko S A 2017 The energy nature of river bends and the problems of cult buildings topography *Historical, philosophical, political and legal sciences, cultural studies and art history. Theory and practice* (**12(86)/1** pp 212-6
[6] Yurchenko S A 2018 Knowledge of ancient civilizations on the accumulation of natural energy in places of construction of cult buildings *Manuscript* **5(91)** pp 61-4
[7] Balezin M, Baryshnikova K V, Kapitanova P, Evlyukhin A B 2018 Electromagnetic propertiesof the Great Pyramid: First multipole resonances and energy concentration *Journal of Applied Physics* JAP(2018)124(03)
[8] Porat N, Sivan D, Zviely D 2008 Late Holocene embayment infill and shoreline migration, Haifa Bay, Eastern Mediterranean *Israel Journal of Earth Sciences*. **57(1)** p 29
[9] Byakov A Yu, Pyatunin M S, Byakov A A 2012 Geophysical studies of dolmens of the Caucasus *Questions processing and interpretation of geophysical observations. Proceedings of the conference dedicated to the 100th anniversary of Alexander Kirillovich Malovichko* (Obninsk: Unified Geophysical Service of the Russian Academy of Sciences) pp 84-7
Gerasimov S A, Lysenko V S, Ogaj I P 2016 About electromagnetic and acoustic radiation of megaliths *Science symbol* **10-3(22)** pp 194-200
[10] Dul'nev G N, Rezunkov A G, Rezunkova O P, Zagustina N A, Golubev S V 2007 Biophysical and energy-information studies of the effect on humans of megalithic structures *Scientific and Technical Journal of Information Technologies, Mechanics and Optics* **35** pp 84-6
[11] Byakov A Yu, Pyatunin M S, Byakov A A 2012 Geophysical studies of dolmens of the *Caucasus Questions processing and interpretation of geophysical observations. Proceedings of the conference dedicated to the 100th anniversary of Alexander Kirillovich Malovichko* (Obninsk: Unified Geophysical Service of the Russian Academy of Sciences) pp 85-7
[12] Mandal D Hypogeum Of Hal Saflieni: The Grand Underground 'Temple' With Acoustic Effects *Realm of History* of September 13, 2015 [Electronic resource]. – Access mode: https://www.realmofhistory.com/2015/09/13/hypogeum-of-hal-saflieni-the-grand-underground-temple-with-acoustic-effects/ (access date: 10.08.2018)
[13] Holodov Yu A 1982 Brain in electromagnetic fields (Moscow: Nauka) 123 p
[14] Balin V D Consciousness as a scientific category? *Bulletin of St. Petersburg University. Psychology. Sociology. Pedagogy* **4** p 98
[15] Koncevich I M 2005 *Optina Pustyn and its time* (St. Petersburg: Biblioteka Frontistes) p 312

IOP Conf. Series: Materials Science and Engineering **483** (2019) 012068 doi:10.1088/1757-899X/483/1/012068

Seasonal changes assessment of motor oil consumption by cars

N S Zakharov, A N Makarova, E I Makarov, G V Abakumov

Industrial University of Tyumen, 38, Volodarskogo St., Tyumen, 625000, Russia

E-mail: makarova-a-n@mail.ru

Annotation. In the article, the question of seasonal conditions influence on motor oil consumption by cars is considered. In the process of research, a hypothesis of a significant impact of seasonal conditions on motor oil consumption by cars is made. The hypothesis is verified by assessing the significance of oil consumption changes during the year. Harmonic analysis is used to solve this problem. The experiment was conducted for cars of six brands and models. The monthly variation of oil change quantity, oil refill numbers, and the amount of refilled oil were analyzed. As a result, it was found that for all the cases under consideration, all the indicators vary significantly over time. The results obtained can be used to improve the motor oil reserve management system.

1. Introduction

Motor oil plays an important role in ensuring durability and efficiency of cars. Engine operating life depends on oil quality and frequency of its change. Therefore, a correct selection and timely delivery of motor oil influence the cost of maintenance and repair [1].

The motor oil consumption planning system currently in use prescribes determination of the planned consumption as a proportion of the amount of fuel consumed. Along with the advantage of the system, which is its simplicity, it has a number of disadvantages. In particular, the impact of operating conditions on oil consumption is not sufficiently taken into account. In addition, some factors have different effects on fuel consumption and oil consumption. As a result, a calculated oil consumption determined by the current method may differ significantly from the actual one. This leads to an increase of oil reserves or their lack. Accordingly, oil acquisition and storage costs increase, or engine durability reduces and repair costs rise [2, 3].

Most of the Russian territory is characterized by a significant change of climatic conditions during the year. Thus, in Western Siberia, the air temperature can vary from – 60 °C to +35 °C [4]. It is established that the ambient temperature affects the coolant temperature and the oil temperature in the sump of the car engine, causing its viscosity change. These processes cause a significant variation of oil consumption rate. The results of this dependence study obtained by different authors are contradictory and sometimes mutually exclusive.

Planning of oil consumption for its change is carried out on the basis of the maintenance production program. It is not taken into account that due to the variation of the car operation rate during the year, the maintenance requirement flow varies over time, causing a change in oil demand.

Thus, there is an urgent problem of planning motor oil requirement, with the variation of operating conditions taken into account, and reducing the cost of revolving fund on that basis.

2. Research method

In the process of research, a hypothesis of a significant impact of seasonal conditions on motor oil consumption by cars is made. The hypothesis is verified by assessing the significance of oil consumption changes during the year, taking into account that operational conditions are changing during the year causing a change of motor oil consumption rate.

Indicators, the values of which change over time, can be represented as a dynamic series. The main hypothesis put forward in the study suggests that oil consumption varies significantly over time, respectively, the indicators characterizing it can be represented as follows [5]:

$$Q = Q_C + Q_T + Q_P,$$

where Q_C is a constant component; Q_T is a periodic component; Q_P is a random component.

The periodic component can be described by a harmonic model of the following type [5]:

$$Q_T = \sum_{k=1}^{g} Q_k \operatorname{Cos}\left(\mathrm{m}(\mathrm{k}T_i - T_{0k})\right),$$

where k is a number of a harmonic; g is harmonic quantity; Q_k is an oscillation half amplitude of the k-th harmonic; T_i is the current time (months); m is the interval between T_i and T_{i+1} (degrees); T_{0k} is the initial oscillation phase (months).

To assess the significance of the seasonal variation of indicators characterizing motor oil consumption, the harmonic model of the periodic component is linearized by replacing the variables [5]. Then we will get:

$$Q_T = \sum_{k=1}^{g} Q_k z_k, \text{ where } z_k = \operatorname{Cos}(\mathrm{m}(\mathrm{k}T_i - T_{0k})).$$

If we assume that the oscillation cycle of motor oil consumption has a 12-month periodicity, the condition of change significance consists in evaluation of the correlation coefficient significance for a linear one-factor equation of the following type

$$Q_T = Q_1 z_1, \text{ where } z_1 = \operatorname{Cos}(30(T_i - T_{0k})).$$

The condition of correlation coefficient significance $r_{Q/z}$ for the first harmonic is [5]:

$$\frac{|r_{Q/z}|}{\sqrt{1-r_{Q/z}^2}}\sqrt{n-2} \geq t_P,$$

where t_P is the table value of Student's criterion for confidence probability P and $n-2$ freedom degrees.

If the indicator change is estimated by monthly averages, then n=12 and $\sqrt{n-2} \approx 3.16$, therefore

$$\frac{3.16\,|r_{Q/z}|}{\sqrt{1-r_{Q/z}^2}} \geq t_P.$$

Thus, the significance condition of seasonal changes of the indicators characterizing motor oil consumption is obtained.

In addition to assessing the significance, it is necessary to assess the magnitude of changes. For this purpose, the value of the first harmonic oscillation half-amplitude can be used. In addition to it, there is one more dimensionless indicator, which is unevenness factor $\mathrm{K_U}$ that shows the ratio of the maximum indicator value in the average annual [5].

3. Experimental research

Data collection on the parameters of the motor oil consumption process was carried out in transport enterprises of OJSC "Surgutneftegaz". The corporate database was used as a source of information. First, cars were selected from the database by brands and models, then the considered technical effects were selected.

To process the database, a program was developed. It allows sorting technical impacts by month, as well as presenting results in a graphical form.

The experiment provides assessment of motor oil consumption for each month during the year. Since the plan matrix is determined by the experiment task, planning is reduced only to assessment of the required number of measurements at each point of the plan, that is, for each month. The minimum required number of measurements N for the normal distribution law of values is determined by the expression [5]:

$$N = t_P^2 V_Q^2 / \Delta^2,$$

where t_P is Student's statistics corresponding to probability P; V_Q is variation coefficient; Δ is relative error.

The results of the calculations carried out using the latest formula are given in table 1.

Table 1. A minimum required number of measurements *N* for the normal distribution law of values

Δ	*N* at *P*=0.95 and various V_Q values			
	0.15	0.20	0.25	0.30
0.10	9	15	24	35
0.05	35	61	96	138

When choosing the number of measurements, it is necessary to test the hypothesis of the distribution normality of the measured values, as well as to determine the value of variation coefficient.

For that purpose, a preliminary experiment was carried out, which consists in selecting from the database the values of the amount of refilled oil for each month. The obtained samples were processed. Verification of the empirical distribution compliance with the normal law was tested by Pearson's criterion [5].

Data were obtained for cars KAMAZ-43118, Ural-4320, GAZ-3302, GAZ-2705, UAZ-3909, Volkswagen Multivan.

The values of variation coefficient vary from 0.15 to 0.24. Accordingly, to ensure relative error of no more than 0.10 with no less than 0.95 probability, it is necessary to carry out minimum 9 ... 24 measurements, and for 0.05 relative error, it is necessary to carry out 35 ... 96 measurements. Calculations show that all empirical distributions are adequately approximated by the normal law with no less than 0.95 probability

4. Results

The experiment involved cars of six brands and models: Ural-4320 – 111 units, KAMAZ-43118 – 33 units, GAZ-3302 – 129 units, GAZ-2705 – 282 units, UAZ-3909 – 332 units, Volkswagen Multivan – 159 units.

The experiment can be described as passive, consisting in collecting information about the actual values of the indicators that form motor oil requirement.

The experiment consisted in processing the database of OJSC "Surgutneftegaz" by selecting information about the moments of the technical impacts on the cars of the above brands and models, including oil refills into the engines.

When processing the experimental data, the average monthly values of the maintenance number, the number of refills, as well as the amount of the refilled oil were calculated. To ensure the possibility of their comparative analysis, the values are translated into relative values (as a proportion of the average monthly values for the year). The results are shown in a graphical form in fig. 1.

Next, a harmonic analysis of time series of values of indicators that form motor oil consumption by cars is performed. A fragment of the results is given in table 2. Consolidated results for the first harmonics are presented in table 3.

The indicators presented in the latest table allow estimating the value and frequency of oscillations of oil consumption for refilling:

Q_k is oscillation half amplitude that shows the maximum relative deviation from the mean value, with the period depending on the harmonic number: one year for the first harmonic; ½ year for the second harmonic; …;

T_i is initial phase which is time in months corresponding to the maximum harmonic value;

r^2 is determination coefficient for the linearized harmonic, showing the influence share of this harmonic in the change of the considered indicator;

r is correlation coefficient for the linearized harmonic characterizing the closeness of the linear relationship of time and changes in the considered parameter with the periodicity corresponding to the harmonic number;

t_r is Student's statistics for correlation coefficient;

$t_{0.95}$ is the table value of Student's statistics for 0.95 probability and 10 freedom degrees.

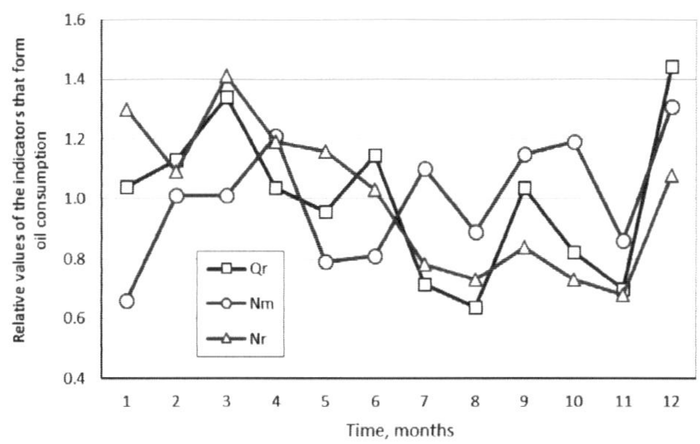

Figure 1. Change of the indicators forming oil consumption by cars KAMAZ-43118 during the year: Nm is maintenance number; Nr is number of operations performed to refill motor oil; Qr is refilled oil amount

Table 2. The harmonic analysis results of consumption changes of oil for refilling the engines of cars KAMAZ-43118 during the year

Harmonic number	Q_k	T_i, months	r^2	r	t_r	$t_{0.95}$
1	0.30	2.96	0.8037	0.8965	6.39	2.23
2	0.02	3.75	0.0044	0.0663	0.21	2.23
3	0.08	2.36	0.0526	0.2293	0.74	2.23
4	0.10	0.97	0.0932	0.3053	1.01	2.23
5	0.06	3.18	0.0331	0.1819	0.58	2.23

Table 3. Parameters numerical values and statistical characteristics of the first harmonic linear model of oil consumption changes for refilling the car engines during the year

Car brand and model	Q_k	T_i, months	r^2	r	t_r	$t_{0.95}$
Ural-4320	0.14	0.38	0.414	0.6434	2.66	2.23
KAMAZ-43118	0.30	2.96	0.8037	0.8965	6.39	2.23
GAZ-3302	0.43	1.91	0.7538	0.8682	5.53	2.23
GAZ-2705	0.19	2.77	0.5697	0.7548	3.64	2.23
UAZ-3909	0.22	0.98	0.5671	0.7531	3.62	2.23
Volkswagen Multivan	0.16	3.87	0.4794	0.6924	3.03	2.23

Graphs of the linearized harmonics and a graph of the harmonic model are shown in fig. 2.

Analysis of the data from the table 2 indicates that for all the cases considered, Student's t-statistic exceeds the table value for 0.95 probability. For example, for cars Ural-4320, correlation coefficient for the first harmonic is 0.6434, t-statistics 2.66 exceeds the table value 2.23 for 0.95 probability.

Similarly, for cars Volkswagen Multivan, r = 0.6924, $t_r = 3.03 > t_{0.95} = 2.23$.

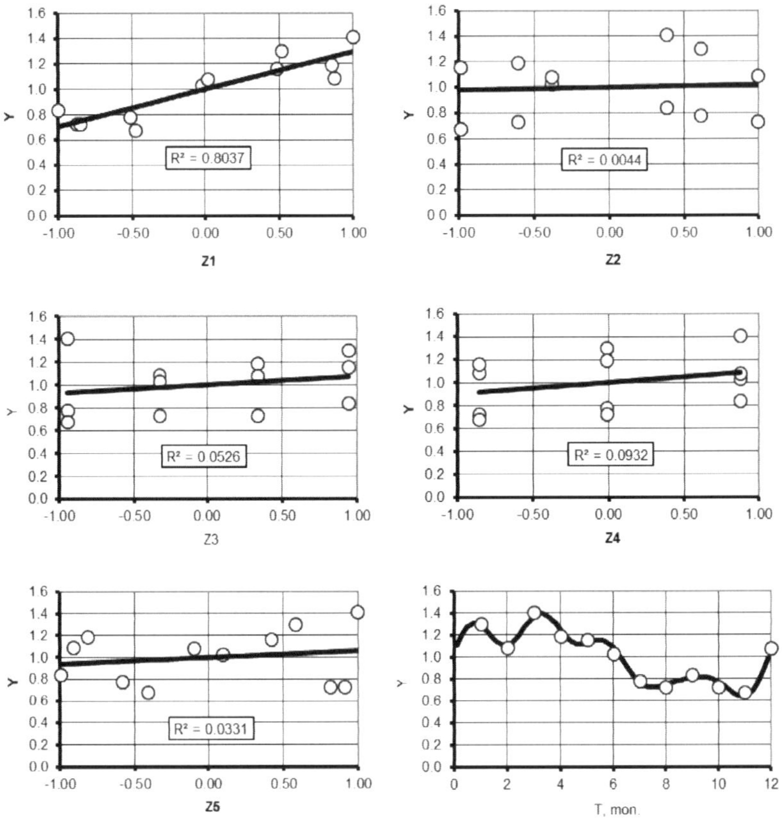

Figure 2. Graphical view of linearized first harmonic of the model of oil consumption change for refilling the engines of cars KAMAZ-43118 over the year

Thus, the hypothesis of a significant change of motor oil consumption during the year is confirmed.

The next issue to be solved in this experiment is determination of the maximum deviation value of the considered parameters from the mean values. For this purpose, for each of the three indicators under consideration, unevenness coefficient equal to the ratio of the maximum value for the year to the average annual value was calculated. The calculation results are given in table 4.

Table 4. Unevenness coefficient values of indicators forming motor oil consumption by cars

Car brand a model	Unevenness coefficient values		
	for oil change number	for oil refills number	for refilled oil amount
Ural-4320	1.54	1.12	1.29
KAMAZ-43118	1.31	1.41	1.44
GAZ-3302	1.27	1.62	1.99
GAZ-2705	1.21	1.36	1.51
UAZ-3909	1.21	1.55	1.74
Volkswagen Multivan	1.23	1.56	1.59

Analysis of the results shows that for the number of changes of oil in the engine, unevenness coefficient varies from 1.21 (GAZ-2705 и UAZ-3909) to 1.54 (Ural-4320).

5. Inference

The hypothesis of a significant change of motor oil consumption during the year is confirmed.

Analysis of the obtained results shows that for the number of changes of oil in the engine, unevenness coefficient varies from 1.21 to 1.54, for the number of refills, it varies from 1.12 to 1.62, and for the amount of the refilled oil, it varies from 1.29 to 1.99.

6. Conclusion

On the basis of the data on the actual motor oil consumption, the statistical significance of the seasonal variation of indicators that form motor oil requirement for cars of six brands and models is proved.

The results obtained can be used to improve the motor oil reserve management system.

References

[1] Bauer V I, Kozin E S, Bazanov A V, Nemkov M V and Mukhortov A A 2014 The methodic of forming a rational structure of a distributed production base of transport divisions in the pipeline industry. *Biosciences Biotechnology Research Asia* **11** 287-295

[2] Zakharov N S, Savin S A, Ivankiv M M and Lushnikov A A 2014 Factors influencing on downtime of transport-technological cars under current repairs. *Oil Industry* **4** 82-84

[3] Zakharov N S, Elesin S V, Novoselov O A, Kichigin S J and Makarova A N 2016. Improving the efficiency of technical operation of the vehicles by optimizing the qualification structure of engineering servicing workers. *International Journal of Applied Engineering Research* **11(3)** 1998-2006

[4] Zakharov N S and Panfilov A A 2015. The influence of seasonal conditions on the heavy metals emissions during vehicle operation. *Research Journal of Pharmaceutical, Biological and Chemical Sciences.* **6(1)** 1838-1851

[5] Zakharov N S 1999 *Simulation of changes in the quality of the cars* (Tyumen: Tyumen industrial university) p 127

IOP Conf. Series: Materials Science and Engineering **483** (2019) 012069 doi:10.1088/1757-899X/483/1/012069

Computer-aided design of technical documentation on the digital product models of Industry 4.0

D A Zakoldaev[1], A V Shukalov[1], I O Zharinov[1], O O Zharinov[2]

[1] Faculty of Information Security and Computer Technologies, St. Petersburg National Research University of Information Technologies, Mechanics and Optics, 49, Kronverksky Av., St. Petersburg, 197101, Russia
[2] Department of Problem-Oriented Computing Complexes, Saint Petersburg State University of Aerospace Instrumentation, 67, Bolshaya Morskaia str., St. Petersburg, 190000, Russia

E-mail: mpbva@mail.ru

Abstract. The task of organizing project activities in a digital factory is considered. Digital Factory is the industry 4.0 enterprise, operating in the digital environment. The components of a digital enterprise at the physical level of technological equipment (cyber-physical systems) and at the cyber level of cloud services are described. A set of tools that provide information support for the process of designing instrument-making products is offered. The scheme of interaction between the components of the Industry 4.0 digital factory and cloud services for project activities is given.

1. Introduction

The development of modern industry is inextricably linked to improvement of the technological base of the project environment of enterprises. In organizations, new information technologies are being introduced [1-3], to substantially automate the design activities of instrument-making product developers and to carry out the activities of manufacturing enterprises in an automatic mode. Such technologies are cloud technologies, technologies of the industrial Internet of Things (IoT), technologies of cyber-physical systems, etc., implemented using cloud services [4, 5].

Enterprises operating with the use of new digital technologies are now commonly considered the Industry 4.0 enterprises. The concept of "Industry 4.0" is a collective one and describes the transition of industry to a new technological order in a historical perspective. According to the accepted terminology, the enterprises of Industry 4.0 are [6-8]: a digital factory (designing enterprise), a smart factory (production enterprise) and a virtual factory implementing the full life cycle of an instrument-making product.

To organize the design activities of the Industry 4.0 enterprises, the components of their infrastructure must be defined both at the physical level of technological equipment and at the cyber level of cloud services, and the properties of these components relevant to the field of instrument engineering [9, 10] should be determined.

2. The Industry 4.0 digital factory Infrastructure

The scheme of the Industry 4.0 digital factory components interaction, specializing in the field of instrumentation, and cloud services that support the project activity, is shown in Figure 1.

Physical level of the Industry 4.0 digital factory is [5, 6]:

- automated workplaces (AWP) of schematic designer for instrument-making products;
- automated workplaces of text document designers for instrument-making products;
- automated workplaces of drawing designers for instrument-making products;
- automated workplaces of software documentation designer for instrument-making products.

In accordance with the types of individual technical documents developed and drawn up for

instrument-making products, the AWP group and the Industry 4.0 digital factory specialists form:
- research and development department, tasked with preparing of electronic version of circuit and text design documentation for instrument-making products;
- design department, tasked with preparing of electronic version of text and drawing design documentation for instrument-making products;
- chief technologist department, tasked with preparing of electronic version of text and drawing process documentation for instrument engineering products;
- software development department tasked with developing of functioning programs codes and preparing program documentation for instrument engineering products;

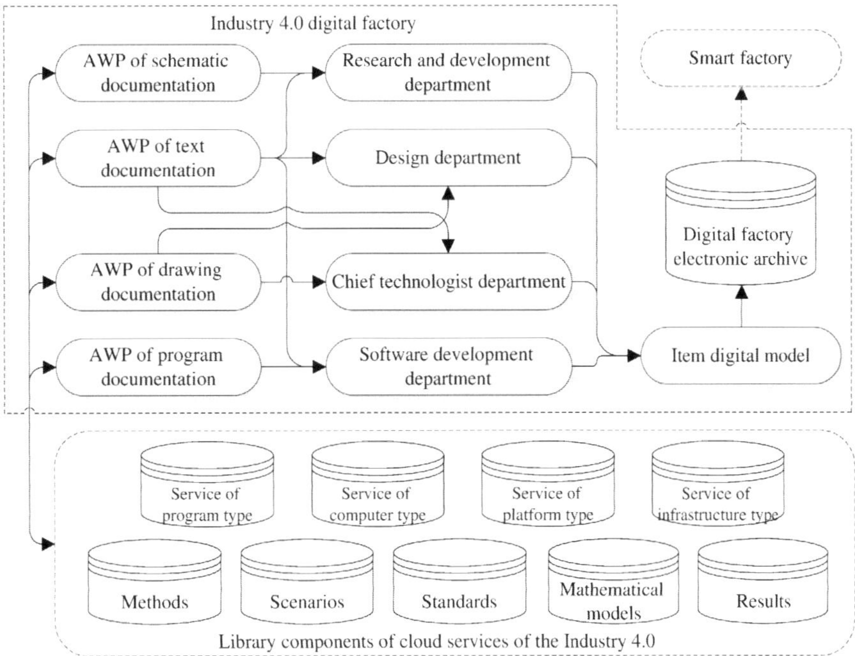

Figure 1. Interaction scheme of the Industry 4.0 digital factory components and cloud services for project activity.

Diagrammatic documentation for instrument-making products includes electrical schematic diagrams, wiring diagrams, connection diagrams, kinematic diagrams, optical circuits, etc. Drawing documentation for instrument-making products includes assembly drawings of parts and assemblies, dimensional drawings, packaging drawings, etc. Software documentation for instrument-making products includes program texts, program data (a component loaded into an instrument-making product), etc.

The entire technical documentation for an instrument-making product prepared at the Industry 4.0 digital factory forms a digital product model. Unlike the electronic model of the product, which requires only the electronic documentation to be made taking into account weight and size characteristics of the product, digital model includes descriptions (at the level of mathematical dependencies) of physical properties of materials and components of the product. The finalized digital model of the product is transferred to the electronic archive of the digital factory for the subsequent transfer of documentation to the smart factory for the series production of the product.

3. Cloud services for the Industry 4.0 simulation modeling

The description of instrument-making products in the form of digital models is related to the need of replacing the traditional, expensive physical tests of the finished product conducted at Industry 3.0 enterprises to confirm the quality of technical documentation with virtual tests conducted at the

Industry 4.0 enterprises by means of simulation modeling.

The AWP scheme of the technical documentation designer for digital models of the Industry 4.0 products is shown in Figure 2. The technical support of AWP is formed by an instrumental computer. The AWP grants an access for the designer to the cloud services based on the Internet of Things protocol [3, 4].

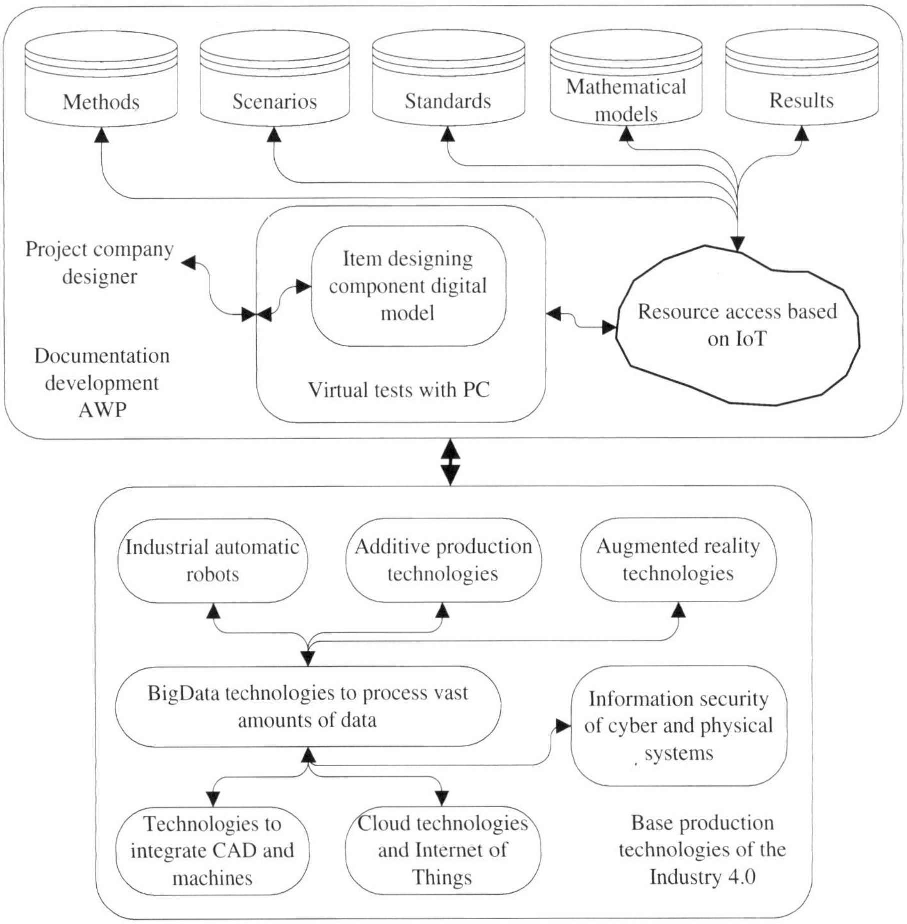

Figure 2. Scheme and components of the technical documentation designer AWP for the Industry 4.0 digital models product.

Simulation modeling is a process of testing a digital model of a product, performed at an automated site of a designer of technical documentation, in which not only a product, but also test equipment, external operating environment, etc., are presented in the form of mathematical models described at the level of variable parameters. Simulation modeling of a digital product model makes it possible to evaluate the behavior of a future product in actual operating conditions at the design stages when a physical sample of the product does not exist yet. The possibility of changing the values of variable parameters with sufficient accuracy for practice allows determining the limits of applicability of an instrument-making product under expected operating conditions.

To perform the simulation of a product digital model, designer must have the tools implemented in software at the cyber level. These tools are:

- cloud services such as software, computer, platform, infrastructure, providing design development environment (design of technical documentation) and testing of a digital model;
- mathematical models of external influencing factors (the environment expected by the product in

operation), models of materials and components used in the product at the level of description of their physical properties, etc.

- scenarios of statistical tests performed by modeling tools, where in certain sequences (or at the same time) there are simulated environmental effects on the digital model of an instrument-making product;
- test methods, with the parameters of external influencing factors determined at the level of numerical values (high and low ambient temperature, pressure, humidity, etc.);
- standards (regulatory and technical documentation) necessary for the preparation of technical documentation for instrument-making products and virtual test methods.

The result of CAD (Computer-Aided Design) simulation of a product digital model are design solutions drawn up by designers in the form of technical documentation (schematic, text, drawing, software), according to which the Industry 4.0 Smart Factory will subsequently manufacture the product. All design solutions obtained in each virtual test scenario are documented in the cloud storage of a digital factory. The final design decision is a variant (design alternative) corresponding to the optimality criteria given in mathematical models.

4. Conclusion

Practical experience in the preparation of technical documentation for instrument-making products shows that the average product design time is up to a year. In the next few months, the manufacture of a product prototype is carried out and the stage of testing a physical sample begins. The test phase can be performed up to six months.

The test results of the product are the basis for its refinement (in the event of errors in technical documentation and manufacturing defects) and for introducing changes in the design, software and technological documentation. In case of changes in documentation and product, tests on the modified sample product are repeated.

Thus, from the beginning of the product designing stage to the moment of drawing up the technical documentation for the product, where most of the errors are identified and eliminated, it can take three to four years. Only then, the product documentation can be transferred to the serial factory for the production of samples. This design procedure is still the same at the Industry 3.0 design enterprises.

The implementation of tools for simulation of digital models and virtualization of design solutions using cloud services in the design procedures of the Industry 4.0 digital factory allows reducing (up to a year) the development time (design) and improve the quality of technical documentation for instrumentation products.

References

[1] Stef I D, Draghici G, Draghici A 2013 *Procedia technology* **9** 451-462
[2] Shariatzadeh N, Lundholm Th, Lindberg L, Sivard G 2016 *Procedia CIRP* **50** 512-517
[3] Gjeldum N, Mladineo M, Veza I 2016 *Procedia CIRP* **54** 158-163
[4] Ruttimann B G, Stockli M T 2016 *Journal of service science and management* **9** 458-500
[5] Kemeny Z, Beregi R J, Erodus G, Nacsa J 2016 *Procedia CIRP* **54** 53-58
[6] Gurjanov A V, Zakoldaev D A, Shukalov A V and Zharinov I O 2018 *Journal of Physics: Conference Series* **1015** 052034
[7] Jeong H-Y, Jeong Y-S, Park J H 2014 *Sustainability* **6** 8510-21
[8] Centobelli P, Cerchione R, Murino T and Gallo M 2016 *International journal of simulation modelling* **15** 223-235
[9] Agyapong-Kodua K, Asare K B, Ceglarek D J 2014 *Procedia CIRP* **17** 744-749
[10] Qian Zh, Yu Y, Fan G 2015 *Applied mathematics & information sciences* **9** 1981-92

IOP Conf. Series: Materials Science and Engineering **483** (2019) 012070 doi:10.1088/1757-899X/483/1/012070

Formation principles of digital twins of Cyber-Physical Systems in the smart factories of Industry 4.0

A V Gurjanov[1], D A Zakoldaev[2], A V Shukalov[2], I O Zharinov[2]

[1] Director, Stock Company «Experimental Design Bureau «Electroavtomatika» named after P A Yefimov, 40, Marshala Govorova St., St. Petersburg, 198095, Russia
[2] Faculty of Information Security and Computer Technologies, St. Petersburg National Research University of Information Technologies, Mechanics and Optics, 49, Kronverksky Av., Saint Petersburg, 197101, Russia

E-mail: mpbva@mail.ru

Abstract. The task of organizing the production of instrument-making enterprises of Industry 4.0 using digital twins of Cyber-Physical Systems (CPS) based on cloud services is considered. Cloud services are a component of SPS, acting in production as active Internet agents for the manufacture of instrumentation products. The combination of CPS, united by a single computerized control system, forms the production infrastructure of a smart factory. Smart Factory is a manufacturing plant of Industry 4.0, operating in automatic mode. A scheme of interaction between the components of the smart factory Industry 4.0 and cloud services for the production activity of the instrument-making enterprise is proposed.

1. Introduction

Modern instrument-making production is focused on the introduction of new digital technologies that provide information support for technological processes of manufacturing instrument-making products. Such digital technologies are [1-3] additive technologies, sensor technologies, BigData technologies for processing large arrays of production data, cloud technologies, etc. Manufacturing enterprises introducing promising digital technologies in the technological processes of manufacturing instrument-making products are the smart factories of Industry 4.0 [4, 5]. Smart factories implement a closed technological cycle of manufacturing products in automatic mode due to the deep automation of production processes. The main production unit of a smart factory is a cyber-physical system [6, 7], which has two components: a component at the level of a physical device and a component at cyber level (a cloud service containing a digital twin of a physical device). The interaction of components is provided by means and protocols of the Industrial Internet of Things (IIoT) [8, 9]. The primary task of building the automatic production of Industry 4.0 is to determine the composition of the components of CPS and the order (scheme) of interaction of these components.

2. Models of CPS at the physical and cyber levels

The models of CPS at the physical and cyber levels are shown in Figure 1.

The physical component is formed [6, 7]:

- digital controller, which is the core of the automatic control system of CPS, and implements the algorithm for performing the technological operation of manufacturing a part (product);
- actor system, which is a set of tools fixed on a single base (rotational-translational motion of the base is carried out by an electric motor under the influence of control signals from the controller);
- working chamber of the technological process, inside which the processing of the material, which is a blank for the manufacture of parts, being a part of the product is performed;
- sensor (set of sensors), which provides measurement of the parameters of the technological

process of manufacturing parts in the working chamber. The sensor measurement results are transmitted via the main feedback line to the controller. In the controller, the sensor measurement results are compared with the expected values (an error value is generated) defined by the reference tool path. The reference trajectory is fed to the controller input from the technological process algorithm library. By comparing the two input values, a control action is generated in the controller, which corrects the actors position in the direction of decreasing the control error value.

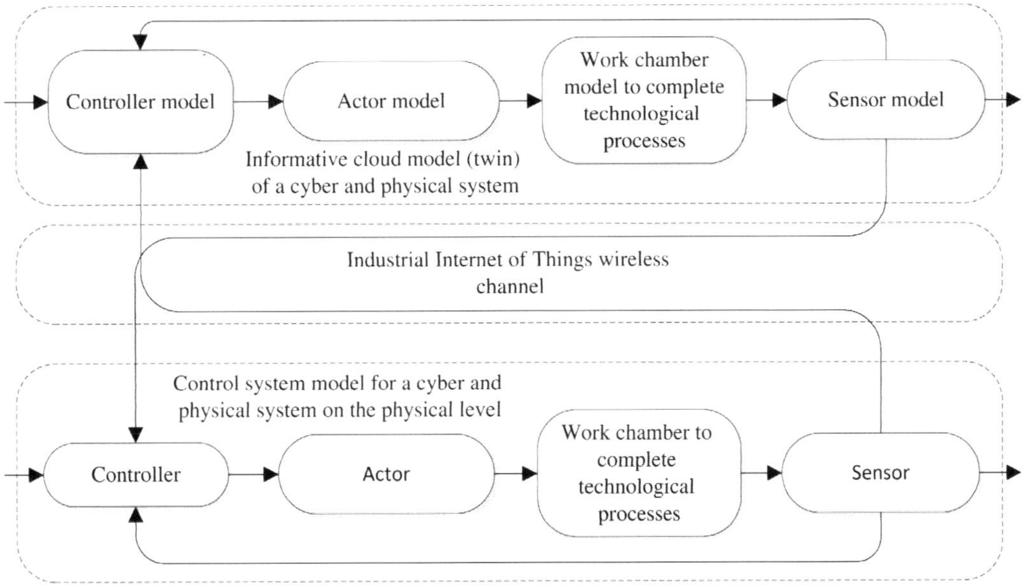

Figure 1. Models of the CPS at the physical and cyber levels.

The cyber world component of the cyber-physical system is located in the cloud service and implements the information model (digital twin) of the physical device. The information model of a cyber-physical device consists of:

- controller model;
- actor model;
- work chamber model of the technological process;
- sensor model (s).

The components models of the physical devices of the cyber-physical system are discrete-time mathematical equations solved by the computing resources of cloud services

The main purpose of the digital twin of the physical device of CPS is to support the pre-production phase. At the preparatory stage of production, the operator by means of modeling evaluates the possibility of manufacturing a part using a digital product model on a certain type of equipment, the properties of which are described mathematically. Such assessment is based on the results of technological operation at the level of mathematical models in the cloud service. Thus, at the stage of preparation of digital production, the routing sheet for the manufacture of instrument-making products within the production complex of an instrument-making enterprise is issued. The simulation results are also used by the computerized control system of a smart factory to ensure the self-organization property of cyber-physical equipment in digital production. Information exchange between the components of the cyber-physical system at the physical and cyber levels is carried out over a wireless communication channel using the protocols of the Industrial Internet of Things. Data exchange on the results of technological operations at the physical and mathematical models levels can also be used to improve the accuracy of the cyber-physical system in the production using the digital Kalman filter.

3. The scheme of components interaction of the smart factory of Industry 4.0

The scheme of interaction between the components of the smart factory of Industry 4.0 and cloud services for the production activity of the instrument-making enterprise is shown in Figure 2.The physical level of the smart factory of Industry 4.0 is represented by a set of cyber-physical systems that are combined into technological lines to perform the route sheet for the manufacture of an instrument-making product. Information exchange between cyber-physical systems is implemented by the Systems-to-Systems (S2S) technology. Physical interaction of cyber-physical systems is implemented by means of a robotic transport system that moves the workpieces and parts between the equipment and the finished goods warehouse. Control (monitoring) of the state of technological operations in production is carried out by an operator using Human-to-Machine technology (H2M) and technical means of an automated workplace [10, 11].

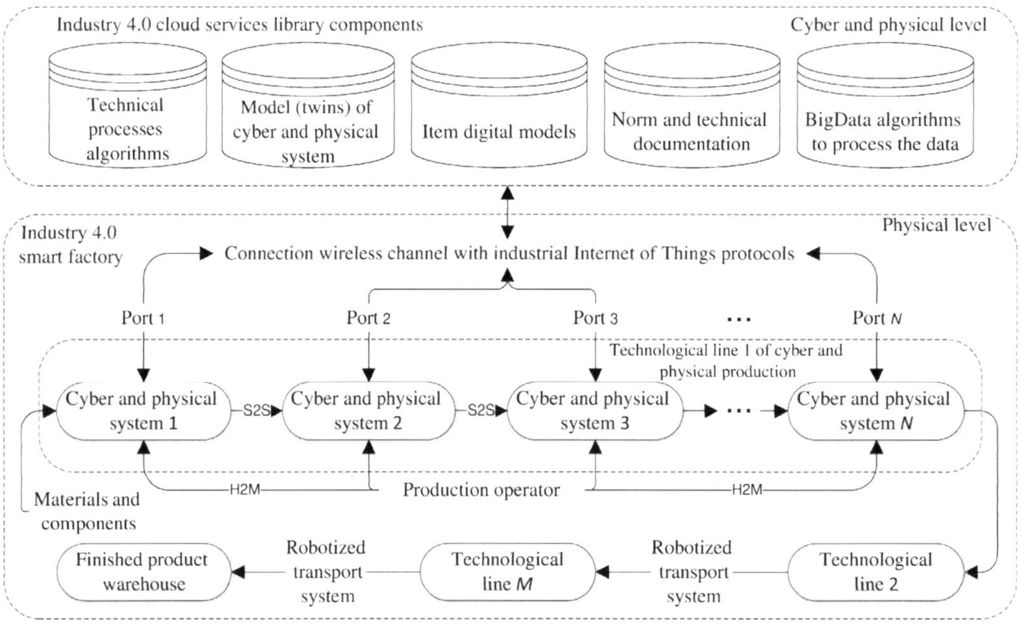

Figure 2. Diagram of interaction between components of the smart factory of Industry 4.0 and cloud services for production activities.

The cyber-physical level of the smart factory of Industry 4.0 is implemented by means of cloud services containing in the form of ready-made libraries [12]:

- technological processes algorithms for the manufacture of instrument-making products, implemented by the technological equipment available at the production;
- digital models (digital twins) of cyber-physical systems installed in production;
- digital models of instrument-making products, technical documentation for which entered the smart factories for the digital factory of Industry 4.0 products manufacture ;
- regulatory and technical documentation regulating the procedure for performing technological operations in production and their compliance with the requirements of state and industry standards;
- BigData algorithms for processing large arrays of production data resulting from measurements from sensors of cyber-physical systems [13].

The interaction of the components of cyber-physical systems with cloud services is carried out over a wireless communication channel using the IIoT protocols. Each cyber-physical system acts as part of a smart factory being an active Internet agent operating in the "self-service on demand" access mode. To organize such a scheme of components interaction, each cyber-physical system is allocated an individual port of information exchange in a single wireless communication channel for a production line (workshop, plant). Technological line components of the smart factory of Industry 4.0 at the physical level and their equivalent components of the digital counterparts of cyber-physical equipment with temporary characteristics of the production operations are shown in Figure 3.

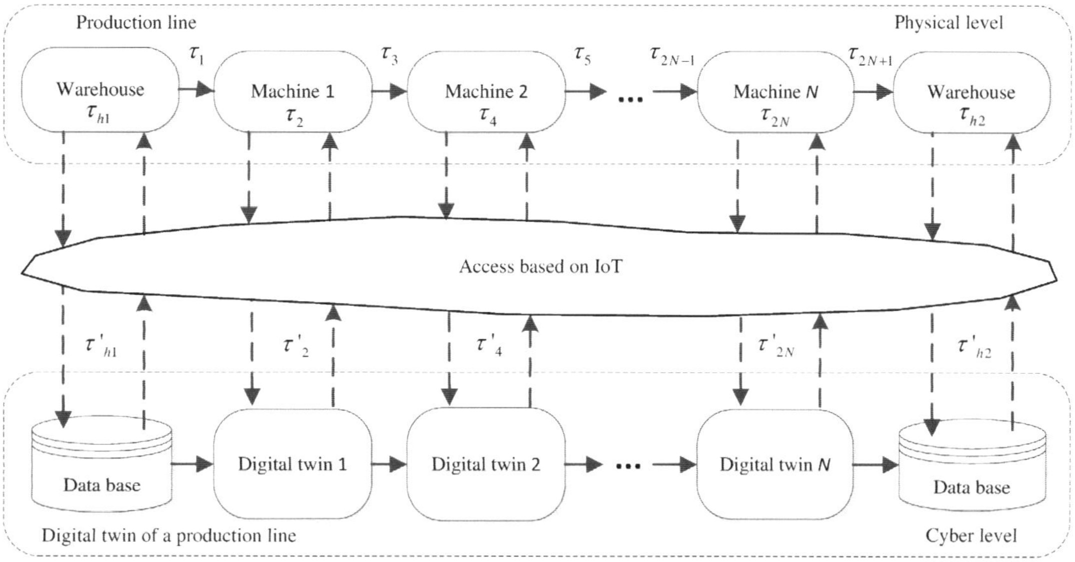

Figure 3. Process line diagram of the smart factory of Industry 4.0.

- τ_i - execution time of the production operation on cyber-physical equipment (physical device) in the process chamber;

- τ'_i - execution time of the production operation, obtained by computer simulation tools in the cloud service (time τ'_i is a random value, since it includes the temporal characteristics of IIoT traffic in the protocol with guaranteed packet delivery).The finished product warehouse and production site for storing materials and blanks in the smart factory are shown in Figure 3 with equivalent databases, the models of which implement cloud storages. Time τ_i with even indices characterizes the execution of production processes performed in the working chambers of cyber-physical systems [14]. Time τ_i with odd indices characterizes the execution of production processes by a robotic transport system. Analysis of the information flows implemented between the components of the smart factory of Industry 4.0 at the physical devices and cyber levels shows that the cyber-physical production control scheme is a closed-loop digital automatic control system. A feature of this automatic control system is the delay in the control channel, due to the temporal characteristics of the IIoT protocol with guaranteed packet delivery. The packet delivery time for the IIoT protocol with a wireless communication channel is a random value, therefore, the control system of the production automation of the smart factory of Industry 4.0 has a random delay.

4. Conclusion

The transition of the existing enterprises of Industry 3.0 to digital manufacturing techniques is linked to a significant modernization of the software and hardware of organizations. It is obvious that such replacement of enterprise infrastructure components will lead to reduction in production time and to an increase in product quality. However, step-by-step modernization of the enterprises of Industry 3.0 will not lead to the emergence of the smart factory of Industry 4.0.To create the smart factory of Industry 4.0, it is advisable to use new production areas, advanced cyber-physical equipment integrated into production lines, and qualified personnel who have been pre-trained in digital production technologies. Priority industries for the introduction of digital production technologies are:

automotive instrumentation, railway instrumentation, aviation instrumentation, rocket and space instrumentation.

References

[1] Liao Y, Deschamps S, Loures E F R and Ramos L F P 2017 *International journal of production research* **55** 3609-29
[2] Jung K, Choi S S, Kulvatunyou B, Cho H and Morris K S 2017 *Production planning & control* **28** 2108-22
[3] Radziwon A, Bilberg A, Bogers M and Madsen E S 2014 *Procedia engineering* **69** 1184-90
[4] Hwang G, Lee J, Park J and Chang T-W 2017 *International journal of production research* **55** 2590-02
[5] Lizunkov V, Politsinskaya E, Malushko E, Kindaev A, Minin M 2018 Population of the world and regions as the principal energy consumer *International journal of energy economics and policy* **8 (3)** 250-257
[6] Ergunova O T, Lizunkov V G, Malushko E Yu, Marchuk V I, Ignatenko A Yu 2017 Forming the system of strategic innovation management at the high-tech engineering enterprises *IOP Conference Series: Materials Science and Engineering* **177 (1)** 012046
[7] Qu T, Thurer M, Wang J, Wang Z, Fu H and Li C 2017 *International journal of production research* **55** 2622-49
[8] Gurjanov A V, Zakoldaev D A, Shukalov A V and Zharinov I O 2018 *Journal of Physics: Conference Series* **1015** 052035
[9] Bradley J M, Atkins E M 2014 *Journal of Aerospace Information Systems* **11** 48-60
[10] Almada-Lobo F 2015 *Journal of innovation management* **3** 16-21
[11] Bligh-Wall Sh 2017 *Cyber security: a peer-reviewed journal* **1** 61-68
[12] Fedotov N G, Moiseev A V, Syemov A A, Lizunkov V G, Kindaev A Y 2017 New Methods of Three-Dimensional Images Recognition Based on Stochastic Geometry and Functional Analysis *IOP Conference Series: Materials Science and Engineering* **177 (1)** 012047
[13] Lee J, Kao H-A, Yang Sh 2014 *Procedia CIRP* **16** 3-8
[14] Zhong R Y, Xu Ch, Chen Ch and Huang G Q 2017 *International journal of production research* **55** 2610-21

IOP Conf. Series: Materials Science and Engineering **483** (2019) 012071 doi:10.1088/1757-899X/483/1/012071

Professional qualifications and standards as a policy instrument aimed at improving the competitiveness of the Russian market for tourist industry

V A Zolotovskiy, A U Bazhenov, A. Yu .Kindaev

Volgograd State University, Prosp. Universitetsky, 100, 400062 Volgograd, Russian Federation
Penza State Technological University, Russian Federation

E-mail: zolotovskiy.azi@volsu.ru

Abstract. The paper discloses the current condition and promising ways of tourism policy aimed at improving the skills of tourist enterprises personnel and enhancing the overall competitiveness of the Russian market for tourist services. The study has defined the goals and methods of the government policy as to improving the skills of tourist enterprises, identified the teleological content of the professional qualifications and standards as instruments of the tourism policy, formulated the concept of integrated development of the professional training in tourism. The material in this paper can be useful both for training institutions and representatives of the tourism and hotel businesses.

Introduction

The worldwide growth in tourist arrivals and the increase of the share of tourism in the global economy are attracting growing attention of businesses, international tourism organizations and national tourist administrations. According to UNWTO forecasts that were pronounced at the 19th session of the General Assembly in 2011, by 2030 the annual international tourist arrivals will amount to approximately 1.8 billion [1]. Judging by the world dynamics of tourist arrivals from 2017 and general statistics of inbound and outbound tourism in Russia we can assume that the Russian market will be about 84 million people. Special importance of tourism for the economy is emphasized by the fact that the tourism sector constitutes about 10% of the world GDP and 8.7% of full employment [2]. It is expected that alongside with the expansion of the tourist market, a qualitative change will take place primarily due to improvement of professional skills of tourist enterprises personnel.

The purpose of the paper is to describe the current state and promising ways of the policy to improve the skills of tourism enterprises personnel and the general competitiveness of the Russian market for tourist services. The objectives are to determine the goals and methods of the state policy on improving skills of tourist enterprises personnel, to identify the teleological content of the professional requirements and standards in tourism as an instrument of the tourism policy, and to formulate a concept of the integrated development of training in tourism.

1. Discussion

The issue of forms, content and methods in tourism education and training has been repeatedly debated in literature. The interest in these questions has been boosted by the special importance of tourism in the global and national economies.

Vanessa A. Amoah and Tom Baum studied the role of the state policy to solve the problem of improving the quality and professionalism in tourism industry through public education activities. As for the institutional aspect of the problem, the authors note that the difficulties of creating a balanced system of interaction between the training and the business come from the multisectoral nature of tourism and the dynamics of its development. Accordingly continuous monitoring and making the training system more flexible are needed.

Experts note that it is the training policy in tourism that can solve such pressing issues as the strategy for introduction of training in the field of tourism, the availability of training as well as promising directions and factors for the sustainable development of tourism. The experts write that this content of training policy in tourism should be based on the adoption of a special general program (curriculum). At the same time, the development of such a program is only possible through consultations between the educational authorities, educational institutions and representatives of the tourism business [3].

Jameson-Charles Madgerie's study focuses on the concept of relationship management that shapes the training in tourism in the Caribbean. Using the methods of interviewing and social survey, the author puts an emphasis on the issue of compliance of the training system with the needs of employers. Defining the role of the training policy in tourism the author concludes that it is necessary to make the whole system more open and to allow businesses to participate in shaping training programs [4].

The group of experts is particularly interested in the issue of training based on the problem method. Using the data of a long-term research of the training practice the authors reveal the specificity of the training content and describe the characteristics of its application in undergraduate training programs. The authors conclude that using the problem-based training and other active training methods for tourism and hotel clusters can improve the quality of advanced training programs. The main conclusion leads to the idea that the problem method does not only allow to improve the approaches to the forming of professional skills, but also to bring students closer to the tourism business in the region [5].

The paper by Tawanda Makuyana and Melville Saayman deals with the matching of the training system in tourism to the needs of the tourism industry. On making a comprehensive analysis of certified training curricula for tourism and hospitality the authors conclude that there is a lack of coherence between the teaching staff and the employers. A similar situation is in the advanced training system [6].

Iryna Skrypchenko stresses the need to purposefully train specialists for youth tourism. She notes the importance of joined efforts of training institutions of all levels, travel agencies and public organizations that are focused on the practical problems of youth tourism in Ukraine. Exchange of experience between these organizations should contribute to the successful training, retraining and advanced training of specialists in tourism. The author also focuses on the prospects of problem-based learning that involves simulation of problem situations in order to develop students' cognitive and creative activities [7].

2. Analysis and Results

A review of the directions and forms of the state policy on development of skills of tourism companies staff reveals two groups of specific instruments:

1 program-and-organizational instruments that set the development priorities, goals and objectives, and mechanism of their achievment;

2 framework and recommendatory instruments that set the qualification standards of tourist enterprises personnel as well as the main goals and characteristics of programs of skills development and retraining in the tourism industry.

Before we study the first group of instruments it must be noted that in Russia tourism is determined as a promising sector of the economy. The case is not unique. In many countries the significance of tourism for the economies is estimated very high. In this regard the development of the state policy strategy for tourism is considered very important [3].

The Russian state authorities know that resources themselves are not enough to develop the national tourism market that has been forming in a rather chaotic manner. This conditioned the development of a specific approach to the state regulation in the Russian tourism industry. It led to adoption of special target and inter-branch state programs in order to ensure a high quality in the tourism industry. We are going to consider their pivotal points.

In 2011 the Russian Government adopted the Federal Target Program "The Development of Domestic and Incoming Tourism in the Russian Federation (2011 – 2018)". It stipulates for creation of the up-to-date, highly efficient and competitive touristic and recreation complex in Russia. It should provide extensive possibilities to meet the demand from Russian and foreign tourists for various touristic services [8]. The program stipulates for improvement of quality of touristic services.

The goal of the program is to increase the competitiveness of the Russian tourist market so that Russian and foreign tourists could enjoy high-quality services. The goal should be achieved through a series of tasks. One of them is improvement of touristic services. To fulfill the task some measures must be taken to develop the tourism specialists training system. The system should include higher education institutions, secondary professional education institutions, improvement of professional skills and retraining programs. Fulfillment of the second task can be greatly promoted by assigning to market participants the rights to control and supervise over tourist organizations including the quality of services they provide. Self-regulatory organizations also must be supported.

Among the principle drawbacks that undermine the competitiveness of Russia in the international tourist market is a low-quality service in every sector of the tourist industry. It is caused by the lack of high-skilled personnel.

In December 2012 the State Program "Culture and Tourism Development" for 2013 – 2020 [9]. Clause 2.2 formulates the main goal of the program. It is the realization of the strategic role of culture as a psychological and moral basis for the individual and state development, integrity of the Russian society and development of tourism for familiarization of people with the world culture and nature heritage. Among the tasks to reach the goal the program states "Development of quality and availability of the domestic and international touristic services".

It must be noted that according to the State Program "Culture and Tourism Development" for 2013 – 2020 the lack of high-skilled personnel and the low quality of service in every sector of the tourist industry are among the principle impediments in the development of tourism in Russia. In this regard the authors of the program note that the state regulation is oriented to introduction of the voluntary unified system for accreditation (classification) of guides and guides-interpreters; introduction of professional standards in tourism; and creation of the legislative basis for the development of organizations for tourism specialist accreditation.

In a draft of the program "Development of domestic and incoming tourism for 2019 – 2025" the task of improvement of professional training in tourism is formulated in more detail [10]. The draft states that the whole range of measures of the future program will contribute to improvement of the quality and competitiveness of Russian tourist services in the home and international markets. It will favor business and public initiatives in tourism including creation of small businesses.

The draft of the program sets the task to develop the system of training specialists and qualification improvement to solve the problem of low-quality services in tourism. It is proposed to continue the work on improvement of the methodic basis of the training system in the tourist industry including the technological basis for development of the tourist industry and introduction of innovative technologies of touristic services via the system of qualification improvement in the tourism infrastructure. Beyond the programs activities programs of qualification improvement and professional adaptation should involve the constituents of the Russian Federation and touristic organizations. Their participation can be in organizational forms and co-financing taking into account the real personnel demand in touristic organizations. The realization of the whole range of measures will remove the problem of low-quality service in every sector of the tourist industry due to personnel qualification improvement.

The principal mechanism of solving the task is the established practice of Government orders for preparation of specialists in tourism on the secondary professional education, bachelor and magistracy levels. The training must be carried out strictly in accordance with the federal educational standards.

Besides under conditions of the established tourist market the real ways of the personnel development are in-service qualification improvement and retraining programs. In this regard the program draft specifies the task of personnel qualification improvement – target indices are planned as the percentage of the specialists who have completed courses of development and retraining including the targets for the constituents of the Russian Federation that are realizing investment projects on tourist clusters.

It should be noted that there have occurred considerable changes in the mechanism of development of the training and qualification improvement system for tourism specialists who are the potential staff for solving the problem of quality of touristic services. In particular, the Russian executives have passed from declarations of intent to direct participation in the process – from government orders for tourism education on the bachelor and magistracy levels to creation of the special on-line project of the Federal Agency for Tourism (the educational platform Turobrazovanie) [11]. On the basis of the latter interactive training courses are introduced by means of regional educational organizations. On the modular system basis programs are run for qualification improvement of heads of means of accommodation; hotel reception and accommodation personnel; hotel room stock management personnel; catering service personnel; travel agencies personnel; guides; up-to-date marketing instruments in tourism industry organizations; tourist services management for disabled persons; raising the efficiency of sales in the services field; effective use of modern marketing instruments for attraction of visitors to national parks; innovative approaches to forming museum exhibitions and conducting excursions; revenues management in a hotel; effective communication with visitors of mass event-driven actions.

Besides in the few last years the practice of professional competitions has been introduced – WorldSkills Russia ("Young Professionals" Union) [12]. WorldSkills Russia is a representative of the non-commercial world organization WorldSkills International whose mission is to develop professional qualifications, raising prestige of professionals, and demonstration of importance of high qualification for economic growth and personal success.

In September 2017 the first inter-university championships started in Russia according to the WSR standards. Due to these championships the public interest to the profession is developing and guidelines are set for development of touristic training programs. The competitions are to demonstrate skills in real working conditions on the basis of the expert-community standards that require high professional skills of competitors. The competitions are organized on the modular principle [13].

Before we analyze the second group of instruments it should be noted that after the liberalization of the market in the beginning of the 21st century the mandatory certification of touristic products was abolished as well as accreditation of tourist agencies and tour operators. In the state regulation of the touristic market the optional method became a priority when tourism organizations can choose their staff strategy themselves. In the last few years special documents have been adopted as main guidelines for formation of the structure of tourism organizations in order to improve commercial and non-commercial organizations management. These documents are "qualification guidebooks", state standards, and professional standards. We are going to review them in more detail.

The basic document is "The Common Qualification Guidebook for Management Positions, Specialists, and Office Workers" [14]. The section "Qualification Characteristics of Tourism Organizations Workers" is designated to solve problems of the regulation of labor relations, providing an efficient system of personnel management in tourism organizations irrespective of their ownership and legal form. The qualification characteristics are applied as normative documents or as a basis for working-out job descriptions with detailed lists of responsibilities taking into account the peculiarities of labor organization and management as well as rights, responsibilities and competency of personnel.

The qualification characteristic for each job has three sections: "Job description" (basic functions), "Must know" (principal requirements as to specific knowledge) and "Qualification requirements" (level of skills necessary for fulfillment of functions). For the purpose of this paper we need to view the qualification requirements of the guidebook.

Section 2 is on the qualification characteristics of personnel of excursion organizations. The higher education degree in tourism or professional training are required for the following positions: heads – director (manager) of an excursion office; specialists – interpreter (in tourism); guide-interpreter (in tourism); guide; excursion organizer. Section 3 sets forth the qualification characteristics of travel agencies personnel. The higher education degree in tourism or professional training are required for the following positions: heads – director (manager) of a travel agency; head of the department of marketing and sales of a tourist agency; specialists – tourism manager (outbound, inbound tourism and domestic tourism); manager of reservations and sales; head of tourist groups; reservation agent; tourism agent (outbound, inbound and domestic tourism). Section 4 sets forth the qualification characteristics of tour operators. The higher educational degree in tourism or professional training are required for the following positions: heads – director (manager) of a tour operator organization; head of the reservation and sales department of a tour operator organization; head of the department of tourist products of a tour operator organization; specialists – manager for forming tourist products; manager for tourist products (outbound, inbound and domestic tourism); manager for exit visas; manager for corporate organizations servicing; adviser on touristic services. Section 5 sets forth the qualification characteristics for hotels personnel. The higher educational degree in tourism or professional training are required for the following positions: heads – director (manager) of a hotel; manager for reception and accommodation.

Professional standards are no less important. Today the standard "Guide" is active in the tourist industry. The professional standard "Specialist on organization and providing tourist services" is also worked out. Both documents contain detailed requirements for tourist organizations personnel. In particular, the professional standard "Guide" determines competencies for travel and transport agencies personnel and for heads of organizations on the basis of functions (accepting and processing orders for excursions; coordination of execution of orders; organization of excursions; working out excursion programs; conducting excursions; determination of the conception and the strategy of an excursion organization development; organization of activities on excursion services; formation and execution of the staff policy of an excursion organization). Besides for each job the professional qualification and work experience are stipulated [15].

Professional skills and educational requirements for jobs in the tourism industry are similarly formulated in the draft of the professional standard "Specialist in organization and delivery of tourist services". The skills are determined on the basis of the following generalized functions: processing orders for formation of excursion (tourist) groups; subsidiary activities on formation, promotion and selling a tourist product; organization of tourists leisure; working out and use of tourist routes; working out, promotion and selling a tourist product; management and servicing of a tourists group; management of a subdivision in a tourist organization [16].

In spite of the detailed elaboration and immediate necessity of the professional standards for tourism organizations still they are not mandatory. According to the interpretation of the Russian Ministry of Labor and Social Protection, the Russian Labor Code stipulates for application of the professional standards if activities are connected with compensations, benefits and restrictions or if the application of the standards is demanded by the Labor Code and other legislation of the Russian Federation. Otherwise the standards are optional [17].

The optional character of the professional standards is a temporary measure needed for adaptation of the market to higher professional requirements. The professional standards will become mandatory in 2020 only for state and municipality-owned organizations as well as for public-private organizations [18]. Modern federal educational standards 3++ are being introduced in the higher education system obviously in connection with these plans. These mandatory acts provide educational organizations with more freedom in formulation of the professional competencies. At the same time the authors of the educational standard direct educational organizations toward the requirements of respective professional standards. In particular, the Federal State Educational Standard for tourism bachelor degree directs authors of educational programs toward the "Guide" professional standard.

IOP Conf. Series: Materials Science and Engineering **483** (2019) 012071 doi:10.1088/1757-899X/483/1/012071

Such approach allows educational organizations to form their educational programs according to the problem-based learning method (PBL) [5] considering the demand from a regional tourist industry market [19]. In this regard it is necessary to point out that the authors of the professional standards focused on the demand of the market [20]. So, we think that it is the professional standards that must become a guideline for the elaboration and realization of programs of qualification improvement and retraining. Our opinion is supported by the list of the programs of qualification improvement and retraining that are being worked out and realized according to the project "Turobrazovanie" (Tourism Education).

Special state standards that set the requirements for skills in separate organizations (sectors) in tourism are particularly important. Particularly we can mention the state standards requirements to personnel of tour operators and travel agencies, the requirements to the programs of qualification development of guides and accommodation personnel [21, 22, 23, 24]. Besides it must be noted that the special state standards that set the requirements as to the basic and supplementary touristic services and to touristic facilities and territories also hold certain requirements to the personnel skills. The formulation of the important tourism services shows the most important tourism sectors that require the special skills. These are accommodation personnel, catering personnel, excursions personnel, sales of tourism products, promotion and sales of tourism services, services in active, cultural, recreational tourism, and so on. The general analysis of the state standards for tourism services and products and for personnel development programs permits to formulate the important requirements as to the duration of training programs, fields of training, and objectives of personnel development programs [25, 26, 27]. As for the latter, the most important requirements refer to foreign languages competency, knowledge of the legislation, providing security, knowledge of the conditions of providing services in accordance with tourist categories, and so on.

3. Conclusion

On the analysis of realization of the policy on qualification improvement and on the general competitiveness of the Russian tourism market improvement we can conclude:

1. The first group of instruments seem important and can solve the problem of qualification improvement of tourism organizations personnel on the strategic level. At the same time the actual system of the higher education that is being transformed to the Standard 3++ can strengthen training programs by the problem-based learning method that is oriented on the local, regional, federal and global problems of development of the tourist industry [28].
2. The second group instruments due to their optional character can formulate the guidelines for market solving of the mentioned problem. Focusing on the recommendations for professional skills using the problem-based learning method in personnel development and retraining we can solve local tactic tasks on raising the quality of tourist organizations activities. The development of the education system with the on-line courses via the "Tourobrazovanie" portal will keep in action the tourism sector.
3. Due to the inter-branch character of tourism we consider the most effective a mixed form of on-line training in the form of the modular system. Particularly prospective we consider forming a system of distant education with a program that is realized as autonomous modules (sub-programs – personnel development programs). Each of them can be devised on the problem-based learning basis [19]. Each module of the program will be oriented on creation of specific skills. Training on the modular system will depend on the demand from customers – the employer or employee of a tourist organization. Some modules of the program can be of a universal character and realized via the "Tourobrazovanie" portal. However, the organizational connection of tourism with a touristic area defines advisability of realization of such training programs in regional educational organizations.

The approach that we propose to develop the State tourist policy as to touristic personnel development and the general competitiveness of the Russian market for touristic services is of a model character. If the similar instruments of the personnel policy are created in other industries, this approach can be adapted to be used in other industries of services and production.

Acknowledgement
The reported study was funded by RFBR and the government of Volgograd region according to the research project № 18-413-342003.

References

[1] UNWTO General Assembly 19th Session 2011 : Tourism towards 2030. Global overview (Republic of Korea) 19

[2] Seselkin A 2017 Sustainable tourism as a priority in sustainable development strategies *Bridges* **10** **(4)** [Electronic resource]: URL: *https://ru.ictsd.org/bridges-news/мосты/news/устойчивый-туризм-как-приоритетное-направление-в-стратегии-устойчивого#_ftn3/*

[3] Amoah V A and Baum T 1997 *International Journal of Contemporary Hospitality Management.* **9(1)** 5-12

[4] Madgerie Jameson-Charles 2012 *Caribbean Curriculum* **19** 145-70

[5] Yumatov K, Kiriyanova L, Yakimova N, Zaitseva N, Larionova A, Korsunova N 2017 *Eurasian Journal of Analytical Chemistry* **12 (5b)** 803-12

[6] Makuyana T and Saayman M 2018 *Conference: Research based STEMitised TVET innovations: Solutions to revive Zimbabwe, At Inyanga-Troutbeck Resort (February)*

[7] Skrypchenko I 2017 *Topical problems of modern science* **5** 30- 40

[8] Decree of the Government of the Russia dated of August 2 2011 № 644 "On the federal target program" Development of domestic and inbound tourism in the Russian Federation (2011 - 2018) *Collection of legislation of RF* **34** *art. 4966*

[9] Decree of the Government of the Russia dated of April 15 2014 № 317 "On the approval of the state program of the Russian Federation" Development of Culture and Tourism " for 2013-2020 *Official Internet portal of legal information* [Electronic resource]: URL: http: //www.pravo.gov.ru

[10] The project of Federal target program "Development of domestic and inbound tourism in the Russian Federation (2019 - 2025)" *Federal portal of projects of regulatory legal acts* [Electronic resource]: URL: http://regulation.gov.ru/projects project

[11] The educational portal Toureducation (Turobrazovaniye) [Electronic resource]: URL: https://tourstudy.ru/

[12] Kandyba O V, Barashok I V, Kushnareva I Yu 2017 *Krasnoyarsk science* **6 (4)** 223-41

[13] Final of the frst National Interuniversity Competition according to the Worldskills standards 2017 [Electronic resource]: URL: http:// worldskills.ru/nashi-proektyi/chempionatyi/mezhvuzovskie-chempionatyi/fnal-pervogo-naczionalnogo-mezhvuzovskogo-chempionata. Html

[14] Order of the Ministry of Culture of Russia dated of March 12 2012 № 202n "On approval of the Unified Qualification Directory of managers, specialists, employees" section "Qualification characteristics of employees of tourism organizations" *Bulletin of normative acts of federal executive bodies* **29**

[15] Order of the Ministry of Culture of Russia dated of August 4 2014 № 539n «On the approval of the professional standard "Guide (Guide)"» *SPS "ConsultantPlus"* [Electronic resource]: URL: http://www.consultant.ru/

[16] Projects of professional standard "Specialist in the organization and provision of tourist services" *Federal portal of projects of regulatory legal acts* [Electronic resource]: URL: http://regulation.gov.ru/

[17] The information by the Ministry of Labor of Russia on the application of professional standards dated of April 5 2016 [Electronic resource]: URL: https://rosmintrud.ru/docs/mintrud/payment/128

[18] Decree of the Government of the Russia dated June 27 2016 № 584 «On the peculiarities of the application of professional standards in terms of the requirements mandatory for use by state

extra-budgetary funds of the Russian Federation, state or municipal institutions, state or municipal unitary enterprises, as well as state corporations companies, more than fifty percent of shares (shares) in the authorized capital of which is in state ownership and or municipal property» *Official Internet portal of legal information* [Electronic resource]: URL: www.pravo.gov.ru June 29 2016

[19] Velikova E, Ianeva M, Gatovski I 2017 *Conference: International Conference The Future of Education* **7**

[20] Order of the Ministry of Labor of Russia of November 2 2015 № 832 "On approval of a directory of new professions in demand in the labor market, including those requiring secondary vocational education"

[21] GOST R 54602-2011 Tourist services. Services instructors guides. General requirements

[22] GOST R 54603-2011 Accommodation facilities. General Service Requirements

[23] GOST R 57807-2017 Tourist services. Requirements for guides (guides)

[24] GOST R EN 15565-2012 Tourist services. Requirements for the provision of professional training of tourist guides and advanced training programs

[25] GOST R 54600-2011 Tourist services. Travel agency services. General requirements

[26] GOST R 57280-2016. National standard of the Russian Federation. Tourist services. General requirements for the provision of ski tourism services "

[27] GOST R ISO 13810-2016 Tourist services. Industrial tourism. Provision of services (with amendment)

[28] Sandeep Malik. (2018), Importance of Training and Development in Tourism Industry *International Refereed Journal of Reviews and Research* **6** 11-19

IOP Conf. Series: Materials Science and Engineering **483** (2019) 012072 doi:10.1088/1757-899X/483/1/012072

Reducing cutting forces at advanced plastic deformation of metals

Abrosimov K.S.[1], Morozova A.V.[2]

[1] Lipetsk State Technical University, Lipetsk, 398055, Russia
[2] Bryansk State Technical University, Bryansk, 241035, Russia

E-mail: ambsk@mail.ru, niotiostu@gmail.com

Abstract. The article carries out the analysis of models for studying the effect of reducing cutting forces when cutting mechanically hardened materials from the standpoint of the theory of dislocation and energy balance. The article presents the perspective model of researching the process of cutting with the advanced deformation allowing calculating the cutting force reduction. The model takes into account the history of damage accumulation considering changes in the stress state at each stage of the combined treatment process.

1. Introduction. Energy model

Modern trends in developing cutting theory [1, 13, 16, 17] affect both the problems of creating new types of machine tools [9] and metal-cutting tools [2, 6, 10, 18, 20], evaluation of the metal cutting quality [8], and the problem of optimal application of cutting fluids [4, 12], study of residual surface stresses in metal cutting [3], as well as modeling the cutting process [11, 14, 15], including the use of modern numerical methods [5, 7, 19].

One of the perspective directions of developing the theory of cutting is studying the effect of cutting force reduction under treating "strengthened" (mechanically strengthened) metals from the standpoint of the plastic deformation balance, which was first explained by Ya.G. Usachev [21]. He wrote that: "before the metal collapses, it goes through all the stages of hardening. The fracture of metal will occur at the moment when strengthening reaches the greatest value. To fracture the metal, already strengthened to some extent, it needs to be less deformed than the unstrengthened metal by the amount corresponding to the preliminary strengthening." And further: "the difference of works will be equal to the work that is required to bring the sample metal to this degree of strengthening."

However, experiments of hardened and annealed samples with broaching to failure, did not confirm this last conclusion. The work spent on rupturing non-strengthened samples was four times as high as the work spent on strenthened samples, and when cutting this excess was from 1.1 to 1.4 times.

2. Dislocation model

According to A.M. Kuznetsov, the theory of dislocations explains the process of improving metal machinability by cutting after applying the cold plastic deformation. [22].

The authors differentiate 3 stages of hardening at applying the deformation stress on the metal. In the first stage of hardening, voltage is applied and the microscopic limit of elasticity is achieved. So, the source of dislocations begins to operate, generating loops of dislocations moving under the action of stress in the sweeping plane [23].

Step occurs in the last loop when the loops achieve the surface of the crystal. The part of the loops is delayed in the sweeping plane and generates a reverse voltage that impedes the action of the source. The sliding line can grow under the influence of constantly applied voltage. And if the voltage is sufficient, the source can send new loops that pass freely through the entire sweeping plane.

Since new obstacles do not arise due to the lack of dislocation forest (sliding occurs only along one system of planes). This process can be repeated at the same voltage level. That's why hardening is little at the first stage under the shear conditions. The height of the step is usually small 50-100 Å, while a sliding line length is 1 mm.

According to Hirsch the transition from stage I to stage II occurs at a time when the distance between the dipole accumulations becomes very small, and the internal stresses caused by heaps of dislocations, together with the applied voltage become sufficient for the occurrence of dislocations in the secondary system. In the second stage, the sliding lines are shorter and less regular than in the first stage. Studying the dislocation structure of copper shows that here there are plexuses of dipoles with the forest dislocation, resulting in an irregular grid of dislocations. The average density of dislocations reaches 1016 cm-2 by the end of the stage. The density of dislocations in the annealed material is on average 106-108 cm-2. The beginning of the second stage corresponds to the catastrophic avalanche-like process of dislocations generated by Frank Reed's sources in the secondary system of sweeping planes. As a result, a large number of sedentary dislocations are formed, which are effective barriers for passing new dislocations, i.e. there appear their piles. The number of dislocations increases in proportion to the strain.

The third stage is characterized by the development of slip bands, the length of which depends on the magnitude of the previous deformation. At a certain stage, developing transverse sliding is observed, in this case, the slip strips expand slightly. As a result of the dislocations accumulation in front of the obstacle, the shift in the original sweeping plane can be slowed down. Screw dislocations can move to the adjacent parallel plane by means of cross-sliding under certain conditions, and this plane forms a new source, which under the action of the applied voltage generates additionally a number of dislocations and thereby provides a plastic shift. At this stage, the coefficient of strenthening decreases in comparison with the observed coefficient at the second stage, as a result, the intensity of the flow resistance growth decreases. At the third stage, the dislocation structure of the crystal depends not only on the magnitude of the stress or strain, but also on the stress state.

The nucleation or presence of dislocations in the crystal does not directly lead to violating the crystal lattice continuity, i.e. to forming a free surface (submicrocrack). However, the atomic layers at the dislocation place are elastically distorted in such a way that there is a local stress concentration.

Dislocation serves as the elastic deformation center. Distancing from this center stress concentration decreases according to the law L/r. Dislocations, in which Burgers vector is located in the sweeping plane, are quite mobile and under the action of the applied stress move in the sweeping plane, so that a plastic shift occurs. Dislocations are able to multiply in the plastic deformation process, the number of them in the metal increases by many orders of magnitude. In the undeformed metal dislocation density is 106-108 cm-2, in the metal strenthened by plastic deformation dislocation density reaches 1011-1012 cm-2.

The free energy of the crystal lattice increases in the dislocation area. Therefore, the dislocation itself is a thermodynamically unstable formation, but because of the balance of forces acting on the dislocation, it will be in some metastable state. To remove it from this state, it is necessary to energize additionally from the outside (for example, to give thermal energy). The energy of the crystal will decrease, and the dislocation will either move to a new metastable state, or leave the crystal.

The convergence of dislocations of different signs leads to their annihilation, and that of one sign leads to increasing the stress concentration, because the fields of one sign repel. The energy of the crystal in the general case increases due to the free energy increase of each dislocation and due to the interaction of dislocation fields and dislocations with crystal interfaces and foreign phases.

Dislocations moving in the crystal, give rise to defects, which arise by the mutual intersection of dislocations and the diffraction of their small inclusions. These defects, dislocated atoms and vacancies, cause the formation of submicrocracks and, ultimately, lead to the material destruction.

There are three types of mechanisms for forming microcracks:

1. Dislocation reaches a critical value in the same sweeping plane
2. Piles of dislocations occur due to their interaction in the intersecting sweeping planes.
3. Cracks are formed as a result of interacting defects in the crystal lattice of vacancies and dislocations.

For example, Stroh and Griffiths suggested that a crack should occur if its formation leads to a decrease in the energy of the lattice distortion, and the fracture should occur when the elastic energy in magnitude exceeds the surface energy required for forming the crack.

Pitch suggested that brittle fracture should occur at a time when theoretical strength exceeds as a result of piling dislocations.

Gilman in the zinc single crystal made an artificial obstacle in the form of a brass bracket, which served as a reliable barrier to advancing dislocations on the shear planes, because the yield point of brass is higher than that of zinc. As a result of the sweeping plane at an obstacle of a heap, there was a weakened layer which opened in a crack under the action of the broaching component which was approximately perpendicular to the sweeping plane.

Not yet proven, but noteworthy, is Price model, in which the origin of the crack occurs due to the local combination of dislocations moving in parallel sweeping planes, which can create areas that combine large internal stresses near the source.

Despite the fact that most of the models leading to various mechanisms of brittle crack nucleation have not yet been confirmed experimentally, it is doubtless that they are the main factor of merging cracks leading to brittle fracture. The possibility of closing and self-healing microcracks, determines rather a low efficiency of the destruction process.

When cutting metals, the energy goes mainly to deforming the adjacent zone, and only a negligible part goes to increasing the system surface energy, i.e. to forming the main crack, leading to separating the material layer. The total energy of destruction exceeds the surface energy by more than 3 orders of magnitude. As the deformation degree increases, the cracks lose their "stability" and the subsequent cutting of such a layer leads to individual microfractures merging into a macrofracture under the action of the cutting tooth.

With the preliminary surface plastic deformation, transverse cracks are gradually moving in the macrocracks, while under the repeated stress they are moving in the longitudinal ones, thereby forming a scaly exfoliation.

The maximum values of the submicrocrack density should, obviously, be in the zone having the maximum value of microhardness, as this is where the maximum density of dislocations is. It should be noted that neither the direction of the microcracks, nor the nature of changes in their depth density relating to the surface plastic deformation has been studied in detail, although this is important to determine the reduction of energy consumption for cutting at advanced deformation.

3. The model considering the depletion of the plasticity resource at different damage accumulation

The accumulation of microcracks in any type of plastic deformation, including cutting, depends on the damage accumulation [24]. Damage accumulation is associated with both the degree of deformation and loading conditions, which are determined by many factors, including the stress state and loading history. The dynamics of the microcrack accumulation is influenced by the sign and the value of the hydrostatic pressure, with the growth of which the intensity of healing the microcracks increases and the efficiency of the destruction process decreases.

In recent years, much attention has been paid to the combined machining with advanced deformation (MAD), such as dimensional combined machining by rolling and deforming-cutting broaching [25, 26]. However, there is still no reliable and valid mathematical apparatus for calculating the reduction of cutting force in MAD. The reason for this is that while the complex deforming action and cutting there are different mechanisms of damage accumulation with different indicators of the stress state, which must be taken into account. In addition, when developing a model of the cutting process with MAD (Fig. 1), it is necessary to lead the damage accumulated during the advanced deformation (curve 1) to the damage accumulated during the cutting (curve 2) [24], through the plasticity diagram of the processed material (curve 3), which determines the degree of shear deformation corresponding to the moment of the material destruction depending on the stress state..

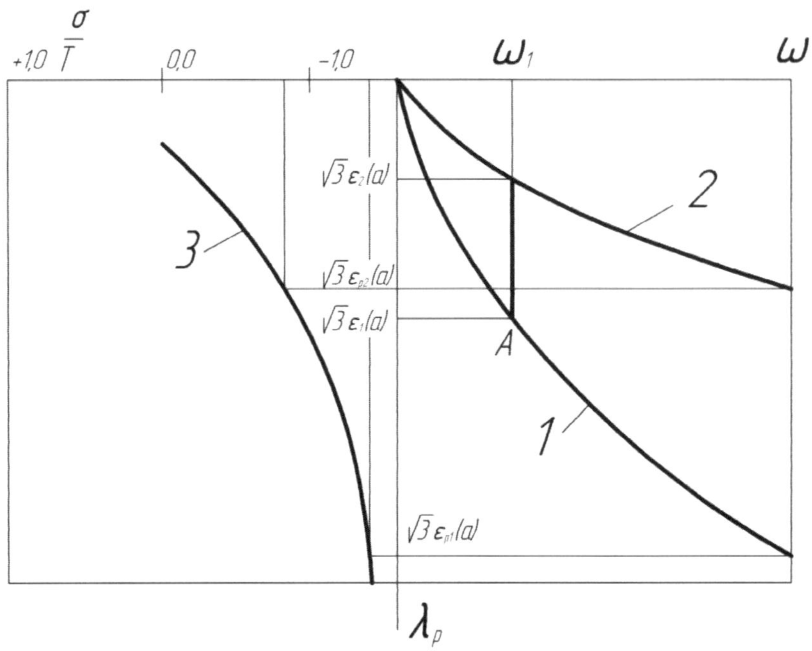

Figure 1. Scheme of 2-stage damage accumulation when cutting with MAD

At first, the process of damage accumulation of the preliminary deformation takes place along curve 1 with the stress state σ/T less than the stress state of the cutting process, and then along curve 2 with a large value of σ/T.

Thus, in the work balance model it is necessary to take into account not all the work of pre-plastic deformation and the corresponding shear deformation, but only a part with a smaller value of shear deformation.

For example, while having deforming-cutting broaching with MAD and elastic-plastic loading (EPL) [26], first there is damage accumulation to the corresponding MAD process with a negative stress state indicator-σ/T, since there is a deforming element before the cutting-deforming tooth in the course of broaching. Further, a deforming-cutting element enters the treatment zone, at the same time, the deformed layer starts broaching in the cutting zone. There is significantly reduced the magnitude of the shear deformation degree of the corresponding damage accumulated earlier in the MAD, rendering into a positive indicator of the stress state characteristic of uniaxial tension. Elastic-plastically broached layer of metal, located between the cutting teeth acquires additional damage under the stress state characteristic of the uniaxial tension process. The cutting tooth, located between the deformers, cuts the broached surface layer, with a new value of σ/T, characteristic of cutting the broached layer. The known index of the stress state of the cutting process is corrected using the principle of superposition, i.e. imposing additional tension on one of the axes. The total adjusted degree of shear strain corresponding to the cutting with MAD and EPL are used in the model of the work balance proposed by Ya.G. Usachev.

Conclusion

From the position of the dislocation theory and plastic deformation balance, theoretical studies do not fully identify and explain the reasons for improving the machinability of materials by cutting while these materials were pre-treated with surface plastic deformation.

Promising models for calculating the machinability of cutting with MAD and EPL should take into account the change in the stress state and the history of damage accumulation considering the change in the stress state and the corresponding change in the shear deformation degree at each stage of the combined action.

References

[1] Binder M, Klocke F, Doebbeler B 2017 *Abrasive wear behavior under metal cutting conditions* (Wear, Elsevier)
[2] Binder M, Klocke F, Doebbeler B 2017 An advanced numerical approach on tool wear simulation for tool and process design in metal cutting *Simulation modelling practice and theory* (Elsevier)
[3] Buchkremer S, Klocke F 2017 Compilation of a thermodynamics based process signature for the formation of residual surface stresses in metal cutting *Wear* (Elsevier)
[4] Gajrani K K, Sankar M R 2017 Past and current status of eco-friendly vegetable oil based metal cutting fluids *Materials Today: Proceedings* (Elsevier)
[5] Gok K, Sari H, Gok A, Neseli S 2017 Three-dimensional finite element modeling of effect on the cutting forces of rake angle and approach angle in milling *Proceedings of the national academy of sciences of the united states of America* (journals.sagepub.com)
[6] Johansson D, Hägglund S, Bushlya V, Ståhl J E 2017 Assessment of Commonly used Tool Life Models in Metal Cutting *Procedia Manufacturing* (Elsevier)
[7] Kaselouris E, Papadoulis T, Variantza E 2017 A study of explicit numerical simulations in orthogonal metal cutting *Solid State Phenomena* (Trans Tech Publ.)
[8] Königs M, Brecher C 2018 Process-parallel virtual quality evaluation for metal cutting in series production *Procedia Manufacturing* (Elsevier)
[9] Krestyaninov P N, Khusainov R M, Yurasov S Y 2017 Selection of metal-cutting machines in operational design by means of PLM systems *Russian Engineering Research* (Springer)
[10] Low I M 2018 *Advances in ceramic matrix composites* (books.google.com)
[11] Malakizadi A, Hosseinkhani K, Mariano E 2017 Influence of friction models on FE simulation results of orthogonal cutting process *The International Journal of Advanced Manufacturing Technology* (Springer)
[12] Mang T, Freiler C, Hörner D 2017 Metalworking fluids *Lubricants and lubrication* (Wiley Online Library)
[13] Razak N H, Rahman M M 2017 Cutting force and chip formation in end milling operation when machining nickel-based superalloy, Hastelloy C-2000 *Journal of Mechanical Engineering Education* (jmes.ump.edu.my)
[14] Rodríguez J M, Jonsén P, Svoboda A 2017 Simulation of metal cutting using the particle finite-element method and a physically based plasticity model *Computational Particle Mechanics* (Springer)
[15] Salehi M, Schmitz T L, Copenhaver R, Haas R 2018 Probabilistic Prediction of Cutting and Ploughing Forces using Extended Kienzle Force Model in Orthogonal Turning Process *Procedia CIRP* (Elsevier)
[16] Shi H 2018 *Metal Cutting Theory: New Perspectives and New Approaches* (books.google.com)
[17] To S S, Wang V H, Lee W B 2018 Single Point Diamond Turning Technology *Materials Characterisation and Mechanism of Micro-Cutting in Ultra-Precision Diamond Turning* (Springer)
[18] Vereschaka A A, Grigoriev S, Sitnikov N N, Bublikov J I 2018 Effect produced by thickness of nanolayers of multilayer composite wear-resistant coating on tool life of metal-cutting tool in turning of steel AISI 321 *Procedia CIRP* (Elsevier)
[19] Zhang C, Lu J, Zhang F, Butt S I 2017 Identification of a new friction model at tool-chip interface in dry orthogonal cutting *The International Journal of Advanced Manufacturing Technology* (Springer)
[20] Zhang T, Jiang F, Yan L, Xu X 2018 Research on the size effect of specific cutting energy based on numerical simulation of single grit scratching *Journal of Manufacturing Science*

and Engineering (asmedigitalcollection.asme.org)

[21] Panchenko K P *Russian scientists – founders of metal cutting science.* (Moscow: MASHGIZ) p 480

[22] Kuznetsov A M 1975 *Technology of basis for creating treatment methods in engineering* (Moscow: MAMI) p 298

[23] Ivanova V C Gordienko L K, Gemenov V N 1965 *Dislocation role in hardening and fracture of metals* **(**Moscow: Nauka**)** p 180

[24] Bogatov A A Mizhiritskii O N, Smirnov S V 1984 *Plastic lifetime of metals in pressure treatment* (Moscow: Metallurgiya) p 144

[25] Ambrosimov S.K. 2016 *New progressive methods and tools of deforming-cutting broaching* Bulletin of the Lipetsk State Technical University vol 1 **(27)** 54-63

[26] Abrosimov S K 2011 *Experimental studies of deforming–cutting broaching, with advanced plastic deformation and elasto-plastic loading, elastic-plastic loading and oblique cutting Effective technology of mandreling, broaching and deforming-cutting treatment* (Moscow: Spektr) p 327

IOP Conf. Series: Materials Science and Engineering **483** (2019) 012073 doi:10.1088/1757-899X/483/1/012073

Using online update of distributional semantics models for decision-making support for concepts extraction in the domain ontology learning task

A Anikin, A Katyshev, M Denisov, V Smirnov, D Litovkin

Volgograd State Technical University, Russia

anton@anikin.name, poas@vstu.ru

Abstract. Most of the information processed by computer systems is presented in the form of text corpuses. The number of such texts (as well as the corpus as a whole) only increases with time, and therefore the word processing tasks remain relevant to this day. Ontology allows to describe semantics using domain concepts and relations between them [1, 2]. In the ontology learning task, the ontology is dependent on quality of corpus which may not be readily available. There are different approaches to creating ontologies (including the use of different tools and frameworks). This paper discusses the use of *word2vec* (group of related models that are used to produce word embeddings) *using online vocabulary update and extension of the original data corpus with additional training* for the domain concepts extraction to automate the domain ontology creation.

1. Introduction

So far, word2vec [3] can't increase the size of vocabulary after initial training. To handle unknown words, not in word2vec vocabulary, the model should be retrained with the updated documents.

Definitely, the quality of the construction depends not only on the parameters of the model itself, but also on the very composition of the body of texts on which the model will be built [4]. Currently, there are a huge number of already constructed markup models, which are based on a huge corps of articles, books, etc. (for example, Google News word2vec model, Russian Distribution Thesaurus (RDT) Word2Vec model [5], etc). They contain ready-marked vectors of words and links between them.

But for a specific domain (for example, object-oriented programming), the data constructed models will be somewhat abstract and fuzzy. For the basic terms of our subject domain discussed in the models above, links with other words will be less clear and meaningful (for example, in Google News completely different sources from different domains are used, so some sense of the words this week will be somewhat blurred and distorted).

In this paper we consider the methods of constructing online dictionaries. It is based on the additional training of the already existing mark-up model with the corpus of texts of a specific domain that the user needs at the moment. We assume that when the already markup model of a huge corpus is completed with the corpus of texts of a specific domain (or with the imposition of its created model of this specific domain on the already prepared, markup model of a huge corpus), the links between the basic concepts of this specific domain will become more meaningful and adequate.

2. Background work

In this paper, we consider the additional training on the following ready-made markup models:

- RDT-model, built on the basis of a collection of books in Russian (150 GB text);
- Wikipedia model built on the basis of Russian Wikipedia articles dated October 1, 2018.

Also in this paper we consider the construction of a specific domain model: object-oriented programming, object-oriented design patterns. The basis of this subject domain is the following books:

- Elementary design patterns [Elementarniye shablony proektirovaniya] (Jason McColm Smith);
- Expressive JavaScript [Vyrazitelniy JavaScript] (Marain Haverbeke);
- Corporate Application Templates [Shablony korporativnyh prilojeniy] (Martin Fowler);
- Programming technologies [Tehnologii programmirovaniya] (Kamaev V. A.);
- Object-oriented programming, analysis and design [Ob"ektno-orientoravannoe programmirovanie, analiz i dizain] (V. Mukhortov, V. Yu. Rylov);
- Dive into design patterns [Pogruzhenie v patterny proektirovaniya] (Alexander Shvets);
- Object-oriented design techniques. Design patterns [Priemy ob"ektno-orientoravannogo proektirovaniya. Patterny proektirovaniya] (E. Gamma, R. Helm, R. Johnson, J. Vlissides);
- Design patterns on the NET platform [Patterny proektirovaniya na platforme NET] (Teplyakov S.);
- Object-oriented thinking [Ob"ektno-orientoravannoe myshlenie] (Matt Weisfeld);
- Object-oriented analysis and design with examples of applications in C++ [Ob"ektno-orientoravanniy analiz i dizain s primerami prilozheniy na C++] (Buch G.);
- Application of design patterns [Primenenie shablonov proektirovaniya] (J. Vlissides).

These books were used in a normalized format (all words are singular, masculine, lowercase: text without punctuation; each sentence starts on a new line).

To build a model of this specific domain based on the above books, the following approaches were used:

- "intersection" of the constructed model of a specific domain with a ready markup huge model (in the RDT-model) using the online dictionary;
- the extension of the original data corpus with additional training (Wikipedia model).

3. State of work

As mentioned above, to improve the quality of connections between terms and to improve the quality of highlighting related subject concepts, two approaches were used:

- online vocabulary update;
- extension of the original corpus data with additional training.

In this work, word2vec was used by Gensim. All ready-made markup word2vec models are stored as a KeyedVectors object - a module which implements word vectors and their similarity look-ups. Table 1 shows the advantages and disadvantages of this module.

First, create a word2vec model of a specific domain based on the above books.

Table 1. Using *KeyedVectors* vs. full model.

Capability	KeyedVectors	Full Model	Note
Continue training vectors	-	+	Need the full model to train or update vectors
Smaller objects	+	-	KeyedVectors are smaller and need less RAM, because they don't need to store the model state that enables training
Save/load from native word2vec format	+	-	Vectors exported by the Facebook and Google tools do not support further training, but you can still load them into KeyedVectors
Append new vectors	+	+	Add new entity-vector entries to the mapping dynamically
Concurrency	+	+	Thread-safe, allows concurrent vector queries
Shared RAM	+	+	Multiple processes can re-use the same data, keeping only a single copy in RAM
Fast load	+	+	Load data from disk instantaneously

```
my_model = Word2Vec(size=500, sg=1, min_count=5, window=7,
iter=5) files = os.listdir(data_path)
for current, file in enumerate(files, start=1): update =
    current != 1
    my_model.build_vocab(LineSentence(data_path + file), update)
    my_model.train(
      LineSentence(data_path + file),
      total_examples=my_model.corpus_count,
      epochs=my_model.epochs)
```

Results of the buiding of the model shows in table 2

Table 2. Specific domain word2vec model.

Compared words	Similarity
klass + ob"ekt	0.5569945351865275
shablon + pattern	0.7296937667116223
izolirovat' + inkapsulirovat'	0.6953800021098984
pattern + stroitel'	0.5999575695010988
pattern + fabrichniy	0.47411129901295646

Consider both of these approaches in more detail and compare the results.

3.1. Online vocabulary update

According to the table 1 *KeyedVectors* does not support additional training of the markup model. To change the weights of the already marked vectors on the ready-made model, we can merge in an input-hidden weight matrix loaded from the original C word2vec-tool format, where it intersects with the current vocabulary. No words are added to the existing vocabulary, but intersecting words adopt the files weights, and non-intersecting words are left alone.

Below is the source code for online vocabulary update:

(1) Load the input-hidden weight matrix from the original C word2vec-tool format and precompute L2-normalized vectors:

```
pretrained_model = KeyedVectors.load_word2vec_format( pretrained_vectors,
  binary=True, unicode_errors='ignore')
pretrained_model.init_sims(replace=True)
```

(2) Build vocabulary from a sentences of each book:

```
my_model = Word2Vec(size=500, sg=1, min_count=5, window=7, iter=5) for
current, book in enumerate(books, start=1):
    is_update = current != 1
    my_model.build_vocab(LineSentence(books_path + book), update=is_update) total_examples =
my_model.corpus_count
```

(3) Update (expand) the created vocabulary with words from the ready-made markup model:

```
my_model.build_vocab([list(pretrained_model.vocab.keys())],   update=True)
```

(4) Merge in an input-hidden weight matrix loaded from the original C word2vec-tool format, where it intersects with the current vocabulary:

```
my_model.intersect_word2vec_format(
  pretrained_vectors, binary=True, lockf=1.0,
  unicode_errors='ignore')
```

(5) Update the models neural weights from the corpus of books of specific domain:

```
for book in books: my_model.train(
        LineSentence(books_path + book),
        total_examples=total_examples, epochs=my_model.epochs)
```

Results of online vocabulary update shows in table 3

Table 3. Online vocabulary update model.

Compared words	Similarity
klass + ob"ekt	0.42248948304496003
shablon + pattern	0.43343555274649304
izolirovat' + inkapsulirovat'	0.39584950908535205
pattern + stroitel'	0.2778668522863343
pattern + fabrichniy	0.28110550019336966

3.2. Extension of the original corpus data

Extension of the original corpus data can be used when the original data corpus is available in the primary version (for example, there is the original Google news corpus, the original Wikipedia articles, the original set of books of artistic literature, etc.).

So, in this work we used the original set of articles of the Russian Wikipedia (version of October 1, 2018), taken from the Wikipedia dump.

All articles were converted to a single text file, on the basis of which the word2vec model was created. Then the constructed model was trained with the corpus of books of a specific domain.

Below is the source code for extension of the original Wikipedia corpus data:

(1) Loading the word2vec Wikipedia model:

```
#wiki_modelismodelwhichsupports
#onlinetrainingandgettingvectorsforvocabularywords
pretrained_model = Word2Vec.load(wiki_model)
```

(2) Update vocabulary model from a corpus of texts of specific domain and update the models neural weights from that:

```
for book in books:
    pretrained_model.build_vocab(LineSentence(books_path + book), update=True) pretrained_model.train(
      LineSentence(books_path + book),
      total_examples=pretrained_model.corpus_count,
      epochs=pretrained_model.epochs)
```

Results of extension of the original corpus data shows in table 4.

Table 4. Extension of the original Wikipedia corpus data model.

Compared words	Similarity
klass + ob"ekt	0.7179621516632567
shablon + pattern	0.6129935413162676
izolirovat' + inkapsulirovat'	0.5621600231928434
pattern + stroitel'	0.46872551920656125
pattern + fabrichniy	0.5412095954734097

4. Comparison of the results

According to the results presented in the tables 2, 3 and 4 it can be concluded that the data obtained during the additional training of the RDT-model showed worse measures of similarity of key terms of the domain than the data obtained when creating a simple model of a specific domain.

This is primarily due to the original corpus of the books that were used to create the RDT- model. This model is based on a huge number of books of different types: encyclopedias, fiction, educational literature, articles, etc. Therefore, the results were less accurate than expected.

Especially worth noting the results obtained in the construction of an extended model of Russian Wikipedia. The similarity of terms such as *klass* and *ob"ekt* has increased significantly. However, the similarity of other pairs of terms (for example, *shablon* and *pattern*) decreased slightly.

We also analyzed the top most similar words from a variety of domain terms (such as *klass, ob"ekt, shablon, pattern, inkapsulaciya, abstrakciya*, etc). Several other terms turned out to be more similar to these words, however, they are synonymous with these word pairs, which were considered in the tables 2, 3 and 4.

The table 5 shows some of the most similar words to term *inkapsulirovat'*.

Table 5. Most similar words with *inkapsulirovat'*.

Similar word	Similarity
skonfigurirovat'	0.8240861892700195
abstragirovat'	0.816191554069519
instancirovat'	0.8075761795043945
specificirovat'	0.8035067915916443
abstragirovat'sya	0.8031325936317444

It is preferable to use Wikipedia models for analyzing a more specific domain and building its ontology on the basis of an already prepared markup model, since this model contains a more accurate description of articles.

5. Conclusions

Methods of additional training of ready-made markup models considered in this paper have both advantages and disadvantages (and in general they are interchangeable).

The main advantages of the online vocabulary update are

- relatively simple using (we only need to build a basic dictionary of the specific domain we need, and make its intersection with the existing ready-made markup model);
- easy connection with any ready-made word2vec model (for online vocabulary update you can use any word2vec model with different parameters).

As the main disadvantage, we can highlight the absence of the necessary ready-made model. In the free access may not be the marked up model, on the basis of which we need to build a model of a more specific domain (for example, we need a ready-made word2vec model built on the corpus of all historical notes of the XVII century).

The main advantages of the extension of the original corpus data are

- freedom in choosing the original corpus (we can independently select the necessary corpus);
- simpler opportunity for additional training (we do not operate with *KeyedVectors*, but with the full model, so we do not need to build basic vocabulary, but we can immediately proceed to train full model with our corpus of texts of the specific domain).

The main disadvantage is the long time for building the original large model because of the huge data corpus based on which this model is created (for example, it took a little less than a day to create a model of Russian Wikipedia articles: 10 hours to create a single text file with all articles of Russian Wikipedia and another 10 hours for direct creation of Russian Wikipedia word2vec model).

The results of the work can be used for the domain ontology learning task to automate the domain concepts extraction for the ontology [1].

6. Acknowledgements

This paper presents the results of research carried out under the RFBR grant 18-07-00032 "Intelligent support of decision making of knowledge management for learning and scientific research based on the collaborative creation and reuse of the domain information space and ontology knowledge representation model".

References

[1] Anikin A, Litovkin D, Kultsova M and Sarkisova E 2016 Ontology-based collaborative development of domain information space for learning and scientific research *Knowledge Engineering and Semantic Web: 7th International Conference, KESW 2016, Prague, Czech Republic, September 21-23, 2016, Proceedings* 301–315

[2] Anikin A, Sychev O and Gurtovoy V 2019 Multi-level modeling of structural elements of natural language texts and its applications *Biologically Inspired Cognitive Architectures 2018* 1–8

[3] Mikolov T, Chen K, Corrado G and Dean J 2013 Efficient estimation of word representations in vector space vol abs/1301.3781 (*Preprint* 1301.3781)

[4] Venu S H, Mohan V, Urkalan K and TV G 2017 Unsupervised domain ontology learning from text 132–143

[5] Panchenko A, Ustalov D, Arefyev N, Paperno D, Konstantinova N, Loukachevitch N, Biemann C 2017 Human and Machine Judgements for Russian Semantic Relatedness *Analysis of Images, Social Networks and Texts: 5th International Conference, AIST 2016* 221–235

IOP Conf. Series: Materials Science and Engineering **483** (2019) 012074 doi:10.1088/1757-899X/483/1/012074

Ontology-based approach to decision-making support of conceptual domain models creating and using in learning and scientific research

Anton Anikin, Dmitry Litovkin, Elena Sarkisova, Tatyana Petrova, Marina Kultsova

Volgograd State Technical University, 28 , Lenina avenue, the city of Volgograd, 400005, Russia

anton@anikin.name, poas@vstu.ru

Abstract. The paper presents an approach to knowledge management in learning and scientific research that allows increasing the availability of expert knowledge and reducing semantic noise during knowledge transfer. The availability of expert knowledge is ensured by transforming them from implicit into explicit form (in the form of Semantic Link Network). The reduction of semantic noise is achieved through the integration of knowledge in different forms and their personalization for different groups of knowledge recipients. In the paper, the tasks of decision-making support are formulated which should be performed by the knowledge management system on the base of the proposed approach.

1. Introduction

In learning and scientific research, knowledge management is the main and most important process. Knowledge management implies creation (identifying as well as stimulating the creation) of new knowledge, storage, transfer (and sharing), and knowledge applying [1]. Explicit knowledge is stored in various types of documents - textbooks, books, presentations, diagrams, audio and video works, etc. Tacit knowledge (for example, a culture of thinking, experience, skills, intuition) is stored in the brain neural structures of teachers and researchers and it is difficult to fix these types of knowledge on the material media. There are also potentially explicit knowledge that is not yet fixed in material form but can be transformed into some explicit form. Usually, for the acquisition of explicit knowledge, the methods of "collecting stuff (content)" are used, meanwhile, for implicit knowledge - the methods of "connecting people". This paper considers the problem of decision-making support in the context of management of explicit and potentially explicit knowledge.

Main participants of the knowledge management process in learning are teacher and student. Teachers usually act as experts, i.e. the owners of domain knowledge. They also act as knowledge managers, who know where knowledge is located and who connect people with knowledge to those who need it. They capture and codify tacit knowledge so as facilitate its reuse, and transform user needs into knowledge requirements. Students usually act as recipients (users) of knowledge. In scientific research, some researchers participating in the knowledge management process act as experts and knowledge managers, while others act as recipients. Each participant of the knowledge management process has an individual cognitive style [2] and current knowledge of the domain, which affect the efficiency of creating and transferring knowledge.

2. Knowledge Management Issues in Learning and Scientific Research

In learning and scientific research, two main knowledge sources are available - document collection and expert knowledge (knowledge of teachers or researchers). Often, a collection consists of a large number of disparate, heterogeneous and contradictory documents. Documents can have different forms of information presentation (text, numeric, graphic, combined, etc.). They are usually presented in a well-known informal language. Its main advantage is high availability but the disadvantage is a rather high semantic noise. Semantic communication noise is a type of disturbance in message transmission that interferes with the message interpretation. It is generated by the content or semantic errors, and message distortions during their encoding/decoding. Semantic noise arises from the cultural factors, incorrect use of the language, and also perceptual filters or psychological deflectors [3].

Expert knowledge is implicit and potentially explicit knowledge. Implicit knowledge can only be gained by connecting people (for example, telling the stories, sharing and discussing the problems and opportunities, discussing the best practices, and talking over the lessons learned, etc.). The main disadvantage of this knowledge source is non-replicability.

Potentially explicit expert knowledge can be transformed into an explicit one, however, this process implies the advantages and disadvantages inherent for explicit knowledge. Also, experts often can not code their knowledge in explicit form effectively.

3. State of the Art: Approaches and Tools for Knowledge Management

There is a wide range of various tools for support of knowledge management [1]:

- electronic libraries (such as ACM Digital Library, IEEE Xplore Digital Library, SpringerLink, ISI Web of Sciences) are widely used in scientific research for access to high- quality peer-reviewed scientific papers, but they provide access only on subscription base and do not support semantic search in the subject domains;
- reference management tools (such as EndNote, Mendeley, RefWorks) record and utilize bibliographic citations for scholars and authors;
- mind mapping and concept mapping software tools for visual models creation (such as MindManager, MindMap, CMap) which allow visualizing the human mental models;
- cloud storage services (such as Dropbox, Google Drive service) offer secure cloud storing, file backup, and file sharing.

The current trend in Knowledge Management is to integrate various software tools in the complex knowledge management system. As a result, the following tools listed below are being created: Organizational Memory Information System (this system functions to provide means by which knowledge from the past is brought to bear on present activities, thus resulting in increased levels of effectiveness for the organization [4]), Semantic Web Portals (such as OntoWeb, Esperonto), Intelligent Information Systems [5]. These systems provide access not only to information resources of a specific domain but also to structured knowledge about the domain itself.

The semantic link network (SLN) has been widely used in the field of Knowledge Management. SLN is a network that represents semantic relations between concepts [6]. SLN can be considered not only as a tool for integrating heterogeneous information from different sources but as a significant knowledge source as such.

Among the applications developed for Knowledge Management, the most widely used method of knowledge representation and operating is an ontology. An ontology formalization is given in a logical (formal) language, which describes a structure of the world that considers all objects involved within the domain of study, their possible states and all relevant relationships between them [7]. Ontology representation in a formal language (such as OWL 2) allows: a) performing an unambiguous description of the ontology; b) proving its consistency; c) creating new knowledge that was not explicitly embedded in the ontology; d) handling various requests to ontology. However, logical language is complicated for human perception, that increases the semantic noise when encoding/decoding knowledge.

To improve the perception of ontology by a human, the ontology can be represented as a visual graph [8]. Such visual mental models use and combine the functions of the left and right brain hemispheres to achieve a holistic and visual presentation of the idea [9]. On the one hand, graphical interpretation of knowledge allows compressing the knowledge, and on the other hand - stimulating

IOP Conf. Series: Materials Science and Engineering **483** (2019) 012074 doi:10.1088/1757-899X/483/1/012074

thinking, that in general increases the efficiency of knowledge decoding, in particular during learning [10].

However, existing visual representations of ontology (such as Graphol, OWLGrEd, VOWL) are either difficult for human perception or limited in their expressive possibilities. One of the ways to improve the visual presentation of ontology is the use of the cognitive frames. The cognitive frame refers to the visualized fragment of ontology, which allows transmitting adequately the knowledge of a target concept to the user [11].

4. Ontology-based Decision-making Support of Knowledge Management in Learning and Scientific Research

4.1. Knowledge Storage

The knowledge memory consists of two knowledge repositories - the collection of documents and SLN collection.

The documents usually represented with well known informal language with large expressive possibilities (text, numerical, graphical, audio, video and combined forms of information presentation). It allows presenting different types of knowledge [10] for a wide range of users (knowledge recipients) with different needs and cognitive styles. On the other hand, the authors of these documents (domain experts) can express their knowledge in the form corresponded to their own cognitive style.

For each user group with a similar profile, an SLN is created for a certain domain. The SLN contains WHAT-knowledge [10] about the domain structured into a single mental model and answers some focus question. The focus question determines the context, main subject, and boundaries of the scope of knowledge being studied. Each SLN fragment (specific concept and/or specific link between specific concepts) is assigned an assessment of the importance of this fragment in terms of sensemaking of the SLN as a whole. In particular, this assessment is used for overview SLN when it is visualized. To reduce the semantic noise associated with encoding and decoding of SLN, the SNL is presented both in the visual language ORM 2 and the formal language OWL 2. The visual representation of the SLN is used by the human, and the formal representation is used by the computer.

The visual language ORM 2 is chosen because it is widely used to create conceptual database models and has great expressive capabilities, ORM 2 is more expressive than the UML.

OWL 2 is intended to describe knowledge in the form of ontologies, that can be interpreted by a special program - reasoner. Reasoner allows validating the consistency of knowledge, deriving new knowledge which was not embedded in the ontology explicitly, and responding variously to formalized queries to the ontology.

To reduce the semantic noise, SLN encoding/decoding using visual and formal languages should take into account the following two aspects. The first one: the coding and decoding of knowledge should be performed in terms of well-recognized structures of WHAT-knowledge, i.e. using WHAT-knowledge patterns. And the second one: because visual representation of the SLN is a strongly connected graph with many nodes, edges, and labels for various purposes, the interactive visualization is necessary. Ben Shneiderman formulated the visual information seeking Mantra: overview first, zoom and filter, then details on demand.

In addition to the knowledge repositories mentioned above, for each SLN a conceptual index is required, which stores the set of links between knowledge fragments in SLN and documents (and documents fragments) in which this knowledge is described. Each link is associated with an assessment that reflects the usefulness of the document as a description of the knowledge piece in terms of the user profile and focus question. The assessment takes into account the quality of knowledge coding for the user group and the semantic noise that occurs when knowledge decoding. As a result, the assessment of the usefulness of the same document will be different for users with different profiles and for different SLNs.

In addition, the fragment of SLN can be associated with the document collection that describes the fragment of knowledge at a sufficient level.

The specific SLN and specific conceptual index are applicable only to the one user group with a certain knowledge level, identical needs, and cognitive styles. The user profile includes: knowledge / competence, which the user already has, and their level; user cognitive style; knowledge / competence with their level required to be acquired. The user profile should be consistent with the focus question.

Thus, for each group of users (knowledge recipients) with the same profile, we have some cognitive information space (CIS) [12] that includes focus question, user profile, SLN, document collection, and conceptual index. The cognitive information space defines the representation of a certain domain from the point of view of this user group. It should be noted that a common document collection is used for the different cognitive information spaces. To search for a specific CIS, all its components (or their fragments) can be used, except the document content.

4.2. Knowledge Creation

Cognitive-information space creation process (Fig. 1, 2) includes the steps below.

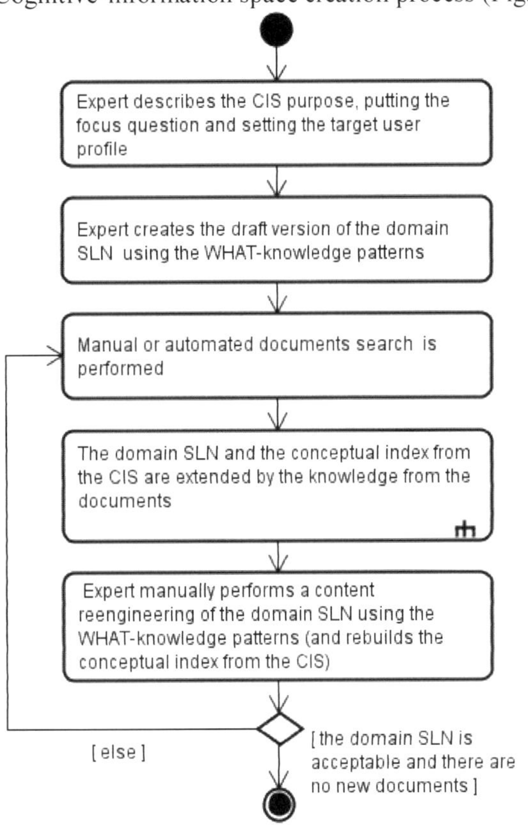

Figure 1. CIS creating

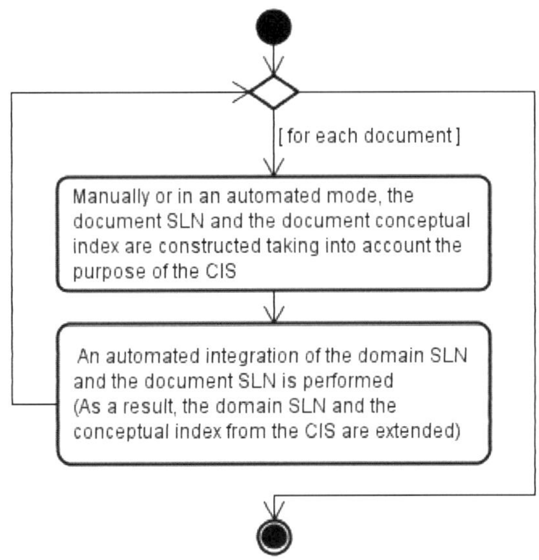

Figure 2. Extending domain SLN with the knowledge from the document

Step 1. Expert describes the CIS purpose, putting the focus question and setting the target user profile.

Step 2. Expert creates the draft version of the domain SLN (this step can be skipped) based on the expert knowledge for a specific user group. The SLN is created in a special editor with the ORM 2 visual language using the WHAT-knowledge patterns. The editor automatically translates the visual representation of SLN into formal OWL 2 ontology. In the process of SLN creation, reasoner automatically validates the ontology consistency, and editor performs partial validation of the knowledge completeness. If possible, detected errors are corrected by an expert, but their complete elimination is not necessary on this step.

Step 3. Manual or automated document search is performed for a more detailed description of knowledge from the SLN. The document search is performed in the already existing document repository, in electronic libraries, the Internet and other sources. This paper does not discuss the document search methods.

Step 4. For each document found and pre-selected manually or in an automated mode, its SLN and conceptual index are constructed taking into account the purpose of the CIS. Automated SLN building for the document is performed using ontology generation methods. For manually and automatically SLN building, WHAT-knowledge patterns are used. Additionally, validation on inconsistency and partly on the completeness of knowledge is carried out automatically. If possible, detected errors are corrected by an expert, but their complete elimination is not necessary on this step.

Step 5. An automated integration of the domain SLN and the document SLN is performed. As a result, the domain SLN and the conceptual index from the CIS are extended. Additionally, validation on inconsistency and partly on the completeness of knowledge is carried out automatically. If possible, detected errors are corrected by an expert, but their complete elimination is not necessary.

Step 6. If necessary, expert manually performs a content reengineering of the domain SLN using the WHAT-knowledge patterns. Additionally, validation on inconsistency and partly on the completeness of knowledge is carried out automatically. If possible, detected errors are corrected by an expert, but their complete elimination is not necessary.

Step 7. The process of building the domain SLN and the conceptual index continues until an acceptable result is obtained in terms of the purpose of the CIS, and all the knowledge inconsistencies are eliminated. After that, the CIS is published for users.

Step 8. After publication of the CIS, the expert continues to improve the domain SLN and the conceptual index by adding new documents. It can also perform content reengineering of the domain SLN.

In order to effectively implement the process of creating the CIS, the Knowledge Management System should provide decision-making support for the following tasks:

- The domain SLN and document SLN creation and content reengineering by an expert. The system should support a) the experts' use of the WHAT-knowledge patterns which are relevant to the encoded knowledge and b) the encoding of complete consistent knowledge in accordance with the WHAT-knowledge patterns.
- Automated document SLN creation. The system should: a) provide the expert with a preliminary set of concepts and relationships, and b) create document SLN based on the concepts and relationships chosen by the expert using WHAT-knowledge patterns. At the same time, the experts labor costs for reengineering of the SLN obtained should not exceed the labor costs for its creation from scratch.
- a) Automated creation of the preliminary document conceptual index without usefulness assessments; and b) automated redefining the conceptual index of the document in the reengineering of the document SLN.
- Automated integration of domain SLN and document SLN. System should: a) determine the alignment between the concepts and relationships of the domain SLN and the document SLN; b) correctly expand the domain SLN on the expert request by including concepts and relationships from the document SLN (according to the WHAT-knowledge patterns) and c) correctly extend the conceptual index of CIS when extending the domain SLN.

4.3. Knowledge Transferring and Sharing

CIS learning process (Fig. 3) includes the steps below.

Step 1. The user automatically searches for a suitable CIS. To search, the user sets his profile and keywords that describe his knowledge needs. As a result, a ranked list of CIS is generated. The CIS assessment takes into account the similarity between the profile of the current user and the profile of the target user of the CIS, the presence of keywords in the CIS and in the focus question. Next, the user reviews the CIS (focus question, target user profile and the SLN) and selects the best one.

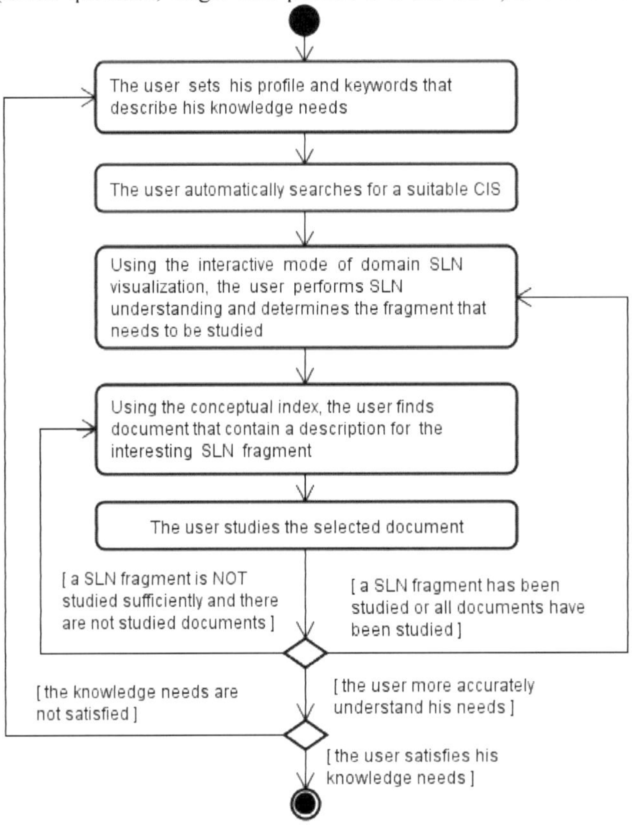

Figure 3. CIS learning

Step 2. Using the interactive mode of domain SLN visualization, the user performs a preliminary SLN understanding and determines the fragment that needs to be studied the first. The interactive visualization mode includes the following mechanisms: a) high-level SLN overview in the form of a graph, when only nodes and edges with the maximum importance assessments are displayed; b) show / hide nodes on the request of the user, taking into account the assessment of their importance; c) filtering nodes and edges of the graph; d) setting the focus point by setting the geometric center of the visible area of the graph; e) changing the visibility of the visual characteristics of nodes and edges.

Step 3. Using the conceptual index the user finds documents that contain a description for the interested SLN fragment. Looking through the usefulness assessments of the documents, and their content, the user decides which document should be studied the first.

Step 4. After studying the selected document, the user decides to return to Step 3 and continue studying other documents for the given SLN fragment, or return to Step 2 for re-understanding domain SLN and find a new fragment for study.

Step 5. The study of the domain SLN and related documents continues until the user satisfies his knowledge needs. On the other hand, after some immersion in the domain, the user can more accurately formulate his needs and return to Step 1.

In order to effectively implement the process of CIS learning Knowledge Management System should provide decision-making support for the following tasks:

- Automatic ranking CIS collection based on the search criteria. When evaluating the CIS, the system should take into account the user profile, his needs (expressed by a set of keywords), the purpose of the CIS (defined as the focus question and the target user profile), and the content of the domain SLN.

- Visualization of the domain SLN taking into account the settings specified by the user (such as level of details, focal point, geometric scale, concepts and relationships filtering), and the assessments of importance of knowledge fragments from the SLN specified by the expert.
- Ranking documents that describe knowledge pieces from the SLN, selected by the user. When evaluating a document, the following criteria should be taken into account: a) the amount of knowledge that describes the document, and b) its usefulness in describing this knowledge.

5. Conclusion

The paper presents an approach to knowledge management in learning and scientific research, which increases the availability of expert knowledge and reduces semantic noise during knowledge transferring.

The availability of expert knowledge is ensured by transforming them from an implicit into an explicit form (in the form of SLN).

The reduction of semantic noise is achieved through the integration of knowledge in different forms and their personalization for different user groups. We propose to integrate the following knowledge sources. The first knowledge source is the structured expert WHAT-knowledge, it is presented in the form of SLN, which is encoded both with the visual language ORM 2 and the formal language OWL 2. The second knowledge source is a document collection that contains various types of knowledge. Each document is presented in a well-known informal language in various forms (text, graphic, etc.). To integrate these knowledge sources, the conceptual index is used and, as a result, the cognitive information space is generated for a certain user group. This CIS is personalized because SLN and conceptual index are created for each user group.

The paper defines the decision-making tasks that the Knowledge Management System should support using the proposed approach.

6. Acknowledgements

This paper presents the results of research carried out under the RFBR grants 18-07-00032 "Intelligent support of decision making of knowledge management for learning and scientific research based on the collaborative creation and reuse of the domain information space and ontology knowledge representation model" and 18-47-340014 "Development of the mechanism of semantic zooming for the ontology geometric OWL graph for increasing the efficiency of decision making in the tasks of studying a new domain, knowledge accumulating and sharing".

References

[1] Gao T, Chai Y and Liu Y 2017 A review of knowledge management and future research trend *Proceedings of the 2Nd International Conference on Crowd Science and Engineering* ICCSE'17 (New York, NY, USA: ACM) 82–92

[2] Gavrilova T A, Leshcheva I A 2016 Conceptual knowledge structures and cognitive style *Psychology. Journal of Higher School of Economics* **13** 154–176

[3] Mironov D 2015 Informacionnyi shum i obrazovatel'nyi process *Vestnik Sankt-Peterburgskogo gosu- darstvennogo universiteta kul'tury i iskusstv* **4** 24–30

[4] Stein E W and Zwass V 1995 Actualizing organizational memory with information systems *Info. Sys. Research* **6** 85–117

[5] Zagorulko Y, Borovikova O and Zagorulko G 2018 Development of ontologies of scientific subject domains using ontology design patterns *Data Analytics and Management in Data Intensive Domains* 141–156

[6] Zhuge H 2011 Semantic linking through spaces for cyber-physical-socio intelligence: A methodology *Artificial Intelligence* **175** 988 – 1019

[7] Sharman R, Kishore R and Ramesh R 2006 *Ontologies: A Handbook of Principles, Concepts and Applications in Information Systems (Integrated Series in Information Systems)*
[8] Anikin A, Litovkin D, Kultsova M, Sarkisova E and Petrova T 2017 Ontology visualization: Approaches and software tools for visual representation of large ontologies in learning *Communications in Computer and Information Science* vol 754
[9] Jonassen D 1998 Designing constructivist learning environments *Instructional design models and strategies* (Mahwah, NJ, US: Lawrence Erlbaum) 215–239
[10] Kudryavtsev D, Gavrilova T 2017 From anarchy to system: A novel classification of visual knowledge codification techniques *Knowledge and Process Management* **24** 3–13
[11] Lomov P, Shishaev M 2014 Creating cognitive frames based on ontology design patterns for ontology visualization *Knowledge Engineering and the Semantic Web* 90–104
[12] Anikin A, Litovkin D, Kultsova M, Sarkisova E 2016 Ontology-based collaborative development of domain information space for learning and scientific research *Knowledge Engineering and Semantic Web: 7th International Conference, KESW 2016* 301–315

Developing intercultural communication competence of operating staff within corporate training program

T N Astafurova[1], O P Kozlova[1]

[1] Volgograd State Technical University, Russia

Abstract. Global integration and globalization processes forced numerous companies in non-English-speaking countries, including Russia, to enter foreign markets and establish joint ventures. Such state of affairs stimulated companies' reliance on English as the language of corporate discourse and international cooperation as concepts, terminology and legal documentation in English have already been shaped and do not cause discrepancies. Therefore joint ventures began to pay special attention to advanced training of production personnel in terms of mastering intercultural professional competence. The current paper deals with the ways of improving efficiency of production personnel within corporate vocational training. The paper is an attempt to present a multi-faceted model of mastering intercultural professional competence and describe its implementation in three modules converging sociocultural knowledge, linguistic skills and ability to cooperate effectively in English. The paper emphasizes the corporate program learning content, technologies of mastering sociocultural, lexico-grammatical and communicative components of intercultural competence. The paper presents the results of experimental teaching and confirms a greater efficiency of the developed model.

1. Introduction

Professional development is improving theoretical and practical knowledge of operating staff, enhancing their skills to meet the ever-increasing requirements. Advanced training can be implemented:

- within the framework of long- and short-term programs with profound study of innovations and modern problems of professional activity of the staff;
- through thematic and problem seminars, distance interactive programs, master classes, individual internships and workshops on scientific, technical and socio-economic problems aimed at accelerating the acquisition of skills, vital for successful performance in new conditions;

Corporate training in joint ventures appeared as a result of integration and internationalization of the world economy in the 21st century, which caused the use of English as the language of international cooperation and contributed to the emergence and development of this area of training in methodology of teaching foreign languages.

The choice of English as language of international cooperation is not accidental, since the global character of English is evidenced by its status in non-English-speaking countries. Hence, English is currently the leader among languages studied being a compulsory subject at all stages of education in more than 100 countries (Russia, Germany, China, Spain, Brazil, Egypt, etc.).

Integration and globalization have forced many companies of non-English-speaking countries, including Russia, to enter the foreign market and establish joint ventures in both English-speaking and non-English-speaking countries, which implies the use of English as "lingua franca", because conceptual and terminological apparatus and legal documentation in English have been already shaped and do not cause discrepancies or disagreements. This explains the fact that joint ventures gradually and voluntarily began to use English as language of "corporate communication", which to some extent proved to be a practical solution to the problem of "language barrier" [1, p. 380].D. Crystal gives a number of reasons of historical, political, practical, intellectual and economic nature contributing to

IOP Conf. Series: Materials Science and Engineering **483** (2019) 012075 doi:10.1088/1757-899X/483/1/012075

the strengthening of the international status of the English language [2, P. 106-107]: the Industrial revolution in England in the 18th century, the country's fast technological development and scientific progress in the 19-20 cc. turned English into the language of international intercourse, which conventions and rules, reflecting the structure, thesaurus, rhetoric and multilevel peculiarities of business language penetrated into the European languages, including Russian.

Pervasiveness of English as a result of world economy and business globalization is confirmed by the following facts:

- record keeping, sales and marketing require close interaction with other subsidiaries or affiliates of joint ventures, so their staff must be fluent in the language;
- technology transfer or exchange in joint ventures is often in English, especially with the development of the global network and electronic communications;
- the growth of joint ventures has spurred the need for hotel services, international transportation and all related services, which imposes certain requirements for the level of language proficiency among the staff of the hotel and tourism business of these countries [3, P.32].

2. The specificity of teaching English within in-company training

In this regard, in joint ventures, special attention is paid to improving the skills of personnel in terms of mastering their English-speaking skills and intercultural professional communication, i.e. intercultural competence. As the efficiency of the advanced training in English has increased significantly due to corporate programs, corporate language training offered mainly by private educational institutions, language schools and centres. Key aspects of the corporate training, its specificity and problems have triggered a response both among specialists in the field of teaching a foreign language, and in the business environment, namely among middle and senior managers responsible for making decisions about personnel advanced training and education, as well as the strategic development of companies and organizations.

The emerging demand for corporate training in a foreign language accelerated the intensive development of the private sector of Russian education. And while in the late nineties corporate foreign language teaching as a direction in the Russian methodology was in its infancy [4], at present, it has become one of the most sought-after and profitable directions of language learning for adults. Starting their work in a joint company, Russian production personnel faces the notion of a corporate culture, which is understood as "the set of norms of behaviour that the organization acquired in the process of adaptation to the external environment and internal integration that have shown their effectiveness and are shared by the majority of the company members"[5].

The undeniable advantage of corporate training is the focus on the needs of its staff, differentiation between the learning process and its regular monitoring. This is achieved by choosing the content of training and its intensity, depending on the specific needs of the joint venture. If it is necessary to improve the general level of language knowledge, the standard course of business English is chosen. If getting highly specialized knowledge in the shortest possible time is a priority, an individual training program is developed for each department. For example, managers are taught the art of making presentations and negotiations in a foreign language, the secretariat is studying the rules for business correspondence and clerical work, accountants master the financial terminology in a foreign language, while the legal department is getting acquainted with the English sublanguage of law.

The business English course is based on modern effective techniques, including interactive programs and language games. This enables not only quick mastering the language, but also strengthening team spirit and boosting the prestige of the enterprise in the eyes of partners, customers and potential employees. Depending on the needs, the most appropriate teaching methodology is selected. If the employees need to develop conversational skills, communicative methodology is advisable, for the purposes of drawing up contracts and conducting business correspondence the grammatical approach is more reasonable to use: "grammar ⍰ vocabulary ⍰ writing". If the goal is to increase the overall level of English, then it is better to combine grammar exercises, simulations, case studies, watching movies, and so on.

The company management can regularly receive reports on the attendance and performance of its operating staff. This makes it possible to take timely measures to improve the professional competence of the staff, since the entire training process is under control. As corporate training is commonly conducted at the company office, it saves time due to flexible schedule caused by heavy workload of staff during the periods of tight deadlines.

However, access to the international market of national enterprises is not only associated with overcoming the "language barrier", but also a "cultural barrier" that presses to be overcome in order to effectively interact in the production sector. This implies the need to acquire information about the differences and similarities of norms, values, behavioural conventions in different cultures, patterns of understanding another discourse and construction of utterances in a foreign language. This leads to the development of ethno-cultural pluralism, adaptation to a foreign culture without abandoning one's own and damaging its values, thus forming and developing intercultural communication [6].

The model of developing intercultural competence of operating staff

Intercultural communication is traditionally considered as "a combination of various forms of relations and communication between individuals and groups belonging to different cultures" [7, p. 142]. The efforts of higher education institutions have been directed to intercultural competence for half a century, but the results in this direction frequently leave much to be desired, since a linguistic competence at universities is formed on the basis of a universal lexico-grammatical code, usually without a cultural component. As a result, cultural lacunae, empathy, ethno-cultural pluralism and the ability to perceive the "alien" are not part of the traditional communicative competence and pose additional challenges for the staff of joint ventures.

This fact confirms the need for advanced training on intercultural communication, which are designed to fill this gap in the framework of personnel training programs to improve professional intercultural competence: courses in English, training in intercultural communication and adapting to another culture, etc. Large joint ventures demonstrate a positive and serious experience of corporate development and training programs: in some of them corporate universities have been established and are actively functioning. Their training programs take into account foreign language needs of personnel in the critical sectors of economy and business in order to increase its competitiveness outside Russia.

The study of the basic parameters that form the lingua-semiotic conceptual sphere of the compared cultures helps to prevent failures in intercultural communication. These parameters are reflected in:

- "communicative grammar" by E. Hall and his classification of cultures according to the time and information parameter (polychronous / monochronous, oriented to the present / past / future, high / low-context cultures, etc.) [8];
- "mental mapping" by G. Hofstede and his classification of cultures according to interpersonal and gender parameters (individualism / collectivism, vertical / horizontal cultures, feminine / masculine cultures, etc.) [9];
- the typology of business cultures by F. Trompenaars according to parameters of universalism / particularism, specificity / diffusion, affectivity / indifference, etc.) [10].

The analysis of the problems encountered in the course of acquiring an "alien" culture, cultural shock and acculturation raises the issue of targeted training of staff for the purposes of intercultural communication in the professional context, for successful performance in joint ventures, transnational companies and even in a foreign culture [11]. For this purpose it is not sufficient to develop only the knowledge of the language in trainees but also their "cross cultural sensitivity / empathy" [12]. The development of curricula for corporate training in English assumes, on the one hand, the preliminary selection and organization of training content, and, on the other hand, constant discussion with trainees and adjustment of the program to the specific needs of trainees [13, p. 83]. In this case, the content of training is considered as a unity of three components: linguistic, intercultural and professional, each of which affects the formation of the corporate training content.

These components are included in the multi-faceted model for acquiring intercultural competence of operating personnel, which is implemented in three modules / stages of training aimed at integrating "alien" cultural knowledge, language competence and ability to cooperate in the professional context:

1) acquiring "alien" sociocultural knowledge (mutual cognition),
2) developing mutual understanding (rapport),
3) mastering professional interaction (cooperation in a foreign language) .

At all stages of training the development of language skills and communicative competence occurs along with the socio-cultural aspects of professional interaction [14].

At the first stage trainees receive primary knowledge of sociocultural universals, according to which cultures are distinguished: the attitude to the environment, time, space, communicative styles, power distance, types of thinking, emotions in social interaction, decision making, strategy implementation, etc.

The second stage of training is aimed at enhancing the lexico-grammatical phenomena necessary for successful information exchange in different situations of professional interaction, discussing the conditions of production activity, summing up the results, agreeing on terms of contracts in English.

The third stage focuses on consolidating language knowledge, mastering communicative skills acquired by personnel at the previous stages of training. Additionally, trainees develop skills of interpreting hidden information. Thus, when approaching an agreement, partners often change the "grammatical tone" of the utterance, namely, an unreal condition changes for the real: *If the details were worked out, we would be prepared to ...* (at the beginning of the discussion) and *If you accept ... we shall ...* (at the end of the discussion). The hidden information in the statement *That is our standard price* is interpreted as *Ask for a discount*; *We do not normally deliver* is interpreted as *We sometimes do*, etc. Particular attention is also paid to wording rhetorical strategies, balance of interests, the comparative analysis of socio-cultural traditions, styles and peculiarities of communicative behavior of a foreign ethnos, that are "as important in teaching a foreign language as other components ⍰ reading, writing, speaking, comprehension and translation "[15: p. 107; 16].

The analysis of the results of the experimental training was carried out at its final stage and consisted of control tests and communicative tasks, the purpose of which was to assess the level of intercultural competence of trainees. The obtained data confirmed the effectiveness of the advanced training model. Qualitative indicators of the formation of intercultural professional competence components in trainees in group G1, where training was implemented within the developed model, is significantly higher than in group G2, where training was carried out according to the traditional methods (see Tables 1-3).

Table 1. Formation of *socio-cultural and behavioral knowledge*, aimed at mutual knowledge of partners.

Criteria for assessing formation of socio-cultural and behavioral knowledge	Placement test		Final test	
(First module)	G_1	G_2	G_1	G_2
Acquisition of knowledge of sociocultural universals relevant for foreign-language interaction	4%	4%	9%	6%
Recognition of language markers of behavioral strategies in cooperation and conflict	2%	1%	8%	2%
Clarification of a partner's intention through metacommunication signals	2%	1%	7%	4%
The general indicator of socio-cultural and behavioral knowledge acquisition	**8%**	**6%**	**24%**	**12%**

Table 2. Development of *language skills* aimed at mutual understanding of partners in the professional sphere.

Criteria for assessing language skills formation necessary for mutual understanding of business partners	Placement test		Final test	
(Second module)	G_1	G_2	G_1	G_2
Ability to correctly apply concepts and terminology of the professional sphere	2%	3%	20%	6%
Skills to correctly structure and share information	4%	5%	7%	6%
Skills to verbalize intentions and strategies of cooperation	2%	4%	5%	4%
The general indicator of the developed language skills aimed at mutual understanding of partners in the professional sphere	**8%**	**12%**	**32%**	**16%**

Table 3. Mastering *integrated skills of intercultural communication* aimed at interaction in the professional sphere.

Criteria for assessing integrated skills of intercultural professional communication (Third module)	Placement test		Final test	
	G_1	G_2	G_1	G_2
Skills to integrate lexico-grammatical skills, communicative and behavioral tactics	6%	4%	10%	7%
The ability to use rhetorical strategies of persuasion and positivity	4%	3%	9%	3%
The ability to achieve communicative and practical goals	4%	3%	9%	4%
The general indicator of developed integrated skills of intercultural professional communication	**14%**	**10%**	**28%**	**14%**

The results of training in each module are presented in the following diagrams reflecting the dynamics of developing components of intercultural, professional communication (Figures 1–4).

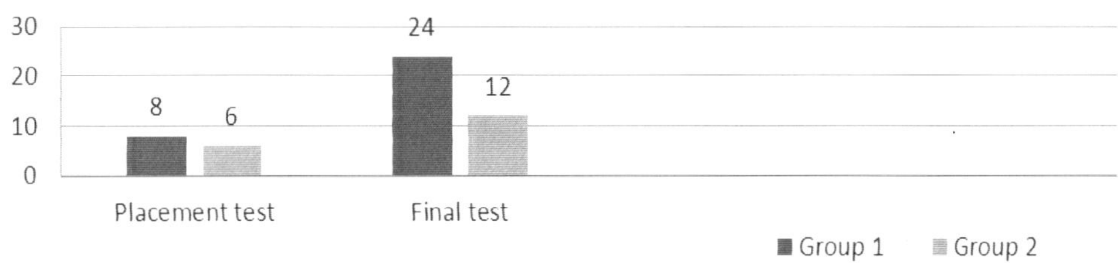

Figure 1. General indicator of acquired socio-cultural knowledge (Module 1)

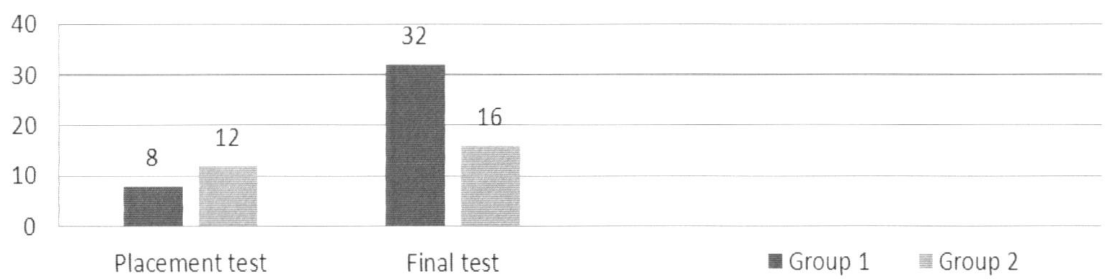

Figure 2. General indicator of acquired language skills (Module 2)

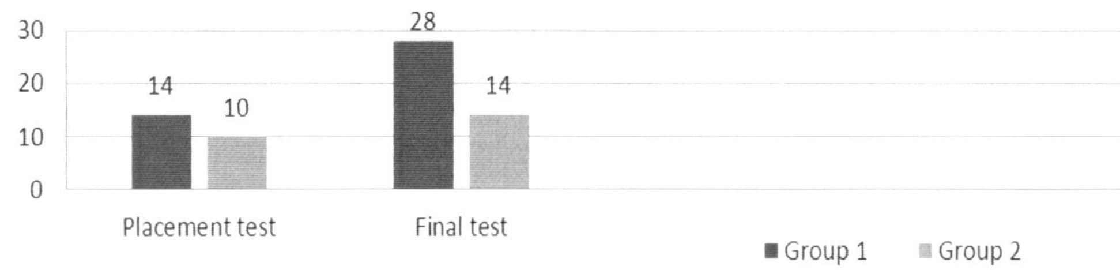

Figure 3. General indicator of acquired integrated communicative skills in intercultural professional communication (Module 3)

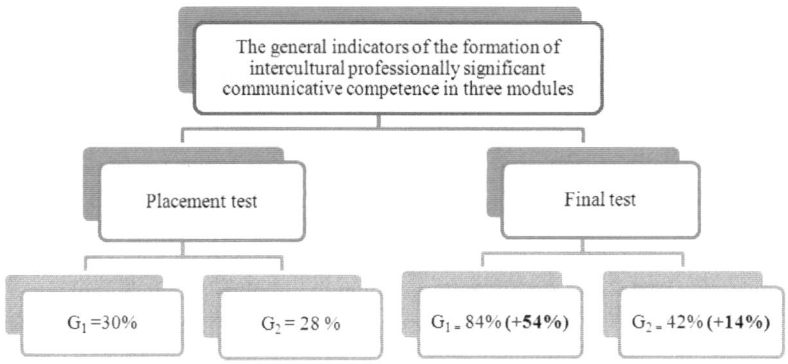

Figure 4. General indicators of acquired communicative competence in three modules

3. Conclusion

Integration and globalization of world economy entailed entering foreign markets and establishing joint ventures in non-English speaking countries, including Russia. This resulted in employing English as language of "corporate communication", since the terminology and legal documentation in English have already been fully formed and do not cause discrepancies or disagreements. Consequently, joint ventures special attention has been paid to mastering English-speaking skills, development of intercultural professional communication of their personnel, i.e. their intercultural competence.

The effectiveness of the educational process has increased significantly due to corporate English training programs aimed at meeting the needs of operating personnel, differentiating between learning process and its regular monitoring. However, development of intercultural professional competence is associated not only with overcoming the "language barrier" of personnel, but also with the "cultural barrier" that hardly creates serious errors in the process of intercultural interaction and is associated with a lack of information of socio-cultural, behavioural and lingua-semiotic character.

In this respect, it is particularly relevant to develop a multi-faceted model of corporate training in intercultural communication, implemented in three modules and focused on convergence of sociocultural knowledge, linguistic competence and ability to communicate effectively in English in professional situations. The analysis of the results of experimental teaching confirmed the advantages of the presented model for further development of intercultural professional communicative competence of operating staff.

References

[1] Marschan-Piekkari R, Welch D, Welch L. Adopting a common corporate language: IHRM implicastions. *The International Journal of Human Resource Management*. Routledge Vol 10 Issue 3 377-390

[2] Crystal D 1997 English as a Global Language. (Cambridge: Cambridge University Press) 150 p
[3] Graddol D 1997 The Future of English (London: The British Council) 64 p
[4] Belyaeva G E, Shuklina Z I 2006 Corporate program of English language teaching (SPb.: European House) 40 p
[5] Boonsong S 2018 Development of Learning Management in Moral Ethics and Code of Ethics of the Teaching Profession Course *IOP Conf. Ser.: Mater. Sci. Eng.* **306**
[6] Astafurova T N 1999 Business communication as an invariant of professional intercultural communication. Teaching a foreign language as a communicative interaction (Moscow) 35-43
[7] Grushevitskaya T G, Popkov V D and Sadokhin A P. 2003 Basics of intercultural communication: textbook for universities (Moscow: YUNITI-DAPA) 352 p
[8] Hall Ed T. 1987 The Hidden Differences: Studies in Intercultural Communication (Tokyo: Bungei Shunju) 155 p
[9] Hofstede G 1999 The Cultural Relativity of Organizational Practices and Theories. (Reader 12) ; in Samovar L A., Porter R E (Intercultural Communication: A Reader. Belmont, CA: Wadsworth Publishing Co) 405 p
[10] Trompenaars F. 1994 Riding the waves of culture: understanding diversity in global business (N.-Y) 216 p
[11] Hastjarjo S, Nuryana A 2018 Intercultural Communication Training in Vocational and Industrial Education Trainin *IOP Conf. Ser.: Mater. Sci. Eng.* **306**
[12] Roth Yu, Kopteltseva G 2001 Meetings on the verge of cultures: games and exercises for intercultural learning. Kaluga: Polygraph-Inform 188 p
[13] Kolesnikova I L, Dolgina O A 2001 English-Russian terminological reference book on methods of teaching foreign languages (Cambridge: Cambridge University Press; SPb Blitz) 223 p
[14] Astafurova T N 1997 Strategies of communicative behavior in professionally significant situations of intercultural communication (linguistic and didactic aspects): the author's abstract of dis. ... Dr. ped. sciences. 41 p
[15] Sternin I A 1996 Communicative behavior in the structure of national culture Ethnocultural specificity of language consciousness (Moscow) 97-112
[16] Liando N V F 2018 Vocational High School Students' Profile and their English Achievement *IOP Conf. Ser.: Mater. Sci. Eng.* 306

IOP Conf. Series: Materials Science and Engineering **483** (2019) 012076 doi:10.1088/1757-899X/483/1/012076

Modeling of object competitiveness based on hybrid method assessment

T L Bezrukova[1], G P Fomin[2], S S Kirillova[1], B A Bezrukov[1]

[1] The Voronezh State University of Forestry and Engineering named after G.F. Morozov, 394087, Russia, Voronezh, Timiriazeva st., 8
[2] Russian University of Economics named after G.V.Plekhanov, 117997, Russia, Moscow, Ctremiannyj pereulok, 36

E-mail: bezrukova_t_l@mail.ru

Abstract. The article covers the aggregation informal and mathematical methods in competitiveness of objects of different origin basing on hybrid assessment method. The given method includes principles of systemic analysis of structurization and quantification that leads to the analysis of the competitiveness problem and complex implementation of multicriteria method of optimization, and also methods of ranging of characteristics of the object to be assessed, methods of dual comparisons, methods of integral programming, Hungarian method, reduction, method of tests and errors. The feature of the method is the advantages integration of different quality methods of analysis with formal methods to obtain the quantitative characteristics of results of competitiveness for different consumer levels. The certain extracts of the big solved task are given together with a lot of single and integrated assessments. The certain parts of a huge task solved are also presented in the article. The result tables with conclusions are also presented here. The results change primary view of consumer, make mental outlook wider and change the attitude of consumer towards the competitiveness in particular. The description of transition from methods and models use to the assessment of object competitiveness is given. That shows the possibility of hybrid method application according to the object competitiveness of different origin.

1. Introduction

At present some organizations are more successful than the others. Some are in deep crises. That is the natural economical process. To provide the competitiveness of the enterprise it is needed to reach a certain level of profit that will help it to survive in future. It is evident, that the mechanism of enterprise adaptation to the changeable market conditions should be created. The mechanism of competitiveness management should be worked out to aggregate measures for systematic development of constituent parts of the object. Methodology of competitiveness assessment for practical use has not been worked out by the present time.

The basis of the competitiveness of the enterprise is the competitiveness of its product, that is its system generated component. All parts of the chain are concentrated on it. That is the basis of the entrepreneurial activity. [1] Goods are the basis of entrepreneurial activity. Business and market policy are defined by product, goods and services. It is evident that competitiveness assessment is advisable to do in complex according to the more accessible indexes and direction of activity. Competitiveness deflection means risks, and deflection value defines the program, content and necessary investment to provide the object competitiveness.

IOP Conf. Series: Materials Science and Engineering **483** (2019) 012076 doi:10.1088/1757-899X/483/1/012076

2.Theoretical, Informational, Empirical, and Methodological Grounds of the Research

Target assessment of competitiveness is to reveal the readiness of staff and individual worker to fulfill the particular kind of activity and also to reveal the level of potential ability for professional and occupational development of individual worker [2].

Competitiveness of enterprise means: sales volume cost and quantity, profit, share of cost of unsold products in store, competitiveness index, profitability, costs. Assessment of competitiveness of object is a multi-complex task that defines a good deal of factors of different origin. However, we need to carry out the given task for fulfillment of a number of activities such as finding of the main strategic ways for goods production, working out the firm staff training program to provide the competitiveness of the staff and the whole organization itself. The methodology of competitiveness has not been worked out properly for practical use by the present time. The complexity of origin of competitiveness specifies a number of methods of its assessment. [3]

Methods of competitiveness assessment are multifarious and are limited in usage. Competitiveness is preferable to assess in complex way according to the most accessible indexes and ways of activity. Deviation of competitiveness defines risk, but the value of deviation defines the program, content and investments needed to provide the competitiveness of the organization. Competitiveness of organization is defined by the following: sales volume cost and quantity, profit, share of cost of unsold products in store, competitiveness index, profitability, costs.

Assessment methods can be divided into three categories, they are: qualitative, quantitative and combined. [4] The most objective methods are quantitative because all the results are obtained in digits with the help of which it's easier to compare them. To get more objective result the combination of qualitative and quantitative methods is used. It's called combined method. [5]

However, the methods mentioned in the article do not allow of getting the objective results for competitiveness assessment of the organization, personnel, group of workers, staff of the division and department. Moreover, the methods are labour intensive at initial data collection. [6] That gave cause for usage of advantages of the existing qualitative and quantitative methods in combining them with the other methods. To eliminate the given faults in the existing models of competitiveness assessment the hybrid method of competitiveness assessment of object, organization, staff or product is proposed to apply. Multitude of qualitative and quantitative indexes of any object is suggested to use as a background. In general we shall work with indexes that define its ability to dominate in the specific sphere of activity, then we shall work with ranging with the calculation of weight next following and estimation of integral characteristics along the all levels of hierarchical indexes tree, that makes the competitiveness of object. All mentioned above made us formulate the hybrid method of competitiveness assessment from the point of systems analysis which is suitable for the objects of different origin. Nowadays problems of competitiveness assessment and conducting the optimization of investments are the most popular discussed among the bosses within international forums. The purpose of competitiveness assessment is to reveal the readiness of the staff and individuals the ability to fulfil exactly that kind of activity which they should do. And also the task is to reveal the potential of the workers for defining the professional and official growth.

3. Results

The following methods are worked out basing on contraction of criteria, indexes or characteristics where one scalar criteria instead of multitude of different origin partial criterion is considered. It is acquired by combination of partial criterion. There are multiplicative and additive methods of criteria contraction. The criteria must be commensurate, for example, standardized and weight criteria defined, that characterizes the importance of every criterion. Then we build a new efficiency function and the task of scalar criterion optimization is fulfilled [7].

In this case, the method of transition from several i.e. multitude indexes $P_1, P_2, \ldots, P_m$ to one specified by a new function $Q = \sum(V_j P_j)$ is called compression or the method of generalized criterion, where V_j are weight coefficient criteria, provided weight sum $\sum V_j = 1.0$ .B It's fairly evident that the more is V_j,the more contribution j index is being contributed to the integrated criterion Q.

In this case, for each quality index $p_j(a_i)$ of the object $a_i \in A$ its weight V_j is calculated defining its meaning, after that the weighed sum of these indexes can be considered as a summarized assessment of the competitiveness of the object ai:

$$K(a_i) = \sum_{i=1}^{n} V_j p_j(a_i) \tag{1}$$

The suggested type of competitiveness assessment can be formulated as a three – level structure and modify the existing methods for competitiveness assessment of product, organization or personnel [8]. As a framework it's necessary to use individual, single, group and integral aggregation indices (table 1). In this case at the beginning a hierarchical, usually a three – level structure of indices of the object being under consideration is been graphed.

Table 1. Analyses of the object characteristics.

	X_1	X_2	X_3	X_4	X_5	X_6	X_7			X_8	X_9
							$X_{7.1}$	$X_{7.2}$	$X_{7.3}$		
A₁	150	1998	589	1580	170	12.8	10.2	6.3	7.7	182	124
A₂	150	1998	591	1601	170	11.2	10.6	6.8	8.2	180	18
A₃	150	1998	564	1469	172	11.2	10.6	6.8	8.2	180	31
A₄	150	1998	463	1455	210	9.8	8.20	5.9	6.7	187	0
A₅	150	1998	415	1455	195	11.9	10.5	6.8	8.1	188	7
A₆	141	1997	410	1454	200	11.3	10.8	6.8	8.2	178	62
A₇	167	2384	420	1750	200	11.0	12.8	7.3	9.3	175	2
A₈	148	1998	410	1580	180	11.5	9.60	6.5	7.6	180	85
A₉	152	1798	405	1505	180	9.0	10.6	6.8	8.0	192	6
A₁₀	150	1390	470	1621	200	9.6	10.1	6.7	8.0	192	3

To assess competitiveness of the object and to choose the preferable one we distinguished some models of vehicles: A_1 – Honda CR-V, A_2 – Hyundai ix 35, A_3 – KIA Sportage, A_4 – Mazda CX-5, A_5 – Mitsubishi ASX; A_6 – Nissan Qashqai; A_7 – Opel Antara; A_8 – Toyota RAV 4; A_9 – Skoda Yeti; A_{10} – Volkswagen Tiguan, their characteristics are given in Table 1 according to the indices: X_1 – engine capacity (h.p.), X_2 – volume of engine (cm^3), X_3 – boot volume (l), X_4 – laden mass (kg), X_5 – clearance (mm), X_6 – acceleration up to 100 km/h, X_7 – petrol consumption (l/100 km): $X_{7.1}$ – town, $X_{7.2}$ – motorway, $X_{7.3}$ – mixed; X_8 – max speed (km/h), X_9 – hijacking (item).

To estimate the weight of each index we use the following formula

$$V_i = \frac{S_i}{n^2} \tag{2}$$

Ranging of the indices to define their meaningfulness we follow the rule due to which the domination of matrix elements is defined

$$b_{ij} = \begin{cases} 1, if\ P_i\ equivalent\ P_j, \\ 0, if\ P_i\ ess\ important\ than\ P_j \\ 2, if\ P_i\ dominates\ over\ P_j \end{cases} \tag{3}$$

Ranging operations and defining the weight of indices we conduct foe each group of indices and we put the down into the table 2, for example, economic, technical, comfort ability, esthetics, safety. [9, 10]

By-turn, each group contains the characteristics. We conduct the ranging and also define the weight – meaningfulness of each index.

The above-mentioned method allows us to conduct an activity of indices convolution converting into a single integral one considering the weight of each and their absolute meanings. It is important to mention that the high index value is not always good, sometimes it's vice versa. For example, fuel consumption of the vehicle, it's necessary to introduce 5 or 10 grade assessment of indices $Б_i$ (p_i) due to the scale "worse-better", where the worst index gets low grade. For instance, vehicle operation costs to be compared with sale proceeds. Indeed, the higher the content of grade the $Б_i(p_i)$ the better index [11, 12]. Table 2 shows interval indices values in grades. So, we transform the model of competitiveness of the object to the following

$$К(a_i) = \sum_{i=1}^{n} V_j Б(p_j(a_i)) \tag{4}$$

Table 2. Interval assessments of indices.

$Б_i$ (P_i)	**1**	**2**	**3**	**4**	**5**
X_1	140.0 – 145.0	146.0 – 151.0	152.0 – 157.0	158.0 – 163.0	164.0 – 170.0
X_2	1390.0 – 1589.0	1590.0 – 1789.0	1790.0 – 1989.0	1990.0 – 2189.0	2190.0 – 2390.0
X_3	400.0 – 439.0	440.0 – 479.0	480.0 – 519.0	520.0 – 559.0	560.0 – 600.0
X_4	170.0 – 177.0	178.0 – 185.0	186.0 – 193.0	194.0 – 201.0	202.0 – 210.0
X_5	13.00 – 12.2	12.10 – 11.40	11.30 – 10.6	10.50 – 9.80	9.70 – 9.00
X_6	13.00 – 12.0	11.90 – 11.00	10.90 – 10.0	9.90 – 9.00	8.90 – 8.00
X_7	175.0 – 179.0	180.0 – 183.0	184.0 – 187.0	188.0 – 191.0	192.0 – 195.0
X_8	124.0 – 100.0	99.00 – 75.00	74.00 – 50.00	49.00 – 25.00	24.00 – 0.00
X_9	140.0– 145.0	146.0 – 151.0	152.0 – 157.0	158.0 – 163.0	164.0 – 170.0

Thus, the method was created on the basis of qualitative and quantitative indexes of object application basing on requirement usage [13]. The hybrid methods comprises aggregation of a few methods of research: interview method, expert judgment, method of paired comparison, rank method, method of grade assessment, method of integer programming, Hungarian method, reduction method, trial and error method.

So we have a decisive rule: object a_i is more preferable than object a_j, if $К(a_i)>К(a_j)$. Then we find integral indices $К_э$,$К_т$, $К_к$,$К_{эс}$,$К_б$ all through the groups, after that – weight – meaningfulness $V_э$,$V_т$,$V_к$,$V_{эс}$,$V_б$ of group indices, and at last, we find the integral index of competitiveness according to the formula:

$$К(a) = \sum V_i K_i \tag{5}$$

Thus, we define competitiveness of the object in the large:

$$К(a) = V_э К_э + V_т К_т + V_к К_к + V_{эс} К_{эс} + V_б К_б \tag{6}$$

In the large the structure of object indices consists of several groups: technical, aesthetic, mechanical, safety, economical. Due to this we define the indices weight of group of characteristics by the method of paired comparison:

$$V_э = 0.36V_T = 0.28V_k = 0.04V_{эс} = 0.12V_в = 0.2К_б. \tag{7}$$

Thereupon we write the following target function:

$$К(a) = V_э К_э + V_т К_т + V_к К_к + V_{эс} К_{эс} + V_б К_б = 0.36V_т = 0.28V_k = 0.04V_{эс} = 0.12V_в = 0.2К_б \tag{8}$$

After that we conduct the ranging of each group of characteristics by the method of paired comparison, involving experts – specialists. We define their weights, grades of characteristics and integral grades due to their aggregates (table 3). It allows us to calculate the following:

$$\text{К}_{\text{э}} = 3.82\text{К}_{\text{т}} = 2.8\text{К}_{\text{к}} = 1.6\text{К}_{\text{эс}} = 3.2\text{К}_{\text{б}} = 1.1. \quad (9)$$

Basing on the mentioned above we define general competitiveness:

$$\text{К}(a) = V_{\text{э}}\text{К}_{\text{э}} + V_{\text{т}}\text{К}_{\text{т}} + V_{\text{к}}\text{К}_{\text{к}} + V_{\text{эс}}\text{К}_{\text{эс}} + V_{\text{б}}\text{К}_{\text{б}} = 0.36 * 3.82 + 0.28 * 2 + 0.04 * 2.6 + 0.12 * 3.2 + 0.2 * 2.2 = 2.822, \quad (10)$$

Table 3. Integral grades of indices.

	X_1	X_2	X_3	X_4	X_5	X_6	X_7	X_8	X_9
A_1	0.188	0.064	0.780	0.156	0.10	0.609	0.094	0.23	2.22
A_2	0.188	0.064	0.780	0.156	0.28	0.609	0.047	1.17	3.30
A_3	0.188	0.064	0.780	0.156	0.28	0.609	0.094	0.94	3.11
A_4	0.188	0.064	0.312	0.780	0.38	1.015	0.188	1.17	4.10
A_5	0.188	0.064	0.156	0.624	0.19	0.609	0.188	1.17	3.19
A_6	0.094	0.064	0.156	0.624	0.28	0.609	0.047	0.70	2.58
A_7	0.470	0.080	0.156	0.624	0.28	0.203	0.047	1.17	3.03
A_8	0.188	0.064	0.156	0.312	0.19	0.812	0.141	0.47	2.33
A_9	0.282	0.048	0.156	0.312	0.47	0.609	0.235	1.17	3.28
A_{10}	0.188	0.016	0. 312	0.624	0.47	0.609	0.235	1.17	3.11

Table 4 shows the following criteria of choice: Y_1 – correlation of price and quality (min), Y_2 – horse power cost (rub.), Y_3 – petrol consumption for 60000 km (rub.), Y_4 – cost of three Technical Checkup (rub.), Y_5 – Insurance (Compulsory Automobile Insurance) within 3 years (rub.), $Y_{6,7}$ – cost reduction within 3 years in% (thousand rub.), Y_8 – quality integral grades (max), Y_9 – total cost of possession.

Table 4. Calculations of choice criteria.

	Y_1	Y_2	Y_3	Y_4	Y_5	Y_6	Y_7	Y_8	Y_9
A_1	549346	8127	192780	30973	16632	31	841	2.219	618275
A_2	324332	7127	200340	18870	16632	42	620	3.296	684822
A_3	356995	7399	200340	26442	16632	43	632	3.109	720671
A_4	304177	8300	154908	26850	16632	30	871	4.093	571590
A_5	332287	7060	198450	23348	16632	25	794	3.187	503108
A_6	399534	7305	204120	29692	14256	38	638	2.578	639468
A_7	374340	6796	241920	27565	19008	31	783	3.032	640043
A_8	474023	7459	181440	26500	16632	31	762	2.329	566812
A_9	313528	6770	200340	18347	19008	40	618	3.282	649295
A_{10}	337837	7000	190890	25867	16632	41	602	3.108	681559

Various criteria given in the table 4 denote the choice A_5 – Mitsubishi ASX.

4. Conclusions and recommendations

The obtained results let us evaluate competitiveness of other objects to be compared. Thus we can choose one object out of multitude or formulate a set of choices in the long run. We can also conduct a comparison evaluation of competitiveness of organizations, enterprises, personnel. We can work out some management measures for various spheres of activity.

It's necessary to point out that the attention must be given to the analysis of risks in the system of organizations for it's a factor of competitiveness raise. Risk analysis is one of the widely spread in spheres of production and trading [14].

The given hybrid method can be used for competitiveness assessment of personnel where the etalon-indicators are the initial data. They are the requirements for every occupation (competence). Then the comparison of individual person of the staff is made both positive features (communicative skills, professional experience) and negative (aggressiveness, contentiousness, irritability). After that the comparison of the list of characteristics, ranging, weight defining is made. After that we proceed to the formulation of the integer-valued programming task which can be solved with the help of Hungarian method available for users in the internet. Consequently, the most complicated, force consuming and important is the collection of the initial data and the task formulation.

To be concluded, the hybrid method of competitiveness has been made basing on the implementation of quantitative and qualitative indices of object of different origin. The hybrid methodology includes a number of research methods and techniques: method of interview, method of expert assessment, method of paired comparisons, ranging method, grading evaluation, method of integer-valued programming, Hungarian method, method of reduction, method of trial and error. The hybrid method of competitiveness assessment has lots of advantages those are the various quality methods of analysis that means result comparison convenience all through the customers levels.

References

[1] Bezrukova T L, Stepanova Yu N, Shanin I I, Busarina Yu V and Nesterov S Yu 2017 *European Research Studies Journal* **20 (3)** 183-92
[2] Lizunkov V, Politsinskaya E, Malushko E, Kindaev A, Minin M 2018 Population of the world and regions as the principal energy consumer *International journal of energy economics and policy* **8 (3)** 250-257
[3] Gashenko I V, Romanov D G, Vokina S G, Bezrukova T L and Kozenko Y A 2016 *Contemporary Economics* **10 (4)** 363-72
[4] Han Y, Jeong J, Ko M H, Lee S and Kim J 2018 *ICIC Express Letters, Part B: Applications* **9 (4)** 339-46
[5] Ergunova O T, Lizunkov V G, Malushko E Yu, Marchuk V I, Ignatenko A Yu 2017 Forming the system of strategic innovation management at the high-tech engineering enterprises *IOP Conference Series: Materials Science and Engineering* **177 (1)** 012046
[6] Lau H C, Ip A, Lee C K M and Ho G T S 2018 *Benchmarking* **25 (7)** 2216-29
[7] Zakeri A, Saberi M, Hussain O K and Chang E 2018 *International Journal of Fuzzy Systems* **20 (4)** 1224-39
[8] Bobel I 2017 *Competitiveness Review* **27 (4)** 433-435
[9] Kim Y 2017 *European Journal of Operational Research* **262 (2)** 771-9
[10] Gornostaeva Z V, Povalyaeve V A, Tepina Y V, Trgulova N G and Kushnaryova I V 2018 *Espacios* **29 (1)** 21
[11] Ferreira R, Lizunkov V G, Politsinskaya E V 2017 Formation of entrepreneurial competencies of university graduates in conditions of transition to the universities of the third generation *Novosibirsk State Pedagogical University Bulletin* **7 (6)** 195-211
[12] Kuhlman C, Ramamurthy K N, Sattigeri P, Lozano F C, Mojsilovic A and Varshney K R 2017 *IBM Journal of Research and Development* **61 (6)** 81673366
[13] Caseiro N and Coelho A 2018 *Competitiveness Review* **28 (2)** 213-26
[14] Fedotov N G, Moiseev A V, Syemov A A, Lizunkov V G, Kindaev A Y 2017 New Methods of Three-Dimensional Images Recognition Based on Stochastic Geometry and Functional Analysis *IOP Conference Series: Materials Science and Engineering* **177 (1)** 012047

Towards a graph model application for automatic text processing in data management

E.G. Grigoryeva, L A Kochetova, Y V Pomelnikov, V V Popov, T V Shtelmakh

Volgograd State University, 100 Universitetsky Prosp.,Volgograd, 400062, Russia

E-mail: kochetova@volsu.ru

Abstract. Based on the two models, "bag-of-words" and graph model, the paper deals with the development of methods for automated text analysis with the purpose to classify natural language texts and randomly generated documents. Within "bag-of-words" model, the authors have found that the primary Zipf's law, which states that given some corpus of natural language utterances, the frequency of any word is inversely proportional to its rank in the frequency table, does not hold continuously true. Modifications to this law have been proposed that enable us to classify texts more efficiently. Using the graph model of the text, which takes into account the occurrence of two random words in a sentence, and the median degree of the vertices of the graph, the authors demonstrate that it can be applied to differentiate meaningless texts from meaningful ones even though the word lists of the two texts are identical.

Introduction

Automated text analysis and text mining methods have received a great deal of attention because of the remarkable increase of digital documents. Typical tasks involved in these two areas include text classification, information extraction, document summarization, text pattern mining etc. Most of them are based on text representation models which are used to represent text content. A common and standard approach to model text document is bag-of-words. This model is suitable for capturing word frequency, however structural and semantic information is ignored. Most existing text classification methods (and text mining methods at large) are based on representing the documents using the traditional vector space model [1]. Recently, graph-based models have emerged as alternatives to text representation model. As compared to the traditional models, graph representation is mathematical constructs that represents not only the content of a document but also the relationship among the keywords, therefore, it can model relationship and structural information effectively. A text can appropriately be represented as Graph using vertex (node) as feature term and edge relation can be significant relation between the feature terms [2]. Text representation using Graph model provides computations related to various operations like term weight, ranking which is helpful in many applications in information retrieval.

The application of Graph based representation of text elements provides processing of the information in various areas like document clustering, document classification, word sense disambiguation, prepositional phrase attachment. However, Graph algorithms or techniques need to be extended in order to capture the requirement and complexities of the applications.

As data can be represented by an array of texts, which requires structured representation, it is required to construct a model of 'useless' texts in order to solve the "garbage" filtering problem. In the article, we focus on two simple models. The first model A is based on randomly selected wordlist and punctuation marks. The second model B is randomly selected wordlist and punctuation marks, but for each element of the text probability of sampling is given. The purpose of the study is to define text characteristics that identify a text as "garbage" by applying the first or second model. Mathematically, this means building a function that calculates the value that is used to discriminate a text. We investigate modifications of the Zipf's laws (G.K. Zipf). In the first part of the article, the "bag of words" text model is used, and in the second part of the article a graph model is proposed.

1. Study based on the model "bag-of-words"

1.1. The primary Zipf's Law "rank-frequency"

Let us have a document D. For every word w from the document we can count the frequency of hits $k(w)$, i.e. the number of occurrences of this word in the document. The probability $p(w)$ of the word w occurrence in a random sentence in the document is calculated as the ratio of the number $k(w)$to the total number of words $N w$ in the document.

The words that occur in the document are ranked as their frequencies decrease and are numbered 1,2,3. The frequency sequence number is called the frequency rank. The most frequently occurring words will have a rank of 1, followed by 2 and so on. If two words have the same frequency, then their ranks are equal. The text word list will be compiled after lemmatization has been conducted.

The first Zipf's law is usually formulated as a hyperbolic dependence between the rank $r(w)$ of the word w and its frequency $k(w)$. This dependence can be presented as follows [3, p. 378]:

$$k(w)=A\cdot(r(w))^{-\gamma}, \tag{1}$$

where A and γ—some constant values. The parameter γ is usually close to 1 and there are good reasons for which it is convenient to assume that for many texts $\gamma = 1$ [3, p. 386]. If equality (1) is divided by the number Nw of all the words in the analyzed text and put $\gamma = 1$, then we arrive at the equality:

$$p(w)\cdot r(w)=C, \tag{2}$$

Where $C=A/Nw$ —constant value. The value of the constant C is dependent on the text topic and genre: stylistic features, language variance [3] and other features. Thus, it has been found that for the Russian language texts on electronics C=0.145 [5, p. 105]. The rationale for the formation of laws of this type can also be found in [6].

We will study the frequency characteristics of the text based on Zipf's law in the form (2), allowing us to reveal any properties of the natural text represented by the "bag of words" model. The experiments performed show that both for the individual texts and for the collection of texts significant deviations from this law are observed.

For all the words w of the rank r, the value obtained from the formula $C(w)=p(w)\cdot r(w)$ is the same and depends only on the value of the rank $r = r(w)$. Figures 1 and 2 show the dependence of $C(w)$ on the rank r graphs for 8 texts of classical Russian fiction. In the first case, word lists of full and short forms of adjectives were compiled, in the second case, word lists of all the used words were built. Figure 1 shows that most authors use adjectives from the first half of the range of ranks. In Figure 2, an increase in the value of the parameter Nw (the total number of words in the collection) and the number of ranks r leads to a decrease in $C(w)$.

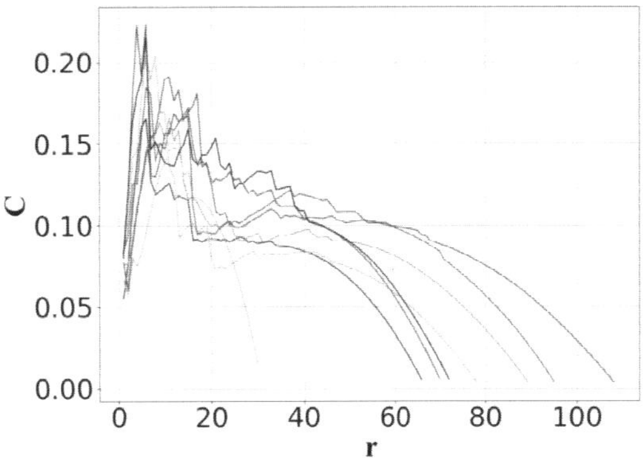

Figure 1. The graph of the function C=p·r for the collection of eight files and their mean value (0.088), the wordlist — adjectives in full and short forms.

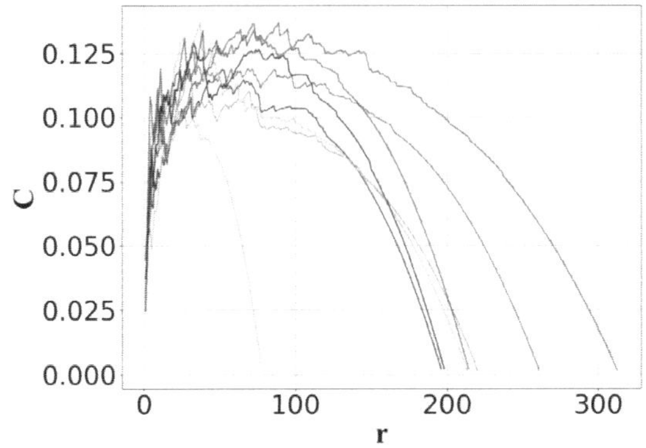

Figure2. The graph of the function *C=p·r* for 8 text files, the wordlist – all the words from the files.

The graphs of the sample files with the respective samples of words are shown below:

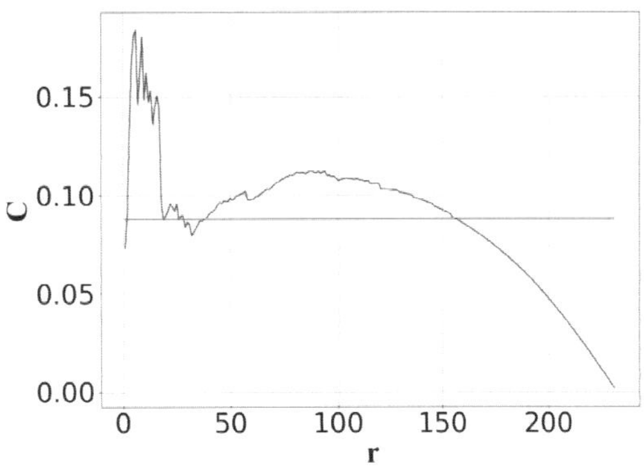

Figure3. The graph of the function *C=p·r* for the eight text files and the mean (0.088), the wordlist — adjectives in full and short forms.

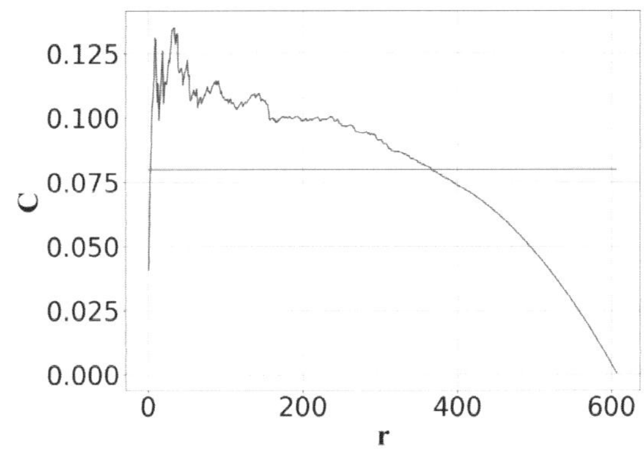

Figure4. The graph of the function *C=p·r* for the eight text files and the mean (0.08), the wordlist – all the words from the files.

We assumed that a significant increase in the number of ranks for the texts represented in Figure 3 indicates significant differences in the sets of adjectives used. Drawing up a collection (Figure 3 and Figure 4) leads to a more uniform distribution of frequencies according to the ranks and "blurring" of copyright differences in the texts. That is, the selection of certain language features from the collection of texts can serve to identify some of their individual characteristics. Rewriting of individual text files with the help of a synonymizer leads to very similar graphs of the function $C(r)$, i.e. the frequency dependence under study does not allow to identify the original text.

We also analyzed a collection of sixty Russian language text files written by Russian and foreign authors of approximately 36Mb.We imposed restrictions on the type of words - only strings of more than two characters were counted. The change in the value of C depending on r will have a character similar to the graphs presented for the samples in Figure 2.

We point out some features of the collection that are found when analyzing it by applying the first Zipf's law. Let us use for analysis $C(r)$ the second Zipf's law, which shows the hyperbolic distribution of the number of words of a certain frequency. The most salient part of the curve is formed by the words that significantly change the frequency characteristics of the text. Formula (2) is scaled by the value of K, equal to the ratio of the total number of words in the text to the one in the dictionary. A comparison of the values considered for the word list of the collection showed that the value $C(r)\cdot K$ would then be in a decreasing range between the mean value and the value of approximately 0.06–0.07.

1.2. Modification of the first Zipf's law parameters

Let us try to estimate numerically the deviation of a document from Zipf's law in the form (2). If we assume that this law is perfectly satisfied for the document D, then $C(w)=p(w)\cdot r(w)=C$ for any word w from D, where C is a constant. In this case, the mean MC of the values of $C(w)$ for all the words w of D is also equal to C. If there are deviations from the law, then the differences $C(w)–MC$ are not always equal to zero. Therefore, you introduce at least three different characteristics of the deviation from the law:

$$\text{Norm0}=\max\{|C(w)-MC|:w\in \text{D}\}, \quad (3)$$

$$\text{Norm1}=\sum\{|C(w)-MC|:w\in \text{D}\}, \quad (4)$$

$$\text{Norm3}=\{\sum|C(w)-MC|^2)^{1/2}: w\in \text{D}\}. \quad (5)$$

For all the nouns from the novel by F.M. Dostoevsky *The Brothers Karamazov* the following values were obtained by formulas (3) - (5): Norm0 = 0.11558, Norm1 = 94.8598, Norm2 = 1.94546.

We introduce other ways to determine the rank of a word, which reduces the total number of ranks for this document and, mostly important, deviations from Zipf's law are reduced.

1.2.1. Method 1. Let us choose an integer $m \geq 1$ and assume that the words w_1 and w_2 have the same rank if the numbers $k(w_1)$ and $k(w_2)$ if divided by m give the same partial numbers (that is, the integral parts of $k(w_1)$ and $k(w_2)$ / m are equal).

We can give another description. We divide the set of non-negative real numbers into half-intervals [0, m), [m, 2m), [2m, 3m), and so on. Then, the words w_1 and w_2 will have the same rank if the numbers $k(w_1)$ and $k(w_2)$ lie in the same half-interval.

If m = 1, then the ranks of the two words are equal only if $k(w_1) = k(w_2)$, that is, we get the usual definition of rank. As m grows, Norm0, Norm1, Norm2 values usually decrease. So, for the above text file, we have:

Table1. Deviations from Zipf's law in relation to the parameters.

	m=1	m=2	m=3	m=10
Norm0	0.11558	0.11105	0.10315	0.09822
Norm1	94.8598	74.6352	61.7326	37.8910
Norm2	1.94546	1.63000	1.38936	0.96081

Method 1 is used when we want to reduce the number of word groups with the same ranks. The way in which the words $w_1$andw_2 are assigned the same rank is not quite correct, if the difference in the numbers $k(w_1)$ and $k(w_2)$ in absolute values does not exceed the previously fixed number m. Let, for example, the frequencies of the words w_1, w_2, w_3 and w_4 are equal to 70, 69, 67 and 66, respectively, and m = 2. Then the adjacent words in this chain, namely the pairs (w_1, w_2), (w_2, w_3), and (w_3, w_4) must have the same rank, therefore the words at the edge of the chain (w_1, w_4) must have the same rank, however the frequency difference 70 –66 is greater than m and, therefore, their ranks must be different.

Let us consider another way of introducing a rank that considers words close, the frequency ratio of which differs little from unity.

1.2.2. Method 2. Let us choose a real number λ greater than one (but close to it) and divide the set of non-negative real numbers into half intervals $[1, \lambda)$, $[\lambda, \lambda^2)$, $[\lambda^2, \lambda^3)$, etc.

We assign the same rank to the words $w_1$and$w_2$if the numbers $k(w_1)$ and $k(w_2)$ lie in the same half-interval. Let us $\lambda = 1.02$. Then, if the frequencies $k(w_1)$ and $k(w_2)$ are not greater than 70, then the same ranks are assigned to the words w_1, $w_2$only if their frequencies coincide. Simultaneously, the same ranks will be assigned to the words, the frequencies of which fall into one of the half-intervals [380, 386), [511, 521], [4431, 4519).

If $\lambda = 1.02$, then for the above-mentioned novel by F.M. Dostoevsky, we have: Norm0 = 0.10098, Norm1 = 75.3674, Norm2 = 1.54230. The deviations obtained are smaller as compared with the deviations found by the traditional method of determining the word rank.

2. Study based on graph text model

To construct a graph model, we introduce the graph $G=(V,E)$, in which the set of vertices is formed by the set of all the words in the text, and the edges indicate the fact that a pair of words is found in one sentence. Moreover, each edge is assigned its weight b - the number of sentences in the text containing this pair of words. Some properties of such a graph were studied in [7]. Note that when the words in the text are mixed, its graph changes its structure. In accordance with the model B, the "garbage" text has the same frequency distribution of words as the source text. Therefore, we set the task of finding the characteristics of a text graph that would not be invariant, regardless of mixing. Among the available characteristics of the graph, we consider:

- average value of the graph edges weights,
- median value of the graph edge weights,
- the average value of the graph vertices degree,
- the median value of the graph vertices degree.

Note that the distribution of the degrees of the vertices of a random graph [8–10] is the subject of a theoretical study and has practical application, for example, when analyzing the graph structure of the Internet. In this case, it is assumed that the degrees have a density described by Pareto distribution: $p(x)=x^{-l}$ [11,12].

To put our model to the test, we took fiction texts with a restricted file size. We considered all the files from the specified directory and its subdirectories with a size limit of 60Kb. The results of the calculations are shown in the figures below.

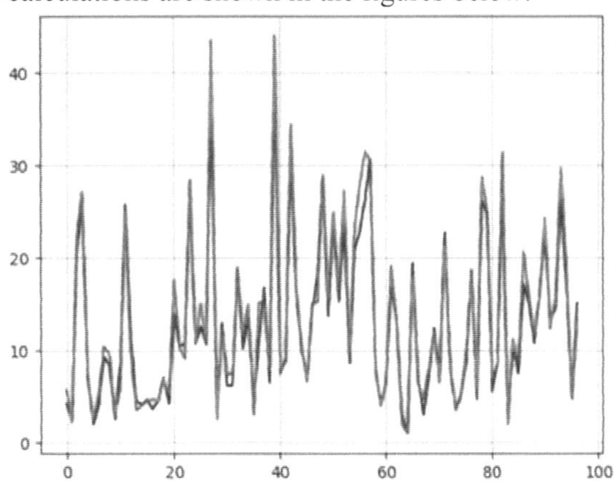

Figure 5. Average weight of the graph edges

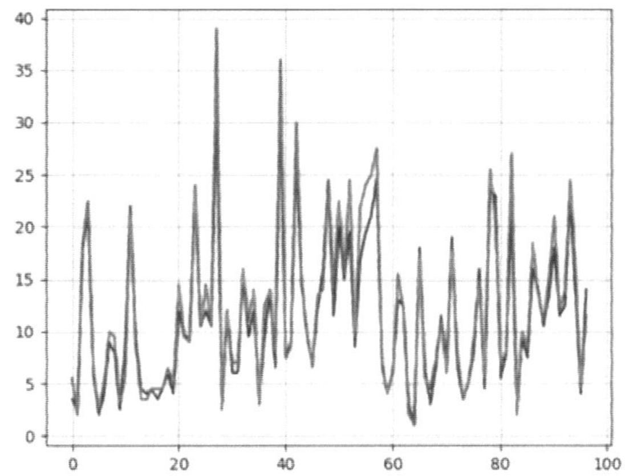

Figure 6. Median weight of the graph edges

In Figure 5, the number of texts in the corpus is plotted on the abscissa axis, and the average weight of the edges is plotted on the ordinate axis. The graph for the source texts is marked blue, the source texts obtained by random mixing are marked red. Calculations show that in 39 out of 97 texts, the first value does not exceed the second.

Figure 6 shows the result of calculating the median value of the edge weights for the same corpus of texts. In this case, in 40 out of 97 texts, the red graph is larger than the blue graph. In both cases considered, the value of the average or median value of the weights of the edges of the graph cannot be an example of the discriminant function of the text.

Figures 7 and 8 show calculations of the average values of the degrees of the vertices of the graph and the median of the degrees of the vertices of the graph.

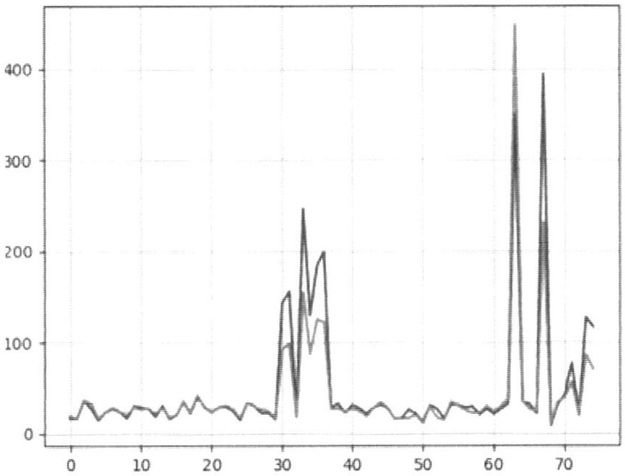

Figure7. The average value of the graph vertices degree.

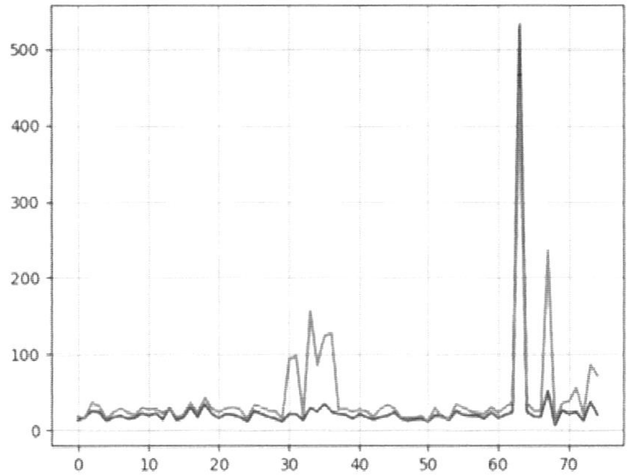

Figure8. The median value of the graph vertices degree.

Calculations show in 41out of 97 texts, the average value of the graph vertices for the source text is less than the same value for the mixed text. As in the case with the previous two characteristics, this value cannot be used as a criterion to classify the text as "garbage".

In Figure 8 the red graph mainly lies above the blue. Calculations show that this does not hold true for only 7 texts out of 97. In other words, the median value of the vertices of the text graph can be used to identify "garbage". In this case, the identification scheme may be as follows. For the source text, the vertices of the graph are calculated. Then, their median value is calculated. Further, the source text is subjected to multiple mixing, and after each such operation the median values of the new graph vertices are calculated. After that, the mean value of the calculated medians is taken and compared with the median value of the source text. If the resulting value is greater, the text is assumed to be real, otherwise it is classified as "garbage" [13].

Thus, the last value is a reliable parameter with which you can separate meaningful texts from the "garbage".

Conclusion

To sum up, computer experiments were conducted to find out the applicability of Zipf's laws to the bag-of-words text model. Using different wordlists based on certain type of words as well as modifying the parameters of Zipf's laws, the study identifies distinctive characteristics of meaningful texts. The constructed graph model of the text, in which links are established for all pairs of words in one piece of text, in this particular case, in one sentence, allowed to effectively investigate the parameters of the model and obtain meaningful information.

Acknowledgement

The reported research was funded by Russian Foundation for Basic Research and the government of Volgograd region, grant № 18-412-340007

References

[1] Mikolov T et. al. Efficient Estimation of Word Representations in Vector Space, arxiv.org. URL: http://arxiv.org/pdf/1301.3781.pdf

[2] Sonawane S S, Kulkarni P A 2014 *International Journal of Computer Applications* **96 (19)** 1-7
[3] Arapov M V 1981 *Systematic Research* (Moscow: Nauka)
[4] Bakulina M P 2007 *Descreet Analysis and Study of Operations* **14 (2)** 3-13
[5] Piotrovskiy R G 1975 *Text, Machine, Human* (Leningrad.: Nauka)
[6] Haitun S D 1983 *Naukometria. Sostoyanie I Perspektivi* (Moscow: Nauka)
[7] Grigoryeva E G., Klyachin V A., Pomelnikov Yu V and Popov V V *Vestnik Volgogradskogo gosudarstvennogo universiteta Seriya 2 Yazykoznanie* **16(2)** 58-67
[8] Fedotov N G, Moiseev A V, Syemov A A, Lizunkov V G, Kindaev A Y 2017 New Methods of Three-Dimensional Images Recognition Based on Stochastic Geometry and Functional Analysis *IOP Conference Series: Materials Science and Engineering* **177 (1)** 012047
[9] Raygorodskiy A M 2009 *Matematika v zadachakh* (Moscow: MTsNMO)
[10] Erdős P, Rényi A. 1959 *Publ. Math. Debrecen* **6** 290- 97
[11] Newman M E, Strogatz S H and Watts D J *Phys. Rev.* **64** 26-118
[12] Pavlov V Yu and Cheplukova I A 2008 Descreet Math **20 (3)** 3–18
[13] Pavlov Yu L 2009 Descreet Math .**21 (3)** 14–23

IOP Conf. Series: Materials Science and Engineering **483** (2019) 012078 doi:10.1088/1757-899X/483/1/012078

Simulation based technologies for Professional Training of Managers in the Arctic Region

M V Druzhinia[1], M V Zakharchenko[1,3], M A Ananina[1], E E Solovyova [2]

[1]Northern (Arctic) Federal University named after M V Lomonosov,17, Severnaya Dvina Emb. Arkhangelsk, 163002 Russia
[2]Arkhangelsk Regional Rescue Service named after Igor Polivany, 10, Leningradskiy prospekt, Arkhangelsk, 163002 Russia

[3]E-mail: marinesko10@gmail.com

Abstract. There are many studies which have arrived at the conclusion that simulation based technology makes a positive impact on learning process. This type of learning ensures active participation and leads to innovative solutions to problems. This study investigates the implementation of simulation based technology with the content in English on the course "Safety Training" for managers of different organizations. The results indicate that using simulation based technology appeared to be a useful tool for creating virtual-reality blended environment and promoting autonomous education. The proposed technology can be used for training and education in a variety of health care situations.

1. Introduction

In 1989, at the initiative of Finland, eight subarctic countries — the USSR (later transferred to Russia), the USA, Canada, Denmark, Norway, Sweden, Finland and Iceland — began to unite efforts to protect the unique nature of the northern polar zone. As a result, the ministers of these countries met in the Finnish city of Rovaniemi and adopted a Declaration on the protection of the environment in the Arctic in June 1991. After that, the Declaration on the establishment of the Arctic Council was adopted on September 19, 1996 in Ottawa (Canada) by representatives of eight Arctic states. The Arctic Council aimed to preserve the integrity of the unique Arctic lands [20]. Russia has successfully introduced work on prevention and preparedness for responding to natural and man-made emergencies. Rescue Centers were created on the territory of northern latitudes, "The Safe City" program complexes were introduced, and the leaders and personnel of the facilities and the public were constantly held in the field of civil defense and emergency protection.

The Arkhangelsk Regional Rescue Service named after Igor Polivany carried out educational courses and professional development of managers and specialists of the near-arctic region in the field of civil defense, protection against emergency situations, and ensuring the safety of people in harsh climatic conditions in the international context. In this regard, it became necessary for training managers capable to organize emergency activity in the process of international communication. Therefore, it was supposed to use simulation based technologies in English content as means of gaining experience to insure safety in risky situations. Simulation based technologies teach managers about realistic decision-making situations involving various changers such as formulating solutions strategies, making decisions by taking into account not only economic but social and psychological considerations and simultaneously improve their language skills.

The main objective of this paper is to identify efficiency of implementation simulation based technologies with the content in English on the course "Safety Training" for managers of international organisations. According the aim, the following issues are to be solved in the article: advances of the simulation based technology with the content in English for managers training, the methods and analysis of the experimental data. The solution of these issues will confirm our presumption about effectiveness of simulation based technologies implementation in "Safety Training" course for managers.

2. Literature review

A number of studies support the effectiveness of simulations for teaching and learning [1, 2, 3, 4, 9, 10, 11, 14, 16, 18]. Simulation based learning approaches aim to imitate a system, entity, phenomenon, or process. They attempt to represent or predict aspects of the behavior of the problem or issue being studied. Simulation can allow experiments to be conducted within a fictitious situation to show the real behaviors and outcomes of possible conditions.

The roots of global simulation technology went back to European reformist pedagogy of the late XIX - early XX century. In the French-speaking countries, the founder of this direction was Selesten Frene. In the pedagogical personality-oriented system of S. Frene, self-development, self-activation, self-realization and autonomy of students, the qualities that can only be achieved through an independent and collaborative process of learning [7]. The main developers of this method are Francis Debuse, Jean-Marc Kar and Francis Yesch [21].

F. Debuse gave the following definition of a global simulation method: "A global simulation is a script that allows a group of up to 30 people to create their own environment, there participants have to use their language skills interacting with each other in this environment and as a result of which they transform a thematic space into a communicative theater" [5].

According to B.E. Avzhanova, educational simulation is a structured scenario with a detailed system of rules, tasks and strategies pursuing a specific goal - the formation of specific competencies that can be transferred directly to the real world [17].

A typology of simulation types derived from the findings of Jonathan Lean, Jonathan Moizer, Michael Towler & Caroline Abbey studies define three types: Computer – based (Gaming situation, Training situation, Modeling situation); None – Computer games (Role Play, Educational Games) [12].Virtual Reality (VR) as a branch of computer simulation, which developed so fast, has been used in various fields such as artistic creation, medical science, entertainment, aviation, space-fight, emergency drill and so on. VR simulation is to generate immersive environments from which users can experience unique insights into the way the real world works [13]. The concept of VR was brought up over fifty years ago when the first immersive human-computer interaction (HCI) named "Man-Machine Graphical Communication System" was invented. The formal term of VR was put up in 1989 [23].VR attempts to replace a user's perception of the surrounding world with a computer-generated artificial 3D environment.

There are many studies which have arrived at the conclusion that VR makes a positive impact on learning process. Xiao Lia, Wen Yib have developed the idea that virtual environment with visual and immersive aids let users feel a "real" sensation. [13]. As reported by Salzmanetal, the higher the sense of presence trainees feel will cause higher motivation which might motivate trainees to work harder to grasp deeper learning [19]. Sh. Pedram, S. Palmisano in their research have concluded that VR technology should be used to train for various operations and dangerous circumstances where it is believed that training objectives cannot be achieved easily or the cost will prohibitive [19]. As stated by Meadows: "When I hear, I forget; when I see, I remember; when I do, I understand" [15]. Therefore, in situations where real life training opportunities are limited, hazardous or impossible, like emergency responses, VR simulators offer the opportunity to emulate a wide range in experiments [20].

In practice Goedert developed a virtual interactive construction education platform which provided game-based safety training through the use of simulation and modeling of education and training methods [8]. It can be named a 3D game technology, which aims to enhance user interactions, refers to computer-based game-like training scenes through integrating visual, interactive, network and multi-user operating technologies and so forth. As well as Goedert, Guo et al. developed a game-based safety training system, which is an online platform that allows trainees to use input devices, such as keyboard, mouse and game controllers and so on, to operate virtual tasks, such as equipment operation and material delivery [9]. Druzinina, Solovyova studied the issues of "Safety Training" and showed that it is necessary to use new teaching technologies [6].

However, very few publications can be found in the scientific literature that discuss the issue of using game–based safety training system in a none-native language (English). This means that students should integrate such skills as reading and listening instructions in English, translating and understanding specific terminology, making research tasks, doing analytical work and formulating recommendations in the language. The methodology and the game-based safety training system which was used in the experiment are presented next.

3. Participants and study context

3.1 Technology in Use

The study was conducted on the basis of the Arkhangelsk Regional Rescue Service named after Igor Polivany. The class of the center is equipped with an educational and technical complex for training the heads of organizations and operational personnel of emergency response services on potentially dangerous objects in the Arctic region.

3.2 Training Scenario

The program is aimed at professional users in the field of radiation monitoring in the field of radiation contamination in conditions of an extremely radiation-related situation. The users of the system are managers and specialists of enterprises, heads of rescue teams participating in emergency response work involving the release of radioactive substances.

The software complex of interactive three-dimensional visualization (simulator of "virtual reality") is a computer system for simulating the work of mobile radiation reconnaissance tea teams at an early stage of an accident.

The computer simulator is designed for training in the following tasks:

- conducting radiation reconnaissance of the territory;
- identification of places of greatest risk for the population;
- detection of a radioactive source;
- assessment of pollution levels and isotopic composition;
- determination of exceeding the permissible levels of intervention;
- transfer of the received data to the control center;
- implementation of continuous monitoring of dose loads.

The computer simulator completely imitates the real means and devices for carrying out radiation reconnaissance, monitoring and search for sources used in the rescue unit.

There are three basic scenarios of radiation exploration group in circumstances of radiation accident for training.

1. Scenario of dispersion of radioactive dust from the cloud made as a result of the explosion of usual charge surrounded with radioactive substance, dispersion of the cloud under the actions of wind metaconditions and subsequent subsidence of radioactive particles on the underlying surface.

In this scenario a big zone of infection is made. Difficulties of localization of the infection zone are connected with difficult geometrical character of the urban development. It does not let the staff designate the zone of radiation infection in a short time.

2. Scenario of not homogeneous dispersion of radioactive substance fragments as a result of explosion or deliberate spread of radioactive substance on the area of the urban development.

In this scenario an infection zone with some local points in spread of power dose is made. The difficulty of infection zone localization is caused by line of sight lack of radioactive substance distribution zone.

3. Scenario of container laying (possibly damaged), containing highly active point source of radiation. In this scenario there is a point source located among objects of city infrastructure. During localization of point sources a "search" object strategy is used. It is recommended to conduct systematic inspection of all development objects.

The participants of the experiments (managers of organisations and rescue formations) in realistic environment of urban development can freely move in three dimensional space taking into account collision with scene elements, can go upstairs, jump from height, follow changes of a relief. According to the rules of simulation game a participant must:

- Choose "virtual" measurement means corresponding the task;
- Do local measurement of radioactive infection degree. It is possible to move the measurement device (dosimeter) freely in space;
- Choose independently a location of signs "infected";
- Make decisions of infected zone way of the round, planning optimum route of dosimetric control group, taking into account available obstacles of transport and urban architecture.
- On completion of the training phase the leader conducts action analysis of a user, point to mistakes and give recommendations.

At the end of the training phase, the head of the exercise analyzes the actions of the user, indicates errors and makes recommendations.

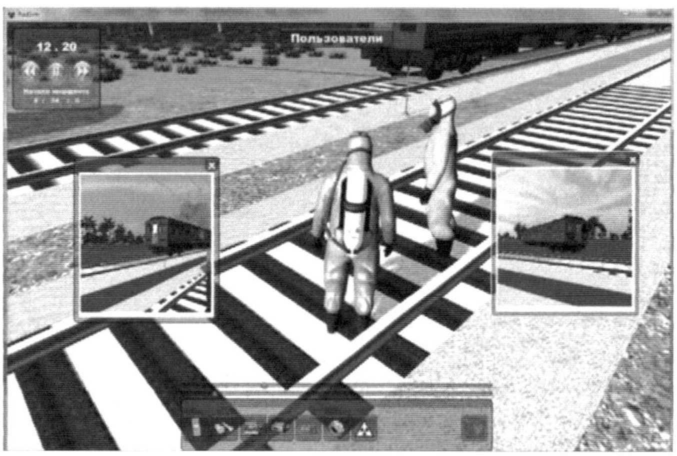

Figure 1. Three-dimensional simulator for training the actions of personnel of rescue units for the detection of radioactive contamination and the conduct of radiation reconnaissance

3.3 Participants

There are 158 people (79 heads of organizations and 79 supervisors of the emergency rescue unit) who took part in the study in the period from 2013 to 2018. In the process of training it was used the following forms of education: lectures, seminars and practical exercises. In addition to the main forms of training, the commanders of rescue formations were trained with the help of technical complex (VR stimulator).

4. Methodology

The purpose of this study is to evaluate the effectiveness of training a control group of facility managers without using VR and an experimental group of emergency response crews using VR to take action in radiation and chemical accidents. To assess the effectiveness of the used questionnaires in which students expressed their opinions about the training. At the beginning of the training was conducted entrance control. The purpose of entrance control is to determine the level of knowledge of participants before the start of training. To compare the degree of learning of students, Student's t-test was used. After training, a final test was conducted to assess the knowledge of students. Learning satisfaction and learning perception were assessed using a Likert score based on questionnaires.

5. Results

At the stage of the experimental training the following methods of the study were used: survey, tests, pedagogical observation, methods of mathematical statistics and data processing. Two groups participated in the experiment. The first control group was taught according to the traditional method without using VR. It consisted of heads of organizations (79 people). Before the beginning of the training an entrance test to diagnose initial level of knowledge was conducted in control and experimental groups. The assessment was carried out with the use of a five point scale. The results of the entrance test can be seen in the Figure 2.

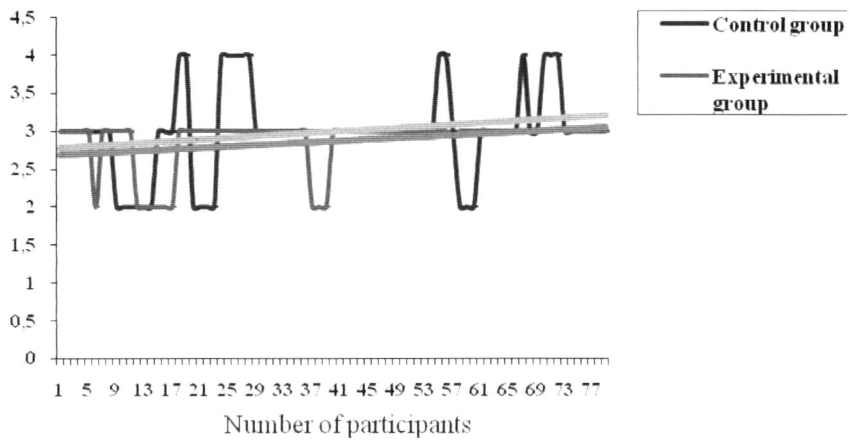

Figure 2. The results of the entrance test in the control and experimental groups

Then the statistical processing of the measurement results was made. Criterion-t by Student was used to assess the reliability of the results, as there were over 50 people who participated in the study. The arithmetic mean, mode and median of the initial level test in the control and experimental groups are presented in Table 1.

Table 1. The arithmetic mean, mode and median of the initial level test in the control and experimental groups.

Group	arithmetic mean X	mode M_o	median Me
Control group	3.0	3	3
Experimental group	2.87	3	3

The analysis of the entrance test (Table 1) shows that the initial knowledge level of the participants in the control group is higher than the initial level of the participants in the control group.

In the control group every value of the level differs from the average value 3 by 0.574. The average value is mode and median. That illustrates the normal distribution of the sample. In the experimental group every value of the level differs from the average value 2.87 by 0.333. The average value is about mode and median, which indicates the normal distribution of the sample.

As follows from the results of the entrance test the initial knowledge level of the participants in the control group is higher than the initial knowledge level of the participants in the experimental group (3.0 > 2.87).

In the experimental group, each value of the series differs from the average value of 2.87 by an average of 0.333. The average value is approximately equal to the mode and the median, which indicates a normal distribution of the sample.

The final test control of knowledge for both groups consists of four tasks with a total of 30 questions. Test control makes it possible to assess:

- knowledge of the features of training in the field of protection from the damaging factors of emergency situations;
- the ability to organize and conduct work to eliminate the consequences of radiation and chemical accidents.

To determine the reliability of the results, the Student's t-test for independent samples was used.

The total number of sample members for both groups is 79.

Calculation of arithmetic mean:

For control group x:

$$x = \frac{\Sigma xi}{n} = \frac{329}{79} = 4.165$$

For experimental group y:

$$x = \frac{\Sigma yi}{n} = \frac{363}{79} = 4.595$$

The results of the final testing are presented in Figure 3.

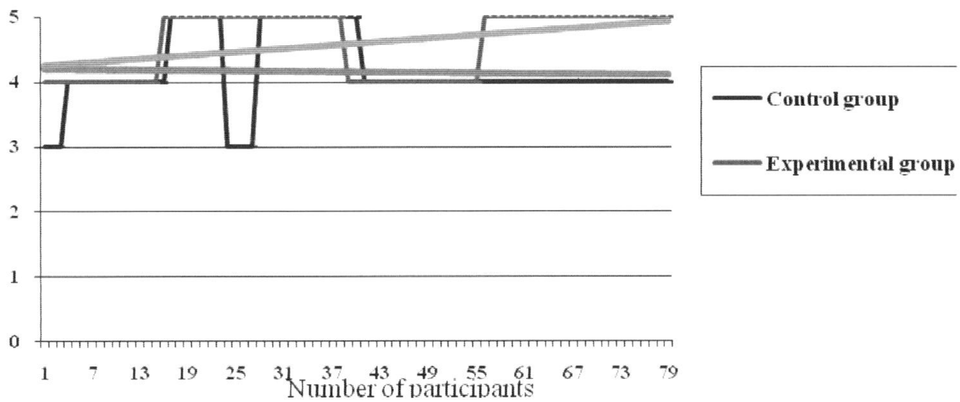

Figure 3. Results of the final control test and experimental group

Standard deviation for the control group:

$$S(x) = \sqrt{S^2(x)} = \sqrt{0.31} = 0.561$$

Standard deviation for the experimental group:

$$S(y) = \sqrt{S^2(y)} = \sqrt{0.24} = 0.491$$

Mean error of arithmetic means G = 0.084

Calculate the temp.

$$\textbf{Temp.} = \mathbf{T}_{\mathbf{emp.}} = \frac{4.595 - 4.165}{0}.084 = 5.11$$

Next, we compare the obtained value of Temp with the theoretical value of T - Student's distribution. In the experiment, Temp = 5.11, tabular Tcr. = 1.99. The obtained in the experiment Temp exceeds the table, therefore, an alternative hypothesis is accepted that the leaders of the rescue formations (EG) showed a higher level of knowledge than the heads of organizations (CG). Temp 5.11> Tcr 1.99, with $\alpha = 0.05$. The differences in the level between the control and experimental groups are statistically significant with an error level of no more than 5%. Thus, a decision is made about the significance of experiential learning and the use of VR.

The level of satisfaction with the learning process was assessed on the Likert scale based on the questionnaires of students [14]. The analysis of the questionnaires showed that the participants satisfaction with the quality of learning in the experimental group using VR (97.55%) is higher than in the control group (CG) trained with the use of traditional methods (65.35) - presented in Figure 4.

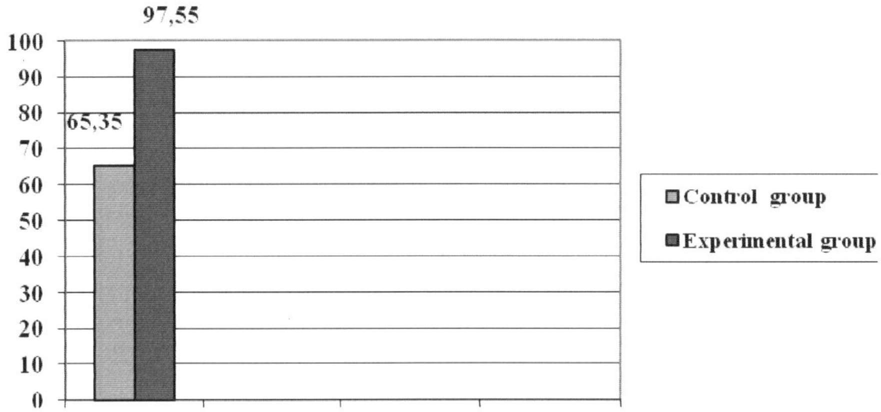

Figure 4. Student satisfaction with the learning process

The assessment of the level of satisfaction with the students' training showed that the learning process using VR is higher than traditional training.

6. Conclusion

In this research, a comprehensive review regarding simulation based technologies in construction the course in "Safety Training" for managers of organisations has been conducted and the experimental methods, application areas and future research directions have been identified. Based on the literature review of stimulation based technologies in particular VR stimulators have enormous potential to increase the quality of professional safety training of managers of the near-arctic region. The data obtained during experiment indicate that our presumption about effectiveness of simulation based technologies implementation in "Safety Training" course for international managers is confirmed. Using VR-based educational program with none-native (English) content didn't cause poor results and on the contrary became an additional stimulus for improving managers' language skills. The design of the courses with the use of simulation based technologies appeared to be a useful tool for creating virtual-reality blended environment and promoting autonomous education. The proposed technology can be used for training and education in a variety of health care situations.

7. Acknowledgement

The authors appreciate the efforts and assistance of the head of the Agency of the State Fire Service and Civil Protection of the Arkhangelsk region A.V. Uvarov for opportunity to conduct research on the base of the Arkhangelsk Regional Rescue Service named after Igor Polivany. The authors wish to acknowledge the assistance and support of the professors from the Northern (Arctic) Federal University named after M.V. Lomonosov.

References

[1] Cavalcanti J F, Soler J L, Alcañiz M, Contero M 2017 Conference: International Technology, Education and Development Conference(*researchgate.net)* 1-6.

[2] Cooper S., Cant R., Porter J., Bogossian F., McKenna L., Brady S., & Fox-Young S 2012 *Women and Birth* **25(2)** 64–78.

[3] Coutts N, Simpson M & Drinkwater R 2001 *Teacher Development* **5(2)** 225-239.

[4] Darragh A R, Lavender S, Polivka B, Sommerich C M, Wills C E, Hittle B A,

Stredney D L 2016 *Clinical Simulation in Nursing* **12(8)** 328–335.
[5] Debyser Fr.1996 *L'immeuble* (P.: Hachett) 95 p.
[6] Druzhinina M V , Solovyova E E 2018 *Sovremennaya nauka: aktual'nye problemy teorii i praktiki. Seriya: Gumanitarnye nauki* **6 (1)** 76-80.
[7] Frene S 2002 Antologiya gumannoj pedagogiki Ed Vul'fson V L (Moscow) p.224
[8] Goedert J D, Rokooei S. 2016 *Int. J. Constr. Educ. Res.* **12** 208–223.
[9] Guo H, Li H, Chan G, Skitmore M 2012 *Accid. Anal. Prev.* **48** 204–213.
[10] Jackson, J. 2003 *System* **31(4)** 457–469.
[11] Ktoridou D, Doukanari E, Epaminonda E, & Karayiannis A 2018 *IEEE Global Engineering Education Conference (EDUCON) 1781-1786.*
[12] Lean J, Moizer J, Towler M & Abbey C *Active learning in higher education* **7(3)** 227–242
[13] Li X, Yi W, Chi H-L, Wang X, & Chan A P C 2018 *Automation in Construction* **86** 150–162.
[14] Lienert G A & Krauth J. 1975 Configural frequency analysis as statistical tool for defining types. **35** 231-238.
[15] Meadows D L 2001 *Simulation &Gaming* **32** 522-536.
[16] Moreno R & Mayer R 2007 *Educational Psychology Review* **19** 309-326.
[17] Odaryuk I V, Kolmakova V V 2016 *Filologicheskie nauki. Voprosy teorii i praktiki* **5-2 (59)** 201-203.
[18] Os'kina S D *Psihopedagogika v pravoohranitel'nyh organah* 2017 **1(68)** 75–78.
[19] Pedram Sh, Perez P, Palmisano S, Farrelly M 2017 *22nd International Congress on Modelling and Simulation (Hobart, Tasmania, Australia)* 361-367.
[20] Seymour N E, Gallagher A G, Roman S A, O'Brien M K, Bansal V K, Andersen D K & Satava R M 2002 *Annals of surgery* **236** 458-464.
[21] Shcherbakova O V 2018 *Pedagogika. Voprosy teorii i praktiki* **2 (10)** 67-70.
[22] Tennberg M 1999 *Arctic Environmental Cooperation. A Study in Governmentality* p. 214
Wang P, Wu P, Wang J, Chi H - L, & Wang X 2018 *International Journal of Environmental Research and Public Health* **15(6)** 2-18

Formation of structural and hierarchical model of needs in production personnel for the purpose of forecasting their professional training

N V Ketko, O E Akimova, S V Zemlyanskaya, I V Dneprovskaya, E M Vitalieva

Volgograd State Technical University, 400005, Volgograd, Lenin avenue,28 Russia
Volgograd State University, 400062, Volgograd, Universitetskiy avenue,100 Russia

zemlyanskayasv@volsu.ru, akimovann25@mail.ru

Abstract: The article deals with the problems of professional training of industrial personnel, namely the issues of coordination of training programs with the existing needs of employers. The authors propose an algorithm for constructing a structural and hierarchical model that allows to visualise the needs of employers in professional staff in the form of a hierarchy, built in descending order of their importance.

1. Introduction

Modernity is characterised by the increasing role of higher education, which is the most important condition for the progress of society. The degree of influence of higher education on the development of the state and the country as a whole in different periods of time is determined primarily by indicators of its quality, and therefore at present, in the education system there is no priority higher than the quality.

In the academic community, the problem of the quality of education is recognised as a sophisticated and complex scientific and historical problem. No educational system in the world can fully satisfy the world academic community in matters of the system of provision (management) and evaluation of the quality of educationx [5, 6].

A common problem in Russia is the recurrent dissatisfaction of state, state and public education management bodies, as well as the general public and employers with the quality of professional training. However, it should be noted that in modern conditions the quality of training determines the future of universities, which, like other economic entities, have entered the path of competition for consumers of their services. These circumstances stimulate the search for more effective approaches to the management of the quality of education, taking into account international trends in this area.

It should be noted that the main trend of improving the quality of vocational education is a purposeful transformation in the activities of educational institutions within a number of characteristics, taking into account the implementation of a set of balanced interests of students, employers, society and the state [7].

The interests of students in terms of the quality of professional training are in the implementation of the constitutional right of a person and a citizen to receive the necessary professional training, used for the purpose of the possibility of conducting in the future not prohibited by law work, as well as improving the personal characteristics of the individual, capable of further favorable impact on his professional career.

The employer, within the framework of his / her interests, after passing the professional training, intends to obtain a competent employee capable of demonstrating high results for the benefit of the enterprise, organisation or institution [8].

The plane of the interests of society in professional training is to prepare a large number of competent professionals to perform the most important social tasks in various areas of life.

It is accepted to allocate two approaches to the solution of a problem of quality assurance in professional training of the personnel conforming to modern requirements:

1. Improving the performance of the University through the introduction of control schemes at all stages of management. To this end, the development of normative documentation of the quality management system of the University (according to international ISO standards).
2. Improvement of all processes in the University aimed at ensuring the quality of professional training by creating objective prerequisites for the involvement of all employees in this process (according to the European standard TQM).

However, ensuring the quality of professional training is not a primary issue, since the main issue is the directions of training, the specialists of which professions to train, in which specialists there is a need for employers [9].

Since the choice of professional direction depends on many factors, so this system, which complicates the process of planning and forecasting, for these purposes it is advisable to apply the methods of fuzzy set theory.

2. Application of mathematical methods for building a structural-hierarchical model

Methods of fuzzy set theory allow not only to investigate the state of the system, but also help the expert to make a certain decision. The advantage of these methods is that they allow the use of not only quantitative but also qualitative indicators, and therefore remove some of the limitations associated with the inability to use indicators with qualitative assessments.

There are methods using different approaches to the study. In the framework of the theory of fuzzy sets, this article uses one method – the Maximin convolution. In order to determine the most demanded professions that act as alternatives, the evaluation criteria are identified, which are the factors on which these alternatives depend, or which allow to assess the significance of an alternative. The significance assessment process is carried out as follows.

The refinement of qualitative parameters and the discussion of the importance of alternatives by experts allow for a compromise in the assessments.

- average assessment:

$$m_j = \overline{m}_{ij} / \sum_{j=1,i=1}^{N,K} \overline{m}_{ij} ,$$

where m_j – average assessment of j-motive;

$\overline{m}_{ij}$ - numerical score «weight» of j-motive, given by i-expert

- the weighted average assessment:

$$m_j = \sum_{i=1}^{N} m_{ij} * a_i / \sum_{i=1}^{N} m_{ij} ,$$

where m_j – average weighted numerical score of j-motive;
m_{ij} – numerical score of the average assessment of j-motive by i-expert;
a_i – weight, reflecting the importance of i-expert.

Convolution of point estimates of alternatives is carried out in accordance with the algorithm presented below, based on the method of fuzzy sets.

In the process of basic calculations, the values of the preference ratio for each pair of alternatives M (a, b) are determined by the formula (1):

$$M(a,b) = \frac{m_a - m_b}{n_j}, \quad \text{if } m_a \rangle m_b \tag{1}$$

0, otherwise,

where ma,b is the significance of the motive a in reference to motive b to the employee;
nj – upper limit of the scale of assessments of the experts of the explanation;
j - ordinal number of the scale developed by the experts;
M (a,b) is a function of the consent that the motive a is better for a worker than motive b.

A preference relation is called fuzzy, as is linguistic variables, reflecting the assessment of the experts.

The calculation results are presented in the form of the following matrix:

$$M(a,b)=\begin{vmatrix} \frac{m_{11}-m_{21}}{n_j} & \frac{m_{12}-m_{22}}{n_j} & \ldots\ldots & \frac{m_{1i}-m_{2i}}{n_j} \\ \frac{m_{11}-m_{31}}{n_j} & \frac{m_{12}-m_{32}}{n_j} & \ldots\ldots & \frac{m_{1i}-m_{3i}}{n_j} \\ \ldots\ldots & \ldots\ldots & \ldots\ldots & \ldots\ldots \\ \frac{m_{11}-m_{k1}}{n_j} & \frac{m_{12}-m_{k2}}{n_j} & \ldots\ldots & \frac{m_{1i}-m_{ki}}{n_j} \end{vmatrix}$$

For the ranking of alternatives, the fuzzy sets of dominance of one motive over another is determined, which is determined by the function μD (a, b) characterising the intensity of its dominance:

$$\mu(a,b)= \quad M(a,b)-M(b,a), \text{ если } M(a,b)\rangle M(b,a) \tag{2}$$
$$0, \text{ otherwise,}$$

where μ (a,b) is a function of preferences, reflecting the intensity of the domination of the motive a over the motive b;
M (a,b) is a function of agreement that motive a is preferable to employee than motive b;
M (b,a) is a function of disagreement that motive a is preferable for an employee than motive B.

The fuzzy set of dominance of one alternative over another is determined by the function μD (k,l). The result of determining the dominance ratio is also presented in the form of a matrix, the elements of which are the difference of the elements of the previous matrix:

$$M(a,b)=\begin{vmatrix} M(1,2)-M(2,1) & \ldots\ldots & M(1,b)-M(a,1) \\ M(2,1)-M(1,2) & \ldots\ldots & M(2,b)-M(a,2) \\ \ldots\ldots & \ldots\ldots & \ldots\ldots \\ M(a,1)-M(1,b) & \ldots\ldots & M(a,b)-M(b,a) \end{vmatrix}$$

To improve the objectivity of the assessment is determined by the ratio of non-dominant μND (k,l).

$$\mu_{ND}(a,b)=1-\mu_{D}(a,b) \tag{3}$$

where μND (a, b) is a preference function reflecting the intensity of non – dominant motive a over motive b.

Calculations of the non-dominant ratio we summarise the resulting matrix as follows:

$$\mu_{ND}(a,b)=\begin{vmatrix} 1-\sup\mu_{ND}(1,1) & 1-\sup\mu_{ND}(1,2) & \ldots\ldots\ldots\ldots & 1-\sup\mu_{ND}(1,b) \\ 1-\sup\mu_{ND}(2,1) & 1-\sup\mu_{ND}(2,2) & \ldots\ldots\ldots\ldots & 1-\sup\mu_{ND}(2,b) \\ \ldots\ldots\ldots\ldots & \ldots\ldots\ldots\ldots & \ldots\ldots\ldots\ldots & \ldots\ldots\ldots\ldots \\ 1-\sup\mu_{ND}(a,1) & 1-\sup\mu_{ND}(a,2) & \ldots\ldots\ldots\ldots & 1-\sup\mu_{ND}(a,b) \end{vmatrix}$$

After determining the non-dominant relations, the obtained alternative series is compared with the already formed domination series [12]. The second row should be opposite to the first. Otherwise, it is necessary to repeat the approval procedure. The following is the calculation of the intensity of dominance of each alternative by the intersection of all motives μND (a,b) for all motives that are not dominated by any motives.

$$\mu^{*}_{D}(M_{a})=\min\mu_{ND}(b,a)=\min_{\substack{b=1,\ldots n\\ b\neq a}}[1-\mu(a,b)]= \\ 1-\max\mu_{D}(b,a)=1-\max[M(b,a)-M(a,b)] \tag{4}$$

where μ*(MA) is a function of the membership of an alternative to a set of non – dominant alternatives. (4)

$$\mu^{*}(M^{*}_{a})=\max_{a=1,\ldots n}\mu^{*}(M_{a})=1-\min\left\{\max\left[\underset{\substack{a=1,\ldots n\\ b=1,\ldots n\\ b\neq a}}{M}(b,a)-M(a,b)\right]\right\} \tag{5}$$

where M*a are alternatives from a variety of non-dominant alternatives.

Next, a series of alternatives is constructed in descending order of the value of the preference ratio, i.e. if A1 > A2 > A3, etc.

After the formation of a number of alternatives arranged in descending order of their importance to build a structural and hierarchical model, presented below, which allows you to clearly see the most popular profession and accordingly adjust the training program.

Thus, planning and forecasting the needs of employers in professional personnel, through the construction of a structural and hierarchical model, will allow more effective development and implementation of the process of training of professional personnel, especially in the business sector [13]. Since the lack of the necessary professional production personnel is one of the main factors hindering the development of the business sector.

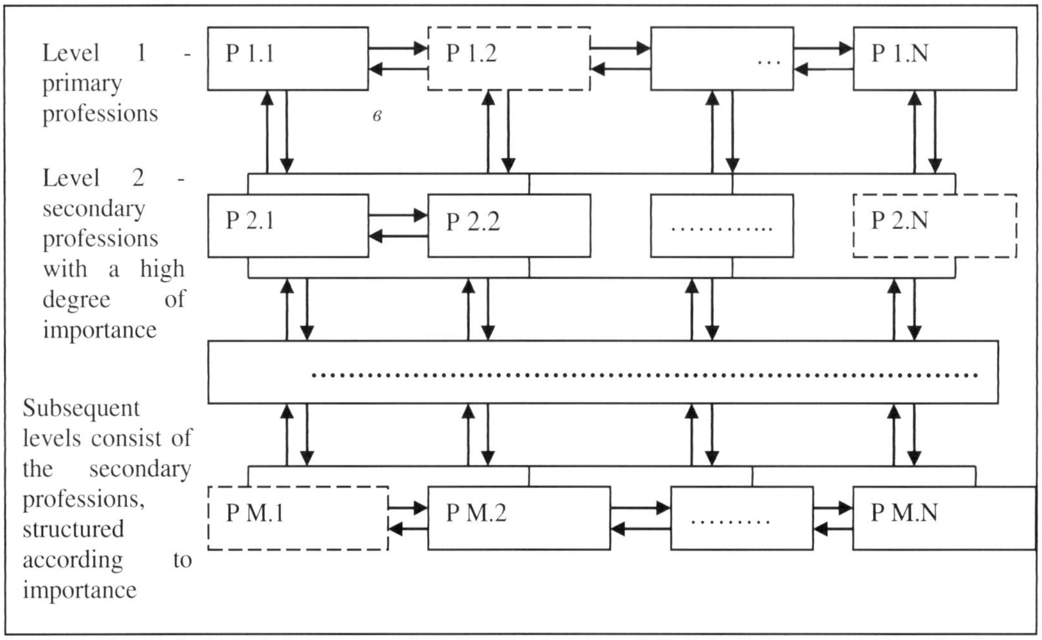

Figure 1. General view of the structural and hierarchical model of the importance of professions

3. Impact on the business sector

According to the Global Entrepreneurship Monitoring for 2016-2017, only 28.4% of respondents believe that they have sufficient knowledge and experience to start their own business. According to this indicator, Russia occupies one of the lowest places among all the countries of the project - 63 in developed countries, the figure is about 50% [10]. In this regard, it is necessary to improve the professionalism of personnel in the field of entrepreneurship through the development of educational programs based on state educational standards aimed at training qualified personnel, the creation of a network of educational institutions that implement educational programs for businesses [13].

According to the well-known Russian scientist Chepurenko, "entrepreneurs are born", not "made", but it does not exclude the need to identify the professional training of young people for entrepreneurial activity. The level of knowledge achieved in society, the modern organisation of the economy create the necessary prerequisites for the development of entrepreneurial abilities. Training of an entrepreneur has a dual direction: firstly, that is training depending on the nature (kind) of activity and, secondly, training in the field of human relations. It is necessary to introduce programs of acquaintance with business activity in secondary schools already, and to teach students of technical universities business skills (unfortunately, a significant part of our engineering universities still trains "generals for the last war"- hired specialists for large state-owned factories, not engineers and technology managers for medium-sized businesses) [11].

The first steps in this direction are just beginning to be made, in many universities began to appear profiles of training in entrepreneurship, however, while it is not fully developed [14].

Since the bulk of entrepreneurs in Russia and the world are representatives of the generation "X" (the so-called generation with a key on the neck, children who were accustomed to early independence, did their own lessons, were able to warm up their meals left on the stove), for them learning throughout life is the motto in life. Given the professionalism, integrity and responsibility of this generation, it is necessary to stimulate their self-employment (SE), as for X work is life.

From this point of view, it is interesting not only the programs within the framework of higher education, but also the programs within the framework of additional professional education. Such programs are implemented today by various organisations. For example, JSC "SME Corporation". It offers several training programs: "ABC of the entrepreneur" (creating a business from scratch) and"

School of Entrepreneurship" (business development). Training programs of JSC "Corporation "SME" can be passed both independently and by visiting trainings in organisations that form the infrastructure to support SMEs implementing training programs of the Corporation [15].

The program "ABC of the entrepreneur" is aimed at training potential and novice entrepreneurs to develop a business plan for the further implementation of the developed business project and the launch of a new business. In order to identify the most appropriate business idea, the Corporation has developed a module "Generation of business ideas".

The program "School of Entrepreneurship" is aimed at training actively functioning entrepreneurs who want to develop, expand or repurpose their business.

In addition, communication between the state and entrepreneurs through social networks is possible. If the materials set out in the training programs of the SME (Small and Medium Entrepreneurship) Corporation are presented in the format of short-duration videos, with the possibility of online assignments, this will receive a positive response.

Targeted training of production personnel in accordance with the needs of not only production, but also other sectors of the economy, not only contributes to the development of the business sector and the economy as a whole, but also acts as a tool to solve the problem of poverty.

4. Training as a tool to overcome poverty

Currently, Russian educational organisations do not smooth social inequality, and in some cases, on the contrary, it is aggravated. [2]. Indeed, the opportunity to get a quality education (starting from the pre-school level) increases for children from wealthy, full and well-to-do families. There is also a big difference in the quality of education between the regions and the Federal center, major cities and the province.

People born in poor families with low cultural and social needs cannot adapt, achieve career growth and high income. As a result, some of these people begin to put up with their situation and perceive poverty as the norm, others become criminals or addicted to bad habits.

It should also be noted that not always people who complete secondary vocational or higher education can find a suitable job with the best salary for them. Often the family even in the adult stage helps to adapt and find a job. At the same time, people from poor families have to adapt on their own. Only a few can overcome poverty and succeed in their careers. If only the economic dimension of poverty is to be understood, it is necessary to consider the number of people living below the subsistence level.

According to Rosstat (Russian Federal State Statistics Service), every seventh resident of Russia lives below the poverty line. According to statistics, in 2016 the number of people whose income was below the subsistence minimum increased by 300 thousand and reached 19.8 million people [3].

In 2018, the number of people in Russia who have an income below the subsistence minimum amounted to 20 million people.

Let us consider the dynamics of the number of people with income below the subsistence minimum from 2015 to 2018.

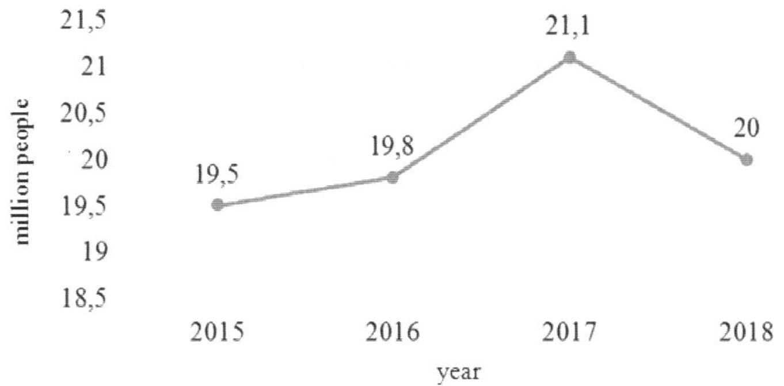

Figure 2. Dynamics of the number of people with income below the subsistence minimum in Russia 2016-2018

Source: author's materials [1] [4]

People who find themselves below the poverty line begin to perceive their standard of living as the norm, resigned to their position. In this regard, it is possible to speak about occurrence of the phenomenon of "institutional poverty".

Institutional poverty is the social and economic norms and rules that form the basis for meeting human needs at the most primitive level.

In order to overcome institutional poverty and enable individuals to improve their quality of life, it is necessary to formulate a set of measures for the state and public structures, as well as educational organisations at all levels.

In May 2018, the decree of the President of Russia declared the national goal to "halve the level of poverty in the Russian Federation" [2].

To implement the decree, Educational Institutions of Secondary Professional and Higher Education on their part can:

- develop a program of measures for the introduction of the profession for students;
- develop training courses and retraining of specialists relevant to each region. When developing courses, educational organisations should cooperate with regional labour and employment committees to determine the demand and relevance of professions and the correlation with the labour market.

The mechanism of interaction is shown in figure 3.

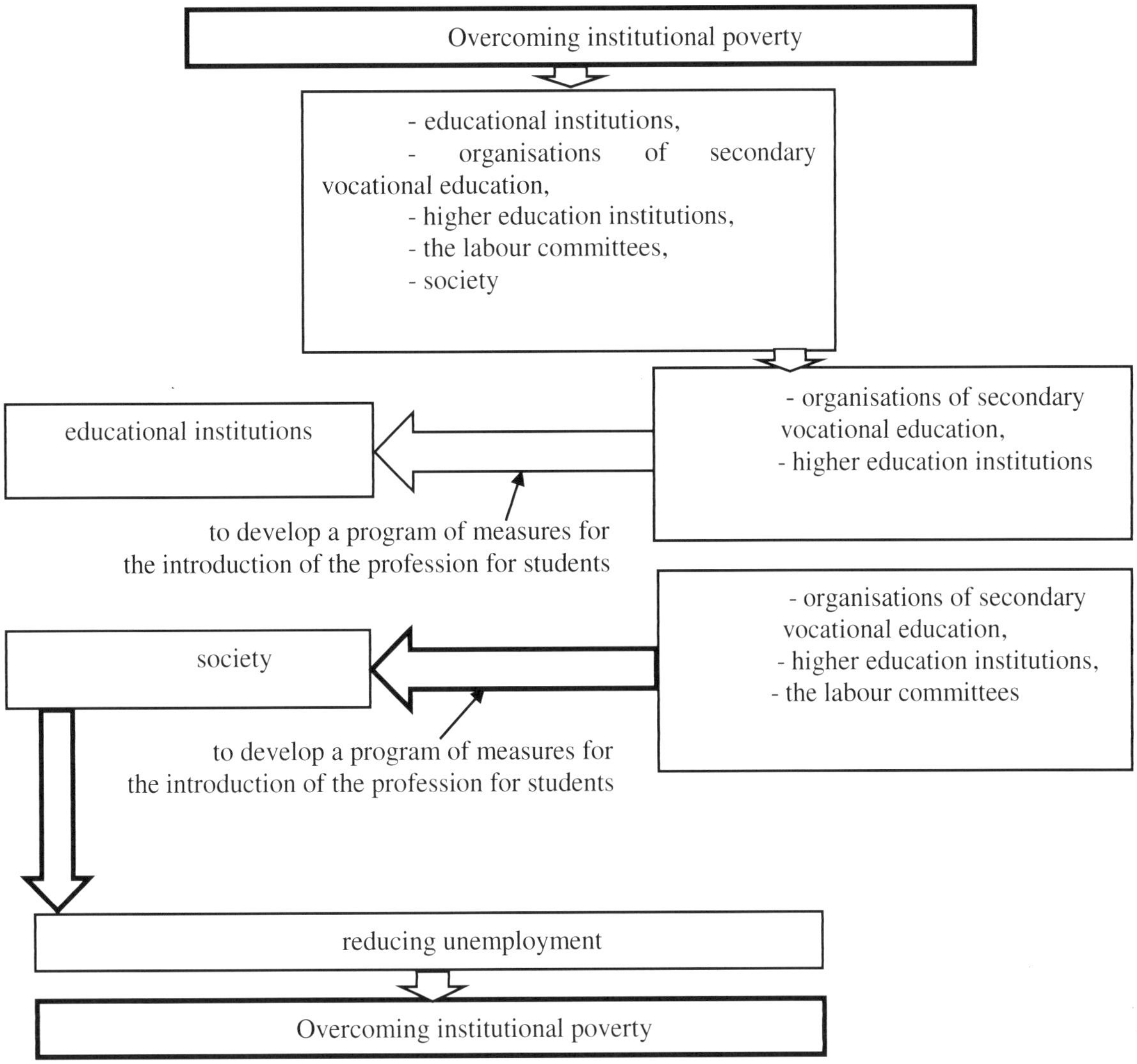

Figure 3. The mechanism of interaction with the labour market

Thus, the mechanism implements the possibility of overcoming institutional poverty. In this mechanism, the leading role is played by educational organisations, since it is in the learning process that the motivation of individuals to work and the desire for career growth should be formed. Periodic training or retraining allow to deepen or master new professions. In turn, work with students will form the desire to work and commitment to the specialty.

References:

[1] The poor counted less. The number of citizens with incomes below the subsistence minimum has decreased for the first time in five years [Electronic resource] Access Mode: https://iz.ru/768023/anna-ivushkina/bednykh-naschitali-menshe
[2] Education should be the main instrument of poverty reduction [Electronic resource] Access Mode: https://www.hse.ru/news/edu/220512836.html
[3] Rosstat. Standard of living [Electronic resource] Access Mode: http://www.gks.ru/wps/wcm/connect/rosstat_main/rosstat/ru/statistics/population/level/
[4] The number of people with incomes below the subsistence minimum in the first half of the year

IOP Conf. Series: Materials Science and Engineering **483** (2019) 012079 doi:10.1088/1757-899X/483/1/012079

decreased by 5.2% [electronic resource] Access Mode: https://rns!online/economy/CHislennost-lyudei-s-dohodami-nizhe-prozhitochnogo-minimuma-v-pervom-polugodii-sokratilas-na-52-2018-10-24/

[5] Andreychikov A V, Andreychikova O N 2000 Analysis, synthesis, planning decisions in Economics (Moscow: Finance and Statistics) 368p

[6] Zade L A 1976 The Notion of linguistic variable and its application to the adoption of approximate solutions (M.: Mir) 165p

[7] Trahtengerts E A 1998 Computerised decision support (M SINTEG) 376p

[8] Spandagos C 2018 Fuzzy model of residential energy decision-making considering behavioural economic concept *Applied Energy* **213** 611-625

[9] Lizunkov V, Politsinskaya E, Malushko E, Kindaev A, Minin M 2018 Population of the world and regions as the principal energy consumer *International journal of energy economics and policy* **8 (3)** 250-257

[10] James H Stock, Watson M W 2013 Variable Trends in Economic Time Series *Journal of Economic Perspectives* **2** (**3**) 147-174

[11] Ferreira R, Lizunkov V G, Politsinskaya E V 2017 Formation of entrepreneurial competencies of university graduates in conditions of transition to the universities of the third generation *Novosibirsk State Pedagogical University Bulletin* **7 (6)** 195-211

[12] Verkhovskaya O, Aleksandrova E, Bogatyreva K A, Dulepova M V, Shmeleva E V 2016/2017 National report Global entrepreneurship monitoring (Graduate School of Management St. Petersburg State University) 15

[13] Ergunova O T, Lizunkov V G, Malushko E Yu, Marchuk V I, Ignatenko A Yu 2017 Forming the system of strategic innovation management at the high-tech engineering enterprises *IOP Conference Series: Materials Science and Engineering* **177 (1)** 012046

[14] Chepurenko A Yu Entrepreneurship and Entrepreneurship Policy in Russia [Electronic resource] Access Mode: URL:http://nisse.ru/business/article/article_1797.html?effort=1

[15] Ergunova O, Lizunkov V, Blaginin V, Politsinskaya E, Shaykina O I 2018 Formation of the price mechanism for energy resources in Russia and the countries of the European union *International Journal of Energy Economics and Policy* **8 (1)** 122-127

Agent-based approach for analysis of electricity distribution technological processes in power systems

S S Kostinskiy, V A Mokhov, T N Kruglova, D V Shaikhutdinov, A S Vlasov

PlatovSouth-RussianStatePolytechnicalUniversity (NPI) 132 Prosveschenia Str., Novocherkassk, Rostov Region, 346428, Russia

d.v.shaykhutdinov@gmail.com

Abstract. The question of minimizing the loss of the active component of power in power transformers of distribution networks is considered.Authors' attention is focused on the reduction of additional losses due to the non-sinusoidal nature of the load. The proposed solution is based on the application of the principle of level decomposition for the system "digital substation - distribution network". There are three logical levels distinguished, which are described in terms of the agent approach within the framework of this system. At the single-phase consumer level the solution of the combinatorial problem of dynamic connection of consumers by means of agent metaheuristics is offered. It is proposed to integrate an intelligent agent into the transformer substation. The main functions of the agent are to control the launch of agent metaheuristics and adaptive filtering of higher harmonic components of the current. Also, the integration of an intelligent agent is offered at the digital substation level. Its task is to synthesize a new configuration of the distribution network, subject to receiving information from agents of transformer substations. For the coordinated work of all agents, the organization of their interaction through a wireless network is proposed.

1. Introduction

Modern electronic equipment in most cases is a non-linear electrical load for the power supply system, that is, it consumes a non-sinusoidal form, which creates distortions in the supply network and, as a result, voltage distortion that affects other equipment that receives electricity from a common source [1, 2]. The situation with power consumers, distorting the quality of electrical energy, is continuously deteriorating [3, 4, 5]. In recent years there has been a significant deterioration in the quality of electrical energy in the networks of non-industrial consumers - shopping complexes, office buildings, educational institutions, residential buildings [6]. Such a load generates high harmonics of current into the electrical network. Thus, every year specific weight of such a load increases in the overall structure of power consumption. [7]. Quantity and characteristics of nonlinear power consumers change, and, consequently, the harmonic composition of currents and voltages change too [3]. Deterioration of the quality of electricity in the distribution networks is typical for most developed countries [8, 9], since there are less and less power consumers with linear current-voltage characteristics [10].

High harmonic components of the current cause the following undesirable consequences:

- distortion of the form of the supply voltage [1, 8, 11, 12];
- occurrence of overheating and destruction of the zero working conductors of supply lines [1, 2, 8, 11 - 13];
- increased wear, swelling and premature destruction of capacitors of reactive power compensation installations [3, 11-14];
- false operation of fuses, circuit breakers, residual current devices, relays and electronic units [11–15] and erroneous operation of measuring devices [2];
- interference in telecommunications networks, provided that power cables and telecommunications cables are laid close to each other [1, 9, 11 - 13, 16];
- resonance phenomena in 0.4 kV electrical installations [12, 15];

- reducing the power factor of electrical installations [12, 15];
- increase in resistance of earthing switches [13];
- increased acoustic noise in electromagnetic equipment [1, 17];
- increase in voltage drop in supply wires due to the skin effect [14].

The continuous growth of the installed power of non-linear, asymmetrical and rapidly alternating loads is not accompanied by the timely implementation of solutions aimed at correcting the quality of electrical energy [23]. Each individual electrical receiver influences the network slightly, but their combination has a significant impact on the electrical distribution network [24], since when the power of a nonlinear load does not exceed 10 - 15% of the power of the power supply system, significant changes in the operating mode of the system do not appear, but with the share of non-linear load exceeding 25%, negative and sometimes accidental consequences occur in the power grids [8, 21].

Recently, there has been a tendency of the prevalence of higher harmonic components of the current of non-linear small receivers over the highest harmonic components of high-power consumers. In connection with the recent increase in the number of used household appliances, a tendency of growth in the influence of higher harmonic components on the part of consumers on the operation of low and medium voltage networks is expected [25].

Considering the intensive distribution of semiconductor converter technique, in recent years, the improvement of the quality of electric power has become a pressing and important technological challenge. This is due to the decrease in the influence of higher harmonic components on the operation of electrical equipment [26, 27]. The current situation in electrical networks requires research aimed at improving the methods for assessing the influence of non-sinusoidal nature of electrical quantities on the loss of active power and electricity in the equipment of electrical networks.

2. An object, subject and research task

The researchobject is the electric power distribution system of electric energy, which is a combination of “digital substation - distribution network” (CP-PC). The distribution network includes distribution transformer substations (TS), power lines connecting them to the CPU, as well as many electricity consumers that receive electricity from the TS (Figure 1).

The subject of research is the technological process of transmission and distribution of electricity to consumers and increasing its efficiency by reducing electricity losses caused by the harmonic components of the current.

The objective of the research is to develop a concept based on the use of multilevel solution approaches to the synthesis of an effective trajectory of the technological process of transmission and distribution of electrical energy from the CPU through the RS to consumers, which will be based on the principle of reducing additional losses of the active component of power in two-phase power transformers of distribution networks caused by current components.

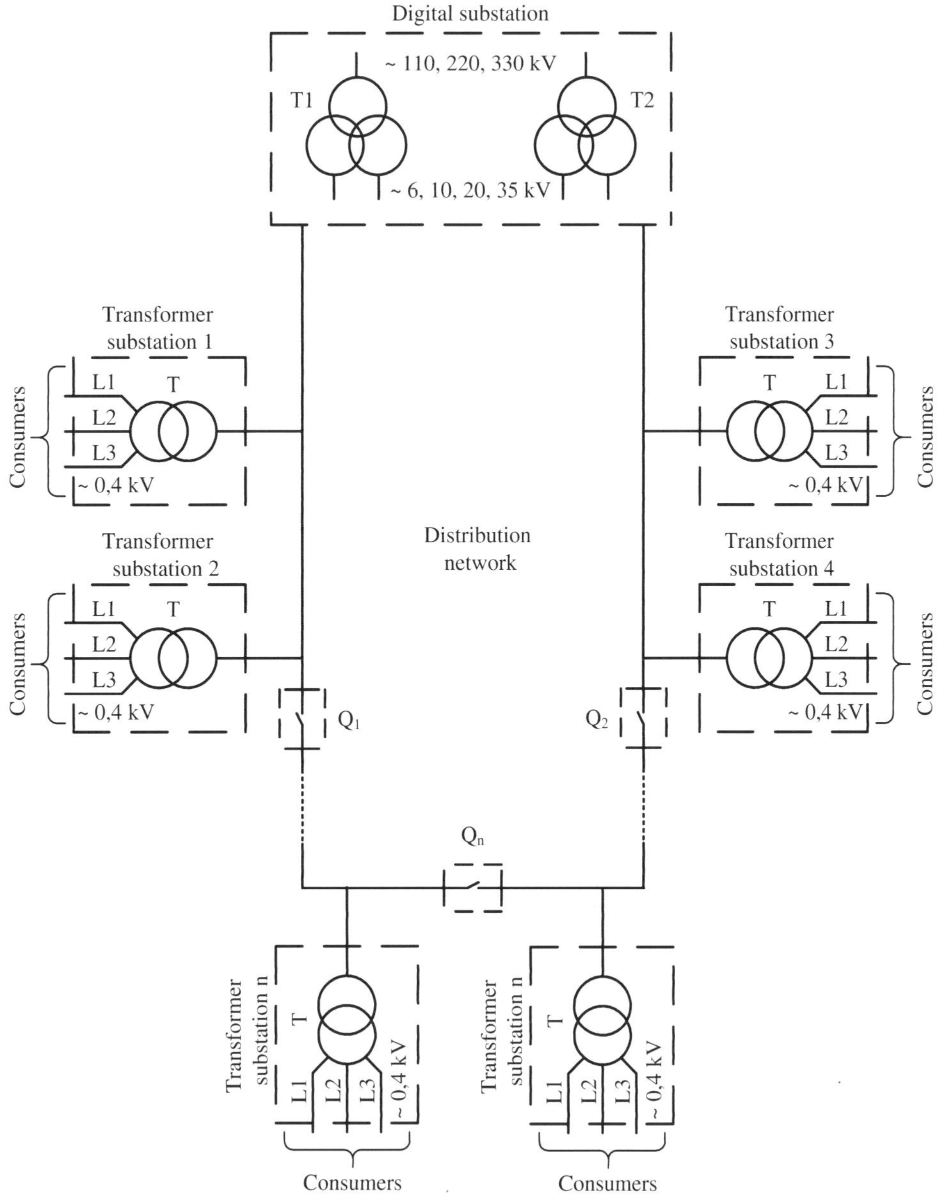

Figure 1. Electric Power Distribution System,representing a set of CP-PC:Q_1, Q_2, Q_n-reclosers

3. Anticipated decision

To solve the problem of minimizing electric power losses in the electric power system CP-PC, caused by non-sinusoidal load, we propose its logical division into a number of subsystems in accordance with the basic principles of level decomposition in computer science and application of agent-specific algorithms that are different for each level.

3.1 Levels decomposition

Within the framework of the CP-PC system, it is logical to distinguish three sublevels/subsystems with the aim of further detailing the solution of the research problem (Figure 2).

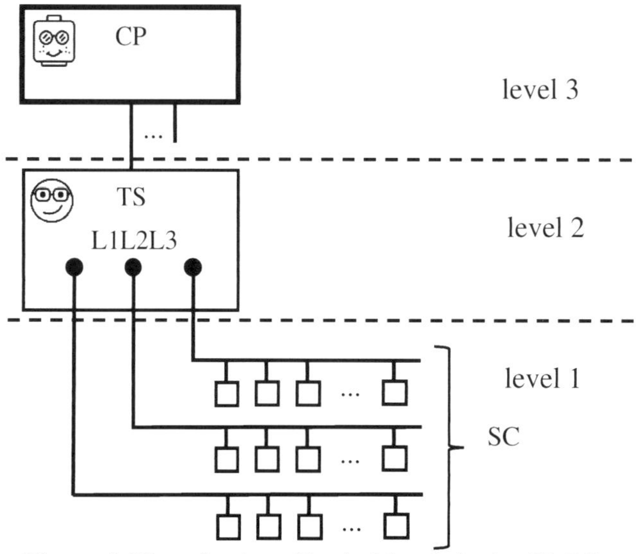

Figure 2. The selection of logical levels in the CP-PC system

The first level of the CP-PC electric power system includes a multitude of single-phase consumers (SC), which are connected to the L1, L2 and L3 phases of the TS (related to level 2) by means of lines rated for 0.4 kV and receiving power supply through them. The third level is formed by a digital substation and a network of distribution power lines designed for one of the medium voltages and intended for power supply of the transformer substation.

At level 1 (Figure 2), it is proposed to apply the distribution of SC to the phases so that their superposition on each of the three phases L1, L2, L3 and on the TS as a whole creates a connection configuration that minimizes the total losses of active electricity in power transformers of the distribution network at the level 2. It is assumed that in the distribution network (for example, an apartment building), the mechanism of dynamic switching of the final SC of electricity between the phases works.

The proposed solution is based on one of the ways to reduce the negative impact of higher harmonic components of current and voltage, which is to ensure the symmetrical operation of the three-phase power transmission system [1, 28]. In order to balance the loads across the phases and at the same time ensure the minimum current in the neutral conductor and the minimum harmonic content in the output current and, as a consequence, voltage, appropriate control and management circuits are necessary [28]. It is also known that harmonic current components different electrical consumers differ in phase, and as a result, when they work together on the same network, the total harmonic components in it can be mutually suppressed [14].

The development of a solution regarding the configuration of the SC connections is planned by reducing the indicated problem to the combinatorial optimization problem and its solution based on agent metaheuristics [29, 30]. Solving such a task will require the organization of a separate computing resource for each TS or a specialized cloud service to service all TSs of the CP-PC system in the complex.

After obtaining a reconfiguration of the SC connections in phases at level 1, the residual harmonic components of the current on the low voltage windings of the transformer substation are minimized at level 2 - using adaptive filtering. This decision is due to the fact that the main means of attenuating higher harmonic components in power supply systems are harmonic filters [26]. One of the most effective examples of such equipment are filter-compensating devices. In turn, passive, active and hybrid filters are distinguished by the presence of active elements [28]. To set up filter-compensating devices, it is necessary to have information about the harmonic composition of the mains current. To this end, before introducing filter-compensating devices in existing power supply systems, one should have information on the levels of the main harmonic components present in them [31]. The use of ac-

IOP Conf. Series: Materials Science and Engineering **483** (2019) 012080 doi:10.1088/1757-899X/483/1/012080

tive filters, sometimes referred to as active harmonics smoothers (active power filter - APF) [32], seems promising. Active harmonic filters in contrast to passive filters contain controls that allow you to change the frequency characteristics of the filter. The development of power electronics: GTO-thyristors and IGBT-transistors, determined the element base, which is the basis for the construction of active harmonic filters. The wide possibilities of the functional properties of the active filter make it possible to apply it not only to reduce the level of harmonic components, but also to compensate for reactive power [28].

It is assumed that it is permissible to reduce the magnitude of the harmonic components of the current at levels 1 and 2 cannot be achieved by economically justified methods. The elimination of the residual harmonic components of the current is assumed at the third level based on the procedures for dynamically changing the configuration of TS connections [33] within the framework of the corresponding distribution network (Figure 1), by changing the network topology using reclosers.

3.2. Ideology of agents

At level 2, it is planned to integrate the software and hardware complex into the TS, which in the aggregate is an intelligent TS agent, indicated in Figure 2 as . This agent is authorized to perform the following functions (Figure 3):

1) «Control» - measurement of the magnitude of the harmonic components of the current and voltage for each phase;

2) «Calculation» - calculation, on the basis of the results of claim 1, of the value of additional losses of the active component of power in two-phase power transformers of the distribution network, caused by harmonic components of the current;

3) «Analysis» analysis of the current situation (each situation is determined by one of the states of the control logic variable F):

- «000» - it is necessary to start reconfiguration of the SC connections by phases;
- «100» - requires the inclusion of adaptive filtering;
- «110» - it is necessary to transfer the command to take measures to minimize additional losses of the active component of power, caused by harmonic components of the current, to level 3;

4) «Reconfiguration» - determination of the need to start reconfiguration of the SC connections in phases and transfer (if necessary such) of the corresponding command to level 1, setting for F the value «100». After that, clauses 1 and 2 are re-executed and the decision is made to turn on the adaptive filter at level 2 («Filtering» command) and then transfer the command to take appropriate measures at level 3 («to level 3» command);

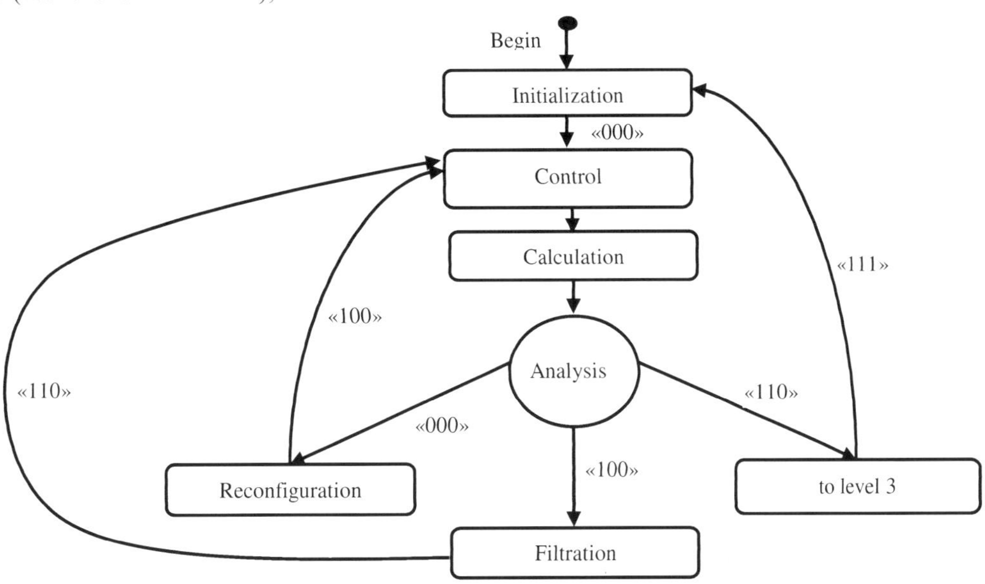

Figure 3. TP agent life cycle diagram

5) «Filtering» - enabling the adaptive filter, setting for F the value «110»;

6) «To level 3» - transfer of a command to level 3 on the need to take measures to combat residual phenomena of the harmonic components of the current (F = «111»);

7) «Initialization» - initialization of the state of the control logic variable F (F =«000»).

It is also supposed to introduce a separate agent in the composition of the CPU, indicated in Figure 2 as . The main function of this agent is the direct synthesis of a new distribution network configuration linking the TS group and the CPU in a circular pattern (Figure 1), subject to receiving commands from all TS agents.

For the coordinated work of all agents, it is assumed that their interaction (exchange of data and commands) is synchronized via a wireless network.

4. Conclusion

1. The current situation in the distribution electric networks, associated with the intensive spread of semiconductor converter technique, requires research aimed at improving the methods for assessing the influence of the non-sinusoidal nature of electrical quantities on the loss of active power and electricity in electrical network equipment.
2. Concept is based on the principle of logical decomposition of the "digital substation - distribution network" system in order to multilevel solve the problem of synthesizing an effective trajectory of the technological process - supplying electricity to consumers, with minimizing the loss of the active component of power in power transformers of distribution networks caused by harmonic components current, using solutions based on agent-based algorithms and metaheuristicsis proposed.
3. The presented work is conceptual in nature and contains a description of the basic ideas for solving this problem.

Acknowledgments

The reported study was funded by RFBR according to the research project № 18-38-20188 "Development and research of tools for intellectual modeling and synthesis of electric power networks technological processes trajectories" using equipment of shared facility "Diagnosis and energy-efficient electrical equipment" (NPI).

References

[1] Prishchepa D N, Zagrutdinov Yu A, Mozhaisky A F 2014 *Proceedings of the Military Space Academy* **644** 199 -203
[2] Cazacu E, Petrescu L 2014 *International Conference on Harmonization and Quality of Power* **6842930** 488-492
[3] ReznichenkoV V, DemidovV P 2016 *Bulletin of civil engineers* **2(55)** 171-173
[4] Kostinskiy S S 2018 *Industrial Energy* 8 29-39
[5] Kostinskiy S S 2018 *Industrial Energy* **9** 25-32
[6] Boyarskaya N P, Temerbaev S A, Dovgun V P, Kabak A L, Kolmakov V O 2013 *Bulletin of the Krasnoyarsk State Agrarian University* **8** 180-184
[7] Yu I Khanin, Yundin M A 2014 *Development and implementation ofresource- and energy-saving technologies and devices* (Sat.ist.VIntern.scientific.practical.conf.) 98-101
[8] Kolmakov V O 2015 *Authoreferat Candidate of technical Sciences*(Krasnoyarsk) p 20
[9] Fuchs E, Masoum M 2008 *Academic press* 638
[10] Leshchinskaya T B,Taranov M M 2009 *Bulletin of the Federal State Educational Institution of Higher Professional Education* (Moscow State Agroengineering University.V.P. Goryachkina) **2** 54-61

IOP Conf. Series: Materials Science and Engineering **483** (2019) 012080 doi:10.1088/1757-899X/483/1/012080

[11] Biryulin I N, Larin O M, Rybalkin O M, Khoroshilov N V 2006 *Electro.Electrical engineering, electric power industry, electrical industry* **2** 11 -14
[12] Petukhov V 2009 *Electrical engineering news* **5(59)**
[13] Chernysheva D V 2012 *Proceedings of the South-West State University.Series: Technique and technology* **2 (1)** 100-105
[14] Mantorski Z 2008 *Light engineering* **2** 30 -33
[15] Guzhov S, Polishchuk A, Turkin A 2009 *Semiconductor lighting engineering* 1 42-46
[16] Halpin S M, Singhvi V 2007 *IEEE Transactions on power delivery* **22 (1)** 270- 276
[17] Cazacu E, Petrescu L, Ionita V 2017*Proceedings of the15thinternational conference on electrical machines, drives and power systems (ELMA)* **7955463** 358-361
[18] Kostinskiy S, Troitskiy A 2017 *Journal of Engineering and Applied Sciences* **12(20)** pp.5058-5068
[19] Singh R, Singh A 2010 *Proceedings14th International Conference on Harmonization and Quality of Power* (ICHQP) **5625347**
[20] Faiz J, Ghazizadeh M, Oraee H 2015 *IET Electric Power Applications* **9(7)** 486-495
[21] Cazacu E, Petrescu L, Ionita V 2017 *Proceedings of the15thinternational conference on electrical machines, drives and power systems* (ELMA) **7955464** 362-365
[22] Kruglova T N, Yaroshenko I V Melnikov M A, Rabotalov N N, 2015 *Electro. Electrical engineering, electric power industry, electrical industry* **2** 37-41
[23] Kobelev A V, Kochergin S V, Bakulin A V, Khrebtov N A, PavlinovV V 2011 *Questions of modern science and practic*e (Vernadsky University) **3** 374-377
[24] Vasilyeva T N,AronovL V 2014 *Young scientist* **6(65)** 148-151
[25] Olhovsky V Ya, Mutiny T V, Nayaksov S Yu 2016 *Reports of the Academy of Sciences of Higher School of the Russian Federation* **1(30)** 84-97
[26] Boyarskaya N P DovgunV P, Shevchenko E S, Barybin P A, Kabak A L 2014 *Bulletin of the Krasnoyarsk State Agrarian University* **6** 243-250
[27] Shklyarsky Ya E Skamyin A N 2013 *Electrical Engineering* **1** 69-71
[28] Averbukh M.A, Limarov D.S.2016 *Monograph*(Belgorod: BGTU publishing house) 185
[29] Mokhov V A, Turovsky F A, Turovskaya E V 2016 *Fundamental research* **9** 279-283
[30] Mokhov V A, Grinchenkov D V, Spiridonova I A *Industrial engineering, and manufacturing applications* (ICIEAM) (IEEE International conference) 1-6.
[31] Belyaev V L, S N Radimov 2012 *Electromechanical and energy saving* systems **3(19)** 469-471
[32] Kolmakov V O, Panteleev V I 2012 *Electricity: from production and distribution to efficient use* (Sat. scientificst. National Research Tomsk Polytechnic University) 87-90.
[33] Kubil V Mokhov V Grinchenkov D 2018 *Proceedings of the 6th international conference*(ICAIIT) 29-35

Cross-cultural professional communication in the context of globalization

N L Shamne, M V Milovanova, E Yu Malushko

Volgograd State University, 100 Universitetskij Ave, Volgograd, 400062, Russia

E-mail: nikolay.shamne@volsu.ru

Abstract. The article provides evidence that business communication and especially at the industrial enterprise depends on a lot of factors. One of such factors, cross-cultural communication, is focused on as a central notion. The author attracts examples and considers exogenous and endogenous factors for effective industrial business communication. One of the widely covered aspects of communication is the role of the individual, his personality and personal traits of behavior in establishing and maintaining mutual understanding. The concept of culture in corporate or business ethics is proven to be an inseparable factor of communication which should guide the communicators in treating contracts and employees.

1. Introduction

Cooperation between institutional actors and different cultural backgrounds, which lasts longer than usual, can be called cross-cultural business relation. It occurs as industrial enterprises are in contact with cultures. A special cross-cultural context is formed by business contacts between countries that exist in different industries and serve different purposes. Finding partners for sales or suppliers abroad, low-cost production in low-wage countries or international cooperation in research are examples of cross-cultural contacts as a part of activities at industrial enterprises.

The end of the 20th and the beginning of the 21st century witness formation of a new picture of the world, which reflected major changes in all spheres of human life. Two paradoxically linked processes – globalization and national identification - have taken an important place [1].

Globalization is associated with unification of life in different countries and creation of a new elite, detached from the national ground, being cosmopolitan in a lifestyle, English-speaking, associated with transnational structures and standard "models of life".

The issue of national identity is connected with the growing desire of peoples to preserve national identity, to strengthen the traditional national identity.

In the future, we can expect that the number of business contacts between industrial enterprises from different countries will only grow. This is due to a sharp rise in the national economy of countries, especially in the Asia-Pacific region, formation of economic unions and, to a large extent, the processes of globalization. This enhances the attractiveness of regional and national markets. But due to the fact that partnership requires communication, coordination and planning, the requirements to the participants are increased, as they often realize the cultural complexity of the situation and tend to misinterpret the facts and make mistakes in their work.

What values affect the long-term relationship between industrial enterprises with different cultural backgrounds? It is not easy to answer this question because cross-cultural business relations are a multifaceted subject of research [2].

(1) The first range of issues concerns the conceptualization of culture. There is still no universally accepted definition, and the notion remains unclear even in anthropology due to the diversity of research interests. Before exploring cross-cultural business relationships, it is necessary to define the essence of culture.

(2) The situation which cultures face is "cross-cultural". The notion of "business relations". Business relations cover a variety of inter-organizational contacts, the latter being of great importance in practice. Creating and maintaining relations with foreign partners become the main problem in management. If we consider culture as an essential determinant of human behavior, it becomes obvious that culture also affects economically motivated relations. To make them fair, it is not enough to have just foreign language skills and mobility, they need to be complemented with cross-cultural awareness and knowledge.

2. Methodology and research

2.1 Definition of the concept: cross-cultural business relations

Business relations are generally understood to be mutual relations between buyers and sellers at the enterprise level.

Business relations are also inter-organizational relations where the object is the industrial enterprises themselves. There may occur a situation that only relations between certain persons make it possible to establish business relations. There is a cross-cultural situation that links social organizations together at different levels, primarily at the level of enterprises and the country [3]. The main importance is given to various national and cultural ideas which are developed in country-specific cultural systems of actors and organizations. They work not only at the enterprise level but also in all business units. Thus, partners are trying to learn above-situational factors of behavior and to build their own behavior based on this. As a result, there is an agreement that is fair to all participants. In the case of international business relations, there appears an understanding of the meaning of national and cultural ideas. Interculture dominating cultural ideas in society. Therefore, business relations should be considered taking into account that there are cultures of different countries.

2.2 Factors affecting cross-cultural communication in the sphere of business relations

In any communicative situation, business relations are affected by a number of factors starting from the current macroeconomic conditions up to the expectations of the communication participants of the specific business cooperation activity. A cross-cultural analysis takes into account, first and foremost, the factors that are both a starting point and a result of cross-cultural interaction, both are most closely related to business relations and also subject to constant changes. However, it is almost impossible to characterize the process and the result of forming business relations solely due to the so-called *endogenous* factors. There also exist many *exogenous* characteristics that are not so closely related to specific business relations, but affect the process of communication from the outside and are hardly changeable in the course of cross-cultural interaction [4].

Exogenous factors include:

- macroeconomic factor
- characteristics of the communication participants
- characteristics of organizations involved in the process of business communication
- own culture and the culture of a business partner (as a central component).

Endogenous factors include:

- economic and sociopolitical factors of influence
- history of developing business relations
- desires and intentions of business partners
- cross-cultural factor (as a central component)

The peculiarity of the presented concept includes consideration of culture as an endogenous or exogenous parameter in the sphere of business relations. Own culture or the culture of a business partner is a cultural background of the participants of cross-cultural interaction. A cross-cultural factor is a factor occurring in specific business situations that can undergo certain alteration only in the part of a key group of individuals participating in this communication. Thus, taking into account certain business relationships, the cross-cultural factor is endogenous while individual cultures of business partners are considered exogenous.

2.3 Consideration of external factors

Any organization is exposed to external factors, so it should take them into account. But industrial enterprises do not often have enough resources to fully analyze all the relationships in the external environment. The external environment is understood as a complex structure consisting of several categories, the priority among which is given to political, legal, economic, social, cultural, physical, technological ones and the category of competitiveness.

For example, the analysis of external environmental factors is an essential task in the framework of strategic management and an important prerequisite for strategic planning. In this regard, international corporations should analyze all geographical areas within which they operate or plan to start work.

2.4 Personality and other individual factors

In business processes of cross-cultural interaction the result of communication is achieved by its participants. Even the decision-making, at first sight having absolutely objective character, is conditioned by personal positions, feelings and principles of communicators. In research on cross-cultural communication, personal qualities of individuals are of great importance. Their absence makes it impossible to obtain reliable and objective knowledge in this field.

2.5 Personality and similarity

Human personality is understood as a psychological structure, expressed in the form of a stable reaction of the individual to the surrounding reality. This is linked to the assumption that all individuals have certain intrinsic qualities on the basis of which they differ from each other. The situational factor of behavior, as a rule, is not taken into account. Supporters of the opposite opinion consider the situational factor decisive in the analysis of human behavior. The results of the research show that even knowing personal qualities of a person, it is impossible to fully predict his/her behavior.

Kale and Barnes offer a typology of individuals including four scales:

- extrovert - introvert
- sensual perception (awareness of details, facts) – intuition (ingenuity, rich imagination)
- thinking (logic and objective decision-making) - subjective perception (subjective reasoning of decision-making)
- perception (growing interest in obtaining information, observation and understanding of what is happening around) - assessment (completion of information retrieval, organizational skills) [5].

In the process of cross-cultural communication (in negotiations) the relationship between extrovert behavior and general attractiveness of the participants of communication was established, which in turn had a positive impact on business relations. Thus, personal traits are considered as potentially basic parameters of communicative behavior, especially the style and content of communication between participants.

The influence of personality on the negotiation process has been the subject of numerous studies. However, there is no single unambiguous opinion on this problem in science. For example, talented communicators can change their external characteristics depending on the situation itself, thereby masking the actual character traits. Therefore, it would be appropriate to talk about the influence of the individual and to take into account the ability of participants to respond to a particular situation in cross-cultural interaction.

Like any form of social organization, industrial enterprises are characterized by specific structures, processes and cultural ideas.

The **culture** of the organization is understood as the standard of shared values and beliefs that help individuals to understand the principle of enterprise functioning and to set standards of internal behavior [6].

It is often assumed that corporate culture is concentrated on management. Culture is a tool of influence and guidance for a manager. For strategic reasons, culture is a competitive advantage. Culture forms an individual's sense of self-awareness and commitment to others, thus being a relevant factor. The influence of culture on industrial enterprise functioning is often compared with the influence of the personality on the individual behavior.

3. Culture parametres

In the study, Reznolds sets the parameters of a culture [7]. For Kale and Barnes, some of these parameters are most relevant because they structure the content and the style of communication.

- Orientation of the industrial enterprise to an external activity is aimed at satisfying the needs of clients and business partners. Internal orientation involves production and activities within the industrial enterprise.
- Within the framework of one type of culture, the priority is given to fulfilling the tasks of the industrial enterprise – the effective achievement of economic growth and profit. Other cultures, on the other hand, involve the personal and social welfare of their employees. Orientation of the enterprise to one or another value system is expressed in the style of cross-cultural communication aimed at the most effective transaction. Interpersonal relations and the interests of all participants are important for socially oriented industrial enterprises in business negotiations.
- The company's culture, which is focused on maintaining its own stability and security in a dynamically developing market, is conservative and cautious. The enterprises that are ready to take risks calmly perceive any changes and even see certain positive chances. The enterprises, focused on stability and safety of own activity, try to record all agreements in writing, take into account a reputation of the business partner and aspire to establish personal relations. Enterprises, ready to take risks, do not hesitate to search for new partners and try some new creative solutions. They accept imperfect terms of contracts.

- A tolerant attitude within the framework of the corporate culture to the behavior of employees, deviating from the generally established norms, emphasizes the recognition of the individual nature of each employee. Such enterprises represent themselves in the business community, observing all the formalities. They offer their partners already proven ready-made solutions. These enterprises are ready to make concessions, adhere to the terms of the contract and are oriented to the ritual character of intercultural interaction. Enterprises that recognize an individual approach to their employees usually offer a wider range of freedom to their representatives when establishing contacts. At the same time, the process of interaction is not carried out in line with the usual unilateral approaches but involves the creative basis of business relations.

3.1 How "superficial" is the business culture?
The scientific opinion in this regard is very contradictory. Some scholars believe that business values are as important as national ones. Others strongly refute this view. According to Schein (1984), the importance of business culture cannot be determined by external behaviour and hidden values. To understand the culture, in his opinion, it is necessary to take into account the basic generally accepted rules and norms that are unconsciously perceived by employees but, nevertheless, affect their perception, thinking and attitude.
The evidence supporting this view is found in the Hofstede business culture study (1990) [8].

Six factors were identified to describe cultural differences between organizations:

- Process priority vs. result of activity
- Priority of social guarantees vs. tasks performed by the enterprise
- Organization itself vs. type of tasks performed
- Open system vs. closed system
- Soft control vs. strict control
- Normative vs. pragmatic orientation

These empirical data are similar to the results of the Reynolds logical analysis (1986) and indirectly confirm his point of view. The most important thing is that the activity of the enterprise as a "superficial cultural phenomenon" in comparison with "hidden values" is the most acceptable for the analysis of business culture. The analysis of the results of the sociological survey held by Hofstede shows that these six factors reflect social relations rather than the structure of business culture [9]. On the basis of available research on business culture, it is impossible to obtain unambiguous conclusions about its essence. The data obtained is a subjective interpretation made by the observers. Culture is a product of thought and thus cannot be characterized by external factors.

3.2 The difference between business culture and business climate
The results of the research are not a cultural phenomenon, but an aspect of the social system, being components of the *business climate*. In this case, we are talking about a relatively stable property of the internal environment of the organization, in which its members reside and which affects their behavior.
Business climate (atmosphere) and business culture are not the same, they are just related. Climate is a pattern by which the daily life of the organization is built. Culture is the basis of the organization, its originality and individuality.

3.3 The Impact of culture on the individual character
It can be assumed that certain personality traits in some countries appear much more often and in more contrast than in others. Even if the concept of personality presupposes the absence of cultural-specific conditioning o human behavior, studies of national stereotypes indicate that each country has its own national character.

The concept of national character is not based on the psychological identity of society. The main task is to characterize the population of a country through relevant socio-cultural concepts. It is necessary to find out which characteristics are inherent in this very society, and which ones are quite universal.

Campbell (1988) points out that cultural-specific characteristics are found in the negotiations on purchase and sale.

In the United States, for example, the course of negotiations depends on decisions taken right at the negotiating table. Therefore, their result will be positive if all participants adhere to the same principle of solving the problem.

In France, on the contrary, the most important factor of commercial success is personal qualities of negotiators and their similarity.

In Germany, the relationship between the commercial interest of the supplier and satisfaction of the customer with the transaction is important in the negotiations.

In the UK, an important role is given to the context of negotiations which guides sellers.

The process of cross-cultural communication can be depicted in the form of a triangle, which consists of cultural differences in tems of different value systems, individual differences concerning the character and personal experience of the individual, as well as intercultural knowledge and experience. This reflects the complexity of the combination of cultural and individual factors in intercultural communication.

Studies of the national character of individuals result in singling out certain traits of their behavior. The differences are interpreted as a product of national cultural specificity without taking into account the ideas that differ countries or cultures from each other.

Some studies are devoted to the content of culture. In one empirical study on the negative attitude of society to innovation, the relationship was identified between the value orientations of the consumer (e.g.. fatalism as an expression of the attitude to the relationship between man and nature) and his attitude to human innovation. Supporters of fatalistic cultures showed the least interest in approbation of new production, nevertheless, taking it as a small risk.

Other studies are mainly aimed at one goal which is to identify cross-cultural differences based on certain cultural concepts. For example, through a special method of assessment, significant differences were established in relation to the values of students from different continents, including students from Germany and the United States.

In enterprise management, different cultural values and postulates affect management functions and manifest themselves in different approaches and ways of management. Leadership values always reflect the values of a culture

Numerous cross-cultural studies of personal values of leaders and the shared professional values in a number of states have revealed both similarities and differences. Differences in cultural values were found in Germany and Switzerland, despite the fact that they border on each other and are characterized by partial kinship of languages. International studies have found that political and role systems, as well as the principles of subordination and leadership, are determined by the country of origin of the leader. The fourth point was assessed through the fact whether it is important for the leader to know exact answers to all professional questions of subordinates. Only ten per cent of Swedish and eighteen per cent of American leaders surveyed responded affirmatively. In Germany, the figure was almost fifty per cent and in Japan eighty per cent. As a result, leaders from different cultures have different values, which affects their management functions. Despite this, traditional cultural values can be conditioned by socio-economic conditions, as it turned out in a comparative analysis of the value system of leaders from the United States, Hong Kong and China.

4. Endogenous factors in business relationships

Endogenous factors differ from exogenous ones with their own specific features in certain human business relationships. Some personal characteristics are preserved regardless of their affiliation to an organization (demographic factor, personal values, personality, etc.) while such parameters as position, role or status are very relative and subject to change. When considering business relations between industrial enterprises, we are talking about "absolutely typical" social organizations, which can be characterized by common universal parameters. Cross-cultural ideas, which express social relations between partners, are of interest in cross-cultural communication. It is important to consider the history of business relations as a whole, as well as to analyze the desires and intentions of the participants of business communication in relation to each other.

Communication is an important factor in business communication. In business relations marketing is always a communicative process and includes all aspects of interaction.

5. Versatility of the communication structure.

When establishing external communication links, the issues of "who communicates with whom", "what", "how", "which way", "when", "with what consequences" are not significantly different from other important decisions within the framework of the communication policy of an industrial enterprise. Unlike traditional marketing the boundaries of communication between industrial enterprises are established jointly by its participants. This is carried out explicitly through treaties or implicitly when the negotiators are kept silent.

As a result of creating and using communication channels, organizations can communicate with each other. Structures and processes of communication reflect structures and processes of social relations between the partners [10]. This concept is supplemented by the observation that the description of the climate of the organization contains aspects of communication and vice versa. The communication climate is largely similar to the business climate of the organization [11], [12]. At the same time, the set of communicative aspects in business relations affects other parameters of the organization. In relations between organizations it is assumed that communication:

- can be used to persuade a partner by reporting information
- facilitates the process of making decisions by its participants
- coordinates the programs of the enterprises
- finds application in realizing the functions of government
- increases the sense of responsibility and loyalty
- interrelated with the atmosphere of trust and mutual support
- related to conflict situations in business relationships

It is necessary to add that a communicative behavior of business partners extends influence of sociopolitical factors on business relations and vice versa it is caused by these factors.

6. Conclusion

Compared with the consideration of communication as a mechanical phenomenon, the psychological factor does not reflect the impact on the communicative structures and processes. Its influence extends just to the internal state of communication participants. The impact of psychological factors on the process of communication consists in correct choosing incentives relevant in the course of cross-cultural interaction by communicators themselves. They also need to select the correct signal processing mode to recognize the value of the sender's information. At the same time, mutual understanding can be improved in case of similarity of cultural perception by business partners. Learning the language and nonverbal communicative behavior of the partner significantly increases the accuracy and speed of communication when establishing business relationships. From a psychological point of view, there is the following relationship between communication and other socio-political factors: the executor strives hard to learn the psychological characteristics of his client and focuses on them in the process of business communication.

An important addition to the above-said is an interpretive and symbolic point of view, as it highlights the important functions of communication – formation of culture and business relations. In contrast to the psychological point of view, the main focus is not given to the internal state of the participants but to the process of interaction. Through interrelated actions partners are able to directly influence processes and structures in business relationships. It is not about creating physical channels of information transmission, but about generating congruent meanings that facilitate understanding in communication. The synthesis of the culture of business relations on the way to cross-cultural understanding is an appropriate basis through which the visual support of the communication attributes, as well as other socio-political factors. Common cultural ideas or standards are created with the help of interpretative and symbolic communication. Their lack or absence would make communication structures and processes, and as a result, relations between organizations quite impossible.

References

[1] S V Serebryakova, A I Milostivaya 2017 Semantic emergence as a translation problem *Vestnik Volgogradskogo Gosudarstvennogo Universiteta-seriya 2-yazykoznanie* **16 (3)** 48-57

[2] Shamne N L 2012 Development of short-term forecasts of the development of the Russian literary language in the Lower Volga region. *Vlast'*. **7**

[3] Milovanova M V, Terent'eva E V 2011 The language space of a multi-ethnic region: environmental problems (Volgograd: Izdatel'stvo VolGU) 292

[4] Omel'chenko E A From the experience of teaching Russian and socio-cultural adaptation of migrant workers: problems and prospects.: Available at: http:// refugee.ru/Events/2013/omelchenko_ehtnosfera.pdf.

[5] Kalenkova O N 2013 Lessons of Russian language. CMD for children of preschool and primary school age. CMD for children of preschool and primary school age. Moscow: Russkij yazyk, Kursy

[6] Sztompka P 1991 Society in Action: The Theory of Social Becoming (Cambridge: Polity Press and University of Chicago Press)

[7] Sztompka P 1993 The Sociology of Social Change (Oxford and Cambridge: Blackwell)

[8] Shamne N, Milovanova M V, Terentyeva E V, Velibekova I M 2016 The polyethnolect phenomenon in the current state of language in Germany and Russia *XLinguae*. T. 9. **4**

IOP Conf. Series: Materials Science and Engineering **483** (2019) 012081 doi:10.1088/1757-899X/483/1/012081

[9] Veber M 2016 Economy and society: essays understanding sociology. Sociology. Moscow: dom Vysshej shkoly ehkonomiki
[10] Murugova E V 2017 Theoretical and applied aspects of corpus based researches, *Vestnik Volgogradskogo Gosudarstvennogo Universiteta-seriya 2-yazykoznanie* **16** **(2)** 197-200
[11] Parsons T 2018 Social system (Moscow: Akademicheskij proekt)
[12] M A Suzdalova, V G Lizunkov, E Yu Malushko, N A Sytina, V E 2017 Medvedev Innovative Forms of Partnership in Development and Implementation of University-Business Cooperation *The European Proceedings of Social & Behavioural Sciences EpSBS* **XIX**450-455

IOP Conf. Series: Materials Science and Engineering **483** (2019) 012082 doi:10.1088/1757-899X/483/1/012082

Self-checking quality of welded structures

V V Miroshnikov[1], A V Morozova[1], D O Vasin[2]

[1] Bryansk State Technical University, Bryansk, 241035, Russia
[2] Bryansk machine-building plant, Bryansk, 241035, Russia

[1]niotiostu@gmail.com

Abstract. The problem of controlling the production processes is considered, the qualitative result of which cannot be confirmed by subsequent monitoring and measurements, and deviations from the established norm become apparent only after using the products. In this regard, there is a need to organize the product quality self-control by the performers of the technological process. The author's technique of self-controlling production quality of mechanical engineering is offered

1. Introduction

The basis of product competitiveness in the modern market is its quality, which is provided by many factors, both technological and marketing. Implementing the methodology of product quality improvement in the production process[6, 7, 13, 27], including methods of defect analysis [21, 31], methods of changing the quality of the technological environment [2, 5, 12, 26], means of increasing the level of interaction between the quality of technology and the quality of production relations [1, 4, 11] create the basis for further improving consumer evaluation of product quality[18, 22, 25, 30], which also includes informing the target audience of the product qualimetric characteristics [9, 10, 14, 19, 23].

In most enterprises, the quality control (from verifying the purchased product to product quality control) is handled by the specialized structural units, for example, the departments (authorities) of Quality Control (DQC). The main purpose of the quality control is to eliminate defects in the output products. However, this goal cannot be achieved in all technological processes.

At modern machine-building enterprises there are such special technological processes as welding, soldering, painting, crimping, gluing. A special technological process is a process, the result of which cannot be confirmed by subsequent monitoring and measurements, the deviation of which becomes apparent only after the beginning of using the products. The proper execution of such processes determines the quality of the final product. The workers of technical control are responsible for controlling the performance of technological operations, they can't provide it as they aren't experts in the field of the particular special process. Thus, the problem of organizing the product quality self-control by the performers of the technological process becomes vital. Currently, it is not sufficiently developed from a methodological point of view. Therefore, the author's team offers its methodological vision of providing quality self-control of special technological processes at the machine-building enterprise.

2. Discussion

At present, the technical control service is only able to separate low-quality products from high-quality ones, in other words – to reject, but it is not able to prevent the production of low-quality products, in this regard, the consumer's risk increases.

Improving the activity of technical control must provide carrying out works on introducing self-control of the main production workers (it also includes forming the list of the technological operations transferred to self-control, and providing the workplaces with necessary monitoring, measuring and test devices, tools, documentation and organizing the workers' special training).

The main task of self-quality control is implementing all the operations of quality control by the contractor directly in the workplace in full compliance with the technical documentation requirements. Thus, the bulk of control operations are transferred to the earliest stages of the product quality formation process, which naturally contributes to the timely detection and rapid correction or isolation of defects.

As an example, let's consider welding as one of the most intensively studied types of such processes [8, 15, 16]. This process is examined and improved both technologically [3, 17, 24, 28] and instrumentally [20, 29]. Since it is impossible to assess the quality after a special process, it is necessary to organize a system of self-quality control of the welding process. The challenge is to encourage the performer to ensure quality, and the performer's knowledge in this area may not be enough. The solution to this problem is to develop a system of welders' certification for self-control suitability. Training the performers of the special process should be focused on ensuring the quality of work performed by them.

Quality performance of welding operations in total is provided by [32]: the quality of welding equipment; the quality of welding materials; the quality of basic materials (products); the requirements for welders and welding operators' qualification; requirements for normative and technical documentation; requirements for labor protection and safety of welding works; production rooms and welders' workplaces needed to be appropriate for performing welding works.

The developed by the authors method of regulating the process of welding self-control is based on analyzing the experience of the product quality of machine-building enterprises (for example, Bryansk machine-building plant) [33]. This method can be presented in the form of the following block diagram (Fig. 1).

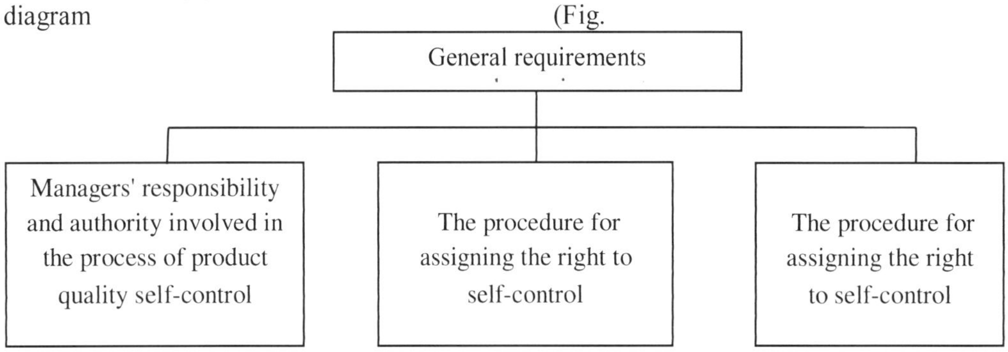

Figure 1. Structure of the welding processes self-control method.

As an "engineering" application to the self-control technique (Fig. 1) the authors propose a method of improving the reliability of the quality self-control process through PFMEA-analysis of a special process that requires "tightening" self-control. The algorithm for this analysis is shown in Fig.2 [33].

PFMEA of the technological process is an analysis of the originally developed and proposed technological process and (or) completion of this technological process in the course of the corresponding PFMEA-teamwork. It allows to prevent the introduction of insufficiently developed technological processes into production and to analyze all design features with respect to the planned technological process.

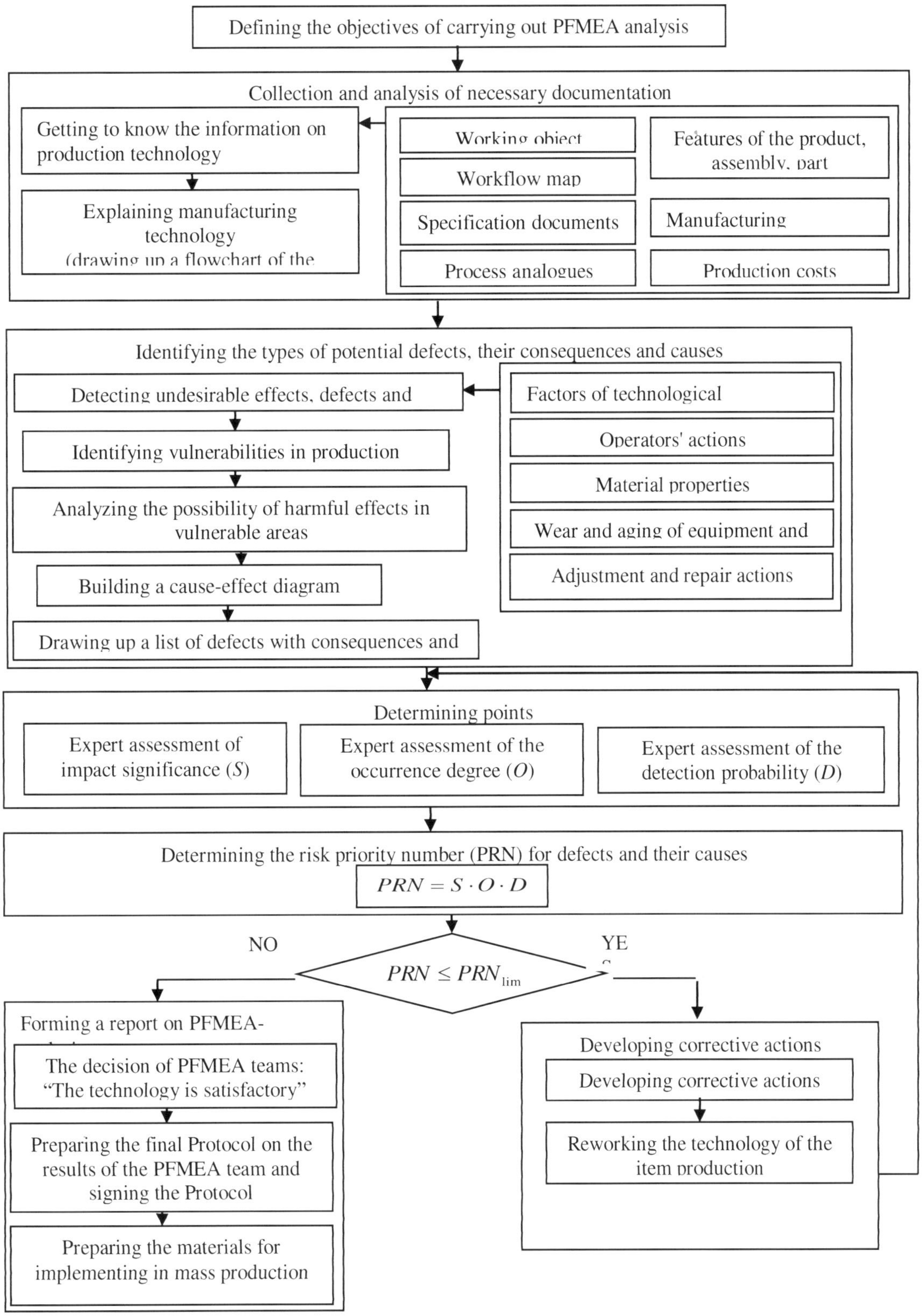

Figure 2. The algorithm of the FMEA team work.

A key question in PFMEA process is: "How can failures in the technological process affect the product, process efficiency or safety?"

PFMEA is conducted by a specially selected FMEA team in the following sequence.

1. Determining the process analysis objectives. The aim of FMEA-analysis of the production process is to ensure the compliance with all requirements for the production processes quality, to insure the assembly by making changes to the process plan for technological actions with an increased risk, as well as by improving the quality of processes and product competitiveness.

2. Collecting and analyzing necessary information, including information on production technologies; on counterparts, technical documentation, etc. PFMEA team should not start the analysis until all the necessary information is collected. After getting to know the information on the production technology, an explanation of the manufacturing technology is carried out on the basis of drawing up a flowchart of the technological process.

3. Determining the types of potential defects, their consequences and causes. At this stage, adverse effects, the technological process mismatches are identified that may occur under the influence of various factors: various factors of technological environment; various properties of materials; operators' actions; wear and tear, aging of equipment, tools; steps for installation and repair.

4. Determining points. The consequences of potential failures are assessed by their significance (S). Significance is assessing the severity of a potential failure for the next component, subsystem, system, or consumer. The degree of occurrence possibility (O) of specific causes or mechanisms is described by a rank number from 1 to 10. The effectiveness of the envisaged control measures (D) is assessed by the ability of the proposed control actions to detect a potential cause (structural weakness) or to detect the next type of failure before the component, subsystem or system are put into production.

5. Determining the risk priority number (PRN). For each identified potential defect and each cause, the risk priority number is determined by the formula: $PRN = S \cdot O \cdot D$.

For risks having a calculated value of *PRN* larger than the predetermined threshold value $PRN_{\lim}$,special steps are taken allowing you to: eliminate the causes of risk; to prevent or reduce the probability of its occurrence; to reduce the impact of the risk; to facilitate and improve the probability of detection and identification.

6. PFMEA concludes with a report on the risk analysis in the form of a Protocol on the results of the PFMEA team.

Experimental verification of the proposed method of the welding process quality self-control is carried out on the example of welding the bottom of the container, AYDMK manufactured in Joint-stock company "Management Company "Bryansk machine-building plant".

The PFMEA-analysis of the technological process of welding the container base, conducted by a multifunctional PFMEA-team reveals a potential defect that can lead to the failure of the structure that is failure of welded joints of the container base frame.

A potential cause of the defect is the welding technology imperfection, in particular, welding large thicknesses of welded parts (25 mm) in one pass, as well as welding joints in a vertical position, which leads to excessive complexity of the welding operation.

While discussing the type of the potential failure, the PFMEA team offered to weld the products by multi-pass welding with overlapping rollers in staggered order. With regard to implementing the weld in the vertical position, it was decided to develop and implement a horizontal marching rotator for the possibility of welding the container frame in the lower horizontal position.

In addition, the PFMEA team identified other potential defects in the process and eliminated their causes.

3. Conclusion

As the result of the carried-out PFMEA analysis, the risks identified in the technological process of welding the container base were reduced by technologists' performing corrective actions and, thus, the reliability of the self-control process of the welding work quality was increased.

References

[1] Aydalot P, Keeble D 2018 High technology industry and innovative environments: the European experience (taylorfrancis.com)
[2] Chapman W L 2018 Engineering modeling and design (taylorfrancis.com)
[3] Cornu J 2013 Advanced Welding Systems: 1 Fundamentals of Fusion Welding Technology (books.google.com)
[4] Davies A 2018 Industrial relations and new technology (taylorfrancis.com)
[5] Ellul J 2018 The technological system (books.google.com)
[6] Fischer KE, Peters K, Stargardt T 2017 Information on product quality, change agency and product adoption Academy of Management Proceedings (journals.aom.org)
[7] Fukuda R 2018 CEDAC: A tool for continuous systematic improvement (taylorfrancis.com)
[8] Górka J, Czupryński A, Adamiak M 2017 Properties and Structure of Nanocrystalline Layers Obtained by Manual Metal Arc Welding (MMA) Archives of Metallurgy and Materials (degruyter.com)
[9] Hallak J C, Schott P K 2011 Estimating cross-country differences in product quality The Quarterly journal of economics (academic.oup.com)
[10] Hallak J C 2006 Product quality and the direction of trade Journal of international Economics (Elsevier)
[11] Hartley J R 2017 Concurrent engineering: shortening lead times, raising quality, and lowering costs (taylorfrancis.com)
[12] Hitomi K 2017 Manufacturing Systems Engineering: A Unified Approach to Manufacturing Technology, Production Management and Industrial Economics (taylorfrancis.com)
[13] Ji-fan Ren S, Fosso Wamba S, Akter S 2017 Modelling quality dynamics, business value and firm performance in a big data analytics environment International Journal of Production Research (Taylor & Francis)
[14] Kahn B K, Strong D M, Wang R Y 2002 Information quality benchmarks: product and service performance Communications of the ACM (dl.acm.org)
[15] Katayama S, Yohei A, Mizutani M, Kawahito Y Development of deep penetration welding technology with high brightness laser under vacuum 2011 Physics Procedia (Elsevier)
[16] Khan M I 2007 Welding science and technology (books.google.com)
[17] Khanna O P, Dhanpat Rai & Sons 2004 A text book of welding technology
[18] Kugler M, Verhoogen E 2011 Prices, plant size, and product quality The Review of Economic Studies (academic.oup.com)
[19] Lawrence X Y 2008 Pharmaceutical quality by design: product and process development, understanding, and control Pharmaceutical research (Springer)
[20] Majeed T, Wahid M A, Ali F 2018 Applications of Robotics in Welding Research in Management and Technology(ermt.net)
[21] Robinson S L 2017 Automated inspection and quality assurance (content.taylorfrancis.com)
[22] Rokonuzzaman M, Paswan A 2017 Effect of Product Return Policy on Consumer's Risk Perception, Store Image, and Store Patronage: A Causal Investigation Creating Marketing Magic and Innovative Future Marketing Trend (Springer)
[23] Schnurr B, Brunner-Sperdin A, Stokburger-Sauer N E 2017 The effect of context attractiveness onproductattractiveness andproduct quality: the moderating role ofproductfamiliarity Marketing Letters (Springer)
[24] Seidlitz H, Fritzsche S, Ambrosio M 2017Advanced Welding Technology for Highly Stressable Multi-Material Designs with Fiber-Reinforced Plastics and Metals Open Journal of Composite Materials (scirp.org)
[25] Tsiotsou R 2006 The role of perceived product quality and overall satisfaction on purchase intentions International journal of consumer studies (Wiley Online Library)
[26] Tukker A, Charter M, Vezzoli C, Stø E, Andersen M M 2017 System innovation for sustainability 1: Perspectives on radical changes to sustainable consumption and production

(books.google.com)
[27] Wei Z, Feng Y, Hong Z, Qu R, Tan J 2017 Product quality improvement method in manufacturing process based on kernel optimisation algorithm International Journal of Production Research (Taylor & Francis)
[28] Xiaoming Z, Longlong S 2017 Summary of the Development of Hydraulic Technology and Welding Technology Fluid Power Transmission & Control (en.cnki.com.cn)
[29] Yang S, Li W 2018 Surface Quality and Finishing Technology Surface Finishing Theory and New Technology (Springer)
[30] Yoo O S, Sarin R 2018 Consumer choice and market outcomes under ambiguity in product quality Marketing Science (pubsonline.informs.org)
[31] Zhang X, Kano M, Tani M, Mori J, Harada K 2018 Defect Data Modeling and Analysis for Improving Product Quality and Productivity in Steel Industry Computer Aided Chemical Engineering (Elsevier)
[32] Gladkov E A 2006 Management processes and equipment for welding (M: Publishing center "Academy")
[33] Miroshnikov V V 2012 Model of complex multilevel FMEA analysis of complicated systems (Bryansk: BSTU)

Modern coatings of metal-cutting tools as a way to control the efficiency of noncorrosive steel treatment

Mokritsky B[1], Morozova A[2], Sebrennikova A[1]

[1] Komsomolsk-on-Amur State University, 27 Lenina str., Komsomolsk-on-Amur, 681013, Russia
[2] Bryansk State Technical University, 7, 50-let Octyabrya str., Bryansk, 241035, Russia

[1] boris@knastu.ru; [2] niotiostu@gmail.com

Abstract. Treating specialized stainless steels is always associated with the need to improve processing performance and efficiency of metal-cutting tools. Functional coatings for domestic metal-cutting tools are proposed, their efficiency is estimated by the efficiency parameters, the analysis of cutting forces is involved to explain the results

1. Introduction

The paper presents the results of developing and applying modern functional coatings on domestic metal-cutting tools for turning viscous and highly corrosion-resistant stainless steel, the grade of the steel is 09Х17Н7Ю (09H17N7YU). As a criterion for evaluating the effectiveness of coatings, the paper applies a performance comparison of the feed-through turning cutters by the value of their wear on the back face for equal operating time under the same operating conditions.

The aim of the work is to identify the most rational coatings without reducing the productivity and treating quality.

Taking into account the actual absence of recommendations for treating such stainless steel, we selected domestic tool hard alloy of the VK8 (ВК8) brand as the basic version of the tool material in the delivery condition from JSC “Kirovgrad hard alloy plant”. The considered coatings were applied to the same material. The time to achieve wear on the back face, equal to 0.5 mm, was taken as the criterion of assessing the tool performance. Wear and tear measurements were carried out every 15 minutes of the tool performance.

2. Results and their discussion

2.1. *Initial condition*

The data are given only on quadrangular square plates (diameter of the described circle is 17,5 mm) with the central fixing aperture and without it, with a chip groove and without a chip groove.

The diameter of the workpiece varied during the treatment from 280 to 60 mm. The cutting speed was maintained within 50-55 m / min by varying the number of spindle revolutions. The screw-cutting lathe of model 16K25 was used. The cutter feed from the requirements of the treated surface roughness is selected equal to 0.21 mm / Rev. The cutting depth was taken equal to 0.5 mm for finishing conditions and 1 mm for roughing conditions. In both cases, the maximum permissible amount of wear on the back face was considered to be 0.5 mm. The cutting tool was compared according to the period of wear resistance, i.e. by the work time of cutting plates under the due roughness to achieve a wear of 0.5 mm for the back face.

Wear measurement was carried out on a multi-sensor measuring center (video measuring machine) of the model Micro Vu Sol 161. The test results were duplicated and documented.

2.2. *Discussing the results*

Fig.1 shows a graph of the experimental dependence of the wear value h3 on the time t of the carbide tool of brand BK8 (VK8) in the delivery condition. This tool produced at JSC "Kirovgrad hard alloy plant" is taken as basic. All other tool materials (substrate + coating) are evaluated in relation to this base and among themselves. Similar graphs are obtained for different instrumental materials.

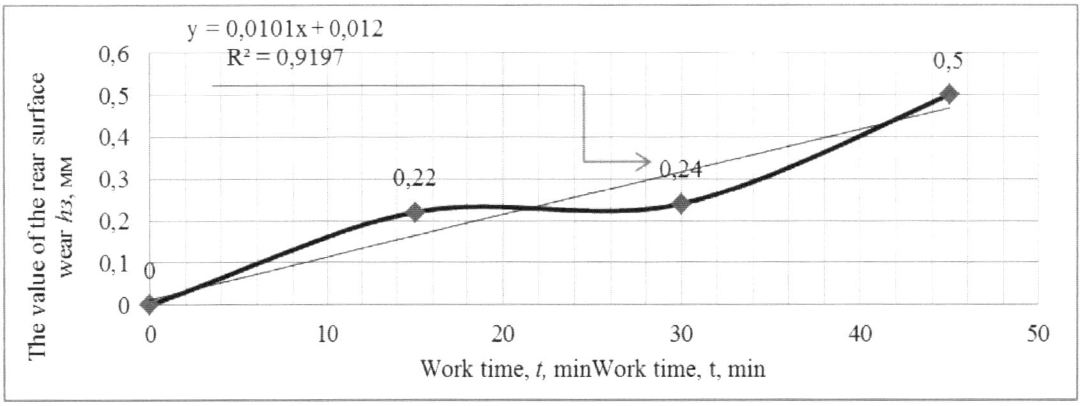

Figure 1. An example of the dependence of the wear value h3 on the work time t of carbide tool (grade VK8) (cutting mode v=55 m/ min, n=160 rpm, s=0.21 mm / rev, t=1 mm)

The visualization of individual results is shown in Fig.2, where: 1 is carbide brand VK8 in the delivery condition; 2 is tool material CCMT 120404-HMP of the company KORLOY; 3 is tool material VK8 coated with Ti (up to 1 μm) + TiN(1 μm) + (TiAl)N(2 μm) + TiN(0.5 μm) [applying coating layers* by the condensation method with ionic bombardment (CIB) using filtered droplet phase and assisting with accelerated ions]. In this case the thicknesses of the layers in the coating are given in parentheses, and the conditions for applying the layers are given in square brackets. Information on filtering the droplet phase and assisting with accelerated electrons is given in the work [1]; 4 is tool material VK8 coated with Ti (up to 1 μm) + TiN (1 μm) + (NbZrTiAl) N (2.5 μm) [applying coating layers by the condensation method with ionic bombardment (CIB) using filtered droplet phase and assisting with accelerated ions].

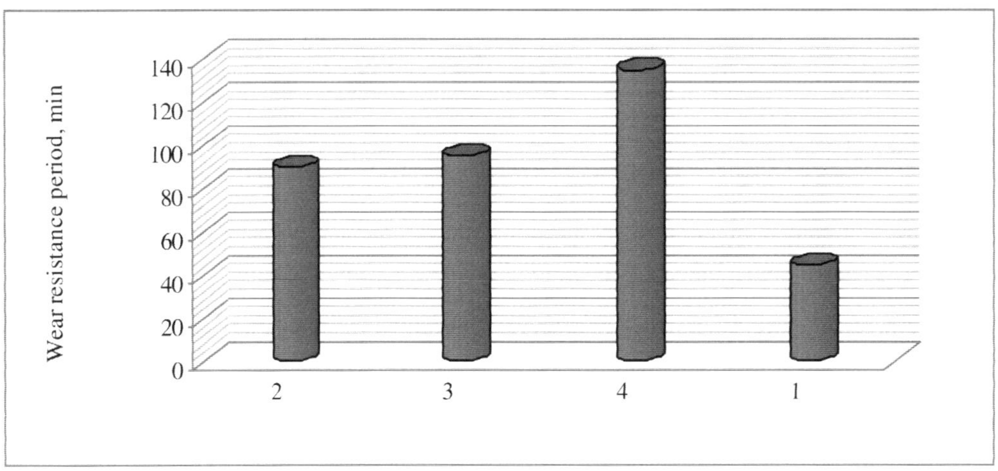

Figure 2. Comparing the period of wear resistance of various tool materials for specified operating conditions

Another example of comparing wear resistance of tool materials is shown in Fig.3, where: 1 is VK8 in the delivery condition; 2 is VK8 + coating No. 1; 3 is VK8 + coating No. 2; 4 is VK8 + coating No. 3; 5 is VK8 + subjected to the diamond thermal sharpening of the front surface and polishing the rear surface of the plate: coating №1 is Ti (up to 1 μm) + TiN(1 μm) + (NbZrTiAl) N (2.5 μm) [applying coating layers by the condensation method with ionic bombardment (CIB) using filtered droplet phase and assisting with accelerated ions]; coating No. 2 is Ti (up to 1 μm) + TiN (1 μm) + (NbZrTiAl) N (2.5 μm) [applying coating layers by the condensation method with ionic bombardment (CIB) using filtered droplet phase and assisting with accelerated ions]; coating No. 3 is (AlCr)N(1.5 μm) + (AlTi)N(2 μm) [applying coating layers by the condensation method with ionic bombardment (CIB) using filtered droplet phase and without assisting with accelerated ions].

The wear resistance here refers to the time of the plate to reach a certain amount of wear on the rear surface, this value is taken to be 0.5 mm.

To explain the identified results and patterns strain gaging was made in all three of its components (Fig.4), in which the cutting mode was the following: cutting speed was 50 m / min, feed was 0.21 mm / rev., depth of cut was 0.5 mm: *a* is a test stand using Stand STD.2012-2 model dynamometer; *b* is visualizing and documenting the process of recording the cutting force components; *c* is a graph sample showing the effect of the tool coating on the vertical component value of the cutting force (1-VK8 in the delivery condition; 2-VK8 + (AlCr)N (1.5 μm) + (AlTi)N (2 μm) [applying coating layers by the condensation method with ionic bombardment (CIB) using filtered droplet phase and without assisting with accelerated ions]; VK8 + Ti (up to 1 μm) + TiN (1 μm) + (NbZrTiAl)N(2.5 μm) [applying coating layers by the condensation method with ionic bombardment (CIB) using filtered droplet phase and assisting with accelerated ions]; 4-VK8 + Ti (up to 1 μm) + TiN (1 μm) + (NbZrTiAl)N (2.5 μm) [applying coating layers by the condensation method with ionic bombardment (CIB) using filtered droplet phase and assisting with accelerated ions]; 5-VK8 + Ti (up to 1 μm) + TiN (1 μm) + (NbZrTiAl)N (2.5 μm) [applying coating layers by the condensation method with ionic bombardment (CIB) using filtered droplet phase and assisting with accelerated ions]; 6 is MC2210).

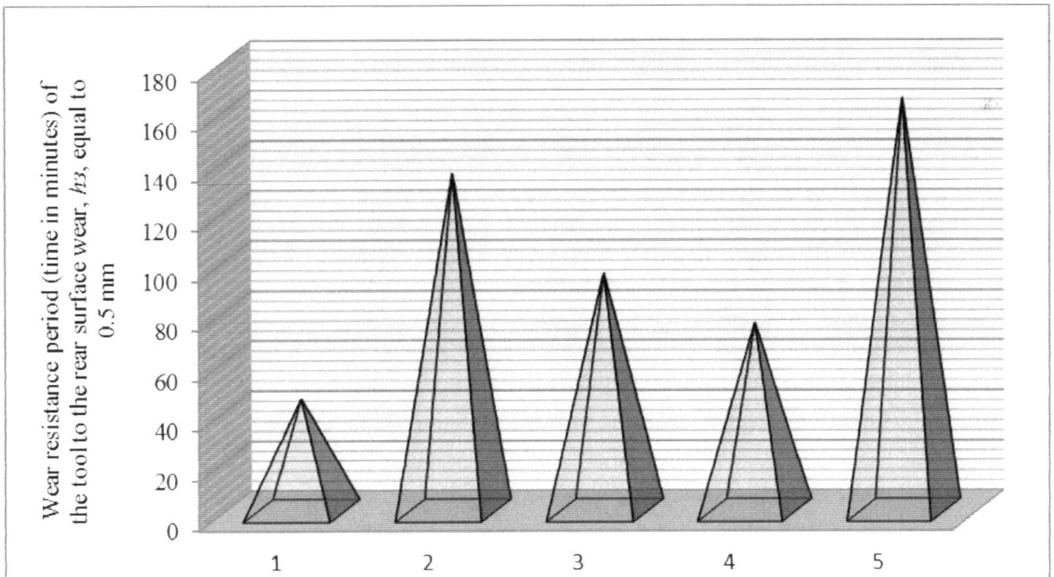

Figure 3. Sample of comparing wear resistance period of different tool materials

a

B

c

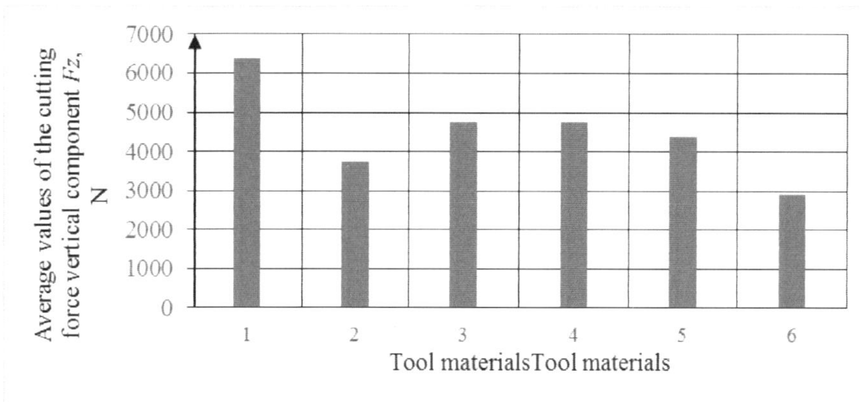

Figure 4. Sample of entering components of cutting force when treated the workpiece made from a specialized stainless steel of grade 09X17H7IO with various tool materials

3. Conclusions

The conducted researches allow drawing the following conclusions.

1. The use of coatings [2, 3, 4, 5] on the domestic hard alloy of brand VK8 significantly increases the performance of the cutting plates.

2. Technological method of refinement (thermal sharpening of cutting edges from the front and rear surfaces) also significantly increases the tool efficiency.

3. The insignificant difference in the size and shape of the medium dimension plates (the circumference diameter is from 16.5 to 18 mm) entails a significant difference in the wear resistance period, the cutting plate wear. The pentagonal shape of the plate is preferred. A rhombic form is a little bit inferior. The type of chip scraps produced is similar.

4. For plates of larger dimensions (the circumference diameter is 21.5 mm), the shape of the plate significantly affects its wear resistance period. In this case, a four-sided square shape is preferred.

5. The findings do not contradict the existing ideas about the cutting processes [5, 6, 7, 8, 9], they are based on the synergetic mechanisms of the blade treatment process of viscous materials and allow us to offer recommendations for selecting and strengthening domestic hard-alloy materials for treating the said stainless steel.

6. Analizing the results of recording the components of the cutting forces showed that their value and the ratio between them depend significantly on the tool material used. It is established that the role of the coating in improving the tool performance depends significantly on the operating conditions, including the cut depth. The coating, which was the most effective (of the considered) at a cutting depth of 0.5 mm, may not be so at a cutting depth of 1.5 mm and may give primacy to the coating,

which was the second or third in efficiency at a depth of 0.5 mm. This does not contradict the existing ideas, but on this specialized steel the coating is displayed most clearly. This is typical not only for turning, but also for end milling [10, 11]. For end mills, the effect of coatings is evaluated by modeling, including using ANSYS software.

*The authors express their acknowledgment to A.A. Vereshchaka and A.A.Moskvitin for assisting in coating and researching..

References

[1] Vereschaka A A, Vereschaka A S, Grigoriev S N and Sladkov D V 2013 Nano-scale multi-layered coatings for cutting tools generated using assisted filtered cathodic vacuum arc deposition (AFCVAD) *Applied Mechanics and Materials* **325-326** 1454-1459
[2] Mokritsky B I, Sablin P A, Vereshchaka A A 2017 Instrumental support of modern machine-building industries (Komsomolsk-on-Amur: Publishing House K/ASTU)
[3] Mokritsii B Ja, Morozova A V and Usova T J 2017 Results in kompostite hard-alloy and milts design based on simulation of their operation conditions *Procedia Engineering* (**206**) 1093-1098
[4] Sitamav E S 2018 Results of a comparative study of carbide tool wear resistance when treating stainless steel Metalloobrabotka **4 106** 7-13
[5] Cselle T 2002 Nanostracturierte Schichten in der Werkstaff Platit AG. *Warkzeugtagung* (Google Scholar)
[6] Vereshchaka A S, Vereshchaka A A 2005 Effectiveness increase of the instrument by controlling the composition, structure and properties of coatings *Uprochnyayushchie tekhnologii i pokritiya* **9** 9-19
[7] Tabakov V P 2008 *Formation of wear-resistant ion-plasma coatings for cutting tools* (Moscow: Mashinostroenie)
[8] Grigoriev S N, Vereschaka A A , Vereschaka A S and Kutin A A 2012 Cutting tools made of layered composite ceramics with nano-scale multilayered coatings *Procedia CIRP* **1** 318 – 323
[9] Vereschaka A S, Grigoriev S N, Tabakov V P, Sotova E S, Vereschaka A A and Kulikov M Yu 2014 Improving the efficiency of the cutting tool made of ceramic when machining hardened steel by applying nano-dispersed multi-layered coatings *Key Engineering Materials* **581** 68-73
[10] Vereschaka A A, Vereschaka A S, Batako A Dl, Aksenenko A Y and Sitnikov N N Improvement of structure and quality of nanoscale multilayered composite coatings, deposited by filtered cathodic vacuum arc deposition method *Nanomaterials and Nanotechnology* **7** 1-13
[11] Vereschaka A A, Vereshchagin V Y, Sitnikov N N, Oganyan G V, Aksenenko A Yu 2017 Study of mechanism of failure and wear of multi-layered composite nano-structured coating based on system Ti-TiN-(ZrNbTi)N deposited on carbide substrates. *Journal of Nano Research* **45** 110-123

Specialised technical translation: Global Village requirements and regional opportunities

E Yu Novikova[1], S R Khairova[2]

[1]Volgograd State University, 100 Universitetskiy Ave., Volgograd, 400062, Russia
[2]Lomonosov Moscow State University, 1/13 Leninskie Gory, Moscow, 119991, Russia

E-mail: nov-elina@volsu.ru

Abstract. Specialised translation is going through major changes as demand is growing for high quality translation that is terminologically accurate and performed at the level of native speakers. At the same time, economic globalisation creates a complex mix of various types of cultures within professional communication, which also exerts influence on the translation process. The increasingly diverse market requirements change the competency profile of specialised translators, but the actual needs may be seen differently by different market actors. This paper looks at the current requirements and challenges in the area of specialised language services, with a particular focus on the technical fields. It discusses translator competencies as seen by translation practitioners and theorists, clients and end users, as well as translation companies. Comparisons are made on the basis of surveys of the target groups and an analysis of the websites of translation agencies. Drawing on the findings of the study, the authors make recommendations on how translator training should be adapted to incorporate the requirements of the global market.

1. Introduction

The training of translators for specialised areas is facing increasingly stringent requirements in the current environment of active cross-cultural interactions, international economic and political cooperation, growing number of professional contacts, as well as the ongoing reform of higher education. All this encourages educators to improve translation didactics and develop new, more effective teaching methods complying with the current professional standards and the European language portfolio criteria. Yet, despite the introduction of common standards and criteria, active exchange of knowledge and practical expertise between translation scholars and practitioners, there are significant differences in the translator's professional profile formed by the academic curricular and the desired profile shaped by the needs of actual market actors. Expectations from professional translators are very high as the quality of translation products has a direct impact on the company's overall performance and even people's lives, as in the case of medical or technical translation – for example, translation of clinical trial protocols or safety manuals. Importantly, a high proportion of vacancies on the modern international market for language services are those in the sphere of industry-specific translation, technology-related areas, so a large number of today's graduates from translation departments will end up in these areas [1,2,3].

Technical translation, which takes the dominant position on the market of today, includes translation of highly specialised documents and localisation of websites and software. The changing character of requirements to modern translators, as compared with the past, stems, among other factors, from the changing character of clients. These include: clients (1) – private clients who are not the end users of the translated texts; and clients (2) – large translation companies who act as intermediaries for clients (1). As Abdallah [4] points out, clients (2) commission large volumes of translation, expect fast execution and try to save on fees for the actual translators. They are less concerned about the quality of the end product, as opposed to clients (1). It means that translators have to constantly shift between the different professional requirements imposed by different market participants. The following section will discuss the current landscape for professional communication in the globalising world, and the competencies that are expected from translators in specialized professional discourses.

2. Results and Discussion

The existing body of literature on specialised translation is fairly large. Yet, the modern translation studies do not offer a dedicated theoretical framework for industry-specific or technical translation as a separate area in its own right. For example, the Dictionary of Translation Studies [5] recognises the existence of area restricted theories of translation, which are defined as partial theories limited to specific languages and/or cultures under consideration, but there are no entries dedicated to specialised translation as a subdivision of translation studies. At the same time, the high concentration of translation activity in this particular area requires a comprehensive theoretical framework to address the emerging issues. The competency profile of an industry-specific translator is one of the many points of disagreement that call for more detailed research.

It should be noted that translation scholars generally project a very positive image of specialised translators describing them as experts in specific subject areas [6, 7].However, if one looks carefully at any professional standard – for example, the German DIN EN 82079-1 – it becomes clear that developers ascribe to translators only the communicative competence, and not that of an expert, as opposed to the related profession of a technical editor.

According to Sandrini [8], industry-specific translators must have the combination of the following competences:

- procedural competence to produce pragmatically adequate translations based on the Scopos theory principles;
- competence to create specialised texts that are terminologically correct and linguistically adequate;
- subject matter competence (knowledge of a specific subject area);
- translation competence (knowledge of the theory and practice of translation);
- language competence;
- general cultural competence;
- special cultural competence, which comprises knowledge of those features of industries that are unique to a particular culture/country.

The importance of the last two components should not be underestimated. Globalisation in various areas of professional activity results in the creation of an industry-specific internal culture, in which the participants share a system of knowledge, perceptions and information exchange methods. One of the crucial elements of such a uniform culture is industry terminology. On the one hand, international standardisation of terminology is facilitated by the relatively large body of specialised vocabulary with one-to-one equivalents in the majority of contacting languages. Furthermore, as Novikova et al. suggest, technical terms of the globally important professional areas (such as economy, politics, advertisement, cinema, sport, tourism, etc.) originate from one source language – English, the major language of international communication [9]. However, in a lot of industries, such as oil production for example, regional, national and transnational companies have to comply with different standards and might use terms that are unique to their countries or regions. It means that translators should be able to apply their knowledge of the global and country-specific professional cultures and choose the appropriate translation strategies, tactics and techniques to ensure smooth communication between the international partners.

Some scholars discuss the competence of specialised translators with a focus on personality aspects. Thus, Gavrilenko [10] argues that "the personality component of the professional competence of an industry-specific translator comprises professionally relevant personal qualities affecting the basic parameters of an effective translation process (productivity, precision, adequacy, reliability, etc.).These qualities are the prerequisites for the professional activity of a translator; they are improved and polished first during the training and further in the course of practical work". Gavrilenko also notes that these personal qualities should be combined with specific types of knowledge:

- subject-related knowledge in certain fields of science and technology;
- communicative competence in the native and foreign languages;
- discourse competence in specific subject areas in the native language;
- sociocultural competence regarding communication in specific industries in the translator's home country;
- knowledge of IT, skills in computer-based information and terminology search;
- ability to interact with specialists from the related sectors [8].

We support Gavrilenko's view that subject-related knowledge is of paramount importance for a translator. The same opinion is expressed by a translation practitioner with an engineering background, Israel Shalyt, who maintains that language proficiency and proper dictionaries are not sufficient to produce a really good translation of a technical text. A translator must also: "1) understand the meaning of the source text (which is not always expressed explicitly and often has to be deciphered by surmising); 2) be able to expound ideas in accordance with the conventions that have developed in the given subject area" [11].

The expert knowledge of the subject matter at a level comparable to that of a specialist is rather an ideal which can rarely be found in real life, except for in-house translators who work for one and the same company for years. Without challenging the above views on the competency profile of industry-specific translators, we still have to admit that this perspective is mostly supported by translation theorists and practitioners.

The question is how far these views are similar or different from those held by employers and real economy actors. In this respect, Canfora [12] cites interesting findings of two different surveys. The first one is an online survey conducted by Britta Görs in February 2012 among technical editors regarding their expectations from technical translators, in which 90% of the respondents pointed out that translators must possess expertise knowledge of the subject area [13]. Importance of specialised knowledge was also highlighted by respondents of another survey carried out by tekome.V., involving 222 specialists of procurement and technical documentation departments [14]. Canfora provides a summary table of the respondents' replies:

Table 1. Expectations of potential clients: from standard to high quality translation.

	Not important	Standard translation	Quality translation	High quality translation
Translation by a non-native translator	22.2 %	37.9 %	34.8 %	5.1 %
Correct spelling up to 95%	5.4 %	46.0 %	34.7 %	13.9 %
Completeness of translation up to 95%	8.0 %	46.0 %	32.0 %	14.0 %
Correct grammar – up to 95%	5.5 %	42.5 %	38.0 %	14.0 %
Translator's self-check	5.2 %	28.8 %	46.2 %	19.8 %
Final formatting	21.7 %	15.6 %	38.2 %	24.5 %
Localisation	8.5 %	22.3 %	42.7 %	26.5 %
Use of style reference books	6.9 %	19.9 %	46.3 %	26.9 %
Correct word order	0 %	23.1 %	49.5 %	27.4 %
Correlation of content between ST and TT	8.0 %	17.0 %	45.8 %	29.2 %
Stylistic adequacy at the level of a native speaker	4.2 %	19.5 %	44.2 %	32.1 %
Use of specialised vocabulary/terminology	0 %	18.1 %	45.1 %	36.7 %
Translator's expertise in the subject area	5.6	15.3	41.4	37.7

Correct spelling up to 100%	2.3	13.1	46.5	38.
Correct grammar up to 100%	1.9	13	47	38.1
Translation by a native translator	4.2	15.5	41.9	38.6
Terminology consistency	0	11.1	49,5	39,5
Terminology correctness	0	14,5	43,9	41,6
Review by an industry expert	8.4	9.3	39.5	42.8
Completeness of translation up to 100%	0.5	16.4	40.2	43.0
Final review by a native speaker	9.9	11.7	30.0	48.4

The surveys suggest that if potential end users expect a standard translation, they focus primarily on linguistic correctness and completeness of translation. If clients expect a quality translation, their requirements will cover practically all the above parameters, with a focus on the subject-related adequacy of the translated text. Expectations from a high-quality translation will include 100% linguistic and, especially, terminological correctness, which also implies high subject-related competence of the translator.

It can be concluded, therefore, that the market of language services expects translators to have extensive knowledge of the subject area and precision in using industry-specific terminology. The second most important parameter emerging from the surveys is linguistic competence, while all other competencies (skills in using TM-systems and machine-based translation, project management, work experience, cross-cultural competence, personal qualities, etc.) are viewed by potential text recipients as less relevant.

These findings are supported by another study conducted by Chinese scholars [15], who surveyed three target groups in their country – translation practitioners, specialists from various industries, and educators – to understand which skills and competencies are needed by technical translators. Interestingly, the Chinese respondents came up with answers similar to those discussed above. Specialised knowledge featured large in their comments, and other important competencies included a good command of the target language and advanced skills in using the computer for translation purposes. Much focus was also placed on translators' personal qualities, such as "initiative, self-discipline, ability to work independently and under pressure, responsibility", and ability to learn fast. As in the two surveys cited above, the respondents did not give a high priority to cultural competence.

Let us now look at the requirements to specialised translators set by translation companies. As part of this study, we analysed the websites of several categories of translation agencies specialising in technical translation: large transnational companies, large national companies and regional companies. On their home pages, translation agencies generally emphasise that their staff includes professional translators, native speakers, specialists with an in-depth knowledge of specific industries. Yet, the list of requirements to specialised translators rarely includes subject-related knowledge. To give an example, Transtech, a large Russian national company, lists the following job responsibilities and requirements for candidates applying for technical translation positions:

Responsibilities:

- Translation of technical texts in the area of military technology (air defense) from Russian into Arabic;
- Ensuring terminology consistency;
- Comparative check against the company glossaries.

Requirements:

- High level of proficiency in the Arabic and Russian languages;
- Minimum 3 years of experience of written technical translation;
- Confident computer user;
- Excellent typing skills;
- Knowledge of SDL Trados Studio is an advantage (if the candidate has no skills in Trados, they will be provided with training and technical support);

- Teamwork skills, high sense of responsibility, high productivity;
- The interview stage will include a mandatory test.

Regional translation agencies are even less demanding. For example, here is the list of general requirements to applicants set by the Unit translation agency:

- Minimum 5 years of relevant work experience;
- High level of language competence;
- High sense of responsibility, accuracy and scrupulousness;
- Advanced user of MS Office;
- Skills in using CAT software;
- Availability for feedback.

As for transnational companies, they rarely detail their requirements to job applicants on their home pages. Normally, they limit these to very general comments concerning high-level professional skills in the sphere of technical translation that they expect to see in the candidates.

This short overview of translation companies leads to a conclusion that their expectations from translators focus primarily on language competence and work experience, as well as skills in machine translation and computer-assisted translation. The emphasis on the latter parameter is understandable as translation agencies demand very high speed of translation, and one of the most effective ways to accelerate the translation process is the use of MT and CAT software [16, 17].

3. Conclusion

This study has revealed that participants in professional communication, including in the area of specialised translation, may hold different perspectives regarding fundamental requirements to professional competence, and these differences should be taken into account both in translation practice and translation training at the university level. Let us compare professional competence profiles as they are seen by practicing translators, end users and translation agencies acting as intermediaries [18].

Translator	*Client/End User*	*Translation Company*
Procedural competence	Subject area competence	Language competence
Subject area competence	Native speaker language competence	Instrumental competence
Cultural competence	Other	Localisation skills
Instrumental competence		Subject area competence
Communicative competence		Other
Language competence		
Other		

Industry-specific translators recognise the need to constantly enhance their knowledge of the subject area they specialise in. End users expect to receive the quality of the translated text and terminology accuracy that can only be provided by a highly-skilled professional. Translation companies, in their turn, try to attract clients by promising to render top quality services of specialised translation and boasting to have highly-qualified translators with expertise in specific industries. Their requirements to job applicants focus mainly on language and instrumental competences (skills in using state-of-the-art TM systems and other CAT software) [19].

There is no doubt that subject-related knowledge is critical for performing specialised translation, independently or in a team as part of a large translation project. However, other important factors, such as cultural differences, should not be overlooked. Professional communication takes place at a juncture of various cultural fields: national culture, transnational culture, industry-specific culture, the culture of a specific enterprise/company. It would be impossible to deny that intensification of international contacts results in the gradual elimination of cultural boundaries and facilitates terminology standardisation. However, cultural differences are still important, and translation process still involves a cultural transfer, in which translators should balance the requirements of the globalising language services industry and the actual practices that are in place in the contact cultures. An important role in building general and special cultural competences belongs to translation training institutions [20].

Teaching professionally oriented translation should be based on the trainer's personal translation experience, creativity, and a careful choice of the teaching material. Classroom work should provide a learning environment that reflects real-life professional situations, which can be achieved, for example, through using authentic technical texts. Other factors to consider include the emergence of new types of translation services, involving complex technical documentation with blue prints and descriptions of technological developments. These new market realities call for a review of didactic models for teaching specialised translation and increase the role of practitioners in the teaching process.

References

[1] Risku H 2016 *Translationsmanagement: Interkulturelle Fachkommunikation im Informationszeitalter* (Tübingen: Narr)

[2] Schmitt P A, Gerstmeyer Land Müller S 2016 *Übersetzer und Dolmetscher. Eine Internationale Umfrage zur Berufspraxis* (Berlin: BDÜ-Fachverlag)

[3] Brence D 2007 Ausbildungs- und Anforderungsprofil der Translatorischen Berufspraxis – eine Empirische Studie *Lebende Sprachen* **52 (2)** 50–64

[4] Ergunova O T, Lizunkov V G, Malushko E Yu, Marchuk V I, Ignatenko A Yu 2017 Forming the system of strategic innovation management at the high-tech engineering enterprises *IOP Conference Series: Materials Science and Engineering* **177 (1)** 012046

[5] Abdallah K 2012 *Translators in Production Networks: Reflections on Agency, Quality and Ethics* (Joensuu: University of Eastern Finland)

[6] Shuttleworth M and Cowie M 1997 *Dictionary of Translation Studies* (Manchester: St. Jerome)

[7] Holz-Mänttäri J 1984 *Translatorisches Handeln. Theorie und Methode* (Helsinki: Suomalainen Tiedeakatemia)

[8] Lizunkov V, Politsinskaya E, Malushko E, Kindaev A, Minin M 2018 Population of the world and regions as the principal energy consumer *International journal of energy economics and policy* **8 (3)** 250-257

[9] Risku H 1998 *Translatorische Kompetenz, Kognitive Grundlagen des Übersetzens als Expertentätigkeit* (Tübingen: Stauffenburg)

[10] Sandrini P 2010 Fachliche Translation *Diskurs und Terminologie beim Fachübersetzen und Dolmetschen.* ed J Maliszewski (Frankfurt a.M.: Peter Lang) 31-51

[11] Novikova E, Mityagina V and Charfavi E 2017 Pragmatika Terminologicheskoy Nominatsii: Turisticheskiy Menedzment [Pragmatics of Terminological Nomination: Tourism Management] *XLinguae* **10 (3)** 285–299

[12] Gavrilenko N 2018 Lichnostno Orientirovanniy Podhod k Formirovaniuy Otraslevogo Perevodchika [Learner-Centered Approach to Industry-Specific Translator Formation] *Vestnik of Moscow State Linguistic University* **798** 199

[13] Shalyt I 2013 Zametki o Tekhnicheskom Perevode [Notes on Technical Translation] Access mode: https://i-shalyt.livejournal.com/11444.html

[14] Ferreira R, Lizunkov V G, Politsinskaya E V 2017 Formation of entrepreneurial competencies of university graduates in conditions of transition to the universities of the third generation *Novosibirsk State Pedagogical University Bulletin* **7 (6)** 195-211

[15] Canfora C 2017 Image und Kompetenz von Fachübersetzern *Fachsprache* **39 (1-2)** 2-19

[16] Görs B 2012 Nachfragen erwünscht Access mode: http://www.brittagoers.de/auswertung-der-februar-umfrage-2/.17.02.2017

[17] Ergunova O, Lizunkov V, Blaginin V, Politsinskaya E, Shaykina O I 2018 Formation of the price mechanism for energy resources in Russia and the countries of the European union *International Journal of Energy Economics and Policy* **8 (1)** 122-127

IOP Conf. Series: Materials Science and Engineering **483** (2019) 012084 doi:10.1088/1757-899X/483/1/012084

[18] Straub D 2012 Studie Einkauf von Übersetzungsdienstleistungen. Hrsg. tekom, Stuttgart, 17.02.2017 Access mode: http://www.tekom.de/upload/alg/tekomStudieUebersetzungsqualitaet.pdf.

[19] Ying Y *et al* 2018 *IOP Conf. Ser.: Earth Environ. Sci.* **126** 012090

[20] Hao Peng 2018 *IOP Conf. Ser.: Mater. Sci .Eng* **322** 052024

IOP Conf. Series: Materials Science and Engineering **483** (2019) 012085 doi:10.1088/1757-899X/483/1/012085

Peculiarities of translation of English technical terms

L S Polyakova, Yu V Yuzakova, E V Suvorova and K E Zharova

Foreign Languages for Engineering Department, Nosov Magnitogorsk State Technical University, 38 Lenin Ave., Magnitogorsk, 455000, Russia

E-mail: lilitmgn@mail.ru

Abstract. Modern linguistics manifests an increased interest to the study of particular sublanguages serving specific fields of professional activity. The attention of linguists to languages for special purposes is primarily due to the growing number of terms in various sciences. The spread of terminological vocabulary in everyday speech outside communicative situations related to professional activities is a distinguishing feature of our time. In current conditions, the use of terms is no longer the prerogative of specialists in a specific field. This trend is also valid in relation to the vocabulary of technical discourse, since technical terminology is regularly used in a variety of contexts and it is continuously developing in the conditions of modern technical progress. The article considers the structural and semantic features of the English terminology used in technical texts and ways of translation English terminology into Russian. The study of terminology peculiarities is based on the material of technical literature due to its extensive use in modern life and production. Furthermore, the growth of production, science and technology is an important source of new terms, not yet recorded in the dictionaries. New terms appear to create significant difficulties if you need to translate them into another language.

1. Introduction

The relevance of the issue under consideration is related to the rapid development of various areas of scientific and technical knowledge, which is accompanied by new notions formation, followed by their naming and coining the so-called term. Today, there are many dictionaries in specific fields of science and technology, containing tens of thousands of words, but none of them is able to provide a translation of all the terminological units existing and newly coined within this branch.

Modern Internet resources allow you to perform a fairly complex translation by means of empirical identification, i.e. by searching for matching pairs of terms in different languages. For example, one can search for matching of German and English versions of a certain term with corresponding notion thus formulating the corresponding Russian equivalent, by means of perceiving the meaning of the term through pictures found in various search bases. These translation operations are usually carried out at the end of the first viewing of the text, take a long time and do not always lead to the desired result. Therefore, again and again, the translator has to turn to the study of the peculiarities of term translation which existed earlier or have appeared recently in connection with new lexical units coining.

2. Literature Review

Before proceeding to the study of term translation issues in scientific and technical texts, we are to consider the definition and linguistic nature of the term. According to O. S. Akhmanova, the term is a word or a word combination of a specific language, coined (accepted, borrowed, etc.) for the exact expression of specific notions and denotation of specific objects [1, p. 474]. S. V. Grinev gives a similar definition. According to the researcher, the term is a nominative specific lexical unit (a word or a phrase) belonging to a specific language and giving an exact specific concept name [2, p. 33]. The aforementioned definitions indicate that the terms are to be unambiguously identified and understood in the context.

According to this requirement, there are two types of terms: 1) general scientific and general technical terms, 2) specific (nomenclature) terms. The former ones serve to denote the general concepts of science and technology, while the latter are part of a certain terminology. As explained by T.M. Dementieva, terminology representing a set of terms is an autonomous part of any national language that is closely related to professional activity. Terms in any branch of science, technology and production form their own systems that have conceptual connections with professional knowledge and express these connections by means of the language [3, p. 61]. If in a common language (beyond the terminology) a word can be multivalent, then, getting into a certain terminology, it acquires unambiguity [6, p. 208].

Thus, the term does not need a context, as an ordinary word, so it 1) is part of a specific terminology that serves as context, 2) can be used in isolation, for example, in the texts of registries or orders in engineering, 3) for this reason it should to be monosemantic not in the language in general, but within the limits of the terminology. Summarizing the available research data, we have identified the following properties inherent to the terms: monosemy, specific meaning; belonging to a certain terminological system; lack of expression.

In lexicological studies, the branch terminological subsystems development issues are of key importance. In the 30s of the twentieth century, terminology studies, a new applied science discipline, the research object of which are terms and terminological systems, emerged from linguistics [4, p. 17]. Everything connected with the notion "term" brings linguistics nearer not only to different fields of scientific knowledge, but also to different fields of professional work. Terminology vocabulary is growing extremely quickly. The significance and proportion of this lexical layer in the general vocabulary are increasing all the time. The terms of specific terminosystems are unified not according to linguistic features, but features external to the language. They represent not a simple family of words, but a system of words or phrases organized in a certain way [5, p. 28].

The most important problem in achieving translation equivalence of scientific and technical texts is the reproduction of the original content of the text using the terminological system of the target language. The difference in the terminological systems of the source language and target language is the cause of the greatest difficulties in translating scientific and technical texts. Hence there is need to study the term systems and find ways to translate partially equivalent and non-equivalent vocabulary. The problem of exploring terminology is one of the key issues in the study of scientific and technical texts. A term or a phrase denotes a concept of a special field of knowledge or activity. Identification of differences in the concept system expressed by the terms of the source language and the target language is an important step on the way of term system interlanguage harmonization, providing a solution to the problems of term translation in the fields of their functioning [1, p. 8].

So, terms are units of linguistic and professional knowledge that ensure the intercultural communication effectiveness. For this reason, equivalent translation of terminology is of great significance in the translation of scientific and technical texts [6, p. 138].

3. Results and Discussion

The widely used methods of translation uncommon elements being present in an original text, such as terms, are: search for an equivalent, use of analogues, calquing, transliteration/ transcription, functional sense, descriptive translation and translator's comments [2-10].

1. The search for an equivalent in the target language, i.e. a word that is fully equivalent to its counterpart in the original language. The equivalent should be as short as possible and convey the full volume of the original term meaning. The following metal names can serve as examples: *zinc*, *magnesium*, *copper*, etc.
2. The translation with the help of an analogue is used when an original word has several correspondent equivalents in the target language. It is the most widely used translation method, as the practical mass of words is polysemic. Moreover, it is of great help if a translator deals with

universal technical terms or units, borrowed from commonly used languages. E.g.: *strenght, fatigue, hardness.*

3. Calquing is the reproduction of a combinatorial (but sound) composition of a word or phrase, when the components of a word (morphemes) or a phrase (lexemes) are translated by the relevant elements of the target language. As a translation method calquing served as a basis for a large number of borrowings in the process of intercommunication in cases, when the transliteration was unacceptable for some aesthetic, semantic or other reasons [2, c. 88]. The essence of calquing is making a new word or an idiom in the target language, which replicates the structure of the original lexical unit: *mechanical wear, martensite structures*. In this case the translation can be accompanied by the word order change *wear resistance, grain size*. When calquing, one can use such types of lexical and grammatical transformations as a word type change or explication. E.g. *titanium alloys* (a noun is changed for an adjective), *high-temperature materials* – materials with a high point of melting (the term meaning is clarified, i.e. explicated.
4. Transliteration/transcription. The transcription is a formal phoneme by phoneme reproduction of an original lexical unit with the help of the target language phonemes, phonetic imitation of the original word. The other method is transliteration, i.e. formal letter by letter reproduction of the original lexical unit with the help of the target language alphabet, a letter imitation of an original form of a word. In this case, the original word in the translated text is presented in the form adapted to the pronunciation features of the target language [2, c. 90]. Theses types of translation are mainly used in the translation of proper nouns and names, e.g. *Super Duplex Stainless Steel* , *Charpy impact testing*. Besides, the method combining transliteration and transcription is most commonly used in practice.
5. Sometimes the term that has no equivalent or an analogue in the target language, that would convey its meaning sufficiently, is translated with the help of the words, reflecting its functional sense. For instance, the word shoulder in the texts on metalworking will be translated as a support collar, and a hollow drawbar as a hollow pull stud. The search for the functional relevance turns out to be especially urgent in case of so called nonequivalent lexicon, i.e. words, that are absent in bilingual dictionaries for some reasons. This category often includes neologisms and words denoting objects or phenomena unfamiliar for the culture of the target language that have not come into the dictionaries (realities or culturonim).
6. Descriptive translation and translator's comments. The descriptive translation, as a rule, is used in parallel with the transcription and is applied when translating terms, culturonims, unique objects, etc., having no lexical equivalents in the target language. The word meaning is presented by means of a more or less common explanation, made as short as possible. If the description as a translation method usually accompanies the word, presented in a simple form, or even is used instead of the word, the translator's comments are given beyond the text, being mentioned either in the footnotes at the same page or into the endnotes of the text. The comments as a translation method suggests a more detailed (in comparison with the description) explanation of what the original word means in a wide context of the original culture. There is also a translation method, by which units of the original language get into the translation. It mainly concerns various acronyms, abbreviations, and terms, such as types of equipment, e.g. I-Stir PDS machine, or materials (alloys X-65, AL-6XN, etc.).

Thus, translating terms, different in lexical composition, is practically difficult, as it requires from a translator both understanding of the original term meaning and the knowledge of its possible counterparts in the target language.

The terms can be polysemic, synonymous and homonumous. Regarding their form all the terms can be divided into: 1. Simple - – *circuit, flexibility*. 2. Complicated – *flywheel, nonlinear, non-waste*. 3. Word combinations – *circuit breaker, nonmetallic inclusions, diverse scale process, heat and mass exchange, wide spectrum properties*. Terms – word combinations are the most difficult to translate, as the words comprising them are independent. The components of such terms can be distinguished as separate units. For instance, *brake gear, electric motor, powder materials*. The most difficult for the translation are: a) the terms, whose meanings come from the integration of two words *load governor*, *brake landing*; b) the terms, whose components are linked by means of grammar elements, e.g. *ways of measurement*, *body of reactor*, *breaking with rocket*; *lid of frame*. Among the other types of the terms, the word combinations of the following constructions are the most difficult: a) noun + noun, e.g. *back – coupling*; *variable – capacitor*; *peak – energy*; *phase transition*, *percussion waves*; b) adjective + noun, e.g. *remote control*; *safety switch*; *direct current*; c) participle I + noun, e.g. *alterating current; calculating theory*; d) participle II + noun, e.g. *estimated cost, unbounded stream.* If the second component or both of them are common in use, their translation is not difficult. For instance, *safety switch, line wire, thermal stress.*

One should mind, that one and the same word form can have different meanings in several branches of one and the same scientific or technological sphere. The word fuel has a meaning of fuel in power engineering and a meaning of atomic fuel in atomic engineering. Thus, when translating technical terms, a translator should stick to the topic of the text to avoid the wrong variants of translating the homonymous term. One should lay a special emphasis on the translation of not only general scientific, but also non-diversified technical terms when working with English scientific and special technical texts. A certain problem lies in identifying the peculiarities of usage not of general vocabulary, but of specific terminology of scientific and technical texts and, consequently, the development of acceptable options for their translation into Russian language as the Russian scientific vocabulary differs significantly from English technical lexical forms. As a rule, Russian translations of the sentences, that illustrate the use of these technical terms in their contexts, reflect the essence of the author's sources, whereas specialized English terminology should be translated taking into account the present general context of the original source.

A comparative analysis of English-Russian translations of scientific and technical literature shows that a considerable number of translation errors can be found in the use of scientific vocabulary, they can be both the errors in choice of the wrong terms in translation, and the cases of wrong interpretation of terminology phrases. The latter case demonstrates the most widespread situation for errors in translation, because it is easier to translate a single term than to understand its original terminology link within the given author's context. It should be mentioned that it is almost impossible to consider all the nuances of translation difficulties that could emerge in the translation process, but we should highlight at least some of them. Without the right use of terms it is impossible to understand the author's ideas, the way the devices or some mechanisms work in some industry. And here, a translator may face a number of unexpected difficulties, even despite the use of specialized dictionaries.

In some cases, the translation is simplified by the fact that there is an English term, and there is a precise Russian translation. Let us take term "*water*", for example. It is clear that the fantasy of an interpreter should be completely excluded in doing the translation, as this term implies a molecule consisting of two atoms of hydrogen and one atom of oxygen.We cannot deny or argue this law. Similar terms can be, for example, *barograph, ozone, centigrade, gyroscope* and others.

But often translation activity is characterised by the occurrence from minor to significant difficulties in translation when a term has in a heterogeneous interpretations, for example *power – strength, electrical energy; line – line, cable line.* This fact makes the translation of English technical terms more complex.

But there can be more complicated cases, when a creative approach is needed from the interpreter, considering his/her scientific and technical training, for example, the word "*motor*" can be translated as a word "*motor*", or as a word *"engine",* or as a word combination *"to operate in the mode of engine".* There may be found many other examples as well, for e.g: a term "*a handle"* can be translated as *"a control lever"* and *"a handgrip",* a verb *"to handle"* can be interpreted as *"to operate, to manipulate, to serve";* a term *generator as "a generator"* and a verb *to generator as "to work in a generator mode";* a term *to increase* as "*to rise, to enhance";* a verb *to record* can be interpreted as *to register, to put down* and a noun *a record* as *a note, registration*; a term *empty* as *hollow* and its verb *to empty* as *to eviscerate*, etc. And there can be found many other cases of different ways for interpretation of one and the same term.

In some cases, the prefix is added to the English term and then the adequate translation may bear the opposite meaning to the original term, for example: *im- (impossible); counter- (to counteract).* Technical English literature abound in terms which were formed by adding the separate words, but being in a terminological unit they may indicate a single concept, for example: *sliding bearing, correction for displacement.*

Different meaning can be found in such examples as: *treatment of water, a water treatment.* So, these two, "the same" combinations of terms have different semantic meanings, and, respectively, they have different understanding of process technology. The same term may occur in technical English texts, but is used in its translation in different semantic meanings depending on the field of technical knowledge which is used: aviation, marine, electrical equipment, etc. That is, a specific term carries an entirely different meaning depending on its specific use. For example: *pocket* has the following special meanings in interpretation: *air pocket* (in aviation); *surroundings* (in military affairs); *the dead zone* (in radio); *a deposit* (in geology); *a cable channel* (in electrical engineering); and there are many similar terms. When working with the text, a translator should avoid so to say his own general vocabulary, but pay attention to the use of terms.

Another problem which an interpreter may face is that there can be various abbreviations in the original which can be seldom found in a Russian-language text, such as: a. c. (*alternating current*), PP *(peripheral processor).* Of course, when dealing with such a text, a translator should give a full meaning of abbreviation. An accurate translation of abbreviations can be found to some extent in specialized dictionaries.

One more difficulty in translation technical terms is the translation of terms which are different in their semantic meaning at the first sight. Here can arise a number of nuances, for example:

- a single term is a part of another one, e.g.: *valve amplifier*,
- both terms are independent, but in their semantic meaning, they present a single notion, for example: *machine frame,*
- a term attributes the second feature to another one, for example: *machine weight, motor vibrations, bronze washer*, etc.

Conclusions

Without a specific understanding of the term it is impossible to understand the message of the idea described by the author of the text. However, if a term and its specific semantic meaning are more common for the Russian language, then the term in the English technical literature can have a rather multifaceted value: from the specific translation to the translation, requiring scientific literacy of the interpreter. Thus, the best technical translation of the English text into Russian is mainly determined by the general polytechnic training of the interpreter, the knowledge of the interpreter's area of expertise. First and foremost, one needs to remember that the translated technical terminology in a specialized text is full of various difficulties we mentioned above. In conclusion it should be noted that for the adequate translation of the authentic scientific and technical texts, containing a large number of special terminology requires accuracy and unambiguous understanding that can provide interpreters with the high level of linguistic and professional skills.

References

[1] Ahmanova O S 2004 *Slovar' lingvisticheskih terminov. Dictionary of Linguistic Terms* (Moscow: Editorial)
[2] Bauer L, Nation I S 1993 Word families. *International Journal of Lexicography,* **6** **(4)** 253-279
[3] Chung T M, Nation I S *Identifying Technical Vocabulary*
[4] Horace B, Ava Champney 1958 A Comprehensive Dictionary of Psychological and Psychoanalytical Terms (New York: Longmans, Green and Co.)
[5] Grinev S V 1991 *Terminology and Scientific Cognition* (Moscow) **1** 15-27

[6] Hanks P 1984 *Collins dictionary of the English language* (London: Collins)
[7] Jastrzembski J Multiple meanings, number of related meanings, frequency of occurrence, and the lexicon. Cognitive Psychology 198 **13(2)** 278–305
[8] Lotte D S 1968 *Kak rabotat' nad terminologiej. Osnovy i metody.* How to work on terminology The rules and methods. (Moscow: Nauka)
[9] Marchuk J N 1992 *The Basis of Machine Translation* (Moscow)
[10] Pearson J 1998 *Terms in Context* (Amsterdam)

Communicative qualities of a person as the component of professional self-consciousness of the Director of the social service

O Golub[1], T Klenova [2], A Ozerina[3], T Timofeeva[4], O Maletina[5]

[1] Volgograd State University, Psychology and Pedagogy Department, Volgograd State Medical University, Volgograd, Russia
[2] Volgograd State University, State and Municipal Management Department, Volgograd, Russia
[3] Volgograd State University, Psychology and Pedagogy Department, Volgograd State Medical University, Volgograd, Russia
[4] Volgograd State University, Psychology and Pedagogy Department Department, Volgograd, Russia
[5] Volgograd State University, German and Romanic Philology Department, Volgograd, Russia

golub@volsu.ru, klenova@volsu.ru, ozerina@volsu.ru, timofeeva@volsu.ru, o.maletina@volsu.ru

Abstract. The article is devoted to the scientific analysis of professional self-consciousness of managers and specialists in different fields of professional activity. Authors consider modern problems connected with communicative qualities of a manager in the organization of the activity of social service. The actual growth of social service is determined by transformation processes occurring in the modern society. There are characteristics of social services from the point of view of peculiarities of their organization and provision. It is determined that their character influences the degree of complexity of personnel management. According to the social and psychological research communicative qualities of a person are a leading component of professional self-consciousness of a manager of social service. In conclusion there are recommendations where the results of the research can be used effectively.

Introduction

The development of the scientific and technical progress, the complication of production tasks, new methods and strategies of management change requirements to managers in the modern society. Nowadays the level of the efficient work of any organization depends directly on the capabilities of its managers to work out and take modern and relevant decisions, to develop positive patterns of behavior and activity of each employee. The increase of his personal efficiency can be reached if there is knowledge of theoretical basis of management, practical experience and skills to use science, theory and practice.

Despite the fact that tasks of discontinuous professional development are constantly actual, not every manager can assess his own potential and realize development perspectives. Such situation may happen if the manager has a developed self-consciousness when he is able to realize himself as the subject of the professional activity that means he has a professional self-consciousness.

The problem of the professional self-consciousness development of managers and specialists of different fields of professional activities is considered in the researches of different scientists (Bragina V.D., Vaskovskiy S.V., Yeldyshova O.A., Kaziev V.N., Klimov Ye.A., Kosheleva S.V., Kryakhtunov M.I., Markova A.K., Matveeva L.G., Mironova T.L., Mitina L.M., Parygin B.G., Fam T.N.).

IOP Conf. Series: Materials Science and Engineering **483** (2019) 012086 doi:10.1088/1757-899X/483/1/012086

However, in psychology till nowadays there is no single and common definition of the professional self-consciousness of director of social service [1, 2, 3, 4].

1. Specific features of social service in the sphere of production and consumption of services

As the organization of social service is formed to provide people with professional assistance in the sphere of social relations and individual problems of clients [5]. The development of market relations caused a sharp stratification of the society where one part being owners of private property began to receive stable profit, and another part became the social group which due to various reasons is not capable to support its living. This transformation of the society led to the growth of the importance of social service.

The content of the social service differs significantly from other activities because the object is different, and it is influenced by the employee in the process of his labour activity and the character of services is also different in this service sphere. Scientists consider service differently, that is why we need to study definitions of service in wide and narrow senses (Figure1.). Services in the social sphere are characterized by specific features in the sphere of their production and consumption.

Firstly, services produced in fields of the social sphere are to have significant external positive effects in the process of their provision (for example, reduction of risks and scales of population diseases, distribution of social values and development of economic sphere). At the same time the process of services' consumption provokes a significant social interest and this fact allows us to consider services as the category of socially meaningful benefits.

Secondly, as the direct recipients of the social service are people, that is why the location of institutions and organizations of social sphere on a definite territory depends directly on such demographic facts as: population and its density, demographic structure and etc.

Wide sense	**Narrow sense**
Service means results of labour for social use which satisfy different needs of other people and enterprises	Service means a special consumer cost

Figure1. Classification of approaches to the definition of service

Thirdly, big expenditures of social service production and high level of complexity of service provision influence service cost.

Fourthly, there is a high individuality of employees of social service and this means that in other words the result of the activity of employees of social service is directly dependent not only on subject-producers, but on capabilities of a person who is the recipient of this service [6,7].

All the above-mentioned features of service influence professional and personal qualities of employees of the social sphere and define the complicated character of this work.

2. Communicative qualities of a person as a leading component of professional self-consciousness of the Director of social service

Professional and personal qualities of employees of social services which are necessary for their successful performance of given tasks are becoming an object of the scientific research. As a rule, the first group of qualities includes high level of professional education of an employee that is employees should have theoretical knowledge of social work, legal aspects of their activity, pedagogy and psychology [8].

The second group of qualities includes personal qualities of employees of social service because namely these qualities are the main reason of success and effectiveness of the performed work.

Honesty, justice, patience and kindness are the main features of professional suitability of employees of social service and they define the success of interaction with clients [9].

Thus there are some objective facts which make us have a new look at the tasks of management in this field and also at requirements of Directors of social service:

1. Nowadays the sphere of social service is one of the developing and demanding spheres;
2. There are specific features of the Director;
3. The growing number of citizens who need social services makes the Director create favourable working conditions for his subordinates.

Considering the above-mentioned the main tasks of a modern Director are not only to consider and justify general perspectives of the organization development but also to realize prospects of his own and his subordinates' professional development. The problem of professional development of self-consciousness of Directors of social services is determined by the growth of people who need social services because the increasing number of clients leads to the situation when a Director has to pay more attention to his subordinates (creation of favourable conditions, increase of labour motivation, stimulation with the help of financial rewards and etc.) than to his professional development.

We can conclude that the leading component of professional self-consciousness of a Director of social service at the modern stage of this system development is communicative qualities of a person. Firstly, these qualities allow a Director to perform all management functions which are necessary for production and provision of social services including adequate orders and control of subordinates' performed tasks. Secondly, they help to form horizontal and vertical connections between subordinated managers. Thirdly, they allow to exclude communicative barriers between employees of the organization from the process of communication because they have a negative effect on the use of information.

3. Empirical research of components of professional self-consciousness of managers of social service

9 managers and 90 employees of 3 social services in Volgograd and Volgograd region (3 Directors and 30 employees from each social service) took part in the research. To measure components of Directors' professional self-consciousness the following techniques were chosen: study of professional self-conception of a person of a Director as a core of the professional self-consciousness [10], method of self-effectiveness (J. Madduks and M. Scheer, A.V. Boyarintseva's adaptation) which is aimed at the research of self-consciousness and self-assessment of a person. Social and psychological climate among colleagues is measured with the help of the method which is called map and scheme of psychological climate assessment (L.N. Lutoshkin).

The results of empirical research: according to the results in Table 1 we can say that managers of Social Service №1 appreciate the feedback with their employees, they try to take into account their employees' opinion. They consider the following qualities important: volitional, communicative and constant interaction with subordinates.

Table 1. Comparative analysis of components of professional self-consciousness of Directors of Social Services № 1, 2, 3.

Social Services №1		Social Services №2		Social Services №3	
Components of professional self-consciousness	Average figure	Components of professional self-consciousness	Average figure	Components of professional self-consciousness	Average figure
Volitional qualities of a Director (Looking-glass Self)	145	Volitional qualities of a Director (Normative Self)	170	Moral and ethical qualities of a Director (Looking-glass Self)	133

Communicative qualities of a Director (Normative Self)	121	Moral and ethical qualities of a Director (Real Self)	130	Moral and ethical qualities of a Director (Real Self)	128
Communicative qualities of a Director (Looking-glass Self)	110	Moral and ethical qualities of a Director (Looking-glass Self)	129	Organizational and administrative qualities of a Director (Real Self)	108
Level of self-effectiveness (in the sphere of subject activity)	40	Organizational and administrative qualities of a Director (Ideal Self)	118	Organizational and administrative qualities of a Director (Looking-glass Self)	109
Level of self-effectiveness (in the sphere of interpersonal communication activity)	12	Organizational and administrative qualities of a Director (Real Self)	104	Level of self-effectiveness (in the sphere of interpersonal communication activity)	14

In Social Service № 2 managers consider that the key factor in their professional activity is the correlation of the manager's image among subordinates with his own ideal. Firstly, they take into account volitional, moral and ethical, organizational and administrative qualities. In Social Service № 3 managers appreciate moral and ethical, organizational and administrative qualities from the point of view of self-assessment and subordinates. But they do not compare themselves with standard features of a Director.

Among managers of Social Service № 1 the following components dominate:

1) Level of self-effectiveness (in the sphere of interpersonal communication activity) means that subjective manager assessment of his interpersonal communication, knowledge and feelings help to achieve results during the interaction with subordinates;

2) Volitional qualities (Looking-glass Self) of professional Self-conception are manager's view of assessment of his capabilities in management from the point of view of subordinates (for example, how his subordinates assess his capability to cope with problems, how he can achieve his goals and perform definite tasks).

3) Communicative qualities (Looking-glass Self) determine views of a Director about how his subordinates assess his flexibility, communicative skills, skills of interpersonal communication.

4) Communicative qualities (Normative Self) show which communicative skills subordinates would like their manager to have and if the manager can assess these qualities himself.

Among managers of Social Service № 2 the following components dominate:

1) Moral and ethical qualities (Real Self) mean that the manager can give a real assessment of himself, and there are the following prevailing qualities: independence, high professionalism, priority of the personal achievement.

2) Moral and ethical qualities (Looking-glass Self) are manager's views of his independence, professionalism according to his subordinates' opinion.

3) Volitional qualities of a Director (Normative Self) show which volitional qualities according to his opinion his subordinates would like their manager to have and if the manager can assess these qualities himself.

4) Organizational and administrative qualities of a Director (Real Self) mean real assessment of his own initiative; independence in solving problems; self-organization, discipline; sense of duty.

5) Organizational and administrative qualities of a Director (Ideal Self) are comparisons of the manager with his own ideal assessing his initiative; independence; self-organization, discipline; sense of duty.

Among managers of Social Service № 3 the following components dominate:

1) Moral and ethical qualities of a Director (Real Self) mean the real assessment and domination of the following qualities: independence, high professionalism, priority of the personal achievement.

2) Moral and ethical qualities of a manager (Looking-glass Self) are manager's views about his independence, professionalism from the point of view of his subordinates.

3) Organizational and administrative qualities of a manager (Real Self) means real manager's assessment of his own initiative; independence in solving problems; self-organization, discipline; sense of duty.

4) Organizational and administrative qualities of a manager (Looking-glass Self) are manager's views of his initiative; independence in solving problems; self-organization, discipline; sense of duty according to his subordinates' opinion.

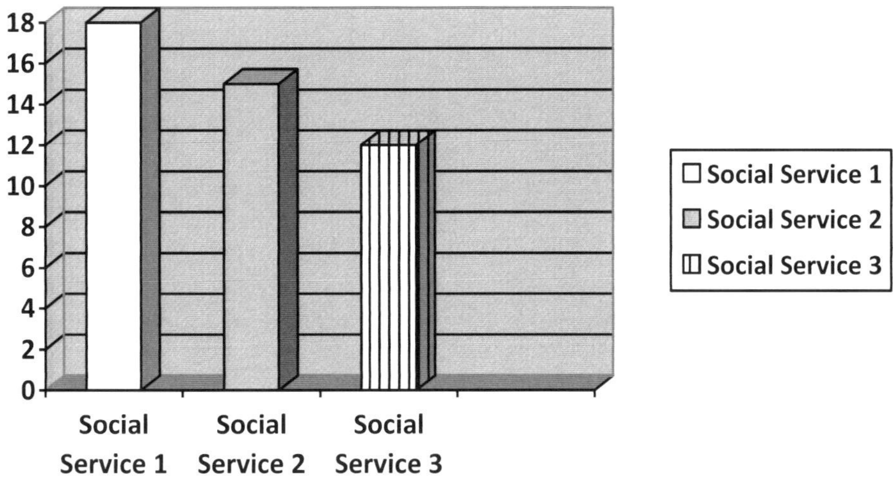

Figure 2. Assessment of social and psychological climate among colleagues

According to the results (Figure 2) all three Social Services have the average degree of favorable conditions, but Social Service № 1 has the highest level of the auspiciousness.

4. Conclusion

Thus, nowadays social work is one of the main mechanisms of social support and population provision. Social services are the final unit in the system of social provision whose work is to realize the process of social services' provision. The character of services differentiates social service from other kinds of activities and determines managers' professional features in these organizations.

According to the analysis communicative skills are the basis of the effective performance of management functions in organizations of social services.

The results of this research can be used in training managers of social services (for example, MBA, President program of training managers for organizations of the national economy).

References

[1] Goloschapov M V 2015 *Scientific Discussions: Tendencies and Perspectives of The Development of Modern Society* 28-32

[2] Kocheshkova L O 2014 *Economic and Social Changes: Facts, Tendencies, Perspectives* **4 (34)** 165-178

IOP Conf. Series: Materials Science and Engineering **483** (2019) 012086 doi:10.1088/1757-899X/483/1/012086

[3] Lubenko V V 2014 *Scientific Community of Students of The XXI Century. Social Sciences: Collection of Material of The XVII International Student Scientific and Practical Conference.* **2(17)** 49-55

[4] Nechaeva Ye S and Nechaev Yu V 2016 *Izvestiya TulSU. Economic and Juridical Sciences.* **3-1** 26-32

[5] Nikitina L Ye 2017 *Questions of Psychological Health of Children and Teenagers.* **317** 170 - 172

[6] Syupova M S and Bondarenko N A 2016 *Scientific Notes TOGU* vol **7 1** 308 – 312

[7] Savinov L I and Ryabova E N 2010 *Servis Plus* **3** pp 14 - 19

[8] Romanina L A, Kurin A Yu 2014 *Vestnik TGU.* **12 (140)** pp 1 – 8

[9] Kholostova Ye I 1993 *Professional, Spiritual and Moral Portrait of a Social Worker* (M., Soyuz) 25-40

[10] Danilova O Yu 2013 *Psychological Sciences.* **13** 75-79

IOP Conf. Series: Materials Science and Engineering **483** (2019) 012087 doi:10.1088/1757-899X/483/1/012087

Efficiency gains for using production capabilities of an oil and gas production enterprise based on innovative process solutions

M I Eliseeva, E I Inyakina

Tyumen Industrial University, Volodarsky Street, 38, Tyumen 625000, Russia

E-mail: m.krasnova@myrambler.ru

Abstract. This article touches upon the issues of efficiency gains for using the production capabilities of an enterprise in the oil and gas industry environment. It is suggested that the most critical elements, including innovative potential would be comprised in the production potential. To assess the degree of utilization of the production capacity of an oil and gas producing enterprise, a rapid assessment methodology based on performance indicators that take into account industry characteristics has been proposed. Implementation of this technique for an oil and gas company has shown that reserves for the fuller employment of its resources are available. Calculations have also been made proving that implementation of innovative process solutions will increase the efficiency of the production potential above the industry average level.

1. Introduction

Currently, most oil and gas fields are characterized by a high reserve recovery and a tendency towards a decrease in oil and gas production. In this regard, to identify ways to improve the efficiency of employment of the production capacity of oil and gas enterprises will be the priority task for the foreseeable future [1]. The issue of the development of scientific and methodological foundations for the building and assessment of the production capacity, as well as the development of specific mechanisms for its implementation, seems relevant.

Production capacity of an enterprise forms a basic prerequisite for its innovation-driven development. There is a relationship between them – the higher the technical and economic level of components of the potential and the greater the utilization efficiency, the stronger is the physical facilities for innovation, the ampler are the opportunities for the implementation of science and technology achievements, improving and increasing the components of an industrial enterprise. This is of particular importance for the oil and gas industry.

2. Theoretical foundations of the production capacity in the oil and gas industry

A review of the scientific literature has shown that almost all researchers believe that the production capacity forms the economic potential of an enterprise [2]. Based on the study of publications, a great variety of approaches to the building of the production potential has been established [3].

For the study of production capacity of industrial enterprises, the major challenge is that all its components perform simultaneously and interdependently, tied together, that is, the production capacity for system, which is a complex and dynamic one, since it is the interconnections between individual constituents of the potential that directly affect its components by changing the initial specifics of their operation, thus contributing to the transition of the potential to a new quality. That is, depending on the extent to which the laws governing the functioning of individual components are studied, it is possible to identify specific features of performance of the production potential of an enterprise as a system as a whole [4].

Development of the production capacity of an enterprise is determined by the influence of innovations, instability of external and internal factors, market conditions, resulting in certain structural changes in the production capacity, as well as in qualitative and quantitative changes in technical and economic indicators that characterize the potential.

Currently, the priority of the economy is the transition to an innovative basis; for this purpose, the majority of enterprises, including oil and gas companies, have developed appropriate innovative programs by creating operational environment for the production. Given this fact, we suggest that capacity for innovations be included in the production capacity as its component. Thus, the production capacity will include four main elements (Figure 1).

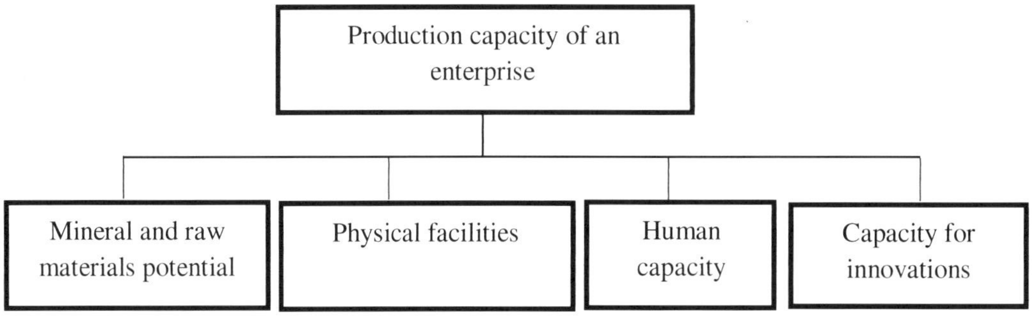

Figure 1. Components of production capacity for an oil and gas enterprise

A constituent element of the production capacity of an enterprise should also be the capacity for innovations that determines the scale, pace, and effectiveness of economic development. Innovative potential is the ability of an enterprise to produce products on a new technical and technological basis. Capacity for innovations is a resource pool available to the company for scientific discoveries, inventions, and technical innovations. Certainly, when studying capacity for innovations, the main attention is paid to the consideration of scientific and technical resources available to the enterprise. In the modern context, innovation-based development is the only option for the development of enterprises that ensures its qualitative growth.

3. Assessment procedure for utilization of production capacity

The issue of assessment of the extent to which production potential of an enterprise is utilized, as well as the formation of a strategy of management of the existing potential is one of the most pressing and critical problems, both in research terms, and in terms of the practical significance for an enterprise.

A rapid assessment technique makes it possible to get a comprehensive view of the extent to which production potential of an enterprise is utilized [5].

For the purpose of rapid assessment of oil and gas companies, we suggest clarifying the assessment technique based on the author's classification of components of the production capacity (Figure 2).

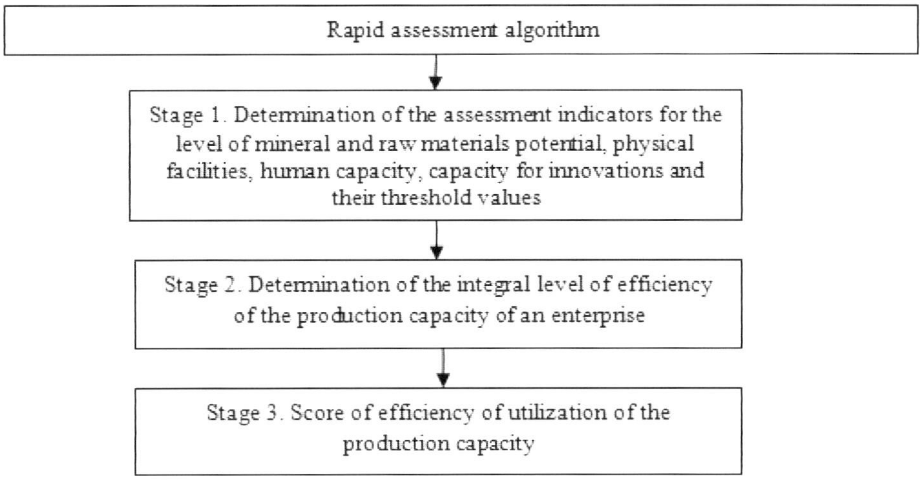

Figure 2. Stages of rapid assessment of efficiency of production capacity of an oil and gas enterprise

Assessment of the efficiency of utilization of the production capacity of an enterprise makes it possible to identify the integral summarized value of this indicator, and to determine the extent to which production capacity is used relative to industry average values.

Stage 1. To assess the utilization of production capacity, it is necessary to determine the components of generalizing and individual indicators for assessing the state, change and efficiency of its utilization and calculate their values for the company under study. Then, the threshold values should be determined, which allow to classify the production capacity by levels. The level B is determined by industry average indicators, while Level A (high) is above average, and Level C (low) is below average. Next, an assessment is made of the indicators that form the production capacity with regard to their threshold values. Given the industry average values, the components of indicators have been provided, taking into account the specifics of the oil and gas industry, and their threshold values have been identified based on the expert estimates (Table 1).

Table 1. Indicators of the assessment of the levels of production capacity of an enterprise and their threshold values

Indicator	Level	Threshold values
1. Mineral and raw materials potential		
Hydrocarbon reserves recovery factor, %	A	> 140
	B	100 - 140
	C	< 100
Hydrocarbon reserves depletion factor, %	A	< 4
	B	4 - 6
	C	> 6
Average petroleum production rate, tons per day	A	> 25
	B	15 - 25
	C	< 15
2. Physical facilities		

Plant and equipment update ratio, %	A B C	> 15 7 - 15 < 7
Return on assets, %	A B C	> 1 0.5 - 1 < 0.5
Return on assets, rubles/rubles	A B C	> 5 3 - 4 < 3
3. Human capacity		
Human employment turnover ratio, %	A B C	< 0.4 0.4 - 1 > 1
Well stock maintenance support, person/well	A B C	< 1 1 - 3 > 3
Output change, %	A B C	> 10 0 - 10 < 0
4. Capacity for innovations		
Scientific and research assets ratio, %	A B C	> 10 5 - 10 < 5
Production gain tempo due to innovation-based technologies, %	A B C	> 20 10 - 20 < 10
Innovative growth ratio, %	A B C	> 1 0.5 - 1 < 0.5

Stage 2. To evaluate the efficiency, a combination of levels for individual indicators shall be considered to assign a level to each component of the production capacity. Next, these indicators reflecting the effectiveness of utilization of individual components of the production capacity for a particular enterprise shall be calculated. Based on the assessments of the levels assigned to the indicators, the expert shall determine their combination to assign the assessment to the level of the relevant component (Table 2).

Table 2. Score of the extent to which the production capacity components are utilized

Extent of utilization of the components	Combination of assessments of the ratio levels			
	Mineral and raw materials potential	Physical facilities	Human capacity	Capacity for innovations
High level (A)	A.A.A. A.A.B. A.A.C. A.B.B.	A.A.A. A.A.B. B.A.A. A.A.C.	A.A.A. B.A.A. A.B.A. A.C.A.	A.A.A. A.A.B. B.A.A. A.A.C.

	A.B.C.	B.A.C.	B.B.A.	B.A.C.
	A.C.A.	C.A.A.	B.C.A.	C.A.A.
	A.C.B.	C.A.B.	C.A.A.	C.A.B.
	A.B.A.	B.A.B.	C.B.A.	B.A.B.
Medium level (B)	A.C.C.	C.A.C.	C.C.A.	C.A.C.
	B.A.A.	A.B.A.	A.A.B.	A.B.A.
	B.B.B.	A.B.B.	A.B.B.	A.B.B.
	B.B.A.	A.B.C.	A.C.B.	A.B.C.
	B.A.B.	B.B.B.	B.B.B.	B.B.B.
	B.A.C.	B.B.A.	C.A.B.	B.B.A.
	B.C.A.	C.B.A.	B.A.B.	C.B.A.
	B.B.C.	B.B.C.	B.C.B.	B.B.C.
	B.C.B.	C.B.C.	C.C.B.	C.B.C.
	B.C.C.	C.B.B.	C.B.B.	C.B.B.
Low level (C)	C.A.A.	A.C.B.	A.A.C.	A.C.B.
	C.A.B.	A.C.A.	A.B.C.	A.C.A.
	C.B.A.	B.C.A.	B.A.C.	B.C.A.
	C.C.C.	A.C.C.	A.C.C.	A.C.C.
	C.C.A.	B.C.B.	B.B.C.	B.C.B.
	C.A.C.	B.C.C.	B.C.C.	B.C.C.
	C.C.B.	C.C.C.	C.A.C.	C.C.C.
	C.B.B.	C.C.A.	C.B.C.	C.C.A.
	C.B.C.	C.C.B.	C.C.C.	C.C.B.

This is to determine the extent of utilization of the production capacity of an enterprise according the same pattern as the level of components (Table 3).

Table 3. Assessment and description of the extent of utilization of the production capacity of an enterprise

General extent	Levels of the components	Description of utilization of the production capacity
High level (A)	A.A.A.A B.A.A.A A.B.A.A A.A.B.A	The company is in a state of absolute equilibrium in almost all components in accordance with all assessment criteria. The physical volume of production has remained unchanged. The equipment and technology used in manufacturing the products are in good condition.
Medium level (B)	A.A.C.B A.B.B.B A.B.C.B A.C.B.B A.C.A.B B.B.B.B B.B.A.B B.A.C.B B.C.A.B C.A.A.B C.A.B.B C.B.A.B	The company's current business is successful; difficulties can be overcome, because adaptation mechanisms do work. Manufactured products are being sold. The physical volume of production has been maintained (although there may be a decrease in certain periods). The equipment and technology are in good condition. Resource efficiency is industry average.
Low level	A.C.C.C B.A.B.C B.B.C.C B.C.B.C B.C.C.C C.A.C.C C.B.C.C C.C.A.C C.C.B.C	Most of the parameters of all functional components are chronically violated, i.e.there are problems with the provision of enterprises with raw and other materials, human resources – these are used inefficiently. Technical and technological support of production activities causes concern. Utilization of basic production assets, labor and material resources is below the industry average.

IOP Conf. Series: Materials Science and Engineering **483** (2019) 012087 doi:10.1088/1757-899X/483/1/012087

C.B.B.C	
C.C.C.C	

Stage 3. To determine efficiency of utilization of the production capacity, it is necessary to provide the scoring. For this, each value of the indicator within the range should be assigned a certain score, which is used afterwards with due regard to the weighting ratio. The highest score should correspond to the most favorable value, while the lowest score – to the most critical one. The level and range of scoring shall be determined by experts (Tables 4).

Table 4. Scoring for utilization of the production capacity of an enterprise

Level	Scoring, points		Threshold values, points
	Indicator	Group of indicators	
A	10	30	26 - 30
B	6	18	14 - 22
C	2	6	6 - 14

Weighting ratio also shall be determined by experts (Tables 5).

Table 5. Weighting ratio

Component of the production capacity	Weighting ratio
1. Minerals and raw materials potential	0.2
2. Physical facilities	0.3
3. Human capacity	0.1
4. Capacity for innovations	0.4
Total	1

The proposed improved procedure for rapid assessment of efficiency of utilization of production capacity was applied to the operational environment of the oil and gas company PJSC Rosneft [6]. Table 6 shows the final values for all elements and the production capacity as a whole.

Table 6. Results of assessment of the indicators for efficiency of utilization of the production capacity of an oil and gas company

Indicator	Value	Level	Score	Weighting ratio	Total score
1. Assessment of minerals and raw materials potential		B	22	0.2	4.4
Hydrocarbon reserves recovery factor, %	186.0	A	10	-	-
Hydrocarbon reserves depletion factor, %	4.2	B	6	-	-
Average petroleum production rate, tons per day	22.30	B	6	-	-
2. Assessment of physical facilities indicators		B	22	0.3	4.4
Plant and equipment update ratio, %	15.0	A	10	-	-
Return on assets, %	1.32	A	10	-	-
Return on assets, rubles/rubles	4.36	B	6	-	-
3. Assessment of human capacity		B	22	0.1	4.4
Human employment turnover ratio, %	4.3	B	6	-	-
Well stock maintenance support, person/well	1.23	B	10	-	-
Output change, %	3.4	B	6	-	-

4. Assessment of capacity for innovations		B	18	0.4	7.2
Scientific and research assets ratio, %	8	C	2	-	-
Production gain tempo due to innovation-based technologies for geological and technical events, %	22.9	A	10	-	-
Innovative growth ratio, %	9	B	6	-	-
Total score				1	20.4

Thus, in accordance with the total score, PJSC Rosneft is at an average level in terms of the efficiency of utilization of the production capacity (which corresponds to the industry average value).

4. Ways of improvement of efficiency of utilization of the production capacity

The key way of raising the efficiency of utilization of the production capacity of the enterprise in general is the development of its innovative component [7]. To this end, PJSC Rosneft has developed a program focused on the achievement of strategic objectives, based on the main priorities, such as efficiency, sustainable growth, transparency, social responsibility and innovation [8].

The objectives can be achieved through a set of activities aimed at the development and implementation of new technologies; design, manufacture and launch of new innovative products and services that meet international standards; promotion of modernization and technological development of the Company through a significant improvement in the main indicators of the efficiency of production processes; increasing the capitalization and competitiveness of the company in the global market.

Capacity for innovations can also be improved on the basis of the introduction of a system of digital field technologies [9]. Information or digital technologies have been already widely used in the global oil complex at all stages of the oil chain. The costs of these technologies are one of the important items of the world leading companies, which allows improvement of the efficiency of field development, increase of production rates, and reduction of expenditures at all stages of the oil cycle [10].

Innovation-based development of the company will lead to increased efficiency in utilization of the production capacity.

The implementation of ways to increase the extent to which the corresponding types of production capacity are utilized by PJSC Rosneft enterprises will allow reaching higher projected values of efficiency indicators.

Based on the rapid assessment, projected values of the efficiency of the production capacity have been calculated with due regard to implementation of the innovative technologies envisaged by the program. The company may reach a high extent of the utilization of almost all constituent components of the capacity, with the score increasing from 20.4 (medium level) to 27.2 (high level).

5. Conclusions

The calculations performed have demonstrated that, in general, for the period concerned, the value of the production capacity of PJSC Rosneft has increased. However, indicators of the efficiency of it utilization, as well as indicators of the efficiency of individual types of capacity, changed diversely.

Based on the improved assessment methods, the efficiency of utilization of the production capacity of the enterprise, the generalized value of this indicator has been identified, and the extent of utilization of Rosneft's production capacity against industry average values has been determined. The calculations showed that, according to the total core, the company is at an average level of efficiency in the production capacity efficiency.

Based on the analysis of the system of objectives of the company's innovation-based development program, the following areas have been identified:

- ensuring the development of PJSC Rosneft as a high-tech energy company;
- ensuring technological leadership in key competencies, namely, oil and gas production, oil processing;
- maintenance of the specific capital and operating costs at the highest global industry level;

- improvement of the energy efficiency of production to the highest global level.

Accomplishment of these areas as well as the introduction of digital technologies of management of field development will let the company improve the efficiency of utilization of the production capacity, provide the competitive edge of the company both in the domestic and global markets.

References

[1] Kryanev D and Zhdanov S 2012 *Drilling and Oil* **8**
[2]] Lepitanova M B, Romanov S I and Kostenko T G 2015 *Methodological Electronic Journal "Concept"*. **8**
[3] Chalenko A. 2012 *Country Capital* **66569**
[4] Avdeenko V I, Kotlov V A 2012 *The Production Potential of Industrial Company* (Moscow) 208
[5] Fomin P A and Starovoitov M K 2013 *Corporate Management* **635**
[6] *PJSC Rosneft* 2017 Official web-site
[7] Biriukova V V, Krainova E A 2016 Features of Formation of Innovative Oil Company Strategy *Problems and Experience of Economic Management of Enterprises* 183-186
[8] *Passport of the Innovation Development Program of JSC Oil Company "Rosneft"* 2015 (Moscow)
[9] Gugulian A G 2017 *Evaluation of the Economic Efficiency of the Use of Digital Field Technologies in Management Decisions in Oil and Gas Production* (Moscow: Russian I.M. Gubkin State University of Oil and Gas)
[10] Ageev N P, Cherniaev M V 2012 *ETAP: Economic Theory, Analysis, Practice* **12** 115-120

The Role of the Regional Support of Technical University in the Training of Professional Personnel for Enterprises of Volgograd region

S P Sazonov, A I Chunakov, E Ekharlamova, A A Polyanskaya, S D Glebov

Volgograd State Technical University, 28, Lenina avenue, the city of Volgograd, 400005, Russia

sazonovsp@mail.ru, helenderbet@mail.ru

Abstract. The article is devoted to assessing the role of the basic technical university in the training of professional production personnel for enterprises of the Volgograd region. The university acts as a communication platform for the implementation of constructive interaction with representatives of business, government, science, education, the public, being the initiator of holding negotiations, meetings, conferences, congresses, symposia at its base.

1. Introduction

Volgograd State Technical University (VSTU) is a supporting technical University with a promising educational infrastructure (scientific and technical library, multimedia classrooms, laboratories with modern equipment).

The role of the supporting University in the region is determined by its target model: to provide advanced training of highly qualified specialists in demand in all spheres of economic activity of the region: design engineers, technologists, economists of production and construction sectors of the economy, builders, programmers, architects, specialists for the branches of engineering, metallurgy, chemical industry, defense industry, housing and communal services, transport. Training for the most part of these specialties, as well as the formation of a number of professional competencies in the region is carried out only in the VSTU. Notably, this group includes even those UGSN (the enlarged groups of specialties and areas of study), which are implemented by other universities in the region. In particular, training specialists in the field of economics and management (UGSN group 38.00.00) is conducted in all educational institutions of the region, but specialists in this field for industrial enterprises (metallurgy, engineering, chemical industry, construction) are prepared only in the Volgograd State University. In addition to training specialists in the basic educational programs of Higher Education, the University provides training of middle managers (technicians, technicians, technologists), retraining and advanced training of specialists and managers of enterprises and organisations in the region, conducts vocational training.

For the regional support University, there are several characteristic areas of training, in which the share of the reduced contingent of students from the total number of students enrolled in the region is more or close to 50%: aviation and rocket and space technology – 100%; weapons and weapons systems – 100%; geology, exploration and development of minerals – 100%; architecture and construction 87.4%; chemical technology and biotechnology -78%; metallurgy, mechanical engineering and material processing – 70%; vehicles – 65%; automation and control – 60%.%; technology of food products and consumer goods – 52%; computer science and technology – 48%; life safety, environmental management and environmental protection – 45%.

The University performs a large amount of applied research in scientific and technical areas for enterprises in the real sector, implements scientific and technological educational chain of "VSTU – VNTK (Volga Science and Technology Complex) – the industrial enterprise", uses the results of research in educational process for development of the corresponding professional competences of future specialists. VNTK platform allows to attract structural units and small innovative enterprises to the production of competitive high-tech products, solving problems of import substitution, promotion of scientific developments of VSTU to the markets of high-tech products.

The University acts as a communication platform for fruitful interaction with representatives of business, government, science, education, the public, being the initiator of the negotiations, meetings, conferences, congresses, symposiums.

The University participates in the development of regional development programs, as well as orients its development strategy in the field of educational and scientific activities under the key areas of development of the region. The University updates its portfolio of educational and scientific products in accordance with the selected program of development of the region segments of key sectors of the economy. These segments include: metallurgical production (production of finished metal products); production of building materials (production of concrete and gypsum products); food industry (conservation and processing of vegetables, production of animal feed production of vegetable oil and fats); production of machinery and equipment (oil and gas equipment, buses, electrical equipment); chemical production (production of rubber products and plastic products); production of petroleum products; transport and communication; production, transmission and distribution of thermal energy; mining [1].

2.Analysis of the results of training of professional personnel for the region

At the moment, the VSTU is a leader in the field of science "Engineering, technology and technical sciences", preparing 75% of the total number of students in this industry in the region.

The structure of demand for educational services determines the development of industrial, construction, agricultural and transport complexes of the Volgograd region. One of the largest consumers of specialists is LLC "EuroChem-Volgakaliy", implementing a project related to the creation of mining and processing plant for the extraction and enrichment of potassium salts.

Table 1. The employment of graduates of VSTU [2]

Year	Number of graduates (pers.)	Employed graduates of the Russian Federation (pers.)	Share of employed graduates (%)	Employed graduates in Volgograd region (pers.)	Employment in Volgograd region / employment of graduates in the Russian Federation (%)
2014	2521	2000	79	1 382	69.1
2015	3873	2987	77	1 883	63.0
2016	1861	1503	81	894	59.5

Based on the table data (see table 1), in 2014 the number of graduates was 2521, of which 79% were employed, which is 2000 people, 1382 of which are employed in Volgograd region (69.1% of all employed graduates of the VSTU. The average amount of payments to graduates 24615 RUB. The number of regions where graduates were employed – 39. Of the total number of employed graduates – 1% - individual entrepreneurs, the rest – employees. Experts of the project "Social Navigator" identified Russian universities, graduates of which are most in demand by future employers, as well as ranked educational institutions according to the criteria of orientation to the real sector of the economy, the demand for scientific works and many others ("Ranking of demand for universities in Russia-2015" - the project " Sociological research of the labour market). Among 140 technical Russian universities, VSTU is on the 59th place [3, 4].

In 2015, the number of graduates amounted to 3873, of which 77% are employed, which is 2987 people, 1883 of which are employed in Volgograd region (63% of all employed graduates of VSTU. The average amount of payments to graduates - 26559 RUB. The number of regions where graduates were employed – 48. Of the total number of employed graduates – 2% - individual entrepreneurs, the rest - employees.

In 2016, the number of graduates amounted to 1861, of which 81% are employed, which is 1503 people, 894 of whom are employed in Volgograd region (59.5% of all employed are graduates of the VSTU. The average amount of payments to graduates 27893 RUB. The number of regions in which graduates are employed – 41. Of the total number of employed graduates – 3% - individual entrepreneurs, the rest - employees. Experts of the project "Social Navigator" identified Russian universities, graduates of which are most in demand by future employers, as well as ranked educational institutions according to the criteria of orientation to the real sector of the economy, the demand for scientific works and many others ("Ranking of demand for universities in RF2016 "– the project" Social Navigator "MIA "Russia today"). Among 132 Russian technical universities, VSTU – 65 places [5, 6].

The General analysis of table 1 allows to draw a conclusion that the majority of students of Volgograd Regional Support Technical University is employed, that proves their demand in the labour market, and also shows the importance of VSTU in preparation of experts for the enterprises of Volgograd region as the majority of graduates is employed there [7].

3. Analysis of the need for qualified personnel in the region

In order to overcome the imbalance between the labour market and the market of educational services, to provide employers, including those involved in the implementation of investment projects, with employees, it is necessary to carry out a research on the needs of the regional market for qualified specialists [8].

According to a survey conducted by the regional labour Committee of Volgograd region, among 1871 employers of Volgograd region, 320 employers (17.1%) showed zero need for personnel.

The largest number of enterprises from the number of respondents has no need for personnel for a seven-year period in the following municipal districts and urban districts of Volgograd region as: Novonikolaevsky district – 63.9%; Nehaevsky district – 57.1%; Oktyabrsky district – 50.9%; Alekseevsky district – 41.2%; Dubovsky district – 39.3%; Mikhailovka-36.3%.

In 11 municipal districts and urban districts of Volgograd region there is no need for personnel from 1.7% to 7.6% of enterprises depending on the number of respondents.

In 7 municipal districts and urban districts of Volgograd region all surveyed enterprises indicated the need for personnel.

As a result of the work carried out, the need for 17.1 thousand skilled workers with higher education for 2017-2023 has been identified (figure 1).

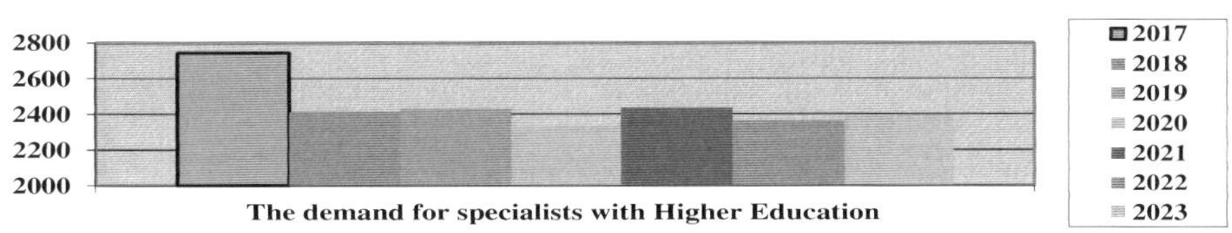

Figure 2. Distribution of demand for qualified workers and specialists by level of education, 2017 – 2023

Specialists with Higher Education are most in demand in the following types of economic activities:

- economy and management - 5,0 thousand people (66,4% of the total demand by type of economic activity);
- education and science - 3.5 thousand people (63.4% of the total need for the type of economic activity);
- health and social services - 2.9 thousand people (57.0% of the total need for the type of economic activity).

IOP Conf. Series: Materials Science and Engineering **483** (2019) 012088 doi:10.1088/1757-899X/483/1/012088

The lowest need for specialists with Higher Education is noted in the following areas:

- provision of public, social and personal services -14 people (0.7% of the total need for the type of economic activity);
- light industry and chemical production – respectively 67 and 89 people (4.1 and 5.2% of the total need for the type of economic activity).

The need for specialists with Higher Education is completely absent in the production of building materials, vehicles, as well as wholesale and retail trade.

The most popular specialties of Higher Professional Education level are presented in table 2.

Table 2. The most popular Higher Education level professions

Name	2017	2018	2019	2020	2021	2022	2023	In total
commodity researcher	359	378	388	408	428	459	475	2895
educator	224	197	195	196	200	198	196	1406
doctor (general medicine)	113	92	97	91	83	83	83	642
technologist of production and organisation of public catering	81	82	81	75	78	78	77	552
agronomist	84	75	76	60	64	66	65	490
builder	59	53	53	53	147	57	54	476
economist	71	61	59	61	60	58	57	427
social worker	55	54	55	54	53	53	53	377
therapist	73	43	47	43	44	42	45	337
Mathematics teacher	55	47	40	51	39	43	39	314

Based on the data presented in table 2, we can conclude that the most popular specialists, who are trained by the VSTU will be: product technology and catering (total need 552 specialists), as well as economists (427 specialists).

4. Conclusions

The formation of the Regional Support University of engineering profile is necessary for advanced training of highly qualified specialists (design engineers, technologists, economists of industrial and construction enterprises, builders, programmers, architects), which creates staffing for the growth of industrial production, construction, reconstruction of housing and communal services, road network and transport [9, 10]. In addition, the Regional Support University will provide a large amount of applied research in natural science and technology areas, demanded by enterprises of the real sector. The possibility of obtaining high-quality and practice-oriented education allows to prevent the leakage of labour and intellectual resources and, moreover, to attract such resources from other regions of the Russian Federation and from abroad.

Volgograd Support Technical University produces highly qualified specialists in demand in all spheres of economic activity of the region: design engineers, technologists, economists of the production and construction sectors of the economy, builders, programmers, architects, specialists for the branches of mechanical engineering, metallurgy, chemical industry, defense industry, housing and communal services, transport [11, 12]. Now, the percentage of employment of VSTU graduates is high and is more than 70-80% for each year of graduation. More than 60 % of employed graduates stay to work in Volgograd region. This fact demonstrates the attractiveness of the region as an employer and the most important role of VSTU in providing new highly qualified specialists of Volgograd region.

5. Recommendations

The development of the Regional Support University will allow the region to create a unique combination of educational programs and research areas that ensure the development of the main elements of the socio-economic system of Volgograd region.

To achieve this strategic goal, the following tasks are planned:

- modernisation of implemented and development of new flexible project-oriented educational programs for bachelors and specialists;
- development and implementation of new programs for training masters and teaching staff in post-graduate school, combining the educational process with the solution of research, design and production and technological problems;
- development of the network of basic departments of the University and training of specialists on target order in the territory of industrial partners and power structures;
- inclusion of distance learning and open online courses in the organisation and content of the educational process of the University;
- diversification of the portfolio of Higher Education services in the direction of specialties with a high level of demand;
- creation of more effective programs for employment of graduates and cooperation with organisations of the region.

Due to the diversity of its functions, the Regional Support Technical University is the most important link in the socio-economic system of the region. Special attention should be paid to the Regional Support University's effective interaction with stakeholders - consumers and customers of educational services, federal, regional and municipal authorities with the regional community [13]. Together with the main subjects of regional labour markets, the leadership of the Regional Support University needs to develop and implement programs of target (profile) training in the field of Higher and Postgraduate Education, training and retraining of various categories of specialists and managers on the basis of the introduction of annual monitoring of the need for qualified personnel in Volgograd region.

References

[1] Sazonov S P, Kharlamova E E, Gorshkova N V, Polyanskaya E A 2016 Competitive advantages of the regional support university and its role in the regional development strategy *Science and Society*) **3 (1)** 180-189

[2] Sazonov S P, Kharlamova E E, Chehovskaya I A, Polyanskaya E A 2015 Evaluating Financial Sustainability of Higher Education Institutions *Asian Social Science* **11 (20)** 34-41

[3] Sazonov S P, Kharlamova E E, Ezangina I A, Gorshkova N, Kovazhenkov M A, Polyanskaya E A 2017 Theory and Methodology of the Financial Management of the Regional Supporting University *Journal of Advanced Research in Law and Economics* **8 (1)** 211-219

[4] Analytical information on the results of the formation of the forecast of the needs of the regional labour market in workers and specialists necessary for the sectors of the economy of Volgograd region (Including investment projects) 2017-2023

[5] Ergunova O T, Lizunkov V G, Malushko E Yu, Marchuk V I, Ignatenko A Yu 2017 Forming the system of strategic innovation management at the high-tech engineering enterprises *IOP Conference Series: Materials Science and Engineering* **177 (1)** 012046

[6] Ferreira R, Lizunkov V G, Politsinskaya E V 2017 Formation of entrepreneurial competencies of university graduates in conditions of transition to the universities of the third generation *Novosibirsk State Pedagogical University Bulletin* **7 (6)** 195-211

[7] Road map of the Regional Support University URL: http://www.vstu.ru/razvitie/programma-razvitiya/programmaazvitiya/ДОРОЖНАЯ%20КАРТА%20ОПОРНОГО%20УНИВЕРСИТЕТА.pdf

[8] Regional Support Universities of Russia: URL: http://flagshipuniversity.ntf.ru/

[9] The program of development of the Federal state budgetary educational institution of higher education "Volgograd State Technical University", aimed at the formation of the Regional

IOP Conf. Series: Materials Science and Engineering **483** (2019) 012088 doi:10.1088/1757-899X/483/1/012088

Support University for the socio-economic development of Volgograd region URL: http://www1.vstu.ru/files/webmaster/2016-09/14573/upload/programma_opornogo_universiteta.pdf

[10] Ranking of demand for universities in Russia URL: https://vid1.ria.ru/ig/ratings/Engen-2016.htm

[11] Labour market and employment in Volgograd region. Official statistics URL: http://volgastat.gks.ru/wps/wcm/connect/rosstat_ts/volgastat/ru/statistics/employment/

[12] Kharlamova E E, Polyanskaya A A 2016 Institutional foundations of the regional support University and its impact on the socio-economic development of the region *Development of territorial socio-economic systems: issues of theory and practice* 100-103

[13] Lizunkov V, Politsinskaya E, Malushko E, Kindaev A, Minin M 2018 Population of the world and regions as the principal energy consumer *International journal of energy economics and policy* **8 (3)** 250-257

[14] Fedotov N G, Moiseev A V, Syemov A A, Lizunkov V G, Kindaev A Y 2017 New Methods of Three-Dimensional Images Recognition Based on Stochastic Geometry and Functional Analysis *IOP Conference Series: Materials Science and Engineering* **177 (1)** 012047

IOP Conf. Series: Materials Science and Engineering **483** (2019) 012089 doi:10.1088/1757-899X/483/1/012089

Professional jargon units in workplace communication (by the example of professional languages of miners, oilmen, railway workers and ambulance doctors)

LN Rebrina, LM Generalova

Volgograd State University

Abstract. This paper explores professional jargon units used by specialists in workplace communication in various fields. Professional jargon, as a means of internal informal workplace communication, is an important element of the organizational culture of any enterprise. Professional jargon units as products of professional and sign activity can be interpreted as verbal manifestations of social symbolism in professional communication. The authors define distinctive features of professional jargon, composition and internal structure of organization of subsets with professional jargon units used by representatives of different professions. The article describes functioning of professional jargon, mechanisms of jargon nomination, its specificity in relation to different professional spheres.

1. Introduction

The study of professional jargon (PJ) units functioning in workplace communication of representatives of different professions enables us to understand actual changes in the modern language within the anthropocentric paradigm in the context of socially determined parameters of its implementation and reflection of professional language picture of the world in it.

Learning any profession and subsequent professional activity includes as a mandatory component mastering a professional vocabulary; and organizational culture of any enterprise is a complex system, where formed skills of professional communication are an important element, its professional jargon plays a significant role, primarily as a means of internal informal workplace communication.

PJ units reflect dynamic aspect of professional (sub)culture [Elistratov 1995], which we interpret as a certain standard determining, among other things, rules of speech behavior in the professional environment, fixed in minds of its representatives as a traditional, "inherited" aspect of normativity, causing individuals' adaptation to the environment through behavior unification, they represent the values of subculture (normative and value aspects determine each other). At the same time, a nationwide language functions as a basis for forming PJ units, it is manifested in the remaining semantic connections of professional jargon units and common lexical units. On the other hand, PJ units are associated with a specific terminological sphere of use, special nature of meaning, and paradigmatic relations with terms.

PJ units as products of professional and sign activity can be interpreted as verbal manifestations of social symbolism in professional communication, reflecting the result of socialization of professional group members, acquiring conventional norms and values, common goals, assimilating the corresponding picture of the world, defining and providing group identity, consciously striving for isolation [Kulpeisova 2009].

For the Russian language, scientists identify four periods of jargonization (1. after the October revolution in 1917, due to "proletarian language" spread; 2. in the middle 30s and post-war 40s of the 20th century, associated with massive repression, increase in crime; 3. 60s-70s of the 20th century, due to the endeavor to form a counterculture; 4. since 90s of the 20th century, associated with restructuring, capitalization, emphasizing the value of civil and linguistic freedom) [Beregovskaya, 2013].

According to YouGov Jargon Survey [https://today.yougov.com/join-community] professional jargons are becoming more widespread, employees of different departments, workers of large and small enterprises have used PJ units actively, but differently. Due to the limited range of carriers of professional subculture, their differentiating and integrating functions and expressive connotations they can foster (with informal oral internal workplace communication on narrowly specialized topics) or hinder (with external or official internal communication) solution of speakers' communicative tasks [MESSAGE (technology of effective communication) 2010. No. 6. http://www.media-factor.ru/img/august/index.html].
Characteristics of professional jargon usually include duplication of terminology, expressiveness, stylistic lowering (there is, however, a point of view according to which PJ units, when used in the workplace communication for its intended purpose, have no stylistic function), relevance of various word-formation models and their core functions: communicative, contact, and expressive [Golovanova, 2008; Morozova, 2006; Putsenko, Eritskaya, 2015; Savinov, 2007; Yakovenko, 2007]. Environment and language can influence each other, which is also true for different pictures of the world, reflected in the nationwide language and its socially determined varieties.
Some professional jargons (for example, ones of soldiers, sailors, programmers, doctors) have been analyzed quite extensively.

2. Results and Discussion

The study explored PJ units used in workplace communication of miners, railway workers, oil workers, ambulance doctors. Our analysis shows that PJ units in different professional languages represent a means of effective brief formation of the concept and corresponding meanings in workplace communication. They reveal the similarity of internal structure for organization of the relevant subsets, namely the same composition of PJ units' subgroups, distinguished on denotative ground: PJ units that nominate items in the relevant industrial sphere (equipment, tools, buildings and facilities, material, products, industrial clothing); subjects or actors of the industrial sphere (employed in the production, customers); professional actions. The first two classes of the mentioned nomination objects are considerably larger in terms of lexical representation.
To illustrate this statement, we present fragments of relevant professional jargons (table 1)

Table 1. Internal organizational structure of professional jargons' lexical subsystems

professional group using jargon	items of professional sphere	professional actions	people (members of professional groups, clients)
miners	(equipment of mines, parts and components of machines, mine sections, specialized facilities, equipment of miners) *баран* (literally meaning 'male sheep') – electric drill (similarity with twisted horns); *дом пенсионеров* ('old folks' home') – operations control (evaluation of operators' job as very easy); *змеёныш* ('young serpent') – miner's belt; *кишка* ('gut') – cable of electric drill; *крот* ('mole') – tunneling machine; *мартышка* ('marmoset') – hand-operated	*басить* – ('to sing bass') – lift the side of something with a lever (possibly from the Kazakh word *'басу'* (push)); *палить* – ('to burn') – explode; *качать* ('swing') – to transport coal by conveyor; *забуриться* ('go off the rails') – go off the rails;	*водолей* ('water-carrier') – miner, pumping, water from the workings; *ежик* ('small hedgehog') – student of the miner; *погонялка* ('goad') - shift supervisor; *проходимец* ('passer-by' or 'con-man') – shaft sinker; *бандерлог* ('bandar log') – operator of mine or conveyor transport at work site; *белые каски / белоголовые* ('white helmets' /

	jack (probably, by characteristics of "silly, useless"); *мышеловка* ('mousetrap') – man-riding cage; *соха* ('plow') – manual scraper; *балда* – ('fool') – large sledgehammer; *борода* ('beard') – chain of scraper conveyor; *верблюд* ('camel') – compressed air operated pump (by characteristic of "low resources"; *лифчик* – ('bra') – supporting system for working face (by function); *маргарин* ('margarine') – gases after explosion; *пчёлка* ('small bee') – air drill; *теща* – ('mother-in-law') – hammer (by function);		'white-headed') – chiefs (orange helmets for workers); *мутнозвон* ('blurred ringing') – operator; *упыры* (pronounced [upy:r], meaning 'undead') – shaft sinker (from abbreviated name of site – UPR);
oilmen	(production tools, facilities, raw materials, products, equipment) *ёлка* ('fir tree') – component of blowout prevention equipment (similarity of appearance); *мартышка* ('marmoset') – hook-wrench for steering wheel; *кривулька* ('something curve') – inclined well; *прялка* ('spinning wheel') – rig; *корка* ('crust') – residual rock in a well (similarity of appearance, associations); *папка-мамка* ('male-female') (*мамка-мамка* ('female-female'), *мамка-папка* ('female-male'), *папка-папка* ('male-male')) – various types of adapters for connecting rock cutting tool and drill pipe; *яйца* ('eggs') – crown block saver, counterweight in the form of two balls (the resemblance)	*ковырять* ('to dig into') – to drill; *куститься* ('to bush') – to drill a number of wells close to each other; *стрелять* ('to shoot') – to do seismic stimulation of oil production, seismic exploration	*пуля* ('bullet') – driller (association by function); *лепила* ('a person bringing freebies') – electric welder; *дневники* ('dnevniki') – a team working in the day-time; *помазок, маслопуп* ('shaving brush') – drill rig equipment mechanic (association by functions);
railway employees	(trains, lines, types/series of trains, train parts and components, railroad equipment) *барыга* ('profiteer')– train Chelyabinsk – Chita; *бешеные* ('crazy') – high-speed trains	*задавить машину* ('to crush machine') – to use locomotive brake; *убить поезд* ('to kill the train') – to disrupt train timetable; *весёлые старты* ('school sports	*болотники* ('from bog', pronounce [balotnıkı]) – locomotive brigades of train depot Bologoe; *бабуины* ('baboons') – locomotive brigades of

	on the Moscow – St. Petersburg line; *вазелин / лизун* ('petroleum jelly' / 'slime') – rail lubricating train / locomotive; *велосипед* (bicycle) – railway trolley; *генерал* ('general') – steam locomotive P-36 (has red stripes on the sides, resembling trouser stripes); *горбатый* – ('humpback') – train with an oversized load; *дупло* ('hole in a tree') – metrodepot; *метельник* ('sweeper' (from "sweep")) – track cleaner; *рогатый* ('horned') – electric locomotive; *крокодил* ('crocodile') – locomotive ZTE10M (long and green); *Серго* (Sergo) – steam locomotive series CO; *сопли* ('snot') – electric cables for rolling locomotive out the depot ditch; *череп* ('skull') – electric locomotive VL8; *Вильгельм* (Wilhelm) – electric locomotive VL-8; *башкировоз* (carrier for Bashkirs)– train Ufa-Sibai	day') – ticketless passengers running over to an already checked car; *дать / дёрнуть петуха* ('take a fake', 'squeaky note') – to apply the emergency brake; *дёрнуть за хвост* ('pull the tail') – to open the end crane of the last car of the train to purge the brake line of the train; *играть в шахматы* ('to play chess') – to provide and remove cars on access roads	train depot Babaevo; *кнут* ('whip') – train dispatcher / station duty officer (association by function); *снегири* ('bullfinches') – track workers (because of orange vests); *внезапник* ('inspector with sudden'/ 'unexpected inspection') – inspector of slow passenger trains; *душманы* ('mujahideen') – children and teenagers on rail ways; *мухомор* ('fly agaric') – operator (because of red cap)
ambulance doctors	*померация* (pomeratsia) – intensive care unit; *буксир* ('tugboat') – sodium oxybutyrate (psychotropic drug) (re-interpretation of sound complex and association by function); *валежник* ('fallen' / 'lying trees') [valezhnik] (from*лежать*) [lezhat'] (to lie) – ward with immobile patients; *галочка с фенечкой* [galochka s fenechkoy] – combination of the drugs haloperidol and phenazepam (sound associations); *гармошка* ('accordion') – lung ventilator manually-operated (similarity of appearance, characteristic "manual"); *дискотека* ('disco') – (turned on) siren and flashing lights of the	*стрельнуть / стукнуть* ('to shoot / to hit') – to use a defibrillator; *загрузить пациента* ('to download patient') – to introduce psychotropic drugs; *сделать скворечник* ('to make a birdhouse') – to make the operation of cranial trepanation;	*самотек* ('natural flow') – a patient who independently came to the substation for help; *скрипач* ('violinist') – pseudo-suicide with cut veins on the wrist; *синий одутловатик* ('blue bloater') – patient with chronic obstructive bronchitis; *самоделкин* ('do-it-yourselfer') – traumatologist; *консервы* ('conserves' or 'canned food') – conservative, i.e. non-surgical treatment patients; *рецидивист* (recidivist) – patient with re-recurrence; *архангел* ('Archangel')

	ambulance; *жаба* ('toad') – angina; *крикаин* [krikain] – insufficient anesthesia (association by function)); *ПМЖ* (PMZh) – clinic by place of residence; *уши* ('ears')– phonendoscope; *будильник* – ('alarm clock') - liquid ammonia solution		– morgue attendant; *банан* ('banana') – patient with yellow disease; *беспилотник* ('drone') – unconscious patient; *дураки* ('fools') – workers of Emergency medical services, part of psychiatric brigade

In the professional jargon of railway workers, there is a significant group (in quantitative terms) formed by 1) names of locomotives and trains; motive for jargon designations' nomination includes similarity of appearance with the everyday object (non-professional sphere, for example, *генерал, крокодил, череп, рогатый*), associations with abbreviations nominating models, series of trains / locomotives / electric locomotives (for example, *Вильгельм, Серго*); reinterpretation on the basis of daily experience of speed characteristics, function of train or passenger characteristics (e.g., *вазелин / лизун; бешенные, башкировоз, барыга, метельник*); 2) names of locomotive crews, their jargon nomination is often based on geographical concepts (e.g., *бабуины, болотники*); 3) designations of workers employed in this professional sphere, based on accentuation of external features similar to non-professional sphere objects (e.g., *снегири, мухомор*) or functions, re-interpreted on the basis of associative features of everyday experience fragments (eg, *кнут, внезапник*). Widespread use of proper names in the jargon nomination formation is typical of railroad workers' professional jargon (that it is due to professional specific nature of this sphere) [see about vocabulary based on personal names of such jargon Feoktistova 2010].

Characteristics of miners' work also affect formation of a professional picture of the world, which is reflected in the unique professional language [see about professional language and professional picture of the world of miners Pribytova 2005]. This professional sphere involves using various tools and special equipment with many components identified by comprehensive and detailed terminology, which explains wide representation of relevant jargon units in miners' professional language, providing a quick and brief exchange of information at their workplace. In this PJ units' subgroup frequent motives for nomination are similarity of appearance, similarity of function, emotional associations; however, extreme nature of the profession, high physical and psychological stress determine the function of emotional release, performed by PJ units (e.g., *баран, кишка, соха, балда, верблюд, лифчик, маргарин, теща, борода*). Denominations of professional sphere subjects constitute the second subgroup (by quantitative parameters); the function performed at work (which is also reinterpreted in a humorous / ironic way), as well as external signs of subjects (e.g., *водолей, погонялка, проходимец, упырь, мутнозвон, бандерлог, белоголовые*) are important for their nomination.

Oil industry employees are different-area specialists: drillers, operators for oil extraction, diesel operators. As D.S. Ermolin noted [Ermolin 2008], jargon of drillers is most developed, which might be associated with a higher (compared to the drilling table) level of production noise in diesel-engine room or near the oil production plant, which led to development and widespread use of sign communication system of diesel mechanics and operators and less significant use of verbal communication. In this subset as in the professional jargon of miners, nomination of equipment and its parts play a crucial role, frequent motives of the nomination include similarity of appearance to everyday objects and associations by functions (for example, *ёлка, кривулька, яйца, мамка-мамка, мамка-папка, папка-папка*), diminutive forms of words are its distinctive feature. The second group (in quantitative terms) includes naming of the production sphere subjects, where the overriding motive for nomination is reinterpreted similarity by function and features of actions (for example, *помазок, маслопуп, лепила, пуля*). Several researchers [Golovanova 2008; Morozova 2006; Suleymanova

2006; Gläser 2000] mentioned high relevance of astrometric parameters (age characteristics, physical signs and condition, emotional experiences, body parts, etc.) for jargon nomination.

Ambulance doctors use jargon to refer to equipment, medicines, medical institutions and facilities, diseases and their external signs, doctors and patients. A common feature for ambulance doctors' PJ units is active use of a laughable metaphor, pun, language game, which is associated with specific psychological conditions of their professional activities [Kuvshinnikov 2011]. Nomination of professional sphere subjects is often based on humorous reinterpretation of sound complex of the corresponding term and association by function (e.g., *померация, буксир*), similarity of appearance (e.g., *гармошка, дискотека*), transfer by function, combined with its ironic evaluation (*крикаин*). Designations of subjects (mainly patients) constitute the second largest group; the most frequent motives determining nomination are external features (e.g., *банан, синий одутловатик*), humorously reinterpreted on the basis of similarity with non-professional sphere objects, the cause of condition or the condition of patient (e.g., *скрипач, безпилотник, рецидивист, консервы*) When denominating colleagues, humorously reinterpreted function becomes a leading motive (e.g., *архангел, самоделкин*).

The analysis proves that common social experience assumes common typified emotions, emotive meanings, which explains typification of PJ units' connotations, their adequate interpretation and identification regarding the objects of thought by representatives of professional groups (these elements of the PJ units' semantics point to the important for workplace situation aspects of the referent).

3. Conclusion

Professional jargon as a socially determined type of language reflects a professional picture of the world of specialists, emphasizing their values and specific nature of semantic images. Specialization of conceptual professional apparatus, importance of generating new knowledge about objects and phenomena of professional spheres based on expressive, imaginative reinterpretation, significance of linking scientific experience with the ordinary, importance of subjectivity in non-professional and professional spheres are the key for professional jargon functioning.

In professional jargons of the spheres characterized by deep specialization in the professional field (all considered jargons), jargon units duplicating formal terms of the professional language or naming terminologically uncovered objects and processes prevail over units that coincide with common lexical items.

PJ units arise as a result of creative reinterpretation taking place at work reference situations or their elements and are characterized by new meanings that do not coincide fully with the professional terms, due to new image associations with nomination objects, as well as different motives of nomination. Semantic images, forming nomination by PJ units and by terms can vary greatly. Deeper, essential features as motives determine terminological nomination (for example, cause-and-effect relations, etc.), while for jargon nomination they are more superficial (for example, similarity of appearance with everyday objects, environment and surroundings, similarity in use). Motivating component is reflected in the internal form of PJ units, which is a point of conjugation of the original and reinterpreted meanings, providing "inheritance" of information and generating new knowledge about nomination object, accentuation of significant for the production situation features of the object that determine associated (subjective, stereotypical) image, which, in turn, provides expressiveness of PJ units, based on the subjectivity of understanding.

References

[1] Beregovskaya E M 2013Youth slang: formation and functioning *Questions of linguistics* **3** 32-41
[2] Elistratov V S 1995 Argo and culture (Moscow)
[3] Ermolin D S 2008 Units of the general and professional vocabulary in the mental lexicon of carriers of literary norms and colloquialisms (on the material of the sublanguage of oilmen)

(Sankt-Peterburg) 119 p
[4] Feoktistova L A 2010 Onomastics and Otomonastic Vocabulary in Railwaymen Slang *Onomastics*.**2 (9)** 69-85
[5] Gläser R 2000 Should LSP Dictionaries also Include Professional Jargon and Slang? *Fifth International Conference of the African Association for Lexicography. Stellenbosch* 86–98
[6] Golovanova E I 2008 Typology of professional communication units: a cognitive-pragmatic aspect *Philology. Art history* **9 (110)** Vol. 19 25-29
[7] Korovushkin V P 2005 Fundamentals of contrastive sociolectology (Cherepovetsk) 283 p
[8] Kulpeisova T S 2009 Manifestations of elements of verbal social symbolism and symbolic interactionism in professional jargon *Vestnik KarGU* https://articlekz.com/article/4543
[9] Kuvshinnikov A I 2011 Medical jargon http://www.kuvschinnikov.ru/2012/07/medicinskij-zhargon.html
[10] Morozova O A 2006 Specificity of interlingual asymmetry in categorizing artifacts, persons and actions in professional sublanguages (based on the vocabulary of the Russian and English versions of the professional oil sublanguage)]: PhD Thesis (Kazan) 154 p
[11]] Pribytova L V 2005 Professional language of miners of the Kemerovo region (Terminology and jargon) PHD Thesis (Kemerovo) http://www.dissercat.com/content/professionalnyi-yazyk-shakhterov-kemerovskoi-oblasti-terminologiya-i-zhargon
[12] Putsenko K N Epritskaya N K 2015 Professional jargon: an approach to the systematization of terms related to the specialty "oil and gas production" *Youth Bulletin of IRSTU* **4** 58
[13] Savinov D A 2007 On the issue of professional jargon *Materials of scientific conference* (Magnitogorsk) 43
[14]] Serdobintseva E N 1998 Speech characteristics of professional vocabulary *Speech characteristics of professional vocabulary* (Moscow)
[15] Suleymanova A K 2006 The terminological system of oil business and its functioning in the professional discourse of a specialist PhD Thesis (Ufa) 459 p
[16] Yakovenko L K 2007 Speech culture of professional communication: stereotypes: norms and behaviors PhD Thesis. (Almaty)

Public organization management: Trade Union Cafeteria Plan of Volgograd State University

N A Archebasova[1, 2, 5], E G Grigorieva[1, 3], S A Linchenko[1, 4]

[1] Volgograd State University, 100 Prospekt Universitetskij, Volgograd, 400062, Russia
[2] Department of General and National History, Archaeology
[3] Department of Germanic and Romance Philology
[4] Department of Social Technologies
[5] Trade Union Organization of Workers of Volgograd State University

nadezhda.archebasova@volsu.ru, e.grigorieva@volsu.ru, linchenko@volsu.ru

Abstract. The article analyses the effectiveness of the current project of systematic social package expansion according to the flexible management system in the Trade Union Organization of Workers of Volgograd State University. The application of such methodologies as customer / product development, Total Quality Management and empathic design has allowed focusing on the customization of social benefits on the Cafeteria Plan menu. The solution to the problem of combining production management and social benefits distribution in a public organization constitutes the novelty of this project. The authors conclude that the use of flexible and situational methods in the system management of an enterprise has enabled creating breakthrough product innovations in the services and benefits sector, extending their action to a wide range of users in the region, as well as applying this practical experience in the work of trade union organizations of other Russian universities.

1. Introduction

The combination of social benefits and corporate events is called the social package of workers (SPW) of an enterprise / organization. The Labor Code of the Russian Federation determines a mandatory part of SPW – state social guarantees. Therefore, SPW should include state social guarantees or expand them. The SPW extent depends on the management policy at an enterprise in relation to its staff, and on its workforce activity. Mutual relations between the management and the workforce are stipulated in the Collective Agreement, one of the main local acts of an enterprise. The representative of the workforce is usually a trade union organization [1]. The Collective Agreement contains the guaranteed part of expanded SPW; furthermore, both parties to the agreement may come to its further supplementing.

The authors of the article explore the effectiveness of the systematic SPW expansion project functioning in the Trade Union Organization of Workers of Volgograd State University (VolSU). The project is developed according to the flexible management system [2].

2. Methodology

The flexible management mechanism – the Cafeteria Plan [3] - adapts well to the activities of a public organization. The study uses the customer / product development methodology which allows orienting different types of social benefit packages on the Cafeteria Plan menu towards their further customization. The term "customization" is understood in this paper as the ideal of interaction between the service provider and the consumer. Such an approach to organizing the production of social services in scientific research is called a service factory [4]. For a successful long-term operation of such a factory, the interplay of the following methodological components is of especial value: Total Quality Management, Teambuilding for social activists, a positive "economy of scale" and empathic design [5] of the production of services.

3. Results and Discussion

The practical experience of introducing the Cafeteria Plan in an enterprise as a working benefits differentiation tool already exists [6]. The Cafeteria Plan in the Trade Union Organization of Workers of VolSU was created in 2015. Its aims were 1) to enhance the protection of social, labor and professional rights of workers; 2) to develop business administration forms and methods in a public educational institution; 3) to modify the emotional and regulatory components of the team loyalty program.

The Cafeteria Plan Project typology [7] is as follows: it is a social program (by level), average regional (by scale), organizationally complex (by complexity); long-term (by terms of its implementation); modular (by quality requirements and methods of its provision); a multiproject (by resource requirements); Russian territorial (by location) and anti-crisis (by goals and objectives). The main reasons for the project include, firstly, the need for structural and functional transformations due to the increasing gap in the life expectancy of employees of the Russian state university and their actual socio-economic position in modern conditions [8]; secondly, the emergence of new opportunities for independent business administration and commercialization of trade union activities under the Federal Laws of Russia "On trade unions, their rights and guarantees of activity" and "On public associations" [9], [10].

It is necessary here to clarify exactly what is meant by the Cafeteria Plan menu. It's the area of implementing various projects where they can be tested, corrected, combined and where they can interact with each other.

The Trade Union Cafeteria creation algorithm includes seven steps, each of which is constantly repeated except for the second one.

1. Analyzing organizational motivation systems, loyalty and corporate culture programs.
2. Making a decision by the Workers' Trade Union Committee on the transition to the Cafeteria Plan.
3. Formation of sections and menu dishes.
4. Approval of the list of effective and efficient benefits.
5. Informing the employees about changes in the social package.
6. Constant analysis of the new benefits system work, their systematic customization.
7. Periodic review of the menu.

The Trade Union Cafeteria menu consists of ten sections / packages of social benefits: 1) optional, 2) corporate, 3) business lunch, 4) "green", 5) fitness, 6) catering, 7) baby/ children, 8) family, 9) seasonal, 10) exotic.

Numerous dishes / service products of these menu sections include a) a formed standard (non-flexible) SPW in the framework of the Collective Agreement; b) a set of elements of the university corporate culture mobilizing the workers to fulfill the university mission; c) a system of privileges for trade union activists; d) labor protection, industrial safety, ergonomics of workplaces; e) the workers' health improvement, treatment at a health resort and recreation activity; f) holding and servicing off-site events; g) social benefits for children and grandchildren of the workers; h) benefits and guarantees extended to the workers' families including several generations of relatives; i) seasonal, daily and holiday social programs; j) inaccessible and organizationally complex services, types of leisure activities and cultural tourism.

The Workers' Trade Union Committee analyzes all completed projects, decides on their further use. The projects of such categories as "new", "upgrade", "popular", "hit of the season" and "best" from each menu section will be on the priority list of the perspective work plan.

Further the paper reveals the peculiarities of applying eight principles of quality management when planning the activities of this public organization under the Cafeteria Plan.

1. Focusing on the customer is important because the workers monthly transfer voluntary membership fees (1% of their salaries) to the Trade Union Organization and expect to receive regular protection of their labor, professional and social rights. Those who are dissatisfied with the activities / inaction of this organization leave it.

2. Leadership. The designer of the introduction of flexible management in the public organization is N.A. Archebasova, the Chairperson of the Trade Union Organization of VolSU, a holder of a Master's degree in Management, the leading auditor of the quality management systems of the "Bureau Veritas" company [11]. The project team includes the Trade Union Committee (20 people) and trade union activists from the university structural units (64 people). The Trade Union Committee headed by the Chairperson represents the interests of the entire workforce (more than 1,500 workers, more than 500 children and family members of the workers) as the second party in the negotiations on concluding the Collective Agreement.

3. The involvement of people in a public organization is based on the awareness of their personal benefit [12]. The vast majority of Volgograd State University workers (83%) are aware of the privileges of their Trade Union Organization membership. In this regard, it is worthwhile to pay tribute to the effective team building of this public organization. A public organization teambuilding is based on the distribution of tasks, taking into account the free time and personal wishes of employees, encouraging innovative activities, the opportunity to learn new forms and methods of social activity. The Cafeteria Plan allows creating a team of efficient trade union activists. They have a simultaneous opportunity to build a public career along with carrying out their professional activities and to realize their creative abilities as a lead user. The lead user method was proposed by E. von Hippel [13] to characterize consumers creating improved versions of products / services. Lead users are potential coordinators of social projects, successful members of a trade union activist team. An important motivator of their activities is the opportunity to realize their own innovative project and get a desired service [14], [15].

4. The strategic planning of production and distribution of social services is based on the SWOT-analysis method (see table 1):

Table 1. SWOT analysis of activity planning of the Trade Union Organization of Workers at VolSU

Strengths	Weaknesses
• a socially oriented Collective Agreement with a wide list of optional compensation benefits and social guarantees, • a positive effect of the main "hygienic factors", • democratic principles of management and grading	• an average team size (up to 2000 employees) - it is difficult to apply the "economy of scale," • predominantly the state formation of the salary fund of the university workers – a modest trade union budget for expanding benefits and compensation, • "obsession" of the team at work - weak motivation for the development of recreational culture
Opportunities	**Threats**
• wide customization of benefits due to the average team size and trade union activists' work, • building a parallel public career or alternative to professional scientific and educational career for trade union activists, • formation of a unique corporate culture, a synergistic effect	• the state "optimization" of higher education, • a long-term decline in the standard of living of Russians and a university worker's social status, • the "sweating system" of an effective contract and of the academic activity of the teaching staff

In the current planning, the process approach is applied. The Trade Union Organization of Workers of Volgograd State University has developed and introduced a technological map of the implementation process of each project (see table 2). This map is designed to monitor the manageable parameters of the service factory: the volumes and types of benefits provided; costs and expenses; timing, duration and relationships; resource provision. This constant monitoring provides, on the one hand, the individualization of the choice of benefits. On the other hand, the public organization has a strict control over the costs of social benefits.

Table 2. Technological map of the implementation process of the VolSU Trade Union Project.

Calendar		Project name/ project member status (C [a] / M [b])	Project costs		Project audience (persons) / amount of financing (RUB [l]) / %			Project PR: announcement / report / photo (A / R / P)		
Month	Date		Total MV [c]/ TU [d]	SF [e]: TU [d] / PF [f] / OSF [g]	Total	TUM [k]	Children	VolSU site	Intra-net	social net-works
Total		C M	MV TU	TU PF OSF: CA [h] VR [i] SP [j]	Pers. RUB %	Pers. RUB %	Pers. RUB %	A R P	A R P	A R P
2017/ 2018		143/ > 170	TU<MV TU= 0-60% MV	Financing from multiple sources	1-2000	all	all	A R	A R	all

[a] Coordinator
[b] Member of the project team
[c] Market value of the project
[d] Trade Union project costs
[e] Sources of financing
[f] Personal funds of the Trade Union Organization members
[g] Other sources of financing
[h] Collective Agreement
[i] Volgograd Regional Trade Union Committee
[j] Social partners
[k] Trade Union Organization members
[l] Rouble

5. Making decisions is based on facts. The own production of social services allows successful implementing both personalized projects (1-3 people) and mass ones (up to 2000 people). All the workers receive information on each completed project / package of social benefits through various communication channels. The employees independently decide which projects to take part in as well as which benefits to use. The Trade Union Committee regularly monitors the effectiveness of projects with the help of technological maps, coordinators' reports and consumer assessment of the services provided.

6. Mutually beneficial cooperation with service providers is also valuable. The situational approach and constant experimentation enable, on the one hand, creating breakthrough production technologies [16] of social benefits and, on the other hand, using the agile-method of product development in collaboration with all the participants of the process on the principles of three social interactions: social partnership, cooperation and co-financing.

7. The system management approach is provided by the Cafeteria Plan and a positive "economy of scale". The cost of production decreases with increasing production volumes. Therefore, the larger the social project, the more benefits can be obtained from the social partners. The optimal scale for

effective social planning is regional (the Volgograd region as a subject of the Russian Federation). On the initiative of N.A. Archebasova, the Chairperson of the Trade Union Organization of VolSU, the chairmen of trade unions of ten state universities of the Volgograd region created an intersectoral coordination council and successfully implement projects at the regional level of interaction (up to 2000 people) using the co-creation method. Joint regional work gives a significant synergistic effect.

8. The perpetual improvement of the quality of social services and the growth in the volume of the provided services on the Trade Union Cafeteria Plan menu have been going on since 2015. The current achievements of the workers of VolSU, on the one hand, and Volgograd State University, on the other hand, in the social sphere are far ahead of those in the previous years. In our view, the progress has been made by introducing democratic innovations [17]. We should mention here, firstly, the improvement of the quality of life of university employees and their families; secondly, their satisfaction with the conditions of work and rest; thirdly, obtaining cultural and spiritual benefits, psychological comfort and other privileges due to implementing numerous projects on the Cafeteria Plan. Each project is a compositionally complete event, and being a "puzzle" it is creating a picture of the lifestyle of the university workers.

As explained in the previous paragraph, it is clear that the Trade Union Cafeteria Plan works to satisfy the workers' existing (perceived) needs. Another undoubted advantage of the Cafeteria Plan is that it develops the consumer of social services. On the other hand, when innovations outpace the needs of the target audience, the production of a service factory is in a situation of uncertainty, waiting for consumers to understand and accept the services of a new product [18]. This entails periodic negative emotional attacks by consumers-traditionalists on a team launching a new product. Moreover, this increases the time for workers to accept a new benefit / service. The rate of introduction of new preferential products at the VolSU Trade Union service factory can be seen in table 3. It shows that over the four years of the Cafeteria Plan implementation, the volume of the provided services has increased eight times. Episodic services have become a permanently diverse menu of benefits.

Table 3. The volume of services / benefits for the Trade Union Organization members at VolSU

Year	The number of social projects for the period		
	year	month	week
2014	21	1.6	0.4
2017	143	11.9	2.9
2018	about 170 (159 for 11 months)	14.2	3.5
2014-2018	The indicators have increased 8 times		

Having discussed the progress made in the sphere of providing the workers with social services at VolSU, the final paragraphs in this section of the paper address the stages of implementation of the Cafeteria Plan.

Stage 1. The 2015-2016 period was that of preparation and implementation of a pilot Cafeteria Plan Project. The pre-project analysis was carried out taking into account the achievements of the VolSU Trade Union Organization of Workers in previous years, the trade-union analytics of the "expectations" of Russians [19] and the best world practices of motivational and personnel management. The concept of the project is based on the need for continuous improvement of socially oriented organizations.

Stage 2. 2017 was the year when the tested successful projects were included into the Collective Agreement for 2017-2020.

Stage 3. The 2018-2019 period is to face constant improving the menu of the expanded SPW as a tool aimed at:

a) improving the university performance as a whole;
b) high-performance labor stimulation;
c) improving the socio-psychological climate in the team;

d) developing corporate culture;
e) increasing the welfare of workers;
f) creating a favorable public opinion about VolSU.

Before proceeding to the conclusions, it will be necessary to briefly dwell on the results. Social projects are potentially characterized by a high degree of novelty. The novelty of this project lies in the compositional solution to the problem of production management and distribution of social benefits in a public organization. Some separate elements of the Cafeteria Plan Project are based on effective programs of the Central Council of the Trade Union of Education and Science Workers of the Russian Federation, primary trade union organizations of a number of Russian universities, but the very concept of the Cafeteria Plan Project is unique because it is adapted to the specific corporate culture of the VolSU Trade Union Organization of Workers. The social significance of this project is confirmed by diplomas of III and I degrees (2015, 2016) of the All-Russian "Trajectory of Success" competition of trade union organizations of workers of Russian universities.

4. Conclusion

The article presents an original way of applying flexible management aimed at creating a service factory which produces benefits in a public organization. The analysis was based on the practical work of the VolSU Trade Union Organization.

The results substantiate the reality of successful long-term operation of a service factory, subject to the skillful use of the interplay of the following methodological components: Total Quality Management, Teambuilding for social activists, a positive "economy of scale" and empathic design of the production of services.

The results obtained confirm the value of flexible and situational methods in the system management of an enterprise. As the analysis shows, applying these methods helps to create breakthrough product innovations in the sector of benefits and services.

The major results of the paper may serve as recommendations for expanding the scope of work of university trade union organizations in the Volgograd region and in the Russian Federation.

5. References

[1] Trade union analytics: linear distribution according to the results of the all-Russian poll on the topic "The role of trade unions in modern Russian society" 2015 *"Solidarity" newspaper supplement* **3** 1-8

[2] Kravets M V 2015 A flexible approach to the formation of employees' social package in an organization *State administration. Electronic messenger* **50** 137-51

[3] Romanyuk D "The Cafeteria principle" is a worthy upgrade of the social package [Electronic resource] URL: http://www.astrosoft.ru/about/press_room/articles/printsip-kafeteriya-dostoynyy-apgreyd-sotsialnogo-paketa/

[4] Vapnyarskaya O I 2014 Genesis and modern approaches to the definition of customization *Service in Russia and abroad* **6**(53) 189-201

[5] Leonard D, Reymport J J 2006 Empathic design as a path to outstanding innovations *Harvard Business Review*

[6] Working tools for differentiating benefits. The "Cafeteria" technique: Practical experience of implementation *Illustrative presentation material for HR & Trainings EXPO'13* [Electronic resource] URL: http: // www.HRpacc.ru

[7] Shapiro V D 1996 *Project management* SPb

[8] Ryzhenkova Yu 2015 By the minimum standard: the satisfaction of basic needs and the position in society are more valuable for Russians than money *Solidarity* **18** 8-9

[9] On trade unions, their rights and guarantees of activity: Federal Law of Russia No. 10-FZ of 04.14.2014 [Electronic resource] URL: http://www.consultant.ru/document/cons_doc_LAW_8840/

[10] On public associations: Federal Law of Russia No. 82-FZ of December 31, 2014 [Electronic

resource] URL: http://www.consultant.ru/document/cons_doc_LAW_6693/
[11] Archebasova N A 2017 Using flexible trade union management to expand the social package of workers: the "Cafeteria" principle *Forming a development strategy for a primary trade union organization in ensuring the competitiveness of higher education: Proc. Int. Trade Union Scientific and Practical Conf. on October 17-20, Novocherkassk, Russia* 56-62
[12] Sardaryan A, Komarova T, Khozhempo V 2008 Motivational function of the social package: how to interest an employee of the 21st century? *Personnel Management* **8** [Electronic resource] URL: http://www.top-personal.ru/issue.html?155
[13] Hippel E von 2005 *Democratizing Innovation* Cambridge MA
[14] Luthjt Ch, Herstatt C 2004 The Lead User Method: an Outline of Empirical Findings and Issues for Future Research *R & D Management* **34**
[15] Eisenberg I 2011 Lead-User Research for Breakthrough Innovation *Research-Technology Management* **54**(1)
[16] Laptev G D, Shaitan D K 2015 Managing the creation of breakthrough product innovations in the established company *Moscow University Bulletin: Series 6: Economics* **3**76-92
[17] Hippel E von 2005 *Democratizing Innovation* Cambridge MA
[18] Laptev G D, Shaitan D K 2015 Managing the creation of breakthrough product innovations in the established company *Moscow University Bulletin: Series 6: Economics* **3**76-92
[19] Trade union analytics: linear distribution according to the results of the all-Russian poll on the topic "The role of trade unions in modern Russian society" 2015 *"Solidarity" newspaper supplement* **3** 1-8

Method of lemmatizer selections in multiplexing lemmatization

O A Sychev, N A Penskoy

Volgograd State Technical University, 28 Lenina Avenue, Volgograd, 400005, Russia

E-mail: oasychev@gmail.com

Abstract. The article describes the method of selecting a particular lemmatizer while multiplexing lemmatization using word's part of speech and suffix. The article also describes the experiment to determine the best suffix (or prefix) length for multiplexing lemmatization. Lemmatizers, that are used to build the multiplexing lemmatizer, show worse results, than the resulting lemmatizer. The developed lemmatizer is compared with the best theoretically possible multiplexing lemmatizer The practical problems of creating multiplexing lemmatizer are discussed.

1. Introduction

Getting a word's lemma is necessary for solving various problems of natural language processing such as extracting knowledge from the natural language texts, texts classification and clustering, information retrieval, text abstracting and annotating and natural language translation, especially from a synthetic language to an analytic language [1]. Lemmatization is a problem of creating a binary relation between word forms and their lemmas, that represent the basic form of words. The problem is made worse because of word homonymy – one word can serve for different word forms of different lemmas (for example a word "lay" can be both past and past participle form of the verb "lie" and nominal form of the verb "lay").

The problem of mapping words to their lemmas in English can be reduced to transforming word's prefix and suffix [2].

There were developed several methods of lemmatization and part-of-speech determining.

The first method uses a dictionary of all word forms where each word form is mapped to its lemma and, possibly, part of speech. That method requires a lot of resources to create and verify the dictionary and big amount of memory to store it. The Prague Dependency Treebank is an example of that method [3]. That method also can be used approximately, while calculating editing distance, such as Levenshtein distance [4] or Damerau-Levenshtein distance. However, such approach could yield wrong results for languages with short words.

The second method is the rule-based approach. It uses a set of grammar rules to form lemmas. Some rules are applicable to certain words (like irregular verbs), others are applicable to thousands of words that are inflected using the same grammar rules [5].

The third method of lemmatization, which is another form of rule-based approach, is based on generating a graph in which nodes are states and edges change word during transitions from one state to others [6, 7].

Different researchers used different methods to compare the lemmatizer efficiency: some use information retrieval on a set of sentences [1, 8], others use a database of words with defined word forms and lemmas [9, 10]. However, the problem of using a particular dictionary for testing lemmatizers is that even if lemmatizer, which uses dictionaries (the first method), shows the best efficiency, that doesn't guarantee it works well with other words, that don't belong to dictionary.

The results of experiments in [10] show that lemmatization efficiency can be increased by multiplexing [11] several lemmatizers, while using each one for those words it performs with best. Multiplexing implies selecting different lemmatizers for different words by some algorithm.

2. Methods

The method of multiplexing lemmatization requires using several lemmatizers and choosing one of them by some criteria to lemmatize a particular word. In [10] it was shown, that generally PHPMorphy, which uses a dictionary-based approach, is the best lemmatizer available. However, it has its flaws and deficiencies and can be improved by multiplexing it with other lemmatizers. To determine the best lemmatizers for multiplexing the authors compared potential gains in the best possible scenario.

Table 1 shows a ratio of words that were correctly lemmatized by different tools while being incorrectly lemmatized by PHPMorphy to all words that were incorrectly lemmatized by PHPMorphy. That is the maximal gain in efficiency that can be obtained by multiplexing PHPMorphy with that tool. English Penn Treebank tagset is used to designate parts of speech.

The ratio is measured separately for each part of speech. If we get the same lemma while lemmatizing two forms of each word, the word is considered correctly lemmatized. Tests for noun lemmatization include infinitive form (NN) and plural form (NNS); tests for adverbs include infinitive (RB), comparative (RBR) and superlative forms (RBS); tests for adjectives include infinitive (JJ), comparative (JJR) and superlative forms (JJS). Verbs are used in their base form (VB), past tense form (VBD), present participle form (VBG), past participle form (VBN), non-3rd person singular present form (VBP), 3rd person singular present form (VBZ).

In table 1 it can be seen, that generally, word_operators lemmatizer showed the best results for multiplexing with PHPMorphy. Word_operatrors lemmatizer was developed using the rule-based approach [10], which efficiently complements PHPMorphy in areas where it performs worse. The best case is the lemmatization of adjectives: word_operators lemmatizer correctly lemmatized 100% of words. However, it can't absolutely replace PHPMorphy. Analysis of noun lemmatization shows that word_operators lemmatizer is better for about 3 thousand words, but there can be seen no clear pattern to use for rule-based multiplexing PHPMorphy with word_operators.

Verb analysis (VBD/VBG, VBN/VBG, VBP/VBZ cases) shows a large group of verbs with suffix "is" that are poorly lemmatized by PHPMorphy. But the analysis of all verbs with that suffix doesn't show any significant difference between PHPMorphy and word_operators. So, the researchers formed a hypothesis that PHPMorphy problems in verb lemmatizing were mainly caused by its incomplete dictionary, not by a word pattern that can be identified and used for multiplexing it with other lemmatizers.

Table 1. The relative improvement of PHPMorphy efficiency by multiplexing it with other lemmatizers (ideal multiplexing).

type/tool	NLTK lemmatizer	SimpleNLG	ruby lemmatizer	skyeng	word operators
JJ/JJR	0.353211	0.117737	0.166667	0.472477	0.467890
JJ/JJS	0.248603	0.069832	0.167598	0.365922	0.332402
JJR/JJS	0.720000	0.272000	0.188000	0.788000	1.000000
NN/NNS	0.204370	0.001490	0.147256	0.207350	0.766327
VBD/VBG	0.531250	0.010417	0.156250	0.531250	0.744792
VBD/VBN	0.409091	0.181818	0.227273	0.409091	0.590909
VBD/VBZ	0.530000	0.010000	0.140000	0.550000	0.440000
VBN/VBG	0.553191	0.010638	0.228723	0.553191	0.750000
VBN/VBZ	0.545455	0.010101	0.116162	0.565657	0.424242
VBP/VBD	0.476923	0.000000	0.174359	0.528205	0.487179
VBP/VBG	0.520408	0.010204	0.204082	0.551020	0.673469
VBP/VBN	0.497382	0.000000	0.120419	0.549738	0.471204
VBP/VBZ	0.685393	0.033708	0.179775	0.685393	0.797753
VBZ/VBG	0.580952	0.023810	0.176190	0.609524	0.547619
Total	0.293603	0.029917	0.154679	0.320178	0.688256

Considering that most English words change their suffix during inflection, it is possible that grouping words by their suffix (of some fixed length) and determining the best lemmatizer for each group will improve the resulting lemmatization efficiency. Since attempts to find specific clusters of words, on which particular lemmatizers perform best, failed, a database with mapping each possible suffix to the relevant lemmatizer is necessary.

As an alternative hypothesis, using prefixes instead of suffixes must be considered.In English language the first letters of a word are often the root of the word, so using fixed length prefixes we can explore how word roots affect words' lemmatization by modern software.

To do that we have to group all words by their parts of speech and inside these groups we have to group words by their prefix or suffix. The best lemmatizer will be determined for each group. That data were stored in a file. We will evaluate the efficiency of a multiplexing lemmatizer using different lengths of prefixes and suffixes to find the best implementation.

A multiplexer lemmatizer should determine the part of speech for the word, find its prefix or suffix of necessary length and find the lemmatizer to use that data and get lemma.

The authors performed the experiment to find the optimal suffix (prefix) length for that algorithm. As shown in [13], in the English language the average word length is 5 characters, so suffixes and prefixes with a length of up to 5 characters were considered. The use of a large prefix (or suffix) brings multiplexing lemmatization close to creating a dictionary containing the best lemmatizer for each word. Too long suffix or prefix will make the multiplexing lemmatizer less efficient when encountering words that weren't used for its training. But the use of too short suffixes doesn't allow enough distinctions between differently inflected words.

The suffix size also affects the number of rules we must record in the file. Not every letter combination of that length will be used (because not all the letter combinations actually occur in words), but the number of possible entries will be growing exponentially. It will affect the lemmatizer performance.

There was used the English dictionary from the Language Tool project to perform the experiment. That dictionary contains about 350 thousand of entries that consist of a lemma, a word form, and a part-of-speech tag. The lemmatizer efficiency with a given suffix (prefix) length was measured in two ways. The first one was for measuring the efficiency of multiplexing lemmatizer that was built using all words in the dictionary for given suffix length. The second lemmatizer was built using 80% of words; the words were chosen randomly from the dictionary; the efficiency of this lemmatizer was measured too. The second measurement allows us to evaluate how efficiently the multiplexing lemmatizer will behave for words that were not used for its building; decrease of that efficiency would signal us that increasing suffix length more isn't effective.

Five lemmatizers were used to build the multiplexing lemmatizer: PHPMorphy, NLTKLemmatizer, SimpleNLG, rubylemmatizer, skyeng, and word_operators lemmatizer.

3. Results

Table 2 shows the number of rules in the multiplexing file, the lemmatization efficiency and the lemmatization efficiency when training lemmatizer on partial data set, depending on suffix length.

Table 2 shows that the optimal suffix length is 3. Suffix lengths' value over 3 letters considerably increases the multiplexing file while decreasing efficiency for words not present in the original dictionary (cf the third and the fourth columns). The efficiency for lemmatizers built using the full dictionary rises insignificantly.

After performing a similar experiment using prefixes instead of suffixes, it was found that the optimal prefix length is 2. Lemmatization efficiency for lemmatizers trained on 80% of dictionary drops faster with prefix length than with suffix length.

Table 3 shows the comparison of the best prefix-based and the best suffix-based multiplexing lemmatizers for different parts of speech. The suffix-based lemmatizer performs better than the prefix-based one; the difference is the most evident for comparative adjectives (JJR) and verbs in past tense form.

Table 2. The number of rules and lemmatization efficiency depending on suffix length.

Suffix length	Number of rules	Lemmatization efficiency	Lemmatization efficiency for the training of lemmatizer using 80% of the dictionary
1	42	0.97488	0.97484
2	569	0.98090	0.98060
3	4230	0.98507	0.98276
4	18765	0.98702	0.97622
5	49947	0.98905	0.95127

Table 3. The efficiency of the prefix-based and the suffix-based multiplexing lemmatizers for different parts of speech.

Part of speech	prefix_tool	suffix_tool
JJ	0.998663	0.998736
JJR	0.605851	0.856156
JJS	0.804255	0.876123
NN	0.999924	0.999975
NNS	0.941226	0.948157
VBD	0.653851	0.974248
VBG	0.966683	0.981184
VBN	0.967657	0.975084
VBP	0.999250	0.999417
VBZ	0.981024	0.981690
Total	0.969263	0.979621

4. Discussion

Table 4 allows us to compare the developed multiplexing lemmatizer (multiplexing_tool) with the two best existing lemmatizers (PHPMorphy and word_operators) and the best theoretically possible multiplexing lemmatizer (i.e. the lemmatizer that performs correct lemmatization for every word for which at least one of the included lemmatizers does it correctly). The multiplexing lemmatizer using 3-letter suffixes performs better than any existing lemmatizer; it performs slightly worse than the theoretically ideal case. It's especially evident for lemmatizing comparative adjectives, where multiplexing lemmatizers have a big advantage over two best individual lemmatizers.

So multiplexing lemmatization is a useful way to efficiently improve lemmatization. The experiment shows us that the best multiplexing lemmatizers could be built using word suffixes with a length of 3 letters and parts of speech to determine which lemmatizer to use.

Practical concerns of using multiplexing lemmatizers should address a problem of getting all multiplexed lemmatizers to work on a particular platform. These lemmatizers are implemented in different programming languages. So, it's better not to just note the most effective lemmatizer for each suffix in the file, but several most effective ones to have a fallback if a particular lemmatizer is unavailable at a particular platform or in particular time. That allows the lemmatizing system to be more reliable to cases when external web services are going down.

Table 4. Comparison of multiplexing lemmatizer efficiency with the best lemmatizers and the theoretically ideal multiplexing lemmatizer.

word_operator/POS	PHPMorphy	word_operators	multiplexing_tool	ideal_multiplexing

JJ	0.793151	0.998718	0.998736	0.999982
JJR	0.709061	0.779358	0.856156	0.905323
JJS	0.803783	0.791962	0.876123	0.892671
NN	0.887252	0.999962	0.999975	1.000000
NNS	0.924569	0.931075	0.948157	0.991306
VBD	0.961792	0.545215	0.974248	0.994610
VBG	0.966683	0.549976	0.981184	0.994487
VBN	0.962746	0.544562	0.975084	0.994729
VBP	0.987286	0.999300	0.999417	0.999767
VBZ	0.948899	0.931067	0.981690	0.998803
Total	0.887812	0.928967	0.979621	0.995122

References

[1] Ozturkmenoglu O and Alpkocak A. 2012 Comparison of different lemmatization approaches for information retrieval on turkish text collection *Innovations in Intelligent Systems and Applications* (Trabzon: IEEE) 1-5

[2] Blevins J P 2006 English Inflection and Derivation The handbook of English linguistics ed Aarts B, McMahon A M (Wiley-Blackwell) pp 507-36

[3] Jan H, Panevová J, Hajičová E, Sgall P, Pajas P, Štěpánek J, Havelka J, Mikulová M, and Žabokrtský Z. 2006 *The Prague Dependency Treebank 2.0* (Philadelphia: Linguistic Data Consortium)

[4] Lyras D P, Sgarbas K N and Fakotakis N D 2007 Using the levenshtein edit distance for automatic lemmatization: a case study for modern greek and english *Tools with Artificial Intelligence* (Patras: IEEE) **2** 428-35

[5] Joel P, Nada L and Dunja M 2004 A rule based approach to word lemmatization 7th International Multi-Conference Information Society IS (Ljubljana: IEEE)

[6] Nicolai G, Cherry C and Kondrak G 2015 Inflection generation as discriminative string transduction *Proceedings of the 2015 conference of the North American chapter of the association for computational linguistics: human language technologies* (Denver: ACL Antology) 922-31

[7] Bhattacharyya P, Bahuguna A, Talukdar L, Phukan B 2014 Facilitating multi-lingual sense annotation: human mediated lemmatizer *Proceedings of the Seventh Global Wordnet Conference* 224-31

[8] Kanis J and Skorkovská L 2010 Comparison of different lemmatization approaches through the means of information retrieval performance International Conference on Text, Speech and Dialogue 93-100

[9] Eger S, Gleim R and Mehler A 2016 *International Conference on Language Resources and Evaluation* (Portorož: European Language Resources Association)

[10] Sychev O, Gurtovoy V and Penskoy N 2017 A study of efficiency of modern inflection and lemmatization software *Proceedings of the IV International research conference "Information Technologies in Science, Management, Social sphere and Medicine"* 308-15

[11] Bruneton E, Coupaye T and Stefani J 2002 Recursive and dynamic software composition with sharing *Proceedings of the 7th ECOOP International Workshop on Component-Oriented Programming*

[12] Marcus M P, Marcinkiewicz M A and Santorini B 1993 Building a large annotated corpus of english: the penn treebank *Computational linguistics 19* **2** 313-30

[13] Sigurd B, Olofsson M, Van Weijer J 2004 Word length, sentence length and frequency–zipf revisited Studia Linguistica 581 37-52

IOP Conf. Series: Materials Science and Engineering **483** (2019) 012092 doi:10.1088/1757-899X/483/1/012092

Hierarchical analysis in optimization of enterprise units

S U Kuznetsov[1], P V Tereliansky[2]

[1]ISE Department, Volgograd State Technical University, 28 Lenin pr., Volgograd, Russia
[2]IS Department, State University of Management, 99 Ryazanskiypr., Moscow, Russia

e-mail: tereliansky@mail.ru

Abstract. In the article the method of hierarchical analysis as applied to research of functioning of the technical services in industrial enterprise. Formed the hierarchy of objectives that realizethe service ofproduction'stechnical preparation.The hierarchical model representsthe main functions of the enterprise unit that implements the planning of production facilities. Significance of the main functions is determined by several important criteria.Analysis used the matrix of pairwise comparisons and hierarchical synthesis. In calculationwe using expert assessments of the individual criteria credibility. Comparison the function's values with the cost of their implementation allows to detect the mismatch and optimize the department work.

1. Introduction. To gain strategic advantages in the markets of engineering products is a permanent task for the industrial enterprise. This taskrelated to the awareness of the priority of scientific and technical preparation services in the enterprisestructure. It is their activity that determines the level of quality and competitiveness of the goods and services offered to the market for industrial purposes. In this regard, an important task is to optimize the organization and management of services of the enterprise on the basis of modern scientific methodologies.

A powerful tool for the implementation of tactical and strategic changes in the organization is the methodology of activity based cost (ABC). It can be used to optimize the various activities and levels of the organization. However, the use of this tool in the field of organization and management is not well developed at the moment. Expediency of this methodology is determined by the possibility of allocation in the management of elementary, repetitive functions, works and operations.Its also important the availability of relevant documentation and regulatory framework [1].

We researchpossibility of using the ABC instruments on the example of the technological preparation service of machine-building enterprises.This service may be called Department of the chief technologist, DCT; or Bureau of technological preparation of production, BTPP) [2, 3]. These units occupy an intermediate position between the design product development service and productionunits.

2. Discussion. The main function of this division is to ensure the technological readiness of the enterprise to produce products in a given quality, a given amount and at a certain cost. The consumer of this function is the subsystem of the main production; quality indicators of performance are the following:

- the volume and composition of the set of technological documentation;
- quality of documentation;
- terms of documentation;
- the cost of work on the TPP.

Another main function, which is external, is the organization and management of TPP processes. The use of a functional approach is aimed at optimizing the activities of the DCT to improve the

IOP Conf. Series: Materials Science and Engineering **483** (2019) 012092 doi:10.1088/1757-899X/483/1/012092

management efficiency of technological preparation of production [5]. Based on the essence of the subject area, and the requirements for this unit in the enterprise, you can build a tree of goals to improve the organization and management of TPP. This takes into account the statement of the chief technologistsdepartment of the machine-building plant.

Based on the study of normative documents that regulate the activities of theDCT, it is possible to build a hierarchy and interconnection of works. We based on the target orientation activities of specialists in organization and management of technological preparation. The most significant and deployed here are the following groups of functions:

- development of organizational and technical measures;
- calculation of production capacity;
- development of calendar schedules at the chamber of technological preparation of production [4].

As an example, to illustrate the capabilities of the functional approach to the analysis, let us consider the second functional group. Its hierarchical model is presented in the corresponding scheme (Fig. 1). According to it, the main function F1 is implemented with the help of several basic functions; the output of process is the control information. Each of the main functions, for its part, implies the execution a number of more specific formalized works [5].

Further implementation of the functional-cost approach requires the construction of a combined functional-structural model of the object. This means the establishment of a correspondence between the functions performed and the structural elements that implement these functions. This will allow us to accurately determine the cost associated with the performance of each function. After that it is possible to determine the existing imbalances in the organization of processes (figure 1).

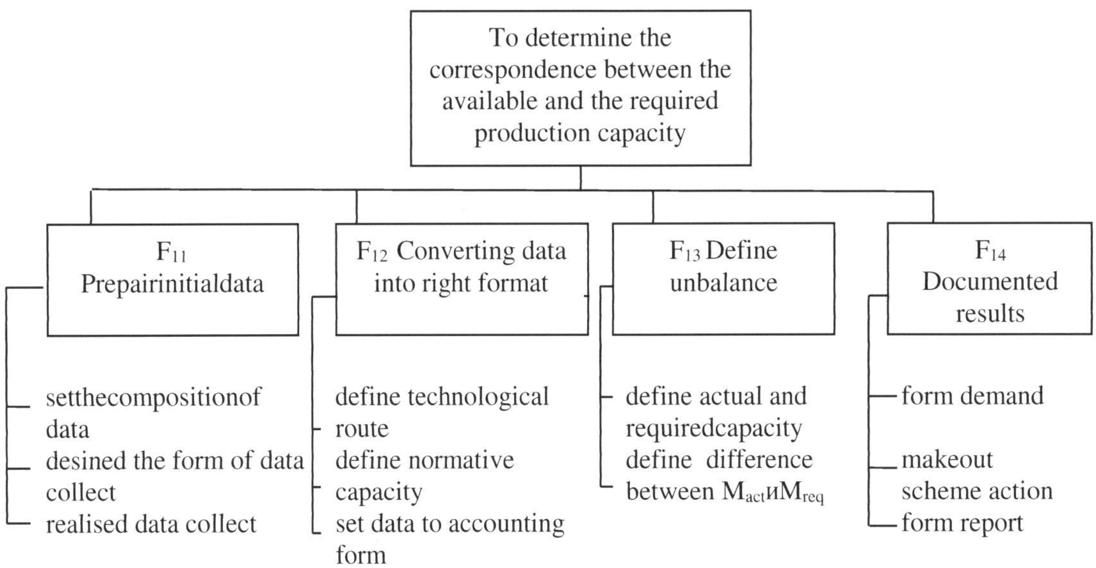

Figure 1. Hierarchical model of object's main function

This feature is implemented by the corresponding function unit which is called Bureau of facilities; that is the part of DCT. The elements that implement those functions are the employees of this division, as well as their computer equipment.

For further calculations, we present the staffing of Bureau (table 1).

Table 1.The staffing of the Bureau of facilities.

Heldpost	Number of people	Monthly salary, rub	Sum, rub
1. HeadofBureau	1	38000.00	38000.00
2. Principalengineer	1	32000.00	32000.00
3. Engineerof 1 category	2	25000.00	50000,00
4. Technician-technologist	2	22000.00	44000.00

Total	6	-	164000.00

Implementation of the considered functions is provided with the use of three automated workstations on the personal computers basis. Their total cost (account with the installed licensed software) is 70000,00 rubles.

The salary of specialists is the main component of the costs for the implementation of the considered functions. To accurately determine the cost of each function, we present a matrix with distribution of functional responsibilities between the employees of this unit. This distribution is characterized by the proportion of time (in percentage) spent by each specialist performing work within each function (table 2):

Table 2. Matrix of the functional distribution of work between specialists.

№	Function / Specialist	Prepair initial data	Converting data	Define unbalance	Documented results
1	Head of Bureau	10.00	20.00	-	50. 00
2	Principal engineer	30.00	20.00	10.00	20.00
3	Engineer of 1 category	20.00	-	30.00	
4	Engineer of 1 category	-	40.00	-	30.00
5	Technician-technologist	40.00	-	30.00	-
6	Technician-technologist	-	20.00	3000	-
	Subtotal	100.00	100.00	100.00	100.00

The time norm for the j work of i object t_{ij} will be defined as the sum of constant part of it t_{cij}, and the variable t_{vij}, which is determined by the complexity of the managed object.

In order to establish a constant part t_{cij} of the time expenditure, determined by the specifics of the field of activity, it is necessary to classify the work into routine, formalized and creative. To do this, we apply a functional approach.

At the most General level, it can be selected the following functions implemented in the course of project activities:

- data entry;
- data conversion to the appropriate format;
- analysis of the original data array;
- search for previously saved solutions in the system;
- find acceptable analogues;
- develop / transform solution;
- build possible solutions;
- choose the optimal solution.

In practice, the implementation of these functions involves a different set of operations and tools, in particular software.

In the course of work, implementing the basic functions, repeated a number of basic thinking techniques. For project activities techniques form a set of identical in nature blocks: "input", "process", "output". They are repeated in the course of work, but differ in the volume and complexity of data processing. In this case the complexity depends on the frequency of repetition of elementary operations in the process.

A set of information about the importance of operations and the distribution of their specific weights in the work can be recorded in a special matrix. Then the complexity of the work, and accordingly the constant part of complexity t_{cij} will be determined by the ratio of combinations of conditions variants on the blocks "input", "process", "output".

Later we will return to the final distribution of the performing functions costs. Now let us turn to the essence of the main functions, and determine their significance relative to the main function [6].

For a more accurate solution of this problem we use a hierarchical approach. In this we determine the significance as a complex criterion, which is concretized through a number of subordinate criteria [7]. The formation of these particular criteria set is an unformalized creative procedure performed by an expert in this subject area, based on the characteristics of a specific task.

In this case, we propose to use three particular characteristics of the functions:

- K_1 – responsibility;
- K_2–difficulty;
- K_3–strategic value of results.

Display the appropriate hierarchy for a more visual representation of the task (figure 2):

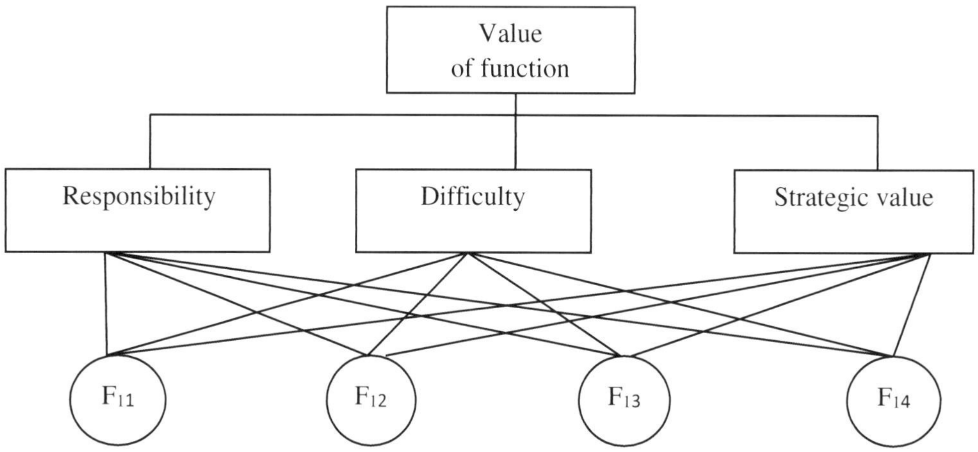

Figure 2.Hierarchy for determination of value function.

According to the methodology for the analysis of hierarchical systems it is necessary to make a pairwise comparison of alternatives – in this case alternatives is the main functions – with respect to each of particular criteria. The comparison is carried out in matrices using a generally accepted nine-digit scale, according to which 1 means the equivalence of the compared alternatives, and 9 – the maximum degree of preference of one alternative over another. Fractional values, respectively less than one, are used to indicate the inverse relationship between the alternatives [8].

We present a matrix of paired comparisons (MPC) filled with expert estimates, with the calculated values of eigenvectors. This values determine the ranking of alternative functions for each of the particular criteria:

K_1	F_{11}	F_{12}	F_{13}	F_{14}	W_1
F_{11}	-	3	3	1	0.381
F_{12}	1/3	-	1/3	1/2	0.103
F_{13}	1/3	3	-	1/2	0.230
F_{14}	1	2	2	-	0.286

K_2	F_{11}	F_{12}	F_{13}	F_{14}	W_2
F_{11}	-	1/7	1/5	1	0.075
F_{12}	7	-	3	5	0.514
F_{13}	5	1/3	-	1/4	0.211
F_{14}	1	1/5	4	-	0.199

K_3	F_{11}	F_{12}	F_{13}	F_{14}	W_3
F_{11}	-	1/3	1/5	1/7	0.054
F_{12}	3	-	4	1/4	0.265
F_{13}	5	1/4	-	1/5	0.207
F_{14}	7	4	5	-	0.546

Next it is necessary to carry out a convolution, or hierarchical synthesis of calculated priority vectors of alternatives, taking into account the different weight of the selected particular criteria as part of the general criterion "the significance of the function". To do this we multiply the matrix composed of the MPC eigenvectors by the vector of particular criteria weights.

In general it can be represented as follows (equation (1)) [9]:

$$\begin{pmatrix} h_1 \\ h_2 \\ .. \\ h_m \end{pmatrix} = \begin{pmatrix} w_{11} & w_{21} & .. & w_{1n} \\ w_{21} & w_{22} & .. & w_{2n} \\ \dots & \dots & .. & \dots \\ w_{m1} & w_{m2} & .. & w_{mn} \end{pmatrix} \times \begin{pmatrix} a_1 \\ a_2 \\ .. \\ a_n \end{pmatrix} \quad (1)$$

where h_i – the integral estimate of the i-th alternative, i = 1...m;

w_{ij} –evaluation of the i-th alternative by the j-th criterion, j = 1...n;

a_j – the weight of the j-th partial criterion in relation to the general evaluation criterion (the vector of weights is normalized, i.e. $\sum a_j=1$).

Expert rate the weights of individual criteria importance of the functions in the following way: A1 = 0.3; A2 = 0.4; A3 = 0.3.

We calculate the ranked vector of integral estimates of the importance of alternatives in MS Excel, using the built-in matrix multiplication function (see figure 3):

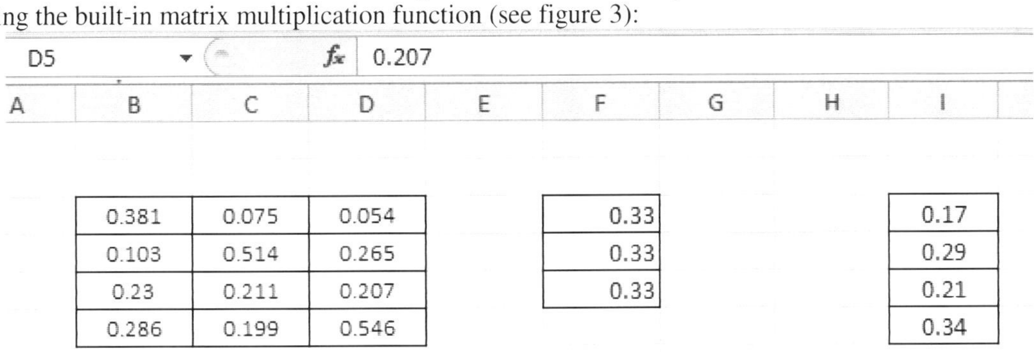

0.381	0.075	0.054
0.103	0.514	0.265
0.23	0.211	0.207
0.286	0.199	0.546

0.33
0.33
0.33

0.17
0.29
0.21
0.34

Figure 3.On-screen calculation form in MS Excel.

We get the ranked vector of integral estimation of alternatives:

H = (0.16; 0.32; 0.22; 0.33)

These values reveal the predominant importance of the second and fourth main functions as part of the main function for the analyzed object.

Next, we must return to the costs of implementing the functions under consideration. Comparison of values of these costs and the significance of the functions will reveal the existing imbalances in the organization of the capacity Bureau work. In the future we will be able formulate measures to optimize it.

In the process of determining labor costs, it is possible to calculate the share of employees in the performance of functions, and the corresponding costs of functions (equation (2)):

$$C_{Fj} = \sum_{1}^{n} C^{i}_{Fj} \qquad C^{i}_{Fj} = \alpha_{ij} \cdot L_i \qquad (2)$$

where C^{i}_{Fj} – cost of the j-th function of the i-th employee;

α_{ij} – share of the i-th employee in the performance of the j-th function;

L_i – average monthly salary of the i-th employee.

Submit the average monthly cost of functions in the following table (table 3) [10]:

Table 3. Calculation of costs for the implementation of basic functions, rub.

Costs elements	Salary	Social	Amortization of	Total costs

Functions		contributions	AWP	
Prepairinitialdata	38600.00	10800.00	220.00	49600.00
Converting data	50700.00	14200.00	460.00	65400.00
Define unbalance	32500.00	9200.00	330.00	41900.00
Documented results	42200.00	14100.00	270.00	54300.00
Subtotal	164000.00	48300.00	1280.00	213580.00

Should now calculate the values of specific relative functional costs and build the chart "importance – costs", which graphically displays the results obtained (equation (3)):

$$z_i = \frac{Z_i}{h_i} \qquad (3)$$

where z_i – specific relative costs to 1 point of the importance of the i-th function;
Z_i – relative total cost of the i-th function, as a proportion of the total cost.
Substituting the previously obtained values, we have the following values:
$z1 = 1.46$; $z2 = 0.96$; $z3 = 0.98$; $z4 = 0.78$.

3. Conclusion. Results are graphically submitted for drawing (see figure 4):

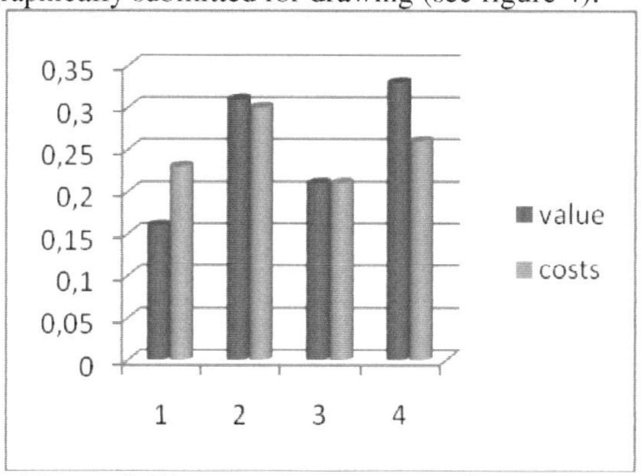

Figure 4. Diagram «Importance – Costs».

Obviously mismatch in the ratio of cost and importance for the first and fourth functions. To solve this problem, it is necessary to optimize the processes of the relevant work, to redistribute responsibilities between employees, to unify the documentation used.

References

[1] Karpunin M G Kuzmin A M and Shaledenkov S V 1990 *Activity-Based Costs in Engineering.*Moskow 75
[2] Sato Y, Kaufman J J 2005 *Value analysis tear-down: a new process for product development and innovation* New York 207
[3] Robert S K, Robin C2008*Cost & Effect: Using Integrated Cost Systems to Drive Profitability and Performance* Moscow 352
[4] Tereliansky P V and Andrejchikov A V 2007 *The Information Technologies of the Prognostication of the Technical Solutions on the Basis of the Fuzzy Sets and Hierarchical Models* Volgograd 204
[5] Andrejchikov A V, Kuznetsov S U, Dekatov D E and Egorova I E 2004 *Models and Stuffs of Conceptual Design of Vibroprotection Systems* Volgograd 144
[6] Andrejchikov A V and Andrejchikova O N 2000 *The Analysis, Synthesis, Planning of Decisions in Economics*Moskow 224
[7] Andrejchikov A V, Kuznetsov S U and Dekatov D E 2005 *Models and Program Stuffs for Support of ABC Procedures of Manufactures: the Monograph* Volgograd 112

IOP Conf. Series: Materials Science and Engineering **483** (2019) 012092 doi:10.1088/1757-899X/483/1/012092

[8] Ivanov V V, Khan O K 2007*Management consideration of effective management* Moscow 208
[9] Gordashnikova O Y 2006 *Functional and cost analysis of product quality and marketing management at a factory* (Moscow) 88
[10] Kuznetsov S U 2015 *Int. Conf. On Economic security of Russia and development strategy of its regions in modern conditions* vol 1 (Volgograd: VSTU/ RFFI) 107-111

Status and perspectives for the use of additive technologies in various branches of Russian industry

I B Teslenko[1], O B Digilina[2], N V Abdullaev[1]

[1]Department of Business Informatics and Economics of Vladimir State University
[2]Department of Political Economy of Peoples' Friendship University of Russia

E-mail: iteslenko@inbox.ru, o.b.digilina@mail.ru, nizamka33@mail.ru

Abstract: The article analyses the state and perspectives of the use of additive technologies in the Russian industry. The authors emphasise that additive technologies not only changed the process of product design, but also modernised the process of value creation. In Russia, additive technologies are developing slowly due to the lack of qualified personnel, the high cost of technology and the underdevelopment of the necessary infrastructure.

Introduction

Global digitalisation and informatisation of all spheres of human life take place in modern conditions. Thanks to this phenomenon, it becomes possible to use a large number of innovations, without which it is increasingly difficult to imagine modern life. So in the industry, the leading sector of the economy, additive technologies that change the algorithm of manufacturing goods/products began to be actively introduced.

Additive manufacturing (referred to as AM - additive manufacturing) is regarded as a class of advanced technologies for the customised production of parts of complex shape on a three-dimensional computer model by successive application of the material (mostly often layer-by-layer) - in contrast to the subtractive production (for example, traditional mechanical processing) [2].

Such production involves the following value creation algorithm:

1. Preparation of CAD-model (computer-aided design model).
2. Creation of an STL-file that contains a 3D model.
3. Virtual separation of the 3D model into layers.
4. Transfer of this file to the AP-system for the layer-by-layer formation of the final product.
5. Finishing treatment.
6. Release of the finished product (for use).

Each company independently decides what kind of additive technologies to use in manufacturing their products. The specifics of the company's business, the available funding, the necessary resources, including staff and other factors, are taken into account.

The aim of this work is to assess the state and perspectives of the additive technology development in the Russian industry, as well as to identify the main obstacles to their spread.

1. Research Materials and Methods

Implementation of research tasks was achieved on the basis of analysis of statistical data and experience of enterprises of the Russian industry, comparative analysis, content analysis and survey of top managers of leading Russian enterprises.

2. Findings

The originator of the additive manufacturing industry is the founder of 3D Systems Charles W. Hull. In 1986, he made a proposal regarding the method of layer-by-layer synthesis (the use of ultraviolet radiation focused on a thin layer of photopolymer resin). It was he who introduced the term "stereolithography". In the same year, the engineer assembled the world's first stereolithographic 3D printer-SLA (Stereolithography Apparatus), which enabled digital technologies to get ahead. Almost at the same time, the founder of the company, Stratasys, invented the first FDM-device. After that, the market of 3D printing began to grow rapidly and replenished with the latest modifications of unique printing equipment [5].

Since 2016, additive technologies have become more actively used in the industry, mainly due to such advantages of AP [2] as: reduction of the time and cost of launching the product into production; the possibility and economic feasibility of small-scale production; timely changes in the project at the production stage; functional optimisation of products; economic feasibility of production of customised products; reduction of costs and production waste; simplification of logistics, reduction of delivery time, reduction of inventory; personalised design.

These technologies have changed the processes of design and construction of products, as a result of which they have become continuous product creation processes. At the moment, modern design and production of products are unthinkable without additive technologies. For example, in the West, 3D printers are common along with conventional computers and are used in the manufacture of fabrics, shoes, food and even the cultivation of living human organs.

Despite the obvious advantages of AP technologies, they have their limitations and drawbacks. For example, the high cost of industrial printers and consumables, the lack of manufacturers of such equipment, the lack of appropriate knowledge of the staff, the problem of protection of intellectual property rights of the patent holder of AM technology, etc. All the above-given factors impede the widespread use of these technologies.

However, this did not prevent the world market of additive technologies from growing. According to the forecasts of the consulting firm Frost & Sullivan, the market volume will increase from $ 5.31 billion in 2018 up to $ 21.5 billion in 2025. By 2025, AP technologies will be globally applied in the following industries: automotive, architecture, aviation and aerospace, consumer electronics, medicine and others [3].

In 2018, the largest consumers of additive technologies in the world are North American countries. The increase in the North American market by 2025 is estimated at $ 7.65 billion. The second largest market is in Europe and the Middle East, which could increase to $ 7.18 billion by 2025.

It is worth highlighting the rapidly growing market in the Asia-Pacific region. It will reach $ 5.56 billion in 2025. At the same time, China will take a large share - 70%.

After analysing the global market for additive technologies, Frost & Sullivan noted such key trends as [3]:

- continuous growth in the number of parts, in the manufacture of which additive technologies were used;

-development of 3D printing technologies at a rapid pace, reducing the time and cost of production through the use of heterogeneous materials;

- growth in the introduction of 3D printing technologies in various industries;
- the use of 3D printing in order to create a rapidly changeable production;
- increase in R & D (Research and Development) financing for additive manufacturing;
- market consolidation through the creation of consortia;
- formation of special institutions to unite the efforts of companies and scientific groups engaged in the development of solutions in the field of additive manufacturing;
- gradual reduction of production costs by reducing the cost of equipment and increasing the availability of technologies.

According to Frost & Sullivan, the leaders of AP-production include: 3D Systems (USA); EOS Gmbh (Germany), SLM Solutions (Germany) Stratasys (US); Objet Geometries (USA-Israel); Envisiontec (USA-Germany (DLP); ExOne (USA); Voxeljet (Germany); Arcam AB (Sweden).

IOP Conf. Series: Materials Science and Engineering **483** (2019) 012093 doi:10.1088/1757-899X/483/1/012093

Up to the present time in Russia the technology of additive manufacturing (design and implementation) developed very slowly. The country almost did not participate in the international organisations that were engaged in the development of AP-technologies in the world; it had neither the appropriate equipment and machinery, providing high-quality products, nor materials and raw materials for 3D-devices. There was also a lack of highly qualified, specially trained staff. In General, Russia has not created the 3D-environment infrastructure necessary for additive manufacturing.

Therefore, now Russia is among the Laggards in such areas as the creation of equipment for 3D printing, production of raw materials and auxiliary materials, etc. As of February 2018, the share of Russia in the structure of the world market of additive production is only 1% [3].

However, domestic companies are expanding the base of production of 3D printers and actively use the additive equipment. They are used in those areas where the need for products of complex geometry is high and allow to establish their production on their own sites without intermediaries, suppliers, delivery risks, reducing production costs.

There are examples of effective implementation of 3D printing in the country. For example, the All-Russian Research Institute of Aviation Materials is engaged in serious research projects, including the production of consumables for three-dimensional equipment. The great interest is demonstrated by the enterprises of metallurgy, aviation, space industry and the military-industrial complex, which use 3D printing mostly for prototyping parts, not end products [7].

Tikhvin Freight Car Building Plant (UWC - United Wagon Company) was the first to implement the project of application of additive technologies in production. At the enterprise with the help of 3D-printer elements of foundry model tooling are created. The technique provides high-quality printing - the thickness of the layer is 0.05-0.15 mm (depending on the settings). And the weight of this product can reach up to 150 kg. This technology gave the opportunity to produce components of complex configuration (with a length of up to 3 metres) for one week with the immediate start of production of pilot castings. Moreover, the equipment has good indicators of energy efficiency and ensures economical consumption of material [7].

STL files for printing are created on computers with standard software. Investments in the project amounted to about 60 million rubles.

Perm Motor Plant "Aviadvigatel" uses selective laser sintering technology (SLS) for the production of cast parts on the burned-out models; for the repair of parts it uses the method of laser cladding (LMD); for the cultivation of metal parts - selective laser melting technology (SLM). The motor plant purchased the equipment within the framework of the program of technical re-equipment of the enterprises of the United Engine-Building Company. Additive technologies are also used in fine-tuning of parts for ground-based engines such as gas turbines and power plants.

At the first conference on additive technologies, the "Luch" Design Bureau presented an unmanned aircraft. The design was carried out on a computer using a special program. Then within 30 hours all the components and parts of the aircraft were printed on the 3D-printer. Assembly of the device was carried out by personnel. The production took two months, and it used to take several years.

Federal State Unitary Enterprise All-Russian Research Institute of Aviation Materials (FSUE VIAM), which is a big powder producer, also uses additive technologies. Activities for the production of powders in VIAM arose from the need to organise the production of solders, which are mainly produced in the form of powders for high-temperature vacuum soldering. The requirements for these powders are the same as the requirements for metal powder compositions in additive technologies. In this regard, the Institute, using the company PSI (England), produced an atomiser, an apparatus that allows obtaining the necessary powders in a fairly narrow range of time. However, since atomisers have two purposes, VIAM began to investigate the method of independent design and production of the apparatus [2]. In addition to its main activities, VIAM takes part in the creation of specialised equipment and technologies for additive manufacturing and software, conducted by the St. Petersburg Polytechnic University.

Another powder producer is the Research Center for Powder Materials Science (RCPM) at Perm Polytechnic University. The research center has a German laboratory unit ALD VIGA-2B, which produces metal powders and alloys by a gas-dynamic spray. The purpose of the AM-machine is to research and obtain small experimental batches. The maximum melting point of the unit is 1.7 thousand C°.

Penza Research and Production Enterprise "Rubin" also joined the development of digital technologies in production. Design engineers of the enterprise create three-dimensional drawings with the help of computer-aided design (CAD), followed by the production of prototypes on 3D printers.

ZAO "Promtractor-Wagon" also included 3D-printers in its arsenal of technical means. At first, the technology was used to print parts of the rail-truck "33", which are copies on a smaller scale, which became useful in the development of assembly algorithms and testing for the collection of innovative rail trucks. In addition to the truck "33", the company, using additive technologies, manufactures components and parts of the forklift [7].

Well-known large corporations such as "Roscosmos", "Rostekh", "Rosatom" are seriously engaged in the promotion of additive technologies. The management of these corporations is aimed at mastering all the elements of "digital production", from the development of materials to the creation of products. For this reason, the program is implemented on additive technologies, consisting of the following sections: technology, raw materials, equipment, standardisation.

In the structure of "Rosatom", additive technologies are developing in the fuel company "TVEL", which actively cooperates with the regional engineering centre established at UrFU, working on the creation of a Russian 3D printer.

In the scientific and educational centre "Modern production technologies" of Tomsk Polytechnic University there is a printer of electron-beam fusion; laser printer; printers that print with reinforced composites, as well as ultrasonic tomography, carrying out at the machine strict control of finished products. Employees of the centre create AM-apparatus and the corresponding software to them and also plan to advance further [7].

The whole production cycle - from the idea to the implementation of the finished product is set up in the centre of additive technologies of Tomsk Polytechnic University. There is an opportunity to create and test details for the covering of spaceships, implants for craniofacial surgery, products of a sophisticated form for the aircraft industry and others, and also to create cutting-edge digital devices.

Thus, although Russia lags behind the leading countries in the field of additive manufacturing, the beginning of the development of AP technologies is laid and already developed unique technologies are proof of this. The innovative products presented above are able to create import-substituting products that are much cheaper than imported analogues and are not inferior to them in quality. So the existing design can make a good competition with foreign manufacturers.

Experts predict the development of additive manufacturing in Russia in the following areas [13]:

- the use of granules and powder materials in 3D printing will help to get rid of the use of triangular and cylindrical shapes in the manufacture of products;
- the use of carbon (graphite) fiber and metal powders will improve the properties of products;
- development of solutions for 3D printing will reduce the error in the manufacture of products and increase the accuracy of production;
- optimisation of performance and development of additive technologies will increase the accuracy, speed and quality of 3D printing;
- service development, that is 3D-printer leasing;
- development of production of 3D-printers will allow creating large-size products with high accuracy;
- the use of the material "graphene" for the production of metal fibers and batteries.

Analysts argue that the stimulation of development in the field of additive manufacturing in Russia should be supported by state subsidies and through direct investment. For example, the Industrial Development Fund provides financial support to projects in this area.

In order to increase the number of qualified specialists in the field of AP-production in the country, experts see the need to develop new professional standards and qualification requirements. Here we are talking about cooperation and the emergence of completely different formats of interaction between Federal Executive bodies, state corporations, high-tech industries, universities and scientific organisations. At the moment the Order of the Ministry of Education and Science of Russia of 22.12.2015 No. 1506 according to which the Federal State Educational Standard of Secondary Professional Education in the specialty 15.02.09 "Additive technologies" is approved. Graduates of this direction will be involved in the organisation and management of the technological process for the manufacture of products on the apparatus for additive production [11].

3. Conclusion

Additive manufacturing marked the beginning of a real industrial revolution in the design and manufacture of certain products. The use of new methods of layer-by-layer synthesis of parts has repeatedly reduced the time to create products. The potential of such production is huge: from the transformation of ways to solve many technological problems that appear in the production process of almost any product, to the creation of new industries. However, there are challenges and difficulties in the implementation and use of these technologies. Recently, there has been a growing interest in the new format technologies from domestic manufacturers, which can be the key to the active development of the AP-technologies market in the future.

References

[1] Averyanova U V, Klementjeva C V. 2016 The expediency of using additive technologies in a single production *Approbation* **12** (51) 130-132

[2] Additive manufacturing: at the peak of inflated expectations [Electronic resource]. – Available at: http://www.umpro.ru/index.php?page_id=17&art_id_1=610&group_id_4=110

[3] Additive manufacturing (AM) [Electronic resource]. – Access mode: https://clck.ru/DBe2q

[4] Additive technologies in Russia: Let's Go! [Electronic resource]. – Access mode: http://www.up-pro.ru/library/modernization/technologies/additive-russia.html

[5] Antonova V. S., Osovskaya I. I. Additive technology: textbook / VSTA SPbgutd. SPb, 2017. - 30 p. [Electronic resource]. – Access mode: http://www.nizrp.narod.ru/metod/kaffizikollchem/25.pdf

[6] Balyakin A V, Smelov V G, Chempinskiy L A 2012 Experience of practical application of additive technologies in the production of aerospace parts *Aircraft industry of Russia. Problems and prospects. Symposium with international participation*. Samara State Aerospace University named after academician S P Korolev (National Research University).. Publisher: Samara National Research University named after academician S P Korolev 57 – 58

[7] Eight Russian enterprises successfully implementing additive technologies [Electronic resource]. – Access mode: http://blog.iqb-tech.ru/am-technologies-russian-experience

[8] Dresvyannikov V A, Bunimovich I D 2017 Status and prospects of the use of additive technologies for the production of consumer goods *Proceedings of higher educational institutions. Volga region. Economics* **2**(6) 35-43

[9] Sannikov D U, Kostygova L A 2016 State and prospects of development of additive technologies in the production of titanium products *Economics and Industry* **1** 8-11

[10] Smirnov V V, Shaihutdinov E F 2013 Introduction of additive manufacturing technologies of parts in mass production *Bulletin of Kazan State Technical University named after A N Tupolev* **2-2** 90-94

[11] Order of the Ministry of Education and Science of 22.12.2015 N 1506 [Electronic resource] – Access mode: https://ktps24.ru/wp-content/uploads/2016/04/FGOS-Additivnye-tehnologii.pdf

[12] Chemodurov A N 2016 The implementation of additive technologies in the production

of engineering products *Proceedings of Tula State University. Technical sciences* **2** 210-217
[13] Frost & Sullivan: additive manufacturing technologies - market, trends and prospects until 2025 [Electronic resource]. – Access mode: http://3dtoday.ru/blogs/news3dtoday/frost-sullivan-additive-manufacturing-technologies-market-trends-and-p/

IOP Conf. Series: Materials Science and Engineering **483** (2019) 012094 doi:10.1088/1757-899X/483/1/012094

Human resources development as the basis of Russian technological breakthrough

G N Tuguzkina, L V Rozhkova, O V Salnikova, S A Vlazneva and S V Taktarova

Penza State University, 40 Krasnaya street, Penza, 440026, Russia

E-mail: galina066@mail.ru

Abstract. The article deals with one of the most significant problems faced by enterprises at the stage of modernization which is staff training in accordance with new organizational and technological requirements. The article discusses the international and Russian experience of enterprise staff retraining and advanced professional training. There is a gap between Russia's quality performance indicators and the highest international standards both in the field of production and vocational education. The authors point out that labour force development is a long-term, future-oriented policy based on investment in human capital. Russian enterprises are aimed at integrating into the world economy and using the most successful experience in the field of vocational training.

1. Introduction

In his presidential address to the Federal Assembly on March 1 2018, Vladimir Putin marked breakthrough development as one of the main strategic courses of Russia. All projects and investments in infrastructure, education, health care and ecology, science and research as well as economic subsidies will be directed to achieve this goal. Special focus was placed on the fact that knowledge, technologies end expertise are the most important competitive advantages [1].

Since the viability of societies, states and position of countries, especially countries as large as Russia, in the world depends on advanced technologies, their efficient development and rapid implementation, scientific and technological breakthrough is among the key national goals and priorities.

Thus the solution of main economic tasks outlined in the Strategy of Scientific and Technological Development and the Decree of the President of the Russian Federation "On National Goals and Strategic Objectives of the Russian Federation for the Period until 2024" is impossible without high-quality human resources of domestic enterprises.

A logical question arises: is it possible to solve the ambitious tasks set by the President with the help of human resources?

In 2010 Russian Public Opinion Research Center and National Research University "Higher School of Economics" conducted a large-scale study "Human resource management barriers in the development of the Russian economy" in order to obtain data on human capital management problems that could act as an obstacle to the intensive growth of Russian companies. The study showed the gap between structure and training of human resources and employers' requirements (employers note insufficient staff qualification); prevalence of "manual" approach to staff management that is inefficient in the context of staff development and skills transfer. The following ways of bridging this gap were named: federal and local government participation (70% of respondents) - changes in the economy structure and improvements in the education system; 11% of company experts saw the need for employee training while 6.6% found specialized training necessary[2].

2. Discussion and results

One of the most significant problems faced by enterprises at the stage of modernization is staff training (second place after financial problems), namely the training of staff in accordance with new organizational and technological requirements. The staff issue is even more urgent for the companies that approach modernization - professional skills of the employees are lower than required. According to a study conducted by the National Research University "Higher School of Economics", 64% of such companies face difficulties in retraining or finding the right staff [3].

By various estimates, the working class in Russia currently accounts for approximately 50% of economically active population with total employment rate of 65%. The largest group of employees is skilled workers of industrial enterprises, construction, transportation, communications, geology and exploration. In 2014 their number was 9.4 million people (the latest available data). The number of mid-level industrial workers that includes operators of machines and installations is 8.7 million. According to the Federal State Statistics Service, there are seven million unskilled labourers. Thus, approximately 25 million people belong to the working class - 35% of all employees in Russia [4].

Ten million employees work in trade, consumer services and housing and communal services and that increases the total number of urban working class to 35 million people. Another 2.35 million wage earners are employed in agriculture, forestry, fishing and fish farming [4].

Generally the working class accounts for 53% of total employed population or 50% of total economically active population (employed and unemployed). Therefore, the Russian working class has not ceased to exist, although it has changed. Economic reforms have significantly adjusted the lifestyle, social consciousness of this social group [4].

Most vacancies on the market are for operation staff- 17% (figure 1) and only 3% for management staff [5].

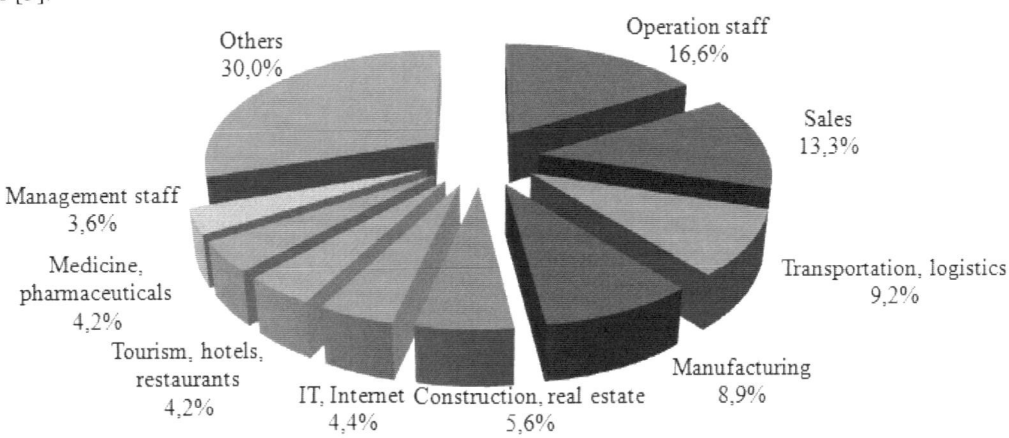

Figure 1. Vacancies in Russia (2018)

Considering foreign experience, it should be noted that at the beginning of the current decade, the structure of the European employment market was dominated by middle-level specialists. They accounted for 17%, the number of high-level specialists exceeded 15%, senior and middle executives accounted for 9% of the total number of employees. Office employees, as well as trade and service workers, accounted for almost 25% of the total. Blue-collar jobs represented 35% of European employment: skilled workers, predominantly manual, workers (13%), skilled machinery operators (8%) and unskilled workers (10%) [6].

According to the forecast, by 2020 the total labor market of European countries will increase by about 3.6%. The biggest growth (13%) is expected for medium-level specialists. The number of highly qualified specialists is projected to increase by 8% [6].

The analysis of the international experience of staff retraining and development shows that its successful implementation is carried out by special institutions that develop and coordinate activities in the area of vocational education. Thus, the International Labour Organization has established an international centre for modern technical and vocational training. EU member states have created the European Centre for the Development of Vocational Training. Major US-based firms have established

CEO-headed organizations that develop permanent programs of retraining and advanced professional training.

German experience in this field deserves attention. Germany is the country with the highest paid workforce and having a greater number of less qualified employees in the enterprise is unprofitable. Therefore, the professional development of production staff is a direct task of production managers.

The most renowned organization, both in and outside Germany that develops methods and carries out employee training is the federal union REFA. One of the main principles of the REFA union is "from practice and for practice", for example, a production manager who has successfully completed training in one of the REFA programs, then not only applies the acquired knowledge in his daily work but also is also able to transfer it to production staff.

Fundamental and comprehensive methods of REFA include training in the workplace [7].

It should be noted that the usual understanding of the German system of training is traditionally associated with "dual training". Each educational program in the German dual system is a network. Its implementation involves an enterprise (one or more), a professional school and a competence centre. Professions regulations (guidelines) in Germany are the official documents for the development and implementation of educational programs. Such documents serve as a basis of professional training programs in enterprises and vocational schools. The enterprise is obliged to form an individual training plan of each student, which includes practical (in the enterprise and/or in the competence centre) and theoretical (in vocational school) training.

Vocational training is most common in industries with active development and introduction of new products and technologies (production of computing equipment, chemical, aerospace industry etc.). However, it is common for even large organizations to join forces. A special fund "10 Cents per Hour" created by three automotive giants ("Chrysler", "General Motors" and "Ford") in cooperation with United Automobile Workers Union is an illustrative example of the integration of corporate resources in the organization of professional training. Financial resources of the fund were invested in the modernization of the training system of automotive industry staff. Japan is one of the recognized leaders in the development and implementation of effective professional training systems. Japanese technical schools, an analogue of vocational education institutions in Russia, are the basis of workers training system. Workers in Japan can receive secondary-level education at part-time and correspondence high schools- an analogue of Russian evening schools. Further training of highly qualified workers is conducted in Junior colleges- an analogue of vocational-technical schools in Russia. It should be noted that the Japanese system of vocational training is not aimed at flexible and continuous educational trajectory. Technical schools and vocational institutions educate workers for specialized corporate tasks.

Usually, blue-collar qualification blocks access to other education levels such as universities. In Finland, vocational training programs are improved in cooperation with labour sector representatives and other interested parties in order to facilitate workers' access to the labour market. Finland is a member of «World Skills International» since 1988. Since 1993, the country has been developing a national competition of professional skills among young people (up to 21 years); it is called "Tajtaja" (artisan) and is supported by the national fund "Skills Finland". Every year up to 1.5 thousand young professionals partake in this competition [8].

The analysis of training systems in the leading countries showed that the main types of vocational education are alternating and continuous training. The first type includes "school-enterprise" approach with varying training duration depending on the student's choice: 2-4 years for persons aged 15-25. Normally training is conducted in vocational centres, where students learn a theoretical course combining it with practical tasks in an enterprise (Denmark, Portugal, Germany, Finland, Switzerland and the Netherlands). Continuous training implies that employees do not stop their work. In some countries, e.g. Portugal, Spain, Finland, Belgium and Norway, vocational training is part of the social policy of the state that allocates funds for its implementation. In other countries, e.g. France, Italy, Luxembourg, Switzerland and the Netherlands, it is conducted with the use of equity capital or payroll tax, i.e. independently from the state institutions. Industrial-technical training in most countries is a form of alternating training and involves theoretical and practical courses. In this case, the student spends 80% of the time in the enterprise, which, after the completion of training, must confer a new qualification to the employee and pay a scholarship (approximately 30-60% of the minimum wage) during the entire period of training. Therefore, labour force development is a long-term, future-oriented policy based on investment in human capital. It is largely due to the consistent and systematic work on the formation of the vocational training system in Germany, Japan, and Finland that a significant increase in productivity in high-tech sectors of the economy was achieved [8].

The priority for the Russian economy and education is not only training of skilled workers, but also the formation of intellectually active employees with broad-based knowledge who are able to see the prospects of technology, product, equipment, production, processes. Consequently, such employees should be able to determine tactical productivity-oriented decisions both individually and collectively [8].

Today most experts share the opinion that the new economy needs a new workforce. Professor Eskindarov, president of the Financial University, reasonably remarked, "...The market is changing, the economy transforming and the process of digitalization is creating new professions. Today the University should not follow the labour market but anticipate the trends of its development..." [9]. Valeriy Fedorov, director of VCIOM, reported on the results of the sociological poll "Professions of the Future". According to respondents, manufacturing occupations will be in great demand in the next 7-10 years and lawyer will become the least attractive profession. Herewith the majority of respondents claimed that they would wish their children or grandchildren to be doctors, members of military or lawyers and not entrepreneurs, pilots or service sector employees. The opinion concerning robotic automation was almost unanimous: most believe that robots will not be able to perform their work duties in the future.

Since 2014, Agency for Strategic Initiatives (ASI) is implementing the program "Human Resources for Industrial Growth" which is a set of measures to train new generation workers and engineers for the Russian industry in the period until 2020. These measures are implemented on the international level (Russia's integration into the WorldSkills movement, Global Education Futures, dialogue at international platforms including Skills Development Working Group under the BRICS Business Council) and on the federal and regional level (WorldSkills Russia, dual education and leadership projects). Various competitions and contests facilitating the development of training systems have been organized in Russia. For instance, the Ministry of Education and Science is responsible for the all-Russian Olympiad of vocational education students. The Ministry of Labour and Social Protection organizes the all-Russian professional expertise contest "Best Professional". The Ministry of Industry and Trade and ASI coordinate sectoral WorldSkills contests, specifically corporate WorldSkills professional championships, WorldSkills Hi-Tech, Junior Skills and Future Skills. Individual state and private enterprises, development and educational institutions are also working to optimize and develop the system of vocational training [10].

Since 2014, ASI has been implementing the "Dual Education" project in 13 pilot regions of Russia. Taking into account the requirements of employers, training of students in new educational programs began. These programs were developed within the framework of the dual model. Dual education is a type of practice-oriented professional education, in which the theoretical part is received in the respective education institution and the practical part is taken in the workplace. Subsequent employment at the enterprise allows reducing time costs for adaptation of the young specialist as he or she is already familiar with the equipment of the concrete enterprise and its staff. Herewith the employer's role is becoming more important: companies request a specified number of specialists, creating on-site jobs for students, and provide mentors from among current employees, as well as participate in the curriculum.

Dual education can be considered as one of the effective formats of interaction between a vocational educational institution and the employer's enterprise, while at the national level this format can become an infrastructure model that provides the ability to predict the needs of production in human resources. Consequently, this process facilitates the implementation of a system of qualification assessment and, ultimately, leads to practical changes in the vocational education system in the country.

These organizations present examples of active and efficient work in this field:

- PJSC "United Aircraft Corporation". PJSC "UAC" is developing and testing a certification system of university graduates and professional audit of aircraft-related education programs in the framework of a joint project with the Ministry of Industry and Trade and RUSNANO;
- JSC "United Engine Corporation". JSC "UEC" implements programs of dual education at corporation enterprises;
- JSC "Russian Railways". The company conducts training in corporate training institutions; in 2013-14 34 professional standards specific to railway transport were developed;
- JSC "Ruselectronics". JSC "Ruselectronics" project staffing requirements and conducts training within the production modernization framework;
- PJSC "RusHydro". PJSC "RusHydro" implements an advanced staff development program and develops professional standards for hydropower industry;
- JSC "ChelPipe". JSC "ChelPipe" conducts on-site training for 11 education plans [10].

In the XXI century, nation's successful development is determined not by the geographical location and the availability of fossil fuels, but by the intellectual potential. Human capital sets key trends in the development of national economies of the world today. In early 2018, an international team of scientists prepared a forecast of scientific and technological development of the world and determined the most relevant professional competencies: creative thinking, planning and implementation of research activities.

3. Conclusion

Thus, every year the development of vocational training and retraining system is becoming more large-scale. Numerous accumulated problems can be solved through close cooperation between employers, education institutions and regulatory authorities. Russia is aimed at integrating into the world community and using the world's most successful experience in the field of vocational training. Russian is open for dialogue on the strategic production goals of the next 10-20 years and determining the competencies of the economy of the future.

However, there is a lack of connection between the system of vocational education and production needs both current and future. Currently, there is no proven mechanism that allows monitoring and forecasting the real needs of the economy, therefore educational institutions fail to train specialists in accordance with such needs. Industrial work placement of students is often of formal character and they do not master the necessary skills.

There is a gap between Russia's quality performance indicators and the highest international standards both in the field of production and vocational education. If a site of industrial work placement does not comply with the world standards, then it is not possible for students to acquire necessary skills. Most vocational institutions in Russia do not meet the needs of the economy and do not provide the opportunity for rapid modernization of facilities and equipment.

References

[1] Presidential Address to the Federal Assembly [Electronic resource]. - Access mode: http://www.kremlin.ru/events/president/news/56957 (access date: 18.09.2018)

[2] Labor market research. *Aging of employees and skills obsolescence are the main HR problems* [Electronic resource]. - Access mode: http://hr-portal.ru/news/starenie-personala-i-nedostatochnyy-uroven-ego-kvalifikacii-osnovnye-kadrovye-problemy (access date: 20.09.2018)

[3] Bondarenko N V, Krasilnikova M D 2014 *Education economics monitoring* **1(75)** 1–64

[4] Razumova T, Chernykh E 2017 *Economist* **1** 71–84

[5] Vacancies distribution by industry. *Time history of the number of vacancies in Russia: October 2017 - September 2018* [Electronic resource]. - Access mode:

https://gorodrabot.ru/salary?l=%D1%80%D0%BE%D1%81%D1%81%D0%B8%D1%8F (access date: 20.09.2018)

[6] Vishnevskaya N T, Zudina A A 2017 *Bulletin of international organizations* **4** 109–129

[7] Manufacturing management. *Training and professional development of production staff: experience of REFA* [Electronic resource]. - Access mode: http://www.up-pro.ru/library/personnel_management/training/podgotovka_i_povyshenie_kvalifikacii_proizvodstvennogo.html (access date: 26.09.2018)

[8] Podverbnyh U S 2015 *Proceedings of Irkutsk State Economic Academy* **4** 621–629

[9] Education for the future. *New workforce for the new economy* [Electronic resource]. - Access mode: http://www.fa.ru/News/2018-03-23-futureedu.aspx (access date: 26.09.2018)

[10] Best staff training practices of the BRICS countries [Electronic resource]. - Access mode: https://asi.ru/upload/0b6/BRICS_ru_int.pdf (access date: 1.10.18)

IOP Conf. Series: Materials Science and Engineering **483** (2019) 012095 doi:10.1088/1757-899X/483/1/012095

On the Essence of Technology (the Case of Collaboration between Chinese and Russian Enterprises)

F. Liu[1], I. M. Zashikhina[1], M. V. Druzhinina[1], Y. Sun[1]

[1]Northern (Arctic) Federal University named after M.V. Lomonosov, Naberezhnaya Severnoy Dviny, 17, Arkhangelsk, 163002, Russia

E-mail: f.lyu@narfu.ru

Abstract. The subjects of technology and communication have been studied widely during the last decades. After Martin Heidegger many researchers have noted connection between these notions. Heidegger himself argued the essence of technology is connected to intentionality. This perspective did not find a due resonance in the later studies on the nature of technology. In search of efficient ways of business communication our research focuses on the phenomena of intentionality, communication and technology. The aim of the study is to discover how the three mentioned components are reflected in interactions for business purposes. With this aim a questionnaire for enterprise senior employees was developed. Answers to the questionnaire revealed communication features, means and barriers of communication between representatives of Chinese and Russian enterprises. Taking into account that most of the collaboration took place via technological means of communication, we had to consider the influence of ICT on the communicative process. Results of the analysis showed that such communication is dependent upon disposition and intentions of the participants.

1. Introduction

In his well-known work M. Heidegger argued that the essence of technology lies beyond the area of technological [1]. The essence of technology derives from its orientation to the world, which M. Heidegger calls "enframing" [2]. In its own turn, enframing emerges from the human need for a precise and scientific knowledge of the world. Such a demand puts forward intention, which constitutes a mental state, like, for example, beliefs, love or hatred exhibiting intentionality [3]. M. Heidegger described enframing in terms of orientation and attitude [2]. Such an approach points out a constructive role of human agency. Thus in search of the nature of the technological, our attention turns to the individual's intentionality as a feature of communication with the use of ICT.

It seems logical to assume that the essence of technology lies in intentionality. This assumption is supported by W. Arthur's view on the nature of technology. In his book "Nature of technology" W. Arthur stated that all technologies, standard or not, belong to "purposed systems" [4]. In a common sense a purposed system presents intentionality. The author suggests that the essence of technology lies in communication, since intentionality is highly integrated into the discussion about communication. In "Beyond the Constitutive-Representational Dichotomy: The Phenomenological Notion of Intentionality" A. Corey argued that "humans are fundamentally suspended in a network of intentional relations by explicating the phenomenological notion of intentionality by Heidegger and Merleau-Ponty" [5]. The fact that the essence of technology lies in communication appears to be grounded, though not definite enough. Portuguese scholars A. Fidalgo et al. analyzed interaction between man, mobile technology and communication in terms of attitude, willingness, intentional expression and acts of speech [6]. According to the authors, relationship between technology and communication are reflected in the interfaces used in social media, like, e.g., Facebook. In unison, O. Yang claims that ICT actively reshapes our understanding of ourselves and of our world [7].

In 2005 J. Budd presented a paper on phenomenology, where he articulated the idea that communication by dialogue can release phenomenology's potential in terms of understanding the nature of technology [8]. N. Arnaud and B. Fauré in their work "Communicative Approach to Sociomateriality: the Agentic Role of Technology at the Operational Level" conducted a qualitative research in French transport industry and reached three conclusions:

1. sociomateriality is the expression of what matters in conversation;
2. communicative practices are at the core of skillful achievement in interorganizational collaboration;
3. communication is at the heart of modern forms of work organization and deserves more researchers' and managers' consideration [9].

Sociomaterial perspective is also recommended as a theoretical framework theorizing computer mediated communication (W.J. Orlikowski, S.V. Scott [10]), W.R. Grace [11], F. Fonseca, A. Bavdaz [12]). Within this approach technology is ascribed with both material and social features, including its relation to routines, humans, organizations. Technology is vastly used for communication in various organizational settings, providing electronic texts with qualities of a particular digital medium within the analysis. As opposed to a synchronous face-to-face communication, computer-mediated practices dismiss the traditional for the context concepts of an author, audience and community. Sociomaterial perspective moves us towards interpretations within the area of material and the embodied intersubjective interactions.

The findings of N. Arnaud and B. Fauré are specially inspiring here since this article is devoted to the discussion of collaboration between Chinese and Russian enterprises in terms of communicative nature of technology. There is a long history of geopolitical relationship, economical common interest and socio-cultural connections between China and Russia, the two ancient and powerful countries. The recent decades have been marked by a significant rise in the intensity of business relationship and intercultural communication [13]. Collaboration between enterprises brought significant mutual benefits in diverse fields, such as Sino-Russian bilateral trade, nuclear energy, electric power, chemical industry, agriculture and mineral development [14]. In order to achieve better collaboration, better communication is critical [9]. Regardless the theory of intercultural communication so far, international business communication presents a serious potential for technological and socio-cultural development in the 21 century [15][16]. This fact states a necessity to further investigate the ties between the two countries in terms of communication, its ways, channels and settings. Taking into account the geographical conditions and vast territories of the two countries, as well as the technological turn in communication of the 21 century, in this research we are analyzing the communicative nature of technology and its implications relevant for the mutual relationship between the nations in business area.

2. Literature review

2.1. Definition of intentionality

Intentionality is the power of minds to be about, to represent, or to stand for, things, properties and states of affairs. "The puzzles of intentionality lie at the interface between the philosophy of mind and the philosophy of language. The word itself, which is of medieval Scholastic origin, derives from the Latin word intentio, which in turn derives from the verb intendere, meaning being directed towards some goal or thing." [3] In the philosophical dictionary [17], it is explained as metaphorical stretching of consciousness to be of something. F. Brentano reintroduced it at the end of the nineteenth century, and held that intentionality is the hallmark of the mental [3]. Both F. Brentano and E. Husserl conceived of intentionality primarily as a conscious phenomenon, and proposed that mental (conscious) phenomena should be defined as 'those that contain an object intentionally within themselves' [17].

E. Husserl suggested that "all consciousness is consciousness of something", and described intentionality as the fundamental characteristic of consciousness [5]. J.P. Sartre in "Being and Nothingness", like E. Husserl, identified intentionality with consciousness, stating that they were indistinguishable from one another. However, he interpreted intentionality in a freer manner, arguing that the "being of the external world is essentially meaningless and so people must create meaning from out of the "nothingness" of consciousness" [18]. In J.P. Sartre's model of intentionality, the core of consciousness is of phenomenological scope; its occurrence is a consciousness-of-an-object, and beneath or behind it lies their "being-in-itself" [19].

Russian scholar, L.V. Mikeshina, referring to Heidegger, claimed intentionality to be the first "discovery" of phenomenology. She also wrote intentionality is not a relation to an external subject that could be endowed with experiences and which would be revealed in some experiences; experiences are intentional in themselves, as such"[20]. Another Russian scholar, N. Blokhina explains that intentionality traditionally refers to the intention of consciousness towards objects or states of affairs in the world, and includes not only ordinary intention to do something, for example, to go to the cinema, but also belief in something, hopes, desires, emotions, perceptions and much more [21].

2.2. Intentionality and communication

It is widely accepted that between intentionality and communication there is a very close relation, concerning language, sociology, culture, cognition, speech act, etc. For example, A. Corey states that in communication research intentionality is used mainly to address whether acts of communication, by definition, require an "intent to communicate" and also when and to what extent we can "know about" those intentions [5]. In the broad sense according to A. Corey, scholars seek answers to why people do or say what they do and at what level of conscious awareness their intentions are placed. In his work A. Corey employed intentionality to demonstrate how to accommodate representational accounts regarding J. Dewey's work on language and the concept of a constitutive character of human speech.

R. Koerber claimed that communication contains three aspects: intentionality, understandability and significance [22]. In his eyes, intentionality relates to higher forms of communication. As he referred to P. Freire, communication is both dialogue and participation, which is a concept of intentionality. Additionally, he claims intentionality and the other two aspects are all contextual in nature, and contextual intentionality is quantifiable, so are the other two.

B. Absher discussed about communication, language, intentionality, intentional states, speech acts etc. by referring to M. Hiedegger, L. Wittgenstein, J. Searle et al. He found the capacity of sentences and marks to represent reality is not intrinsic but is derived from the intentionality of the mind [23]. It is also found that there is a double level of intentionality in the performance of the speech act which derives its capacity to represent the world from direct and inherent relation of intentional mental states to their conditions of satisfaction. In the process of an utterance formation, first of all, the intentional state is expressed; and then comes the intention.

D. Zoller combined M. Heidegger, culture, and intentionality, and found many interesting facts concerning communication, language, sociological perspectives, intention, mental/intentional content, etc. For example, it is written that "a system like an economy or the government communicates only with itself in its own language and by its own rules" when "it is not even clear that the level of individual intentionality has anything to do with the makeup of the social world as it is today" with the reference to N. Luhmann [24].

L. Mikeshina wrote that taking into account the communicative nature of activities and knowledge of the subject in society, it should be borne in mind that intentionality is always collective, even if it belongs to an individual subject, which should be taken into account when analyzing the nature of a social fact [20]. J. Searle defined collective intentionality as a human capability of collaboration [21].

N. Blokhina additionally specified that collective intentionality is usually needed when people compete and even conflict.

Collaboration, competition, conflict are three common types of relationship between enterprises. In leadership terms (micro view) according to R. Campbell [25], intentionality refers to identifying a specific outcome for a communication interaction, recognizing how one should or does appear to one's audience, and adapting one's behaviors, tone of voice, word choice, etc., to convey the desired motivation in pursuit of a goal.

3. Results and discussion

3.1. Communicative nature of technology: perspective of revealing and enframing

M. Heidegger in "The Question Concerning Technology" states the essence of technology lies in revealing and enframing [1]. Revealing is described as a process through which "bringing-forth comes to pass ... something concealed... into unconcealment". And then: "...every bringing-forth is grounded in revealing." As for enframing, Heidegger wrote: "Enframing, as a challenging-forth into ordering, sends into a way of revealing … Enframing means that way of revealing which holds sway in the essence of modern technology and which is itself nothing technological … It is the way in which the real reveals itself as standing-reserve…" Revealing and enframing have direction, i.e. they are purposed. These two phenomena interact with the environment, happen in man, through man, within human doing. They are expressed in a claim (an utterance). So, revealing and enframings, being technological, are manifested through communication. In other words, communication is the core of the notions, that present the nature of technology. Thus the essence of technology lies in communication.

After M. Heidegger the idea that the essence of technology belongs to the area of communication has been supported by many scholars. Following S. George [26], enframing equals to the total understanding of reality as compliantly manipulable resource. Understanding of reality is a part of communication discourse. K. Gülenç and H. Arıtürk [27] explained enframing as gathering of human beings within the order, and enframing is an epoch where human beings claim to be the masters of Earth. It is also clearly stated that "the essence of technology means the way in which intelligibility happens for us these days, that is, as 'enframing'". Intelligibility proves that enframing means a process of communicating, thus once more it becomes evident that the essence of technology is of communicative origin.

3.2. Communicative nature of technology: perspective of intentionality

Intentionality is treated as being particularly relevant to the philosophy of technology [28]. A. Corey wrote the world in its various profiles is always disclosed by and according to multiple and diverse intentional relations [5], while treating intentionality from M. Heidegger' perspective, i.e. close to revealing. S. Tala [29] acknowledged that technology is humanity's way of being in the world. It is not just a means to an end, but rather a mode of human existence, which very much fits with understanding of intentionality explained before.

Studying the subject of philosophy of technology, A. Jacob used D. Ihde's human-technology relations patterns [28] to demonstrate where intentionality is placed in the relations between I, technology and world. To sum up in relations patterns (embodiment relations, hermeneutic relations, alterity relations), intentionality stays close to technology as showed in figure 1 (see Figure 1 "Simplified D. Ihde's Human-technology Relations").

Figure 1. Simplified D. Ihde's Human-technology Relations

It has been acknowledged that intentionality and communication are so closely related that we may assert communication and intentionality can be used interchangeably. At the same time the essence of technology can be revealed or enframed in the same context following the logic of argumentation above. There are also researchers like J. Zuern and H. Ikaheimo suggesting the essence of technology lies in intentionality directly. H. Ikaheimo proposed recognition as a cognitive technology for communication [30]. As a cognitive function, recognition emerges in three forms: identification, acknowledgement and a human sense of recognition. The latter manifests as an altruistic and intrinsic motivation to understand intentionality, in which, as it has been implied, the essence of technology lies.

3.3. Communication and collaboration between Chinese and Russian enterprises

Since we have stated essence of technology lies in communication, we would like to introduce a recent empirical study on international communication and collaboration between Chinese and Russian enterprises as an example to demonstrate how the understanding works. Initially we draw from the authors' professional experience of joint international projects between Chinese and Russian enterprises. Secondly, we have conducted a qualitative research survey, which consists of a set of questions in the Russian language to study the aspects of communication and cooperation between enterprises.

Participants of the research are sales/purchasing managers, directors of factories, professors at universities, representatives of organizations based in Arkhangelsk, Russia, taking part in Sino-Russian collaboration among enterprises. Universities are also included within the group of enterprises, since in the era of academic capitalism they can also be treated as such [31]. Three of the observed enterprises belong to timber industry: Region Les Corporation, LLC Torgmaster, China Forestry Group Corporation; two in education: Northern (Arctic) Federal University (NArFU), Office of Chinese Language Council International (Hanban); one in pulp and paper industry: LLC Tramp. Five of the participants are Russian, and two are Chinese, who know Russian well enough to fill in the survey form.

We found that Region Les Corporation has been working with China and is constantly looking for Chinese partners. It has worked with Russian Forest, a Shanghai company, which arrived to sign a new contract, while Region Les Corporation was opening a new modernized workshop [32]. The Corporation is also working with a few other Chinese partners. LLC Torgmaster is a middle size enterprise, actively seeking new partners from China, because the market is seen as appropriate in terms of requirements and prices. China Forestry Group Corporation has entered Arkhangelsk market half a year ago, and is now operating in the northern market successfully. It is partially moving out of Arkhangelsk to the Republic of Komi for its own interest. NArFU and Hanban have been cooperating with each other for about two years, which boosted Chinese study in Arkhangelsk in terms of both language and culture to a large extent. LLC Tramp, now named as Arkhangelsk Carton, is planning for a new base of pulp production from wasted paper by means of a new machine line produced in China .

3.4. Features of communication at business meetings

It has been found out, that 83.3% (5) of enterprises often communicate in Russian, 50% (3) in English and 33.3% (2) in Chinese; 33.3% (2) of them have business deals with their partners around twice per quarter, 33.3% (2) have business operations around five times per quarter, one company has multiple business operations. 33.3% (2) of them plan to meet their partners from one to five times in a quarter, 33.3% (2) plan "multiple meetings", 16.7% (1) plan six to ten meetings a quarter, and 16.7% (1) mentioned distance communication once a year. Face to face meetings are most preferred. Although assistance of agents is described as especially useful in terms of overcoming language barriers by the participants, 83.3% (5) of them would like to attempt at direct contacts. In terms of communication means, email as the most popular and effective way got 83.3% (5), while direct conversation is anyway twice more preferred and phone calls are also frequently used. Social networks got 66.7% (4), among which Wechat dominates with 66.7% (4), comparing to Whatsapp/Viber with

16.7% (1). VK, a Russian domestic social site, was mentioned among other social media names. As for the communication barriers, language as a hindrance to communication was considered by 66.7% (4) of the participants, whereas cultural peculiarity was chosen by 50% (3) of them.

All respondents thought of the collaboration between Chinese and Russian companies in positive ways. It was mentioned that the participants set a goal to preserve agreeable dispositions and enjoy their collaboration with the intent to team up and cooperate in professional areas. The answers to the survey show high willingness of enterprises to communicate with each other through varied communication means in spite of barriers. The main communication objectives are to build contacts, sign business deals and operate in business area for profitability. Overall, communication between enterprises was aimed at getting mutual understanding in order to achieve best possible material and financial outcomes.

In the world of technology communication also happens without human language to express mental or intentional content by sense and reference. To illustrate the process of non-verbal communication, we can mention the color of cables used in the process in different with different functions. European standards are accepted and used in Russia which are different from Chinese standards. For instance, black stands usually for the negative in China, while in EU stands for signals. In China red is a positive sign, whereas in EU brown is used for the purpose. As has been found out in the course of observation and interviews, Chinese engineers are aware of the difference and use tapes to mark wires without additional explanation to the workers, who are competent to deduce meanings with the use of abductive logic. Generally, motivation for success, entrepreneurial spirit, common sense, technical and technological knowledge, life experience as well as good intentions are listed as the methods of effective communication by the participants of the survey.

4. Conclusion

The role of communication between Russian and Chinese enterprises is continuously growing. Communication has brought two nations closer, and, more importantly, triggered economic and technological growth in diverse fields. This research clarifies understanding of intentionality, essence of technology, and the relation between them and communication. Based on the analysis of the Heidegger's philosophical concept of enframing, the fact of the essence of technology lying in the area of communication has been justified. Besides, the conducted survey illustrates features of communication between Chinese and Russian enterprises, including frequent contact, varied means for communication and intercultural barriers between them. Through the analysis of their collaboration, it can be observed that enterprises are striving for the highest profit, the safest production, the easiest and most authentic negotiations, which matches the concept of intentionality with its focus at goal and enframing. The tendency to pursue efficiency and succeed in business brings about humans' positive and active commitment. Cases of miscommunication may occur, but better understanding of the essence of technology and communication makes sustainable and efficient collaboration between Chinese and Russian enterprises highly possible.

References

[1] 1977 The Question Concerning Technology [Electronic resource]. - Access mode: http://www.psyp.org/question_concerning_technology.pdf
[2] Heidegger: The Question Concerning Technology [Electronic resource]. - Access mode: http://www.english.hawaii.edu/criticalink/heidegger/index.html
[3] 2014 Intentionality [Electronic resource]. - Access mode: https://plato.stanford.edu/entries/intentionality/
[4] Arthur B W 2014 *The essence of technology* (Zhejiang: Zhejiang People's Publishing House) p 285
[5] Corey A 1999 *Communication Theory* **9(1)** 26–57
[6] Fidalgo A and Tellería A S Carvalheiro J Canavilhas J and Correia J 2013 *Revista Latina de Comunicación Social* **68**

[7] Utesheva A and Boell K S 2017 *Database for Advances in Information Systems* **47(4)** 106–10
[8] Budd MJ 2005 *Journal of Documentation* **61(1)** 44–59
[9] Arnaud N and Fauré B 2016 *Communication Research and Practice* **2(3)** 290–310
[10] Orlikowski W J and Scott S V 2009 *The Academy of Management Annals* **2(1)** 433–74
[11] Grace W R and Fonseca F 2014 *iConference 2014 Proceedings* 791–7
[12] Bavdaz A 2018 *Athens Journal of Social Sciences* **5(1)** 51–78
[13] 2016 Economic Relations between Russia and China – Increasing inter-dependency? [Electronic resource]. - Access mode: https://helda.helsinki.fi/bof/bitstream/handle/123456789/14346/bpb0616.pdf?sequenc e=1
[14] 2016 Ambassador, Li Hui An interview with Interfax: Embassy of the People's Republic of China in Russia [Electronic resource]. - Access mode: http://ru.china-embassy.org/chn/jmhz/zejmhz/t1425688.htm (access date: 11.11.2018)
[15] 2017 Why Russian, Chinese Economies Have 'Large Potential to Complement Each Other [Electronic resource]. - Access mode: https://sputniknews.com/business/201706031054271724-russia-china-economies
[16] 2018 Shaolei Feng The Evolution of Sino-Russian Relationship: A Perspective from the Meeting Between Xi Jinping and Vladimir Putin [Electronic resource]. - Access mode: http://valdaiclub.com/a/highlights/evolution-of-sino-russian-relations/
[17] Adler J M 1995 *Adler's Phiosophical Dictionary* (New York: Scribner) p 224
[18] 2018 Intentionality [Electronic resource]. - Access mode: http://www.newworldencyclopedia.org/entry/Intentionality (access date: 11.11.2018)
[19] 2013 Phenomenology [Electronic resource]. - Access mode: https://plato.stanford.edu/entries/phenomenology/
[20] Mickeshina A L 2016 *Voprosy Filosofii* **2**
[21] 2014 Ontology philosophy [Electronic resource]. - Access mode: https://www.rsu.edu.ru/wp-content/uploads/2014/10/constr_realnosti.pdf
[22] Koerber C R 2011 *Determining the value of communication The development of a theory and three-dimensional scalar model for evaluating and assigning values to relating communications relating to adult education* (The University of Memphis) p 159
[23] Absher D B 2010 *Situating language: Language, practice and meaning in the thought of Ludwig Wittgenstein and Martin Heidegger* (University of Kentucky) p 236
[24] 2009 Heidegger Culture, and Intentionality [Electronic resource]. - Access mode: https://www.academia.edu/8620642/Heidegger_Culture_and_Intentionality
[25] Campbell D R 2009 *The awareness and practice of international leadership* (University of southern California) p 140
[26] George K S 2017 *AI & Soc* **32** 191–9
[27] Gülenç K and Arıtürk M H 2016 *Synthesis philosophica* **61** 83–103
[28] Jacob P A 2015 *The technological mediation of psychopathologies Depth psychology and the philosophy of technology* (Pacifica Graduate Institute) p 280
[29] Clough M P, Olson J K and Niederhauser D S 2013 *The Nature of Technology- Implications for Learning and Teaching* (Rotterdam: Sense Publishers) p 51–84
[30] Ikaheimo H 2010 *ASCS09: Proceedings of the 9th Conference of the Australasian Society for Cognitive Science* 145–153
[31] Sharonova S, Nazarova E and Erokhova N 2015 *Sociology and Anthropology* **3(12)** 642–8
[32] 2018 In Shenkursk sawmill was restored and modernized [Electronic resource]. – Access mode: https://region29.ru/2018/04/04/5ac4d9ac12f17bf0362ea232.html

Application of the theory of inventive problem solving to amplifying creativity of employees

V N Tsygankova

Volgograd State Technical University, Volgograd, Russia

verats1@mail.ru

Abstract. The development of staff creativity becomes an important task in an innovative economy, as it's necessary to analyze market trends, develop strategies, participate in innovative projects and offer new solutions. TRIZ is one of the most accepted methods to construct systematic innovations (TRIZ is the internationally acknowledged Russian abbreviation for Teorija Resenija Isobretatelskih Zadac, which can be translated as the Theory of Inventive Problem Solving). TRIZ can be combined with traditional methods of strategic planning and allows for systematic training of staff. In this article are analyzed the results of TRIZ application to the training the creativity and forecasting and reengineering projects.

1. Introduction

Innovative economy is based on new ideas and their rapid implementation, and today we need employees, who can work in conditions of constant changes, analyze it and offer new solutions. Thus, it is necessary to develop new competencies of staff based on creativity and ingenuity.

The founder of the study of creativity Guilford J [1] showed a wide scope of this feature, and since the 40s creativity has been studied in pedagogy, management, personnel management, social psychology, knowledge management and human capital. Mednick S A [2] considered that creativity based on associations and as a process of redesigning elements in new combinations. Torrence E.P developed tests for diagnosing creativity [3]. Nonaka I and Takeuchi H [4], Senge P [5] examined the impact of knowledge on the development of organization.

Amabile T M [6] considers that daily creativity, developed by fixing the progress of achievements, providing social and emotional support, promoting creative results in related projects. Also she considers it necessary to keep up group creativity, which is essential in the conditions of distributed innovative projects, and the individual creativity often insufficient to promote innovative project within the organization and outside it [7]. Accordingly, the development of group skills may be trained on the base of the combination the methods of generating ideas, internal motivation, intellectual resources management, the forecast of market development.

2. Methods and discussion

TRIZ was developed by Altshuller G S [11, 12]. It's a complex method that takes into account the achievements of the previous methods (for example, William Gordon's Sinectic, method of test questions, mind storming), using 39 inventive principles, Algorithm of Solving Inventive Problems (ARIZ), contradictions matrix, laws of the development of technical systems, 9- Screen Model. Substance-field (Su-field) Analysis. Petrov V continued to develop the theory [13] and proposed new options for its application.

TRIZ differs from other methods in its attempt to reconstruct the process of thinking and create tools for solving inventive problems. TRIZ theory had developed in the USSR, and since the 80's began to spread around the world. Although TRIZ was originally developed for engineering problems, it was found quite wide application in marketing, design, quality management, which are also distinguished by sufficient complexity, as well as the high cost of the consequences of the decisions made.

TRIZ tools are widely used in various fields:

- conceptual development of new products, processes, business strategies [14];
- the prediction of the evolution of technical and economic objects and processes, functional cost analysis;
- comprehensive search for solutions and protection of the company with the help of patents;
- assessment of hidden desires and needs of consumers, market segmentation based on consumer qualities, development of new marketing projects [15];
- preliminary analysis of deficiencies and their elimination in new and existing products;
- additional opportunities for generating ideas, the development of personal and group creativity, new forms of motivation [16];
- QFD.

The figure 1 shows a diagram of methods and techniques used in TRIZ.

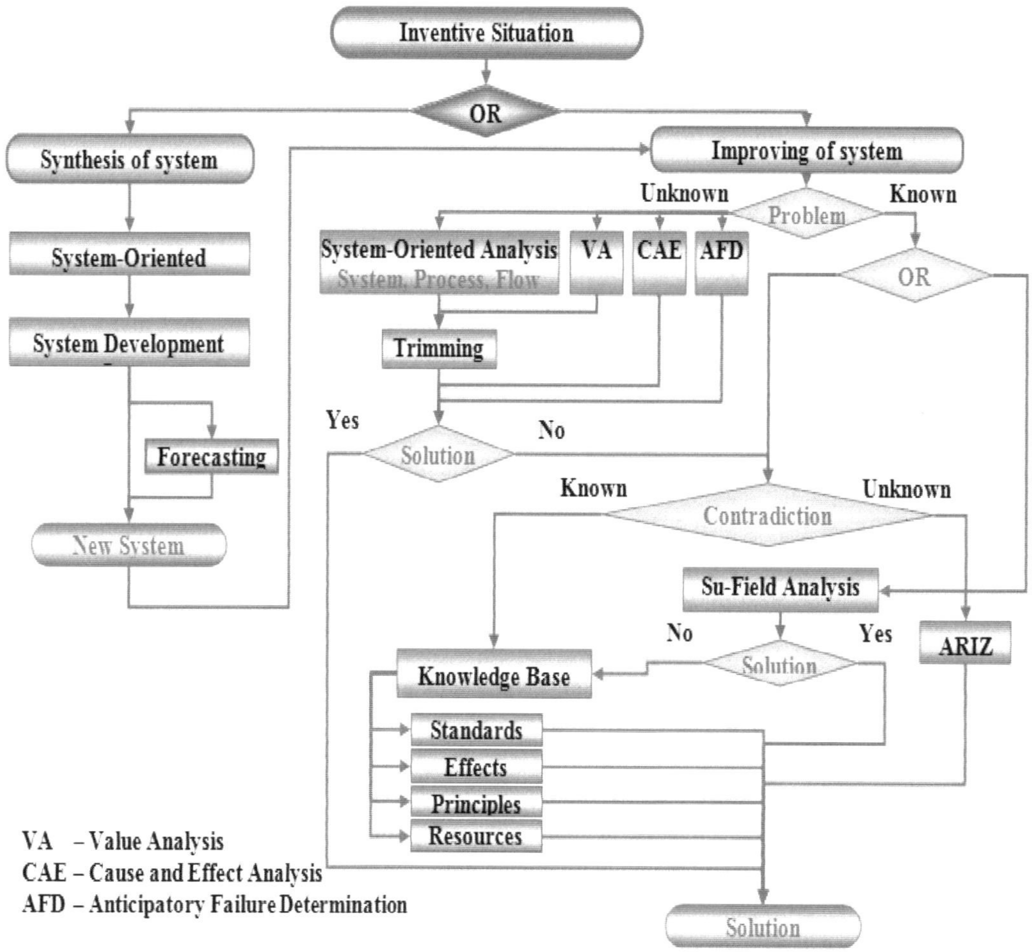

Figure 1. Algorithm for selecting TRIZ tools [13].

The use of TRIZ requires [16]:

- All members of the organization should know and use TRIZ.
- Trainings and coaching should be made possible for all involved.
- Additional training time.
- The advantages of the TRIZ methodologies should be explained.

TRIZ as a method can be applied in different formats - both group and individual training (commercial and educational).

During the training, there are several options for the distribution of roles:

- free distribution of roles;
- choice of roles by the participants;
- assign role (mostly for cases).

Training allows create new competencies that are important both for the individual and for the team. Different opinions create higher requirements for the product and make it more efficient. Using TRIZ allows you to develop and use a unified scheme of work; an important advantage of the methods of generating ideas is gamification, which allows you to develop additional scenarios and options for training [17, 18, 19].

TRIZ can be used to forecast the development of the organization and individual products (on the base of 9- Screen Model), that can be used for developing new products, planning carrier.

Table 1 presents a forecast of the development of education and training, while cognitive abilities must be maintained throughout life, and TRIZ here can be used as a regular training and a set of techniques for analyzing inventive tasks. At the same time, the set of competencies and their professional interactions become more complicated.

Table 1. Forecast of development of system of motivation.

	Past	Present	Future Scenario 1	Future Scenario 2
Super system	Training and Control System External motivation Team	The combination of internal and external motivation	Self-learning Organization Motivational environment	Self-learning Organization Self-learning environment MOOC
System	Basic Education	Basic Education + Continuing Professional Development	Lifelong Education Combination of professions	Lifelong education Professional interactions Emotional intelligence
Sub-system	employee	employee	employee	employee

Thus TRIZ has a number of advantages: adapted techniques for solving economic problems and the system of cases, focus on the formation of a creative personality on the basis of systematic training, management of the direction of the search, wide practice of application, different formats of training, the formation of new relationships in the group, the definition of the rules of preparation and sessions.

Table 2 shows the evolution of motivation systems, careers, sources of creativity, education. Forecasting based on the transition to the super-system shows the transformation of social systems and increasing demands on employees, but at the same time the growth of resources that can be used to achieve the aim.

For economic tasks, the structure of the algorithm is simplified; often 4 main stages are identified [20]:

- Task description, technical and economic constraints, desired target.
- Systems analysis (highlighting relationships and components, transition to a super-system, building an ideal end result).
- Synthesis (application of methods for eliminating contradictions, material-field resources, system operator, laws of systems evolution, S-curve).
- Analysis of options, their selection (if the problem is not solved, return to the first stage).

Table 2. Forecast of development of system of education.

	Subsystem	System	Super system
Motivation Modes	Internal motivation Self-realization	Internal motivation Self-realization External motivation (material and non-material) Competition Self-Actualization Needs	Corporate social programs participation in the company's capital Self-Actualization Needs Intra-organizational marketing
Career	Freelancer Outsourcing Downshifting	Horizontal vertical Step career	Inter-organizational career
Sources of creativity	Individual training Insight	Group trainings Foresight Delphi method Synectics Cooperation	Online Foresight, Online trainings On-line contests for solving non-standard tasks (project contests)
Education	Online Courses mutual learning self-education	self-learning organization quality circles	MOOC, Mentoring Forums Social networks Corporate universities

TRIZ can be applied in the reengineering of business processes, in this case an analogue of the model (as is - to be), see Figure 2. In this case, it can be combined with other methods (for example, foresight, SWOT – analysis, functional-cost analysis), which allow concentrating efforts on the chosen direction.Direct and reverse planning process should be carried out in parallel with business modeling to verify the proposed changes.

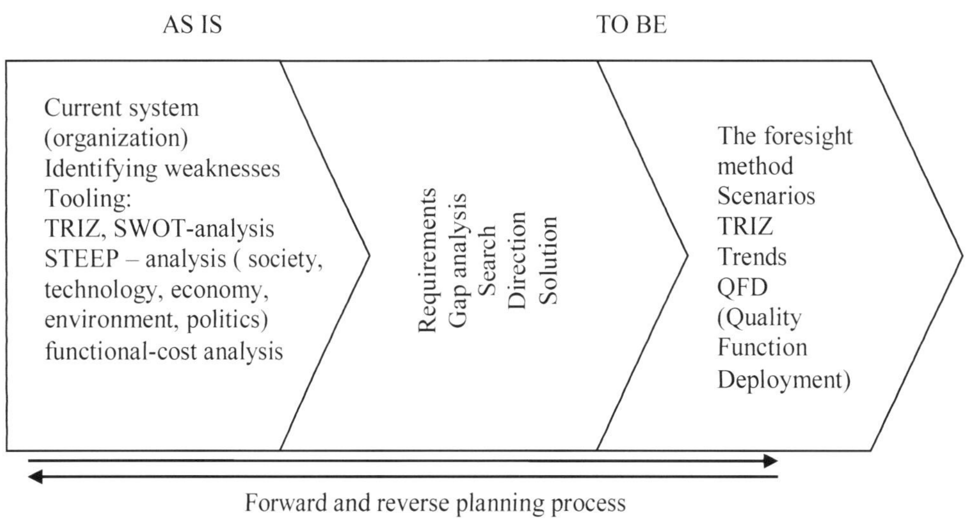

Figure 2. TRIZ and organization transformation.

Thus, TRIZ can integrate with other methods and helps to find the right solution in the field of innovation and marketing, because it helps broaden the horizons, properly analyze complex tasks, conduct patent searches, carry out business planning [21].

2. Conclusions Modern conditions require the integration of competencies, tracking of market trends and quick response. The development of new directions requires systematic efforts to develop the creativity of the staff. Although there are many methods of generating ideas, TRIZ has significant advantages:

- The main techniques are published and available for study, TRIZ achieved results in various industries.
- The use of algorithm for solving inventive tasks, integration with traditional methods of strategic and marketing analysis.
- Developed inventive techniques for economic systems.
- Teamwork contributes to the joint promotion of ideas, the development of emotional intelligence, the development of scientific research skills, team building, cooperation.

References

[1] Guilford J P 1967 *The Journal of Creative Behavior* (**1**) 3-14
[2] Mednick S 1962 *Psychological Review* (**69**) 3 220-232
[3] Torrance E P 1966 *Torrance tests of creative thinking: Directions manual and scoring.* (Princeton: Personnel Press)
[4] Nonaka I, Takeuchi H 2008 *The Knowledge-creating company: How Japanese companies create the dynamics of innovation* (Oxford, Oxford University Press)
[5] Peter Senge 1992 *The fifth discipline* (Century Business)
[6] Amabile T M, Kramer S.J. 2011 *The Progress Principle: Using Small Wins to Ignite Joy, Engagement, and Creativity at Work* (Harvard Business Review Press)
[7] Amabile T M, Kramer S J 2010 *Harvard Business Review* (**88**) 1-2 pp 44–45
[8] Kirton M J 2003 *Adaptation and innovation in the analysis of the procedure of reinventing, the course context of diversity and change* (Routledge, London)
[9] Belbin R M 1981. *Management Teams: Why They Succeed or Fail.* (Oxford, U.K.: Butterworth Heinemann)
[10] Aritzeta A, Swailes S and Senior B 2007 *Journal of Management Studies*, (**44**) 1 96-118
[11] Altshuller G 1996 *And suddenly the inventer appeared: TRIZ, the theory of inventive problem solving* (Technical Innovation Center)
[12] Altshuller G, Shulyak L and Rodman S 2005 *40 Principles Extended Edition: TRIZ Keys to Technical Innovation* (1st ed.) Worcester, United States: Technical Innovation Center, Inc
[13] Petrov V 2016 *TRIZ Journal* https://triz-journal.com/algorithm-for-selecting-triz-tools/
[14] Livotov P 2015 *Procedia Engineering* (**131**) 767–775
[15] Gazem N A J, Rahman A A 2013 *Advances in information Sciences and Service Sciences,* **5(11)** 273-282.
[16] Gronauer B, Naehler H 2016 *Procedia CIRP* (**39**) 185-190
[17] Mann D A 2001 *Creativity and Innovation Management* (**10**) 2 123-125
[18] Vaneker T, van Diepen T *Procedia CIRP* (**39**) pp 67-72
[19] Rubina N V 2014: *Coordination and integration of TRIZ tools, TRIZ developers SummitLibrary* (**6**) 175-198
[20] Rantanen K, Conley D W, Domb E R 2017 *Simplified TRIZ: New problem-solving applications for technical and business professionals* (CRC Press,Productivity Press) 282.
[21] Tsygankova V N 2013 *WASJ* (**27**) **1** 83-86

Organization of structured interaction on the base of psychographic characteristics within the model of personality traits DISC

[1]E A Chigova, [2]I V Plyushch, [3]I V Leskova

[1]Reshetnev Siberian State University of Science and Technology, 31 "Krasnoyarskiy Rabochiy" pr., Krasnoyarsk, 660037, Russian Federation.
[2]Federal State Educational Institution of Higher Professional Education Siberian Federal University, 79 Svobodny pr., Krasnoyarsk, 660041, Russian Federation.
[3]Russian state social University 4 Wilhelm Pieck St., bld. 1, Moscow, 129226, Russian Federation

E-Mail: chigovaea@yandex.ru, reft@inbox.ru, leskova.i@yandex.ru

Abstract. The paper discusses impact factors assuring the change in the efficiency of structured interaction at enterprises and in organizations. Methodological tools included a pilot study based on the model of different personality traits (DISC). Participants of the study were 47 representatives of small work teams in companies of Siberian region. The paper interprets a correlation between the efficiency of structured interaction in work teams and social personality traits. The use of training forms aimed at adjusting a social model of work teams is said to be restricted. An interrelation between team building based on psychographic traits and application of latest organization principles of structured interaction was ascertained according to analysis data.

1. Introduction

In the present day society, improvement of manageability is recognized to be a burning issue in all activity fields. As more parties concerned are getting involved in the process, management tends to change.

Management techniques have been studied for a long time so far. As soon as mankind is aware of its needs, managing attempts are made, although an understanding of management is a quite new phenomenon. For two recent decades new approaches to management have been established in science. Human recourses management experiences the most dramatic changes. Management has been considered a process towards the goal. Subjectivity of management is becoming a more important factor in research of the time. A modern understanding of management implies projecting and maintaining of such an environment, where individuals co-working in groups achieve certain goals in an effective way [1]. Therefore, management is found in all spheres with intended transformation of properties or states in order to increase the efficiency of a system.

According to a common belief in the present day society a social factor is vital for production management. Individuals are involved in the most production phases, making processes uncertain. Several attempts have been made to improve this uncertainty through taking into consideration more parameters of human behavior, making behavior more controllable. But these attempts were complicated.

First, people management is impossible without an ethic factor. Therefore, social management relied on thoroughly developed formalized regulations, agreements, laws etc. However, a complex nature of human behavior impedes its regulation on the base of formalized rules. Moreover, work activities with a great number of rules and strict control is low efficient [2]. To date, individuals have a more subject role in the society, becoming an active force instead of being just a tool or one of recourses. Specifics of people management enjoy recognition in management.

Social management is thought to be the process of going towards goals in a company; it has two moving forces – economic and administrative motivation of human behavior, and transformation of social maturity in staff, its active involvement into strategic planning of the company [3].

Second, it is hardly possible to balance or consider all uncertainty factors even if IT-technologies are used. As accepted, a complex system is to be self-balanced in an uncertain situation, which is not beyond approved system facilities. As a consequence, qualifications of human recourses have to meet much higher standards. Any work team or social group is a complex system and is to be capable for working more efficiently in ordinary situations.

2 Problem statement

Management implies implementation of preliminary realized needs of subjects. A new challenge facing management in the present day society is the use of realized needs both of managers and executives; that is based on employees' interest in work product and strategic development of the enterprise [4]. Manageability turns into a more required feature of organizations, groups, manufacturing in any activity field.

The involvement of staff into decision making process with the purpose to increase manageability has been gaining in importance in the society [5]. Participation in management and decision making, in its turn, necessitate certain social characteristics and states. For instance, it is a psychological climate in staff. The psychological climate in staff depends in most aspects on the social competence of employees. First of all, such characteristics as ability to be responsible for decisions made, a skill of making precise statements, being polite and reasonable in arguments, willing to discuss and accept ideas of other people [6]. Furthermore, in some activity fields social competences are strongly interrelated with professional skills and are close to them. For example, social competence is an important constituent of professional competence in the field «person-to-person» [7]. Introducing IT-technologies, people still participate in processes requiring a complex interaction between individuals. A majority of managers try to develop work teams. The most widespread practice in the Russian society is training to build local social groups (work teams).

Training as a form to influence a group dynamics is popular with managers due to certain characteristics. They include the development of a certain skill, willingness to practical application of this skill, time limit of preparation [8]. Trainings are used to change the dynamics of work groups.

However, training can't solve all problems of group dynamics for two reasons. First, managers underestimate the phase of problem statement; they are more focused on trainings with a certain subject area. For example, team building, negotiating skills, public speech, time management etc. The reason for conflicts or low efficiency in staff might not agree with the subject area of trainings. Second, training is far from being a universal tool to solve all problems existing in staff. Efficiency of training as a tool increasing social competence is quite low. Mainly, because skills of a social competence imply external reaction of an individual to the surroundings, to him/herself, it is a part of emotional intelligence, that can't be trained, only developed [9]. For example, communication training can promote personal development of an employee but not initiate it. The development of certain skills can't develop the personality of an employee; it provides him/her only with more sophisticated tools. A question is stated – what conditions are unfavorable for application of training as a universal tool for team development. Identification of these conditions makes it possible to determine potential application of trainings within the certain group dynamics. To answer this question a study was carried out on the base of psychographic characteristics.

3. A pilot study on a model of individual personality traits in DISC

A model of individual personality traits DISC is frequently used to analyze groups. Methodically, this model is based on a four-factor model of the same name and centers on test data in 4 different behavior types (rational thinking, feelings, emotions, intuition) [10]. In Russia this methods is also applied now but mostly for selection of highly skilled personal.

Method description. In September 2018 a study was carried out in Krasnoyarsk with the purpose to pilot the method and assess its validity for Russian conditions. The study was conducted in Krasnoyarsk (Siberia, Russia). Employees of 4 companies, aged 20-50, participated in this study. Work teams have no functional connections and were tested according to DISC method without interrupting their work activity.

The study included double DISC-testing of respondents: in the evening of an active week day and during the day time at the weekend. Staff and managers of 4 structural (2 business and 2 governmental service) sub-divisions were tested.

Two business structures:
- pizza delivery service in the city (one company, 7 people),
- shop warehouse «Mebeldesign» (four subdivisions, 26 people)

Two structures of governmental services:
- PR-department of the territorial administration (one sub-division, 3 people),
- kids' clinic (one sub-division, 11 people).

Therefore, forty seven male and female respondents were questioned, including 7 managers and 40 ordinary employees.

As findings of tests were determined, respondents were also interviewed. The interview included three questions. They were: «Do you agree with DISC-test results?», «What is your position in the company?», «Do you like your job and your position?». The majority of respondents (except for 2) agreed with the study outcomes. Being emotional and using certain examples from their private life, they confirmed validity of results. As the second question «What is your position?» was asked, the data of testing and position of a respondent were compared.

4. Results

The results of DISC-test data obtained in the evening of a weekday and at the weekend are quite similar. They differ in one aspect only: tired respondents (after a working day) had more points in the scale «passivity». For instance, results of behavior style "compliance" exceeded slightly those of behavior style "dominance", whereas "steadiness" was above "inducement". And, vise-versa, results of the scale «activity» were higher at the weekend. However, the principal result, indicating the dominant DISC behavior style, was stable (89%). Therefore, a time factor is shown to be quite irrelevant for results of DISC-testing.

As for correlation between the position in a company and dominance, there is an unambiguous regularity. For instance, five of seven managers demonstrated a dominant behavior style. Two managers indicated a lower dominant style, according to assessments one of them was inducement (head of the PR-department), the other one demonstrated the style of compliance (head of accounting department). All respondents had minimum 6 months of working experience in the managerial position. All of them emphasized to feel comfortable in their managerial positions; and all their tasks were completely performed.

Analyzing the data provided by executive employees, entirely different conclusions were drawn. Among ordinary employees staff of kids' clinic (5 of 10 respondents) and shop warehouse «Mebeldesign» (4 of 22 people) demonstrated the dominant behavior style. Therefore, managerial positions are held more frequently by people with a high dominance factor, that is, active, purposeful, strong-willed, capable for problem solving, and authoritarian ones. In other cases these are positions with relative operational autonomy, where self-reliant decisions are supposed to be made (clinic). Modest, conservative, reserved, piece-loving, slow in decision making and easily-agreeing employees are executives. Here, the question «Do you like your job and your position?» was answered with "yes" by two thirds of executive respondents. In most cases they represented the style steadiness, who didn't feel comfortable with a dominant and conflicting manager.

5. Discussion

Efficient organization of any structured interaction implies a communication environment with certain rules of structural and personal communication, goals and content of manufacturing or business processes. Principally, efficient organization of structured interaction presupposes low proneness to interpersonal conflict on all levels of the manufacturing process, i.d., between managers, structural sub-divisions and executives. The role of emotions employees have in the process of work has been emphasized by experts of different research fields for almost fifty years [11]. Optimally, regular checking of negative emotions is recommended, the number of employees is not important in this regard. Using special techniques, the matter of difficulties should be disclosed in order to resolve conflicts arising in a work team. Otherwise, a conflict like any psychological problem caused by communication might deteriorate professional activity, lead to neglecting of functional tasks, and worsen attention focusing, as a consequence, employees make mistakes in their work activity.

Methods of staff diagnostics were not applied for recruitment in any organization under consideration. Employees were recruited according to formal characteristics (level of education, work experience). When recruiting social competence was not taken into account, as well as adaptability, orientation towards rules accepted in the organization. This is quite typical for the majority of organizations in Russia; as a result, they face such problems as labor fluctuations, low work efficiency and inefficiency of the organization. Poor-efficient methods of labor optimization are used to resolve these problems: introduction of an updated norm system, change of work schedule, training aimed at the development of certain skills. These methods can hardly change the efficiency of a work team as whole.

At the same time it is quite impossible at this stage to reveal reasons for certain psychographic characteristic that employees have. Several explanations are found. First, executives might have prevailing average values of certain characteristics due to random selection according to personality traits formed before. In this case people tend to choose unconsciously a professional role with more psychological comfort. Second, the tested might not demonstrate strongly-pronounced personality traits; those are seen as a consequence of some professional deformation. In this case the accepted passive role presupposes a leveling effect of psychographic characteristics with regard to the chosen role position. Third, it might be an all-Russian general tendency of low social competence. I.V. Plyushch emphasizes that misbalancing formation and development of certain personality traits cause later maturation, and a longer period of education is not the only reason for it. More infantilism manifestations, emotional immaturity, unformed social competences of adults are registered [12]. Present day infrastructural and social life conditions slow down the development of some social competences, sometimes making their formation even impossible.

In recent decades recruitment based on certain social and psychological characteristics has been gaining in importance. Different methods of recruitment or assembling work teams are used in this case; these methods can be divided into qualitative (relying on psychological and sociological theories) and quantitative (including formalization and modeling of logical deductions by decision making individuals, binary choice models and artificial neuron nets applied by HR agencies) [13]. As a result, all methods of recruitment and assessment of personal single out psychographic characteristics revealing a life style of an employee, his/her goals and values. The data on psychographic characteristics make it possible to choose a work team according to a set model, with necessary/sufficient level of personality development and orientation of this development. A work team formed this way operates maximally efficiently and is changeable by many up to date managerial methods, including training forms.

6. Conclusion

To date, management of a work team is an important part in management in any organization. Qualitative characteristics of personal are in focus of research. Two trends have been formed in personal management; their implementation relies on principles accepted in two different managerial schools.

One approach pays attention to the development of personal hired in a random way. This approach is dominant in countries with paternalistic managerial and cultural tradition, e.g. Russia. The advantage of this approach is higher security of employees. A low efficient employee is still employed. It is important in a society with a low level of social and professional mobility. Using short termed forms of training, a manager makes attempts to develop certain skills of employees. The efficiency, in this case, increases insignificantly and doesn't depend on made effort. Performing work activity for a long period of time, these employees develop certain tactic models of their profession and improve the efficiency of work under stable conditions or in certain professional fields. Therefore, work efficiency increases in a long run provided that a work team is constant. This approach is in line with management principles stated by A. Fayol.

The second approach to personal management relies on rigid diagnostics procedures of recruitment at the initial stage. The recruitment is carried out via determining psychographic personality traits of individuals. This approach makes it possible to assemble a work team with a high level of compatibility immediately at the beginning of professional activity. Short-termed training in this work group would result in a long-termed effect, increasing, in this way, the work efficiency, that is in line with management principles, described by the founder of personal manager, F.W. Taylor [14]. Nowadays, scholars also agree with F.W. Taylor's ideas and suggest four principles of personal management, which are efficient in a highly compatible work team [15]. In general, management is aimed at making the activity more rational and supporting the rational way of thinking. The rational activity is result-orientated; however, different groups of a work team might not always accept a certain work result as a main one. The efficiency of work activity is regulated both by management and social norms [16]; that's why we admit the development of any work team should be initiated in a managerial group. The most managers expect changes provided by employees, although the change of a managerial group is needed.

Therefore, organization of structured interaction on the base of psychographic characteristics makes it possible to reveal psychographic characteristics of potential employees at the stage of recruitment, and also use up to date principles in organization of structured interaction. For instance, a deep professional specialization of employees would be maximally efficient under conditions of appropriately formed work team, where managers and employees share work tasks and responsibilities, and both material and non-material remuneration for work done is important. Otherwise, no managerial methods, both formalized and communicative trainings, can guarantee high efficiency of organization.

7. Acknowledgements

We appreciate deeply collaboration of all participants of the study. They were willing to learn more about themselves with the purpose to become better. These are participants listed in groups: employees of pizza delivery service «Al Pacino», shop warehouse "Mebeldesign", PR-department of Regional Administration, kids' clinic №2.

References

[1] Koontz H & Weihrich H 1990 *Essentials of management* **5** 13-22
[2] Vogt J, Leonhardt J, Koper B and Pennig S 2010 *Ergonomics* **53**
[3] Mazurkiewicz A. and Pizło W 2006. *Podstawy Marketingu* (Warsaw: PWSBiA) p 165
[4] Chizhova E A and Oleshkevich Yu V 2017 *Theory of journalism* **2** 218–227
[5] Maksimov B I 2014 *Sociological Studies* **3** 107-113
[6] Urbach N. and Ahlemann F 2016 *HMD-Praxis der Wirtschaftsinformatik* **53(1)** 16-18
[7] Ahlemann F and Eckl C 2013 *Strategisches Projektmanagement* (Springer-Verlag Berlin Heileierg) p 163
[8] Plotnikov A V and Skuridina I V 2014 *Life Science Journal* **11** 548–553
[9] Segundo Vito Aliaga Araujo and Scott N Taylor 2012 *Cross Cultural Management: An International Journal* **19** 19-29

[10] Nikishina A 2016 *Modern research and innovation* **16** 7
[11] Marston William Moulton 1928 *The Emotions of Normal People Emotions Of Normal People* (Osmania University: Publisher Kegan Paul Trench Trubner And Company) 443
[12] Plyushch I V 2017 *Pedagogy* **3** 64-69
[13] Zinchenko A A 2010 *Personnel management* **3(15)** 70-75
[14] Taylor F W 1911 *The Principles of Scientific Management* (New York: Harper & Brothers) 156
[15] Barabash M A and Chizhankova I V 2017 *Materials of Ivanovo readings* **3-2** p 6-12
[16] Elster Jon 1989 Journal of Economic Perspectives **3(4)** 89–117

IOP Conf. Series: Materials Science and Engineering **483** (2019) 012098 doi:10.1088/1757-899X/483/1/012098

Social aspects of industrial safety: current challenges

I V Plyushch[1,2], V I Ruzanov[2], L M Safiulina[3], N V Vysotckaya, T N Yudina

[1]Reshetnev Siberian State University of Science and Technology, 31 "Krasnoyarskiy Rabochiy" pr., Krasnoyarsk, 660037, Russian Federation
[2]Federal State Educational Institution of Higher Professional Education Siberian Federal University 660041, 79 Svobodny pr., Krasnoyarsk, 660041, Russian Federation
[3] Russian state social University 4 Wilhelm Pieck St., bld. 1, Moscow, 129226, Russian Federation

reft@inbox.ru, ruvich@yandex.ru, luzia@inbox.ru, vysotskaya25@yandex.ru, ioudinatn@mail.ru

Аннотация. The paper discusses the importance of social aspects in industrial safety. Specifics of the Russian society, manifested in culture and set of population's values cause underestimation of social constituents when breaking labor safety rules. At the same time, industrial infrastructure differs significantly from the environment, so safety in these conditions tends to change its essential characteristics. Governmental measures aimed at the improvement of industrial safety are insufficient. According to the study it is important to consider a technological environment as a field of human activity.

1. Introduction

Anthropogenic environment came into being as a protection tool aimed at the improvement of human safety. People started confronting with new hazards because of technological advancements. These days, the issue of human interrelation with the anthropogenic environment is thought to be the major area of human fears. As the society develops, an individual tends to realize deeper the specific character of human interaction with the anthropogenic environment. Here, this interaction can be characterized from an entirely new point of view, furthering, as a result, changes in worldview concepts of the mankind. In the 20th century issues of ecology had a breakthrough effect on the worldview, in the early 21st century problems of safety and interrelation with the anthropogenic environment are of high importance. Undoubtedly, various kinds of hazards and risks were an integral part of the entire evolution of human civilization. However, understanding of essential characteristics of hazards and identification of system safety characteristics have been center-stage only in recent years.

Safety is supposed to be one of characteristics and criteria of function and development in social, economic, technical, environmental and biological systems. To date, safety and security are key issues for an individual and any spheres of the society – organization, institution, state. Safety arrangements are vital for efficient management.

This problem should be of key importance in the Russian Federation, since security and safety have been permanently undervalued in the Russian society. Any crisis has complex consequences for the system itself and for all individuals, in particular. In the present day social conditions the issues of safety have been gaining in importance, accentuating contradictions between general theoretical fundamentals of safety and standards of their practical implementation.

The issue is becoming even more complicated because of fast historical transformations. Social conditions and characteristics are changing extremely, impeding the traditional and natural theoretical development of social safety. To date, the evolution of theory through accumulation and systematization of facts and their further interpretation and generalization, as well practical model building, are parallel processes, taking place in the same time continuum. This feature in the development of social safety emphasizes the relevance of expansion and transfer of systematic research results in the society. Nowadays, a high level of safety can be guaranteed in the society, provided that the gap between science and management, theory and practice is as closed as possible.

2. Problem statement

Social aspects of safety imply a multidiscipline research domain on the edge of sociology, psychology, law and social philosophy. This recently established field of research is hardly thought to be a new level in reality investigation or an object, type of activity field unknown before. It is likely a new level of understanding of common and well-known facts in our life. This is a domain discourse conditioning activities of groups and individuals in the process of manufacturing. Within this domain some efforts are made to determine appropriate strategies of activity in order to reduce risks and estimate a risk tolerance. In fact, these are limits, guaranteeing certain safety of our manufacturing activities (being safe in a broad sense of this word - for living beings, personalities, representatives of certain cultures and ethnic groups, elements in the entire life system in the world).

Safety is a complex phenomenon, depending on stability and (or) repetition of the situation, therefore, on its predictability. As the development of a situation has expected or predictable results a working system can save energy, resources, so it can be considered as a positive characteristic. In simple words, safety is seen as a lack of hazards, protection against hazards [1]. Industrial safety can't rely on the lack of hazards only, because any manufacturing activity is implemented in different conditions, therefore, this definition is not a complete one for current industrial safety.

Anthropogenic and natural environment are two principally different phenomena. There are no entirely positive or negative phenomena in nature, since there are only constituents in the common process of planet existence. However, elements of the manufacturing system function on the base of other principles. Each anthropogenic system is a separate structure with a preset purpose of its activity, for which other systems, e.g. human beings, are irrelevant. Therefore, industrial safety arrangements can be viewed as a secondary function of activity in this anthropogenic system. An anthropogenic system can't respond flexibly, so human life safety in manufacturing conditions depends on two factors, first, preset restrictions of an anthropogenic system, second, human behavior (directly or indirectly).

Some employees don't obey industrial safety rules; as a result, it is a statistically frequent cause of diseases, disabilities and even death of people. According to the statistical data almost 37.6 thousand industrial injures were registered in Russia in 2017, including 5.96 thousand serious or fatal accidents; 1.6 thousand people died [2]. The data over a longer period confirm this tendency: «above 75 % accidents are caused by organizational issues and the so-called «human factor» (insufficient organization of the manufacturing process, violation of labor safety rules, gaps in training, violation of labor discipline)» [3]. The main tendency is also found in various industries. According to monitoring carried out by the Ministry of Energy of the Russian Federation: «there is a slight change only in basic industrial injures in 2018 in comparison with 2016 and 2017» [4]. In most cases employees are responsible for injures: no labor safety trainings are carried out, any protection equipment isn't used, labor safety regulations are violated, and jobs are performed without being registered. The data above emphasize a low level of industrial safety in the Russian society. Findings of sociological surveys confirm the dimension of this problem: health care is not important «in the system of life values Russian population; in the scale of relevance it has third or fourth position» [5]. This feature of culture even worsens the issue of safety, including industrial safety.

As for human safety parameters of the anthropogenic environment are modified and tested, stated in several technical documents and standards, can be checked in a formal way, therefore, easily controllable. First of all, it is relevant for technologically up to date manufacturing processes. To date, parameters of the anthropogenic environment correlate with the capacities available in the society for safety protection, so they are harmless for infrastructures.

However, a behavioral factor depends on numeral conditions, has a multiplicative effect, therefore, it is hard predictable, requiring other forms of control. A multifactor nature of a behavioral effect on human safety in industry is complicated, being an interesting research domain, consequently. An individual is a main risk for human safety in a complex industrial system. Two options (scenarios) are possible and are realized:

1 development of a person, his/her social and psychical qualities, e. g. self-control, discipline, intended obeying of safety regulations etc.

2 creation of closed anthropogenic systems capable for functioning without a human interference in certain situations.

The second scenario guarantees a wider range of safety; as a consequence, the development of manufacturing processes is oriented towards this model. The implementation of this scenario is limited: it is more cost-intensive, science-demanding, correlates with scientific and technological advances, so its introduction can be facilitated.

The first option is less efficient, but more affordable at a current level of manufacturing technologies and infrastructure. As a result, this scenario was the first developed in history. Then these two scenarios were combined, in future the first scenario will be gradually substituted by the second one. However, the first scenario can't be stopped in certain industries. It is important for an employer to realize which efforts and manufacturing conditions can provide a high level of safety, which behavioral skills and personality traits of employees, and also their correlation are required.

3. Discussion

Safety is interrelated with any activity. It can be neither absolute nor absolutely objective. In fact, safety is a relative phenomenon, caused by historical, cultural and situational factors. Safety of a living system always comprises objective and subjective constituents [6]; it implies the absence of danger and behavior identical to a situation. Taking into consideration current manufacturing conditions, human safety depends on a person's behavior. A certain level of safety in industry and manufacturing processes can be guaranteed, provided that various regulations and laws are obeyed, contributing, therefore, to human safety in industry. To sum up, observance of industrial labor safety regulations is a principal mechanism of human safety. Inconsistency and ambivalence of this condition is caused by a social character of this managerial mechanism. The observance of safety regulations depends on human behavior only. Numerous studies confirm this fact, indicating that a current manufacturing process is «safe work, which presupposes physical and psychological willingness of employees with certain professional skills and motivation for observance of rules and functions according to labor safety rules» [7].

All employees of a company are informed about the system of safety regulations, since recruitment in the Russian Federation is based on a certain procedure. This procedure comprises two obligatory phases:

- familiarization of an employee with labor safety regulations accepted at this enterprise;
- registration of an employee in a training log;
- signature of an employee;
- date of training.

This procedure is carried out on a regular basis (period varies for different manufacturing processes).

Procedures subject to formal control (registration, person's signature, date of training) are usually registered in a log and checked regularly. However, it is difficult to guarantee socially-caused characteristics of labor safety training, their regular control is low efficient and might be more cost-intensive than a monitoring procedure on the whole. Socially-caused characteristics include guidelines of training, level of its implementation, awareness and responsibility of employees, motivation of an employee for observance of industrial safety regulations, importance and relevance of these regulations in the value system of an employee, etc.

Experts claim, that almost all accidents have numerous causes, result from accidental coincidence of events, but happen basically due to violations of industrial safety regulations [8, 9].

The observance of these regulations should be provided by certain organizational measures taken to prevent industrial injures. They include:

— «training and testing of knowledge on labor safety regulations, safe methods and work procedures;
— all types of training on labor safety and fire safety briefing;
— traineeships and shadowing;
— emergency prevention measures and fire safety trainings;
— further training courses» [10].

However, the majority of violations are registered in organizational measures. All efforts to raise the control level have a short-termed effect. At the same time, improvement of technical, technological and sanitary and medical aspects of industrial safety is in focus of the society and state. The reason might be also worsening physical health of population, which, in its turn, deteriorates psychological abilities to withstand workloads. «In the foreground there are issues of psychological and physiological stress employees have, improvement of work conditions dependent on physical factors is less important. It is possible due to dropping physical stress alongside with growing psychological and physiological stresses, this combination causes chronic fatigue, mental overstrain, bad relations with other employees and managers» [11]. That is why a mainstream direction of current management is motivation of employees for safe work. Managers have a need to identify the effect of motivating influence, and form a positive attitude of workers to safe work.

Safe behavior in industry relies both on professional knowledge, skills and abilities and motivation of an employee. Therefore, the control of human activities is possible exclusively through the control of his/her inner motivation. In the field of industrial safety motivation aims basically at making personal more interested into strict obeying of regulations.

Safety culture, importance of its implementation and development has been in the focus of discussion in recent years, since safety depends both on the external control and on activities of all employees. The principal object of protection is an individual; the secondary important ones are infrastructure and manufacturing processes. In fact, the problem is how to protect a person against him/herself in the manufacturing process. A manufacturing infrastructure reflects a technological essence of the present day human life. Urban conditions also cause numerous problems in interaction with the infrastructure [12]. Therefore, social aspects of industrial safety are integral constituents of a more general problem of social development. To a certain extent this issue can be solved via introduction of breakthrough educational technologies, and infrastructural advances. To date, digital technologies are to be used in Krasnoyarsk («Digital city, a pilot project of Rosatom) as an urban space with a high safety level. However, the principal issue is still a complex character of the internal motivation for safe behavior.

4. Conclusion

The life of each individual comprises several occupational spheres. Labor is one of the most important constituents. Industrial safety of a person is a mainstream direction in the development of a social state. «In present a system of management in Russia has been modernized from the standpoint of labor safety, a cost-based model, based on compensation for industrial injures should be replaced by a new model for assessment and control of professional risks, supporting preventive measures to protect life and health of people in industry. The statistics indicate a general drop of accidents in industry. However, this trend is incomparable with other technologically developed countries. In total, these data demonstrate a worsening tendency of industrial injures as compared with countries Germany, Japan» [9]. At the same time the problem of underestimated social aspects of industrial safety is still sharp. A high level of control can't be guaranteed all the time. A great number of disasters and accidents caused by a social factor emphasize the importance of this problem. The development of safety culture is supposed to become a priority in the Russian society. The implementation of this approach is to be carried out in two spheres: transformation of public values and active use of advanced technologies in education and control. For instance, enterprises of nuclear industry managed to achieve a high level of industrial safety. To date, safety can gain in importance as the value of human life is reconsidered. A level of industrial safety correlates with the level of industrial development and can be viewed as a universal integral criterion of economical or social development of the company.

5. Acknowledgements

Statements of this paper have been discussed with practicing University colleagues, because issues of safety culture are relevant for everybody, irrespectively to the occupational field. We highly appreciate a possibility to get an external feedback. This feedback makes it possible to get an insight into another point of view, make our own attitude more precise.

We are grateful to the Committee of the 1st International Scientific and Practical Conference «Breakthrough technologies and communications in industry» for kindness and professionalism, publication of this paper would be impossible without their competent assistance.

References

[1] Prohozhev A A 2005 General theory of national security // http://uchebnik-online.com/129/173.html
[2] Industrial injuries in the Russian Federation 2018 //http://www.gks.ru/wps/wcm/connect/rosstat_main/rosstat/ru/statistics/wages/working_conditions/
[3] Types and causes of occupational accidents 2016 // *https*://rosmintrud.ru/labour/safety/206
[4] The Ministry of energy of the Russian Federation: the office site 16 10 2018 // https://minenergo.gov.ru/node/4497
[5] Safina L 2011 *Economics 11 p 89-92* // http://ecsocman.hse.ru/text/50315452/
[6] Plyushch I V 2018 *Problems of scientific thought* 12 vol **1** 15-18
[7] Koshechkin Y V, Barabanova S N 2015 *Bulletin of rural development and social policy 12* p 38-40
[8] Social status and standard of living of the Russian population: 2004-2017 http://www.gks.ru/wps/wcm/connect/rosstat_main/rosstat/ru/statistics/publications/catalog/doc_1138698314188 /
[9] Buhtiarov I V, Izmerov N F, Tihonova G I, Churanova A N 2017 *Problems of forecasting* **5 (164)** 140-149
[10] Schennikov N I, Pachurin G V, `Kuragina T I, Mejenin N A 2013 http://window.edu.ru/resource/152/79152/files/Мон%20%20Щен%20и%20др%202013.pdf
[11] Morgunov E B 2010 *Personnel management* **6** 26-30
[12] Plyushch I V 2017 *Pedagogy* **3** 64-69

IOP Conf. Series: Materials Science and Engineering **483** (2019) 012099 doi:10.1088/1757-899X/483/1/012099

Comparative legal analysis of the employment of graduate production personnel in relation to the labor law of Russia and Germany

A N Shamne

Volgograd State University, 400062, Prosp. Universitetsky, 100, Volgograd, Russia

E-mail: antonsh97@gmail.com

Abstract. The article deals with the problem of employing young technical professionals as an important part of a more general problem of providing quality training for professional production personnel. The lack of practical experience is cited as the main reason for the impossibility of employment. Interpretation of such concepts as "business qualities of the worker", "young specialist" is given. Comparative legal analysis of the employment of graduate engineers is made in relation to the labor law of Russia and Germany There is a clearer structuring of the concept of experience in employment and a more developed system of practices during training in German labor law. Possible ways of solving the problem of presence/absence of experience are proposed when employing graduates (young professionals) in relation to the Russian labor law. An effective way to improve the efficiency of professional production personnel in the Russian education system is professional practice/training during university studying in the framework of production capabilities of the regions.

1. Introduction

High-quality training of professional production personnel, meeting the requirements of the time, has been and remains an urgent issue for modern society. This problem should be considered not only in relation to the training of future specialists but primarily within the framework of the dichotomy "college/university – technical division/production department". It is how close the relationship is between college/university – production (in a broad sense – both in terms of technical specialties and humanitarian activity) which influences the effectiveness of training professional production personnel.

One of the most acute issues in this area is the problem of employment because some issues in the field of employment remain not fully resolved despite the fairly complete regulation of labor relations within the legislation. In particular, we are talking about such a concept as "experience" in relation to a certain field of activity.

The materials of the relevant Internet sites offering various vacancies in the field of employment contain a number of requirements for applicants. One of them is the experience in this field of activity or (which is much less common) almost any practical experience in general. Moreover, this pattern is typical for both the Russian and foreign labor markets [1].

Accordingly, the lack of work experience can become the reason and sometimes the main one for refusal in employment. This problem is particularly relevant for young professionals who have just graduated from an educational institution and have not yet worked in the chosen field of activity.

We find it interesting to present a comparative description of this problem on the example of Russia and Germany, since approaches to its solution in labor law are different taking into account a more global problem of training professional production personnel in these countries, especially since this problem has not received proper coverage in the works of legal scholars.

According to the research conducted by the center for economic research of the University "Synergy", 73.7% of employers in Russia pay attention to the presence of practical experience but not to the name and prestige of the university when hiring graduates [2].

Surveys conducted by the portal for young professionals Career.ru showed that Russian companies are reluctant to hire specialists (graduates) without experience. In particular, every fourth company said that it does not accept young professionals without experience. Possible ignorance of the relevant production (possible errors in work), unwillingness to attract additional costs for professional training within the company, etc. [3] are indicated as the reasons.

2. Materials and research

Let us turn to the other side of the dichotomy "college/university – production" which is related to graduates (young professionals). Research centre Superjob cites a survey of graduates in 2015-2017. The difficulty of employing graduates is estimated 7.1 points out of 10. The most important reason for arising difficulties is the lack of work experience which was indicated by 67% of respondents [4].

Let us consier the materials of Internet sites containing information about vacancies for job seekers and forming the empirical base of the study. For Volgograd as an example, such well-known in the labor market sites offering employment are sites "Superjob" (https://volgograd.superjob.ru/) and "Avito" (https://www.avito.ru/volgograd). An analysis of about 200 vacancy announcements in various fields of activity showed that the requirement of work experience is 87%. Moreover, the time range of available experience ranges from "more than 1 year" to "more than 3 years". In some cases, the time characteristics of the experience are not given at all. Here is an example:

Manager on dealing with corporate clients (VTB Leasing Volgograd), experience of more than 1 year; Senior cook ("Carousel", Volgograd), experience in leadership positions; Law Counsel (LLC "Zirax", Volgograd), experience at a manufacturing plant is required; Regional representative ("Simple financial solutions North", Uryupinsk), experience in sales/insurance/NPF is compulsory; Expert at managing services for fixed business in the mass market ("Mobilnyetelesistemy", Volgograd), experience in marketing/sales/management of at least 2 years is required; Seller in hypermarket/stationary merchandiser ("Health", Voroshilovskiy district, Volgograd), experience in food sales and/or as a merchandiser (https://volgograd.superjob.ru/).

Assistant Manager for recruitment (Volgograd, Voroshilovsky district, Academicheskaya st. 22, office 417), experience of at least 1 year; administrator in the guest house "Empire" , experience of more than 1 year; ventilation engineer (NTO UNION, Krasnoarmeysky district), experience of more than 3 years; hairdresser, (Fitness, beauty salons "Hermes", Central district), experience of more than 1 year (https://www.avito.ru/volgograd/vakansii).

Based on the examples of advertisements in which experience is often accompanied by the imperative "necessary/required/obligatory", it can be said that lack of experience is one of the main reasons for refusal in employment.

In the Russian labor legislation there is no official regulation on the need for work experience in employment. Article 64 of the Labour Code states the following: "Any direct or indirect restriction of rights or establishment of direct or indirect advantages at the conclusion of the labour contract depending on sex, race, colour, nationality, language, origin, property, family, social and official position, age, place of residence (including the presence or absence of registration at the place of residence or stay), attitude to religion, convictions, belonging or not belonging to public associations or any social groups, as well as other circumstances unrelated to business qualities of workers, are not allowed with the exception of cases in which the right or obligation to impose such restrictions or benefits is provided by Federal laws" [5].

In this case, only "business qualities of employees" are mentioned in the list as the reason for refusal (we do not take the given exceptions). Let us dwell on this concept. The labour Code does not explain what is meant under business qualities. But Resolution of Supreme Court Plenum of the Russian Federation of 17.03.2004 N 2 (ed. 24.11.2015) "On application of the Labor Code of the Russian Federation by courts of the Russian Federation" provides the following interpretation in the section "*Conclusion of the employment contract*" paragraph 10: "Under the business qualities of the employee should, in particular, be understood the ability of an individual to perform a certain job function, taking into account his professional qualifications (for example, the presence of a certain profession, specialty, qualification), personal qualities of the employee (for example, health, the presence of a certain level of education, experience in this specialty, in this industry)" [6].

And here we find that the concept of "work experience" is one of the personal qualities of the employee. In this case, in our opinion, there is a certain conceptual inaccuracy: on what basis is professional experience related to personal qualities? The meaning of the linguistic unit "personal" is defined in the explanatory dictionaries as "characteristic of the personality" [7]. The concept of personality, in turn, includes the stable properties of human individuality, which are significant for society [8].

Based on these definitions, in relation to personal qualities, it can concern, for example, the discipline, ability to contact people, emotional lability, etc. As for the phenomenon of "work experience", its presence can be included in professional qualities, but not personal. In connection with the above, we believe that the concept "business qualities of the employee" needs significant clarification.

Since we are talking about the problem of employment of graduates, the concept of "young specialist" requires a separate consideration. There is no definition of this concept in the Russian legislation. However, guided by various programs, grants, competitions offered by the state within the framework of both the university (for university employees) and production company, we can make the following conclusion: a young specialist is a professionally trained personnel under the age of 35 years (in some areas – up to 30 years or up to 39 years) having acquired secondary vocational or higher education. A young specialist has the same rights and duties as other employees. As a special condition, a kind of limiting the employment of young professionals, it is possible to introduce a trial period proposed in some areas and industries (for example, some jobs presented on these websites contained this condition), which usually takes a time period of three months.

However, article 70 of the Labor Code of the Russian Federation states that the trial period is not set for "persons who have received secondary vocational education or higher education on state-accredited educational programs and for the first time coming to work in their specialty within one year from the date of receipt of vocational education of the appropriate level" [9], i.e. in this case certain benefits are stipulated for graduates (young professionals).

As for young professionals (graduates of colleges) under 18, the Labor legislation provides a number of benefits (Chapter 42, Labor Code) for them: benefits for employment or dismissal both during production activities and rest.

In the Russian state at the present time, there is also targeted support for young professionals in some separate industries to attract professional personnel (when lacking them). For example, in order to attract young specialists to rural areas, the RF Government Decree No. 598 of 15.07.2013 "On the Federal target program "Sustainable development of rural areas for 2014-2017 and for the period up to 2020" maintains the provision of social payments for the construction (purchase) of housing to Russian citizens living in rural areas, including young families and young professionals living and working in rural areas or who have expressed a desire to move to rural areas and work there [10]. At the level of subjects, certain privileges may also be established for individual categories of young professionals. For example, according to the Law of Moscow of 20.06.2001 N 25 "On the development of education in Moscow" young teachers of Moscow public educational institutions are provided with additional payments of 40% to the salary for being a young specialist and about 50% for having a diploma with honors during the first three years of work [11].

All of the above facts concerning graduates (young professionals) are the so-called "de jure", but "de facto" graduates face the actual problem of employment.

Let us turn to the representation of this issue in the labor law of Germany. Here the presence of experience at employment is also rather an important condition. According to statistics, 67% of surveyed employers prefer their employees to have professional experience (https://www.absolventa.de/karriereguide/berufseinsteiger-wissen/berufserfahrung-fundierte-einschlaegige). However, if we compare the Russian and German practices of employment, we find that the problem of presence/absence of experience is developed in the latter in more detail. In particular, the types of experience meaningful for the employer are distinguished and, accordingly, it is possible to distinguish a kind of level of having certain experience: 1) erste Praxiserfahrung (primary practical experience – *hereafter we give our translation*); 2) einschlägige Berufserfahrung (relevant professional experience); 3) fundierte Berufserfahrung (profound professional experience) [12].

The first and thus the lowest level – erste Praxiserfahrung - assumes the presence of so-called primary experience, usually obtained during a professional practice or university training. This experience is in demand primarily for graduates and other young professionals who just start working in the profession. As such experience the candidate can mention a certain type of practice, some part-time work, and it is not necessarily important that the experience is obtained in the field of activity where the candidate is going to work.

The second or intermediate stage – einschlägige Berufserfahrung - means that there is already some relevant experience obtained, i.e. practical skills in the field of professional activity. Moreover, a young specialist may have such experience since similar practices can be carried out for a shorter or longer time or even permanently while studying (Werkstudentätigkeit) in the chosen field.

The third or the highest stage - fundierte Berufserfahrung - indicates a profound professional experience. In this case, the employer is looking for an employee with long-term, extensive professional experience in the very field of activity. It is no longer possible to indicate practice as a sourse for obtaining experience.

Thus, in German employment practice, the above-mentioned levels of professional experience are generally accepted. This specification of the volumetric concept of "professional experience" allows you to organize the problem of employment for some categories including young professionals.

There is no Labour code in German law. Labour law is regulated by various legal acts, including those at the level of Land law, which is very well developed in Germany. The most complete employment relationship is reflected in the Social Code (Sozialgesetzbuch) as well as in the jurisprudence of the Federal Labour Tribunal (Bundesarbeitsgericht). As in the Russian labor legislation, in German one there is no definition of a young specialist. However, the graduates' categories of secondary and higher educational institutions are more clearly structured. In particular, the "Statistics" section of the official website of the Federal Labour Agency of Germany mentions the following qualifications of graduates: Fachkräfte - workers, Experten - experts, Spezialisten - specialists. Workers are those who have studied for at least two years and have a degree and a relevant qualification; experts are graduates having studied at university for at least four years and they also have the above-mentioned; specialists graduated from magistracy in a higher educational institution or have higher technical education with the appropriate qualification [13]. This structure allows employers to be more clearly oriented towards young professionals.

Let us turn to the well-known German Internet sites that publish information about vacancies. A similar number of vacancy announcements (about 200) in various fields of activity were analysed. As a result, it was noted that work experience is required in 91%. In contrast to the actual indication of the number of years in relation to the experience on Russian websites, here are the qualitative characteristics of the experience (primary, thorough, etc.), or an indefinite amount of time (long-term). Quite rarely we can find that the time characteristics of the experience are not given at all. Here is an example:

Pädagogen, Praxixiserfahrung erste in der Gestaltung von Integrationskursen und der Begleitung von Flüchtlingen und Migranten – Teacher, ***primary practical experience*** (preferably, but not obligatory – in the development of integration courses and support for refugees and migrants); Pädagogen mit einschlägigen beruflichen Erfahrungen im Bereich frühkindlicher sprachlicher Bildung und Bildungsarbeit - Teacher, ***appropriate professional experience*** in the field of training early-aged children in language teaching and learning (https://www.stellenanzeigen.de/job/); Dipl. Pflegefachperson HF (Demenzabteilung), fundierte Erfahrung im Langzeitbereich – Certified specialist in providing care for those who suffer from dementia, ***thorough many years of experience*** (in this area); Leitung Hotellerie, fundierte Berufs- und Führungserfahrung – hotel manager, a

thorough professional and management ***experience*** (http://stellenanzeige.monster.ch/); Data-Analyst / Data-Scientist, fundierte Erfahrung mit Data Analytics – specialist in data processing, ***profound experience*** in the field of data processing and analysis; Servicetechniker (m/f) Standort HamburgHYDRO Systems, mindestens 3 Jahre einschlägige Berufserfahrung – technician (m/f) to service the machines, at least three years of ***relevant professional experience*** (http://stellenanzeige.jobpilot.de/); Business Development Manager (f/m), einschlägige Projektleitungserfahrungen - manager (m/f) in business development, ***relevant experience*** in project management; Rezeptentwickler / Koch (m/f), Fundierte Erfahrung bei der Entwicklung von neuen und Weiterentwicklung von bestehenden Rezepten der deutschen und internationalen Küche – recipe developer (m/f), ***profound experience*** in developing new and improving existing recipes of German and international dishes; **Bustouristik/Reiseveranstaltung** eine/n Reiseverkehrsveranstalter/in erste Praxiserfahrung – specialist (m/f) in the field bus tourism /travel organizer, ***primary practical experience***; Außendienstmitarbeiter/in mehrjährige fundierte Berufserfahrung – sales representative (m/f), years of ***thorough experience***; Lehrer (Deutsch als Fremdsprache / Zweitsprache) (m/f) erste Praxiserfahrung – teacher of German as a foreign /second language (m/f), ***primary practical experience*** (http://stellenanzeige.jobpilot.de/).

Thus, in this case, there are more opportunities for job seekers in the field of employment, compared with the Russian vacancy database (of course, with the appropriate diploma, certificate and other evidence of qualification) because it is possible to get a job, having only in-study practices mentioned as experience.

The analysis of various legal sources has shown that the concept of practice/internship during training is of great importance in German labour law. In general, the possibility of obtaining professional experience during training is given great attention to. The Social code and other legislative sources present the concept of *Werkstudent*, there is no equivalent in the Russian or English language. It means literally a student working in the field of study, a *student who is both studying and working to gain* practical experience [14]. According to the Social code, they have pension insurance under the General rules but have no necessity to pay for health insurance until 25 years. At the same time, the weekly working hours include not more than 20 hours per week during the lecture period of the semester [15].

For employers, such students are an opportunity to find new staff, train them for work in their enterprises, as well as save on social contributions. That is why many large enterprises in Germany take *Werkstudenten*.

The very concept of "practice" within the framework of the considered "College/University-production" relationship means mastering the acquired knowledge and skills in a certain period of time or obtaining the necessary knowledge in order to apply them in the enterprise. A trainee in Germany is a person who undertakes a contractual obligation to enter into a professional activity within a certain period of time in order to acquire practical knowledge and skills for his/her professional activity. This time we consider vocational training in relation to the Act on education or any other comparable practical training (see Minimum wage act § 22 Abs. 1 P. 3 MiLoG [16]); however, the Intern does not have the status of an employee (see: materials for judicial practice of the Federal labor court) [17].

The practices in Germany are a part of the professional development assistance programme developed by the Federal labor Agency in accordance with the Social Code.

3. Conclusion

Thus, in this case we are talking about the possibility of the so-called "accumulation" of professional experience. Here the following practices are taken into account: various practices or internships in a foreign country (if any), work during studying (so-called "working students"), or mini-work, course work, including applied materials, public work, social work, work as support staff in research laboratories at the University [18]. In addition, even social networks (an own blog, an Instagram profile, etc.) can be taken into account as professional experience in certain cases [19].

In our opinion, the Russian labour legislation and professional educational standards do not pay due attention to the system of practices. As it is noted by the graduates of Russian universities, it was the practice/internship during their studies that contributed to their further rapid employment [20].

The comparative characteristics of the concept of "professional experience" or "work experience in the field of study" indicate a more detailed representation of this concept in the Labor law of Germany. We believe that the Russian labour law should clarify the concept of professional experience at the legislative level, and perhaps take into account the allocation of different types (levels) of professional experience represented in German labour law.

As possible ways of solving the problem of presence/absence of experience when employing graduates (young professionals), we propose the following o the system of higher education:

1. to assign the status of official professional activity (as a kind of primary professional experience) to various types of practices (in-study), foreign internships, courses, final qualifying works, significant social projects of the all-Russian and international levels.
2. to officially establish the applied nature of the listed activities as a requirement.
3. to develop mandatory criteria for the professional selection of practices.
4. to establish the obligatory applied character of all subject programs in the block of professional disciplines.

The proposed measures will, to some extent, address the problem of the availability/lack of experience to obtain employment and increase the employment opportunities for young professionals.

In the framework of the considered common problem of professional training production personnel practices/internship should be given priority. In order to develop and widely implement practices, it is necessary to solve the tasks in a single unity, in close cooperation of universities, regional authorities, the production potential of the region. After all, practices in Russian universities are often formal and do not go beyond the University; practice databases are not fixed, or their selection is quite random.

Thus, it is the consistent implementation of the system of in-study practices that will contribute to the effectiveness of professional training of production personnel and closer interaction between a College/University and a production department/organization.

References

[1] https://volgograd.superjob.ru/; https://www.avito.ru/volgograd; https://www.stellenanzeigen.de/job/; http://stellenanzeige.jobpilot.de/

[2] N L Shamne, E V Shishkina 2017 German popular scientific medical online media: structural and functional aspects *Vestnik Volgogradskogo Gosudarstvennogo Universiteta* **16 (2)** 143-151

[3] Career.ru. Russian companies have become less interested in young professionals [Electronic resource] – Access mode: https://career.ru/article/21007

[4] Superjob Research center: 6 out of 10 internships end with employment [Electronic resource] – Access mode: https://www.superjob.ru/research/articles/112085/6-iz-10-stazhirovok-zakanchivayutsya-trudoustrojstvom/

[5] Labour code of the Russian Federation of 30.12.2001 N 197-FZ (as amended on 03.07.2016) (rev. and ext., put in force on 01.01.2017) [Electronic resource]. – Access mode: http://www.consultant.ru/document/cons_doc_LAW_34683/f9ccec223c774c4895b03311bcd7eb355ef9d78f/

[6] Resolution of Supreme Court Plenum of the Russian Federation of 17.03.2004 N 2 (ed. 24.11.2015) "On application of the Labor Code of the Russian Federation by courts of the Russian Federation" conclusion of the employment contract [Electronic resource] – Access mode: http://www.consultant.ru/document/cons_doc_LAW_47257/307b9638d24d24fa83f0937c6a7f80ffd1a4baa5/

[7] Efremova T F 2000 New dictionary of the Russian language. *Explanatory and derivational* **2**

[8] Ergunova O, Lizunkov V G, Malushko E Yu, Marchuk V I, Ignatenko A Yu 2016 Forming

system of strategic innovation management at high-tech engineering enterprises *IOP Conference Series-Materials Science and Engineering* **177**

[9] Labour code of the Russian Federation of 30.12.2001 N 197-FZ (as amended on 03.07.2016) (rev. and ext., put in force on 01.01.2017) [Electronic resource]. – Access mode: http://www.consultant.ru/document/cons_doc_LAW_34683/cc065f088bd92d312b5fae714377b25f557ac44a/

[10] Resolution of the RF Government dated 15.07.2013 N 598 (ed. 02.08.2017, rev. 12.10.2017) "On the Federal target program "Sustainable development of rural areas for 2014-2017 and for the period up to 2020" [Electronic resource] – Access mode: http://www.garant.ru/products/ipo/prime/doc/70319016/

[11] Law of Moscow dated 20.06.2001, No. 25 (ed. 24.06.2015) "On the development of education in Moscow" [Electronic resource] – Access mode: http://docs.cntd.ru/document/3627457

[12] Berufserfahrung: Einschlägig, fundiert, erwünscht – Das solltest du wissen. [Electronic resource.] – Access mode: https://www.absolventa.de/karriereguide/berufseinsteiger-wissen/berufserfahrung-fundierte-einschlaegige

[13] Bundesagentur für Arbeit. Statistik [Electronic resource]. – Access mode: https://statistik.arbeitsagentur.de/Navigation/Footer/Top-Produkte/Fachkraefteengpassanalyse-Nav.html)

[14] Werkstudent Gabler Wirtschaftlexikon [Electronic resource]. – Access mode: http://wirtschaftslexikon.gabler.de/Archiv/85658/werkstudent-v7.html

[15] Murugova E V 2017 Theoretical and applied aspects of corpus based researches *Vestnik Volgogradskogo Gosudarstvennogo Universiteta* **16 (2)** 197-200

[16] Mindestlohnregelung für (echte) Praktika nach § 22 Abs. 1 Satz 2 Mindestlohngesetz (MiLoG) [Electronic resource]. – Access mode: https://www.gesetze-im-internet.de/milog/BJNR134810014.html

[17] Bundesarbeitsgericht, Urteil vom 5. August 1965, 2 AZR 439/64 Electronic resource] Access mode: https://dejure.org/dienste/vernetzung/rechtsprechung?Gericht=BAG&Datum

[18] Bender Ute, Müller Sabine 2010 Berufserfahrung in Deutschland - mit PFIFF : Leitfaden für ein erfolgreiches Praktikum / Bonn [u.a.] : (InWEnt, Internationale Weiterbildung und Entwicklung GmbH)71

[19] Staufenbiel Institut: Erste Berufserfahrung: Was bedeutet das für Bewerber? [Electronic resource.] – Access mode: https://www.staufenbiel.de/magazin/jobsuche/erste-berufserfahrung.html

[20] Suzdalova M A, Lizunkov V G, Malushko E Yu, Sytina N A, Medvedev V E 2017 Innovative Forms of Partnership in Development and Implementation of University-Business Cooperation *The European Proceedings of Social & Behavioural Sciences EpSBS* **XIX** 450-455

European trends in providing food safety in training of technology students in conditions of the development of digital technologies

N S Limareva, T S Shaltumaev, T V Shchedrina, V N Orobinskaya

FSAEI HE (Branch) "North Caucasus Federal University" in Pyatigorsk, the city of Pyatigorsk

orobinskaya.val@yandex.ru

Abstract. The article deals with the problems associated with the use of European experience in ensuring the safety of consumers of food products. The need to ensure the safety of consumers is an urgent modern task for the food industry. Search and analytical work were carried out to identify the main regulatory documents regulating the quality and safety of products and services in catering enterprises in the Russian Federation and abroad. The article presents some results of Erasmus+ Jean Monnet Module grant realisation on development of educational module "Food Safety Regulation in the European Union" for students of 19.03.04 "Production Technology and Organisation of Public Catering". The module includes the study of some aspects of the European and Customs Union legislation in the field of food safety. The necessity of training future technologists of public catering with the use of distance education technologies is substantiated.

Introduction

The new direction outlined by the President of the Russian Federation Vladimir Putin in his address to the Federal Assembly on December 4, 2014, the National Technology Initiative (NTI) FoodNet (system of personal production and delivery of food and water) is one of the priorities of state policy. When forming the FoodNet roadmap, special attention is paid to ensuring the quality and safety of food raw materials and food products.

One of the fundamental tasks of this direction is to provide consumers with high-quality food products. The lack of systematic control in this area of production associated with the presence of a large number of small businesses and small forms of innovative enterprises in the Russian Federation, leading to negative consequences for the consumer and for the food producer. [1]

For this reason, much attention is paid to food safety.

Safety control in the EU countries is regulated by national sanitary and hygienic standards. Taking into account the globalisation processes in order to improve mutual understanding, simplify procedures in the field of management and trade in the world community, it is necessary to develop a generally accepted model of the food security system. [2]

The study of international experience in the implementation of food technology quality systems based on the principles of HACCP (Hazard Analysis Critical Control Point), to create a modern corporate culture of production and sales of products, not only leads to improved food safety for human health, but also to a significant increase in the efficiency and productivity of processes. [3,5]

In order to ensure a unified approach to assessing the implementation of risk analysis at all stages of the life cycle of food products for manufacturers, distributors and persons performing the functions of foreign manufacturers of food products in Russia, standards regulating the safety of food products, similar to international standards ISO 22000 are developed. [4]

Methods and discussion

Search and analytical work were carried out to identify the main regulatory documents regulating the quality and safety of products and services in catering enterprises in the Russian Federation and abroad. (table 1)

Table 1.The main regulatory documents regulating the quality and safety of products and services in catering

The name of the document	Brief description	Application
GOST R ISO 22000-2007 "Food Safety Management Systems. Requirements for Organisations Involved in the Food Chain". [6]	This standard is identical to the international standard ISO 22000: 2005 "Food safety management systems - Requirements for any organization in the food chain"). [6]	This standard is intended for organisations (enterprises) in the field of food industry, developing and implementing a system of management of food safety based on the requirements of GOST R ISO 22000. [7]
GOST R 54762-2011 / ISO / TS22002-1:2009 "Prerequisite programs on food safety - Part 1: Food manufacturing". [8]	This standard is identical to the international document ISO / TS 22002-1:2009 "Prerequisite programs on food safety - Part 1: Food manufacturing". [8]	Establishes requirements for organisations that are involved in the chain of food production. [8]
ISO 22000: 2018 "Food safety management systems. Requirements for all organisations in the food production and consumption chain". [9]	A new version of the standard was developed by the International Organization for Standardization ISO instead of ISO 22000:2005. The transition period for enterprises certified according to the international standard is 3 years from the date of its publication. [9]	As in the previous version of 2005, the requirements of the new document are system-wide and can be applied to all organisations in the food chain, regardless of their type, complexity and size. [9]
GOST R 53755-2009 - "Food safety management systems". Requirements for bodies providing audit and certification of food safety management systems" - analogue of ISO / TS 22003: 2007. [10]	This standard is identical to the international standard ISO / TU 22003: 2007 "Food safety management systems. ISO, carrying out audit and certification of food safety management systems" (ISO/TS 22003:2007 "Food safety management systems - requirements for bodies that carry out audit and certification of food safety management systems"). [10]	This standard is designed to assist in the certification of SMBPS that meet the requirements of ISO 22000 "Food safety management system. Requirements for all organisations in the food production and consumption chain". The content of this standard may also be used to assist in the certification of SMBS, which are based on other specified requirements. [10]
GOST R 51705.1-2001 "Quality Systems. Food quality management based on HACCP principles. General requirements" [11]	This standard establishes the basic requirements for the system of food quality and safety management based on the principles of HACCP - Hazard Analysis and Critical Control Points, as set out in the Directive of the Council of the European Community 93/43. [11]	This standard can be applied to all organisations, regardless of their size or complexity, that participate in the food production chain and intend to implement a program of mandatory measures for food safety in such a way as to meet the requirements set out in section 7 of ISO 22000:2005. [11]
GOST R 55889-2013. "Catering service. Food and catering safety management system" **Recommendations for use GOST R ISO 22000-2007 for the food industry.**	This standard is developed taking into account the main regulations of the British profile BIP 2127: 2007 "Catering services - development of recommendations for the use of ISO 22000 for the food industry", NEQ). [12]	This standard is intended for organisations (enterprises) in the field of the food industry, developing and implementing a system of food safety management based on the requirements of GOST R ISO 22000, as well as for

[12]		personnel responsible for the safety of food products. [12]
GOST R 57621-2017 "Trade Services. Sale of perishable food products through vending machines. Requirements" [13]	The standard establishes requirements for the sale of perishable food products through vending machines, including hygienic, which create conditions for the sale of perishable food products through vending machines by self-service in accordance with the rules of sale. The document for the first time defines the requirements for the operation, cleaning and disinfection of vending machines into which perishable food products are loaded, and also contains special requirements for cooling and storage of food products. [13]	The developed standard is **identical to the German national standard DIN 10527: 2004.** In order to select the best practices of standardisation in this area, a large amount of information was analysed, including European, German and English standards, and detailed requirements for trading services using vending machines were identified in the German version. [13]
GOST R 54762-2011. "Prerequisite programmes on food safety - Part 1: Food manufacturing"). **[14]**	This standard is identical to the international document **ISO/TS 22002-1:2009** "Prerequisite programmes on food safety - Part 1: Food manufacturing"). [14]	This standard establishes requirements for the development, implementation and enforcement of mandatory preliminary programs (PRP)* to assist in the management of food safety risks. [14]
GOST R 56746-2015/ISO/TS 22002-2:2013 "Prerequisite programmes on food safety - Part 2: Catering", IDT). [15]	This standard is identical to the international document ISO/TS 22002-2:2013 "Prerequisite programmes on food safety - Part 2: Catering", IDT). [15]	This standard does not duplicate the requirements of ISO 22000 and is intended to be used in conjunction with ISO 22000 in the development, implementation and maintenance of PRP programs for specific organisations, as well as for the control of basic hygiene requirements in the provision of catering services. [15]
GOST 30390-2013 "Catering service. Public catering products sold to the population. General specifications" [16]	This standard establishes general technical requirements for catering products sold to the public, including the classification of products, general requirements for the safety of catering products and safety procedures, acceptance rules, control methods, requirements for the sale of products, packaging, labelling, conformity assessment of catering products. [16]	The provisions of this standard are applied to products manufactured and sold by catering enterprises (food enterprises) of various forms of ownership and individual entrepreneurs. [16]
GOST 30389-2013. "Catering service. Public catering enterprises. Classification and general requirements". [17]	This standard establishes the general requirements and classification of enterprises (objects) of public catering of various types. [17]	This standard is applied to enterprises (objects) of catering legal entities and individual entrepreneurs. [17]
GOST 30524-2013. "Catering service. Personnel requirements".	This standard establishes general minimum requirements for personnel of catering establishments of different types of all	The provisions of this standard can be used in the selection and placement of personnel,

[18]	legal forms and forms of ownership. [18]	certification, qualification, development of job descriptions and standards of work of personnel of catering services. [18]
Methodical recommendations MP 5.1.00096-14. [19]	Methodical approaches to the organisation of assessment of processes of production (manufacturing) of food products on the basis of HACCP principles. [19]	The materials contain approaches to assess compliance with the requirements of sanitary legislation and technical regulations of the Customs Union for producers (manufacturers) of food products based on the principles of risk management and analysis. [19]
GOST 32692-2014. "Catering services. General requirements for methods and forms of service in catering". [20]	This standard establishes the general requirements to methods and forms of service at the enterprises (objects) of public catering. [20]	The standard is intended for use by legal entities and individual entrepreneurs in the field of catering services. [20]
GOST 32691-2014. "Catering services. The procedure for the development of brand and new dishes and products in catering" [21]	This standard establishes general requirements to the procedure of the development of brand and new dishes and products at the enterprises (objects) of public catering. [21]	The standard applies to the development of branded and new dishes and products manufactured by catering enterprises (objects) of various forms of ownership, organisational and management structure and individual entrepreneurs. [21]
GOST R 54607.3-2014. "Catering services. Methods of laboratory control of public catering products. Part 3. Methods of control of production of public catering observance of processes". [22]	This standard is applied to methods of monitoring of compliance with the processes of catering products manufacturing. [22]	This standard includes methods for controlling the quality of deep-fried fats, a method for determining the efficiency of heat treatment of meat and fish culinary products, a method for determining the mass fraction of residual sulfur dioxide in semi-finished products from sulfited raw peeled potatoes, a method for determining the content of eggs in culinary products. [22]
GOST R 55323-2012. "Catering services. Identification of catering products. Generalities" [23]	This standard is applied to products manufactured at the enterprises of public catering and sold in the rendering of services of public catering, and also at the enterprises of retail trade in the territory of the Russian Federation. [23]	This standard establishes the general provisions for the work and registration of the results of the identification of public catering products. [23]
GOST 31984-2012. "Catering services. General requirements" [24]	This standard establishes a list of catering services, general requirements for services, methods of assessing the quality of services. [24]	-
GOST 31989-2012. "Catering services. General requirements for primary food processing company of public catering". [25]	This standard establishes general requirements for primary food processing companies of catering. [25]	-

GOST 31986-2012. "Catering services. The method of organoleptic evaluation of the quality of food service products". [26]	This standard is applied to public catering products of mass production and establishes a method of organoleptic evaluation of its quality. [26]	-
GOST R 54609-2011. "Catering services. The nomenclature of indicators of quality of products of public catering". [27]	This standard establishes the nomenclature of indicators of quality of public catering production and the order of the choice of indicators of production quality. [27]	This standard is applied to products of public catering, manufactured by catering enterprises of various forms of ownership, organisational and management structure and individual entrepreneurs. [27]

The analysis of information on the creation and stages of formation of management systems and food safety of the world community is presented in table 2.

Table 2. Stages of creation of food safety and management system

Year	Organisation	Document	Purpose of adoption
1997	Codex Alimentarius Commission	ALINORM 97 /13A	Approval of the draft guidelines for the application of the HACCP system
1997	Codex Alimentarius Commission	ALINORM 97 /31A	Approval of the draft guidelines on the application of the HACCP system as analysed.
1997	Codex Alimentarius Commission	CAC / RCP 1-1969 , Rev . 3 (1997)	Establishing the basis for food hygiene, improving food safety through the use of an approach based on the principles of HACCP.
2001	International Organisation of Standardisation	ISO 15161:2001	Revision of the manual on the application of the HACCP system in order to detail the explanations given.
2003	Codex Alimentarius Commission	CAC / RCP 1-1969 , Rev . 4 (2003)	
2005	International Organisation of Standardisation	ISO 22000:2005	Model of food safety management system associated with hazards to the final customer at the time of consumption.
2005	International Organisation of Standardisation	ISO/TU 22004:2005	Publication of official guidelines for the application of ISO 22000: 2005

Work to create a new version of the standard in the field of food safety on the terms of integration of ISO 22000:2018 with other international standards for management systems, such as ISO 9001 "Quality Management Systems" and ISO 14001 "Environmental Management Systems" is carried out.

The new version of the standard includes 10 sections and introduces a new approach to risks as a vital concept for the food industry. The standard identifies risk at the operational level, applying HACCP principles and risk at the strategic level of the management system (business risk), by defining which the company will be able to find opportunities to achieve specific business goals. This ensures connection with the documents of the Codex Alimentarius Commission, which develops common international guidelines on food safety. [1]

Food safety involves the prevention, elimination of potential hazards that may arise at any time from the place of production and to the place of consumption of products. Since food safety hazards can be introduced at any stage of the process, each enterprise in the food supply chain must exercise appropriate risk control. In fact, food security can only be supported by the joint efforts of all parties involved: governments, producers, trade organisations, food enterprises and final customer.

In July 2018, the teachers of the Institute of Service, Tourism and Design of the North Caucasus Federal University (NCFU) won the European grant Erasmus+ Jean Monnet Module. Jean Monnet aims to increase knowledge of European integration processes through teaching, research and debate on topics related to the history, politics, economy and legislation of the European Union (EU), as well as EU relations with other regions of the world. Jean Monnet was launched in 1989 and it got its name in honour of the famous French statesman, one of the "architects" of the European Union [1]. The main goal of the program is to bring the European experience of education in the higher education system of the world. The participation of NCFU teachers in this project will give a real opportunity to develop the export potential of the Russian education system of European countries.

The Jean Monnet module is a curriculum or course in European Union research at a higher education institution. The duration of each module is at least 40 teaching hours per year. Training hours include not only the time of direct contact with the teacher in the form of traditional lectures, practical, laboratory classes and seminars, but also can be implemented in the format of distance learning. The modules can be focused on one particular discipline in European studies or be multidisciplinary. [2]

Erasmus + supports Jean Monnet modules with the aim of:

- promoting research and teaching experience for researchers, scientists and practitioners in matters relating to the European Union;
- collaborating in dissemination of research results;
- creating interest in the European Union;
- development of educational courses on a wide range of research activities in the European Union, relevant for graduates in their professional lives [2].

Module Jean Monnet reinforces and actualises teaching EU Affairs in the curriculum which still included the content associated with the EU only to a limited extent. They also provide facts and knowledge about the European Union to a wide range of students and those who are interested.

In the framework of the European grant of the September 2018 at the Department of Food Technology and Commodity Research of NCFU for students of direction of preparation 19.03.04 "Production Technology and Organisation of Public Catering" the educational module "Regulation of Food Safety in the European Union" is conducted. The developed module is focused on the formation of knowledge and skills to ensure food safety according to European standards for students, future specialists in the field of organisation of production and operation of catering. Implementation of the educational module enables students and teachers of NCFU to use the European experience in the field of food legislation, to master innovative methods of laboratory control of food safety of the European Union, which will improve the educational environment of the University through the introduction of innovative European learning technologies.

Anthropogenic factors of modern civilisation have led to a significant deterioration not only in the overall environmental situation in almost all regions of the world, but also to a change in the quality of food raw materials, food structure and food preferences of modern man. [3].

According to the Federal law "On Quality and Safety of Food Products" food safety is a state of reasonable confidence that food products under normal conditions of their use are not harmful and do not pose a threat to the health of present and future generations [4].

Safety of food raw materials and food products is evaluated by quantitative or qualitative content of anti-nutritional substances of microbiological, chemical and biological nature.

At an early stage of the European Union, the countries that made up it up had thousands of national food safety standards. EU legislators have been trying for many years to harmonise food safety standards. Currently, the attention of EU regulators is focused on food safety from the point of view of consumer health and protection of their interests in cases where there is a falsification, or the consumer cannot make an informed choice.

After the establishment of the Customs Union between the Russian Federation, Belarus and Kazakhstan, there was also a need for harmonisation of standards. The customs Union is based on compliance with the requirements that ensure quality control of the final product. Technical regulations and standards for the common economic space have been developed.

As for the European Union, it relies on preventive measures and minimisation of the risks associated with each process at each stage of the full food chain. The differences between the two systems have a significant impact on the content of legislation, including control, testing, labelling, approaches to food quality and the state's obligations to ensure safe food for consumers. The European Union is called upon to ensure the supply of safe food in the world and the standards by which these goals will be achieved.

The materials of the module compare the systems of the European and Customs Unions on food safety, which ensure the right of consumers to safe food all over the world.

The developed educational module includes the study of the following aspects of food legislation of the European and Customs Unions: requirements for food control and the method of its implementation, requirements for food quality, requirements for food labeling, new foods with related products and materials that require a special permit, food safety criteria and requirements for laboratory control.

The developed educational module is designed for 120 hours and consists of 3 disciplines, which include lectures and seminars:

1. "Food safety legislation". The discipline includes the following sections:
   - ✓ Comparative analysis of the legislation of the European Union and the Customs Union;
   - ✓ Trade networks in the world and technical regulations as an obstacle to the development of trade;
   - ✓ Management system of food control and the responsibility of government and industry.
2. "Food safety criteria and laboratory control". The discipline includes the following sections:
   - ✓ Laboratory control and food safety criteria in the European Union and the Customs Union;
   - ✓ Risk-based approach of the European Union food control system.
3. "Food additives and labelling for food safety." The discipline includes the following sections:
   - ✓ Requirements for food labelling;
   - ✓ Comparative analysis of the use of additives in food products in two systems.

The developed educational module is implemented in the educational process of the FSAEI HE “The North Caucasus Federal University” in Pyatigorsk, on the basis of modern educational technologies. Currently, the share of resource-saving technologies in education is insignificant and their implementation in the educational process is an urgent task.

Distance learning is considered from the perspective of resource saving [5]. The introduction and use of distance learning systems within the framework of resource saving can be determined from the perspective of positive aspects, as it allows to:

- ✓ reduce the cost of training (no rental costs, travel to the place of study of both students and teachers, etc.).);
- ✓ reduce training time (collection, travel time);

- ✓ the participant is free to plan the time, place and duration of classes;
- ✓ conduct training of a large number of people;
- ✓ improve the quality of education through the use of modern tools, large electronic libraries, etc.
- ✓ create a single educational environment (especially relevant for corporate training) [5].

Hypertext technology and multimedia are used to organise individual educational trajectory of students. The distance course of the educational module "Regulation of Food Safety in the European Union" is based on the requirements for the structure of distance learning courses. In addition to theoretical materials, tasks for practical work and control issues, the course includes reference materials, video materials and video lectures. Completed practical tasks are sent and evaluated based on the established schedule. The course provides a system of forums for group communication and webinars to discuss various issues with the teacher.

Conclusion

Thus, the study of the European system of monitoring and safety requirements at all stages of the food chain from raw materials to consumers using digital technologies can be considered as a means of improving the level of knowledge of Russian students to ensure food security and the real prospect of increasing the competitive advantages of domestic food production enterprises.

Acknowledgments

The reported research was funded by within the grant Erasmus+ Jean Monnet Mdule (600244-EPP-1-2018-1-RU-EPPJMO-MODULE "Food Safety Regulation in the European Union").

References

[1] N S Valgina, 2001 “Active Processes in Modern Russian Language: textbook for University students,” (M.: Logos) 304 p
[2] Versan V G 2007 Food Safety: from HACCP to ISO 22000 *Certification* **3** 32-34
[3] Erasmus+Programme Guide Electronic resource http://ec.europa.eu/programmes/erasmus-plus/programme-guide/part-b/three-key-actions/jean-monnet-activities/modules_en
[4] Donchenko L V, Nadykta V D 2006 Food Safety (Krasnodar: KSAU) 371 p
[5] Federal law of January 2, 2000 N 29-FZ "On quality and safety of food products" (2018)
[6] Konopko E A 2018 E-learning system as an element of resource-saving technologies Informatization of education and methods of e-learning *Proceedings of the II International Scientific Conference* 183-187
[7] GOST R ISO 22000-2007 Food safety management systems. Requirements for organisations involved in the food chain
[8] GOST R 55889-2013 Catering Services. System of management of safety of public catering products. Recommendations for the use of GOST R ISO 22000-2007 for the food industry
[9] GOST R 54762-2011 Prerequisite programs on food safety - Part 1: Food manufacturing
[10] ISO 22000: 2018 Food safety management systems. Requirements for all organisations in the food production and consumption chain
[11] GOST R 53755-2009 Food safety management systems. Requirements for bodies carrying out audit and certification of food safety management systems
[12] GOST R 51705.1-2001 Quality Systems. Food quality management based on HACCP principles. General requirements
[13] GOST R 55889-2013 Catering services. System of management of safety of public catering products
[14] GOST R 57621-2017 Trade Services. Sale of perishable food products through vending machines. Requirements
[15] GOST R 54762-2011 Prerequisite programs on food safety - Part 1 Food manufacturing
[16] GOST R 56746-2015/ISO/TS 22002-2:2013 Prerequisite programs on food safety Part 1 Part 2

Food manufacturing
[17] GOST 30390-2013 Catering Services. Public catering products sold to the population. General specifications
[18] GOST 30389-2013 Catering service Public catering enterprises. Classification and General requirements
[19] GOST 30524-2013 Catering service. Personnel requirements
[20] GOST 32692-2014 Catering service. General requirements for methods and forms of service in catering
[21] GOST 32691-2014 Catering service The procedure for the development of brand and new dishes and products in catering
[22] GOST R 54607.3-2014 Catering services. Methods of laboratory control of public catering products. Part 3. Methods of control of observance of processes of production of public catering
[23] GOST R 55323-2012 Catering services. Identification of food service products. Generalities
[24] GOST 31984-2012 Catering services. General requirements
[25] GOST 31989-2012 Catering services. General requirements for primary food processing companies of public catering
[26] GOST 31986-2012 Catering services. The method of organoleptic evaluation of the quality of food service products
[27] GOST R 54609-2011 Catering services. The nomenclature of indicators of quality of products of public catering

IOP Conf. Series: Materials Science and Engineering **483** (2019) 012101 doi:10.1088/1757-899X/483/1/012101

The Schengen Information System: legal support for the automated cross-border business management

A O Inshakova[1], S Y Kochetkova[2], E A Serbina[3]

[1] Volgograd State University, the Department of Civil and International Private Law, prosp. Universitetsky, 100, 400062, Volgograd, Russia. ORCID 0000-0001-8255-8160

[2] Volgograd State University, the Department of Germanic and Romanic Philology, prosp. Universitetsky, 100, 400062, Volgograd, Russia.

[3] Volgograd State University, the Department of Civil and International Private Law, prosp. Universitetsky, 100, 400062, Volgograd, Russia. ORCID 0000-0002-2366-4446

E-mail: gimchp@volsu.ru .

Abstract. This study examines the impact of the development of the information and communication technologies on the social, economic and political systems of the member countries of the united Europe, as well as the countries entering into business and production relations with them. The authors reveal the issues of using the Internet as a tool of the loyal border management of the united countries, ensuring the liberalization of the movement of factors of production through the creation of a single space of freedom, security and justice. The infocommunication technologies are researched as a management mechanism with regard to their influence on modern geopolitical, economic, and specifically industrial processes. The indicated approach is implemented simultaneously both as the management of the content of political relations and the management of their form. The Schengen border management information system is regarded as one of the most striking examples of the information technologies use in the political, economic, and specifically industrial sphere. The structure and legal mechanisms of this system operation are analyzed, on the basis of which the advantages and disadvantages of technological and legal nature are revealed. Special attention is paid to the analysis of the draft laws under development aimed at the improvement of the Schengen information system as an electronic technology for the automated border management of the pan-European space of freedom, security and justice. Among other things the authors have taken into account the latest changes in the legal regulation of the Schengen information system, which took place in October 2018.Taking into account the identified shortcomings in the functioning of the Schengen information system, the recommendations for their elimination are formulated. It is concluded that it is necessary to create a single European information portal in order to optimize the processing of data received from the different agencies. The general trends of the SIS II reform are the simplification and acceleration of the procedures, as well as the strengthening of the personal data protection.

1. Introduction

Production as a process of creating economic wealth and accompanying resource conversion has been experiencing a real revolution in recent decades. The growth of high-tech industries, the increase in gross expenditures on software and hardware, communication equipment and related services, the increase in the government funding of research in nanotechnologies are becoming a stable trend in the development of the economic system of society. Of course, the processes aimed at building and developing the telecommunications infrastructure of the industry and business cannot but have an impact on the main production factors. It should be emphasized, however, that three of the four factors of production - labor, entrepreneurship and capital – are cross-border.

Work as the most important factor of production is an appropriate, conscious human activity aimed at meeting the needs of the individual and society. The market of the subjects of labor activity possessing a set of personal qualities, knowledge, skills, physical forces applied in the industry becomes unusually competitive environment. With the development of integration relations in the economy, a person gets the opportunity to realize himself in the industry (both as an employee and as an entrepreneur) not only on the territory of his country, but also abroad, if this state seems to him more attractive in terms of technology development, creation of conditions for work and creativity and other factors. Cross-border movement of people, as well as interethnic interaction of business entities invariably generates cross-border movement of capital. In this connection, the management of migration and financial flows has a direct impact on the intensification and efficiency of production processes.

The integration activities of international organizations, which have been actively developed in recent decades, are aimed at the liberalization of social relations and all factors of production. The interstate communication in the sphere of production in the context of globalization of the world economy involves the free movement of financial and human capital. One of the manifestations of such liberalization is the reform of the management of migration flows in the direction of simplification and acceleration of the administrative procedures.

In view of the increasing development of the information and communication technologies and communication systems in recent decades, particularly, in the light of the ubiquity of the broadband Internet access, mankind has acquired the tremendous opportunities for communication, as well as for the receipt, dissemination and transmission of information. The information culture has received the status of the most important factor and integral indicator of the society development. [1] In virtual communication, the need for a personal contact with the interlocutor disappears, which can significantly reduce the time of exchange of significant information and neglect the distance between the subjects of communication. Of course, such a powerful instrument of interaction could not be ignored by the political systems of the world.

The information and telecommunication network "Internet" is characterized by the cross-border nature. At the same time, the clearly defined state borders, which are subject to the sovereign jurisdiction of the state power, are being eroded. In this regard, the actions of the subjects of virtual communication should be considered through the prism of international law. [2] The modern political scientists note that the political stability of the current political systems becomes directly dependent on the infocommunication potential of the socio-political relations and the effectiveness of its implementation by the state and civil subjects in a current political practice. [3] At the same time, to date, the main focus of the subjects of interaction is transferred not so much to the transmitted information, but to the development and use of the accessible, functional mechanisms, as well as the mass communication channels.

The impact of the information and communication technologies on modern geopolitical processes is reduced to the two main areas: the management of the content of political relations and the management of their form.

The management of the content of political relations with the help of the information and communication technologies is actively expressed in the western doctrine. In particular, Ronald Deibert emphasizes that any changes in the methods of communication lead to the global consequences in the form of redistribution of power in society, the changes in individual and social consciousness, and the axiological changes. [4]

It should be noted that by the information management of the form of political relations the authors mean the creation of new public institutions, models and mechanisms of interaction between the government and society, which operate on the basis of the modern high-tech means of communication. This view correlates with some doctrinal statements. Thus, L. Smorgunov expresses the idea that the possibilities of information and communication technologies are now understood much wider than before. While the concept of e-government, i.e. a static state in which the government information is open and accessible and public services are provided online, was initially dominant in public administration, the idea of e-government is now embedded in electronic interaction. Thus, there is an extension of the form and sphere of influence of citizens on the process of making and implementing political decisions. [5] In other words, the emphasis of the dominant subject in the public legal relationship shifts from the state body to the citizen.

The undoubted positive aspects of such management are that the electronic administration in the process of public administration significantly increases the speed of feedback between the public authorities and citizens, significantly reduces the financial costs associated with the maintenance of bureaucracy, which is replaced by an automated system; paperwork disappears, as well as the range of entities that have the opportunity to interact with the public authorities (in particular, we are talking about persons with disabilities) is enlarged. [6]

One of the clearest examples of the implementation of e-government is the European Union. The analysis of the European Union's policy on research and technology shows that providing an innovative way of economic development is only possible through the creation of new innovative institutions with active state participation. [7]A distinctive feature of most European programs and the strategies for their implementation is that all the used online technologies – e-commerce, e-government, e-business, etc. - are not considered as isolated activity areas, but as a global environment, which is the basis for transition to the digital economy and the information society. [8]

2. Research

The Schengen information system (SIS), which emerged on March 26, 1995, is a unique phenomenon of the information society. It is the Unified Information System, whose main material and technological base, including the main data bank, is located in Strasbourg. [9] Having passed 3 stages in 23 years of its existence (SIS, SIS1+, SIS II), the Schengen information system performs the most important function of the legal border management of the European Union, maintaining the balance between the interests of the national security and the respect for human rights. The Schengen rules provide for the abolition of border and customs control at the internal borders between its member countries, introduce the unified rules of entry/exit for all external borders of the contracting countries, declare the cooperation of the law enforcement agencies to combat the international crime, as well as introduce the unified requirements of the rules of extradition and registration of all border crossings in the database. Being organically integrated into the centralized information system, including such elements as VIS, EURODAC, ECRIS, the Schengen information system is attentive to data protection. This is expressly provided for in article 102 of the Schengen agreement, which expressly prohibits copying data from one national section of the SIS to other national files (paragraph 2) except in cases of prevention of the highest serious threat to public order and security (paragraph 3).

The mechanism of functioning this system is as follows. The elements of the Schengen information system are the national sections. Each contracting party shall designate a specialized body to act as an information aggregator, whose officials shall be responsible for the proper functioning of the national section. (art. 108, para. 1-2). The interaction of aggregators is carried out through the depository. The access of persons to the information contained in the SIS is controlled by the national supervisory authority, which determines the circle of persons and the way of the information transmission. They are united by the international monitoring body – the European Agency for the management of the large-scale IT systems in the field of freedom, security and justice - consisting of two representatives from each national monitoring body and headed by the European data protection inspector. [10] The Control is carried out in the following areas: the control over the device access, database, data input, use, access, and transport.

Any person has the right to request verification of the data concerning him, to correct the factual inaccuracies about him or her and to initiate proceedings before a court for the purpose of correcting, destroying or sending up a report about him or receiving compensation in this regard. The damage caused to a person in connection with the use of the national SIS data file is borne by the specific information aggregator. The period for information storage depends on the specific type of data and can range from 1 to 10 years.

IOP Conf. Series: Materials Science and Engineering **483** (2019) 012101 doi:10.1088/1757-899X/483/1/012101

The development of the information and communication technologies makes it possible to improve the quality of interaction between the services that ensure the security of European countries by storing a new type of data-biometric information (fingerprints registered through the automatic fingerprint identification system (AFIS), photos for visas and documents). The reform that the Schengen information system underwent in the spring of 2013 also made it possible to provide the intelligence services of the contracting countries with information on hijacked aircrafts, ships, stolen containers and banknotes, arrest warrants, residence permits and means of payment.

Despite these innovative changes, the literature notes the weakness of SIS II and its inability to fully meet its challenges. [11]

First of all, the freedom of movement within the Schengen States enjoyed by the European citizens has worsened the terrorist situation, as the practice has shown. [12] In our opinion, the measures taken by the leadership of the European Union, in particular, the conduct of random checks of the EU citizens on certain categories of databases, are very resource-intensive and ineffective, since the full coverage of persons who are potentially dangerous to the terrorist situation is impossible. A vivid proof of this is the formation of the areas inhabited by former refugees from the Eastern countries, who do not want to assimilate and pose a threat to the national security because of the opposite radical European ideological attitudes.

In addition, the information management architecture for the borders and security is fragmented, because the information is stored separately in the unrelated systems: the EES, VIS, ETIAS, EURODAC, SIS II, ECRIS-TCN, SIRENE and EUROPOL. [13] In this regard, the end – users – border guards, police, customs officials, judges, consulates, as well as offices responsible for vehicle registration – in some cases are unable to quickly access the information that is necessary in their professional activities to ensure security.[14] At the same time, the certain restrictions related to the competence of the office should certainly be retained in order to prevent the right abuse.

We see the solution to this problem in the integration of all the above databases. The integration processes have led to the emergence of an initiative to create a unified European Search Portal (ESP). In our opinion, the concept of a platform for sending requests to the existing information resources is very convenient, because it allows you to get all the necessary information in a single window. At the same time, the radical restructuring of the formed information systems and, as a result, serious financial costs are not required, and the data security remains protected.

Thus, the creation of the ESP is a way out of the current situation of the information isolation within the various departments and in the future will allow the competent authorities and officials more effectively and in a shorter time to carry out the documents verification through a single search and obtain the results from all the systems to which they have the right of access.[15] This will simplify the immigration and asylum procedure, since the distortion of the identity data will be much more complicated.

A new round of the gradual reform of the Schengen information system began in April 2016, when the imperfection of SIS II after three years of operation was officially confirmed. In December 2016, the European Commission officially presented a legislative package aimed at improving the SIS system from a technical point of view and a number of measures to counter the development of some forms of serious crimes, including terrorism. [16] Since then, there has been an active legislative process. After years of refinement and approval, The General data protection regulation (GDPR) came into force on May 2, 2018. In general, this regulation is intended to increase the level of control over the personal data of users and legal entities. The right of the user and the company to information about what data are available on the network and how this information is used. Although this is a general law, it also applies to the SIS.

On the technical side, the development of SIS II can be traced in three main areas: the maximization of benefits of the functioning information systems, the development of new additional systems for the purpose of dealing with the gaps in information and, finally, the convergence of the information systems and increasing their interoperability. [17]

A number of important changes were discussed in Brussels on June 12 this year. The draft reform contains very innovative developments, such as the inclusion of a DNA profile to facilitate the identification of missing persons in cases where fingerprint or photo data are not available or suitable for the identification. [16]

The draft regulation introduces additional information categories in SIS II. The warnings are issued for the investigative purposes and are an intermediate step to a preliminary investigation that allows people to be interviewed. The alerts also help to prevent the abduction of children by one of the parents as well as children and persons in a helpless state, in respect of whom there is a risk of deprivation of freedom of movement (for example, in cases where the exit might lead to the conclusion of a forced marriage, human trafficking, etc.). It will also be possible to send alerts about unknown suspects and wanted persons, which include the introduction into SIS II the fingerprints or palms of a perpetrator found at the scene of a crime or terrorist act.

In addition, it is planned to include a new European service, the border and coast guard agency, in the circle of entities with the access to the SIS II data within its competence. It is also planned to open the access to all categories of data and exchange additional information for Europol. All Schengen countries will be obliged to inform Europol of any attacks where a person is wanted in connection with a terrorist offence.

It is also necessary to note the problems of protection of personal data from distortion, theft and illegal use. Considering the problem a priority, the European Commission supports the development of informatization in the sphere of control not only through the legislation, but also financially. Thus, 1.7 million euros were allocated to co-finance the training of data protection specialists in 2018, another 2 million-to support the national data protection authorities. [18]

The latest changes in the legal regulation of the Schengen information system took place on October 24, 2018. The European Parliament adopted a law that introduces a new type of notification-the notification of the decisions of the national authorities in respect of foreign citizens illegally staying on the territory of the European Union. This innovation is certainly important because the lack of inter-state cooperation on this issue has allowed third-country nationals against whom the extradition or deportation was ordered to move to another European state and to avoid the enforcement of such a decision. [19]

3. Results

As a result of the study, it can be concluded that the development of modern political management methods is largely associated with the development of the information and communication technologies and communication systems. The way in which the information is transmitted, the circle of subjects and the speed with which it is transmitted, may determine the adoption of the most important state decisions that shape its policy.

The electronic administration on the example of the Schengen information system has proved its convenience and efficiency. A well-designed three-stage system of interaction of the elements of SIS II allows processing and storing a significant amount of information and performing the functions related to the security of the member states of the Schengen agreement in a short time.

Despite a number of advantages of SIS II, it is impossible not to note the shortcomings of this system, which were particularly clearly revealed in the period from 2015 to 2017. The main ones are the lack of interconnection between different European information systems and the complication of access to the information stored in different databases of the intelligence services.

The solution to this problem is seen in the development of a unified European search portal, which is the link for all the European information systems. The result of the creation of such a portal will be a significant reduction in the time to process significant amounts of information.

Currently, the Schengen information system is one of the most frequently used information systems. The statistics show more than 5 billion uses by the member states, with more than 240,000 foreign alerts sent and viewed. [20] Therefore, the reform of SIS II should take place in the direction of simplification and acceleration of the procedures, as well as in the direction of strengthening the protection of personal data.

References

[1] Drescher U N, Sultanova E R 2017 The information culture of society as a major factor and the integral indicator of its development *Vestnik KazGUKI* **2** 24-6

[2] Belovatykh A V 2014 The information factor in the dynamics of modern geopolitical processes *Sravnitelnaja politika* **4 (17)** 64-6

[3] Volodenkov S V 2018 The role of information and communication technologies in modern politics *Nauchnij ezhegodnik IFIP Uro RAN* **2** pp 69-86

[4] Deibert R J 1997 *Parchment, Printing and Hypermedia: Communications in World Order Transformation* (New York) 334

[5] Smorgunov L V *E-government, knowledge management and administrative reforms* [Electronic resource]. – Available at: http:// politex.info/content/view/59/40 (reference date 23.10.2018)

[6] Alekseeva L N The transformation of public administration under the influence of information technologies: essence, forms, vectors *The journal "U". Economika. Upravlenije. Finansy* **1** 13-27

[7] Vasilenko N D 2013 The innovation policy of the EU: The theoretical and legal aspect of technoparks and technopolises activities *Vestnik Volgu. Serija 5. Jurisprudentsija* **2(19)** 166-73

[8] Petrova E A 2007 The foreign experience of informatization and the peculiarities of its implementation in Russia *Fundamentalnyje issledovanija* **11** 31-5

[9] Inshakova A O, Kiseleva E V 2017 The Schengen information system. *European Union law: tutorial and workshop for bachelor's and master's degree program* (Moscow) 341-5

[10] *Regulation (EU) No 1077/2011 of the European Parliament and of the Council of 25 October 2011 establishing a European Agency for the operational management of large-scale IT systems in the area of freedom, security and justice//* EUR-lex. URL: https://eur-lex.europa.eu/legal-content/EN/TXT/?uri=celex:32011R1077 (reference date: 28.10.2018)

[11] Nigmatullin R V 2016 On combating the threats and challenges from the perspective of modern international law *Juridicheskij mir* **11** p 56

[12] Proposal for a Regulation of the European Parliament and of the Council on establishing a framework for interoperability between EU information systems (police and judicial cooperation, asylum and migration). Date of document: 12/12/2017)// EUR-lex. URL: https://eur-lex.europa.eu/legal-content/EN/TXT/?qid=1522305554245&uri=CELEX:52017PC0794 (reference date: 28.10.2018)

[13] Inshakova A O 2012 The European area of freedom, security and justice: managing the boundaries *Nauka i obrazovanije khozjaistvo i economika predprinimatelstvo pravo i upravlenije* **8(27)** 38-45

[14] *The Council on establishing a framework for interoperability between EU information systems (borders and visa) and amend Council Decision 2004/512/EC, Regulation (EC) No 767/2008, Council Decision 2008/633/JHA, Regulation (EU) 2016/399 and Regulation (EU) 2017/2226 and Proposal for a regulation of the European Parliament and the Council on establishing a framework for interoperability between EU information systems (police and judicial cooperation, asylum and migration)//*EUR-Lex. URL: http://data.consilium.europa.eu/doc/document/ST-15729-2017-ADD-3/EN/pdf (reference date: 28.10.2018).

[15] *Communication from the commission to the Parliament and the council "Adapting the common visa policy to new challenges".* Date of document: 14/03/2018// EUR-lex. https://eur-lex.europa.eu/legal-content/EN/ALL/?uri=CELEX:52018DC0251 (reference date: 28.10.2018)

[16] Cerdeira V H, Sadet R 12/06/2018 *Schengen information system: agreement between the Council Presidency and the European Parliament* URL: https://www.consilium.europa.eu/en/press/press-releases/2018/06/12/schengen-information-system-agreement-between-the-council-presidency-and-the-european-parliament/ (reference date 28.10.2018)

[17] *Communication from the Comission to the European Parliament, the European Council and the Council twelve progress report towards an effective and genuine Security Union.* COM/2017/ 0779 final/ / EUR-Lex. URL: http://eur-lex.europa.eu/legal-content/EN/TXT/?qid=1522850887595&uri=CELEX:52017DC0779 (reference date: 28.10.2018).
[18] *Statement by Vice-President Ansip and Commissioner Jourova ahead of the entry into application of the General Data Protection Regulation 24/05/2018* URL: http://europa.eu/rapid/press-release_STATEMENT-18-3889_en.htm (reference date 28.10.2018)
[19] *The European Parliament approved new rules for the Schengen system* URL: https://ria.ru/world/20181024/1531420048.html (reference date 28.10.2018).
[20] *The SIS II annual statistics for 2017 02/2018* URL: http://data.consilium.europa.eu/doc/document/ST-8279-2018-INIT/en/pdf (reference date: 28.10.2018)

IOP Conf. Series: Materials Science and Engineering **483** (2019) 012102 doi:10.1088/1757-899X/483/1/012102

On the problem of preservation of inundated ecosystems of the planet

S N Lega[1], I N Tikhonova[2], V N Orobinskaya[3*], M F Marshalkin[4]

[1, 2] Pyatigorsk Medical Pharmaceutical Institute (PMPHI) branch of Volgograd Medical State University, Russia, Pyatigorsk, Stavropol region 11, Kalinin Prospect
[3,4] Federal State Autonomous Educational Institution of Higher Professional Education "North Caucasus Federal University" Institute of Service, Tourism and Design, branch of NCFU in Pyatigorsk 56, str. Ukrainian
orobinskaya.val@yndex.ru, stasjairustix@mail.ru

Abstract. In article the question of value of inundated ecosystems of the planet for the biosphere, importance of preservation of their steady state in the conditions of the escalating pressure of an anthropogenic factor is considered. Authors within several years made monitoring of the vegetable communities of the floodplain of the Yutsa River subject to pollution by an unauthorized dump of solid household and construction waste. Change of a specific variety of grassy communities of the studied territory subject to anthropogenic influence has been as a result registered. The group of the plants which are most adapted (tolerant) for observed changes of the surrounding environment which is probably the so-called destructive link which is a basis for the subsequent restoration of the studied grassy community in these changed conditions is found. It is revealed that the changes in the specific structure of vegetable community which have led to loss of stability of the damaged territory continued some time and after elimination of anthropogenic influence, becoming populated by invasive types.

Introduction

The problem of sustainability of natural ecosystems in the modern world due to global environmental pollution is very important and has great practical value. However, natural formations, like other living systems of the biosphere, are very complex structures, represented by a huge number of interrelated and interdependent heterogeneous elements, often not yet sufficiently studied. Any change in one of the links of this organism inevitably affects the whole system. Moreover, functionally different types of natural communities on the scale of the whole biosphere, as if complementing each other, determine its stability and integrity. The loss of any of its constituent elements can cause serious, often irreversible, negative changes in the cycle of substances, reduce its stability.

Thus, the study of the structure of various natural communities, mechanisms of their sustainability, as well as monitoring and protection are imperative and urgent tasks of our time. The understanding of the insufficiently studied question of the implementation of immunity of plant organisms, the mechanism of their resistance to adverse environmental factors, especially in conditions of increasing anthropogenic pressure, which, along with natural causes, leads to a rapid change of ecosystems (succession), can help to solve this problem, along with the solution of other problems in this area.

Now it is established that development of natural formations in many respects is similar to development of individuals. In ecosystems, as well as in a single organism of an animal or plant, metabolism is detected, due to the opposite processes of catabolism and anabolism. Anabolism processes are manifested in natural ecosystems in the synthesis of biomass producers and its accumulation in ecosystems. The processes of catabolism are provided by the life activity of the reducents. This leads to the sustainability of natural communities of any complexity, ie. their equilibrium state with the natural environment.

However, from year to year the number of sustainable natural systems is decreasing, which negatively affects the global cycle of substances, leads to the destruction of ecosystems and adversely affects climate change throughout the planet. So today, about 63.8% of biogeocenoses on land are more or less susceptible to degradation.

It is known that at the present scale of anthropogenic impact on nature, when changes in environmental factors occur very quickly, living organisms do not have time to adapt to them, which leads to their extinction. The disappearance of species sensitive to various pollutants leads to the settlement of the territory by other forms (invasive) that are more resistant to the created conditions. At the same time, invasive species, and most often in herbaceous communities are ruderal plants, create an unfavourable ecological background, since they often have poisonous pollen, and sometimes poisonous vegetative organs, a more powerful root system, have a more powerful seed fund. All this leads to an increased competitive ability of the minerals in comparison with local forms. They quickly colonise nearby areas, displacing less adapted species.

In this regard, river and floodplain ecosystems deserve great attention. They are today considered as a repository of the planet's biodiversity. Each of them is unique and therefore makes an invaluable contribution to the overall integrity and stability of the biosphere.

1. Goals and Objectives

The task of our work was to study the biodiversity and stability of plant communities in the floodplain of the Yutsa river, to study the process of their transformation under the influence of anthropogenic factors, as well as the mechanisms of recovery after the cessation of this pressure.

The relevance of this work lies in the absence, as far as we know, of data on the study of floodplain ecosystems of the Yutsa river, which is the right tributary of the Podkumok river flowing through the territory of the Caucasian Mineral Waters.

It is known that the plant communities of the floodplain are significantly different from the flora of other areas. This is due to the influence of the river, which during floods, brings suspended matters on their territory. They form a layer of deposit of spring floods, mixed with the fall of vegetation, causing thus soil fertility and creating conditions for the appearance of characteristic plant communities, which are significantly different from each other in different river floodplains.

The floodplain of the river Yutsa is also unique. In the floodplain of the river there is a small lake, with its own biota, increasing the biodiversity of the area and causing its uniqueness. The study and monitoring of natural ecosystems of the recreational region of the CMW (Causacian Mineral Waters) in order to preserve the stability and biodiversity of its unique natural complexes is particularly relevant. The status of the resort of CMW requires great responsibility for the quality of the natural environment of the region, which, as in the modern world as a whole, has certain environmental problems and requires its early solution.

The creation of a tourism and recreation cluster not only develops infrastructure, but also contributes to the creation of unauthorised landfills that have a negative impact on the environment due to the rapid growth of construction institutions nearby.

The objectives of solving global problems related to environmental pollution in the construction, operation and dismantling of "end-of-life" buildings are reflected in the international standards ISO 14040-14044.

Three licensed landfills of solid household waste (SHW) are officially operating in the region: in the vicinity of Essentuki, Mineral Waters and in Lermontov. Only 9 companies out of 23 are licensed for the collection and disposal of solid household waste [1]. The problem of collection, utilisation of solid household waste (SHW) in the urban system, associated with the preservation of the environment takes the second place.

In the CMW region, observations have shown that from time to time unauthorised landfills are formed, which are monitored by the city administration and activists in order to eliminate them. The landfill consisting of SHW (Solid Household Waste) and construction debris was discovered by the authors in the floodplain of the river Yutsa in the area of the settlement Goryachevodsky. After its discovery, having existed for another 2 years, it was liquidated.

In this paper, the dynamics of changes in the species diversity of the floodplain vegetation community of the Yutsa river, transformed by an unauthorised landfill, was studied. The monitoring was carried out for 6 years near the settlement of Goryachevodsky.

2. Methods of Study

The main method of our study was the monitoring of changes in the species diversity of ecosystems located in the immediate vicinity of the landfill in comparison with the control ecosystems located at a distance of 100 m. The object of monitoring was randomly selected areas in three replications of 1m2. With the aim of identifying species by standard methods, herbarisation of plants was carried out, using determinants: Keller B. A., Lubimenko V. N., Maltsev A. I., Fedchenko B. A., 1934-1935; Rycina, 1952; Grossheim, 1949; Galushko, 1980.

The criteria of ecosystem quality was the calculation of biomass of plant communities and their species diversity.

Results

As a result of monitoring, 27 species of plants were found in the study area. These are: blue devil (Echium vulgare), cardaria nibs (Cardaria drada), common yarrow (Achilléa millefólium), sphagnum moss (Sphagnum), white dead-nettle (lamium album), cocksfoot (Dactylis glomerataL.), oatmeal high (Festuca altissima), common thistle (Cirsium vulgare (Savi).), clover field (Amoria Trifolium campestris L.), horse sorrel (Rumex confertus Wilid), common cardaria (Cardaria vulgare), common falcaria (Falcaria vulgaris), tufted vetch (Vicia crassa), hedge vetch (Vicia sapium), small crane (Geranium pussilum), cleavers (Galium aparine), sow-thistle (Sonchus palustris L.), bird's-foot trefoil (Lotus corniculatus L.), burdock (Arctium lappa), common winter cress (Barbarea vulgaris) and plantain (Plantago major L.), field horsetail (Equisetum arvense L.), yellow sweet clover (Melilótus officinális), wormwood (Artemísia absínthium), burdock (Xánthium strumárium), bluegrass (Poa annua L), stinging nettle (Urtícaúrens) (table 1).

Table 1. Species diversity of herbal communities of the Yutsa river floodplain during the monitoring period

Year of study	Control area	Experimental area
1	1. Blue devil - Echium vulgare 2.Cardaria nibs - Cardaria drada 3.The ordinary cardaria - Cardaria vulgare 4. Common yarrow – Achilléa millefólium 5. Sphagnum Moss-Sphagnum 6.Common fascaria - Falcaria vulgaris 7. Tufted vetch - Vicia crassa 8. Hedge vetch -Vicia septum 9. Small crane - Geranium pussilum 10. Cleavers – Galium aparine 11. Sow-thistle – Sonchus palustris L. 12.The common Thistle - Cirsium vulgare (Savi) 13.White dead-nettle – lamium album 14. Oatmeal high - - Festuca altissima 15. Bird's-foot trefoil – Lotus corniculatus L. 16.Cocksfoot – Dactylis glomerata L.	1.Blue devil - Echium vulgare 2.Cardaria nibs - Cardaria drada 3.The horse sorrel – Rumex confertus Wilid 4.Common yarrow - Achilléa millefólium 5.White dead-nettle – lamium album 6, Cocksfoot - Dactylis glomerata L. 7. Tall fescue - Festuca altissima All 8.Field clover-Amoria (Trifolium) campestris.L. 9.Sphagnum Moss-Sphagnum 10.The common thistle - Cirsium vulgare (Savi)

2	1. Blue devil - Echium vulgare 2.Cardaria nibs - Cardaria drada 3.Horse sorrel – Rumex confertus Wilid 4. Common yarrow – Achilléa millefólium 5.White dead-nettle – Lamium album 6. Cocksfoot – Dactylis glomerata L. 7.Oatmeal high – Festuca altissima 8.Field clover – Amoria (Trifolium) campestris L. 9. Sphagnum moss - Sphagnum 10. Common thistle - Cirsium vulgare (Savi)	1. Blue devil - Echium vulgare 2.Cardaria nibs - Cardaria drada 3. Common cardaria - Cardaria vulgare 4. Common yarrow – Achilléa millefólium 5.Sphagnum Moss-Sphagnum 6. Common Fascaria - Falcaria vulgaris 7. Tufted vetch - Vicia crassa 8. Hedge vetch - Vicia septum 9. Small crane - Geranium pussilum 10.Cleavers – Galium aparine 11. Sow-Thistle – Sonchus palustris L. 12.The common Thistle - Cirsium vulgare (Savi) 13. White dead-nettle – lamium album 14. Oatmeal high - - Festuca altissima 15. Bird's foot – Lotus corniculatus L.
3	1. Blue devil - Echium vulgare 2.Cardaria nibs - Cardaria drada 3.Horse sorrel – Rumex confertus Wilid 4. Common yarrow – Achilléa millefólium 5.White dead-nettle – lamium album 6. Cocksfoot – Dactylis glomerata L. 7.Oatmeal high – Festuca altissima 8. Field clover – Amoria (Trifolium) campestris L. 9. Sphagnum moss - Sphagnum 10. Common thistle - Cirsium vulgare (Savi) 11.Wheatgrass janalski – Elytrigiad.shinalica Sablina 12.Hordelymus eureurropaeus 13.Plantain – Plantago major L.	1. Blue devil - Echium vulgare 2.Cardaria nibs - Cardaria drada 3. Common cardaria - - vulgare Cardaria 4. Common yarrow – Achilléa millefólium 5.Sphagnum Moss - Sphagnum 6.Common fascaria - Falcaria vulgaris 6. Tufted vetch - Vicia crassa 7. Hedge vetch - Vicia septum 8. Small crane - Geranium pussilum 9. Cleavers – Galium aparine 10. Sow-Thistle – Sonchus palustris L. 11.The common thistle - Cirsium vulgare (Savi) 12. White dead nettle - lamium album 13.Oatmeal high - Festuca altissima 14. Bird's foot - Lotus corniculatus L. 15.Cocksfoot – Dactylis glomerata L.

4	1. Blue devil - Echium vulgare 2.Cardaria nibs - Cardaria drada 3. Horse sorrel – Rumex confertus Wilid 4. Common yarrow – Achilléa millefólium 5.White dead-nettle – lamium album 6.Cocksfoot – Dactylis glomerata L. 7. Tall fescue - Festuca altissima 8. Field clover-Amoria (Trifolium) campestris L. 9. Sphagnum Moss-Sphagnum 10. The common Thistle - Cirsium vulgare (Savi) 11. Wheatgrass jamalski – Elytrigiadshinalica Sablina 12. Hordelymus European - Hordelymus europaeus 13.Plantain – Plantago major L.	1. Blue devil - Echium vulgare 2.Cardaria nibs - Cardaria drada 3. Common cardaria - - vulgare Cardaria 4. Common yarrow – Achilléa millefólium 5. Sphagnum Moss - Sphagnum 6. Common fascaria - Falcaria vulgaris 7. Tufted vetch -Vicia crassa 8. Hedge vetch - Vicia septum 9. Small crane - Geranium pussilum 10. Cleavers – Galium aparine 11. Sow-thistle – Sonchus palustris L. 12.The common Thistle - Cirsium vulgare (Savi) 13. White dead nettle - lamium album 14. Oatmeal high - - Festuca altissima 15. Bird's foot - Lotus corniculatus L. 16.Cocksfoot – Dactylis glomerata L.
5	1. Blue devil - Echiumv ulgare 2. Cardaria nibs - Cardariadrada 3. Common yarrow – Achilléa millefólium 4. The common Thistle - Cirsiumvulgare (Savi) 5. Wheatgrass jamalski – ElytrigiadshinalicaSablina 6. Tufted vetch - Vicia crassa 7. Hedge vetch - Vicia septum 8.Wild carrots – Daucus carota L. 9. Yellow sweet clover – Melilotus officinalis 10.Hop clover – Medica golupulina L. 11. Cleavers – Galium aparine 12. Flattened meadow grass – Poacompressa L. 13.Wild chicory - Cichorium inthybus L. 14.Celandine – Chelidonium majus L. 15.Acanthus bristlethistle – Carduus acanthoides L. 16.Sea-side lavender - kerlik	1. Blue devil - Echiumv ulgare 2.Cardaria nibs – Cardaria drada 3. Common cardaria - Cardaria vulgare 4. Common yarrow - Achilléa millefólium 5.Sphagnum Moss-Sphagnum 6. Common falcaria - Falcaria vulgaris 7. Tufted vetch - Vicia crassa 8. Hedge vetch - Vicia septum 9. Small crane - Geranium pussilum 10. Cleavers – Galium aparine 11. Sow-Thistle – Sonchus palustris L. 12.The common Thistle - Cirsium vulgare (Savi) 13. White dead-nettle – lamium album 14. Oatmeal high - - Festuca altissima 15. Bird's foot –Lotus corniculatus L. 16.Cocksfoot – Dactylis glomerata L.
6	1.Blue devil - Echium vulgare 2.Cardaria nibs - Cardaria drada 3. Common yarrow – Achilléa millefólium 4.The common Thistle - Cirsium vulgare (Savi) 5.Wheatgrass of janal – Elytrigia dshinalica Sablina 6.Tufted vetch - Vicia crassa 7. Hedge vetch - Vicia septum 8.Wild carrot - Daucus carota L. 9.The yellow sweet clover – Melilotus officinalis 10. Wild chicory - Cichorium inthybus L. 11. Celandine – Chelidonium majus L.	1. Blue devil - Echium vulgare 2.Cardaria nibs - Cardaria drada 3. Common cardaria - - vulgare Cardaria 4. Common yarrow – Achilléa millefólium 5.Sphagnum Moss - Sphagnum 6.Common fascaria - Falcaria vulgaris 7. Tufted vetch - Vicia crassa 8. Hedge vetch - Vicia septum 9. Small crane - Geranium pussilum 10. Cleavers – Galium aparine 11. Sow-Thistle – Sonchus palustris L.

	12.Prickly Thistle – Carduus acanthoides L. 13.Burweed - Xánthium strumárium 14.Wormwood - Artemisia 15. Orache –Atriplex sagittata	12.The common thistle - Cirsium vulgare (Savi) 13. White dead nettle - lamium album 14.Oatmeal high - Festuca altissima 15. Bird's foot - Lotus corniculatus L. 16.Cocksfoot – Dactylis glomerata L.

During the monitoring period, the authors registered a change in the species composition of the plant communities transformed by the landfill, which was reflected both in the disappearance of some species, probably the most sensitive to the changed environmental factors, and also the coverage of the study area with new invasive species for the area.

Among the dropped types were noted: sphagnum moss (Sphagnum), white dead-nettle (lamium album), cocksfoot (Dactylis glomerata L.), tall fescue (Festuca altissima) and identified invasive plants: field horsetail (Equisetum arvense L.), yellow sweet clover (Melilótus officinális), wormwood (Artemísia absínthium), burweed (Xánthium strumárium), bluegrass (Poa annua L), stinging nettle (Urtícaúrens), burdock (Arctium lappa), common winter cress (Barbarea vulgaris) and plantain (Plantago major L.).At the same time, the species diversity of the control area remained unchanged.

The disappearance of the mentioned species brought the system out of a stable state, as evidenced by a significant decrease in plant biomass in comparison with the control area. Later, during the restoration of the disturbed plant community, the difference in the amount of biomass between the control and experimental areas tended to decrease (table 2).

The results of the monitoring showed that if the difference between the biomass of the disturbed community and the control area in the first year of monitoring was 33%, then by the sixth year of observation, after the elimination of the unauthorised landfill, it decreased to 10.8%, which can be explained by the coverage of the territory by invasive species instead of the disappeared local species.

Table 2. Changes in biomass of the plant community disturbed by an unauthorised landfill during the monitoring period

year	biomass gr/м^2		The difference between the amount of biomass in control and experimental area	
	Experimental area	Control area	gr/м^2	%
1	466	700	234	33
2	500	703	203	28
3	510	709	199	28
4	550	710	160	23
5	600	720	120	16.7
6	650	729	79	10.8

It should be noted that for the entire period of observation the plant community of the control area was stable, i.e. we did not record the emergence of new or disappearance of local plant species.

Besides, this work has allowed to identify the plant species growing in the floodplain of the Yutsa river which are most resistant to the prevailing conditions.They are: field horsetail (Equisetum arvense L.), wormwood bitter (Artemísia absínthium), common cocklebur (Xánthium strumárium), stinging nettle (Urtícaúrens), burdock (Arctium lappa), common winter cress (Barbarea vulgaris).

These species were found both in the area disturbed by the landfill and in the control area during the entire observation period. Probably, they provided elasticity and stability of plant community in the created conditions, ensuring its stability and possibility of the subsequent restoration.+

Conclusions

As a result of the monitoring, the change in the species diversity of the herbal community of the landfilled community was recorded, which resulted in its initial impoverishment, accompanied by a simultaneous decrease in biomass.

During the research such plants as sphagnum moss (Sphagnum), white dead-nettle (Lamium album), cocksfoot (Dactylis glomerata L.), tall fescue (Festuca altissima All.) have disappeared from the area transformed by the landfill. Grass coverage biomass decreased by 33%.

Later on, after the elimination of the pollution source, there has been a tendency to recovery of the biomass, due to the appearance of invasive species such as field horsetail (Equisetum arvense L.), yellow sweet clover (Melilótus officinális), wormwood bitter (Artemísia absínthium), common cocklebur (Xánthium strumárium), annual bluegrass (Poa annua L), stinging nettle (Urtícaúrens), burdock (Arctium lappa), common winter cress (Barbarea vulgaris) and common plantain (Plantago major L.).

However, observations showed that some of the species of the original community remained unchanged, as it was detected during the entire period of monitoring both at the experimental site and at the control site. These are: field horsetail (Equisetum arvense L.), wormwood (Artemísia absínthium), common cocklebur (Xánthium strumárium), stinging nettle (Urtícaúrens), burdock (Arctium lappa), common winter cress (Barbarea vulgaris). This allows us to consider this group of plants as the most adapted (tolerant) to the newly established environmental conditions. Probably, it was this group which was the destructive link that served as the basis for the subsequent restoration of the disturbed ecosystem.

An important result of the monitoring, in our opinion, was the observation that after the elimination of the landfill, during the studied period of time, there was no restoration of the former species diversity of this herbaceous community, although plant biomass tended to recover.

Thus, studies have shown that unauthorised landfills and construction waste, so often found in many regions of the country, are not harmless. They have a negative impact on the species composition of ecosystems, disrupting their sustainable state. Even their subsequent elimination does not lead to a complete restoration of biogeocenosis parameters in a short period of time (monitoring time).

References

[1] Tikhonova I. N, Lega S N 2017 Environmental problems of solid waste and construction waste generation and utilisation in the region of CMW on line scientific & educational *Bulletin "Health and Education Millennium"* 19 **7**

[2] Aramisov Yu H Methods of comparative floristics in the study of florruderal habitats 2008 *Natural-geographical research: scientific almanac* **6** 30-32

[3] Hainy S, Kopecky K, Kropach Z 1987 Anthropogenic vegetation – how to assess its role? (International Yearbook "Science and humanity", Ed. "Knowledge" Moscow) 115 – 123

[4] Grossheim A A 1948 Vegetative cover of the Caucasus (publishing house of the Moscow Society of naturalists) 267p

[5] Grossheim A A Manual of the plants of Caucasus 1949 (Publisher: Soviet science) 376p

[6] Galushko A I 1978 Flora of the Northern Caucasus. (Manual: Publishing House of Rostov University) 328p

[7] Ivanov A L 1998 The flora of the Pre-Caucasian region and its genesis (Stavropol: Publishing House of SSU) 204p

[8] Keller B A, Lubimenko V N, Maltsev A I, Fedchenko B A 1934-1935 Weed plants of the USSR vol **1 - 4** (Publisher: Academy of Sciences of the USSR) 1500 p

[9] Marshalkin M F, Lega S N, Tikhonova I N 2014 The role of ruderal plants in the restoration of natural plant communities disturbed by unauthorised landfills *Fundamental study* issue 9 (**2**) 329-332

[10] Maremchuk Yu A 2007 Flora anthropophyta – threat to natural landscapes of the North Caucasus *Successes of Modern Natural Sciences* **12** 44-45

[11] Nikitin V V 1983 Weed plants of the USSR flora (L.: Science) 454 p
[12] Rychina Yu V 1952 Manual of weed and garbage plants (Publishing house Moscow) 280 p
[13] Braslavskaya T Yu Biodiversity and vegetation dynamics in the floodplain of a small river South of Non-Black Soil Zone (on example of the river Nerussa, the Bryansk region), diss. PhD. in. biol. sciences. (Moscow) 2001
[14] Bukvareva E N 2010 The role of terrestrial ecosystems in climate regulation and Russia's place in the post-Kyoto process (KMK Scientific Press Ltd, Moscow)
[15] Lega S N, Tikhonova I N 2016 Formation and implementation of adaptive mechanisms of plants in viral pathogenesis *Modern problems of science and education* **5**

IOP Conf. Series: Materials Science and Engineering **483** (2019) 012103 doi:10.1088/1757-899X/483/1/012103

Optimal technological process planning approach based on the state of mechatronic systems

V A Mokhov, T N Kruglova, D V Shaykhutdinov, A S Vlasov

Platov South Russia State+ Polytechnic University (NPI), 132, Prosveschenija St., Novocherkassk, 346400, Russia

E-mail: d.v.shaykhutdinov@gmail.com

Abstract. The article presents the problem formulation for diagnostics and optimal planning of the trajectory of the technological process taking into account the wear of the mechatronic systems and its electric motors. As an example, we consider a mobile robot with a differential drive, which must successively pass a field of various shapes. The trajectory of movement is chosen taking into account the wear of its electric motors. To solve this problem, it is supposed to use graph theory and probabilistic algorithms. Examples of solutions to the problem are given without taking into account the properties of the soil in the working space of the robot, used by the type of drive and its angle of rotation. The power of sets of feasible solutions will increase significantly with increasing number of technological operations, which, in turn, will lead to inefficiency of using brute force methods for solving the formulated problem and the need to solve it based on probabilistic algorithms agent metaheuristics.

1. Introduction

The reliability of industrial enterprises largely depends on the state of electric power equipment, ensuring long-term uninterrupted operation of the technological chain, an unplanned shutdown of which can lead to serious accidents, resulting in long downtime and significant financial losses. The main consumer of electricity in the world is electric motors, ensuring the reliability and safety of operation of which dictates the need to estimate the spent resource of responsible structural components and to analyze the rate of damage accumulation at these nodes during operation.

The main types of faults that occur in electric motors during their operation are damage to the insulation of the armature windings (50%). This can be: breakdown of insulation (30%), overheating (26%) and mechanical damage (20%). The causes of accidents are caused by normal physical aging and wear (34%), poor lubrication (15%), increased humidity (10%), and faults detected during the inspection (30%) [1]. Also, a significant impact on the life of electric motors have loading conditions during operation, the management of which will significantly extend the uptime of the electric drive and the equipment as a whole [2,3]. Therefore, to solve the problem of improving the reliability of the electric motor separately from the object on which it is installed without taking into account the operation of all other mechanisms is not possible. Therefore, this task should be solved for a group of drives of an object operating in a certain technological process.

An example of the object of research is a mobile robot with four independent electric drives, which are a mechatronic module "motor-wheel" [4]. The robot is used for the needs of agriculture and moves over rough terrain. The task of the robot is the implementation of the technological process / detour of the field of a given geometry [5,6]. In this case, the choice of the trajectory of the robot (the sequence of technological operations of the movement, which determine the technological process of the field processing) should be carried out in such a way that the wear of the drives of its wheels is minimal.

2. Robot wear model

In figure 1 shown model of a four-wheeled mobile robot with a differential drive. It is assumed that the robot is driven in one of three ways:

1) using the wheels D^1 and D^2 (front wheel drive);
2) using the wheels D^3 and D^4 (tasks drive);
3) using all four wheels simultaneously (all-wheel drive).

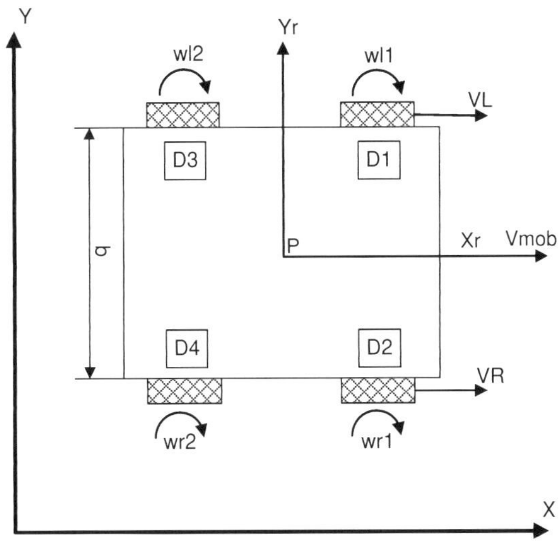

Figure 1. Model of a four-wheeled mobile robot with differential drive

The preferred option for controlling the robot is an all-wheel drive, since in this case the life of the wheel drive is consumed evenly. However, in the event of a failure of one of the engines, the control system can be retuned to front or rear wheel drive.

A complete failure of a mobile robot will be considered to be options for working on the right or left pair of drives (D^1 and D^3 or D^2 and D^4 in Figure 1) or on any one drive, as this leads to a situation of uncontrolled movement of the robot and unreasonable wear of working engines.

The degree of wear of the robot is the sum of the degrees of wear of the drives of each wheel when performing a single technological operation t_i (1).

$$P(t_i) = p^1 + p^2 + p^3 + p^4, \qquad (1)$$

where p^1, p^2, p^3, p^4 is the degree of wear when performing a single technological operation of the drives D^1, D^2, D^3 and D^4, respectively.

At the same time, the wear of the electric drive of each of the wheels for an individual technological operation is determined as a function of three arguments (2):

k_{ground} – variable that determines the properties of the soil (takes on the values: 1 - smooth hard surface, 2 - smooth loose surface, 3 - uneven loose surface);

t_{fun} – A variable defining the mode of operation of the engine (0 - passive, 1- in the “full drive” mode, 2 - in the active front or rear drive mode);

M_{turn} –variable that determines the additional load on the motor when performing a turn (for each drive, turns are taken into account as follows: for drives D^1 and D^3 - to the right, for D^2 and D^4 - to the left).

$$\begin{cases} p^1 = f(k^1_{ground}, t^1_{fun}, M^1_{turn}), \\ p^2 = f(k^2_{ground}, t^2_{fun}, M^2_{turn}), \\ p^3 = f(k^3_{ground}, t^3_{fun}, M^3_{turn}), \\ p^4 = f(k^4_{ground}, t^4_{fun}, M^4_{turn}). \end{cases} \quad (2)$$

3. Reasoning

For experiments in the proposed model of wear, the following restrictions are introduced:

1) the robot moves on a horizontal surface and the Z coordinate is taken constant (Z=const);

2) the robot moves at approximately constant speed when performing all technological operations that determine a separate technological process (V=const).

For the organization of the technological process a specific field is created (the working space of the robot) that has a certain flat geometry / configuration (some of the possible options are shown in Figure 3.a (n = 3), 3.b (n = 4), 3.c (n = 15) and 3.d (n = 14). Processing such a field is a separate process.

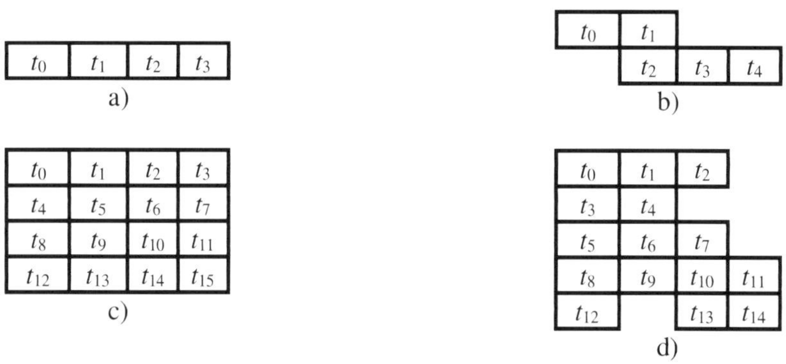

Figure 2. Possible options for the working space of the robot

For the convenience of a mathematical description, the field space being processed is divided into n segments so that each of them uniquely defines the limits of a single technological operation t_i (i = 1..n) for a robot.

In accordance with the variant of the working space of the robot, shown in Figure 2.a, the robot performs a uniform rectilinear motion consistently passing through all the operations of the technological process.

In Figure 2.b, in addition to rectilinear movement, there is a turn to the right, which contributes to uneven wear of the right wheels as compared to the left ones.

In Figure 2.c and Figure 2.d the robot needs to go through a more complex trajectory, the choice of which is advisable to do in view of the wear of its engines.

At the same time, the task of the robot is to fully implement the technological process (performing all its technological stages) so that the total wear of P_0 electric drives is minimal (3):

$$P_0 = \sum_{i=1}^{n} P(t_i) \to \min \ . \quad (3)$$

Additionally, the condition that the wear P_0 is assumed to be minimal when it is evenly distributed between the drives D^1, D^2, D^3 and D^4 is accepted. It is possible to reach the latter with the minimum difference between the perfect $M_{turn(R)}$ and $M_{turn(L)}$ left rotations (in other words, their general approximate equality) (4):

$$\sum M_{turn(R)} - \sum M_{turn(L)} \rightarrow \min . \quad (4)$$

If for a particular field all its n segments (which define a certain technological process in the aggregate) are represented as the corresponding graph $T(V, E)$, the set of vertices of which (V) uniquely correspond to the field segments, and the set of edges (E) logically define possible transitions between adjacent segments (technological operations), the mathematical problem of the synthesis of an efficient trajectory $G(P_0)$ for the technological process of field processing can be represented as (1), (2), (4) and follows (5) [7,8]:

$$G(P_0) = \left\{ t_i \middle| f(t) = \sum_{i=1}^{n} P(t_i) \rightarrow \min,\ i = 1...n \right\}, Z = \text{const}, V = \text{const} . \quad (5)$$

4. Solution

Variants of graph representation for partitions of working space of robot (fig. 2.a, 2.b, 2.c and 2.d) are shown in figures 3.a, 3.b, 3.c 3.d respectively.

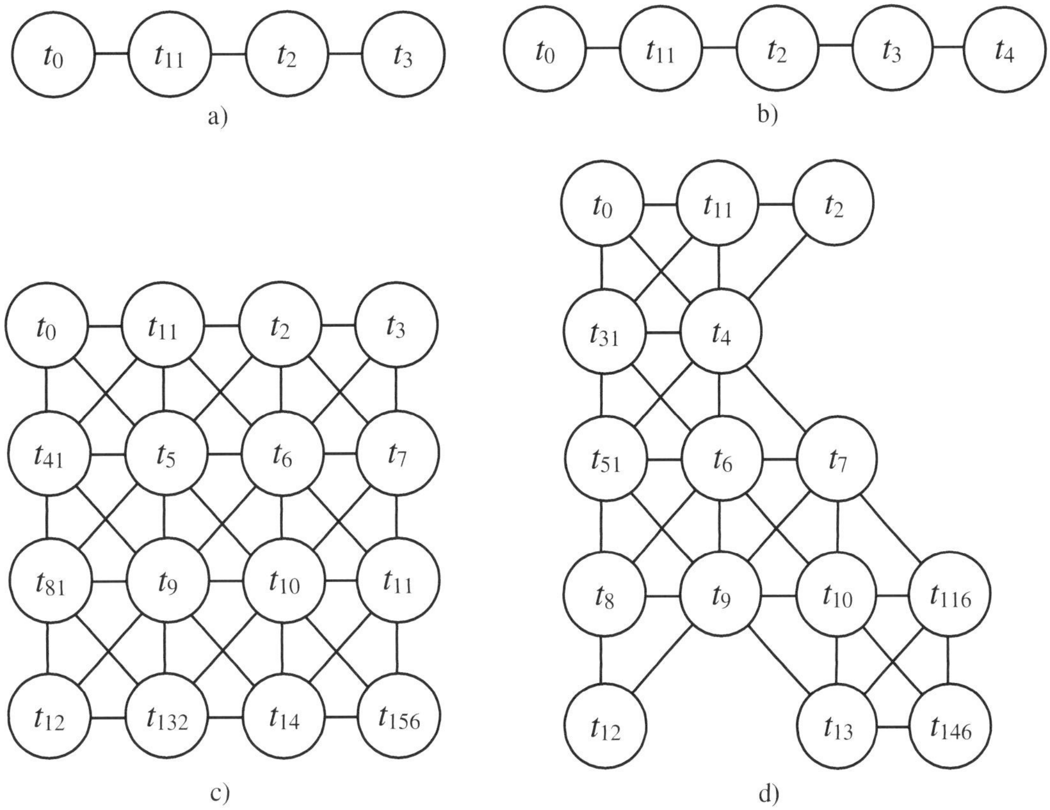

Figure 3. Variants of the graph representation of the workspace of the robot

Possible solutions for these cases shown in figure 3.a ($G(P_0) = \{t_0, t_1, t_2, t_3\}$) and 3.b ($G(P_0) = \{t_0, t_1, t_2, t_3, t_4\}$) are obvious. For more difficult case as in figure 3.c, solution will have form, for example ($G(P_0) = \{t_0, t_1, t_2, t_3, t_4, t_5, t_6, t_7, t_8, t_9, t_{10}, t_{11}, t_{12}\}$), and for the most difficult case as in 3.d, solution will have form, for example ($G(P_0) = \{t_0, t_1, t_2, t_3, t_4, t_5, t_6, t_7, t_8, t_9, t_{10}, t_{11}, t_{11}\}$).

To demonstrate the turns within the selected trajectories, we define the turn matrixes $G(M_{turn})$. To do this, we use the following notation: 0 - turn within t_i (not performed), 1 - as part of the technological operation, turn right $(M_{turn(R)})$, 2 - turn left $(M_{turn(L)})$. Then the matrix $G(M_{turn})$ for each of the four previously defined trajectories will take the following form:

for the 1st trajectory- $G(M_{turn}) = \{0,0,0,0\}$;

for the 2nd trajectory - $G(M_{turn}) = \{0,0,1,0,0\}$;

for the third trajectory - $G(M_{turn}) = \{0,0,0,1,1,0,0,2,2,0,0,1,1,0,0,0\}$;

for the 4th trajectory - $G(M_{turn}) = \{0,0,1,1,2,0,0,2,2,1,1,0,2,2,0\}$).

The selected trajectory can be considered optimal from the standpoint of minimum engine wear if the number of left and right turns for the indicated options of trajectories $G(P_0)$ approximately/conditionally coincides.

It is important to note that the presented examples of solutions to the problem (5) are given without taking into account the properties of the soil in the working space of the robot, used by the type of drive and its angle of rotation. The solution of these problems will be carried out in the framework of further research.

5. Conclusion

Obviously, in this case, the power of sets of feasible solutions will increase significantly with increasing number of technological operations, which, in turn, will lead to inefficiency of using brute force methods for solving the formulated problem and the need to solve it based on probabilistic algorithms (for example, agent metaheuristics). [9, 10, 11].

Acknowledgments

This article is based on results of the project No. SP-359.2018.1, implemented under the program “Scholarship provided by the President of the Russian Federation for young scientists and graduate students engaged in advanced research and development in priority areas of modernization of the Russian economy” using equipment of shared facility “Diagnosis and energy-efficient electrical equipment” (NPI).

References

[1] Moiseeva R R, Samigullina B C, Bazhenov N G and Zatsarinnyi N 2014 *Herald to Zan and Tech. Univ.* **3(17)** 117-119
[2] Bulgakov A G, Kruglova T N 2017 *J. of Appl. Ing. Sci.* **15(4)** 426-432
[3] Kruglova T N, Bulgakov A G, Vlasov A S, Shmelev I A 2017 *MATEC Web of Conferences* **132**
[4] Bishop R H 2002 *The mechatronics handbook: The University of Texas at Austin Austin, Texas* 1229
[5] Vlasov A S, Kruglova T N 2017 *Intellectual energy systems: materials V Intern. youth Forum* **1** 162-165
[6] Vlasov A S, Kruglova T N 2017 *Intellectual energy systems: materials V Intern. youth Forum* **1** 159-162
[7] Mokhov V A, Turovsky F A, Turovskaya E V 2016 *Basic Research* **9** 279-283
[8] Mokhov V A , Grinchenkov D V , Spiridonova I A 2016 *ICIEAM, IEEE* 1-6
[9] Kubil V, Mokhov V, Grinchenkov D 2018 *Proc. of the 6th Int. Conf. on App. Innovations in IT* **6(1)** 29-36
[10] Kubil V N., Mokhov V A 2014 *Prob. of Modernization of Eng. Educ. in Russia* 140-144
[11] Turovskaya E V, Mokhov V A, Kuznetsova A V 2014 *Don Engineering Gazette* **28(1)** 56

Modeling of object competitiveness based on hybrid method assessment

T L Bezrukova[1], G P Fomin[2], S S Kirillova[1], B A Bezrukov[1]

[1] The Voronezh State University of Forestry and Engineering named after G.F. Morozov, 394087, Russia, Voronezh, Timiriazeva st., 8
[2] Russian University of Economics named after G.V.Plekhanov, 117997, Russia, Moscow, Ctremiannyj pereulok, 36

E-mail: bezrukova_t_l@mail.ru

Abstract. The article covers the aggregation informal and mathematical methods in competitiveness of objects of different origin basing on hybrid assessment method. The given method includes principles of systemic analysis of structurization and quantification that leads to the analysis of the competitiveness problem and complex implementation of multicriteria method of optimization, and also methods of ranging of characteristics of the object to be assessed, methods of dual comparisons, methods of integral programming, Hungarian method, reduction, method of tests and errors. The feature of the method is the advantages integration of different quality methods of analysis with formal methods to obtain the quantitative characteristics of results of competitiveness for different consumer levels. The certain extracts of the big solved task are given together with a lot of single and integrated assessments. The certain parts of a huge task solved are also presented in the article. The result tables with conclusions are also presented here. The results change primary view of consumer, make mental outlook wider and change the attitude of consumer towards the competitiveness in particular. The description of transition from methods and models use to the assessment of object competitiveness is given. That shows the possibility of hybrid method application according to the object competitiveness of different origin.

1. Introduction

At present some organizations are more successful than the others. Some are in deep crises. That is the natural economical process. To provide the competitiveness of the enterprise it is needed to reach a certain level of profit that will help it to survive in future. It is evident, that the mechanism of enterprise adaptation to the changeable market conditions should be created. The mechanism of competitiveness management should be worked out to aggregate measures for systematic development of constituent parts of the object. Methodology of competitiveness assessment for practical use has not been worked out by the present time.

The basis of the competitiveness of the enterprise is the competitiveness of its product, that is its system generated component. All parts of the chain are concentrated on it. That is the basis of the entrepreneurial activity. [1] Goods are the basis of entrepreneurial activity. Business and market policy are defined by product, goods and services. It is evident that competitiveness assessment is advisable to do in complex according to the more accessible indexes and direction of activity. Competitiveness deflection means risks, and deflection value defines the program, content and necessary investment to provide the object competitiveness.

2.Theoretical, Informational, Empirical, and Methodological Grounds of the Research

Target assessment of competitiveness is to reveal the readiness of staff and individual worker to fulfill the particular kind of activity and also to reveal the level of potential ability for professional and occupational development of individual worker [2].

IOP Conf. Series: Materials Science and Engineering **483** (2019) 012104 doi:10.1088/1757-899X/483/1/012104

Competitiveness of enterprise means: sales volume cost and quantity, profit, share of cost of unsold products in store, competitiveness index, profitability, costs. Assessment of competitiveness of object is a multi complex task that defines a good deal of factors of different origin. However, we need to carry out the given task for fulfillment of a number of activities such as finding of the main strategic ways for goods production, working out the firm staff training program to provide the competitiveness of the staff and the whole organization itself. The methodology of competitiveness has not been worked out properly for practical use by the present time. The complexity of origin of competitiveness specifies a number of methods of its assessment. [3]

Methods of competitiveness assessment are multifarious and are limited in usage. Competitiveness is preferable to assess in complex way according to the most accessible indexes and ways of activity. Deviation of competitiveness defines risk, but the value of deviation defines the program, content and investments needed to provide the competitiveness of the organization. Competitiveness of organization is defined by the following: sales volume cost and quantity, profit, share of cost of unsold products in store, competitiveness index, profitability, costs.

Assessment methods can be divided into three categories, they are: qualitative, quantitative and combined. [4] The most objective methods are quantitative because all the results are obtained in digits with the help of which it's easier to compare them. To get more objective result the combination of qualitative and quantitative methods is used. It's called combined method. [5]

However, the methods mentioned in the article do not allow of getting the objective results for competitiveness assessment of the organization, personnel, group of workers, staff of the division and department. Moreover, the methods are labor intensive at initial data collection. [6] That gave cause for usage of advantages of the existing qualitative and quantitative methods in combining them with the other methods. To eliminate the given faults in the existing models of competitiveness assessment the hybrid method of competitiveness assessment of object, organization, staff or product is proposed to apply. Multitude of qualitative and quantitative indexes of any object is suggested to use as a background. In general we shall work with indexes that define its ability to dominate in the specific sphere of activity, then we shall work with ranging with the calculation of weight next following and estimation of integral characteristics along the all levels of hierarchical indexes tree, that makes the competitiveness of object. All mentioned above made us formulate the hybrid method of competitiveness assessment from the point of systems analysis which is suitable for the objects of different origin. Nowadays problems of competitiveness assessment and conducting the optimization of investments are the most popular discussed among the bosses within international forums. The purpose of competitiveness assessment is to reveal the readiness of the staff and individuals the ability to fulfill exactly that kind of activity which they should do. And also the task is to reveal the potential of the workers for defining the professional and official growth.

3. Results

The following methods are worked out basing on contraction of criteria, indexes or characteristics where one scalar criteria instead of multitude of different origin partial criterion is considered. It is acquired by combination of partial criterion. There are multiplicative and additive methods of criteria contraction. The criteria must be commensurate, for example, standardized and weight criteria defined, that characterizes the importance of every criterion. Then we build a new efficiency function and the task of scalar criterion optimization is fulfilled [7].

In this case, the method of transition from several i.e. multitude indexes P_1, P_2, ..., P_m to one specified by a new function $\boldsymbol{Q = \sum(V_j P_j)}$ is called compression or the method of generalized criterion, where $\boldsymbol{V_j}$ are weight coefficient criteria, provided weight sum $\boldsymbol{\sum V_j = 1.0}$ .B It's fairly evident that the more is $\boldsymbol{V_j}$,the more contribution j index is being contributed to the integrated criterion Q.

In this case, for each quality index $\boldsymbol{p_j(a_i)}$ of the object $\boldsymbol{a_i} \in \boldsymbol{A}$ its weight $\boldsymbol{V_j}$ is calculated defining its meaning, after that the weighed sum of these indexes can be considered as a summarized assessment of the competitiveness of the object ai:

$$K(a_i) = \sum_{i=1}^{n} V_j p_j(a_i) \quad (1)$$

The suggested type of competitiveness assessment can be formulated as a three – level structure and modify the existing methods for competitiveness assessment of product, organization or personnel [8]. As a framework it's necessary to use individual, single, group and integral aggregation indices (table 1). In this case at the beginning a hierarchical, usually a three – level structure of indices of the object being under consideration is been graphed.

Table 1. Analyses of the object characteristics.

	X_1	X_2	X_3	X_4	X_5	X_6	X_7			X_8	X_9
							$X_{7.1}$	$X_{7.2}$	$X_{7.3}$		
A_1	150	1998	589	1580	170	12.8	10.2	6.3	7.7	182	124
A_2	150	1998	591	1601	170	11.2	10.6	6.8	8.2	180	18
A_3	150	1998	564	1469	172	11.2	10.6	6.8	8.2	180	31
A_4	150	1998	463	1455	210	9.8	8.20	5.9	6.7	187	0
A_5	150	1998	415	1455	195	11.9	10.5	6.8	8.1	188	7
A_6	141	1997	410	1454	200	11.3	10.8	6.8	8.2	178	62
A_7	167	2384	420	1750	200	11.0	12.8	7.3	9.3	175	2
A_8	148	1998	410	1580	180	11.5	9.60	6.5	7.6	180	85
A_9	152	1798	405	1505	180	9.0	10.6	6.8	8.0	192	6
A_{10}	150	1390	470	1621	200	9.6	10.1	6.7	8.0	192	3

To assess competitiveness of the object and to choose the preferable one we distinguished some models of vehicles: A_1 – Honda CR-V, A_2 – Hyundai ix 35, A_3 – KIA Sportage, A_4 – Mazda CX-5, A_5 – Mitsubishi ASX; A_6 – Nissan Qashqai; A_7 – Opel Antara; A_8 – Toyota RAV 4; A_9 – Skoda Yeti; A_{10} – Volkswagen Tiguan, their characteristics are given in Table 1 according to the indices: X_1 – engine capacity (h.p.), X_2 – volume of engine (cm^3), X_3 – boot volume (l), X_4 – laden mass (kg), X_5 – clearance (mm), X_6 – acceleration up to 100 km/h, X_7 – petrol consumption (l/100 km): $X_{7.1}$ – town, $X_{7.2}$ – motorway, $X_{7.3}$ – mixed; X_8 – max speed (km/h), X_9 – hijacking (item).

To estimate the weight of each index we use the following formula

$$V_i = \frac{S_i}{n^2} \quad (2)$$

Ranging of the indices to define their meaningfulness we follow the rule due to which the domination of matrix elements is defined

$$b_{ij} = \begin{cases} 1, if\ P_i\ equivalent\ P_j, \\ 0, if\ P_i\ ess\ important\ than\ P_j \\ 2, if\ P_i\ dominates\ over\ P_j \end{cases} \quad (3)$$

Ranging operations and defining the weight of indices we conduct foe each group of indices and we put the down into the table 2, for example, economic, technical, comfort ability, esthetics, safety. [9]

By-turn, each group contains the characteristics. We conduct the ranging and also define the weight – meaningfulness of each index.

The above-mentioned method allows us to conduct an activity of indices convolution converting into a single integral one considering the weight of each and their absolute meanings. Its important to mention that the high index value is not always good, sometimes it's vice versa. For example, fuel consumption of the vehicle, it's necessary to introduce 5 or 10 grade assessment of indices Б$_i$ (p$_i$) due to the scale "worse-better", where the worst index gets low grade. For instance, vehicle operation costs to be compared with sale proceeds. Indeed, the higher the content of grade the Б$_i$(p$_i$) the better index [10]. Table 2 shows interval indices values in grades. So, we transform the model of competitiveness of the object to the following

$$К(a_i) = \sum_{i=1}^{n} V_j Б(p_j(a_i)) \tag{4}$$

Table 2. Interval assessments of indices.

Б$_i$ (P$_i$)	**1**	**2**	**3**	**4**	**5**
X$_1$	140.0 – 145.0	146.0 – 151.0	152.0 – 157.0	158.0 – 163.0	164.0 – 170.0
X$_2$	1390.0 – 1589.0	1590.0 – 1789.0	1790.0 – 1989.0	1990.0 – 2189.0	2190.0 – 2390.0
X$_3$	400.0 – 439.0	440.0 – 479.0	480.0 – 519.0	520.0 – 559.0	560.0 – 600.0
X$_4$	170.0 – 177.0	178.0 – 185.0	186.0 – 193.0	194.0 – 201.0	202.0 – 210.0
X$_5$	13.00 – 12.2	12.10 – 11.40	11.30 – 10.6	10.50 – 9.80	9.70 – 9.00
X$_6$	13.00 – 12.0	11.90 – 11.00	10.90 – 10.0	9.90 – 9.00	8.90 – 8.00
X$_7$	175.0 – 179.0	180.0 – 183.0	184.0 – 187.0	188.0 – 191.0	192.0 – 195.0
X$_8$	124.0 – 100.0	99.00 – 75.00	74.00 – 50.00	49.00 – 25.00	24.00 – 0.00
X$_9$	140.0– 145.0	146.0 – 151.0	152.0 – 157.0	158.0 – 163.0	164.0 – 170.0

Thus, the method was created on the basis of qualitative and quantitative indexes of object application basing on requirement usage. The hybrid methods comprises aggregation of a few methods of research: interview method, expert judgment, method of paired comparison, rank method, method of grade assessment, method of integer programming, Hungarian method, reduction method, trial and error method.

So we have a decisive rule: object a$_i$ is more preferable than object a$_j$, if К(a$_i$)>К(a$_j$). Then we find integral indices К$_э$,К$_т$, К$_к$,К$_{эс}$,К$_б$ all through the groups, after that – weight – meaningfulness V$_э$,V$_т$,V$_к$,V$_{эс}$,V$_б$ of group indices, and at last, we find the integral index of competitiveness according to the formula:

$$К(a) = \sum V_i K_i. \tag{5}$$

Thus, we define competitiveness of the object in the large:

$$К(a) = V_э К_э + V_т К_т + V_к К_к + V_{\prime c} К_{эс} + V_б К_б, \tag{6}$$

In the large the structure of object indices consists of several groups: technical, aesthetic, mechanical, safety, economical. Due to this we define the indices weight of group of characteristics by the method of paired comparison:

$$V_э = 0.36V_T = 0.28V_k = 0.04V_{эс} = 0.12V_в = 0.2К_б. \tag{7}$$

Thereupon we write the following target function:

$$К(a) = V_э К_э + V_т К_т + V_к К_к + V_{\prime c} К_{эс} + V_б К_б = 0.36V_т = 0.28V_k = 0.04V_{\prime c} = 0.12V_в = 0.2К_б, \tag{8}$$

After that we conduct the ranging of each group of characteristics by the method of paired comparison, involving experts – specialists. We define their weights, grades of characteristics and integral grades due to their aggregates (table 3). It allows us to calculate the following:

$$К_э = 3.82К_т = 2.8К_k = 1.6К_{эс} = 3.2К_б = 1.1. \quad (9)$$

Basing on the mentioned above we define general competitiveness:

$$К(a) = V_э К_э + V_т К_т + V_к К_к + V_{эс} К_{эс} + V_б К_б = 0.36 * 3.82 + 0.28 * 2 + 0.04 * 2.6 + 0.12 * 3.2 + 0.2 * 2.2 = 2.822, \quad (10)$$

Table 3. Integral grades of indices.

	X_1	X_2	X_3	X_4	X_5	X_6	X_7	X_8	X_9
A_1	0.188	0.064	0.780	0.156	0.10	0.609	0.094	0.23	2.22
A_2	0.188	0.064	0.780	0.156	0.28	0.609	0.047	1.17	3.30
A_3	0.188	0.064	0.780	0.156	0.28	0.609	0.094	0.94	3.11
A_4	0.188	0.064	0.312	0.780	0.38	1.015	0.188	1.17	4.10
A_5	0.188	0.064	0.156	0.624	0.19	0.609	0.188	1.17	3.19
A_6	0.094	0.064	0.156	0.624	0.28	0.609	0.047	0.70	2.58
A_7	0.470	0.080	0.156	0.624	0.28	0.203	0.047	1.17	3.03
A_8	0.188	0.064	0.156	0.312	0.19	0.812	0.141	0.47	2.33
A_9	0.282	0.048	0.156	0.312	0.47	0.609	0.235	1.17	3.28
A_{10}	0.188	0.016	0. 312	0.624	0.47	0.609	0.235	1.17	3.11

Table 4 shows the following criteria of choice: Y_1 – correlation of price and quality (min), Y_2 – horse power cost (rub.), Y_3 – petrol consumption for 60000 km (rub.), Y_4 – cost of three Technical Checkup (rub.), Y_5 – Insurance (Compulsory Automobile Insurance) within 3 years (rub.), $Y_{6,7}$ – cost reduction within 3 years in% (thousand rub.), Y_8 – quality integral grades (max), Y_9 – total cost of possession.

Table 4. Calculations of choice criteria.

	Y_1	Y_2	Y_3	Y_4	Y_5	Y_6	Y_7	Y_8	Y_9
A_1	549346	8127	192780	30973	16632	31	841	2.219	618275
A_2	324332	7127	200340	18870	16632	42	620	3.296	684822
A_3	356995	7399	200340	26442	16632	43	632	3.109	720671
A_4	304177	8300	154908	26850	16632	30	871	4.093	571590
A_5	332287	7060	198450	23348	16632	25	794	3.187	503108
A_6	399534	7305	204120	29692	14256	38	638	2.578	639468
A_7	374340	6796	241920	27565	19008	31	783	3.032	640043
A_8	474023	7459	181440	26500	16632	31	762	2.329	566812
A_9	313528	6770	200340	18347	19008	40	618	3.282	649295
A_{10}	337837	7000	190890	25867	16632	41	602	3.108	681559

Various criteria given in the table 4 denote the choice A_5 – Mitsubishi ASX.

IOP Conf. Series: Materials Science and Engineering **483** (2019) 012104 doi:10.1088/1757-899X/483/1/012104

4. Conclusions and recommendations

The obtained results let us evaluate competitiveness of other objects to be compared. Thus we can choose one object out of multitude or formulate a set of choices in the long run. We can also conduct a comparison evaluation of competitiveness of organizations, enterprises, personnel. We can work out some management measures for various spheres of activity.

It's necessary to point out that the attention must be given to the analysis of risks in the system of organizations for it's a factor of competitiveness raise. Risk analysis is on of the widely spread in spheres of production and trading.

The given hybrid method can be used for competitiveness assessment of personnel where the etalon-indicators are the initial data. They are the requirements for every occupation (competence). Then the comparison of individual person of the staff is made both positive features (communicative skills, professional experience) and negative (aggressiveness, contentiousness, irritability). After that the comparison of the list of characteristics, ranging, weight defining is made. After that we proceed to the formulation of the integer-valued programming task which can be solved with the help of Hungarian method available for users in the internet. Consequently, the most complicated, force consuming and important is the collection of the initial data and the task formulation.

To be concluded, the hybrid method of competitiveness has bee made basing on the implementation of quantitative and qualitative indices of object of different origin. The hybrid methodology includes a number of research methods and techniques: method of interview, method of expert assessment, method of paired comparisons, ranging method, grading evaluation, method of integer-valued programming, Hungarian method, method of reduction, method of trial and error. The hybrid method of competitiveness assessment has lots of advantages those are the various quality methods of analysis that means result comparison convenience all through the customers levels.

References

[1] Bezrukova T L, Stepanova Yu N, Shanin I I, Busarina Yu V and Nesterov S Yu 2017 *European Research Studies Journal* **20 №3** 183-92

[2] Gashenko I V, Romanov D G, Vokina S G, Bezrukova T L and Kozenko Y A 2016 *Contemporary Economics* **10 №4** 363-72

[3] Han Y, Jeong J, Ko M H, Lee S and Kim J 2018 *ICIC Express Letters, Part B: Applications* **9(4)** 339-46

[4] Lau H C, Ip A, Lee C K M and Ho G T S 2018 *Benchmarking* **25(7)** 2216-29

[5] Zakeri A, Saberi M, Hussain O K and Chang E 2018 *International Journal of Fuzzy Systems* **20(4)** 1224-39

[6] Bobel I 2017 *Competitiveness Review* **27(4)** 433-435

[7] Kim Y 2017 *European Journal of Operational Research* **262(2)** 771-9

[8] Gornostaeva Z V, Povalyaeve V A, Tepina Y V, Trgulova N G and Kushnaryova I V 2018 *Espacios* **29(1)** 21

[9] Kuhlman C, Ramamurthy K N, Sattigeri P, Lozano F C, Mojsilovic A and Varshney K R 2017 *IBM Journal of Research and Development* **61(6)** 81673366

[10] Caseiro N and Coelho A 2018 *Competitiveness Review* **28(2)** 213-26

IOP Conf. Series: Materials Science and Engineering **483** (2019) 012105 doi:10.1088/1757-899X/483/1/012105

Reclamation of disturbed lands in the municipal district of Yekaterinburg city for sustainable development

E Wegner-Kozlova[1], O Guman[2]

[1] Federal State Autonomous Educational Institution of Higher Education «Ural Federal University named after the first President of Russia B.N. Yeltsin», 620014, Ekaterinburg, Lenin str., 13B; Institute of Economics The Ural Branch of Russian Academy of Sciences, Russia, 620014, Ekaterinburg, Moscow str., 29.
[2] Ural State Mining University

e-mail: katya.human@mail.ru, guman2007@mail.ru

Abstract. The current issues of sustainability are determined by a vital task to combine economic interests with preservation of natural balance of the ecological system. In this context, the key measures to solve a problem of disturbed lands in the municipal district of Yekaterinburg city as one of the factors of unsustainability were proposed in the article.

1. Introduction

Standards of social and economic sustainability of lands aimed at preserving its natural capital make it possible to provide such a quality of life that meets the high requirements of the international community. However, the disturbed lands are one of the signs of unsustainability (its area in the municipal district of Yekaterinburg city is 1,337.64 Ha). It is noteworthy that the modern process of achieving sustainability is not adequately supported by the efficient methodological approaches and practical tools to meet the current ecological and economical challenges. It includes a legal mechanism of public regulation in the spheres of natural resources use, environment protection, and ecological safety

2. The urgent character of sustainability

It is difficult to define the subject of management in terms of a category of natural resources. On one side, the natural resources may be subject to private, state, municipal, and other forms of ownership (par. 2 article 9 of the Constitution of the Russian Federation), i.e. it may be not only state property. On the other, the constitutional law contemplates that it is a duty of the state to protect the natural resources as the basis of the life and activity (par. 1 article 9). That is, the state considers all resources as a national asset and undertakes to protect it as an element necessary for life in the state territory, regardless of the form of ownership of these resources.

The urgent character of sustainability-related issues is caused by the vital task to combine economic interests with preserved natural balance of the ecosystem. The efficient integration of Russia into the global economy and international market depends on a competitiveness level of its markets. In terms of this task, the technological leadership is one of the key factors of enhancing the efficiency of the markets. At first seem, the task to preserve the ecosystem stability is not supposed to be the critical one. However, it is obvious today that the unaddressed ecological problems can have a long-term effect, e.g. they can be reflected by the economic indicators. Since the negative environmental impact compromises a quality of natural and human resources, it assumes taking the coordinated actions in all the spheres of life and conducting structural reorientation of social, economic, and environmental institutions. It is worth pointing out that the task to develop sustainability is urgent not only for Russia, but globally. It is supported by the United Nations appeal to achieve sustainability standards in 2030.

Guidelines for sustainability under market conditions were documented by an order of the President of the Russian Federation dd. February 4, 1994 No. 236. The order provides the realization of rights established by the Constitution of the Russian Federation, namely the right of citizens to a favourable environment, the rights of future generations to enjoy the natural resources potential in order to maintain sustainability. It also provides a solution to current social and economic challenges in strong connection with the relevant measures to protect and develop environment, and to preserve and restore natural resources. They include:

- environmentally sound siting of labour forces
- environmentally safe development of industry, power production, transport, and public services
- environmentally safe development of agricultural sector
- sustainable management of renewable natural resources
- sustainable use of non-renewable natural resources
- increased use of secondary resources; recovery, pro-cessing, and disposal of wastes
- management development in a sphere of environmental protection, natural resource management, emergency prevention and control.

From a methodological point of view, the sustainability concept is considered as a target paradigm of management of economic systems, and it supposes the whole natural and economic complex of large territories (regions and municipal entities) with its traditional economic features to be the object of management. The Urals is one of the oldest mining regions in Russia, and it is one of the Russian regions with the largest volume of known and anticipated mineral reserves. It predetermined the extensive growth of ferrous and nonferrous metallurgy, construction, chemical production, extraction of mineral resources, including gold extraction. Broadly for this reason, there is a stable growing problem of environmental deterioration and degradation of living conditions for critical ecological indicators in the region. The land allocation for transport infrastructure facilities, urban development, waste landfills, tailing and waste dumps - all of these have a negative impact on the environment; the transport networks linking the localities increase the degradation of natural landscape [5] and the area of disturbed lands. The high percentage of disturbed lands is one of the factors of unsustainable regional development.

3. The disturbed lands of the municipal district of Yekaterinburg city

The current global estimates suggest that almost 3.7 billion people are living in cities today and the number is going to increase even further.[8],[1] In fact, for the very first time, cities have been given priority in the global development agenda, due to the progressive rise in the number of people living in urban areas (viz. by the end of 2050, two out of every three people will be living in cities), and because cities can play a defining leadership role in the management of issues like climate change, public health, and food security.[8],[1],[2] Over the next 15 years till 2030, the global community and the member states have committed to accomplish the proposed 17 sustainable development goals (SDGs)[8] and the one of them is SDG-11 which implies making cities and human settlements inclusive, safe, resilient, and sustainable.

The disturbed lands are the lands that lost its economic value, or the lands that are a source of negative impact on the environment due to disturbance of topsoil and hydrological regime, and formation of man-made relief as a result of production activities. In terms of the area of disturbed lands, Sverdlovsk region has occupied one of the leading positions among Russian regions during the last decade. According to the statistics, the area of disturbed lands in Sverdlovsk region was equal to 62,000 Ha in 2016, that is a 400 Ha (0.65%) increase compared to 2014 [6].

Distribution of the disturbed lands of Sverdlovsk region by land categories:

- Reserve lands 1364 Ha
- Forestry fund lands 27156 Ha
- Lands of specially protected territories and objects 248 Ha
- Agricultural lands 26722 Ha
- Residential areas 2790 Ha

– Lands of industrial and other special designation 3720 Ha

It is linked with the increasing area of disturbed lands within the area of lands of industry, power production, transport, communications, radio broadcasting, television, and information, and lands for ensuring outer space activity, lands of defense and security, and lands of other special designation. It can be explained by the reassignment of land plots to this category with the purpose of exploitation, as well as by the commencement of such works in the land plots granted previously (Revdinsky, Rezhevskoy, Nizhneserginsky, and Neviansky districts). Alongside with the exploitation, the land users perform the remedial actions to reclamate the previously disturbed lands [3]. The department of the Federal Service for Supervision of Natural Resources for Ural federal district (Report No. 2-TP (Reclamation of lands) reports that the area of 56,160 Ha of disturbed lands was reclamated in 2015 [3]. However, the reports on form No. 2-TP (Reclamation of lands) for the year of 2015 in Ural federal district were presented only by 237 enterprises [6].

The disturbance of lands within the municipal district of Yekaterinburg city is connected with the exploitation of construction materials, peat mining, terrain relief changes caused by unauthorized excavation, disturbance of land runoff, bogging, accumulation of construction, industrial, and domestic wastes.

The disturbance of lands is a result of natural and man-made processes (the latter prevails in Yekaterinburg city). These processes lead to the deterioration and/or destruction of soil and plant cover, and other quality changes in the state of lands. The man-made disturbance of lands is caused by mining operations, extraction of minerals, geological exploration, surveying, construction, and other works, as well as by operation of city services for residential area development. The category of disturbed lands includes also the degraded lands, where there is a stable decrease in quality and quantity of soil composition and its characteristics. This decrease is induced by the anthropogenic and natural-anthropogenic factors.

According to the data of Uralgeoproekt, LLC in 2011 the total area of disturbed lands in the municipal district of Yekaterinburg city was 1,337.64 Ha, that is equal to 1% of the new area of the municipal district of Yekaterinburg city. The stated area is 6.5 times larger than the lands of specially protected territories and objects. Relative to the land categories, these figures are:

– 8% of the agricultural lands
– 18% of the lands of industry, power production, transport
– 5% of the forestry fund lands
– 30% of the water fund lands

The total area of disturbed lands includes:

– 621.4 Ha of cuts
– 402.9 Ha of fills and piles
– 34.9 Ha of settlers
– 25 Ha of solid waste landfills
– 193.45 Ha of boggy areas
– 60 Ha of sludge reservoirs

The disturbance extent of lands of the municipal district of Yekaterinburg city can be expressed in terms of Disturbance coefficient that is equal to 1.17% (the coefficient is calculated as the ratio of the total area of disturbed lands to the area of the municipal district of Yekaterinburg city).

The monitoring of the disturbed lands of Yekaterinburg city shows that the disturbance extent of city districts varies from 0.98% to 1.52%. The disturbance extent is:

- 0.98% in Oktyabrsky district
- 1.3% in Ordzhonikidzevsky district
- 1.32% in Chkalovsky district
- 1.4% in Verkh-Isetsky district
- 0.41% in Zheleznodorozhny district
- 1.5% in Leninsky district

IOP Conf. Series: Materials Science and Engineering **483** (2019) 012105 doi:10.1088/1757-899X/483/1/012105

- 1.52% in Kirovsky district.

The key problem of land protection is the non-compliance with legal requirements that regulate the restoration of lands disturbed as a result of operational activity. A way the disturbers compensate damage caused to the lands as to the object of environment protection, it is an important tool that prevents and represses violations of the land legislation, and it characterizes the efficiency of state land supervision [5]. Article 13 of the Land Code of the Russian Federation dd. October 25, 2001 No. 136-FZ (rev. dd. July 3, 2016) (as amended , effective as of January 1, 2017) sets forth the responsibility of owners, landlords, tenants, and other land users to reclamate the disturbed lands, to restore the topsoil, to rehabilitate the lands in timely manner.

4. The key measures for solving a problem of disturbed lands

It should be noted that the legislation of the Russian Federation was deeply amended for the last several years, and these changes help to gradually solve the problem of disturbed lands. Firstly, the administrative fines for non-fulfillment or untimely fulfillment of obligations to reclamate lands were increased. According to art. 8.7 of the Administrative Offenses Code of the Russian Federation (as amended by Federal law dd. March 8, 2015 No. 46-FZ), such nonfulfillment or untimely fulfillment of obligations leads to an administrative fine ranging from twenty thousand to fifty thousand rubles for citizens, from fifty thousand to one hundred thousand for officials, and from four hundred thousand to seven hundred thou-sand for legal entities. Until recently, the administrative penalty for nonfulfillment of obligations to reclamate lands was not so strict (a fine was from one thousand to one thousand five hundred rubles for citizens, from two thousand to three thousand for officials, and from forty thousand to fifty thousand for legal entities), that contributed significantly to the negligence in fulfillment of the obligation set forth in art. 13 of the Land Code of the Russian Federation. Secondly, Federal law dd. December 28, 2016 No. 488-FZ On Introducing Amendments to Certain Legislative Acts of the Russian Federation also contributes to the tightening of penalties for a breach of obligations to reclamate lands, since it allows the creditors to make the subsidiary liability for entity's debts be hold by a person, authorized by law or other legal or constituent instrument to act on behalf of a debtor. Therefore, the liquidation of a legal entity is not a probable cause to avoid responsibility. Thirdly, there appeared a possibility to use strip mining, host rock, waste from ferrous metals production of hazard classes IV and V in course of the liquidation of mine workings according to a project of mine workings liquidation (art. 11 of Federal law On Production and Consumption Waste dd. June 24, 1998 No. 89-FZ).

However, the use, allocation, and reallocation of land resources on a city scale have its own significant features. The urban land use is multifunctional; it is a clash of land and property interests of different economic sectors, the interests of municipal services, individual land owners, landlords, land users, tenants of land plots. The regulations on use of lands of different types are applicable simultaneously in the cities, but the urban development regulations are considered as of primary importance. On a city scale, there should be a balance between the citizens' needs for comfortable life conditions and natural resources preservation [7].

These issues are current for the city of Yekaterinburg. The projection of monitoring of the disturbed lands in 2008-2011 shows that the area of disturbed lands in the municipal district of Yekaterinburg city was reduced slightly. In general it is caused by:

- unauthorized waste disposal or soil excavation
- formation of waterlogged and boggy areas due to the "patch" construction with no complex approach to the surface runoff drainage
- growth of the area of the city in 2008-2011
- and abandoned deposits, since at the end of an agreed period of works the license holders usually go bankrupt or disappear or sell a deposit to the firms that have no funds to reclamate it.

Furthermore, the current legislation does not consider the growth of city area. For example, par. 5, art. 12 of Federal law On Production and Consumption Waste dd. June 24, 1998 No. 89-FZ prohibits the waste disposal within the inhabited localities. But the growth of city area leads the new territories

IOP Conf. Series: Materials Science and Engineering **483** (2019) 012105 doi:10.1088/1757-899X/483/1/012105

to be developed for the periods much shorter than the lands need to be reclamated, so the life conditions become worse thereafter.

The role of the municipal authorities in the reclamation of lands is restricted in an explanation of ruling of the Constitutional Court of the Russian Federation dd. October 13, 2015 No. 26-P. The position of the Constitutional Court of Russia on the responsibility for disturbed lands is defined in par. 24, part 1, art. 16 of Federal law On the General Principles of the Organization of Local Government in the Russian Federation. It does not provide the liability of the local government authorities of urban districts to liquidate an unauthorized disposal of domestic and industrial waste, left by unidentified person in the land plots, at the expense of local budget, and it does not authorize the local government authorities with the relevant state powers.

Thus, the subjects of business and other activities that have a negative impact on the environment must bear the full costs of its reclamation. If a person guilty of damaging is unidentified, the reclamation of the relevant land plots is performed by its proprietors. The registered owners have both a right and an obligation to preserve lands that is stipulated by law.

The disturbed lands of the municipal district of Yekaterinburg city are subject to the different forms of owner-ship: private lands, municipal lands, and lands of the Russian Federation. The abovementioned ruling is sup-posed to explain the low figures of target performance indicator of a municipal program, regulating the land resources management in the municipal district of Yekaterinburg city in 2017-2020. This target performance indicator reflects the area of disturbed lands, and for the indicated period it should be equal to 24 Ha, i.e. 6 Ha per year or approximately 1.8% of the total area of disturbed lands in the city.

– To sum up, considering the need to achieve sustainability the following measures can be recommended to solve the problem of disturbed lands:

– To formalize by law the audit of the disturbed lands of regions, cities, and towns
– To create a unified data base, the use of which should also be reflected in law
– To capture in law a supervision of reclamation of disturbed lands and an extension of monitoring after a completion of project of its reclamation
– To tighten penalties for failure to submit a report of form No. 2-TP (Reclamation of lands)
– To determine a list of waste, non-toxic and of hazard class 5, for reclamation of disturbed lands and to justify its possible use in the inhabited localities.

The following recommendations for the municipal authorities should be indicated:

– To enhance the general urban plan of Yekaterinburg city with the zones of disturbed areas that need to be reclamated
– To reclamate disturbed lands and rehabilitate it according to the urban planning zones of the municipal district of Yekaterinburg city, to begin with disturbed areas in the territories intended for building
– To organize monetary funds for the reclamation of disturbed lands with the local government authorities, in compliance with the projects of deposit mining and waste disposal. Then these funds should be used for its intended purpose to reclamate the disturbed areas under super-vision of the local government authorities
– To prepare the lands, that is to drain the peat bog areas and to backfill peat long before selling these lands to the developers. The costs of these activities should be then included in a selling price.

owners, landlords, tenants, and other land users to reclamate the disturbed lands, to restore the topsoil, to rehabilitate the lands in timely manner.

5. Conclusion

Today the objectives of sustainability define the present and future quality of life, thus the economic development of territories should be long-term, environment-friendly, and multi-tasking. To achieve sustainability is the most critical challenge for the entire world, including the Russian Federation. Its solution will be facilitated by driving attention and interest to the indicated problems of the state and municipal authorities, and the citizens of the state.

References

[1] N S Valgina, 2001 "Active Processes in Modern Russian Language: textbook for University students," (M.: Logos) 304 p

[2] Caiaffa W T, Friche A A, Danielle C 2015 Urban health: landmarks, dilemmas, prospects, and challenges *Cad Saude Publica* **31** 5-6
[3] Hodson R 2016 Urban health and well-being *Nature* **531** 49
[4] Report On Condition and Use of the Lands in Sverdlovsk Region in 2015 p 38 Available: https://rosreestr.ru/site/open-service/statistika-i-analitika/zemleustroystvo-i-monitoring-zemel66/
[5] Shrivastava, Saurabh R, Shrivastava, Prateek, Ramasatny, Jegadeesh 2017 Improving urban health standards: Promotion of equity and development of healthier cities for sustainable development *Annals of tropical medicine and public health* Vol 10 **2** 301-302
[6] State report On Condition and on Protection of the Environment in the Russian Federation in 2015. Moscow, Ministry of Natural Resources and Environment of the Russian Federation 2016 (NIA-Priroda) p 109
[7] State report On Condition and on Protection of the Environment in Sverdlovsk Region in 2015, pp.65 Available: http://www.minprir.midural.ru/gosudarstvennye-doklady
[8] V G Bryzhko 2016 Reclamation of Disturbed Lands on a City Scale *Fundamental Investigations* **6** 134-138
[9] World Health Organization. Global report on urban health: equitable healthier cities for sustainable development 2016 (Geneva WHO Press) 1-26

IOP Conf. Series: Materials Science and Engineering **483** (2019) 012106 doi:10.1088/1757-899X/483/1/012106

Parcelled phrases in the aspect of business communication

O V Nikolenko, A V Belozerova, N V Sumina, E Yu Shapovalova

Don State University Technical University Rostov-on-Don, Russia

Olganikolenko@mail.ru, belochka-04.80@mail.ru, Natalia.sumina@rambler.ru, elena19982002@gmail.com

Abstract. The article is devoted to the study of the causes of communicative barriers in the process of dialogue at manufacturing. Special attention is given to parcelled statements as they function in an oral text, having a double semantic load and expressing in addition to external substantive contents the author's implied meaning, which is often a priority in the information perspective, eludes the addressee. The relevance of the proposed work is due to the need for a detailed description of this area of business communication (alongside with the expressive syntax) and identify the causes of misunderstanding of the very nature of the designated language phenomenon by recipients entering into speech cooperation.The expediency of the consideration of the parcelling in the designated key is justified by the fact that it, representing a multilevel semantic structure in structural terms, in addition to the visual shell endowed with a deep subtext interpreting the linguistic picture of the social environment, acts in the business style as an auxiliary element serving as a certain universal code for the allocation of the information segment in the speech flow and the strengthening of suggestion. On the other hand, the incorrect division of phrases in the speech flow not only undermines the literary norms, but also can cause communicative difficulties for the negotiators. Consequently, the strategy of conducting business conversations should be carried out taking into account the complex knowledge of psycholinguistics, culture of speech and office work in order to both recognise the parcelled units in the lexical and grammatical complex and take into consideration their structural and semantic features in the production of the text. This, in our opinion, forms a successful personality in terms of communication.

1. Introduction

The anthropocentric paradigm of General linguistics, psycholinguistics and office-work determines the public interest in expressive phenomena in language in order to use them as a tool of suggestion in all areas of speech interaction, including business communication. Scientists state the fact that any modern text tends to move from the "continuity of the chain <...>, the bright expression of subordinate relations <...> to the actualised constructions with implicitly represented syntactic connections (without special signals of these connections), to the dissection of grammatical structures, the ultimate independence of their individual components" [1], functioning in the form of parcelled syntaxemes, i.e. "division of the sentence, in which the content of the statement is implemented not in one, but in two or more intonation-semantic speech units, following one after another, after the separation pause" [2]. This emphasises the " belonging of the parcelling as a means of semantic division of the message content to the communicative expressive syntax "[3], endowed with "great opportunities, both to strengthen and, on the contrary, to weaken the expressiveness of the statement" [4], to clothe it in an expressive shell , which is "an instrument of knowledge of the inner world of man, his external manifestation, < ... > the text characteristic, which is formed in the process of generating the text for entry into artistic " [2] and manufacture communication. In addition, the parcels create a stylistic effect, emphasise an information segment in the speech flow.

2. Problem Statement

This syntactic versatility of parcelling raises a number of debating problems, the solution of which is the subject of numerous works of scientists-linguists (A. A. Dobrycheva, O. P. Carcoshco, R. I. Zelepukina, A. Z. Khaimursinova, A. N. Kalinina, etc.). Identifying piculiarities of the indicated device from the related syntactic phenomena, the above mentioned linguists argue that parcelled constructions, possessing all the features of the text (cohesion, consistency, modality, informativeness, etc.), emphasise the dynamic status of the parcelled phrase (with the possibility of syntactic unification the parcellates and basic parts in a single structure) to concretise its denotative plan and stir up its expressiveness and emotionality. However, the implementation of semantic relations between the parcellates remains an insufficiently studied aspect in their works.

3. Main Part

The parcelling divides the statement in a special way: on the one hand, a structurally fragmented syntaxeme conveys diversity in unity; on the other hand, the events described sequentially are presented in a separate narrative frame:

I am a communicative person ... Different creative ideas ... And also, quite a good artist.

This example clearly shows that the division of the monological phrase contributes to the transfering of both explicit and implicit information (about additional professional qualities).

Typically, the parcelling disrupts the rhythm of the narrative, creating a pause to further enhance the expression of the unexpected development of events:

But maybe the situation is not going to be like that ... Maybe, something terrible is going to happen ... either to the firm or to the employees.

Another example:

I started my career as a manager in a small company ... Worked there for 2 years. Then I moved to another organisation where I work at the moment / as an accountant.

The intonationally indicated segment *worked there for 2 years* helps to convince the listener of the professional experience of the candidate and pause at the word *accountant* indicates a certain production status.

As a rule, with simultaneous parcelling and emphasis, the sentence is divided into two logical parts, the first of which acts as a certain signal of information supply in order to attract the attention of the addressee of speech, and the second develops the idea of what was said:

- *Why did you apply for a job in our company?*

- *You have a great reputation, and I hope to gain more opportunities ... and the prospect of growth in your company.*

Linear parcelling of the narrative is based on the compositional intention to highlight the details of the picture in general, figuratively to stress a significant section of the speech.

Linguists have repeatedly observed that any part of the sentence can act as a parcellate. In linguistics, "a number of permanent characteristics inherent to parcelled structures has been identified in linguistics: predominance of parcelled adverbial modifiers over the other parts of the sentence, relatively stable parcelling of indirect object, the predominance of parcelling of compound sentences over complex sentences, etc.," [5]. However, particularly interesting to us are the examples involving the parcelling of the lexemes, linguistic nature of which is designed to perform incidental functions (introductory words, participles, particles, etc.).

Our firm was not involved in this incident... Naturally.

In this case, thanks to the parcelling, the introductory word naturally appears in the position of the actualised rheme, which conveys not only confidence in the existence of the fact, but also forms a deep subtext: the speaker's desire to emphasise the exclusivity of his morality, as well as an invitation to the addressee to share this opinion.

Parcelling enhances the psychological contrast in any speech:

The old and so true saying: those who have hard times in life are responsive to the misfortunes of others ... Tatyana Dmitrievna has had tough life – conversational-everyday style;

"I Am Hamlet...a burden line ...
And there is no rhyme ... no plot ... " - poetic speech;

Recently at the company there has been a sharply increased turnover. If you look at the contingent of personnel by age, we basically lose our main strength, that is, those employees who already had a fairly high qualification, experience and level of professional knowledge. Simply put, we lost those who in a few years had to take a higher position ... Future chief specialists ... heads of departments and their deputies. - official business style.

As you can see, structurally, the second phrase serves as a predicate, an indicator of which is the possible introduction of the lexeme is between the parts: *"I am Hamlet" – is a burden line... in which there is no rhyme and no plot.* Semantics of the parts testifies that their semantic relations are based on the certain segment partially verbalised by the word burden. The parcelettes ... *there is no rhyme ....and no plot ...* are cut off from the main sentence, which gives the whole phrase an intermittent character. Moreover, the second syntaxeme *...and no plot ...* serves as additional joining, filling all the statements with psychology and creating the impression that it (the statement) is born at the moment of pronouncing. Singling out syntaxemes "*The future chief specialists ... heads of departments and their deputies*"deprives the phrase of the enumerative tone and opens the veil of the intriguing beginning. All this strengthens logical and semantic accents, dynamism, stylistic tension.

In formal speech parcelled constructions with attributive meaning, in which one semantic node is in relation of submission to another node, having by-word or adnominal-attributive character, are quite frequent:

First of all, we propose to change the shift schedule and the rules and regulations ... inner rules and regulations. And this will require serious analysis and training.

In the example, the sign contained in the parcel definition is subjected to emotional and expressive underlining: the parcel performs a pictorial function, participating in the creation of an image, the nature of which affects the intonation pattern of the phrase.

It is thought that the presence of attributive meaning is another feature of parcelling that distinguishes it from a number of other artistic means of expression used in business communication.

Semantic shades in such constructions can be caught only in the sounding speech due to its rhythmic-intonational design, as well as on the basis of knowledge of the situation reflected in the statement:

And the participants ... were given accommodation on the lower floors! (from the speech).

Compare: *And the participants who were at the meeting were given accommodation on the lower floors;*

And participants, if they were at the meeting, will be given accommodation on the lower floors.

In the written speech of colloquial subjects there are parcelled constructions, in which the identification of the basic typical relationship between the semantic segments is difficult. It is often impossible to determine which part is the priority in semantic perception:

So many things to say ... Need to finish.

Compare: *I want to say a lot, but we have to finish (opposition);*

So many things to say, although it is necessary to finish. (concessive meaning).

Structural changes of such statements represent a semantic variation line and make no specification in the phrase. In this regard, in such constructions the hidden mediating meaning cannot be fully restored without knowledge of the situation and the corresponding intonation.

Another feature of the structural and semantic organisation of the parcelling is the presence of mediated semantic links between its parts in the development of the second syntagma of the word, which is absent in the first part (zero sign), which is replenished by the context, or in the introduction of a hidden mediating meaning, not expressed verbally. It is thought that such constructions are close to the constructions, called homonymous in the Russian language, i.e. having an ambiguous structure, "for which several grammatical meanings are fixed, one of which is actualised in speech" [6]:

Dear ladies and gentlemen! Turn your heads to the right ... This is a mockup of our new building!

Compare: *Dear ladies and gentlemen! Turn your heads to the right and look at the layout of our new building!*

Dear ladies and gentlemen! Turn your heads to the right. This is a mockup of our new building!

It is seen that the content of the original construction can be interpreted in two directions: with the introduction of the zero lexeme, look without division into speech segments, the construction will be a complex sentence of expository character; without the material embodiment of the reference for the second part of the word, the statement turns into a microtext describing the situation in different ways, without grammatical dependence between the components of the statement.

The analysis of the parcelled constructions in the pragmatic aspect with an emphasis on the addressee of the statement is particularly relevant in the light of the anthropological setting of modern linguistics. It seems that with the perception of the parcelling in the text, the reader alongside with the author is following the same path of structuring and restructuring of linguistic units and independently filling in the semantic gaps between parcellates [7], so understanding the nature of the verbal and cogitative processes in the construction and perception of parcelled texts is especially necessary:

In just a month you will be able to understand Chinese well. Or you won't. It all depends on what decision you make now.

The information is supplied by the copywriter taking into account the perception of the text by possible clients: the first basic part of the content is an ordinary advertising proposal; the second part is a verbalised possible objection of the client, to which the author of the statement immediately gives an answer with the third parcellete (and only one of several possible content – beneficial for the author of the multicomponent construction).

The use of the parcelled units in advertising and in psychological practices is explained by their ability to strengthen the plan of suggestion, so their study is promising "to create automatic systems for assessing the strength and quality of the suggestive effect of advertising and other texts of influence" [8]:

"Luxury gate": Protect your home. Around the clock.

The parcellete placed in a separate sentence is loaded semantically and emotionally and receives the position of the actualised rheme (compare: *Protect your home around the clock),* so the reader restores implicitly represented semantic links by the semantic context: just like the gates protect the house around the clock so as the firm producing them provides one hundred percent (valid hours) warranty.

Communicative task of the speaker is a leading factor in the construction of the real text in accordance with the theme and the situation of communication. This provides an indissoluble unity of the external structure of the statement, its internal content and functional purpose.

To attract the attention of a business partner to specific information or to strengthen the meaning of the word, a person resorts to the use of emphases.

Emphatic intonation is a complex rhythmic pattern, timbre bright colors and a dot accent:

Today / the negotiations were successful – the tone of comparison;

Today / negotiations were / successful – a statement of fact.

As you can see, the emphasis, focusing on certain lexemes, reflects the communicative intention of the speaker to emphasise the basic semantic components of the message.

Obviously, due to its auditory nature, the emphasis is realised in the oral statement to a bigger extend; in written texts (especially artistic), the concentration of attention on a particular speech area is achieved through the prism of its perception by the reader or the introduction of the author's signs.

"Calm Down, Mary. I'm / / Dubrovsky," - said the hero of Pushkin. Setting dash after a personal pronoun "forces" the reader, on the one hand, to pause, and on the other hand – to discover new personal qualities of the character.

With the help of punctuation marks speakers in the non-audial text clearly convey important, in their opinion, information and " I-position", which are not difficult for the recipient to read":

The cause has been dealt with ... We have real competitors. Let's now look at how to increase the interest of our employees, what we will motivate them by, at least at the time when we will conquer the market. Who would like to speak on this issue?

Or in a literary text:

And if happens once in a century HISTORY, remember long, reverently transfer from mouth to mouth with the hidden pride – as it happened to us, not at neighbours from Markushev, that only were well-known for on the district that the terrible ensign Savelyev Slavka, the trouble-free repairman of personal officer equipment ... [16].

From the point of view of grammar, culture of speech and graphics, the above excerpts do not correspond to any punctuation, no speech, no graphic standards, but we can say that it is a deviation from them that allows the author of the statement to remove information veils and convey to the reader significant shades of meaning.

The emphasis on individual elements of the sentence can be carried out not only by means of a special intonation, but also by varying the length of the sounds:

You have / b-i-i-i-g problems!

If we consider the nature of language, including the consideration of the data of phonology, we can see that vowels, as well as sibilant and explosive consonants are the most frequently used elements in the emphatic elongation, because they are able to increasingly clothe the word in a melodic shell, as well as to make a hidden meaning: for example, " o " provokes the image of something holistic, three-dimensional; "u" conveys the depth of thought:

Thi-i-i-ink about that!

"Sh" lulls the consciousness:

I'm yours!

This kind of emphatic models are not considered as a deviation from the norm or its violation: they serve as an additional source of giving expression and emotionality to the phrase.

The absence of the author's distinctive semantic signs transfers the statement into a zone of "black square", where its addressee is "offered" to stir imagination and independently interpret the content component. We will illustrate this idea by the example of the (CEO) Chief Executive Ofiicer's speech to the workforce:

After all, we are already ahead now on this indicator and we will fail to significantly raise a salary now ... to us it will be simply economically not profitable. - a direct causal result at parcelling;

After all, we are already ahead now/ on this indicator and we will fail to significantly raise a salary now /... to us it will be simply economically not profitable – temporal priority;

After all, we are already ahead now on this indicator and we will fail to significantly raise a salary now ... to us it will be simply economically not profitable – subject comparison;

After all, we are already on this / parameter, and to significantly raise wages now, we will fail... We just it is economically not profitable – defining-indicating uniqueness.

The meaningful interpretation of such parcelled emphatic constructions depends on the correlation of the accentuated element with the expression of the speaker's communicative intention and on the conditions of its semantic and syntactic design.

The distinctive dominance of certain lexemes serves as an indicator of their communicative significance: they not only actively realise their possibilities of accentuation, but also are characterised by the greatest use.

On the other hand, in our opinion, it is necessary to approach the use in the language of parcelled constructions structurally close to the complex sentence with caution, since it is this layer of syntaxemes that generates grammatical mistakes in the statement due to the wrong understanding of the semantic relations between the parts of the parcelled utterance:

Thus, our employees are deprived of such an important incentive factor as moving to work in Astana. So we need to master the market of Astana.

This kind of parcelled statements are characterised by uncertainty of semantic content and syntactic ambiguities of form that brings them closer to weakly structured units acceptable in oral conversation.

In the scientific literature, such constructions are regarded as structurally incorrect, since the subordinate part cannot be extracted from the limits of dependence in the area of independence. In addition, the business philosophy of the businessman requires from the chief the demonstration of speaking skills to create coherent, logical, "classic" text that contains syntactically deployed phrases with explicitly expressed subordinating relations.

Attention should also be drawn to the fact that the semantic relations in the parcelled constructions found in the official texts are deeper in the denotative subtext than in similar statements built on commonly used colloquial models.

4. Conclusion

The semantic and syntactical features of the parcelled constructions considered within the framework of this research allow to speak about their wide use in the oral official and business statement. The fact of identity of functioning of the described syntaxes in other styles confirms the idea of significant influence of the oral form of speech production on the formal-semantic side of the statement. Adequate perception of the semantic side of the parcelled units is provided by a number of factors, which include super-segmental means, context and knowledge of the situation of speech. These factors, capable of creating a variability of semantic relations, should be taken into account in the selection and presentation of the official text.

References

[1] N S Valgina, 2001 “Active Processes in Modern Russian Language: textbook for University students,” (M.: Logos) 304 p
[2] D E Rosenthal, M A Telenkov, 1976 “Dictionary of Linguistic Terms” (M. Education) 399 p
[3] Linguistics. Great Encyclopedic Dictionary 1998 (M.: Great Russian Encyclopedia) 685 p
[4] V V Vinogradov 1958 From the History of the Study of Russian Syntax (textbook, M. Moscow State University Publ) 174 p
[5] Yu V Bogoyavlenskiy 2013 Parcelling the Headlines: a Dynamic Aspect (on the material of the newspaper "Kommersant") *Vestnik of Tomsk State University* **373** 7-13
[6] O V Nikolenko 2007 Semantic Variation of Complex Sentences with Homogeneous Subordination of Subordinate Parts in the Russian language dis. PhD. Philol. Sciences (Rostov on / D) 158 p
[7] R I Zelepukin Parcelling in V Tokareva's prose 2007 Structure, Semantics, Text-forming Functions extended abstract /dis....PhD Philol (Sciences)
[8] O I Ulanovich 2013 Suggestive function of language sound-producing symbols: the problem of transmission of effects in the translation of advertising texts *Bulletin of Tomsk State University. Philology* **4 (24)** 43-52
[9] A A Dobrycheva 2012 Parceling in the prose of Sergei Dovlatov: from sentences to text Extended abstract/dis... PhD Philol. (Sciences)
[10] O P Carcoshko 2011 Parceling: Structure, Semantics, Function Extended abstract/dis. PhD. Philol (Sciences)
[11] A N Kalinin 2005 Syntagmatics and Syntax of Parcelled Constructions in Modern English extended abstract/dis PhD. Phil (Sciences)
[12] O V Nikolenko 2014 Expressive Function of Complex Sentences with Homogeneous Subordination of subordinate clauses *Bulletin of Moscow State University of Culture and Arts* **5 (61)** 277-280
[13] L G Ramazanova 2004 Metaphor in the Journalistic Text: on the material of the works of A. N. Tolstoy, extended abstract/ dis. PhD. Philol. (Sciences)
[14] A Z Khaymurzina 2008 The Pragmatic Aspects of the Short Story in the Literary Text on the material of the Short Story extended abstract/dis. PhD. Philol. (Science)
[15] E V Shelestyuk 2014 Speech Influence: Ontology and Methodology of Research (M.: flint: Science)344 p

[16] A A Zumarev 2003 Parceling in Modern Newspaper Speech extended abstract./dis PhD. Philol. (Sciences)

Energy saving of hydraulic drives of machines due to increase of effectiveness of hydraulic cylinders cuffs according to the results of simulation modeling

V I Posmetyev[1], I M Bartenev[1], M A Malyukova[1] and S V Malyukov[1]

[1] Voronezh State University of Forestry and Technologies named after G.F. Morozov, 8 Timiryazeva Street, Voronezh, 394087, Russia

malyukovsergey@yandex.ru

Abstract. Analysis of the results of studies of traditional sealing dynamic connections has revealed shortcomings and peculiarities of their design, which must be taken into account when selecting and justifying the design operating parameters of promising sealing dynamic connections. The proposed design of sealing dynamic connections allows increasing the efficiency of a hydraulic cylinder by 8-12%, facilitating the launch of hydraulic actuators of forest machines at freezing temperatures, increasing the durability of a hydraulic cylinder and reducing the complexity of maintenance of a hydraulic cylinder by reducing the number of assembly and disassembly work associated with unscheduled replacement of sealing dynamic connections.

1. Introduction

The efficiency of forestry units is significantly influenced by the use of advanced designs of sealing dynamic connections (SDC) further cuffs, in the hydraulic drives and energy recovery systems. At present, mainly sealing elements made of rubber in the form of rings and cuffs are used in mobile connections of hydraulic drives of forest and agricultural machines at present. In this case, rubbers are made, as a rule, of SKN (MRTU 38-5-1166-64) and SKF (MRTU-38-5-6056-65) raw rubber. The designs of these compounds are regulated by 6969-54, 9833-73 and 14896-74 State Standards. However, the lack of reliability of such movable seals leads to numerous failures of hydraulic drives of machines and equipment. The main drawbacks of standard rubber sealing elements are the high coefficient of friction (0.1-0.4), as well as the high dependence of the change in the elastic properties and elasticity of such elements on the temperature differential. These drawbacks are the main cause of low efficiency, and, in some cases, complete loss of hydraulic actuators efficiency of FM.

2. Materials and methods

Analysis of research results of traditional SDCs has revealed the shortcomings and peculiarities of their structures, which must be taken into account when selecting and justifying the design and operational parameters of prospective SDCs. In this regard, basic requirements for prospective SDCs structures for HC of forest machines have been developed at the initial stage of the study. Such SDCs must:

Minimize the effect of varying pressure of actuating fluid on the tightness of the connection, as well as on the amount of force on the rod;

Ensure reliable tightness when using low-viscosity actuating fluids, as well as with long-term nominal (15-20 MPa) and short-term maximum (25-40 MPa) pressure values in HC;

Reduce the effort of the pistons and rods moving to minimum values;

Prevent significant increase in the resistance force when moving HC pistons at speeds up to 5-10 m/s;

Work stably at low and high ambient temperatures (–40 - + 60 ° C) [3];

Ensure leak tightness in HC up to 0.008 MPa;

Have geometrical parameters not exceeding the similar values laid down in the State standards and standard specifications;

Be compact, technological in design, inexpensive to manufacture and easy to operate.

In this regard, using the described ADS model of hydraulic drives of forest machines, a study has been carried out on the basis of simulation modeling and computer experiments to evaluate the effectiveness of standard and new sealings for various operating modes [1-3].

3. Results of research

The results of computer experiments has showed that if the cuff is rigid and nonmovable (figure 1, a), then the working fluid gradually enters the gap between the cuff and the inner wall of the hydraulic cylinder. However, since the cuff is elastic and movable (figure 1, b), then when the fluid pressure increases in the left part of the cuff, it deforms and rises, resting against the inner wall of the hydraulic cylinder (HC); it increases tightness of dynamic application and reduces the possibility of leakage of working fluid.

Analysis of the simulation results has also showed that the shape of the cuff and, accordingly, the compaction efficiency, significantly depends on the pressure difference between the working fluid on the left and on the right sides. The cuff is in substantially symmetrical state about the vertical axis at equal pressures on its both sides of the cuff (figure 2, *a*). With a small pressure difference (less than 1 MPa) the cuff is pressed against the inner surface of the HC and prevents penetration of the working

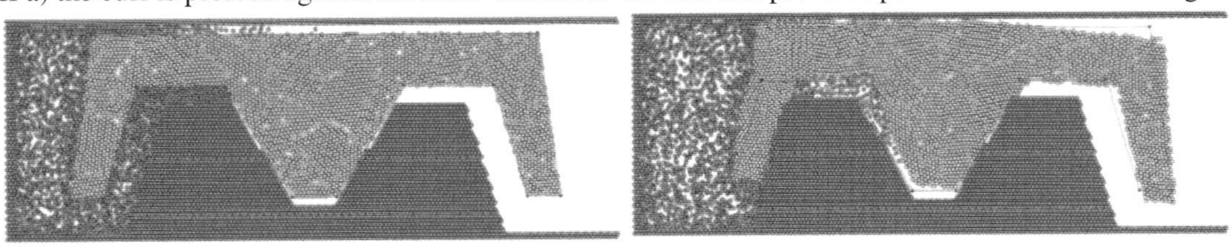

a *b*

Figure 1. Visual comparison of the effectiveness of sealing with tight and elastic cuff: *a* – leakage of the working fluid through tight cuff; *b* - effective compaction due to the bending of the cuff of elastic material

fluid into the low-pressure cavity of HC (figure 2, *b*).

Although the degree of compression is small, the pressure of the working fluid is also small for the occurrence of significant throttling into the remaining gap. With a large pressure difference (about 10-20 MPa), the cuff not only presses stronger against the inner surface of HC, but it simultaneously presses against the skewed wall of the piston with the same force, which causes wedging of the middle part of the cuff and reliable sealing both in its upper and lower parts (figure 2, *c*).

At very high pressures in the hydraulic drive (about 40-60 MPa), the elasticity of the seal (cuff), made of traditional materials, can lead to its crush and (or) destruction (figure 2, *d*). Therefore, cuffs for high-pressure HC must be made of elastic materials with a high modulus of elasticity and protected from destruction with the help of additional elements of highly durable polymeric materials. In addition, it is possible to exclude the crush by changing the geometrical parameters of the cuff (more massive cuff in a whole, or a smaller gap size between the thrust metallic shaped ring of the piston and the inner surface of HC [4-6].

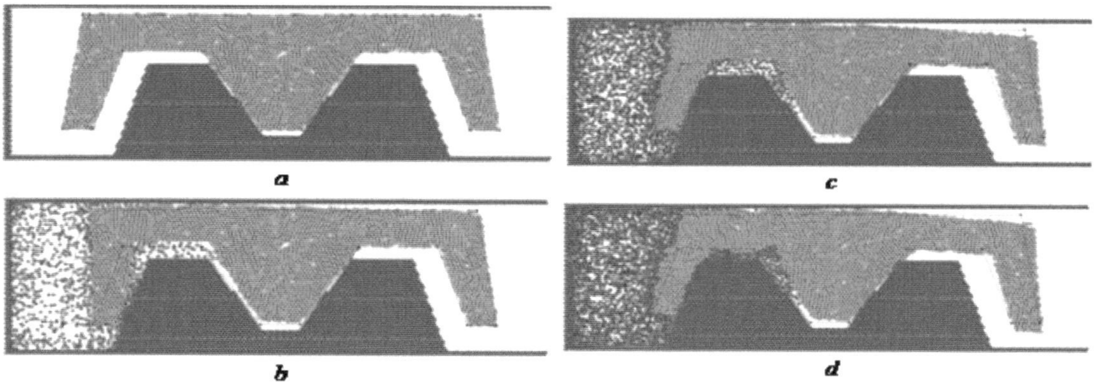

Figure 2. The condition of the cuff at different pressures of the working fluid: *a* – the main state of the cuff in case of equality of pressure on both sides of the piston; state of the cuff when the pressure in the piston cavity changes; *b* – 1 MPa; *c* – start of cuff deformation at a pressure of 10 MPa; *d* – crush of the cuff at a pressure of 50 MPa

The results of simulation and laboratory-field experiments of the investigated SDC, obtained for the three main parameters - linear leakage Q_l of the working fluid, inertia of the response t_r and overall efficiency of HC η_o are presented in figure 3. An analysis of the obtained data indicates the following [7-8].

The values of all three fixed parameters largely depend on changes in the ambient temperature and, especially, in freezing 0 … –10 and low 0 … +10 ° C temperatures. At the same time, the intensity of reducing the values of the parameters in the direction of performance deterioration of SDC is approximately 1.5-2.0 times higher than at T>10 ^{0}C. This difficult drawback of both SDCs, typical for all types of SDCs, is especially evident in spring and autumn periods, which should be taken into account by machine operators, working with forest machines on forest objects at low ambient temperatures.

The leakage Q_l of the working fluid of the new SDCs does not exceed the leakage of the standard seal (figure 3, a), which indicates that it is sufficiently tight over the whole range of possible ambient temperatures.

Delayed action t_r of HC with new SDC is 1.6 times less than that of the delayed action standard SDC (figure 3, *b*). Due to this, both hydraulic actuators themselves and working bodies of the FTT are much less susceptible to dynamic loads during operation of forest machines on cuttings. The efficiency of the new

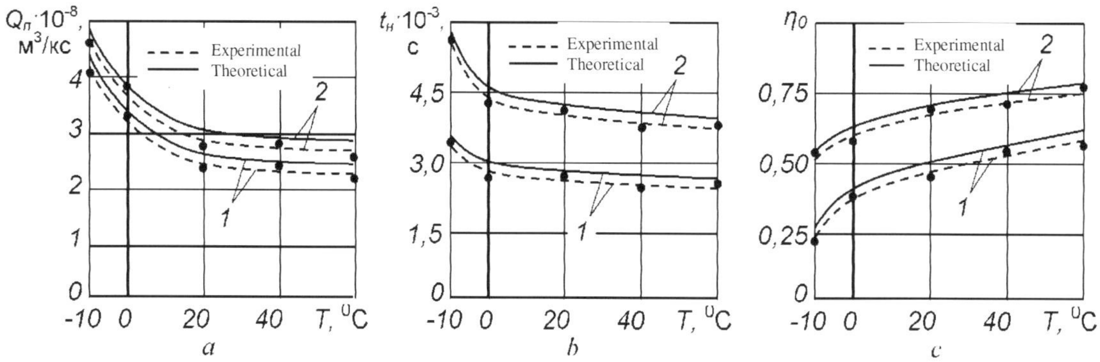

Figure 3. Dependences of the specific linear leakage Q_l of the working fluid, inertia of the response t_r and overall efficiency of HC η_o of the hydraulic cylinder on the working fluid temperature T for the studied sealing elements: *1* and *2* – curves of new and standard SDCs, respectively

SDC has increased two times due to decayed action of the operation of HC, and also due to the use of a polyurethane cuff in the new SDC, which has a much lower coefficient of friction in comparison with the standard rubber cuff (figure 3, *c*) [13-14].

4. Discussion and conclusions

The failures associated with hydraulic systems of tractors are mainly caused by the loss of movable seals tightness of the hydraulic cylinders rods and pistons of the mounted devices and working machines, the wear of spools and valves. The reason for these failures, as a rule, is untimely replacement of worn seals in hydraulic drives, actuating fluid and flushing of tractor hydraulic systems. Unjustified downtime for the repair of tractors adversely affects the intermittency of the mounted system and the quality of works performed at forest sites.

Thus, the results of the analysis show that the use of SDC new design ensures an increase in the efficiency of hydraulic actuators and forest machines by reducing the leakage of the working fluid and the inertia of HC operation [9-12].

In addition to the noted ones, an important positive quality of the proposed SDC structures is the ability of their cuffs to slightly rotate with the pistons in the cylinder during the operation of HC. Presumably, this should ensure more uniform wear of SDC sealing elements and antifriction layers of pistons during long-term operation of HC and, consequently, increase durability and stability of HCs in general. The proposed SDC designs make it possible to increase the efficiency of HCs approximately by 8-12% in comparicson with traditional ones, to facilitate the launch of hydraulic drives of forest machines at low temperatures, to increase the durability of HCs and to reduce the complexity of HC maintenance by reducing the number of installation and dismantling works associated with unscheduled SDA replacement.

5. Acknowledgements

"The reported study was funded by Council for Grants President Russian Federation according to the research project № MK-6621.2018.8".

References

[1] Posmetyev V I, Latysheva M A, Tarasov E A and Zelikov V A 2006 *Patent Russian Federation* no 2269050
[2] Bashta T M 1972 *Hydraulic drive and hydraulic and pneumatic control systems* p 328
[3] State Standard 14896-84. 2003 *Rubber sealing cuffs for hydraulic devices* p 54
[4] Kaverzin S V, Lebedev V P and Sorokin E A 1998 *Ensuring the efficiency of a hydraulic drive at low temperatures* (Krasnoyarsk) p 240
[5] Kondakov L A 1972 *Sealings in hydraulic systems* p 240
[6] Posmetyev V I, Zelikov V A and Posmetyev V V 2015 *Improving the efficiency of forest tillage units based on the implementation of promising scientific and technical solutions* (Voronezh) p 270
[7] Zelikov V A, Posmetiev V I and Latysheva M A 2014 *World Applied Sciences Journal.* **30 (4)** p 486-92 DOI: 10.5829/idosi.wasj.2014.30.04.14053
[8] Posmetev V I, Stanchev D I and Uskov I A 1999 *Collection of scientific works of the All-Russian scientific-practical conference* (Voronezh) p 97-100
[9] Bartenev I M, Malyukov S V, Gnusov M A and Stupnikov D S 2018 *International Journal of Mechanical Engineering and Technology* **9 (4)** p 1008-18
[10] Liu W, Wang S, ZhouY, Wang L, Zhu J and Wang F 2016 *Natural Hazards* **81 (1)** p 347-63. DOI: 10.1007/s11069-015-2083-1
[11] Afzaal Hamra, Zafar Nazir Ahmad 2016 *2nd International Conference on Robotics and Artificial Intelligence (ICRAI)* (Natl Univ Sci & Technol Coll E&ME Pakistan, Islamabad, Pakistan) p 112-17

[12] Kasymov D P, Fateyev V N and Zima V P 2017 *Proceedings of SPIE - The International Society for Optical Engineering* 10466
[13] Posmetev V I Latysheva M A 2005 *Forest. The science. Youth - 2004: a collection of materials on the results of the scientific research work of young scientists in 2004, dedicated to the 75th anniversary of the birth of Professor A. V. Veretennikov* (Voronezh) p 157-60
[14] Posmetev V I Latysheva M A and Kukharev V S 2005 *Science and images to the service of the forest complex (on the 75th anniversary of the All-Russian State Medical University): materials International. Scientific and Practical Conference* (Voronezh) p 79-82

Legal realisation of information technology activities of the government-owned corporations: automation of management and control

A O Inshakova[1], L A Chegovadze [2], T V Deryugina [3]

[1] Volgograd state university, Prospect Universitetsky, 100, 400062, Volgograd, Russian Federation. ORCID 0000-0001-8255-8160
[2] Moscow University of the Ministry of Internal Affairs of Russia named after V.Ya. Kikoty, ul. Academician Volgin, 12, 117437, Moscow, Russian Federation. ORCID 0000-0003-0079-1917
[3] Moscow University of the Ministry of Internal Affairs of Russia named after V.Ya. Kikoty, ul. Academician Volgin, 12, 117437, Moscow, Russian Federation. ORCID 0000-0003-2978-0234

E-mail: gimchp@volsu.ru

Abstract. The article touches upon the problem of digitalization of management and control over government-owned corporation activities. The authors develop the architecture of a unified electronic information system for property management and control of property turnover. The paper reveals a range of issues related to the use of information technologies in management and control over government-owned corporation activities. The researchers analyze the causes of arising problems in legal regulation, general issues being common for all elements of the legal regulation system, as well as private ones relating directly to management and control of the government-owned (public) corporation. The authors use general and special methodological approaches in their research to solve the tasks stated in the paper. This research paper represents the analysis of normative legal acts containing the legal basis for creating an effective system of management and control over government-owned corporation activities. The results of the authors' study indicate general trends of building a system of a legal response by means of established prohibitions, limits and restrictions. The research points out positive and negative aspects of this approach. The authors identify types of activities of public corporation being non-subject to the control of the founder and monitoring agencies due to definite lacks in the current legislation and suggest legal means to fill in such gaps. Analyzing this doctrine in relation to the use of different approaches to the management and control of government-owned corporation the authors determine positive and negative characteristics of the proposed methods. The researchers suggest using a set of legal instruments in legislative regulation of management and control over government-owned corporation activities. Along with the prohibitions, limits and restrictions the article proposes to digitalize the processes of management of a government-owned corporation and control its activities by means of creating and maintaining a unified electronic information system of property management and control over the turnover of property. The authors characterize features of such a system distinguishing it from existing analogs and specify the structure of the subjects having access to such systems. The paper determinates databases necessary for inclusion in its structure.

1. Introduction

Active development of information technology and the formation of new sectors of the digital economy have determined the need to create an effective system of legal regulation reflecting the specifics of the electronic format of civil turnover.

The constant introduction of digital innovations, digitalization of modern processes and use of information technologies in all spheres of life predetermine the rapid obsolescence of existing legal methods and means. At the international level there is a lack of clear approaches to the law necessary to resolve emerging conflicts. [1]

Consequently, we suppose, that the changing architecture of the modern economy creates the nesessaty to change not only the structure of law but also the concept of new mechanisms of legal regulation [2].

For numerous fields and institutions of the economy a legal framework is currently being created to solve problems concerning the area of regulation and control over legal subjects activity. However, there are some institutions which do not use the information technology necessary for the effective management and monitoring of modern organisations up to now.

Thus, Russian law is aware of the non-specific organisational and legal form of a legal entity created as a result of the transfer of state property as a contribution to the authorised capital of a non-profit organization – public corporation.

Debates concerning the activities of the subject of law have not subsided since the moment of incorporation of this legal form in the Russian legislation and the creation of the first public corporation in scientific field. The results of the study indicate the need for an effective system of management of the Corporation and the monitoring of its operations. There are various opinions among researchers about the control of public corporation's property transferred to it by the government, which can be described as a particularly valuable and strategically important property. Similar to the last point, the problem of control over the use and disposal of such property by a government-owned corporation is particularly highlighted.

Legislators made numerous attempts to solve the problem of property management and control over the turnover of public corporation's property. Thus, in order to solve this problem, the federal law "on public corporation on atomic energy "Rosatom" underwent amendments more than sixteen times. In respect of the public corporation "Rostech" legislators have adopted seven federal laws changing its legal status. Numerous changes have been made in the regulations regarding other corporations. It is, however, evident that without fundamental research in this area, it seems impossible to creat an adequate system of legal control of legal relations. The facts considered predetermined the need for this scientific analysis.

The legislators attempts to solve the problem of control over the property management system and civil turnover of public corporation's property, as a rule, are reduced to the establishment of limits and restrictions on the implementation of public corporation's property rights. However, solving the problem by establishing a large number of bans is not entirely efficient, as these bans limit the targeted activities of a government-owned corporation preventing the achievement of its designated objectives.

Specific problems established by the legislator's limits and restrictions, are unconditionally, solved. It is, however, necessary to look at this problem from another angle identifying solutions through the creation of a single electronic information system of property management and control over the turnover of property of public corporation.

2. Results

The analysis of the scientific literature shows that most authors believe that the management of a public corporation is out of state control, [3] which is unacceptable since the creation of a municipal corporation is primarily predetermined by the need to solve domestic and international problems of our country. So, by art. 7.1 of the Federal Law "On Non-Profit Organizations" of the Rostech, Rosatom, Vnesheconombank and other state corporations are not subject to the rules on state control related to their activities as non-profit organizations: control bodies without the consent of a state corporation are not entitled to request statistics on the economic and other activities of the state corporation; carry out inspections, including the expenditure of funds and so on.

Of course, one cannot say that the activities of a state corporation is non-subject of control. Unique control mechanisms have been created to meet the goals that the government decides through the creation of public corporations. Unfortunately, we have to state that they are not sufficient to solve the indicated problems.

Moreover, we should note that the mechanisms created can, as a rule, be qualified as public control (the obligation to provide public reporting has emerged, the Audit Chamber of the Russian Federation can verify activities [4] "On the Accounts Chamber of the Russian Federation"), although the interpretation of the law is ambiguous regarding the possibility of such verification and the subject of the audit, which led to additional discussions in the literature and practice, etc.). However, at present, the problem of control is extremely acute, both for the state corporation management system and for the civil turnover of its property, which, of course, necessitates the development of additional control and monitoring tools.

In the research literature, certain civil legal means of control over transactions by state corporations are proposed, but, as a rule, such proposals are reduced to the establishment of prohibitions on the Commission of certain types of operations; [5] or prohibitions on transactions with certain property of a state Corporation; [3] the establishment of control over transactions by the state; [6] a change in the civil legal standing of a state Corporation. [7]

Some authors see the problem of corporate governance in the activities of governing bodies. [8]

Of course, these points of view are worthy of attention and, to a certain extent, can help solve the problems of managing the property of a state corporation and controlling its turnover, including control over the circulation of funds. However, we consider that absolute significance in determining the possibility of exercising continuous control over the making transactions state corporations is the establishment of duty for it to disclose information about transactions.

Considering the transactions of a state corporation, one can single out some deals in respect of which there are currently no adequate controls, including on the part of the founder.

Giving public corporations the right to contribute funds received, including from the federal budget, to the authorized capital of subsidiaries and affiliates, the legislator did not take into account the fact that individuals can act as co-founders of such companies, which may result in redistribution budget funds in favor of such founders (for example, clause 3, clause 4, article 17 of the Federal Law of December 1, 2007 No. 317-ФЗ (as amended on December 31, 2017) "on public corporation on atomic energy "Rosatom"). [9]

Permission given to government-owned corporations to consent to large transactions, including those related to the alienation of property of unitary enterprises under their jurisdiction either cannot be considered indisputable (paragraphs 16, 17, paragraph 1, article 8.1. Of the Federal Law of December 1, 2007 N 317-ФЗ (ed. of December 31, 2017) "on public corporation on atomic energy "Rosatom"), as well as the right to act as a guarantor (taking into account the possible consequences of imposing a material burden on the guarantor) (clause 1 of Article 7 of the Federal Law of November 23, 2007 N 270- Federal Law (as amended on 04.23.2018) "On public corporation on the Promotion of the Development, Production and the export of high-tech industrial products "Rostech").

We also consider the opportunity given to a government-owned corporation to issue loans on concessional terms. The absence of a ban on the transfer to the sublet of property received by public corporation from the state as a constituent contribution and leased on concessional terms is also a matter of censure. The possibility of gratuitous financing of commercial legal entities also creates opportunities for specific abuses of the right, a similar remark can be made regarding the possibility of investment in Russian and foreign organizations (paragraph 5, clause 5, article 7 of the Federal Law of November 23, 2007 N 270-FZ (Ed. 04/23/2018) "On public corporation on the Promotion of the Development, Production and Export of High-Tech Industrial Products Rostech").

Thus, it should be noted that there is a possibility of making various transactions with property transferred by a state corporation to the state, including the possibility of introducing into public circulation funds allocated from the federal budget. Along with the establishment of legislative prohibitions and restrictions, as well as the use of legal means proposed by various authors and cited above, the unified electronic information system for property management and control over the turnover of property of a state corporation can solve the problems of power.

Analogues of such an information system currently exist in various sectors of the Russian economy — for example, the Unified State Register of Real Estate (Federal Law of July 13, 2015 N 218-FZ (as amended on 03.08.2018) "On State Registration of Real Estate"), the Unified State Information System in the field of health (Resolution of the Government of the Russian Federation of 05 May 2018 N 555 On the Unified State Information System in the Field of Healthcare"), Unified State Register of Rights to Aircraft (Federal Law of March 14, 2009 N 31-FZ (as amended on 13.07.2015)" On State Registration of Aircraft Rights and Transactions with Them") and etc.

There are numerous examples of the use of electronic information systems for international trade (for example, the Decision of the Board of the Eurasian Economic Commission of September 30, 2014 N 180). Moreover, as it is rightly noted in the scientific literature, despite some problems related to the legal nature of electronic transactions, as well as the definition of the moment of their conclusion, there are legitimate bases for creating electronic registers and electronic transactions in the Russian Federation. [10, 11]

However, unlike existing in legislation analogues of unified information systems to provide information to a specific or a wide range of subjects, the centralised electronic information system for property management and control over the turnover of state corporation property should not have a public sign. This feature is associating with the specifics of the activities of state corporations, some of which are carriers of state secrets. This system should be of a closed nature and can be available only to certain subjects strictly defined in law. The unified electronic information management system and control over the turnover of property of a state corporation are intended for information support of the activities of the control bodies and the founder.[12]

The unified electronic information system for property management and control over the turnover of property of a state corporation should be a federal state information system, which includes: information databases on any transactions with property, including money, made by state corporations; subject and object of sales; information about the subjects entering into deals with state corporations; information on the rights to property arising from the parties of the legal relationship at the conclusion of such a transaction.

3. Conclusion, recommendations

1. Crucial in determining the possibility of effective management and continuous control over the turnover of property of the state Corporation is to establish for it the obligation to disclose information about the transactions.

2. Disclosure of information should be carried out using the creation of a unified electronic information system of property management and control over the turnover of property of the State Corporation. The purpose of the system – information support of the control bodies and the founder of the state Corporation to determine the level of management efficiency.

3. The unified electronic information system of property management and control over the turnover of property of a state Corporation should not have a sign of publicity. Access to this system is possible only to the entities defined in the law to obtain information, the scope and content of which is necessary for the performance of their functions.

4. The unified electronic information system of property management and control over the turnover of property of the state Corporation should contain information databases on transactions with the property of state corporations; on the subject and object of transactions; on the subjects of transactions; on property rights arising from the parties to the legal jural relations.

References

[1] Wieland K M 2017 Key Issues for Digital Transformation in the G20. Report prepared for a joint G20 German Presidency (OECD conference Berlin, Germany)

[2] A O Inshakova, A I Goncharov, O A Mineev, M V Sevostyanov 2017 Amendments to the Civil Code of the Russian Federation: Contradictions of Theory and *Practice Contributions to Economics. Russia and the European Union Development and Perspectives Number* **XIII** 147-153

[3] Kucherov N A, Povetkina N E, Abramova 2016 Control in the financial and budgetary sphere:

scientific and practical guide II (M : IZISP, CONTRACT) 320 p
[4] Federal Law 2015, About state registration of real estate (approved by the Federation Council of July 13, 2015 no. 218-FL (as amended on 08/03/2018) (Russian)
[5] Kiselev A G 2012 Improving rationality and using the potential of state corporations in the socio-economic development of regions: Author. diss. .. on the competition (Cand. economical sciences. M.) 156 p
[6] Vinnitskiy A V 2013 Public property, (Prospekt, Moscow) 606 p
[7] Kuzminov A A 2009 Civil legal personality of state corporations: (diss. ... Cand. legal sciences. M) 192 p
[8] Zubarev A V 2013 Compliance with the principle of disposition in determining the competence of the general meeting as the highest governing body in corporations, *Bulletin of the Volgograd State University* Series 5 **1 (18)** 43-46
[9] Federal Law 2007, About the State Atomic Energy Corporation "Rosatom" (approved by the Federation Council of December 1, 2007 no. 317-FL (as amended on 12/31/2017) (Russian)
[10] Kozlovskaya S V 2018, Problems of civil law regulation of transactions concluded within the framework of electronic interaction Experiments of civilized research: a collection of articles AE Ageenko, II Akimova, V A Volgina, etc.; hands auth. count and rep. ed. AM Shirvindt, NB Shcherbakov. (M.: Statute) **2** 368 p
[11] Carayannis E G, Alexander J, Geraghty J 2001 Service sector productivity: B2B electronic commerce as strategic driver *Jornal of Texnology transfer* **26** 337-350
[12] Choudhury B, Petrin M 2018 Corporate governance that 'works for everyone': Promoting public policies through corporate governance mechanisms. Faculty of Laws (University College London) 18 **2** 381-415

Social and psychological features of career strategies mobility of the college students

O V Lyusova[1], A S Seredintseva[2], E M Glukhova[2]

[1]Department of Psychology, Federal State Educational Budget Institution of Higher Education «Volgograd State Physical Education Academy», 400005, Lenin Avenue - 78, Volgograd, Volgograd region, Russia
[2]Department of Secondary Professional Education, Volzhskiy branch of Federal State Autonomous Educational Institution of Higher Education "Volgograd State University", 404130, 40-let Pobedy Street -11, Volzhskiy, Volgograd region, Russia

oxy180171@mail.ru, anna_seredinceva@mail.ru, zvezda130573@yandex.ru

Abstract. The modern world sets high standards for competitiveness, productivity, leadership qualities and commitment of young people who have just received secondary professional education. In this regard, the article deals with the problem of building a career and forming effective career strategies. The authors study different approaches to defining "career" and "career strategies", analyze how students understand what a career means, their attitude to building a career, the appropriate traits for a successful career and the role, which a career plays in people's lives. To conduct the research and gather the data the authors used Solomin's method of colour metaphors, Schein's "Career Anchors" method and a questionnaire survey. On the basis of the survey results the authors proposed a program of development of effective career strategies "My Success" that is aimed at optimizing the process of training qualified specialists for main industrial companies of the region.

Introduction

In the modern constantly changing world a person must be a competitive, productive and successful leader, who is able to achieve goals, resist obstacles, step over barriers, that is primarily associated with the choice of the appropriate career strategy. Experts in the field of economics and psychology accentuate the problems of professional socialization of young people: employment difficulties, the problems of holding the employment position, the growth of youth unemployment. Along with the economic reasons for these phenomena (economic instability, low salaries, lack of work experience for graduates of educational institutions), there is a psychological aspect of the problem associated with the internal uncertainty of the professional way, the inexpressiveness of career orientations or the discrepancy between the abilities and real professional requirements. Today career planning is an important task not only for young people and their parents. There are more parties interested in the appropriate choice: employers, the workers themselves, state employment services, the state itself. Thus, the relevant career and professional orientation is the task which is to be solved not only at the local level by the youth, psychologists and teachers but also at the state level by the authorities. The aim of the research is to highlight the features of career strategies of modern provincial youth, substantiate scientifically and develop a program of designing the effective career strategies. The hypothesis of the research states that career strategies of the modern youth have the following features: students rely on such values as job stability, humanism, material prosperity and personal growth; have high career mobility; female students are more likely to associate a career with creativity than male ones do.

Having analyzed the approaches [1, 2, 3, 4] to understanding career in modern scientific research, we relied on the definition of a career in the broad sense as professional advancement and in the narrow sense as a movement towards the top of our personal achievements. The following personal factors, which largely determine the peculiarity of a career, were identified: anxiety, emotional stability (or instability). Sociability, openness and willpower, which facilitate promotion, are equally important. And, on the contrary, a career can be affected by suspiciousness or suspicion. Maximum results can be achieved when professionalism is combined with high personal qualities [5].

Donald Super identifies four categories to explain career patterns: 1) *Stable pattern.* Highly skilled and educated individuals make a permanent career choice directly after school or college and follow it till the retirement time. 2) *Conventional pattern.* Individuals try several careers before taking up a stable one which is more satisfying to their needs. 3) *Unstable pattern.* Individuals move from one job to another following the sequence trial-stable-trial. Different careers can be considered before making the final choice. 4) *Multiple-trial pattern.* Individuals change employment rather frequently and usually take entry-level positions without establishing in a career [6].

We define career strategies as a specific system of actions designed for adaptation or development of the career and the person in it and aimed at transforming a person himself, other people, changing the conditions and the content of the professional activity in order to achieve career goals [7, 8].

There are two conditions particularly relevant for young students in the process of formation of life strategies [9]: 1) age identification, 2) students as a social group do not stand ready for the certain practical actions and social subjectivity, that is necessary for the life strategies formation, since youth positions adequate to their social and age opportunities in the social space are either functionally limited or absent.

As a result of the analysis, we can conclude that all typological models of life strategies of young people are divided into three groups [10, 11, 12]:

- by the content and form of expressing subjectivity of the individual, characterized by the fact that the life strategy is formed by the individual and the social and economic conditions affect the content of strategies, but are not dominant;
- by the nature of the influence of institutional living conditions when the external socio-structural and economic conditions (type of settlement, social and ethnic origin, region of residence) have a dominant influence on constructing life strategies;
- by the degree of normativity of social stereotypes and behavior in which the content of young people's life strategies can relate to the normative values and attitudes prevalent in the society.

The students of the 1st and 3^{d} (final) year of the Secondary Professional Education Department (Volzhsky branch of Volgograd State University) were chosen as the participants of the experiment. At the ascertaining stage of the research we studied the specific features of career consideration among female and male students.

On the basis of the theoretical analysis we can determine the methods used for conducting the research: questionnaire survey, I S Solomin's method of colour metaphors, Schein's Career Anchors method. The students were offered to answer how they understand the concept of career; to express their attitude to the career; to define personality traits to build a successful career; to describe the role of a career in their lives; to distinguish the peculiarities of career mobility. The questionnaire was processed through the content analysis [13].

In contrast to traditional psychological tests and questionnaires Solomin's method of color metaphors [14] makes it possible to find out what a person really thinks and feels. It is focused on identifying the content and structure of consciousness and unconsciousness. Subjects are asked to associate the things from the proposed list with specific color (a set of M Luscher color cards). The "Anchor of a Career" questionnaire by E Shein contains 41 statements that measure 5 patterns, or anchors explaining the career decisions. The respondent rates on a 10-point scale how the given statements apply to them. The items are grouped into 8-point scales, representing 8 non-mutually exclusive career value orientations ("career anchors"): general managerial competence, technical / functional competence, autonomy / independence, security / stability, entrepreneurial creativity, service / dedication to a cause, pure challenge, lifestyle [15].

Results and Discussion

The diagnostics, run on the basis of E Shein's method "Career anchors" [16], revealed the following results. The most significant value (anchor) of a career is the employment stability. Students tend to long-term employment without any job changes, leading to stress and fear of losing it, what reflects the current economic state and social instability of the society.

The next important value is serving: "working with people", "serving humanity", "helping people", "wanting to make the world better", etc. The students see their future careers in accordance with humanism, the mission of benefiting the society, helping people. They do not plan to neglect their values in relation to other members of the society, and if the place of work (organization) is opposed to these values, they will choose being fired, rather than a compromise.

The third most important value for students is the integration of lifestyles: focusing on a combination of different aspects of life (family, career and self-development), students strive for a balance between these areas. In this case career is considered a means of self-expression in life. Life is much more important than a career in particular, and a career is inseparable from life in general, but not identical to it.

As it is indicated by the students in the questionnaires, the least preferred anchors are: stability of residence, that reflects the mobility of modern young people, and professional competence, that can be associated with inexperience in finding a job, career advancement and orientation towards other opportunities while moving up the career ladder (commitment, workability, arrogance, sociability, "the right contacts").

The survey on the basis of I S Solomin's method of colour metaphors shows that students of the final year are more optimistic about their future career than first-year students. This is due to their overall greater personality and age maturity, the experience of communicating with professionals at master classes, teachers-practitioners, and a special system of educational work aimed at students as future competitive specialists, better adaptation to their professional life.

Among female and male students, there was no difference in the subjective attitude towards their future career. However, they significantly differ in subjective reflection on the connection between career and creativity, career and interesting occupation. While associating with a career connected with creativity or creative attitude to work, terms of "career" and "creativity" have the same emotional coloring among female students, and it is more than among male ones. There is also some difference in their attitude towards the connection of career and success. More female students than male ones associate success with building a career.

Having conducted a study questionnaire about the attitude to building a career among the students it is possible to present the given results:

1. Understanding the concept of career. The students perceive a career through the values of material prosperity (35%); values of power and influence (21%), personal growth as a goal, reaching a new level (14%).

2. Attitude to career. Modern students consider their future careers in the context of values: personal growth, the desire to go forward in life, get over themselves – 28%, to achieve material prosperity, to improve their financial level, financial well-being – 17%; to gain authority, to manage and lead people – 12%.

3. Personality traits to build a successful career. The students distinguish the traits which are essential for a successful professional career: dedication – 15 people, responsibility – 11 people, communication skills, leadership, tact, ability to work in a team – 12 people, traits related to the work quality (diligence, professionalism) – 10 people, traits associated with education (intelligence, education, advanced training) – 4 people, a separate group of negative traits (arrogance, cynicism) – 7 people. The students also mentioned some other traits such as activity, perseverance, willpower, flexibility and so on.

4. The role of a career in a person's life. In the survey 65% of the students are willing to take risks for their careers, and 35% are not ready to take risks, 26% are ready to choose a career at the expense of personal happiness, and 74% choose personal happiness. Among their priorities, students average career an average of 6-7 out of 10.

5. Peculiarities of career mobility: in case of job loss – 16 % of the students will look for a job within their specialty; take any job – 40%; learn a new profession – 30%; set up their own business – 28%; look for a temporary, but well-paid job – 5%; look for a stable job – 16%; hope to have a chance and take no action – 0.

As a part of our research, we proposed a program to develop an effective career strategy "My Success" which contains three forms of work: 1) group counselling, which includes informing on the problems of understanding the essence of a career and career strategies, types of career strategies, qualities and abilities necessary for a successful career and career mobility; 2) participation in the development work of the program "My Success", which includes the awareness of their own career strategies, a career place in the hierarchy of their own values, typical models of behavior in situations of professional activities, especially goal-setting; 3) individual counseling for students on developing their own career strategies.

The aim of the program is to increase the level of students' psychological competence for effective career strategies formation. To achieve the aim it is necessary to take the following objectives: • to promote formation of effective career strategies; • to inform about the psychological foundations and patterns of career building; • to present the methods of self-regulation in difficult situations; • to expand the repertoire of professional behavior in the professional career development; • to promote the students' awareness of the career place in the hierarchy of their values.

The program includes the system of interconnected psychological exercises in the form of group counselling, also based on the diagnostics results, trainings on the reflection development, time management, and self-regulation. It is also connected with developing the system of homework tasks (both individual and for all participants). The program involves the use of the following techniques and methods: inputs (short information messages aimed at introducing the problem with which students will work later); card games of L I Pryazhnikov, art therapy techniques and methods (metaphorical associative cards). The purpose of the mentioned above methods is to promote the students' awareness of their values, motives, the hierarchy, which allows them to make an important life decision, take responsibility for it and simulate the consequences of decision and choice making.

The conducted survey and the launch of the given program at the Department of Secondary Professional Education can contribute to optimizing the educational process in general and improving the preparation of specialists for industrial companies, thereby satisfying the social demand of major manufacturing companies of Volgograd region such as JSC "Volzhsky Pipe Plant", JSC "SunInBev", JSC "Volgomyasmoltorg", JSC "Gazprom khimvolokno" and etc.

Conclusion

Summarizing the results of the empirical research, we have made an attempt to give a descriptive characteristic of the main features of career strategies of modern young people: understanding the concept of career modern students rely on such values as employment stability, humanism, integration of lifestyles (some harmonious balance); building their own future career they rely on the value of material prosperity and personal growth; among the important traits which are necessary for a successful career students choose commitment, responsibility, interpersonal skills; career mobility is expressed in the ability to take any job in case of its loss. Thus, we can state that the aim of the Department of Secondary Professional Education is to prepare future specialists of industrial companies both in the local region and in the whole country in such a way that the students are focused on stable employment, developing their personality and being harmoniously included in the environment. They understand the importance of universal human traits for being a qualified specialist and are ready for being mobile when the external life and professional circumstances change, that emphasizes the relevance of the modern competence-based approach in education to the conditions of social and economic development of the Russian society.

References

[1] Afanasenko I V 2005 Strategies of building a career as a component of professional I-concept *Proc. Int. Conf. of Psychologists* (SPb.: Pushkin Leningrad State University Press) pp 15-20
[2] Mogilyovkin E A 2007 *Career development: diagnostics, techniques, training* (SPb.: Rech) p 336
[3] Mordvincev L 2007 *Career planning* (Moscow: Alpina Business Books) p 208
[4] Prjazhnikov N S 2017 Guidance methods in psychological «spaces» of self-determination *National Psychological Journal* **3** pp 121-135
[5] Vishnyakova M 2006 Career management: a strategic plan of career development *Top personal* **11** pp 28-34
[6] Super D E 1990 A life-span, life-space approach to career development *Journal of Vocational Behavior* **16(30)**
[7] Sorokin P S 2013 Modern studies of career as a social phenomenon: theoretical and methodological problems and ways to overcome them *RUDN journal of* **3** pp 40-47
[8] Efremkina I N 2018 The study of the dynamics of career orientations, educational motivation, their interrelations among bachelor students as the conditions for forming readiness for professional mobility *Perspectives of Science & Education* **2 (32)** pp 175-180
[9] Didkovskaya Ya V 2012 Professional strategies of university graduates *Higher Education on Russia* **8-9** pp 90-96
[10] Muftahova Yu D 2013 Professional self-determination of students *Proc. Conf. of the science publishing center «Sociosfera»* **16** pp 69-70
[11] Reznik S D, Igoshina I A and Reznik V S 2005 *Personal career management* (Moscow: Logos) p 288
[12] Petrushin V I 2004 *Psychology of a career* (Moscow: academic project; Al'ma Mater) p 208
[13] Lyusova OV, Solyankina L Ye and Mazur E Yu 2018 Psychological Features of Career Strategies of the Youth in Russian Province *Revista Espacios* vol 39 pp 20-21.
[14] Solomin I L 2006 *Modern methods of psychological express diagnostics and professional consulting* (SPb.: Rech) p 280
[15] Bigliardi B, Dormio A Ivo 2009 R&D Personnel Career Routes: An Exploratory Study *Journal of Technology Management & Innovation* vol 4 **1** pp 8-21
[16] Schein E H 1993 *Career Anchors: discovering your real values* (Pfeiffer & Company) p 87

IOP Conf. Series: Materials Science and Engineering **483** (2019) 012110 doi:10.1088/1757-899X/483/1/012110

Prospects for the development of virtual logistics on the territory of the EAEU

E A Koryakina, N P Sheveleva, N S Kulakova

Tyumen industrial university, 625000, Tyumen, St. Volodarsky, d. 38

katnasonya@yandex.ru, shnaps@inbox.ru, northstar_nadin@mail.ru

Abstract: The article analyses the state of the e-commerce and virtual logistics market on the territory of the EAEU. The portrait of the participants of the virtual trading market in different segments has been considered. The influence of various indicators on the logistics efficiency index and the impact of each component on it has been studied. The comparison of the leading country on the territory of the EAEU with the world leader in the logistics efficiency index was carried out and the potential of its development was noted. The impact of the index on the country's GDP and the volume of e-commerce is considered. The classification of levels of integration of logistics activities is presented – the influence of integration of organisations providing services in the field of logistics on the level of logistics service is determined. The problems of logistics development in the EAEU are identified by the authors. The study provides the insight into the main promising areas of development of Internet trade and reflects the impact of logistics providers on the development of the EAEU Member States.

1. Introduction

Globalisation of the world market contributes to a radical modification of the system of production management and distribution of products through the introduction of the platform economy while transforming the existing socio-economic, political, communication and other relations. At the same time, globalisation contributes to the emergence of economic unions, such as, for example, the EU, the EAEU.

The development of any territory is impossible without the unification of the information space, built on the strengthening of the importance of various platforms and business development. Further digitalisation of the economy will strengthen the role of the consumer in trade relations, which will lead not only to the expansion of the range of services, but also to the emergence of criteria such as speed and quality of delivery (time and place of delivery, integrity and quality of goods, culture of communication of representatives of the courier service, etc.); unconditional replacement of defective goods; the possibility of return and exchange of quality goods; the breadth of multi-vector services.

2. Findings and discussion

Participation in the virtual logistics system contributes to the competitiveness of the organisation on the territory of the EAEU. Economic growth in the EAEU is impossible without the development of e-commerce and modernisation of the economy through digital technologies. Taking into consideration the large area of the countries belonging to the EAEU grouping - Russia, Belarus, Armenia, Kazakhstan and Kyrgyzstan, it is necessary to pay special attention to the transport and logistics infrastructure of the territory, that is the length and quality of roads and railways, as well as the efficiency of their maintenance. The prolonged socio-economic relations of the EAEU countries allow to identify the population of similar features, which gives the opportunity to target market segments for the formation of a unique selling proposition. At the same time, there are also differences between the countries, primarily in the level of income and purchasing power – the average monthly nominal wage in 2017 amounted to $671 in Russia, $ 463 in Kazakhstan, $ 426 in Belarus, $ 228 in Kyrgyzstan, $ 228 in Armenia (taking into account the employer's payments for compulsory social insurance) – $ 404 [5,8].

The volume of mutual trade between the EAEU countries in 2017 amounted to 54 697.9 million dollars. Mineral products (27.5%) and machinery, equipment and vehicles (18.6%) accounted for the largest share in the structure of mutual trade. The volume of foreign trade of the EAEU States with third countries amounted to 634221.3 million dollars. (imports - $ 247271.1 million), exports-386950.2 million dollars.). Russia's share in the EAEU imports is 85 %, in exports – 84 % [5].

The volume of e-commerce in Russia in the B2B segment in 2015 was estimated at $ 550 billion, and in the B2C segment at $ 14 billion, due to the implementation of state and municipal procurement in electronic form. The dynamics of the volume of e-Commerce in the EAEU countries indicates that the B2C segment is growing at the most significant pace and belongs to potentially attractive areas of business development.

Over the past five years, the Russian e-commerce market has had an average annual growth of 19%, the volume of cross-border trade with an annual growth for the same period – 32.6% and occupies 37% in the structure of trade, the remaining 63% of the market volume falls on domestic Internet trade, with an average paycheck on the local market - 5370 rubles according to the Association of Internet Trade Companies (AKIT); according to Data Insight in the C2C segment - 3250 rubles, in B2C - 4200 rubles [6,12].

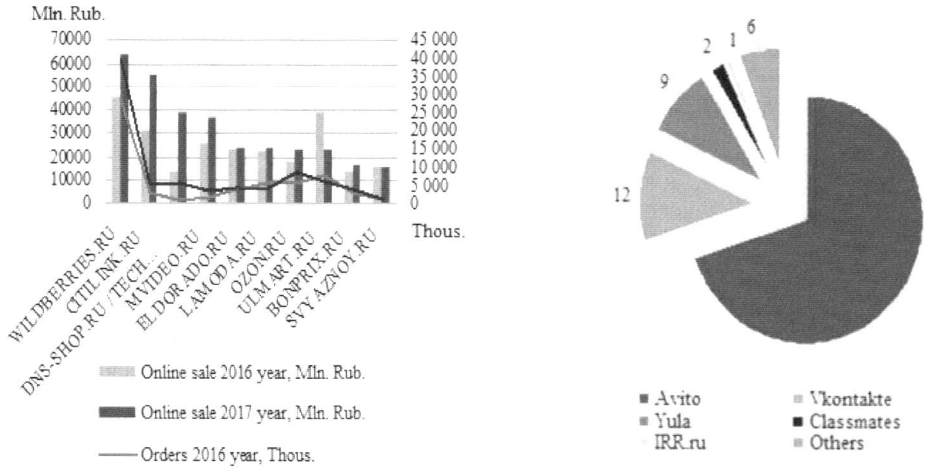

a) b)

Figure 1. Russia's largest Internet participants a) online stores B2C; b) C2C-sales, %

The rating of online stores in terms of online sales according to Data Insight research Agency is shown in Fig. 1 (a). In 2017, Wildberries - the Russian international online store of clothing, footwear, household goods and other items - was recognised as the largest participant on the online platform in the B2C segment. In addition to the Russian Federation, the company represents its interests in Belarus, Kazakhstan and Kyrgyzstan. In second place with the growth of online sales by 75% is the Russian chain of stores "CITILINK" positioning itself as an electronic discounter [6]. The leader in Russia in the share of transactions on C2C sites is Avito (70 %). 21 % of transactions are accounted for two also leading sites - VK.com and Yula (12 and 9 %, respectively) (see Fig. 1 b).

According to the online survey for the first half of 2017, the portrait of C2C market participants showed almost complete identity of characteristics of buyers and sellers (Fig. 2). A typical representative of the users of this model of marketplaces is an individual aged 25 - 34 years old, with an income above average and experience on the Internet for more than 5 years. The predominant share of online purchases by persons under 34 years of age is made with the use of smartphones. The use of various gadgets and high activity are characteristic features of the category of buyers called "extreme shopper" (in Russia, 61 % of the audience of online stores account for "extreme" buyers [10]).

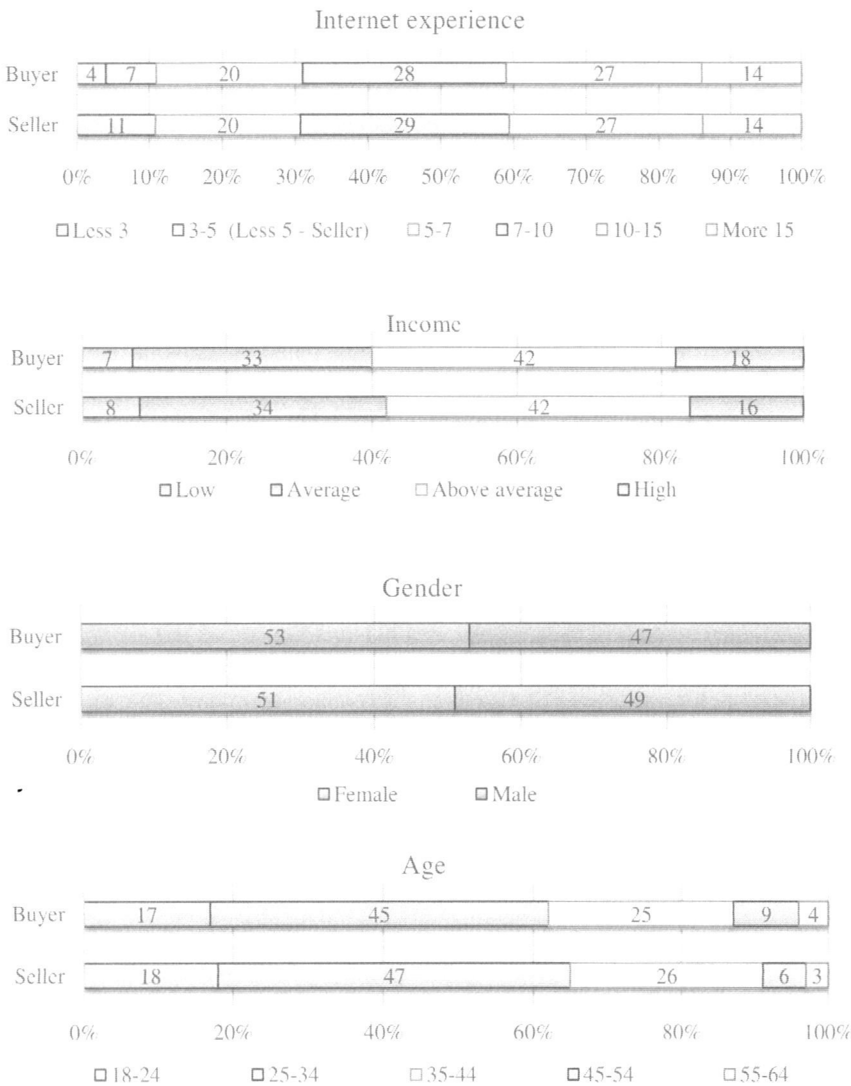

Figure 2. A portrait of the participants in the C2C segment [6]

At the same time, in the B2C segment of e-commerce, along with attractive time and financial savings, there are constraints: psychological (lack of tactile effect; distrust in forms of payment), technical (technical failures and fraud; weak information protection), logistics (unacceptable delivery terms), imperfection of legislation, etc. The organization of e-commerce assumes first of all that the buyer has access to the Internet, the quality of which is uneven in the territory of the EAEU countries.

The research conducted by VCIOM (Russian Public Opinion Research Centre) shows that the share of Internet users in Russia is 81 %, however, the use of the Internet is primarily caused by work or educational purposes (44 %), while Internet purchases are not a significant motive for the use of the Internet [11].

Another problematic factor hindering the development of Internet trade in the EAEU is the level of development of warehousing. The total area of warehouses in Russia is 24 million m2, while the Moscow and Leningrad regions accumulate 70% of them [3]. Only a quarter of warehouses of the highest category A are located outside the Central and North-Western Federal districts of Russia. In the

Republic of Belarus, the area of warehouses, classes A and B was about 800 thousand m2 at the end of 2017. Kazakhstan built two warehouses of class A, in the territory of Armenia there are only category B warehouses and below, and in Kyrgyzstan – the warehouses of hangar type are presented. Consequently, the development of e-commerce and virtual logistics requires a continuous growth of high-quality warehouses of A and A+ categories in all the above mentioned countries.

Despite the existing shortcomings, the volume of e-commerce in the EAEU is constantly growing and will continue to expand due to the expansion of the use of agile marketing - the manufacturer's continuous response to changes in market needs.

One of the most important components of e-commerce is effective logistics (logistics providers), formed on the basis of quality services of the transportation and storage of goods at an adequate cost. Virtual logistics becomes the basis of competitive advantage of the organisation. At the same time, there is a growing trend in the global market for organisations to outsource part of their functions. Virtual logistics ceases to be a part of the organisation, turning into services of third-party organisations for the transportation of goods.

Logistics efficiency (productivity) index is a six-component index, the overall assessment of which reflects the perception of the country's logistics on the basis of:

1. Efficiency of the customs clearance process.
2. Quality of services and infrastructure related to transport.
3. Ease of organisation of deliveries at competitive prices.
4. Competence and quality of logistics services.
5. Cargo tracking and control capabilities.
6. The frequency with which shipments reach consignee within scheduled or expected time.

The index value ranges from 1 to 5, and efficiency increases as the value increases. The analysis uses World Bank information in partnership with academic and international institutions, private companies and individuals involved in international logistics. The dynamics of logistics efficiency index for 2007 – 2016, calculated every two years is presented in Fig. 3 [7].

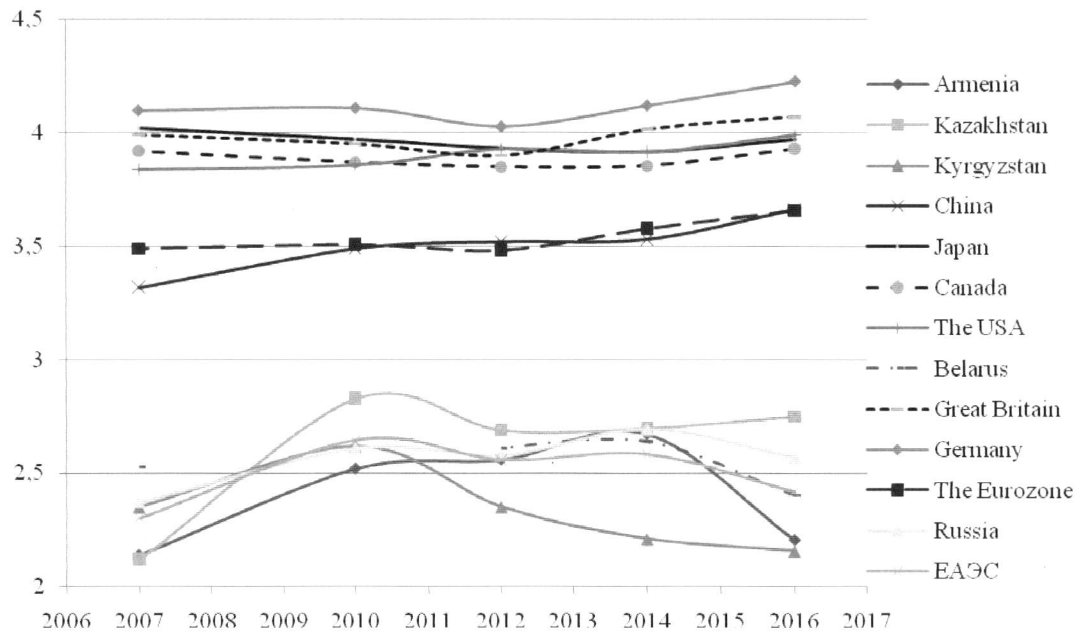

Figure 3. Logistics performance index

The leader in the studied indicator is Germany - 4.23 points. In general, it can be noted that in recent years, most countries have seen an increase in the indicator (the growth rate of the logistics efficiency index: China – 1.04, Japan – 1.01, the US – 1.02, the Eurozone – 1.02, the EAEU – 0.94, Russia – 0.95), while among the EAEU countries the only positive dynamics was observed in the Republic of Kazakhstan – the growth rate of the index 1.93%. The average decrease in the EAEU was 6.4 % in 2016 compared to 2014. This may be partly due to subjective geopolitical reasons, and partly due to objective economic factors. Component analysis of the index (Fig. 4) allows to state that Kazakhstan is the leader among the EAEU countries in 4 indicators, and Russia is the leader in quality of services and frequency of deliveries. Kyrgyzstan is lagging behind in most indicators, and the strong point of logistics in Belarus is the simplicity of the organisation of supplies, the level of prices for logistics services and the frequency of receipt of goods. The effectiveness of customs procedures for the majority of the EAEU members is the most problematic point. International analysts emphasise the abundance of documents for application, preservation of the procedure of physical inspection of goods, the duration of customs clearance procedures [4, 9, 15].

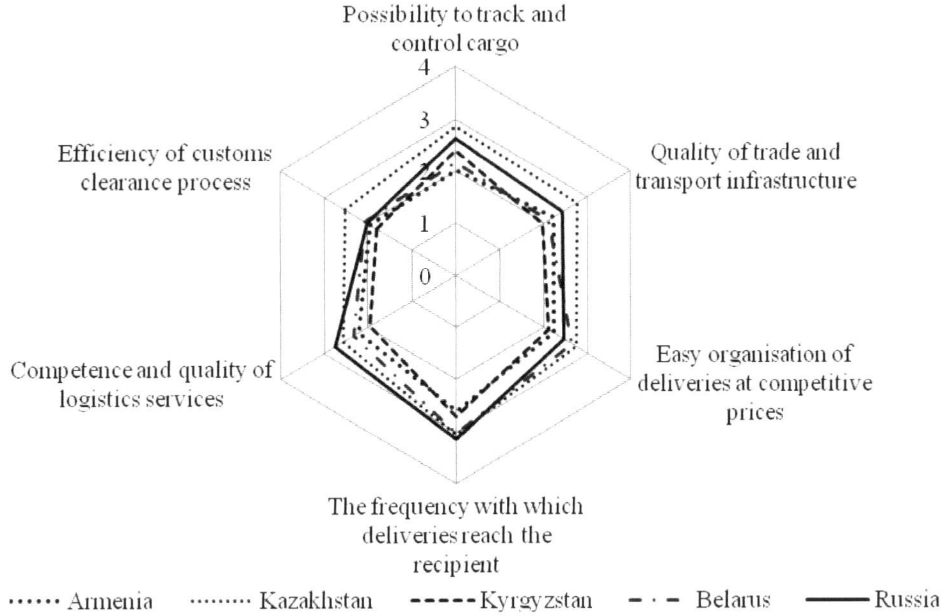

Figure 4. Components of the logistics performance index for the EAEU countries, 2016

A component-by-component comparison of the logistics efficiency index of the leading country in this indicator of Germany and Russia is presented in Fig. 5. The strongest side of the logistics leader is the quality of trade and transport infrastructure, the weakest - the quality of supply and price competitiveness [14].

In the course of the study, the authors established a close correlation between research and development costs as % of GDP and logistics efficiency, which indicates that international logistics is currently a high-tech industry, namely, high technology, primarily digital, that determines its further development. Germany's costs are respectively 2.88 % of GDP, Eurozone 2.14, Russia 1.13, the average for the EAEU - 0.44 % of GDP, which is less than the Eurozone 4.9 times. Therefore, the development of logistics in the EAEU will be largely determined by the volume of investment in research and development.

At the moment, the index of e-commerce development in the EAEU is significantly behind the leading states, with Russia in the 50th place out of 137 countries in 2015, Belarus – 62, Armenia – 87, Kazakhstan – 88, Kyrgyzstan-109 [1].

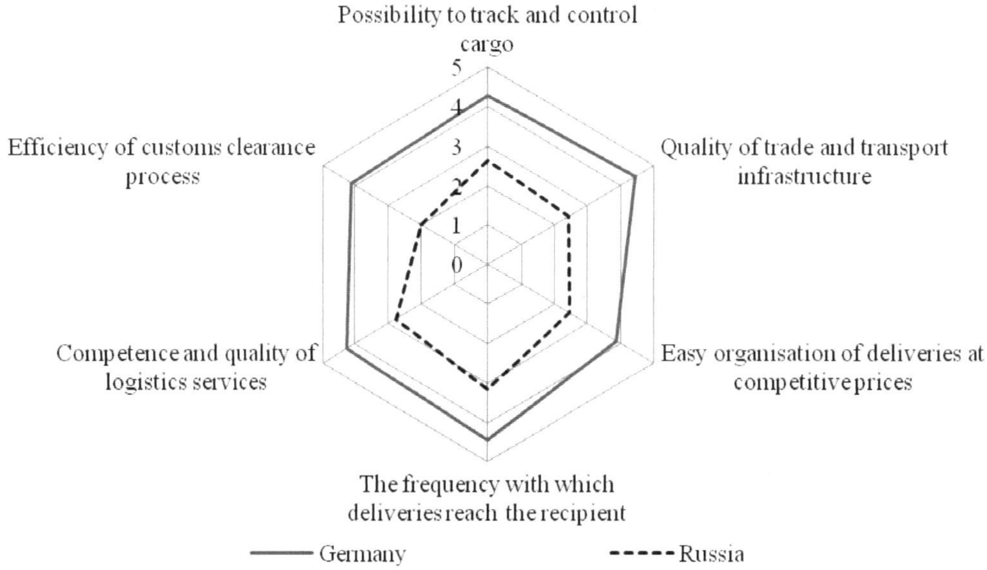

Figure 5. Comparative analysis of the logistics performance index of Russia and Germany, 2016

The coefficient of determination between the costs of research and development and the logistics performance index in 2016 amounted to 0.8372, the dependence is reflected in the formula (1). At the same time, the impact of the logistics efficiency index on real GDP per capita in 2016 R2= 0.9124 was expressed by the formula (2).

$$y=0.7389*x+2.1762 \tag{1}$$

$$y=57292\ln(x)-35091 \tag{2}$$

By elevating the quality of logistics to the level of Germany - the country-leader, Russia's GDP will be able to reach the level of 48012.97 $/person. Paradoxically, the authors could not establish a correlation between the level of the logistics efficiency index and the volume of trade of countries (correlation coefficient in the analysis of 132 countries was 0.05), however, a strong correlation with the volume of e-commerce is identified (Fig. 8).

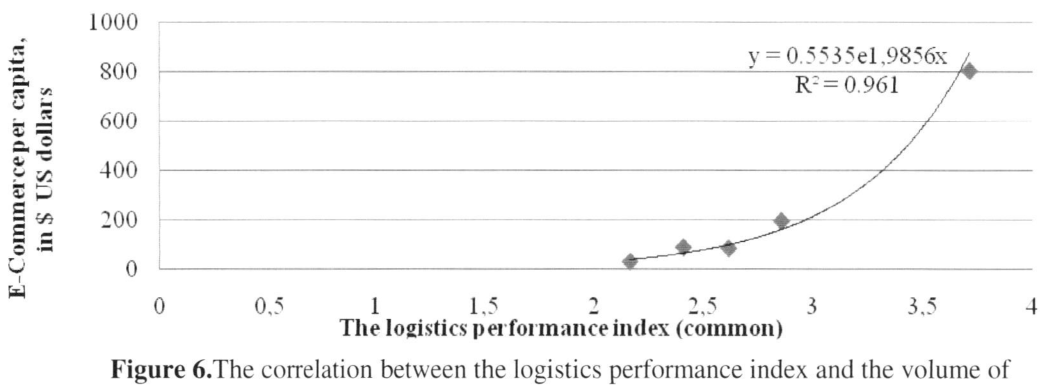

Figure 6. The correlation between the logistics performance index and the volume of e-commerce in 2016

The presence of this dependence indicates that the electronic commerce to a greater extent than trading as a whole depends on effective logistics. Given the continuous expansion of the volume of trade operations in the B2C and C2C segments, we can talk about the need to improve virtual logistics in the EAEU. The nearest prospect of online trading-neuromarketing - the formation of predictive proposals based on the offer of goods to customers before they thought about the need to purchase [2, 13]. The integration of logistics companies will increase the level of services, reduce time costs and shipping costs, which will increase the competitiveness of e-commerce companies in the EAEU and the world market (table.1).

Table 1. Classification of the level of integration of logistics activities

Logistics service levels (LP – Party Logistics)	Participants	Integration levels			
1 PL – logistics insourcing (autonomous logistics)	Cargo owners				
2 PL – partial logistics outsourcing (traditional logistics)	Hauliers	Transportation, warehousing and storage			
3 PL – comprehensive outsourcing of logistics services (logistics for third parties): - international freight forwarder; - providers of warehouse services, transportation, optimisation of transport services, software.	Logistics operators		Forwarding, organisation of customs clearance, services for cargo handling: small-party and multimodal transportation; express courier delivery; door-to-door transportation; cross-docking.		
4 PL – integrated logistics outsourcing	System logistics integrators			IT - integration and coordination of the activities of the focal company and key counterparties in the supply chain; commodity-transport monitoring; logistics capacity leasing; SCM consulting.	

5 PL – "virtual" logistics	"Virtual" logistics integrator				Network business services with the use of the global information technology space	
... n PL	«...» IT – integrator					The integration of information interaction with the ultimate consumer in the field of planned requirements into logistics business processes.

Any company from the Top 100 IT companies in Russia, geographically represented by Moscow, St. Petersburg, Kazan, Ufa, Ulyanovsk, Yekaterinburg, Novosibirsk, Nizhny Novgorod and other cities can act as a virtual logistics integrator. Any of these companies, having experience in the field of IT and management experience, can diversify its activities, acting as a 5 PL logistics provider.

3. Conclusion

E-commerce, as well as manufacturing, in the near future will be built on a scrupulous analysis of big data, which will allow to personalise the offers in the EAEU market [16]. Electronic and platform economy will allow to predict demand more accurately on the basis of digital footprint and collaborative filtering, the volume of product design for the consumer with constant cost will increase.

Thus, one of the main problems of the EAEU logistics development is slow customs clearance procedures, especially veterinary and phytosanitary control, the quality of infrastructure and ease of organisation of supplies. The first and last problems can be solved by virtual logistics capabilities, partly combining the capacities and investment opportunities of logistics providers, which will improve the situation in the second direction as well. The authors believe that the emergence of national logistics providers united in a single transnational provider on the territory of the EAEU is promising, which will allow:

1) to simplify the customs clearance procedure by expanding the possibilities of preliminary information and consulting;

2) to make better and more efficient use of the existing logistics infrastructure;

3) to increase the efficiency of investments in the development of virtual logistics;

4) to reduce the costs of enterprises for the organisation of logistics processes;

5) to improve the efficiency of logistics operations and economic activities in general, by improving the speed and quality of supply. This, in turn, will expand the possibilities of e-commerce as one of the promising areas of development of the integrated Eurasian economic space.

4.References

[1] Sheveleva N P, Luzin D A, Savkina N G Development of International Electronic Trading *Proceedings of the 31st International Business Information Management Association Conference (IBIMA)* 6554-6557

[2] Ergunova O T, Lizunkov V G, Malushko E Yu, Marchuk V I, Ignatenko A Yu 2017 Forming the system of strategic innovation management at the high-tech engineering enterprises *IOP*

Conference Series: Materials Science and Engineering **177 (1)** 012046
[3] Akutkina U M, Karpova O E, Konyavskiy V A, Yasinovskaya E D 2017 Digital economics: conceptual architecture of the digital industry ecosystem *Business Informatics: Higher School of Economics* **4 (40)** 17–28
[4] Analysis of the warehouse market in all regions of Russia 2018. https://zdanie.info/2393/2421/news/12550
[5] Balandina G, Ponomarev Y, Sinelnikov-Murylev S, Tochin A 2018 Customs administration in Russia: areas of improvement *Economic policy*. - St. Petersburg: ANO "Editorial Board of the journal "Economic policy"13 **3** 106-131
[6] Eurasian Economic Union. Facts and Figures. http://www.eurasiancommission.org/
[7] Online trading in Russia 2018. http://datainsight.ru/sites/default/files/DI-RIF2018.pdf
[8] Data directory Trend Economy http://data.trendeconomy.ru/dataviewer/oecd/National_accounts/
[9] Koryakina E A, Kulakova N S 2015 Problems of improving the living standards of the population of the region. Theory and practice of social development **24** 80-83
[10] Ferreira R, Lizunkov V G, Politsinskaya E V 2017 Formation of entrepreneurial competencies of university graduates in conditions of transition to the universities of the third generation *Novosibirsk State Pedagogical University Bulletin* **7 (6)** 195-211
[11] Ergunova O, Lizunkov V, Blaginin V, Politsinskaya E, Shaykina O I 2018 Formation of the price mechanism for energy resources in Russia and the countries of the European union *International Journal of Energy Economics and Policy* **8 (1)** 122-127
[12] Koshanov A 2017 Eurasian Economic Union: opportunities and limitations *Society and Economy* **9** 40-56
[13] Prospects and threats of e-commerce in 2018 https://www.insales.ru/blogs/university/
[14] The Great Interwebs: for work or entertainment? https://wciom.ru/index.php?id=236&uid=9322
[15] E-Commerce market in Russia. Results of the 1st half of 2017 http://www.akit.ru
[16] Lizunkov V, Politsinskaya E, Malushko E, Kindaev A, Minin M 2018 Population of the world and regions as the principal energy consumer *International journal of energy economics and policy* **8 (3)** 250-257

IOP Conf. Series: Materials Science and Engineering **483** (2019) 012111 doi:10.1088/1757-899X/483/1/012111

Methodological basics of assessing ecological safety in the zones where waterworks units influence natural habitats with the use of water resources

V.L. Bondarenko, E.A. Semenova, A.V. Aliferov

Novocherkassk Engineering and Land Reclamation Institute named after A.K. Kortunov, federal state-financed educational institution of higher education "Don State Agrarian University" (Novocherkassk)
North Caucasus Federal University (Branch in Pyatigorsk)

aliferov_92@mail.ru, dekannn@mail.ru

Abstract. Relying on the results of the long-term studies on the interrelation, interaction and correlation of waterworks units, called an "activity object", as a part of natural engineering systems "natural habitat-activity object-population" with the environment within a basin geosystem, the article reports on the methodological framework of evaluating the quantitative and qualitative indicators for the zones of activity object influence as important factors of environment safety assurance.

1. Introduction

The current period of social development both within the terrestrial biosphere of the global system "nature-society-man" and at the local hierarchical levels of river basin geosystems in which almost all types of economic and other activities are conducted is substantially linked with the resource-consuming technologies and water resources.

As the reality shows, the resource consuming technologies under systemic consideration contribute to the growth of the gap between the natural processes of interrelation, interaction and correlation in the systems of various hierarchical levels and the types of economic activity. According to the specialists' research, about 5-6 billion tons of the living matter disappears annually in the biosphere of the Earth because of such human economic activity (Bondarenko et al. 2009; Bondarenko et al. 2012). Such existing opposition of the biosphere and technosphere causes the urgent need to create basically new technologies for using natural resources including water resources.

The transition to the level of essentially new technologies of natural resource utilization cannot be implemented within the existing framework of the traditional development paradigm by transforming even radically the technological systems that are now in use. Consequently, there is a necessity for essentially new methodological approaches of the creation of technological basis that is integrally interrelated with scientific, productive, social and vital needs of the society. In a general view, this problem was formulated by V. I. Vernadskiy and his followers at the beginning of the XXth century, introducing the concept "noosphere" in which reasonable human activities become a development determinant (Bondarenko and Diachenko 2005).

The utilization of water resources in the manifold kinds of economic activity (industrial and agricultural productions, power generation at nuclear power plants (NPPs), thermal power plants (TPPs), hydroelectric power plants (HPPs), and pumped storage hydroelectric power plants (PSHPPs), public water supply, water disposal, etc.) is integrally connected with operating water systems and the building of new water systems in which various types of hydraulic structures are the main technogenic components. Water systems function within the basin geosystems where water resources develop as a superficial and underground flow. So, the use of water resources in the considered basin geosystems of the Kuban River, the Terek River and the Lower Don River is related to intra-basin and inter-basin regulation and river flow transfer that assigns the spatial borders of very considerable influence zones of hydraulic structure complex, being a part of a water system (Bondarenko et al. 2007).

The functioning of a hydraulic structure as a part of a water system makes certain changes in the natural processes of interrelation, interaction and correlation between natural (biotic and abiotic) components within the boundaries of the influence zones of the considered hydraulic structure complex, hereinafter referred to as "activity object", functioning as a part of the natural engineering system (NES) "natural habitat-activity object-population" (NH-AO-P). As the research results prove, the changes in the natural habitats within the influence zones of the activity object in the quantitative and qualitative sense are characterized by the changes in the movement intensity of streams of substance, energy and information, determining all the processes of life activity in the considered spatial limits (Vernadskiy 1988). Under a systematic consideration, the movement of streams of substance, energy and information determines the ecological state in habitats (atmosphere, hydrosphere, top layers of lithosphere, soil layer of pedosphere and sub-soil) that affects all aspects of the vital activity of the living matter including a man. In terms of the water resource utilization, the findings of the long-term studies on operating and being under construction natural engineering systems "NH-AO-P" show that for example, the ecological state in the zones of the activity object influence in the region of the North Caucasus defines the environmental safety as an important factor in the processes of the people's life activity. Thus, the assurance of environmental safety is the most significant and necessary requirement in the technological processes of using water resources (Saling et al. 2002; Saling and Hafer 2009).

2. Materials and methods

The problem of Water as renewable natural resource determines its use in the economic activity and not less important conditions of nature and types of activity object influence on natural habitats, the spatial limits of basin geosystems where the quantitative and quality indicators of water resources are formed as the determinant of environment safety, depending on the ecological state of the area under study.

The processes of interrelation, interaction and correlation of the activity object with natural habitats were studied within the basin geosystems of the Kuban, the Terek and the Lower Don where over 23 million people live (16.3% of the Russian Federation population). In terms of the spatial-temporal consideration, it was found out that the zones of activity object influence in the interaction with natural habitats are different in their nature, borders and time of influence on the movement processes of the flows of substance, energy and information.

Proceeding from the unity of actions of the Nature and economic activity, the systematic consideration of interrelation, interaction and correlation of the activity object with natural habitats identified the need for reasoning a new methodological approach to the delimitation of influence zones and specification of the nature of influence and activity of the population, living in these zones,with the use of the system energy-entropy approach (Davis 1990). In view of the concept, the research of the processes of activity object interaction with the biotic (flora and fauna) and abiotic (atmosphere, river hydrographic network and underground waters, the upper layers of lithosphere in the basis of hydraulic structures and soil cover) components of the environment is considered according to the fundamental laws of the Nature (energy conservation, irreversibility of processes, the second law of thermodynamics, etc.) where the Sun is the main source of energy. The study of issues on reasoning the zones borders of activity object influence is discussed by the example of the hydraulic structure complex and buildings of Zelenchuk Hydroelectric Power Plant-Pumped Storage Hydroelectric Power Plant (Zelenchuk HPP-PSHPP/HPP-PSHPP) which is located in the basin geosystem of the Upper Kuban in the territory of the Karachay-Circassian Republic (KCR).

Considering the Nature actions within the Earth's biosphere and engineering processes of using water resources for the production of electrical energy by the HPP-PSHPP, the interrelation between the processes of transformation of energy forms should be emphasized; in this case it is referred to the transformation of the solar radiant energy into the potential energy of a water flow and then into electric one with a certain coefficient of performance (COP)according to the laws of thermodynamics.Itisimportanttonotethat20%(135.600 TW) of the solar energy, reaching the surface of the Earth's land (149.1×10^6km^2) and World Ocean (361.1×10^6km^2) flows into the global hydrological processes of the water cycle and less than 1% (1.780 TW) – into the photosynthesis processes of the Earth's flora (Kovalchuk 2010).

The interrelation, interaction and correlation of the activity object in the regulation and intra-basin transfer of the river flow from the local areas of the basin geosystems of the BolshoyZelenchuk River (F_{wat}=779 km^2, W_{atm}=7.790 km^2, W_{lit}=233.7 km^3), the Marukha River (F_{wat}=336 km^2, W_{atm}=3.360km^3, W_{lit}=100.8 km^3), and the Aksaut River (F_{wat}=509 km^2, W_{atm}=5.090 km^3, W_{lit}=152.7km^3=15) occurs within the boundaries of the basin geosystem of the Upper Kuban River (F_{wat}=11.0×10^3km^2, W_{atm}=110×10^3 km^3, W_{lit}=3.3×10^3 km^3) that is a basic part of the Earth's biosphere where the global processes of the water cycle run under the influence of the solar energy.

Theformationofthewatersupplyinthelocalareasofthenaturalengineeringsystem "NH-AO-P"in the basin geosystems of the Bolshoy Zelenchuk, the Marukha and the Aksaut requires the solar energy in the amount of 273.429×(10^6×kW/h) 117.136×10^6kW/h, and 178.659×(10^6×kW/h) for the generation of water volumes of 1.232 mln m^3, 290.0 mln m^3, and 446 mln m^3(according to the long-term annual average data)correspondingly, 26% of which is taken (on average, 512×[10](6)m^3)(by the technological system of Zelenchuk HPP-PSHHP to produce the electric power. It is worthmentioning that in the technological processes the solar energy (E_{free} =148206×(10^6×kW/h)) is transformed into the object-oriented water flow of 512×10^6m^{3}for the annual production of electric power in the amount of E_{HPP}=400×(10^6×kW/h)at the HPP-PSHPP with In plants the COP is about 0.1 [5].

At the level of the basin geosystem of the Upper Kuban within which the activity object functions as a part of the natural engineering system "NH-AO-P", its integrity is supported by the continuous processes of self-organization and transformations in the intra-system formations (hydrographic network, flora and fauna, soil cover, top layers of lithosphere, atmospheric boundary layers, etc.) at all hierarchical levels of the basin geosystem. The introduction of the activity object to natural habitats results in the complication of inter-structural relations in the processes of self-organization between natural and technogenic structural formations in the zones of its influence. The irreversibility in the processes of the activity object interaction with the habitats in the zones of influence causes the emergence of new phenomena in the geological environment, hydrographs of stream bottom part of upper and lower pound locks of reservoir hydrosystem, bank areas of inundated parts of river channels, atmospheric boundary layers and basin geosystem as a whole (Kovalchuk and Naraikin 2011).

3. Results and discussion

Analyzing the results of the studies on the types of activity object influence on habitats, the specific zones of influence were identified with the classification characteristics: **low-active, active and highly active**(see Fig. 1).

Low-active zone of influence I (see Fig.1)is defined by the intra-basin transfer of water flow in the basin geosystem of the Upper Kuban and is characterized by minor changes of hydrographs on the tributary streams of the first (the Bolshoy Zelenchuk and the Malyi Zelenchuk) and second (the Marukha and the Aksaut) order.

Active zone of influence II (see Fig.1) is determined by the influence of water bodies and water surface on the microclimate parameters in the atmospheric boundary layers with the growing sum of temperatures over 70 and consequently the duration of the frostless period of 10-15 days. This zone of influence is located on the water storage reservoirs and water catchment areas of hydro-systems, along the route of open areas of the derivational canal "Zelenchuk-Kuban" and in the areas of the upper and lower basins of balance storage (see Fig. 1). It should be pointed out that it is necessary to take into account in the operation period that the air humidity increases with fogging when temperatures go beyond certain limits.

Highly active zone of influence III (see Fig.1) is defined by the considerable changes in natural hydrological, river channel forming and hydraulic processes and the life activity of land and water biocenoses lower the route of the derivational canal "Zelenchuk-Kuban" on the Bolshoy Zelenchuk and the Malyi Zelenchuk. The major factor of influence is 50% of water catchment during the flood season (April-October) from the Bolshoy Zelenchuk, the Marukha and the Aksaut. To reason the possibility of 50% of floodwater catchment, the complex studies (2005-2006) were carried out, engaging the professionals from various scientific fields (biology, ichthyology, zoology, botany, soil science, sociology, hydraulic engineering, etc.) that enabled to assess the nature and level of influence and changes introduced to the natural interaction of natural components between themselves and the population, by reference to the vital needs in the NES"NH-AO-P" under creation.

Highly active zone of influence IV (see Fig.1) is formed by the change of natural hydrograph of the Kuban River in the part (15 km) from the dumping by the hydraulic units of Zelenchuk HPP-PSHHP ($70m^3$/sec) to the facilities of the waterfront of Ust-Dzhegutinsky storage reservoir from which water is pumped to the Great Stavropol Canal (GSC). The factors of influence on the natural hydrograph of the Kuban River is the consumption growth of water discharged by the hydraulic units of the plant, forming the unbalanced hydraulic mode of channel flow movement on the river part under consideration (15 km). The unbalanced changes of water levels in Ust-Dzhegutinsky storage reservoir affect the safety of hydraulic structures of the waterfront, functioning of the water catchment structure of the GSC and the hydraulic operating mode of the canal.

Active zone of influence V (see Fig.1) is specified by the processes of the interaction between the activity object and the upper layers of the lithosphere - geological environment in which the composition, condition and level of natural soil waters change under the influence of seepage waters from the structures (reservoirs, derivational canal "Zelenchuk-Kuban", upper and lower basins of balance storage).The laboratory research of the seepage processes in the geological environment of the zones of the active activity object's influence, conducted by the filtration laboratory of Novocherkassk Engineering and Land Reclamation Institute named after A.K. Kortunov set the quantitative indicators that were used to define the borders of influence zones in the geological environment.

The generalized quantitative indicators of zones of activity object influence are characterized by spatial values of local volumes (W) which specify the changes in the natural habitats: atmosphere (W_{at}), upper layers of the lithosphere (W_{lit}), the bank areas of river channels below site of water catchment hydro-system on the Bolshoy Zelenchuk, the Marukha and the Aksaut (tributaries of Malyi Zelenchuk) and in the part of the Kuban River from the site of Zelenchuk HPP-PSHHP to Ust-Dzhegutinsky storage reservoir and in the local areas of the land surface as well (F_{ZP}) (Bondarenko et al. 2009).

The research data of assessment of influence of Zelenchuk HPP-PSHHP objects as a part of the NES"NH-AO-P" on the natural habitats in the basin geosystem of the Upper Kuban were applied to obtain the following quantitative indicators:

So, the draw of the design water flow in the flood period (April-October) from the Bolshoy Zelenchuk and Malyi Zelenchuk makes from 25.9% to 26.5% and from 39.5% to 43.4% correspondingly, causes the negative phenomena in the natural habitats of the influence zones and reduces the zones of waterlogging and flooding beyond the area of the site of water catchment systems; the area of the NES "NH-AO-P" under discussion amounts to F_{NES}=117, 3 km^2, the volume of the atmospheric boundary layers – $W_{NES(at)}$=1.173 3 km^3, the lithosphere volume $W_{NES(lit)}$=35, 2$km^3$and the total amount makes W_{NES} =1.212 km^3.

The quantitative ratio of the spatial influence zone, formed by the NES "NH-AO-P" (W_{NES} =1.212 km^3) to the special boundaries of the basin geosystem (BG) of the Upper Kuban (W_{BG} =113.3 $thou_{sandkm}$ km^3) is W_{NES}/ W_{BG}=0,011 that means 1.1% and makes the conditions for a considerable dominance of the natural processes in the movement formation of the flows of substance, energy and information within the spatial borders of the zones of activity object influence.

The study results enable to formulate the following concept-based statements on the ecological state in the zones of activity object influence (I-V) and environment safety:

The environment safety in the zones of activity object influence is interrelated with the processes of the life activity of population and proceeding processes in natural habitats.

– The environment safety is directly interlinked with the energy consumption, providing the generation, storage and transformation of energy forms (thermal, mechanical, chemical, electric, etc.).

– The threat of environment safety damage is activated by the uncontrolled emission of energy and substance, accumulated by the activity object, into the external environment in relation to the activity object, for example, the destruction of waterfront of reservoir hydrosystem.

– The uncontrolled emission of energy and substance into the external environment results in the extreme negative change in the processes of the life activity of population in natural habitats and is accompanied by harm (material and ecological).

– Under systematic consideration, the environment safety damage in the zones of activity object influence is a result of the expression of cause-effect chain of prerequisites that develop at the stages of designing, building and operating of activity object.

– The links of the cause-effect chain in the environment safety damage within the system "protection object-source of environmental hazard-protective measures" ("PO-SEH-PM") is the activity object that does not meet the environmental and designing requirements.

The complex assessment of the environment safety in the zones of activity object influence (I-V) was carried out by applying the results of different studies (geochemical, nuclear chemical, lithochemical, biogeochemical indications, the pollution of hydrosphere (river network), atmosphere, upper layers of lithosphere, pedosphere, the energy flows, emitting by vehicles, mechanisms and equipment) and the analysis of structure and principles of flora and fauna distribution. The assessment results were used to define the qualitative (I_i) and quantitative (I_j) component indicators of the ecological state in the zones of activity object influence as factors of the environment safety.

In terms of quantitative and qualitative parameters, the indicators (I_i and I_j) of component group are characterized by maximum permissible concentration, maximum permissible level and maximum permissible emission in relation to original (background) ecological state.

The first component group of indicators I_i involves the main kinds of the activity object influence on the natural habitats: $I_{at(i)}$ – atmospheric boundary layers; $I_{surf(i)}$ – surface and $I_{gr.(i)}$ – ground waters; $I_{soil(i)}$ – soil cover and sub-soil; $I_{flora(i)}$ – flora and $I_{fauna(i)}$ – fauna. Using these indicators, it was found out that the ecological state in the zones of activity object influence features minor and quite permissible pollution level. It should be mentioned that the dynamics of the ecological state within the low active zone of activity object influence (I) is dominantly affected by the other types of economic activity, unrelated to the utilization of water resources at Zelenchuk HPP-PSHPP.

The second group consists of quantitative indicators (I_j) that characterize the changes in the natural processes of water resource formation (hydrological, hydrogeological and in river channels) with in the basin geosystem of the Upper Kuban under study: $I_{MAC(j)}$ – indicator of maximum allowed inverse catchment of water at the HPP-PSHPP from the river network; $I_{IBPDF(j)}$ – indicator of the intra-basin partial diversion of flow with the change of hydrographs in river channels; $I_{RC(j)}$ – indicator that forms river channels and is related to the processes of data transport and suspended sediments in a stream flow in upper and lower canal pounds; $I_{SS(j)}$ – indicator of the seepage security in the basements and bank abutments in the interaction of hydraulic structures with the geological environment; $I_{MC(j)}$ – indicator of microclimate change in the zones of interaction of water surface of water storage reservoirs, upper and lower basins of balance storage and water supply structures (canals and aqueducts) with the atmospheric boundary layers.

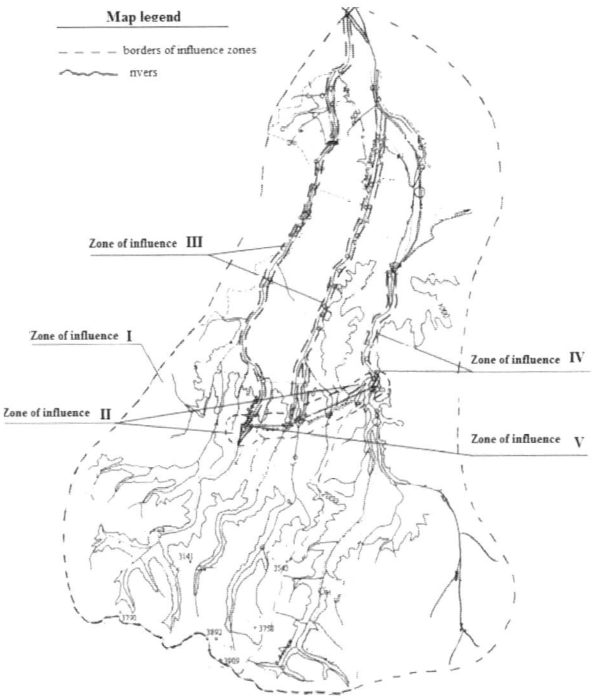

Figure 1. The zones where the complex of hydraulic engineering structures of Zelenchuk HPP influences on the environment of the basin geosystem of the Upper Kuban: I – low-active zone of influence within the territory of the water collecting territory on the Upper Kuban (11,000 km^2); II – active zone of influence of the complex of hydraulic structures on the microclimate indicators (I_{MC}); III – highly active zone of influence of the complex of hydraulic structures on the Bolshoy Zelenchuk and Malyi Zelenchuk (I_{MAC}); IV – highly active zone of influence in the fifteen kilometers part of the Kuban (I_{BPDF}); V – active zone of influence of water storage reservoirs and the complex of hydraulic structures on the geological environment(I_{BSR})

The systematic study of the interaction of activity object with the natural habitats showed that the indicators (I_j) in quantitative terms are defined by the ratio of the spatial borders of the determined influence zones (W_I-W_V) to the spatial borders of local basin geosystems, the Bolshoy Zelenchuk (I_{BZ}), the Marukha (W_{BM}) and the Aksaut (W_{BA}), and the basin geosystem of the Upper Kuban as a whole (W_{BG}) where water resources develop.

So, $I_{MAC(j)}=W_{III}/W_{BG}$= 0.0065(0.03%); $I_{IBPDF(j)}=W_{IV}/W_{BG}$= 0.0003 (0.03%); 0.5≥ $I_{RC(j)BZ}$; $I_{RC(j)(M)}$; $I_{RC(j)(A)}$≥ 1.0 м; $I_{SS(j)}=W_V/W_{BG}$= 0.00003 (0.003%); $I_{MC(j)}=W_{II}/W_{BG}$= 0.0057 (0.57%).

According to the local elements of the basin geosystem $I_{BZ(j)}=W_{II}/W_{BZ}$= 0,0079(0.079%); $I_{M(j)}=W_{II}/W_{BM}$= 0.055 (5.5%); $I_{A(j)}=W_{II}/W_{BA}$= 0.024 (2.4%); $I_{BZ(j)}=W_{III}/W_{BZ}$= 0.0092(0.92%); $P_{M(j)}=W_{III}/W_{BM}$= 0.0063(0.63%); $I_{A(j)}=W_{III}/W_{BA}$= 0.0012(0.12%); $I_{MZ(j)}=W_{III}/W_{BM}$= 0.015 (1.5%); $I_{C(V)}=W_{IV}/W_C$=0.0008/0.08%; $I_{BZ(V)}=W_V/W_{BZ}$= 0.00003/0.003%; $I_{M(V)}=W_V/W_M$= 0.002/0.02%; $I_{A(V)}=W_V/W_A$=0.0000/0.008%; $I_{NES}=W_{NES}/W_{BG}$= 0.125/1.25%.

The analysis of the quantitative indicators I_j suggests the domination of natural processes of interaction between biotic and abiotic components over technogenic processes of interaction between the activity object and natural components in the form of the changes introduced to the natural habitats. The domination of natural processes over technogenic changes in the natural habitats within the zones of activity object influence specifies the ecological state in the considered areas which are characterized by sufficient sustainability. The ecological sustainability in the influence zones in relation to the original background state in the natural habitats is characterized by the ratio of the spatial borders of NES"NH-AO-P" ($W_{NES(at)}$ =1,422.94 km^3)where the technogenic processes take place to the spatial borders of the basin geosystem of the Upper Kuban (W_{BG} =113.300 km^3) where the natural processes proceed and the flow is formed (on average 3000 mln m^3 annually). So, the ratio of W_{NES}/W_{BG}=0.0125 (1.25 %) reflects the quantitative domination of natural processes with the probability rate of 98.5% - 99.003% that determines the ecological state and environmental safety in the zones of activity object influence.

The results of the ecological state monitoring in the influence zones of the activity object "Zelenchuk HPP-PSHPP" that was implemented in the period from 2006 to 2014 prove the facts mentioned in the previous paragraph.

IOP Conf. Series: Materials Science and Engineering **483** (2019) 012111 doi:10.1088/1757-899X/483/1/012111

4. Conclusion

The conceptual principles of methodology for the definition of quantitative (I_j) and qualitative (I_i) component indicators of the ecological state as an important factor of providing ecological safety in the determined zones of influence of functioning activity object or activity object under construction as a part of the NES "NH-AO-P" were scientifically grounded, relyingon the results of the systemic studies of assessing the zones of waterworks unit influence on the natural habitats within the spatial limits of river basin geosystems.

The obtained results were tested under operational conditions by the activity object "Zelenchuk HPP-PSHPP", the Irrigation Water Distribution System of Stavropol Krai and Rostov Region that confirmed the accuracy of methodological approach for assessing the zones of activity object influence on the natural habitats in the spatial limits of the considered basin geosystems where water resources develop.

The qualitative relations of special borders of influence zones(W_I-W_V) to the spatial borders of local basin geosystems (W_{BG}) and NES "NH-AO-P" (W_{NES}) reflect the changes, introduced by the activity object into the natural systems on the background of natural interaction processes. Thus, the ratio W_{NES}/W_{BG}=0.0125 (1.25%) means the dominance of the natural processes over technogenic processes that defines ecological state and environment safety in the zones of activity object influence.

References

[1] Bondarenko V L,Diachenko V B 2005 Assessment of ecological state of basin geosystems in the processes of using water resources *Regional Environmental* **2** 86-92

[2] Bondarenko V L, Gutenev V V, Privalenko V V, Poliakov E S 2007 Assessment of the influence on the environment during the design of the water resources utilization system "Zelenchuk HPP" *Theoretical and Applied Ecology* **1** 47-54

[3] Bondarenko V L, Privalenko V V, Kuvalkin A V 2009 Solution of ecological problems in design of hydraulic structures in terms of the basin geosystem of the Upper Kuban (Rostov-on-Dom: publishing house of Southern Scientific Center of the RAS)

[4] Bondarenko V L, Privalenko V V, Skibin G M, Azarov V N 2012 Environmental safety in environmental engineering, water management and building industry: ecological infrastructure of basin geosystems *Monograph* (Novocherkassk: SRSTU (NPI))

[5] Davis G R 1990 Energy for planet Earth, *V Mire Nauki* **11** 7-16

[6] Kovalchuk M V, Naraikin O S 2011 Constructor for the future *V Mire Nauki* **9** 24-31

[7] Kovalchuk M V 2010 From synthesis in science to convergence in education, *Educational Policy* **11-12 (49-50)** 1-12

[8] P Saling, A Kicherer, B Dittrich-Kramer, R Wittlinger, W Zombik, I Schmidt, W Schrott, S Schmidt 2002 Life cycle management, Eco-efficiency Analysis by BASF: The Method (BASF Germany)

[9] Saling P, Hofer R 2009 "Metrics for Sustainability" *RSC Green Chemistry* **4** 25-37

[10] Vernadskiy V I 1988 *The philosophical ideas of a naturalist*, (Moscow: Nauka)

IOP Conf. Series: Materials Science and Engineering **483** (2019) 012112 doi:10.1088/1757-899X/483/1/012112

Efficient use of Russia's natural resources: agricultural production, export and import

S V Senotrusova[1], V G Svinukhov[2], E V Vavilova[2], V V Zhoglitcheva[2]

[1]Lomonosov Moscow State University,GSP-1, Leninskie Gory, Moscow, 119991, Russia
[2]Plekhanov Russian University of Economics, Stremyanny lane 36, 117997 Moscow, Russia

svetlsen@mail.ru, customs_fgu@mail.ru, alena.elena86@mail.ru, inmar2000@mail.ru, alena.elena86@mail.ru

Abstract. The paper provides assessment of the geographical focus, volumes and cost of import of vegetable products under 0702 Foreign Economic Activity Commodity Nomenclature (FEACN) code under the embargo conditions. To make such assessment, the foreign trade statistics and data, provided by the RF Customs Service were used. At the end of 2017, the import of tomatoes of 0702 FEACN commodity code reduced relative to 2015 in monetary terms by 17.4%, and in weight terms - by 22.5%, and versus 2013 import of tomatoes reduced by 49.5% (down to 558.7 million U.S. Dollars), on a weight basis import reduced by 40.8% (down to 515.8 thousand tons). The completed analysis shows that geographic focus and supply volumes have changed significantly. If in 2013, the main suppliers were Turkey (450,8 million U.S. Dollars), China (105,5 million U.S. Dollars), the Netherlands (101,9 million U.S. Dollars), Morocco (101,3 million U.S. Dollars), Spain (109,9 million U.S. Dollars), Belarus (42,1 million U.S. Dollars), Ukraine (41,2 million U.S. Dollars), Azerbaijan (41,0 million U.S. Dollars), in 2017, the main suppliers became such countries as Azerbaijan (155.1 million U.S. Dollars), China (135,1 million U.S. Dollars), Morocco (132,4 million U.S. Dollars).

1. Introduction

Currently import of some food items to Russia is regulated by the Decree of the Russian Federation President dated August 6, 2014 № 560 "On the application of certain special economic measures to ensure the security of the Russian Federation." The decree prohibits import of some agricultural commodities. The list of agricultural products, prohibited for import into the Russian Federation, includes commodities under 07 FEACN EAEU (Eurasian Economic Union) (vegetables, edible roots and tuber crops) (table.1).

Table 1. Product item group 07 FEACN EAEU

FEACN of the EAEU	Product Name
0701	potatoes, fresh Or chilled
0702	Tomatoes, fresh or chilled
0703	onions, shallots, garlic, leek and other bulbous vegetables, fresh or chilled
0704	Cabbages, cauliflower, kohlrabi, Kale and similar edible brassicas, fresh or chilled
FEACN of the EAEU	Product Name
0705	lettuce (lactucasativa) and chicory (cichoriumspp), fresh or chilled
0706	Carrots, turnips, salad beetroot, salsify, celeriac, radishes and other similar edible roots, fresh or chilled

0707	Cucumbers and gherkins, fresh or chilled
0708	Leguminous vegetables, shelled or unshelled, fresh or chilled
0709	Other vegetables, fresh or chilled
0710	Frozen (raw or boiled in water or steamed) vegetables
0711	Canned for short-term storage (e.g., sulfur dioxide, in brine, sulfur water or other temporarily preserving solution), but in this form unsuitable for direct consumption vegetables
0712	Dried, whole, cut, sliced, broken or in powder, but not further prepared vegetables
0713	leguminous vegetables dried, hulled, peeled or unpeeled, chopped or uncooked
0714	Manioc, arrowroot, salep, artichokes or topinambour, sweet potatoes or yams, and similar roots and tubers with high starch or inulin, fresh, chilled, frozen or dried, whole or sliced or in the form of pellets; sago

2. Materials, methods and organization of research

Commodities of 07 FEACN group (vegetables, edible roots and tuber crops) are one of the significant items of the Russian food import, together with dairy and fish products [1-11]. In 2014 the country ranked 5th in the list of major importers of vegetables and some edible roots and tubers (2.95 billion U.S. dollars), in 2015 - 10th place (1.89 billion U.S. dollars), and in 2016 Russia was located on the 12th place (1.39 billion U.S. dollars). However, in 2017, Russia ranked 9th among the countries of the world in terms of the value of import of the group 07 (1.8 billion U.S. dollars) and in terms of the value of its exports - 16 place (0.5 billion U.S. dollars). The first positions of this products import in 2017 were taken by such countries as the USA (14.2% of world imports), Germany (9.0%), the UK (6.5%), Canada (5.6%) and India (4.7%) (table. 2).

Table 2. The cost of import and export of 07 FEACN EAEU group in the context of the world's leading countries at the end of 2017

Import, billion U.S. dollars		Export, billion U.S. dollars	
USA	10.3	China	10.5
Germany	5.9	Netherlands	7.3
United Kingdom	4.3	Spain	6.8
India	4.0	Mexico	6.6
France	3.3	USA	4.8
Canada	3.1	Canada	4.5
Netherlands	2.6	Belgium	2.6
Japan	2.5	Australia	2.4
Russia	1.8	Russia	0.5

If global import of 07 FEACN group products is analyzed, it may be noted that in all countries there has been a slight decrease in imports of vegetables since 2014. However, Russia from 2014 to 2016 reduced the import of vegetables in value terms by 52%, but later in 2017 slightly increased import. Among the main counteragents of Russian vegetable import in 2017 were China, Turkey, Israel, Egypt and Morocco, their share was up to 72.7% [10]. It should be noted that the agricultural sector is the most socially significant and climate-dependent, vegetable production is quite labor-intensive, and prime cost is affected by temperature, humidity, soil structure and a number of other natural factors, so few countries can become large and efficient exporters of large volumes of vegetable products. In 2017, the top five global exporters were China (about 13% of world exports), Spain (13.2%), the Netherlands (12.2%), Mexico (10.1%) and Canada (7.2%). These countries account for a large share of the global food export.

3. Results and their consideration

Analyzing foreign trade statistic data, we should note that Russia is the largest importer of agricultural produce. However, recently the structure of countries-exporters in Russia substantially changed, in connection with "food" embargo. The question is which countries came to the Russian market of agricultural commodities and what is the volume of their supplied products. To analyze import of vegetable produce in the Russian Federation and export from it, foreign trade customs statistics data, and international trade statistics data was used. At the same time, it is impossible not to investigate the issue of Russian export of vegetable products. Among the countries that import Russian vegetables are Turkey, India, Ukraine and Pakistan. Their share is up to 77% (cost) and about 70% (volume) of Russian goods 07 FEACN codes. In 2015 versus 2014, there was a dramatic growth of vegetable export from the Russian Federation to the Ukraine. According to foreign trade statistics data, export from Russia increased by 51 times (in terms of value) (table.3).

Table 3. Importers of Russian vegetables (group 07 FEACN EAEU) (mln. U.S. dollars /thousand tons)

Countries	2014	2015	2016	2017
Total	246.2/679	402.9/1196	478.1/1332	404.5/1562
Turkey	113.9/344.2	173.0/516.1	200.6/513.0	191.2/585.8
India	-	55.3/139.9	51.8/122.3	54.6/220.3
Ukraine	0.4/0.2	25.6/197.1	25.5/216.7	35.6/227.2
Pakistan	8.6/24.8	22.1/59.7	35.9/79.8	29.3/83.5
Belarus	3.0/2.6	3.3/3.6	4.1/5.3	5.5/8.9
Italy	22.1/70.7	-	-	10.3/42.1

Dynamics of the Russian export of vegetables and tuber crops is uneven. Until 2012, there was a moderate stable growth. Then, in 2013 and 2014, the cost of vegetable produce export reduced versus 2012 by approximately 92 million U.S. dollars. Nevertheless, in the following years the cost of export products and their volumes started growing. In 2016 versus 2014 the cost of export increased by 232 million U.S. dollars. In 2017 versus the previous year, the export of vegetables increased on a weight basis by another 15%. At the same time, Russia ranked 16th in the list of countries exporting goods of 0701-0714 FEACN codes.

Among the main products of the vegetable group of Russian exports are the following:
* dried, shelled, peeled or unpeeled, chopped or uncooked legumes (FEACN code -0713), which amounted to 80% of the export of vegetables and root crops in 2017. As the analysis showed, under

this commodity code is exported most of the vegetables from Russia, such as beans, chickpeas, peas, lentils and other legumes;
* other fresh and chilled vegetables (item code -0709), which amounted to 9% of the export of vegetables and root crops in 2017;
* fresh or chilled potatoes (item code – 0701), which amounted to 4% of the export of vegetables and root crops in 2017;
* exports of the remaining goods of the group of 07 amounted to about 1% or less of the total exports of vegetables and root crops.

Meanwhile, it should be noted that import of vegetable products in Russia since 2013 has reduced by approximately 1.1 billion U.S. dollars, which is about 44 %. The largest drop was in 2015-2016. One of the factors for import cost reduction is ruble depreciation (table. 4).

Table 4. Exporters of vegetable products to Russia (group 07 FEACN EAEU) (mln. U.S. dollars / thousand tons)

Countries	2014	2015	2016	2017
TOTAL	2959/3380	1913/2618	1401/1816	1794/2470
China	466.8/556.6	441.3/503.8	380.6/386.8	488.1/543.7
Morocco	131.2/107.0	105.2/81.9	174.5/133.2	143.7/109.5
Israel	277.1/334.7	189.4/246.6	156.3/183.8	181.1/195.9
Belarus	154.5/233.2	155.4/376.9	145.3/327.6	175.6/373.4
Azerbaijan	104.9/125.5	91.3/119.3	125.6/156.3	207/4/256/7
Iran	117.4/120.3	114.2/128.9	104.0/99.1	119/9/117/4
Egypt	213.2/379.7	165.9/380.2	91.1/229.4	179/6/454/9
Turkey	597.3/557.5	434.6/468.4	29.1/36.7	55/6/97/8
Uzbekistan	23.1/19.5	18.5/27.2	27.4/35.7	34/7/51/1
India	37.5/63.6	27.1/40.3	23.2/34.8	26/7/39/9
Armenia	5.8/6.6	9.4/14.1	21.8/38.5	15/2/27/9

For the last several years, the structure of exporting countries has changed dramatically. Supplies of goods were reoriented from EU countries to EEU. For example, Belarus increased import of vegetable produce in the Russian Federation by 4.3 times; Armenia increased supplies by 25 times. The volume of vegetables import from Azerbaijan has nearly doubled. However, export from France reduced considerably, by 53 times, Poland reduced export from 234 million U.S. dollars down to 106 thousand U.S. dollars, i.e. by 2210 times. The Netherlands, which was one of the main suppliers of vegetables until 2014, actually stopped export. Thus, the main suppliers of vegetable products are China (22% of the volume of supplies, and 27% of the cost), Egypt (18% of the volume, 10% of the cost), Belarus (15% of the volume, 10% of the cost), Azerbaijan (10% of the volume, 12% of the cost).

Currently Russia cannot fully provide population with certain types of its vegetable products; therefore, import still plays an important role. In Russia, there is a lack of equipped warehouses to store perishable goods. Annually some part of a crop becomes spoiled. Climatic conditions of Russia do not allow sustainable growth of produce all year round, therefore, since November to May one can see drastic drop in offers on the counters.

For detailed investigation of embargo impact at formation of the Russian vegetable market, a certain type of products was chosen under 0702 FEACN EEU commodity code (fresh or chilled tomatoes). The market of tomatoes (0702 FEACN) is one of the main by volume of vegetable sales in the Russian Federation, in 2014 Russia was the third in the list of the largest importers (11.4% of the global import), inferior only to the USA (22%) and Germany (15.9%).

On the Russian market, the volume of domestic tomatoes production is continuously increasing. While in 2013 the production was at the level of 199.2 thousand tons, in 2016 production reached the value of 290.2 thousand tons. This is approximately 56% of the volume of imported goods. Meanwhile, for the last years the volume of tomato import into Russia has been demonstrating a sustainable trend towards reduction (table 5). In 2013 855.9 thousand tons were imported for the amount of 1104.4 million U.S. dollars, while in 2017 - only 515.8 thousand tons for the amount of 558.7 million U.S. dollars. The reason for such situation is the course laid towards import substitution, which is also supported by depreciation of the ruble in respect to other currencies. On the other hand, Russian manufacturers of agricultural goods are unable to provide greenhouse produce to the country in the off-season. As a result, imports still play a significant role in the tomato market in Russia.

Prior to embargo introduction in 2013 the products under this code were imported in different volumes (thousand tons) by Turkey (39.2%), China (9.4 %), the Netherlands (8.6 %), Morocco (8.3 %), Spain (8.2 %), Belarus (5.5%), the Ukraine (5.5%), Azerbaijan (5.5%) and other countries in insignificant volumes. In 2013 Russia imported 855.9 thousand tons tomatoes for 1104.4 million U.S. dollars. Following the results of 2017, import of tomatoes of 0702 FEACN commodity code reduced versus 2015 in monetary terms by 17.4%, and on a weight basis - reduced by 22.5%, and versus 2013 import of tomatoes reduced by 49.5% (down to 558.7 million U.S. dollars), on a weight basis import reduced by 40.8% (down to 515.8 thousand tons). Geographical focus of supplies has changed. Whereas in 2013 the main suppliers were Turkey (450.8 million U.S. dollars), China (105.5 million U.S. Dollars), the Netherlands (101.9 million U.S. dollars), Morocco (101.3 million U.S. Dollars), Spain (109.9 million U.S. Dollars). Belarus (42.1 million U.S. Dollars), Ukraine (41.2 million U.S. Dollars), Azerbaijan (41.0 million U.S. Dollars). In 2017, the main suppliers were such countries as Azerbaijan (155.1 million U.S. Dollars), China (135.1 million U.S. Dollars), Morocco (132.4 million U.S. Dollars) (table.5).

Table 5. Dynamics of import of vegetables (group 07 FEACN EAEU) to Russia (million U.S. Dollars/thousand tons)

Country	2013	2014	2015	2016	2017
Uzbekistan	5.8/3.6	1.7/0.9	1.4/1.5	3.9/5.1	7.1/14.2
Armenia	0.3/0.2	0.2/0.1	0.6/0.7	17.6/30.9	12.2/23.1
Ukraine	41.2/46.8	42,.0/43.3	0	0	0
Turkey	450.8/337.0	439.2/365.5	333.7/346.7	0	2.2/1.6
Poland	52.5/46.2	25.1/21.3	0	0	0
Netherlands	101.9/73.6	73.1/53.7	0	0	0
Morocco	101.3/71.0	111.3/78.1	97.6/66.5	164.9/126.4	132.4/107.5
China	105.5/81.0	119.8/90.8	113.7/86.4	105.5/86.4	135.6/109.0
Spain	109.9/70.3	73.0/45.8	0	0	0
Belarus	42.1/47.1	66.1/68.4	38.9/92.0	64.0/75.9	55.2/51.1

Azerbaijan	41.0/46.4	52.8/55.2	57.3/47.1	91.9/98.1	155.1/157.0
Iran	0.6/0.5	0.4/0.3	2.7/2.3	8.9/6.9	12.5/9.2
Total	1104.4/855.9	1044.7/847.1	676.4/665.5	490.4/461.5	558.7/515.8

Therefore, after Turkish tomatoes had left the Russian market in the beginning of 2016, their share was divided among traditional importers: Azerbaijan - the country became the leader by volume of tomatoes import in 2017, having almost tripled import versus 2013 up to 157.0 thousand tons; the second place by volumes of import was taken by China, having increased import to 109.0 thousand tons; and the third place was occupied by Morocco with import volumes of 107.5 thousand tons. It should be noted that Armenia has significantly increased supplies, being one of the top five tomatoes suppliers, however, and its export volumes equal to around 23.1 thousand tons.

According to forecasts of the Ministry of Agriculture of the Russian Federation, in order to fully satisfy the demand of Russia for greenhouse vegetables, it is necessary to launch more than 2 thousand hectares of greenhouses, to increase production up to 1.7-1.8 million tons, which can be achieved by 2020.

If we analyze the cost of imported tomatoes from non-CIS countries, the cheapest are the ones supplied from Morocco and China. Their cost in 2013 equaled to 1.43 and 1.30 U.S. dollars per kg, and in 2017 - 1.23 and 1.24 U.S. Dollars per kg, accordingly. For the last years the cost of 1 kg of tomatoes from Uzbekistan and Armenia was at the level of 0.90-0.50 U.S. Dollars per kg (table.6).

Table 6. The declared value of 1 kg of tomatoes imported to Russia (U.S. Dollars per kg)

Country	2013	2014	2015	2016	2017
Uzbekistan	0.87	1.73	0.90	0.75	0.50
Armenia	1.27	1.62	0.81	0.57	0.53
Ukraine	0.88	0.97	0	0	0
Turkey	1.35	1.20	0.96	0	1.31
Poland	1.14	1.18	0	0	0
Netherlands	1.38	1.36	0	0	0
Morocco	1.43	1.42	1.46	1.30	1.23
China	1.30	1.32	1.32	1.22	1.24
Spain	1.56	1.59	0	0	0
Belarus	0.89	0.97	0.42	0.84	1.08
Azerbaijan	0.88	0.96	1.21	0.94	0.99
Iran	1.25	1.22	1.13	1.28	1.36
Total	1.29	1.23	1.02	1.06	1.08

Source: authors ' calculations

4. Conclusions

Therefore, the Russian "economic" embargo noticeably influenced volumes and cost of vegetable supplies to Russia. The volume of deliveries in 2017 decreased significantly compared to 2013. Prior to the embargo in 2013, the import of tomatoes in different volumes (thousand tons) was carried out by the Turkey (39.2%), China (9.4 %), Netherlands (8.6 %), Morocco (8.3 %), Spain (8.2 %), Belarus

(5.5%), Ukraine (5.5%), Azerbaijan (5.5 %). In 2017, the main suppliers were such countries as Azerbaijan (us $ 155.1 million), China (135.1 million U.S. Dollars), Morocco (132.4 million U.S. Dollars).

References

[1] Senotrusova S V, Svinukhov V G, Makarova I G 2017 Import of dairy products of position 0403 TN VED to the Russian Federation *Modern science: actual problems of theory and practice. Series: Economics and law* **11** 30-33

[2] Senotrusova S V, Svinukhov V G 2017 Russian fish market: export, import and customs payments *International trade and trade policy* **4 (12)** 75-83

[3] Baldina N A, Amosova N I 2017 Formation of an effective system for managing the sale of milk. *Scientific papers of the IV International Scientific Practical Conference. Samara State Agricultural Academy, collection: The modern economy: problems, solutions, prospects.* 213-216

[4] Gavrilova K A 2017 Food embargo and import substitution. *Materials of the International scientific-practical conference. Plekhanov Russian University of Economics, collection: Import Substitution and the Development of Non-Commodity Exports - Priorities of Modern Industrial Policy* 91-98

[5] Demenko O G 2017 On the dairy market in terms of import substitution. *Materials of the International scientific-practical conference Plekhanov Russian University of Economics, collection: Import Substitution and the Development of Non-Commodity Exports - Priorities of Modern Industrial Policy* 116-119

[6] Svinukhov V G, Senotrusova S V, Zhoglichyova V V, Makarova IG 2017 The value of vegetable production in terms of import substitution: a regional aspect *Proceedings of the X Congress of Peter the Great Academy of Sciences and Arts, collection: Science. Industry. Education. Culture Formation of the spiritual, moral and physical health of the nation* 8-10

[7] Goncharova N Z 2016 Methods of analysis of the conjuncture of the meat market 2016 *Moscow Economic Journal* **2** 37

[8] Glotova I S, Svinukhov V G, Senotrusova S V 2017 Import of milk and dairy products in terms of food sanctions *Eurasian Union of Scientists* **6-1 (39)** 72-76

[9] Bazanova M D 2017 The impact of sanctions from Western countries on the Russian economy *Materials of the International scientific-practical conference, collection: Import Substitution and Development of Non-Commodity Exports are priorities of modern industrial policy of Russia* 167-178

[10] Mazurova PS 2017 Impact of import substitution policy on Russian meat producers *Materials of the International scientific-practical conference Plekhanov Russian University of Economics, collection: Import Substitution and the Development of Non-Commodity Exports - Priorities of Modern Industrial Policy* 180-187

[11] Donskaya N P, Zoloyeva Z B 2017 Priorities of livestock in the light of structural reorganization of the agro-industrial complex S*cientific papers of Innovation center for the development of education and science, collection: Actual problems of management and economics in Russia and abroad, (Novosibirsk)*. 27-30

Some approaches to management and provision of industrial and environmental safety at regional enterprises

A A Matveeva[1], E Yu Ladonina[1], O Yu Popova[1]

[1]Volgograd State Technical University, 100 Universitetskij Ave, Volgograd, 400062, Russia

matveeva@volsu.ru, e.ladonina@volsu.ru, popova@volsu.ru

Abstract. The article investigates the challenges of industrial and environmental safety at regional enterprises. It examines the strategies for ensuring these safety areas and identifies the main goals, objectives and implementation priorities for production facilities. The article studies the economic activities of key enterprises of the Volgograd region. The oil and gas, chemical, construction and food production enterprises of different categories of danger were selected as objects of the study. The analysis was based on the following criteria: the sources and level of impact on environmental components, protection measures, environmental policy, industrial and environmental safety. According to the requirements of the current legislation and production controlling the authors proposed the measures for the effective management of the labor protection system and implementation of environmental policies at the industrial facilities.

1. Introduction

For the effective development of any production in regional markets, it is necessary to ensure both industrial and environmental safety. Without sufficient attention to security issues, it is impossible to build a controlling system at an enterprise, make effective management decisions. It is necessary to separate the objectives and component content of the strategic and operational controlling. Using a certain type of controlling in the production process it should be noted that a system of urgent measures for preventing these or other threats is required.

The management of any production process is to be based not only on obtaining sufficient volumes of goods produced and services provided that is crucial for the national economy, but also be safe for the workers, consumers and environment as a whole (Brown, 1999; Grachev, Pliamina, 2018; Mikhailov, Karasev, 2018).

2. Main Part

In the Russian Federation, new industrial and environmental safety strategies that are interrelated with each other will be used up to 2025 (Table 1).

Table 1: Comparative analysis of the Strategies of industrial and environmental safety in relation to enterprises-users of natural resources

No.	Name of Strategy	Key parameters for the implementation of state policy in the field of industrial and environmental safety		
		Aims	Main objectives	Priority areas
1	Industrial safety	- prevention of accidents and	- ensuring the actualization of	- development and implementation of uniform

		incidents at industrial facilities; - solution of legal, economic and social tasks aimed at ensuring the growth of industrial production; - implementation of the constitutional rights of citizens to work in conditions that meet safety requirements for a favorable environment	industrial safety requirements taking into account the development of technologies used at industrial facilities; - cancellation of obsolete, redundant and duplicate industrial safety requirements	criteria for assessing the risk of accidents at industrial facilities and categorizing such facilities; - development of industrial safety culture, awareness of personal responsibility for industrial safety and formation of an intolerant attitude towards violations of industrial safety requirements
2	Ecological safety	- preservation and restoration of nature; - ensuring the quality of the environment necessary for the sustainable development of economy; - elimination of the accumulated harm to the environment	- prevention of water and air pollution, land and soil degradation; - efficient use of natural resources	- introduction of innovative and environmentally friendly technologies, development of environmentally friendly industries; - construction and modernization of treatment facilities, as well as introduction of technologies aimed at reducing the volume of air and water pollution; - minimization (reduction to the established standards) of risks of accidents at the production facilities and other emergencies

The considered Strategies are designed to provide the conditions for the maximum reduction of the negative environmental impact of the industrial facilities. It is necessary preliminarily to assess the current situation dealt with an enterprise activity control according to the regulatory requirements concerning the environmental protection and preservation of the sanitary and epidemiological well-being of population [1-29].

To determine the current environmental safety at the industrial facilities, it is necessary to study the dynamics of pollutant inflow into the environmental components. As a result we can speak about the effectiveness of environmental safety at the industrial enterprises (Matveeva, 2016: 4-19).

Taking into account the dynamics of emissions, discharges and waste in the Russian Federation as a whole, it is impossible to make a relevant conclusion towards improving or deteriorating the environmental situation. The impact of production processes on the environment is constantly increasing, and to minimize the negative effect each company must take a set of appropriate measures. They are used by each enterprise individually based on its specialization, types of harmful factors, climatic and geographical features of the area, and other indicators.

The problem of ensuring industrial and environmental safety at an enterprise is quite acute and requires an immediate solution in a short-term period. This is due to the type of economic growth of the country, poorly developed informational portrait of environmental monitoring, deficiencies in the methodological basis for determining the indicators of environmental safety of some environmental components, as well as a fairly low level of public awareness in this area. The traditional methods of environmental safety management applied at enterprises are sometimes duplicate and not sometimes systematic (Matveeva, 2017: 246-249).

The introduction of industrial environmental monitoring is able to solve partially the problem, however, the state authorities as an administrative resource are still unable to cope with ensuring environmental safety at the industrial facilities. This is due to the fact that there are a number of serious issues: technical and technological resources of enterprises are close to exhaustion, there are inadequate staff qualifications, non-compliance with standards in the design and construction of technical devices and environmental protection equipment, administrative barriers in obtaining relevant permits.

The environmental situation in the regions is unfavorable. The Volgograd region, including the city of Volgograd, can be used as a vivid example. The region belongs to the industrialized regions with large enterprises of the fuel and energy, metallurgical (ferrous and non-ferrous), petrochemical and chemical, agricultural and housing and communal complexes and enterprises for production of building materials. The region has a well-developed transport infrastructure and a network of oil and gas pipelines. Volgograd is a diversified industrial center with a predominance of manufacturing industries.

According to the state registration data, as of January 1, 2018, there are about 46,631 organizations in the Volgograd region, which are distributed by the type of economic activity. At the same time processing industries account for 3119 organizations (6.7%), the degree of depreciation of basic production assets is 41.7% in terms of the aggregate regional industrial production index which is equal to 99.7% compared to the current period of 2017 (Official statistics of the portal Territorial body of the Federal State Statistics Service of the Volgograd region (Volgogradstat)).

Analyzing the economic activities of some city-forming enterprises of the city and region, it should be noted that a systematic approach for ensuring industrial and environmental safety is urgent. It must be based on the classical principles: environmental and economic responsibility, comprehensiveness and economic accounting [1-3].

The assessment of the degree of ensuring environmental and industrial safety during production environmental control at enterprises was carried out according to the following criteria (Matveeva, 2017):

1. The degree of environmental threat in accordance with the category of object of negative environmental impact (NEI).
2. The development of a system of environmental measures to reduce the negative impact on environmental components.
3. The presence / absence of environmental policies.
4. The strengths / weaknesses of an enterprise, as well as threats / opportunities when using the SWOT-analysis method.
5. The development of measures to prevent and eliminate the risk of emergency situations and accidents at work.
6. The presence / absence of an effective labor protection policy at an enterprise in accordance with the established requirements and criteria.

The objects of research were not chosen by chance, since it was important to show different industries and their reference to different objects of negative environmental impact (NEI). A summary of the activity analysis of the selected enterprises is presented in Table 2.

Table 2: Final assessment of the degree of ensuring environmental and industrial safety at the studied enterprises of the region

Name of enterprise / type of activity	Category of object of NEI	Key threats / opportunities	Environmental policy	Control system of labor protection	Safety level on a scale from 1 to 3 (1 - low, 2 - average, 3 - high)	
					Environmental	Industrial
Open Joint Stock Company Zenzevyatskiy Elevator /grain processing	II	The lack of resource-saving technologies in the process of grain production / Implementation of developed environmental project documentation in the field of waste management	Present	Present	3	3
Closed Joint Stock Company Gazpromhimvolokno / production of cord and technical fabrics	II	Increasing emissions, discharges and waste due to the commissioning of new production facilities and pre-sale cleaning of equipment in workshops / Phased introduction of an environmental management system	Partially present	Partially present	2	2
Open Joint Stock Company Volgograd Ceramic Plant / production of building materials and household products	I	The lack of environmental service at the enterprise, no resource-saving technologies for the use of water resources, no closed water supply system / Developed system of nature-conservative measures for organizing and landscaping of the sanitary protection zone	Partially present	Present	2	3
LLC LUKOIL-Volgogradneftepererabotka /	I	The increase in the volume of pollutants with the constant intensification of	Present	Present	3	3

production of petroleum products		production capacity / The presence of developed and implemented environmental and industrial safety programs.				

3. Conclusion

A complex analysis of the selected production objects shows that each enterprise has its own problems associated with insufficient funding. For example, the Volgograd Ceramic Factory is characterized by the lack of its own environmental laboratory and environmental specialists and the inability to implement the best available technologies in the production process at the current stage. The economic activities of LLC LUKOIL-Volgogradneftepererabotka have the classical characteristics of successful production process intensification with an annual increase in gross volumes of pollutants into the environment while the strategic objectives in the field of environmental and industrial safety are sufficiently effective. CJSC Gazpromhimvolokno is an industrial object of federal significance which belongs to the category of high-tech enterprises in Russia. At the same time the company has outdated purification technologies that are to be replaced by more modern and highly efficient ones, which is especially necessary due to the introduction of new industries. There are no resource-saving technologies at the grain processing industry, which, as a rule, will be cost-effective at the first stage of their implementation. Most enterprises of the food sector can not afford such technologies because of the low cost of products which are sold on domestic markets.

The controlling system will quantitatively reflect the state of safety at an enterprise and can be used as effective information support in evaluating investment activities in the form of developing measures to improve the state of industrial and environmental safety as well as the labor protection (Eljack, Kazi, 2016; Evtushenko, Siryk, 2016; Idrisova, Myasnikov, 2018).

However, ensuring industrial and environmental safety at regional enterprises is impossible without a systematic approach, which is aimed at developing constructive relationships between the company's management and legislative and executive authorities.

References

[1] Application "On the formulation of a facility that has a negative impact on the environment on the state accounting" of JSC "VKZ" 2016 (Volgograd) 74 p
[2] Brown A 1999 Risk analysis: An investment in engineering *Process Safety Progress* vol 18 **2**, 121-125
[3] Declaration on payment for the negative impact on the environment of CJSC Gazpromhimvolokno 2016 (Volzhsky) 19 p
[4] Draft standards for the generation of waste and limits on their location of JSC "VKZ". 2014, vol. 1- 2. Volgograd, 102 p
[5] Egorova O 2011 Environmental safety management in the context of innovative development of enterprises, *Annals of DAAAM and Proceedings of the International DAAAM Symposium* 106302 p 1169
[6] Eljack F, Kazi M 2016 Process safety and abnormal situation management, *Current Opinion in Chemical Engineering* **14** 35-41
[7] Evtushenko O, Siryk A, Porodko P 2016 Development of the occupational safety in the food industry with regard for the risk-based approach, *Ukrainian food journal* vol. 5 **1** 174-185
[8] Draft standards for maximum permissible emissions of pollutants into the atmosphere of JSC "VKZ". 2013 (Volgograd) 156 p
[9] Grachev V, Polosin A, Pozdnyakov A, Pliamina O 2018 Environmental policy of industry and manufacturing company *International Journal of Mechanical Engineering and Technology*

vol 9 **8** 213-219
[10] Grachev V, Pliamina O 2018 Industrial and environmental safety in the integrated security policy of the industry, a corporation and an enterprise *International Journal of Mechanical Engineering and Technology* vol. 9 **7** 802-812
[11] Idrisova J I., Myasnikov V N, Uljanov A I, Belina N V 2018 Increasing the efficiency of labor protection in the enterprise *International Conference on Information Networking* 586
[12] Information on the air protection for 2016 of JSC "VKS" 2017 (Volgograd,) 5 p
[13] Information on the current costs of environmental protection and environmental payments for 2015 of JSC "VKZ" 2016 (Volgograd) 4 p
[14] Information on the formation, use, disposal, transportation and disposal of production and consumption waste for 2016 of JSC "VKZ". 2017. Volgograd, 22 p
[15] Koptseva N P, Sitnikova A A 2018 Environmental Management in Central Siberia (Based on Analysis of the Krasnoyarsk Territory), *IOP Conference Series: Earth and Environmental Science* vol **164** Issue **1**
[16] Mikhailov V, Karasev V, Mikhailov G 2018 The Study of the Main Indicators of the Local Environmental and Economic System «industrial Enterprise-Environment» *E3S Web of Conferences*. Conf. Ser. **137557**
[17] Matveeva A A 2017 A systematic approach to ensuring environmental safety at the regional enterprises *Environmental management: traditions and innovations* .246 – 249
[18] Matveeva A A 2016, Ecological safety: theoretical aspects (Section 1) *Ecology and environmental management* (Volgograd: Volgograd State University Publishing House) pp 4 – 19
[19] Official statistics of the portal Territorial body of the Federal State Statistics Service of the Volgograd region (Volgogradstat) http://volgastat.gks.ru/wps/wcm/connect/rosstat_ts/volgastat/ru/statistics/
[20] Presidential Decree of 06.05.2018 N 198 "On the Fundamentals of the RF state policy in the field of industrial safety for the period up to 2025 and the future" 2018 Available at: ConsultantPlus: inform. System
[21] Report on the inventory of emissions of harmful (polluting) substances into the air and their sources for LLC LUKOIL 2016 **1** (Volgograd) 398 p
[22] Report on the inventory of emissions of harmful (polluting) substances into the air and their sources for LLC LUKOIL 2016 vol. 2 (Volgograd) 374 p
[23] Report on the inventory of emissions of harmful (polluting) substances into the air and their sources for LLC LUKOIL 2016 vol. **3** (Volgograd) 431 p
[24] Report on the inventory of emissions of harmful (polluting) substances into the air and their sources for LLC LUKOIL 2016 **4** (Volgograd) 531 p
[25] Stanik J, Kiedrowicz M, Waszkowski R 2018 Security and risk as a primary feature of the production process, 2nd International *Conference on Intelligent Systems in Production Engineering and Maintenance, ISPEM 2018* **835** 701-702
[26] Tang E, Liu F, Zhang J, Yu J 2014 A model to analyze the environmental policy of resource reallocation and pollution control based on firms' heterogeneity *Resources Policy* vol. 39 **1** 88-91
[27] The official portal of the RF Ministry of Natural Resources and Environment. The project "The RF Strategies for the environmental safety for the period until 2025" is undergoing a public discussion procedure. Available at: http://www.mnr.gov.ru/news/detail.php?ID=142853
[28] The plan of environmental protection measures of JSC "VKZ" for 2017 - 2020 on reducing the negative impact on the environment 2016 (Volgograd) 5 p
[29] The program of industrial environmental control of CJSC Gazpromhimvolokno 2016 (Volzhsky) 77 p

IOP Conf. Series: Materials Science and Engineering **483** (2019) 012114 doi:10.1088/1757-899X/483/1/012114

The use of modern management technology to improve business efficiency

A A Ksenofontov[1], A S Ksenofontov[2], M A Kirpicheva[1], P V Trifonov[1]

[1]Financial University under the Government of the Russian Federation, 49 Leningrad Prospect, Moscow, 125993, Russia
[2]Kabardino-Balkarian State University after H M Berbekov, 173 Chernyshevsky St., Nalchik, 360004, Russia

a.ksenofontov@mail.ru

Abstract. In the dynamic world, any company needs tools and methods for improving business efficiency. High competition in the global market dictates the need to find ways to be competitive. For this purpose, it is necessary to use advanced achievements and technologies in business management. New management methods, including a process approach, are required. The purpose of the article is to prove the expediency of the process approach as a method for improving the company's management system and developing recommendations for optimizing key business processes.

1.Introduction

Currently, there is an accelerated increase in requirements for improving business management efficiency. This is due to numerous internal and external factors affecting the company. Top-level managers take into account external factors affecting the company since it operates in the unstable external economic environment. These include globalization of the world economy and production globalization, an increase in interdependence of producers, a high rate of development of scientific and technological progresses and innovation, high expenditures on thematic research and development, sanction trade restrictions.

For successful survival in the market and implementation of a development strategy, a company should be flexible and dynamically developing, since the key factor in increasing competitiveness is time. In addition, the external business environment is becoming increasingly complex and uncertain. The modern company should be able to adapt and improve its financial and economic sustainability.

2. Methods

The basis of the research is scientific and practical works by leading scientists who identified patterns of the evolution of management as a science, principles of scientific leadership and application of the process approach to company management [1, p. 473-486], classical [2, p. 40-47] and modern [3], [4] models of industrial innovation. Special attention is paid to the work on reengineering of business processes as a whole [5, p. 405-414], [6, p. 63-87], and in enterprises belonging to various economic sectors [7, p. 63-66], [8, p. 46-68], [9] in particular. The works on modeling business processes [10, 105-129], the Business Process Management (BPM) System [11], the role of the context in BPM [12] were analyzed. The studies used a wide range of general scientific methods, including a system approach, economic and statistical methods, comparison, classification, scientific abstraction, etc.

3. Research results

In the modern world, many companies are low efficient. To increase business efficiency, let us consider business process management. Most organizational tasks are cross-functional in nature. It is necessary to implement the concept of process management to improve company efficiency. There are three solutions: description and management of processes, creation of a process organizational structure and cross-functional problem solving.

The easiest way is to use business processes to identify certain problems cross-functional groups will work on. One can use formalization. Based on a broad definition of business processes, company executives can explore the workflow and its cross-functional aspects causing difficulties. Then, priorities are set and cross-functional groups are formed. From a methodological point of view, there are two alternative ways for providing group assignments.

First, it is possible to use the formal structure of an organization for setting a task, analyzing and collecting data on its activities, developing and testing possible alternative options in order to make the most preferable management decision. The second way is based on the system reengineering of business processes. The first method based on creation of a formal troubleshooting team is relatively safe for the company. The second method is more radical and may cause a number of problems, especially if it is used as a tool for developing the company as a whole or for comprehensive management process improvement.

The second way is more streamlined. First of all, we are talking about the most important business processes in accordance with unified and rigorous principles. Different approaches can be used, but the most successful one is the process quality management (PQM) method. The PQM describes the mission of the organization (its fundamental goal), units or divisions, a list of key success factors that are necessary and sufficient to accomplish the mission. There are eight key success factors.

The next step is description of daily activities aimed at carrying out company's business operations and new activities which are required for successful accomplishment of the company's mission. As a rule, there are about 20-30 business processes in the company. At this stage, the processes are ranked by their importance, current effectiveness and status of the owners. It is necessary to understand which of the company's employees wins the most if the business process is effective, and who loses if it is ineffective. Selected employees are members of the team describing business processes. They are executives and top managers of the company.

There are a lot of ways to improve management system efficiency, one of which is to use the business process reengineering (BPR) [13, p. 125-134]. Its use requires synchronization, since the decision of the process owner to use the BPR without understanding and support from other top managers can cause chaos in the company. Conversely, if all process owners decide to use this approach at the same time, the result will be the same. Some of the methods used during business process improvement are relatively safe. Even if they turn out to be wrong, the damage will not be significant.

The third way to use our knowledge of business processes in practice allows us to describe the organizational structure of the company and manage the organization itself. The method involves understanding that the organizational structures influence business efficiency. Most organizational structures are designed to impede improvement of the company. Departments of the company deal with one functional area, feel like self-sufficient structures, do not understand the company mission. It is possible to reduce the number of structures influencing business processes by creating work groups. These groups will include specialists responsible for the entire business process rather than for its specific areas.

The organizational structure of the company should comply with its mission. In most cases, reorganizers define development of the organizational structure as placement of the names of company's functional units in squares of the organigram. Consequently, improvement of the organizational structure should be synchronized with existing business processes.

Let us analyze modern approaches to business management. In highly competitive markets, there are two groups of counterparties:

- irregular (one-time) counterparties who can easily contact another supplier of the product or service;
- permanent (loyal) counterparties who constantly work together and form special requirements for the product or service.

The latter group are partners who participate in product cost management. In this case, creation of a business network is one of the key success factors. If business partners cannot reconcile their interests, it will be difficult for them to have loyal customers. Top managers of the company are interested in streamlining their relationships with suppliers and consumers.

The following approaches can be used for structuring relationships in the business network: first, to choose a "key supplier" - a strategic partner that provides key products and services to the client company; secondly, to conclude long-term strategic partnership agreements between the supplier and the customer-consumer. The company's client has to monitor the market of consumed products and services, compare the strategic partner and their competitors to make sure that the products and services you purchase are optimal in price and quality. It is also necessary to integrate interaction of the provider and the customer (including information interaction) based on modern e-business technologies in order to provide products and services of the required quality at the required time.

It is necessary to manage your client portfolio, try to maximize the value created by this portfolio. At the same time, management of the client portfolio excludes two components: financial management of client relations and operational management of client relations (interaction of the company with key partners). Transition from one-time or even regular deals to strong relationships with counterparties will allow participants to gain a stable financial advantage over competitors.

The service-profit chain is a powerful tool for understanding how the integration of partners in a business network can be a lever for increasing business efficiency. There is a direct link between good services and a steady profit growth. Consequently, it is necessary to provide high quality products and services to customers. The service-profit chain emphasizes the importance of quality management of the sales chain, in which effective partner relationships increase the business activity index.

The key to managing the service-profit chain is interfirm relations based on a quick response to changing needs of customers and attentive after-sale attitude towards. The inter-company relations are possible only if managers of the partner companies can make these decisions and are motivated to manage problems. Encouraging the initiative of partners, their proposals for improving products and services, the business network uses its own potential for business development.

Studies have shown that the classical structure of business management where the company consists of a number of functional departments (marketing, financial, marketing and other services), has a number of drawbacks. They do not allow the company to successfully compete with organizations using a process management structure. In a traditional company, hierarchical subordination is more important than horizontal interaction in a business process. In this regard, each employee works for his immediate boss rather than for internal or external clients or for achieving the company mission.

Modern business requires organizational structures that support the system of flexible production, keep up with its short life cycle. Consequently, at the present stage, the organizational structure of the company must be directly opposite to the functional one. It must combine complex tasks carried out by highly qualified personnel and simple systems of coordination and control existing at a low level of bureaucracy. The process structure of the company should include a hierarchy of cross-functional business processes headed by coordinators (process "owners") and implemented by process management groups. The process structure meets the requirements of modern businesses for the following reasons:

- the efficiency of the company significantly increases, as there is an owner of each business process [14, 153-167]. He is responsible for implementation and management decisions;
- along with the exchange of end products of the business process, it is possible to build a system for calculating costs and levels of corporate entrepreneurship;
- it is possible to manage the objectives of the company and develop an effective remuneration system based on personal achievements. Combined with the cost management approach, it allows the staff to be more efficient
- relations between departments (both vertical and horizontal) are based on the "customer-contractor" system which eliminates interaction barriers.

4. Conclusion

Thus, the process-oriented company management system sets the tasks based on company goals. The system of material remuneration makes it possible to implement the tasks set for the staff. The system of restrictions for employees is created. The management system allows managers to intervene when problem situations arise in order to prevent them and minimize financial losses.

Analysis of business processes and their cost models allows us to aim internal corporate relations at creating the main business goal - to increase the consumer value of the product. This business goal is achieved by transferring resources and responsibility to business process owners and their implementers which contributes to timely management decisions. The active use of business process technology will help improve management of inter-company relationships.

To conclude, the dynamic and competitive business world of the twenty-first century creates new problems for the company. Problems should be solved promptly. New opportunities need to be evaluated and used for company development. These tasks are extremely complex, but modern management concepts and technologies can help managers. The company's focus on business process management creates potential opportunities and provides stable benefits contributing to sustainable development.

References

[1] Benner M J 2009 Dynamic or Static Capabilities? Process Management Practices and Response to Technological Changes *Journal of product innovation management* **26(5)**
[2] Abernathy W J, Utterback J M 1978 Patterns of Industrial Innovation *Technology Review* **80(7)**
[3] Parmar R, Mackenzie I, Cohn D, Gann D 2014 The New Patterns of Innovation *Harvard Business Review* **92(1-2)**
[4] Trachuk A V, Linder N V 2018 Innovation and Performance: An Empirical Study of Russian Industrial Companies *International Journal of Innovation and Technology Management* **15 (3)**
[5] Kiran D R 2017 Business Process Reengineering *Total Quality Management Key Concepts and Case Studies*
[6] Bhaskar H L 2018 Business process reengineering: a process-based management tool *Serbian Journal of Management* **13(1)**
[7] Gareeva G, Grigoreva D 2018 Reengineering of Business Processes and Improve the Sales System by 1C In Pharmacies *National Academy of Managerial staff of Culture and Arts Herald* **2**
[8] Turra M, Juliani LI, Salla NMDC 2018 Business Process Management - BPM: A Bibliometric Study on National Scientific Production *Revista Administracao Em Dialogo* **20 (3)**
[9] Nekrasov R Y, Tempel Y A, Tempel O A, Barbyshev B V 2017 Innovative Model of Business Process Reengineering at Machine Building Enterprises *International Conf. on Modern Trends in Manufacturing Technologies and Equipment (Icmtmte 2017)* **129**
[10] Melao N, Pidd M 2000 A conceptual framework for understanding business processes and business process modelling *Information Systems Journal* **10 (2)**
[11] Rialti R, Marzi G, Silic M, Ciappei C 2018 Ambidextrous organization and agility in big data era: The role of business process management systems *Business Process Management Journal* **24 (5)**
[12] Brocke J V, Zelt S, Schmiedel T 2016 On the role of context in business process management *International Journal of Information Management* 36(3)
[13] Trkman P 2010 The critical success factors of business process managemen *International Journal of Information Management* **30(2)**
[14] Danilova K B 2018 Making process ownership work: Evidence from a global Delphi study *Knowledge and process management* **25 (3)**

Opportunities and emerging trends for digitalization of Russian economy

O Ergunova, A Skuratov, O Pozdeeva, A. Yu .Kindaev

Department of Department of Tourist Business and Hospitality Ural State Economic University Yekaterinburg, Russian Federation
Penza State Technological University, Russian Federation

o.t.ergunova @urfu.ru, Alexey.Skuratov@urfu.ru, ogp@usue.ru

Abstract. This working paper gives an overview of the new possibilities opened up by the digital economy and tackles some specific questions in relation to its effects on the regional institutional structure in the world and the rate of its development in the Russian Federation. The digital economy fundamentally transforms modern production, thanks to new technological achievements, including robotization, artificial intelligence and the Internet of things, new materials and technology. Due to these changes, the development of the digital economy has been revealed that a number of countries, including Russia, objectively have the preconditions for the transition to the way of digitalization of the economy and development of their potential as the development potential is exhausted through non-innovative tools and increased market competition. In this paper, the authors aimed at providing a better understanding for the theoretical and legal peculiarities of the “digital economy” and its application benefits for Russia. The main problem is how the Russian Federation acts against the economically developed countries, which are the creators of the digital economy. This paper mainly focuses on presenting the authors views on how to sustain and increase competitive advantage of the Russian Federation by catching and implementing the “digital economy”. The degree of development of the digital economy in the Russian Federation is considered and a conclusion is drawn that the domestic digital economy is lagging behind the world leaders.

1. Introduction

Rapid growth and expansion of new information technologies, development of social networks, smartphone markets, broadband Internet access, artificial intelligence technologies and Internet of Things result in changes which transform the way companies, cities, regions and countries function – they have to fall in line with digitalization of economy by means of either expanding their niches or transforming the ones that they already have.

Today there is no common understanding of the Digital Economy phenomenon; other terms such as E-Economy, New Techno-economic Paradigm, API Economy, Application Economy and Creative Economy are widely used. Europeans tend to use the term Digital Economy whereas Americans, namely Deloitte, IBM and some other companies, prefer a more technological API Economy.

When building the platforms of the digital economy, it is necessary to focus efforts on key areas: transport, telecommunications, energy and data processing. The development of these areas will allow creating an infrastructure and technological basis, replicating which to other areas, Russia will be able to rapidly develop the mature digital economy. A focused construction of a number of industrial digital platforms with a unified architecture and standards will allow in the future to build a single digital space that will unite all industries. Such an approach will significantly increase the transparency, manageability and flexibility of the country's economy. This approach seems most appropriate for Russia today, but it still has its shortcomings. To form the concept of a digital economy, which will serve as a basis for corresponding strategy, it is necessary to take into account both the risks of the proposed strategy and the risks of the digital economy itself.

IOP Conf. Series: Materials Science and Engineering **483** (2019) 012115 doi:10.1088/1757-899X/483/1/012115

2. Theoretical and methodological basis of study

First terms as well as concept of modern digital economy appeared at the end of the XX century with Don Tapscott in his 1995 book "The Digital Economy" defining Digital Economy as the economy based on digital computing technologies [1]. At the same time American computer scientist Nicholas Negroponte used the metaphor explaining the process of moving from a physical world based on atoms to a digital world based on bits [2]. According to Oxford dictionary, Digital Economy is "an economy which functions primarily by means of digital technology, especially electronic transactions made using the Internet".

According to IBM's definition, Digital Economy is an economy capable of providing high-quality ICT infrastructure and mobilizing ICT opportunities for the benefit of consumers, business and the state. BCG specialists give the following definition of the Digital Economy: the use of online and innovative digital technologies by all participants in the economic system, from individuals to large companies and states.

We refer below the definitions of the Digital Economy given by national governments of several countries and international organizations.

Table 1. Definitions of the digital economy approved at the governmental level in several countries of the world

No	Institution	Definition
1.	Government of Australia	"Global network of economic and social activities that are enabled by information and communications technologies, such as the Internet, mobile and sensor networks"
2.	BCS, Great Britain	"Doing business in markets based on the Internet and / or the World Wide Web" [4]
3.	OECD	"markets based on digital technologies that facilitate the trade of goods and services through e-commerce"
4.	UK Government	"Manufacture of computer and digital equipment, Publishing activities, Media production, Computer programming activities"
5.	World Bank	"System of economic, social and cultural relations based on the use of digital information and communication technologies"
6.	Russian Federation Government	"Activities where digital data are key factors of production while their processing and use in large volumes, including directly at the time of their formation, allow to significantly increase the efficiency, quality and productivity in various types of production, storage, sale, delivery and consumption of goods and services in comparison with traditional forms of economy management"

The basic elements of the digital economy are usually as follows:

— infrastructure, including technical facilities, data storage, processing and transformation centers, information transfer centers, software, telecommunications facilities, etc.;

— electronic services of bodies of legislative and executive state power and administration;

— business processes of business entities through computer networks in the context of virtual interactions between market entities;

— e-commerce, which is currently one of the largest segments of the digital economy [5].

3. The practice of legal regulation of the economy digitalization in the Russian Federation and foreign countries

IOP Conf. Series: Materials Science and Engineering **483** (2019) 012115 doi:10.1088/1757-899X/483/1/012115

Consider the practice of legal regulation of the economy digitalization in the Russian Federation and foreign countries. The G20 Ministerial Declaration, adopted in July 2017, supported monitoring activities by the WTO, UNCTAD and OECD, as well as the World Bank and the IMF activities to strengthen trade and investment cooperation. In their view, digital transformation is the driving force behind global, innovative and sustainable growth, contributing to reducing inequality and reaching sustainable development by 2030. Therefore, they committed themselves to providing all their citizens with "digital communication" by 2025, especially while welcoming the development of digital economy infrastructure in low-income countriesConsider the practice of legal regulation of the economy digitalization in the Russian Federation and foreign countries. The G20 Ministerial Declaration, adopted in July 2017, supported monitoring activities by the WTO, UNCTAD and OECD, as well as the World Bank and the IMF activities to strengthen trade and investment cooperation. In their view, digital transformation is the driving force behind global, innovative and sustainable growth, contributing to reducing inequality and reaching sustainable development by 2030. Therefore, they committed themselves to providing all their citizens with "digital communication" by 2025, especially while welcoming the development of digital economy infrastructure in low-income countries [6].

In 2015, the World Economic Forum (WEF) announced the launch of a special long-term program Digital Transformation Initiative (DTI), which, according to its creators, should serve as the "focal point for new opportunities and themes arising from latest developments and trends from the digitalization of business and society". At the same time, DTI is initially regarded as an important part of the broader activity around the "theme of the Fourth Industrial Revolution".

The plan "Digital Europe 2020" (eGovernment Action Plan 2016–2020) was approved in the European Union. According to this plan, if the state actively adapts principles of digitalization, the EU countries will be able to save up to $5 billion annually. Leaders in the level of digitalization of public systems are the countries of Scandinavia, the Netherlands and Austria.

The Presidential Decree No. 203 of May 9, 2017 "On the Strategy for the Information Society Development in the Russian Federation for 2017–2030" states that that the development of the digital economy is a strategically important issue for Russia as it determines its competitiveness in the world arena. In accordance with the resolution of the Government of the Russian Federation No. 1632-r of July 28, 2017, the Digital Economy of the Russian Federation Program was approved, aimed at creating "the necessary conditions for the development of the digital economy of the Russian Federation, in which digital data is a key factor in production in all spheres of social and economic activity, enhancing the country's competitiveness, the quality of life of citizens, ensuring economic growth and national sovereignty" [7].

4. Development of the digital economy in the world

According to a study conducted by Boston Consulting Group in 2016, the digital economy is estimated at $2.9 trillion [8]. The research conducted by BCG on the level of development of the digital economy, taking into account all its sectors, singled out five groups of digital economies for the relative level of development of digital operations and GDP per capita. The group of leaders includes countries with the highest rate of "digitization" of economic operations and high level of technologies used for such "digitization": South Korea, Denmark, Great Britain, Sweden, Norway and the Netherlands. The second group of countries includes the majority of the developed economies of the world, for example, Germany, the United States, Japan, the EU countries. The third group covers countries with a high level of well-being (GDP per capita), but with relatively lower rates of "digitization" of operations. These are the countries of the Middle East, primarily the UAE and Saudi Arabia. The fourth group is "beginning leaders". The level of development of digital operations in these countries is higher than the level of economic development: China and Russia. Experts of BCG refer all the other countries to the group of "lagging behind" in the development of the digital economy.

5. Evaluation of Russia's position in the world's ratings of the economy's digitization

We will evaluate Russia's position in the world's ratings of the economy's digitization (see Table 2).

Table 2. Evaluation of Russia's position in the world's ratings of the digitalization of the economy

Rating	Rank of Russia	Number of countries in the rating
Networked Readiness Index, NRI, World Economic Forum [9]	41	139
E-government development rank, EGDI, UN [10]	35	193
ICT Development Index, IDI, ITU [11]	43	175

Concerning the development of ICT, Russia occupies an average position among other countries. At the same time geopolitical confrontation and international sanctions that have affected Russia have not been sufficient to undermine the effect of technology development, as well as massive use of smartphones and social networks among the population of these countries. Despite the relatively low overall level of digitalization, Russia demonstrates stable growth rates and has the potential to take the lead. According to the McKinsey report "Digital Russia: a New Reality", Russia ranks 1st in Europe and 6th in the world (87 million people) in the number of Internet users. According to Rosstat statistics, from 2010 to 2016, the number of households with access to the Internet increased from 48.4 % to 74.8 %. Russia ranks 2nd in the world in the lowest prices for Internet and mobile communications. And the number of users of state and municipal services portals doubled only in 2016 and reached 40 million people.

The OSP Data survey "The Present and Future of Digital Transformation in Russia" in 2017 showed that about half of Russian enterprises are already transforming their business, and within two years the share of companies that somehow started the process of digital transformation will exceed 80 %. Among the factors hampering the changes, respondents noted mainly various organizational issues. However, they also noted the shortcomings of the technological platform on the basis of which business changes will be made — for example, a weak readiness for deep analysis of data in order to turn them into information useful for business. Nevertheless, IT managers interviewed consider their companies to be basically ready for digital transformation projects.

6. Results

Russian regions considerably differ in the level of infrastructure development. These differences have historic roots and they are associated with characteristic features of Russia's socioeconomic development over the past decades.

Russia features great differences in transport infrastructure in different regions. In 2016, the greatest density of public railroads was 3.076 km of tracks per 10.000 km^2 in St. Petersburg. The lowest density of public railways was 2 km of roads per 10.000 km^2 in the Republic of Sakha (Yakutia) which is 1.538 times less than the greatest density (see Figure 1). By 2016, in comparison with 2006, the differentiation increased by 133 %. This growth was due to the construction of railroads mainly in the European part of the country.

IOP Conf. Series: Materials Science and Engineering **483** (2019) 012115 doi:10.1088/1757-899X/483/1/012115

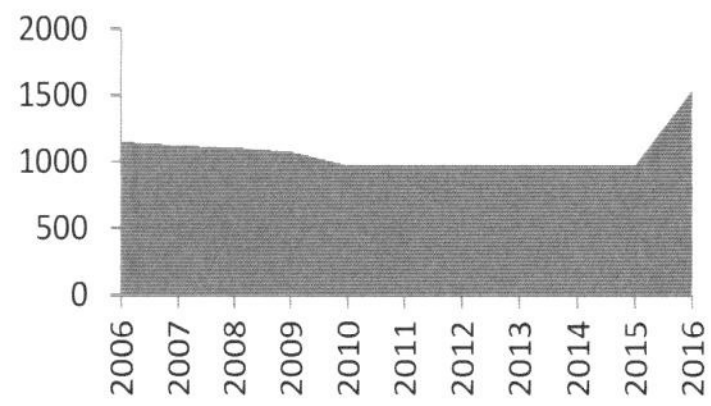

Figure 1. Dynamics of the coefficient of differentiation in the density of public railways in the regions of the Russian Federation [12]. Calculated by authors

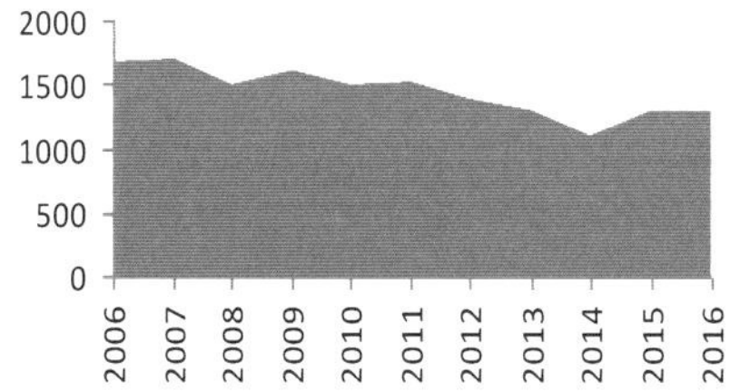

Figure 2. Dynamics of the coefficient of differentiation in the density of highways in the regions of the Russian Federation [12]. Calculated by authors

Regions also have different level of highway availability, with coefficient of differentiation equaling to 1.312 times in 2016. St. Petersburg had the highest density of highways – 2.492 km per 1000 km^2, and the Nenets Autonomous District had the lowest density of highways — 1.9 km per 1000 km^2 (see Figure 2). In 2016, the coefficient of differentiation decreased by 22 % compared to 2006. At the same time, in 2016 as compared with 2006, the density of motor roads had more than doubled in 46 regions out of 85. The pace of construction of new highways had increased significantly since 2012. The reason was intensive road construction in the European part of the country.

Excessive differentiation in terms of the level of road infrastructure development in the regions of the country is an obstacle to the existence of a single economic space and can lead to limiting the opportunities for both the development of the country's economy as a whole and the consideration of regional features and regional interests [13].

Huge territories of the Russian Federation, difficult climatic conditions in most of the country predetermine the need to reduce the spatial costs in the development of the economy. It is impossible to solve these problems through the development of basic infrastructure elements in the foreseeable future. Overcoming such a huge differentiation in the development of road infrastructure between regions is a lengthy process and requires significant investment.

Transition to the digital economy, announced by the Government of the Russian Federation, can contribute to reducing spatial costs between production sites and places of consumption, thus creating the basis for economic relations and interaction between business entities through the creation of new infrastructure elements.

Infrastructure facilities that enable the use of information and communication technologies help to solve scientific and technical problems, contribute to information availability, optimize interregional economic ties, which ultimately leads to a steady increase in labor productivity and, therefore, raises the standard of living of the population [14].

Consider the potential of the regions of the country for the transition to a digital economy. Organizations in the regions had different level of availability of technologies that use microelectronics to collect, store, process, search, transmit and present data, texts, images and sound in 2016. In general, the highest indicator was the total number of organizations using personal computers which amounted to 92.4 % of the total number of organizations surveyed. The lowest indicator was the share of organizations that had a website (45.6 %) (see Figure 3).

However, it is worth mentioning that the level of differentiation between regions in terms of indicators characterizing the development of information and communication technologies is not so vast as in terms of road infrastructure. The share of organizations using personal computers in 81 regions out of 85 was more than 85 %, the share of organizations using servers in 42 regions was more than 50 %, the share of organizations using local area networks in 73 regions was more than 50 %, the share of organizations using global information networks in all 85 regions was more than 60 %, the share of organizations that used the Internet in 84 regions was more than 75 %, the share of organizations that had a website in 43 regions was more than 45 %, the share of organizations using special software in 83 regions was more than 70 %.

All the information and communication technologies considered in all regions for the period from 2006 to 2016 showed the growth rate of 200 or more percent.

Thus, it can be noted that there is a potential for the transition to the digital economy as the growth rate of this area is high. Further development of information and communication infrastructure will make it possible to change the spatial development of the country, to reduce differentiation between regions, to increase labor productivity and production efficiency in all types of economic activity.

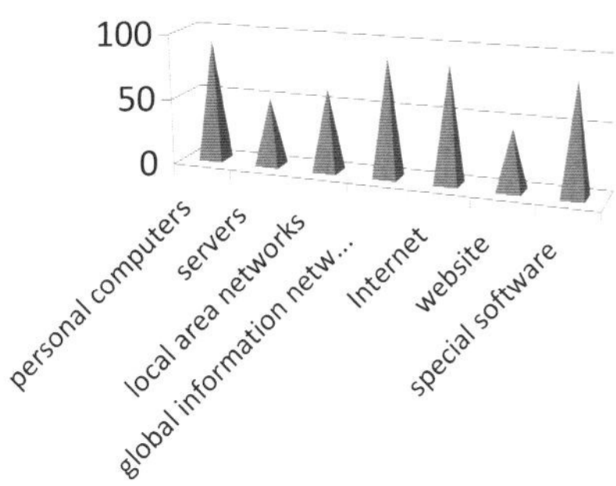

Figure 3. Total number of organizations in the Russian Federation, which used information and telecommunication technologies in 2016 [12]. Calculated by authors

7. Conclusion

In this paper, the authors aimed at providing a better understanding for Digital economy concept and its application benefits for Russia. The main problem is how the Russian Federation acts against the economically developed countries, which are the creators of the Digital economy. This paper mainly focuses on presenting the authors views on how to sustain and increase competitive advantage of the Russian Federation by catching and implementing. With Digital economy, Russian Federation get a bigger share of the world manufacturing value chain. Digital economy "is challenging, since it forces to rethink the way goods are produced, distributed and sold. To fully apprehend the potential Digital economy and anticipate the coming challenges, it is of the utmost importance to consider, and understand, the specific economic nature of digital goods. Countries are divided into four categories:

leaders, countries with a slowing growth rate, promising and problem countries. Some countries are located on the borders of these categories [15,16].

Leading countries have an extremely developed digital economy and a powerful development dynamic. They stimulate the introduction of innovations, effectively using their advantageous position. In order not to lose their positions, these countries must create new demand, and the development of innovative solutions must go at full speed. Otherwise, they risk becoming countries with a slowing growth rate.

Countries with a slowing growth rate have a developed digital economy, but they lose their development dynamics. Top five countries in our ranking (Norway, Sweden, Switzerland, Denmark and Finland) fall into this category, thus showing how difficult it is to maintain growth rates. To overcome the digital plateau, these countries will have to make conscious efforts to rethink their economic model, put everything on those digital technologies and technological areas in which they lead, and remove any obstacles to innovation.

Promising countries currently have a low rate of digitalization, but are developing rapidly. Significant dynamics of development and great potential can make them very attractive for investors. They are hampered by the undeveloped infrastructure and poor quality of the institutional environment. Promising countries have the potential to become leaders of the future. This group is headed by China, Malaysia, Bolivia, Kenya and Russia.

Problem countries are at a low level of digitalization and have low dynamics, which creates great difficulties for them. In some of them, the pace of digital development is declining. The most reliable way to increase the dynamics of development for them will be improved access of the population to the Internet by reducing the gap in the use of mobile Internet, that is, the difference between the number of mobile phones and mobile phones connected to the Internet.

We have to recognize that Russia today has resources for the formation of a mature digital economy, while the distinctive feature of the Russian economy is that the “lion's share” of GDP is created by state corporations (or companies with significant state participation).

References

[1] D Tapscott 1995 The Digital Economy: promise and peril in the age of networked intelligence (McGraw-Hill)
[2] N Negroponte 1995 Being digital (New York: Alfred A Knopf)
[3] UK Digital Strategy 2017. Department for Digital, Culture Media & Sport. 1 March, 2017. URL: https://www.gov.uk/government/publications/uk-digital-strategy/ukdigital-strategy (date of access: June 23, 2018)
[4] A V Babkin, T Yu Khvatova 2010 “Model of the national innovation system based on the knowledge economy”, Economics and management 12 170–176
[5] G20 Digital Economy Ministerial Declaration [G20 Ministerial Conference on Digital Economics Dusseldorf 6–7 April 2017]. URL: http://www.eurasiancommission.org
[6] Digital Economy of the Russian Federation. Official site of the Government of the Russian Federation. URL: http://static.government.ru/media/files/ (date of access: June 29, 2018)
[7] K Gada “Digital Economy in 5 minutes”, Forbes, Jun 16, 2016. URL: https://www.forbes.com/sites/koshagada/2016/06/16/ (date of access: June 23, 2018)
[8] The World Economic Forum. URL: http://reports.weforum.org/global-information-technology-report-2016/networked-readiness-index (date of access: June 23, 2018)
[9] Public Institutions and Digital Government. Department of Economic and Social Affairs. URL: https://publicadministration.un.org/en/Research/UN-e-Government-Surveys (date of access: June 23, 2018)
[10] International Telecommunication Union. URL: http://www.itu.int/en/ITU-D/Statistics/Pages/publications/mis2016.aspx (date of access: June 23, 2018)
[11] Socio-economic indicators 2017 Federal state statistics service. URL: http://www.gks.ru

[12] T Rayna 2008 Understanding the challenges of the digital economy: The nature of digital goods Communication & strategies, **71** 34
[13] Lizunkov V, Politsinskaya E, Malushko E, Kindaev A, Minin M 2018 Population of the world and regions as the principal energy consumer *International Journal of Energy Economics and Policy* **8(3)** 250-257
[14] Ergunova, O, Lizunkov V, Blaginin V, Politsinskaya, E, Shaykina O I 2018 Formation of the price mechanism for energy resources in Russia and the countries of the European union *International Journal of Energy Economics and Policy* 8(1) 122-127.
[15] Fedotov N G, Moiseev, A V, Syemov A A, Lizunkov V G, Kindaev, A.Y. 2017 New Methods of Three-Dimensional Images Recognition Based on Stochastic Geometry and Functional Analysis *IOP Conference Series: Materials Science and Engineering* **177(1)**
[16] Demyanenko, O V, Kopanitsa, NO, Sarkisov Y S,Ikonnikova, K.V., Ikonnikova L F *IOP Conference Series: Earth and Environmental Science* **87(9)**